Polymeric Nanomaterials

Edited by
Challa S. S. R. Kumar

Related Titles

Kumar, C. S. S. R. (ed.)

Nanotechnologies for the Life Sciences

10 Volume Set

ISBN: 978-3-527-31301-3

Kumar, C. S. S. R. (Ed.)

Nanomaterials for the Life Sciences (NmLS)

Book Series, 10 Volumes

Vol. 6

Semiconductor Nanomaterials

2010

ISBN: 978-3-527-32166-7

Vol. 7

Biomimetic and Bioinspired Nanomaterials

2010

ISBN: 978-3-527-32167-4

Vol. 8

Nanocomposites

2010

ISBN: 978-3-527-32168-1

Vol. 9

Carbon Nanomaterials

2011

ISBN: 978-3-527-32169-8

Vol. 10

Polymeric Nanomaterials

2011

ISBN: 978-3-527-32170-4

*Nanomaterials for the
Life Sciences
Volume 10*

Polymeric Nanomaterials

*Edited by
Challa S. S. R. Kumar*

WILEY-VCH Verlag GmbH & Co. KGaA

The Editor

Dr. Challa S. S. R. Kumar
CAMD
Louisiana State University
6980 Jefferson Highway
Baton Rouge, LA 70806
USA

■ All books published by **Wiley-VCH** are carefully produced. Nevertheless, authors, editors, and publisher do not warrant the information contained in these books, including this book, to be free of errors. Readers are advised to keep in mind that statements, data, illustrations, procedural details or other items may inadvertently be inaccurate.

Library of Congress Card No.: applied for

British Library Cataloguing-in-Publication Data
A catalogue record for this book is available from the British Library.

Bibliographic information published by the Deutsche Nationalbibliothek
The Deutsche Nationalbibliothek lists this publication in the Deutsche Nationalbibliografie; detailed bibliographic data are available on the Internet at <http://dnb.d-nb.de>.

© 2011 Wiley-VCH Verlag & Co. KGaA, Boschstr. 12, 69469 Weinheim, Germany

All rights reserved (including those of translation into other languages). No part of this book may be reproduced in any form – by photoprinting, microfilm, or any other means – nor transmitted or translated into a machine language without written permission from the publishers. Registered names, trademarks, etc. used in this book, even when not specifically marked as such, are not to be considered unprotected by law.

Composition   Toppan Best-set Premedia Ltd., Hong Kong
Printing and Binding   betz-druck GmbH, Darmstadt
Cover Design   Schulz Grafik-Design, Fußgönheim

Printed in the Federal Republic of Germany
Printed on acid-free paper

ISBN: 978-3-527-32170-4

Contents

Preface *XV*
List of Contributors *XIX*

Part One   Nanogels, Interfaces, Carriers, and Polymersomes   *1*

1   Towards Self-Healing Organic Nanogels: A Computational Approach   *3*
German V. Kolmakov, Solomon F. Duki, Victor V. Yashin and Anna C. Balazs
1.1   Introduction   *3*
1.2   Methodology   *5*
1.3   Towards Self-Healing Organic Nanogels   *8*
1.3.1   Response of Samples to Tensile Deformation   *8*
1.3.2   Stress–Strain Curve   *10*
1.3.3   Tensile Strength of Nanogel Samples   *14*
1.3.4   Modeling Viscoelastic Nanogel Particles   *15*
1.4   Conclusions   *24*
          Acknowledgments   *24*
          References   *25*

2   Synthesis and Characterization of Polymeric Nanogels   *27*
Yoshifumi Amamoto, Hideyuki Otsuka and Atsushi Takahara
2.1   Introduction   *27*
2.2   Synthesis of Polymeric Nanogels   *28*
2.2.1   Chemical Nanogels   *29*
2.2.1.1   Polymerization in Templates   *29*
2.2.1.2   Crosslinked Micelles and Star Polymers   *36*
2.2.1.3   Polymer Reaction of Linear Chain, Nanoparticles, and Nanocomplexes   *40*
2.2.2   Physical Nanogels   *42*
2.2.2.1   Hydrogen Bonding   *42*
2.2.2.2   Ionic Bonding and Coordination Bonding   *43*
2.2.2.3   Hydrophobic Interaction   *44*
2.2.2.4   Protein Denaturation   *46*

2.2.3	Degradable Chemical Nanogels	47
2.3	Characterization of Polymeric Nanogels	48
	References 50	

3 Stimulus-Responsive Polymers at Nanointerfaces 59
Roshan Vasani, Martin Cole, Amanda V. Ellis and Nicolas H. Voelcker

3.1	Introduction 59	
3.2	Types of Stimulus-Responsive Polymer 60	
3.2.1	Thermoresponsive Polymers 60	
3.2.1.1	LCST Polymers 61	
3.2.1.2	UCST Polymers 62	
3.2.2	pH-Responsive Polymers 62	
3.2.2.1	Polycations 62	
3.2.2.2	Polyanions 64	
3.2.2.3	Polyzwitterions 65	
3.2.3	Photoresponsive Polymers 66	
3.2.4	Dual and Multi-Stimulus-Responsive Polymers 67	
3.3	Generating Stimulus-Responsive Interfaces 69	
3.3.1	Covalent Routes 69	
3.3.1.1	"Grafting-From" Techniques 69	
3.3.1.2	"Grafting-To" Techniques 69	
3.3.2	Noncovalent Routes 70	
3.3.3	Interfacing SRPs with Flat and Porous Substrates 71	
3.3.4	Nanoparticles and Nanotubes 72	
3.3.4.1	Silica, Metal, and Metal Oxide Nanoparticles 72	
3.3.4.2	Quantum Dots 73	
3.3.4.3	Micelles 74	
3.3.4.4	Liposomes 75	
3.3.4.5	Nanotubes 76	
3.4	Applications of Stimulus-Responsive Polymers at Interfaces 77	
3.4.1	Controlled Drug Delivery 77	
3.4.2	Control of Biointerfacial Interactions 79	
3.4.3	Microfluidic Valves 82	
3.4.4	Molecular Separation 85	
3.5	Summary and Future Perspectives 86	
	List of Abbreviations 88	
	References 90	

4 Self-Assembled Peptide Nanostructures and Their Controlled Positioning on Surfaces 105
Maria Farsari and Anna Mitraki

4.1	Introduction 105	
4.2	Vertical and Horizontal Alignment on Surfaces 106	
4.3	Printing Using Inkjet Technology 107	
4.4	Vapor Deposition Methods 107	

4.5	Positioning Using Dielectrophoresis 110
4.6	Laser Patterning 111
4.6.1	Laser-Induced Forward Transfer 111
4.6.1.1	Direct Transfer 112
4.6.1.2	Self-Assembly of Peptides on LIFT-Patterned Biotin 113
4.6.2	Nonlinear Lithography 114
4.7	Summary and Perspectives 117
	Acknowledgments 117
	References 117

5	**Multifunctional Pharmaceutical Nanocarriers: Promises and Problems** 121
	Vladimir P. Torchilin
5.1	Introduction 121
5.2	Established Paradigms: Longevity and Targetability 122
5.2.1	Polymers for Longevity 123
5.2.2	Long-Circulating Liposomes 124
5.2.3	Nonliposomal Long-Circulating DDSs 126
5.2.4	Combination of Longevity and Targeting 127
5.3	Stimuli-Sensitivity and Intracellular Targeting 130
5.3.1	pH-Sensitive Systems 130
5.3.2	Temperature-Sensitive Systems 133
5.3.3	Magnetically Sensitive Systems 133
5.3.4	Ultrasound-Sensitive Systems 134
5.3.5	Redox-Sensitive Systems 135
5.3.6	Intracellular Delivery of Pharmaceutical Nanocarriers 135
5.4	A New Challenge: Theranostics 138
	References 141

6	**Polymersomes and Their Biomedical Applications** 157
	Giuseppe Battaglia
6.1	Introduction 157
6.2	The Chemistry of Polymersomes 158
6.3	Polymersomes: Physico-Chemical Properties 160
6.3.1	Membrane Conformations 160
6.3.2	Responsive Polymersomes 163
6.3.3	Surface Chemistry of the Polymersomes 164
6.4	Polymersomes Formation and Preparation 167
6.5	Biomedical Applications 169
6.5.1	Medical Imaging 169
6.5.2	Cancer Therapy 172
6.5.3	Polymersomes as Delivery Vectors 172
6.5.4	Nanoreactors 172
6.5.5	Artificial Cells and Organelles 173
6.5.6	Gene Therapy 174

| 6.6 | Conclusions 175 |
| | References 175 |

Part Two Nanoparticles 185

7 Synthetic Approaches to Organic Nanoparticles 187
Stefan Köstler and Volker Ribitsch
7.1 Introduction 187
7.1.1 Types of Organic Particle and Scope of This Chapter 188
7.1.2 Characteristics of Organic Nanoparticles 189
7.2 Methods of Organic Nanoparticle Preparation 190
7.2.1 Top-Down Approaches to Organic Nanoparticles 190
7.2.1.1 Reduction of Particle Size by Mechanical Forces 190
7.2.1.2 Lithographic Methods 193
7.2.2 Bottom-Up Approaches to Organic Nanoparticles 194
7.2.2.1 Solution-Based Bottom-Up Methods 194
7.2.3 Vapor Condensation-Based Methods 206
7.3 Application of Organic Nanoparticles 207
7.4 Summary and Future Perspectives 208
References 210

8 Organic Nanoparticles Using Microfluidic Technology for Drug-Delivery Applications 221
Wei Cheng, Lorenzo Capretto, Martyn Hill and Xunli Zhang
8.1 Introduction 221
8.1.1 Batch Synthesis of Organic Nanoparticles 222
8.1.2 Specifications of Reactors: Macroscale versus Microscale Syntheses 223
8.1.3 Properties and Application of Organic Nanoparticles for Drug Delivery 226
8.2 Microfluidic Synthesis of Organic Nanoparticles 227
8.2.1 Overview: Unique Features of Microfluidic Reactors for the Controlled Synthesis of Organic Nanoparticles 227
8.2.2 Microfluidic Reactors for Organic Nanoparticles 228
8.2.2.1 Emulsions 228
8.2.2.2 Nanoprecipitation 230
8.2.2.3 Liposomes 233
8.2.3 Controlled Operating Parameters of Microfluidic Reactors 234
8.2.3.1 Flow Velocity, Microfluidic Dimension, and Mixing Time 235
8.2.3.2 Mixing Time, Aggression Time, and the Damkohler Number 236
8.2.4 Synthetic Operations 238
8.2.4.1 Micromixing 238
8.2.4.2 Online Process of Various Reactants 239
8.2.4.3 Thermal Control and Heat Transfer 240
8.2.4.4 Spatial and Temporal Kinetic Control 242

8.2.4.5	Self-Assembly Mechanism and Competitive Reaction	243
8.3	Microfluidic-Related Organic Nanoparticles for Drug Delivery	244
8.3.1	Drug Encapsulation and Release	245
8.3.2	Stimuli-Responsive Release	246
8.3.3	Nanomedicine Delivery to Target Cells	248
8.4	Conclusions and Prospective Study	250
8.4.1	Materials, Design, and Fabrication	250
8.4.2	High-Throughput Microfluidic Processes	251
8.4.3	Controlled Synthesis of Organic Nanoparticles	252
8.4.4	Spatial and Temporal Kinetics Investigation of Nanoparticles	253
	References 253	

9 Lipid–Polymer Nanomaterials 259
Corbin Clawson, Sadik Esener and Liangfang Zhang

9.1	Introduction	259
9.2	Lipopolymers	260
9.2.1	Synthesis and Fabrication of Lipopolymers	261
9.2.2	Properties and Characterization of Lipopolymers	263
9.2.3	Applications of Lipopolymers in the Life Sciences	263
9.3	Lipid–Polymer Hybrid Nanoparticles	264
9.3.1	Synthesis and Fabrication of Lipid–Polymer Hybrid Nanoparticles	266
9.3.1.1	Synthesis of Polymer-Protected Lipidic Nanoparticles	266
9.3.1.2	Synthesis of Lipid-Coated Polymeric Nanoparticles	269
9.3.2	Characterization of Lipid–Polymer Hybrid Nanoparticles	271
9.3.3	Applications of Lipid–Polymer Hybrid Nanoparticles	275
9.4	Lipid–Polymer Films and Coatings	277
9.4.1	Synthesis of Lipid–Polymer Films and Coatings	277
9.4.2	Characterization of Lipid–Polymer Films and Coatings	278
9.4.3	Applications of Lipid–Polymer Films and Coatings	279
9.5	Summary and Future Perspective	280
	References 281	

10 Core–Shell Polymeric Nanomaterials and Their Biomedical Applications 285
Ziyad S. Haidar and Maryam Tabrizian

10.1	Introduction	285
10.2	Core–Shell Nanomaterials of Biomedical Interest	286
10.3	Core–Shell Polymeric Nanoparticles	287
10.3.1	Biodegradable Core–Shell Polymeric Nanoparticles	288
10.3.2	Core–Shell Colloids	290
10.3.3	Surface Properties and Modification Techniques	292
10.4	Biomedical Applications of Core–Shell Polymeric Nanostructures	293
10.4.1	Bioimaging, Biological, and Cellular Labeling	293
10.4.1.1	Cytotoxicity of QDs	295
10.4.2	Glucose Monitoring and Biosensing in Diabetes Mellitus	295

10.4.3	Drug Delivery 296	
10.4.3.1	Natural Polymer-Based Drug-Delivery Systems: Layer-by-Layer Self-Assembly 297	
10.4.3.2	Synthetic and Composite Drug-Delivery Systems: Functionalized Nanoshells 298	
10.4.4	Cancer 300	
10.4.5	Miscellaneous Applications 302	
10.5	Future Prospects 302	
	Acknowledgments 303	
	Referencesm 303	

11 Polymer Nanoparticles and Their Cellular Interactions 311
Volker Mailänder and Katharina Landfester

- 11.1 Introduction 311
- 11.1.1 Nanoparticle Synthesis by Miniemulsion 313
- 11.2 Nanoparticles as Labeling Agents for Cellular Therapeutics 314
- 11.2.1 Experimental Polystyrene Nanoparticles with Iron Oxide (Magnetite) and a Fluorescent Dye as Reporter 314
- 11.2.2 Gadolinium as a Reporter in Nanoparticles 316
- 11.3 Uptake of Polymeric Nanoparticles into Cells 317
- 11.3.1 Influence of Polymer on Uptake 318
- 11.3.1.1 Polystyrene 318
- 11.3.1.2 Polyisoprene 318
- 11.3.1.3 Nanoparticles Composed of Different Proportions of Polystyrene and Polyisoprene 320
- 11.3.1.4 Poly (n-butylcyanoacrylate) (PBCA) Nanoparticles 320
- 11.3.1.5 Polyester Nanoparticles: Poly(ε-Caprolactone), Poly(D,L-Lactide), Poly(D,L-lactide-co-Glycolide) 321
- 11.3.2 Influence of Transfection Agents: Surface Modifications of Nanoparticles by Covalently Linked Groups 324
- 11.3.2.1 Functionalization of Nanoparticle Surfaces by Carboxylic and Amino Side Groups 324
- 11.3.3 Size Dependency of Cellular Uptake 328
- 11.4 Influence of Nanoparticles on (Stem) Cell Differentiation 329
- 11.5 Endocytosis 331
- 11.6 Summary 334
- References 335

12 Radiopaque Polymeric Nanoparticles for X-Ray Medical Imaging 343
Shlomo Margel, Anna Galperin, Hagit Aviv, Soenke Bartling and Fabian Kiessling

- 12.1 Introduction 343
- 12.2 Synthesis of the Monomer MAOETIB 345
- 12.3 Radiopaque Iodinated P(MAOETIB) Nanoparticles 346

12.3.1	Synthesis of the P(MAOETIB) Nanoparticles *346*	
12.3.2	Characterization of the P(MAOETIB) Nanoparticles *346*	
12.3.2.1	Effect of the MAOETIB Concentration *347*	
12.3.2.2	Effect of the Initiator Concentration *348*	
12.3.2.3	Effect of the Surfactant Concentration *349*	
12.3.3	*In Vitro* X-Ray Visibility of the P(MAOETIB) Nanoparticles *349*	
12.3.4	*In Vivo* X-Ray Visibility of the P(MAOETIB) Nanoparticles: Preliminary Studies *351*	
12.4	Radiopaqe Iodinated P(MAOETIB–GMA) Copolymeric Nanoparticles *351*	
12.4.1	Synthesis of P(MAOETIB–GMA) Copolymeric Nanoparticles *353*	
12.4.2	Influence of the Weight Ratio [MAOETIB]/[GMA] on the Iodine Content and Size and Size Distribution of the Copolymeric Nanoparticles *353*	
12.4.3	Characterization of the P(MAOETIB-GMA) Nanoparticles Prepared at a Weight Ratio [MAOETIB]/[GMA] of 99/1 *355*	
12.4.4	*In Vivo* X-Ray Visibility of the P(MAOETIB-GMA) Nanoparticles *358*	
12.5	Summary *361*	
	References *362*	
13	**Solid Lipid Nanoparticles to Improve Brain Drug Delivery** *365*	
	Paolo Blasi, Aurélie Schoubben, Stefano Giovagnoli, Carlo Rossi and Maurizio Ricci	
13.1	Introduction *365*	
13.2	The General Problem of Brain Drug Delivery *366*	
13.2.1	Basic Brain Physiology *366*	
13.2.2	Brain Drug-Delivery Strategies *368*	
13.3	Solid Lipid Nanoparticles for Brain Drug Delivery *369*	
13.3.1	General Information *369*	
13.3.2	Physico-Chemical Aspects of Lipid Packing and SLN Structure *371*	
13.3.3	Surfactants and SLNs for Brain Targeting *375*	
13.3.4	Evidence of Brain Drug Accumulation *376*	
13.3.5	SLN Potentiality in Brain Imaging *380*	
13.3.6	Toxicity Issues *382*	
13.4	Concluding Remarks *384*	
	References *384*	

Part Three   Nanoscaffolds, Nanotubes, and Nanowires *395*

14   **Architectural and Surface Modification of Nanofibrous Scaffolds for Tissue Engineering** *397*
Jerani T.S. Pettikiriarachchi, Clare L. Parish, David R. Nisbet and John S. Forsythe
14.1   Introduction *397*

14.2	Tissue Engineering Scaffolds	397
14.3	Nanofibrous Scaffolds	399
14.4	Electrospinning	399
14.5	Cellular Interactions with Polymeric Nanofibers	401
14.6	Optimizing Fiber and Scaffold Architecture	403
14.6.1	Fiber Diameter	403
14.6.2	Fiber Orientation	404
14.6.3	Core–Shell Fibers	405
14.6.4	Pore Size	407
14.6.5	Layering Fibers	408
14.7	Optimizing the Fiber Surface	410
14.7.1	Polymer Blending	410
14.7.2	Coating Fibers	411
14.7.3	Chemical Modification of Fibers	412
14.7.4	Polymer Grafting onto Surfaces	413
14.7.5	Plasma Treatment of Fibers	414
14.7.6	Biomolecule Attachment	416
14.8	Challenges with Fibrous Scaffolds in Tissue Engineering	417
14.9	Summary	419
14.10	Future Perspectives	419
	References	420

15 Controlling the Shape of Organic Nanostructures: Fabrication and Properties *429*
Rabih O. Al-Kaysi and Christopher J. Bardeen

15.1	Introduction	429
15.2	Milling, Soft-Templating, and Other Methods for Preparing Organic Nanostructures	431
15.2.1	Mechanical Milling	431
15.2.2	Vapor Growth	431
15.2.3	Supramolecular Self-Assembly	432
15.2.4	Reprecipitation	432
15.2.5	Electrospinning	434
15.2.6	Emulsification	434
15.3	Hard-Templating Methods for Preparing Organic Nanostructures	434
15.3.1	AAO Templates	435
15.3.2	Template-Assisted Synthesis of Polymer Nanowires, Nanotubes, and Nanorods	436
15.3.3	Template-Assisted Synthesis of Small-Molecule Nanowires, Nanotubes, and Nanorods	437
15.3.3.1	Solution-Based Template Wetting Method	437
15.3.3.2	Melting–Recrystallization Template Wetting Method	437
15.3.3.3	Sublimation Method	438
15.3.3.4	Electrophoretic Deposition Method	438
15.3.3.5	Solvent-annealing Method	439

15.4	Applications of Noncovalent Organic Nanostructures	444
15.4.1	Electronic and Optical Devices	444
15.4.2	Chemical Sensing	445
15.4.3	Photomechanical Actuation	446
15.5	Future Challenges and Outlook	448
	Acknowledgments	449
	References	449

16 Conducting Polymer Nanowires and Their Biomedical Applications 455
Robert Lee and Adam K. Wanekaya

16.1	Introduction	455
16.2	Fabrication of Conducting Polymer Nanowires	457
16.3	Surface Modification of Conducting Polymer Nanowires	458
16.4	Assembly/Alignment of Conducting Polymer Nanowires	461
16.5	Biomedical Applications of Conducting Polymer Nanowires	462
16.5.1	Sensing and Detection	462
16.5.1.1	Proteins and Disease Markers	462
16.5.1.2	Detection of Bacteria	464
16.5.1.3	Detection of Small Molecules	466
16.5.1.4	Detection of Heavy-Metal Ions and Pesticides	466
16.5.2	Drug Delivery and DNA Carriers	467
16.6	Summary and Future Perspectives	468
	References	468

17 Organic Nanowires and Nanotubes for Biomedical Applications 473
Keunsoo Jeong and Chong Rae Park

17.1	Introduction	473
17.2	Fabrication of Organic Nanowires and/or Nanotubes	474
17.2.1	Self-Assembly Processes	474
17.2.2	Template-Based Synthetic Processes	476
17.2.3	Nanotubes Based on Modified CNTs	478
17.3	Biomedical Applications of Nanowires and/or Nanotubes	480
17.3.1	Biosensors	480
17.3.2	Cancer Therapy	482
17.3.2.1	CNTs for Photothermal Therapy	482
17.3.2.2	CNTs for Radiofrequency Ablation	483
17.3.3	Optical Bioimaging	484
17.4	Summary	487
	References	487

18 Rosette Nanotubes for Targeted Drug Delivery 493
Sarabjeet Singh Suri, Hicham Fenniri and Baljit Singh

18.1	Introduction	493

18.2	Peptide-Based Nanotubes	494
18.3	Self-Assembling Rosette Nanotubes	495
18.3.1	Self-Assembly Peptides	495
18.3.2	G^C Motif Self-Assembly: Novel Helical Rosette Nanotubes	495
18.3.2.1	Novelty	495
18.3.2.2	G^C Motif Self-Assembly Process	496
18.3.2.3	Built-In Strategy for Manipulating the Properties of RNTs	496
18.3.3	Biological Functions of RNTs	498
18.4	Stability Issues	502
18.5	Nanomaterials for Receptor-Mediated Targeting	502
18.5.1	Human Epidermal Growth Factor Receptor (EGFR)	503
18.5.2	Vasoactive Pituitary Adenylate Cyclase (VPAC)-Activating Peptide Receptors	503
18.5.3	Transferrin Receptor (TfR)	503
18.5.4	Folate Receptor (FR)	504
18.6	Ethical Issues and Future Directions	504
18.7	Conclusions	505
	References	505

Index 509

Preface

What an extraordinary ending to a twenty volume combined series on nanotechnologies and nanomaterials for the life sciences? When the first ten volumes of the NtLS series (Nanotechnologies for the life sciences) were completed in 2007, we assumed that we have covered every aspect of nanotechnologies with respect to their applications in life sciences. When we began work on the second series, Nanomaterials for the life sciences (NmLS), little did we realize that synergic research investigations between nanomaterials and nanotechnologies and life sciences will outpace our ability to capture all the highlights in NmLS series. As I present the final volume in this series, I am overwhelmed with the amount of information that we have been able to capture and I have no doubt that we have created truly a sub discipline within the field of nanotechnologies and nanomaterials-Applications for Life Sciences. This is a long and successful journey made possible only through extraordinary efforts from close to 700 authors from around the globe. I am humbled by their high quality and timely contributions in addition to keeping in tune with the vision for all the twenty volumes. I have had great pleasure in working with the team from Wiley-VCH and in my view one the top rated publishers. Thank you Wiley-VCH!

In this volume, I introduce you to the applications of organic nanomaterials in life sciences. For the sake of clarity, the book is divided into three different sections. The first section provides an in-depth analysis of nanostructured gels, interfaces, carriers and polymerosomes created using organic molecules. While the second section focuses on more traditional organic nanoparticles including polymeric nanoparticles, the final section introduces readers to more sophisticated nanoscaffolds, nanotubes and nanowires.

The very first chapter provides a platform for addressing one of the grand challenges in materials science- to design "smart" synthetic systems with the ability for self-repair. In this chapter, authors demonstrate an important role the computational modeling plays in designing self-healing materials and coatings taking nanoscopic gel particles as an example. Continuing the focus on nanogels, but from an experimental point of view, the next chapter summarizes synthetic approaches to nanogels followed by the tools utilized in their charactreization. Highlighting the importance of nano interfaces in desiging „smart" materials, the third chapter reviews key types of organic nano interfaces that have the ability to repsond to external stimuli. In addition, the chapter provides excellent examples for their application in drug delivery, microfluidics, molecular separation and tissue culture. There is a renewed interest in peptide-based nanostructures and the ability to manipulate their assembly for a number of applications ranging from materials science and bio-nanotechnology to biomedicine and biotechnology. In tune with this, the fourth chapter contains the most recent review on "soft" methods targeted for positioning of self-assembled peptide nanowires and nanotubes. The fifth chapter in this section introduces to the reader the most recent developments in the field of polymeric, multifunctional pharmaceutical nanocarriers. The chapter provides a window to a broad selection of multifunctional systems with combined features ranging from longevity, targetability, stimuli-sensitivity, cell-penetrating ability and contrast properties. In the final chapter, readers learn about polymerosomes, a realtively new class of organic nanostructures that form amphiphilic membranes, bio mimetic analogues of natural phospholipid vesicles, obtained based on block copolymers. In this chapter, one can find not only the chemistry but also the physics and structure-dependent biomedical applications of polymerosomes.

In the second section of the book, a variety of organic nanoparticles made from organic polymers, solid-lipid polymers, core-shell type polymers are covered. It also includes a number of synthetic approaches, characterization methods and applications in biomedicine ranging from medical imaging to drug delivery. The first four chapters in this section bring out the most important tools for their synthesis including the micro fluidic technologies. The fourth chapter in the section (Chapter 11 in the book) is unique due to its focus on bringing out salient features in interactions between different polymeric materials with cells. Lessons from this chapter are the key to their application in biology and medicine. Two specific applications of organic nanoparticles have been included in the last two chapters. The first one is related to the development of radiopaque polymeric particles as contrast agents for x-ray imaging and the second one is on solid lipid nanoparticles (SLN) for brain drug delivery. Overall, this section is central to the theme o f the book.

The final section in the book deals with relatively newer concepts in organic nanomaterials with repsect to their applications in life sciences. For example, chapter 14 refelcts on the most recent strategies for synthesis and modification of polymeric nanofibers and opportunities to enhance their biological functionality especially as scaffolds in tissue engineering. Continuing to reiterate the importance of non-spherical organic nanomaterials, the next chapter surveys recent methods for controlled synthesis of organic nanorods, nanotubes, nanowires with the main focus being on the use of hard templates for the fabrication of organic nanostructures with well-defined shapes and dimensions.

Of the different types of organic nanomaterials conducting organic nanomaterials, which are polymers with spatially extended π-bonding systems due to alternating single and double carbon-carbon bonds, are important as sensing devices. Prominent among the conducting nanopoymeric materials are conducting polymeric nanowires. The chapters 16 and 17 reviews conducting polymer nanowires and nanotubes, which are expected to have increased sensitivities compared to the bulk materials due to their small size, high aspect ratios high surface-to-volume ratios and unusual target binding properties. From the combination of these two chapters, one would learn about all aspects related to various fabrication technologies including self-assembly, template synthesis, and their application in biomedical areas including biosensors, cancer-therapy via photo-thermal and/or radio-frequency ablation, and optical bio imaging. The final chapter brings out the latest information on peptide-based bio-friendly nanotubes including novel rosette nanotubes for drug delivery with special reference to receptor-mediated delivery.

With the publication of this book, the combined twenty volume book series on nanotechnologies/nanomaterials has come to an end. We have made attempts to put together 271 chapters contributed by close to 700 experts and encompassed most of the information related to nanotechnologies and nanomaterials with reference to life sciences. I am aware that knowldge base, new ideas and new perspectives we created will continue to grow at speeds difficult to keep up with. I would like to end this preface with exact words that I used when I concluded the NtLS series. *On behalf of all the authors who have made contributions to this exciting series, it is my privilege to play the role of a catalyst in inculcating new thinking by providing a multi pronged base of knowledge in nanotechnologies/nanomaterials for life sciences. It is my hope that this book series will help in stretching the limits of thinking in all those who come in contact with it.*

Challa Kumar
Baton Rouge, USA.

List of Contributors

Rabih O. Al-Kaysi
King Saud bin Abdulaziz
University for Health Sciences-
National Guard Health Affairs
Department of Basic Sciences
Building Mail Code 3124
Riyadh 11423
Kingdom of Saudi Arabia

Yoshifumi Amamoto
Kyushu University
Institute for Materials Chemistry
and Engineering
744 Motooka, Nishi-ku
Fukuoka 819-0395
Japan

Hagit Aviv
Bar-Ilan University
Department of Chemistry
Ramat-Gan, 52900
Israel

Anna C. Balazs
University of Pittsburgh
Chemical Engineering
Department
Pittsburgh, PA 15261
USA

Christopher J. Bardeen
University of California
Department of Chemistry
Riverside, CA 92521
USA

Soenke Bartling
German Cancer Research Center
Department of Medical Physics in
Radiology
Heidelberg, 69120
Germany

Giuseppe Battaglia
The University of Sheffield
Department of Biomedical
Science
The Krebs Institute
Firth Court
Western Bank
Sheffield S10 2TN
UK

Paolo Blasi
Università degli Studi di Perugia
Dipartimento di Chimica e
Tecnologia del Farmaco
via del Liceo 1
06123 Perugia
Italy

Lorenzo Capretto
University of Southampton
School of Engineering Sciences
Southampton SO17 1BJ
UK

Wei Cheng
University of Southampton
School of Engineering Sciences
Southampton SO17 1BJ
UK

Corbin Clawson
University of California San Diego
Department of Bioengineering
La Jolla, CA 92130
USA
University of California San Diego
Moores Cancer Center
La Jolla, CA 92130
USA

Martin Cole
Aarhus University
The Interdisciplinary Nanoscience Center (iNANO)
Aarhus, 8000
Denmark

Solomon F. Duki
University of Pittsburgh
Chemical Engineering Department
Pittsburgh, PA 15261
USA

Amanda V. Ellis
Flinders University
School of Chemical and Physical Sciences
Adelaide, SA 5042
Australia

Sadik Esener
University of California San Diego
Moores Cancer Center
La Jolla, CA 92130
USA
University of California San Diego
Department of Nanoengineering
La Jolla, CA 92130
USA

Maria Farsari
Foundation for Research and Technology-Hellas (FO.R.T.H.)
Institute of Electronic Structure and Laser (I.E.S.L.)
P.O. Box 1527
Vassilika Vouton
71110 Heraklion
Crete

Hicham Fenniri
University of Alberta
National Institute for Nanotechnology and Department of Chemistry
11421 Saskatchewan Drive
Edmonton, AB T6G2M9
Canada

John S. Forsythe
Monash University
Clayton 3800
Australia

Anna Galperin
University of Washington
Department of Bioengineering
Seattle, WA 98195
USA

Stefano Giovagnoli
Università degli Studi di Perugia
Dipartimento di Chimica e
Tecnologia del Farmaco
via del Liceo 1
06123 Perugia
Italy

Ziyad S. Haidar
McGill University
Faculties of Medicine and
Dentistry
Department of Biomedical
Engineering
Montréal, QC
Canada H3A 1A4

Martyn Hill
University of Southampton
School of Engineering Sciences
Southampton SO17 1BJ
UK

Keunsoo Jeong
Seoul National University
Carbon Nanomaterials Design
Laboratory
Global Research Laboratory
Research Institute of Advanced
Materials
Department of Materials Science
and Engineering
Seoul 151-744
Korea

Fabian Kiessling
Aachen University
Department of Experimental
Molecular Imaging
Aachen, 52074
Germany

German V. Kolmakov
University of Pittsburgh
Chemical Engineering
Department
Pittsburgh, PA 15261
USA

Stefan Köstler
Joanneum Research
Forschungsgesellschaft mbH
MATERIALS Institute of Surface
Technologies and Photonics
Steyrergasse 17
8010 Graz
Austria

Katharina Landfester
Max Planck Institute for Polymer
Research
Ackermannweg 10
55128 Mainz
Germany

Robert Lee
Missouri State University
Chemistry Department
Springfield, MO 65897
USA

Volker Mailänder
Max Planck Institute for Polymer
Research
Ackermannweg 10
55128 Mainz
Germany
University Medicine of the
Johannes Gutenberg University
III. Medical Clinic
Langenbeckstr. 1
55131 Mainz
Germany

Shlomo Margel
Bar-Ilan University
Department of Chemistry
Ramat-Gan, 52900
Israel

Anna Mitraki
Department of Materials Science
and Technology University of
Crete
Vassilika Vouton
710 03 Heraklion
Crete

David R. Nisbet
Monash University
Clayton 3800
Australia

Hideyuki Otsuka
Kyushu University
Institute for Materials Chemistry
and Engineering
744 Motooka, Nishi-ku
Fukuoka 819-0395
Japan

Clare L. Parish
Howard Florey Institute
Parkville 3010
Australia

Chong Rae Park
Seoul National University
Carbon Nanomaterials Design
Laboratory
Global Research Laboratory
Research Institute of Advanced
Materials
Department of Materials Science
and Engineering
Seoul 151-744
Korea

Jerani T.S. Pettikiriarachchi
Monash University
Clayton 3800
Australia

Volker Ribitsch
Joanneum Research
Forschungsgesellschaft mbH
MATERIALS Institute of Surface
Technologies and Photonics
Steyrergasse 17
8010 Graz
Austria

Maurizio Ricci
Università degli Studi di Perugia
Dipartimento di Chimica e
Tecnologia del Farmaco
via del Liceo 1
06123 Perugia
Italy

Carlo Rossi
Università degli Studi di Perugia
Dipartimento di Chimica e
Tecnologia del Farmaco
via del Liceo 1
06123 Perugia
Italy

Aurélie Schoubben
Università degli Studi di Perugia
Dipartimento di Chimica e
Tecnologia del Farmaco
via del Liceo 1
06123 Perugia
Italy

Baljit Singh
University of Saskatchewan
Department of Veterinary
Biomedical Sciences
52 Campus Drive
Saskatoon
Canada SK S7N 5B4

Sarabjeet Singh Suri
University of Saskatchewan
Department of Veterinary
Biomedical Sciences
52 Campus Drive
Saskatoon
Canada SK S7N 5B4

Maryam Tabrizian
McGill University
Faculties of Medicine and
Dentistry
Department of Biomedical
Engineering
Montréal, QC
Canada H3A 1A4

Atsushi Takahara
Kyushu University
Institute for Materials Chemistry
and Engineering
744 Motooka, Nishi-ku
Fukuoka 819-0395
Japan

Vladimir P. Torchilin
Northeastern University
Department of Pharmaceutical
Sciences and Center for
Pharmaceutical Biotechnology
and Nanomedicine
Mugar Building
Room 312
360 Huntington Avenue
Boston, MA 02115
USA

Roshan Vasani
Flinders University
School of Chemical and Physical
Sciences
Adelaide, SA 5042
Australia

Nicolas H. Voelcker
Flinders University
School of Chemical and Physical
Sciences
Adelaide, SA 5042
Australia

Adam K. Wanekaya
Missouri State University
Chemistry Department
Springfield, MO 65897
USA

Victor V. Yashin
University of Pittsburgh
Chemical Engineering
Department
Pittsburgh, PA 15261
USA

Liangfang Zhang
University of California San
Diego
Moores Cancer Center
La Jolla, CA 92130
USA
University of California San
Diego
Department of Nanoengineering
La Jolla, CA 92130
USA

Xunli Zhang
University of Southampton
School of Engineering Sciences
Southampton SO17 1BJ
UK

Part One
Nanogels, Interfaces, Carriers, and Polymersomes

1
Towards Self-Healing Organic Nanogels: A Computational Approach
German V. Kolmakov, Solomon F. Duki, Victor V. Yashin and Anna C. Balazs

1.1
Introduction

The ability to heal wounds is one of the truly remarkable properties of biological systems. A grand challenge in materials science is to design "smart" synthetic systems that can mimic this behavior by not only "sensing" the presence of a "wound" or defect, but also actively re-establishing the continuity and integrity of the damaged area. Such materials would significantly extend the lifetime and utility of a vast array of manufactured items. Nanotechnology is particularly relevant to both the utility and fabrication of self-healing materials. For example, as devices reach nanoscale dimensions, it becomes critical to establish a means of promoting repair at these length scales. Whilst operating and directing minute tools to carry out this operation is still far from trivial, an optimal solution would be to design a system that could recognize the appearance of a nanoscopic crack or fissure, and then direct the agents of repair specifically to that site. Even in the manufacture of various macroscopic components, nanoscale damage is a critical issue. For instance, nanoscopic notches and scratches can appear on the surface of materials during the manufacturing process. Because of the small size of these defects they are difficult to detect and, consequently, difficult to repair. Such defects, however, can have a substantial effect on the mechanical properties of the system. For example, significant stress concentrations can occur at the tip of notches in the surface; such regions of high stress can ultimately lead to the propagation of cracks through the system and the degradation of mechanical behavior.

Thus, one of the driving forces for creating self-healing materials [1–9] is in fact the need to effect repair on the nanoscale. On the positive side, advances in nanotechnology could also provide routes for realizing the creation of these materials. In particular, scientists can now produce a stunning array of nanoscopic particles, and have become highly adept at tailoring the surface chemistry of the particles. In this chapter, recent computational studies on the design of self-healing materials that exploit the unique properties of *soft* nanoscopic particles are reviewed. As noted further below, these studies take their inspiration from biological systems that show remarkable resilience in response to mechanical deformation.

Nanomaterials for the Life Sciences Vol.10: Polymeric Nanomaterials. Edited by Challa S. S. R. Kumar
Copyright © 2011 WILEY-VCH Verlag GmbH & Co. KGaA, Weinheim
ISBN: 978-3-527-32170-4

In a recent study [10], attention was focused on nanoscopic polymer gel particles, or "nanogels" [11], as the primary building blocks in the system. New methodologies have recently enabled the well-controlled synthesis of such colloids [12]. Furthermore, the surface of these particles can be functionalized with various reactive groups, which allow the individual nanogel particles to be cross-linked into a macroscopic material [11]. By using a coarse-grained computational model, it was possible to examine systems of such crosslinked, soft nanogel particles, and to design a coating that would undergo structural rearrangement in response to mechanical stress, and thus prevents any catastrophic failure of the material [10].

It was assumed that the particles were connected via a fraction of labile bonds (e.g., thiol, disulfide, or hydrogen bonds) [3]; the particles were also interconnected by stronger, less-reactive bonds (e.g., C–C, bonds) – referred to as "permanent" bonds – and thus, the system exhibits a so-called "dual crosslinking." Within this system, the stable, "permanent" bonds between the nanogels play an essential role by imparting structural integrity. As discussed below, it is the reactive, labile bonds, however, that improve the strength of the material. In particular, when the material is strained, the labile bonds break before the stronger connections; these broken bonds then allow the particles to slip and slide, to come into contact with new neighbors, and to make new connections that maintain the continuity of the film. In this manner, the labile bonds can postpone catastrophic failure and, thereby, impart self-healing properties to the material. Through computer simulations, the parameter range was pinpointed for optimizing this self-healing behavior. In fact, it was found that only a relatively small volume fraction of labile bonds within the material could cause a dramatic increase in the ability of the network to resist catastrophic failure [10].

The above behavior is conceptually analogous to the properties that contribute to the strength of the abalone shell *nacre*, where brittle inorganic layers are interconnected by a layer of crosslinked polymers [13]. Under a tensile deformation, the weak crosslinks or "sacrificial bonds" are the first to break. These ruptures dissipate energy and thus mitigate the effects of the mechanical deformation. Consequently, the breakage of these sacrificial bonds helps to maintain the structural integrity of the material.

It should be mentioned that, in another recent study [14], inspiration was taken from nature; namely, the functionality of biological leukocytes, which localize at a wound and thereby facilitate the repair process. In the synthetic system, the "leukocyte" represents a polymeric microcapsule, the healing agents represent encapsulated solid nanoparticles, and the "wound" is a microscopic crack on a surface. In the simulation, the nanoparticle-filled microcapsules are driven by an imposed fluid flow to move along the cracked substrate. The goal was to determine how the release of the encapsulated nanoparticles could be harnessed to repair damage on the underlying surface. The simulations revealed that these capsules could deliver the encapsulated materials to specific sites on the substrate, thus effectively generating an alternate route to repairing surface defects. Once the healing nanoparticles had been deposited on the desired sites, the fluid-driven capsules could move further along the surface, and for this reason the strategy was termed "repair-and-go." The latter strategy might be particularly advantageous as it would have a negligible impact on the precision of the nondefective regions, and involve minimal amounts of the repair materials.

It is noteworthy that micron-sized capsules filled with dissolved particles can encompass very high payloads, allowing them very rapidly to carry and deliver large amounts of nanoparticles to a desired location. Furthermore, the continued, flow-driven motion of these micro-carriers, at least potentially, would allow multiple damaged regions to be healed by the capsules.

As the introduction of a synthetic microvasculature [15] into structural materials becomes more developed, the use of such microcapsules as cellular mimics could expand the efficiency of the artificial circulatory systems. In addition to supplying healing reagents in the channels, it could be advantageous to encapsulate "damage markers" within the microcapsules. The microcapsules would then continue to circulate in a "healthy," undamaged system, but become trapped or localized at a damaged site and thus deliver a chemical "marker" (i.e., a visible or fluorescent dye) through its porous shell. Such markers would enable the nondestructive location and tracking of the damaged regions over time.

Below, attention is focused on describing the present authors' studies with nanogels. In previous investigations [10], the lattice spring model (LSM) was utilized, which was adopted from atomistic models of solid-state and molecular physics [16]. The LSM involves a network of interconnected "springs," which describe the interactions between neighboring units. The large-scale behavior of the resultant system can be mapped onto continuum elasticity theory [17]. Advantage was taken of the LSM to formulate new techniques for modeling the interactions between surface-functionalized, soft nanogels. Following Section 1.2, the findings on the behavior of dual crosslinked nanogel particles under strain are discussed, after which new calculations are described that allow modeling of the viscoelastic behavior of the individual nanogel particles.

1.2 Methodology

In this section, the computational approaches are described that were used to examine healing at the nanoscale. Specifically, we address the challenge of modeling deformable nanoscopic particles that are interconnected into a macroscopic network by both reactive and relatively nonreactive bonds. As noted above, the term "dual crosslinking" is used when referring to this material; here, the labile bonds allow the system to undergo significant structural rearrangement, while the strong "permanent" bonds provide an important "backbone." It is the dynamic interplay between these different components that gives rise to the novel and distinctive characteristics of the materials.

A network of associated colloidal particles is commonly referred to a "particle gel." To the best of the present authors' knowledge, there have been no prior

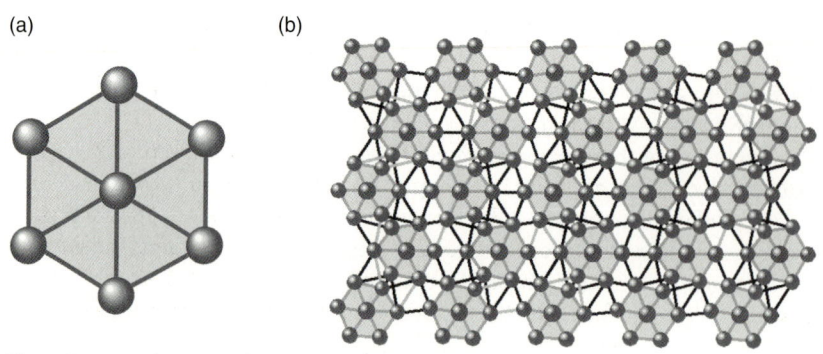

Figure 1.1 (a) Schematic of a deformable gel particle; each particle consists of seven nodes (points) connected by spring-like bonds (lines); (b) Fragment of an undeformed nanogel layer for sample with $P = 0.8$. The dark lines between units (shaded in gray) mark stable bonds, while light gray lines indicate labile bonds.

computational studies on particle gels where each individual particle is itself a deformable gel. Thus, these studies [10] represent the first simulations of a deformable network where each unit can itself undergo deformation.

The approach for simulating the nanogel network is based on the LSM [17, 18], where point-like masses (nodes) are interconnected by Hookean springs, which represent bonds. Figure 1.1a shows the seven-node model that represents the individual gel unit. These units are then interconnected into an extended material by both permanent and labile bonds.

Within a single gel unit, the nodes interact through a potential $U(r)$ that involves an attractive Hookean spring interaction, and a repulsive force, which mimics an excluded volume around the node:

$$U(r) = \frac{\kappa}{2}\left(r^2 + \frac{a}{r}\right) \tag{1.1}$$

with a cut-off distance r_c. Here, κ is the spring stiffness constant, r is the distance between the nodes, and a is the repulsion parameter. The equilibrium distance between the nodes is equal to $\Delta = (a/2)^{1/3}$. In the simulations, the cut-off distance r_c is set equal to 2Δ. Within each gel unit, the bonds do not break during the course of the simulations.

To model bonds between gel units, the same interaction potential is used, which emanates from each of the surface nodes on the gel pieces. Now, however, the spring constant κ for the inter-gel interactions is taken to be sixfold weaker than that for intra-gel bonds. (While different values for the latter spring constants could be chosen, it must be noted that for the large number of nodes considered here – in excess of 1000 for large samples – significant differences between the inter- and intra-gel spring constants can give rise to numerical instabilities.) Additionally, κ has the same value for stable and labile bonds. (The latter choice allowed attention to be focused specifically on isolating effects arising from the dual crosslinking.) In the case of a broken bond, the interaction potential is only given by the repulsive part (i.e., by the term $\propto 1/r$ in Equation 1.1).

The dynamical behavior of the system is taken to be in the overdamped limit, where the inertial terms in the equation of motion for the nodes is neglected. Thus, the velocity of a node is taken to be proportional to the net force acting on it (where the net force is the sum of forces from neighboring nodes and from an external tensile force). It must be noted that this assumption is commonly made in studies on gel dynamics [19, 20]. Specifically, each gel node obeys the following dynamical equation: $\frac{d\mathbf{r}_i}{dt} = \mu \mathbf{F}_i$, where μ is the mobility and $\mathbf{F}_i$ is the force acting on node i. Here, μ is taken to be a constant, and thus the dependence of the mobility on the polymer density is neglected. The force acting on the node i is defined as: $\mathbf{F}_i = -\frac{\partial U}{\partial \mathbf{r}_i} + \mathbf{F}_i^{ext}$, where the elastic energy U is equal to $U = \frac{1}{2}\sum_{m,n}' U_1(|\mathbf{r}_m - \mathbf{r}_n|)$; here, the prime denotes that the summation is for $m \neq n$. The term $\mathbf{F}_i^{ext}$ is the external force acting on particle i. In these simulations, $\mathbf{F}_i^{ext}$ is the tensile force applied to the nodes at the vertical edges of a rectangular sample. These equations of motion are then numerically integrated, using the fourth-order Runge–Kutta algorithm.

As explained above, in response to the applied deformation, the bonds between the gels units can rupture and reconnect. Thus, the Bell model [21] was adopted to describe the rupture and reformation of bonds. Recently, the Bell model has served as a useful framework for describing the relationship between bond dissociation and stress [22], and has also been widely used to describe the reversible bonds formed in proteins [23], between biological cells, or between cells and surfaces [24–26]. In accordance with the model [21, 25], the rupture rate, K_r, is an exponential function of the force applied to the bond:

$$K_r^{(s,l)} = \nu^{(s,l)} \exp\left[\frac{r_0 F - U_0^{(s,l)}}{k_B T}\right]. \qquad (1.2)$$

Here, $U_0^{(s,l)}$ is the potential well depth at zero mechanical stress, F is the applied force, r_0 is a parameter that characterizes the change in the reactivity of the bond under stress, k_B is the Boltzmann constant, and T is the temperature. In the simulations, we set $r_0 = 0.2\Delta$, which is a representative value for chemical bonds [23]. The parameter $\nu^{(s,l)}$ is an intrinsic frequency of an unstressed bond; in the LSM, its value is equal to $\nu^{(s,l)} = \sqrt{\kappa/m}$, where κ is the bond stiffness and m is the reduced mass of the nodes attached to the bond (in the simulations, m was set to 1). The superscripts s and l label the stable and labile bonds, respectively. Taking representative values into consideration, the potential well depth was set equal to $U_0^{(l)} = 100 k_B T$ for labile bonds, and to $U_0^{(s)} = 140 k_B T$ for strong bonds [27].

The reforming rate, K_f, for a broken bond was calculated directly from the detailed balance principle [24, 26]:

$$\frac{K_r^{(s,l)}}{K_f^{(s,l)}} = \frac{K_{r0}^{(s,l)}}{K_{f0}^{(s,l)}} \exp\left(\frac{\Delta U^{(s,l)}}{k_B T}\right),$$

where $\Delta U^{(s,l)}$ is a difference in the potential energies of a connected and broken bond, and $K_{r0}^{(s,l)}$ and $K_{f0}^{(s,l)}$ are the rupture and reforming rates for an unstressed bond. For the Hookean spring interaction described by Equation 1.1, this gives [24, 26]:

$$K_f^{(s,l)} = K_{f0}^{(s,l)} \exp\left\{\kappa(r-\Delta)\left(r_0^{(s,l)} - (r-\Delta)/2\right)(k_B T)^{-1}\right\} \tag{1.3}$$

The probability for a connected bond to break and the probability for a broken bond to reform within a numerical time step Δt were taken to be of the following forms:

$$w_r^{(s,l)} = 1 - \exp\left[-K_r^{(l,s)}\Delta t\right],$$
$$w_f^{(s,l)} = 1 - \exp\left[-K_f^{(l,s)}\Delta t\right]. \tag{1.4}$$

At each simulation time step, the probability of bond rupturing or reforming is computed according to Equation 1.4, where $\Delta t = 10^{-2}$ is the time step of integration.

Gel samples of three different sizes were considered: (i) five rows, with 10 particles in each row; (ii) 10 rows with 10 particles in each row; and (iii) 12 rows with 15 particles in each row. To prepare dual crosslinked materials with different distributions of labile and permanent bonds, these samples were constructed in two steps. In the first step, the layers were arranged into a regular pattern with a lattice spacing of 3Δ between the centers of the gel units, where Δ is the equilibrium distance between the nodes (2Δ is the horizontal size of a gel unit). The vertical spacing between the layers was equal to 1.3Δ. At this step, all possible bonds within the cut-off radius were established, and each node was allowed to subtend at most five interactions. All of these interactions were marked as labile bonds. The sample was then equilibrated for 100 time steps (for the smallest sample), or 1000 time steps (for larger samples). During the equilibration, the initial mechanical stresses undergo relaxation, and the most stressed bonds were ruptured in accordance with the probability in Equation 1.4. In the second step, the characteristics of each interparticle bond were specified, assigning stable bonds with a probability P and labile bonds with a probability $(1-P)$. Thus, even for a fixed value of P, each simulation has a different, independent distribution of stable and labile bonds.

1.3
Towards Self-Healing Organic Nanogels

1.3.1
Response of Samples to Tensile Deformation

In the simulations, the interconnected gel particles form a two-dimensional (2-D) network (see Figure 1.1b). Given that N_{sta} and N_{lab} are the respective average number of stable and labile bonds in the system, the ratio $P = N_{sta}/(N_{sta} + N_{lab})$ is used to characterize the interconnections in the network. For example, for $P = 1$, the nanogel particles are interconnected solely by the stable bonds, whereas for $P = 0.8$, the material encompasses 20% labile bonds. In this case, P is referred to as the dual crosslinking ratio.

Below are described the findings for larger samples; namely, those that encompassed eight rows of gel particles, with 10 particles in each row, and a sample with 12 rows, where each row consisted of 15 particles. In order to characterize the behavior of the system, a Weibull statistical analysis [28] was carried out on these samples. In the relatively thin samples, a certain fraction of the labile bonds are located in the outer surface of the film. When the tensile deformation was applied [10], some of these bonds were readily broken (as they had fewer neighbors to bind them), and this process effectively nucleated a small surface crack which then initiated the ensuing dynamic processes. As the width of the sample was increased, however, the relative fraction of surface bonds decreased. To ensure that the simulations are run in realistic time scales, for the larger samples, a small notch was initially introduced at a random site at the surface, after which the analysis was carried out, as described below.

Following the introduction of a crack at the lower surface (as shown by the vertical arrows in Figure 1.2), the sample was stretched by a tensile force. It was then determined whether the sample fractured after being stretched at a given stress, σ, or not The simulation was repeated eight times, with different initial positions for the crack. The probability of rupture, p_b, was calculated as a ratio of the number

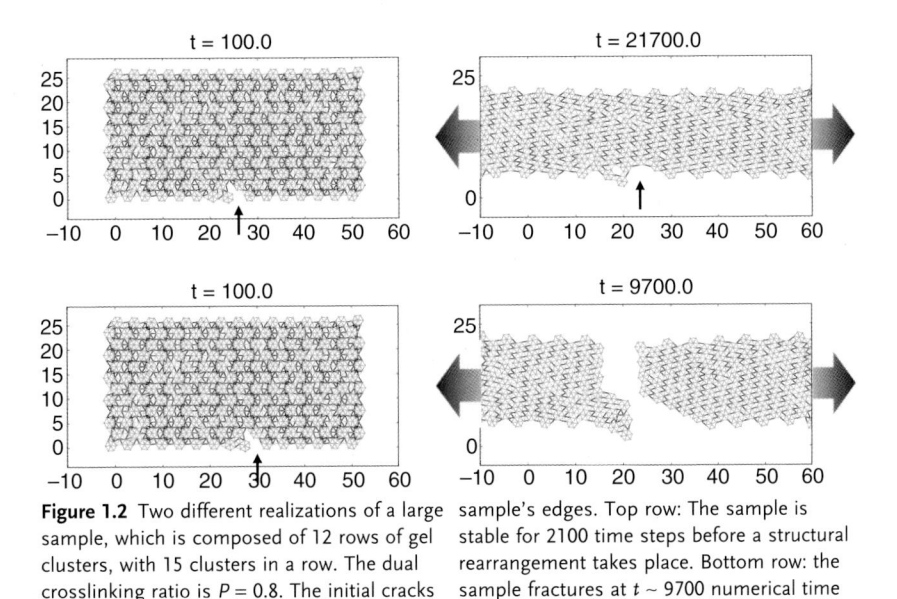

Figure 1.2 Two different realizations of a large sample, which is composed of 12 rows of gel clusters, with 15 clusters in a row. The dual crosslinking ratio is $P = 0.8$. The initial cracks are marked by vertical arrows. Large horizontal arrows indicate the direction of the stretching forces that are applied to the sample's edges. Top row: The sample is stable for 2100 time steps before a structural rearrangement takes place. Bottom row: the sample fractures at $t \sim 9700$ numerical time steps. It is clearly seen that fracture is initiated at the crack.

of times that the sample was completely fractured, n_b, compared to the total number of attempts, n_{tot} ($p_b = n_b/n_{tot}$).

In the Weibull statistical analyses, the probability of a sample breaking is described by the two-parameter cumulative distribution function:

$$p_b(\sigma) = 1 - \exp\left[-(\sigma/\sigma_b)^m\right], \tag{1.5}$$

where σ_b and m are the fitting parameters characterizing the distribution. The parameter σ_b is the characteristic stress at which the sample fractures, and the exponent m characterizes the brittleness of the sample.

The results of the analysis for the two larger samples are quantitatively similar; thus, the data for the largest sample are presented in Figure 1.3. The plot in Figure 1.3a shows the dependence of the probability of rupture, p_b, on the applied stress σ for a sample with $P = 0.8$; each point represents an average of eight independent simulations. The stress σ is normalized by the stiffness of the bond, κ (see Section 1.2), which has the same dimensionality as σ in two dimensions (so that the ratio σ/κ is dimensionless). The curve shows the result of fitting the numerical data to the function $p_b(\sigma)$ in Equation 1.5. The fitting parameters were determined with the aid of the least-squares method. The values for the relevant parameters were $\sigma_b/\kappa = 0.83$ (± 0.04) and $m = 5.15$ (± 0.04) for $P = 0.8$. Thus, the statistical error in the determination of the fitting parameters was ~5% for the characteristic rupture stress σ_b, and less than 1% for the exponent m. It can be seen that, for $P = 0.8$, the curve characterizing p_b exhibits a gentle slope.

By generating plots similar to Figure 1.3a for different values of P, the curve in Figure 1.3b is obtained, which shows the dependence of the characteristic rupture stress σ_b on the dual crosslinking ratio P. The error bars in this plot show the standard deviations for σ_b obtained via the fitting procedure. The maximum in the plot at $P \sim 0.7$ clearly shows that the stress needed to fracture a material with a small fraction of reactive bonds is greater than that required to fracture a material composed entirely of the stable bonds.

1.3.2
Stress–Strain Curve

In order to more completely characterize the behavior of this dual-crosslinked material under tensile deformation, and to demonstrate its self-healing properties, the stress–strain curves were also determined. In contrast to the simulations described above, where a constant stress was applied to a sample, in this case the sample was stretched at a constant velocity, and the tensile stress was computed as a function of the strain, $\varepsilon = (L - L_0)/L_0$. (In particular, the right edge of the sample was held fixed, while the left edge was displaced along the horizontal axes with a speed V_t.) This type of measurement is widely used in the characterization of crosslinked polymers [29, 30]. It should be noted that the engineering stress [31], which is defined as the ratio of the tensile force to the cross-section (in the Y direction) of the unperturbed layer, is calculated. At regions of high strain, where

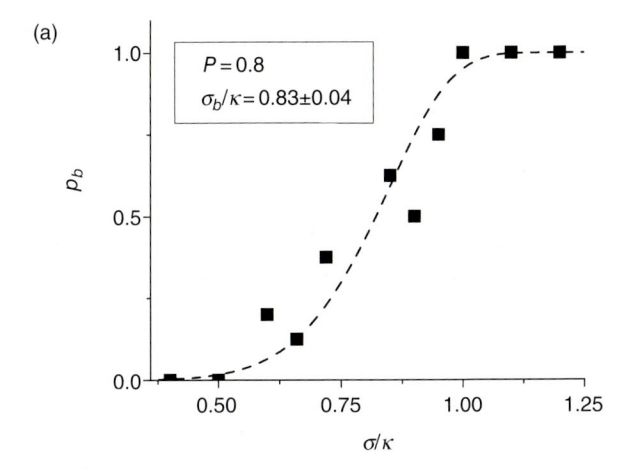

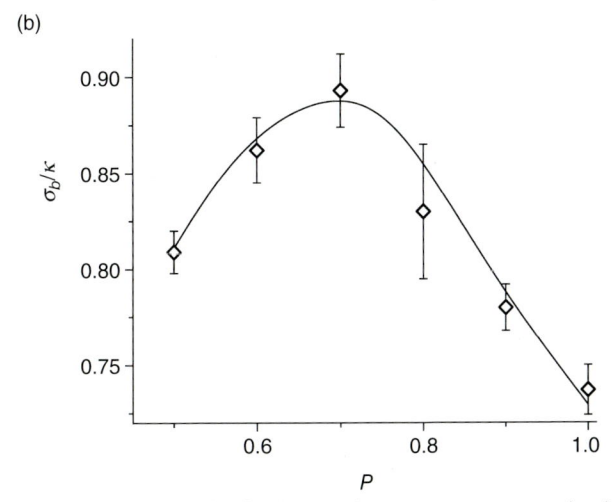

Figure 1.3 (a) Probability for the sample to break, p_b, plotted as a function of the applied tensile stress, σ. The solid squares show the results from simulations for dual crosslinked samples with $P = 0.8$. The dashed curve shows the results of fitting of the data by the Weibull probability distribution function. The stress is normalized by the bond stiffness constant, κ; (b) The diamonds show the dependence of σ_b on the dual crosslinking ratio P, as calculated through Weibull statistical analyses for the largest sample. The full curve is plotted as a guide for the eye.

structural rearrangement takes place, the true stress is higher than the calculated engineering stress, due to the decrease in the sample's cross-section during the course of the rearrangement. Thus, these calculations provide an estimate from below for the stability region of materials encompassing labile bonds ($P < 1$).

Figure 1.4 shows the stress–strain curves calculated for the largest sample (with 12 rows with 15 particles in each row) for $P = 1$ and $P = 0.8$. Here, the tensile

Figure 1.4 Upper diagram: Stress–strain curve calculated for the largest sample. Stress is normalized by the bond stiffness constant, κ. The open squares mark the results for dual crosslinked samples $P(0) = 0.8$; filled triangles indicate the results for the permanently crosslinked $P = 1$ samples. The dashed curves are plotted to guide the eye. The inset shows jumps in the stress–strain curve on an enlarged scale. Lower diagrams: Panels (1–3) showing the evolution of a portion of the sample with increasing strain, ε. The moments in time at which the panels are plotted are labeled in the upper plot by vertical arrows. Panels (1) and (2) show the respective images of the sample just before and after the formation of holes between the gel particles because of the bond rupture. Panel (3) shows the same sample after the holes have collapsed, at a later time. A cluster positioned near these structural rearrangements is marked by slanted arrow.

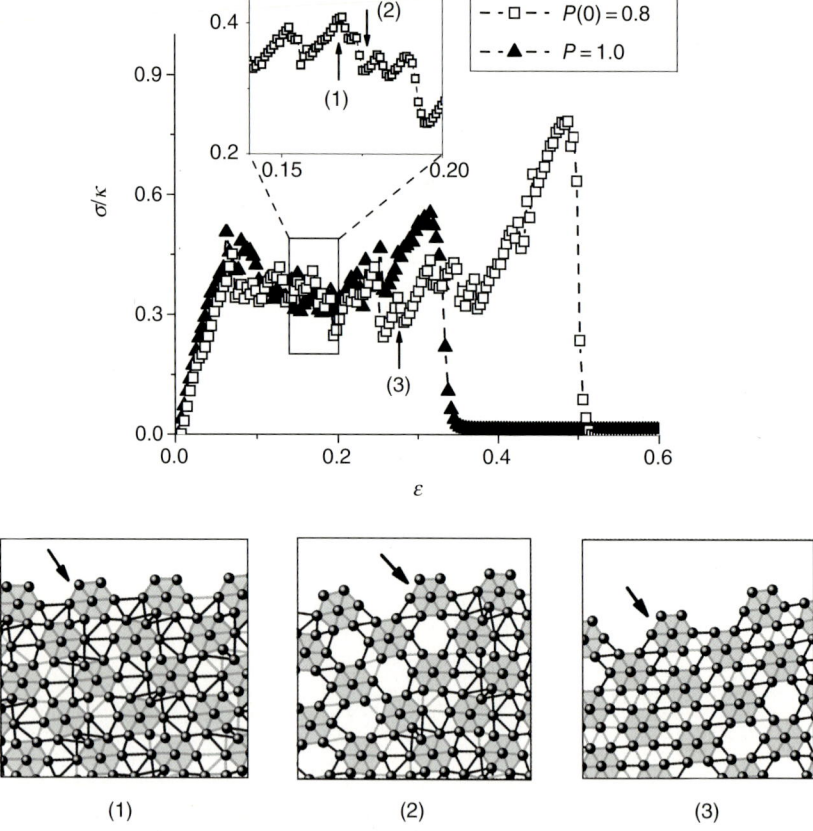

speed was equal to $V_t = 10^{-3}\, d/\tau$, where $d = 2\Delta$ is the characteristic size of the gel particle. The parameter $\tau = 1/\mu\kappa$ is the elastic response time for a bond, where μ is the mobility of the nodes and κ is the stiffness of the bonds (see below and Section 1.2). The first peak at $\varepsilon \approx 0.08$ provides the yield stress, while at stresses below the yield stress the curves for the permanent and dual crosslinked samples coincide with each other. The latter behavior arises because the stiffness constants for the strong and labile bonds are chosen to be equal. The lower panels (1–3) in Figure 1.4 illustrate the mechanism of structural rearrangement (plastic elongation) of the sample for $\varepsilon > 0.08$. As is apparent from panels (1) and (2), elongation of the sample at $\varepsilon > 0.08$ is accompanied by rupture of bonds and the formation of cavities in the sample bulk. (The formation of cavities during the plastic deformation of solid samples was also observed in molecular dynamics simulations in crosslinked polymers [29, 30, 32].) The bond rupture is also responsible for the saw-tooth-shaped fluctuations seen in the stress–strain curves, and is clearly visible in the inset. In the case of the dual crosslinked sample, the cavities collapse at later times due to the formation of new labile bonds between the clusters, as is evident from panel (3) (see also Figure 1.2). In effect, the dual crosslinking allows the particles to move relative to each other, without compromising the structural integrity of the sample and thereby, to decrease the strain energy.

It is clear from Figure 1.4 that the strain at which the $P = 0.8$ dual crosslinked sample fractures ($\varepsilon_b \approx 0.5$) is approximately 1.5-fold greater than that for the sample with $P = 1.0$ ($\varepsilon_b \approx 0.33$). Furthermore, the stress needed to fracture the $P = 0.8$ sample is greater than σ^*, which is the stress necessary to fracture the $P = 1$ material. The latter finding is in agreement with the plot in Figure 1.3, which was obtained from simulations involving constant applied stress. These observations support the conclusion that the introduction of labile bonds leads to an increase in the mechanical stability of the nanogel material.

Figure 1.5a reveals how the total number of bonds in the sample, $N_{tot} = N_{lab} + N_{sta}$, vary with the applied strain for the tensile deformation shown in Figure 1.4. The data are plotted up to the point where the samples undergo fracture: $\varepsilon \approx 0.5$ for $P = 0.8$ and $\varepsilon \approx 0.33$ for $P = 1.0$. While the total number of bonds is decreased during the deformation for both $P = 0.8$ and $P = 1.0$ samples, the total number of bonds for the dual crosslinked sample is always higher than that for the permanently crosslinked sample in the plastic deformation region $\varepsilon > 0.08$. This difference is due to the reformation of ruptured labile bonds during the structural rearrangement.

To further characterize changes in the network during rearrangement in the dual crosslinked sample, the saturation parameter $s = N_{lab} / N_{lab}^{(max)}$ is defined; this is the ratio of the number of formed labile bonds to the maximally permitted number of labile bonds in the sample. The total number of labile bonds is limited in the model by the total number of nodes on the surface of the gel particles (see Section 1.2). The dependence of s on ε for the $P = 0.8$ sample is plotted in the inset in Figure 1.5b. Initially, only approximately 28% of all possible labile bonds were formed in the sample; all the other labile bonds were sufficiently stressed that they ruptured, in accordance with the probability in Equation 1.2. During rearrangement of the sample, the number of labile bonds was gradually increased, and reached $s \approx 0.5$ before the sample fractured.

On the other hand, the less-reactive, stable bonds mostly simply rupture (without reforming) during rearrangement for the $P = 1$ sample (see Figure 1.5a). As a consequence, the value of P (the ratio of the number of stable bonds to the total number of bonds) is decreased from its initial value of $P = 0.8$ to $P \approx 0.53$ during the course of deformation (as shown in Figure 1.5b). These data support the contention that reforming of the labile bonds plays a crucial role in maintaining the stability of dual crosslinked samples.

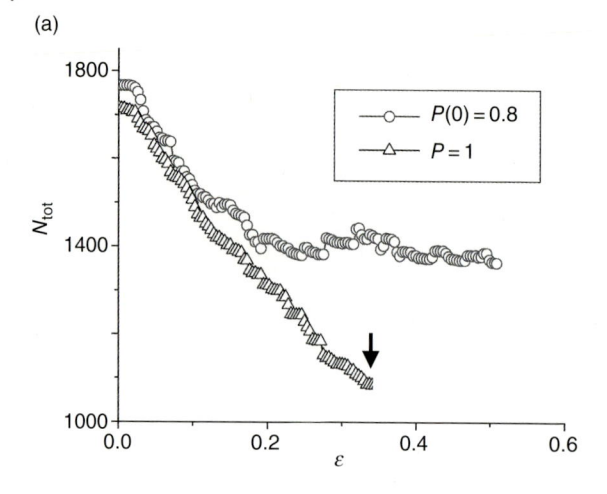

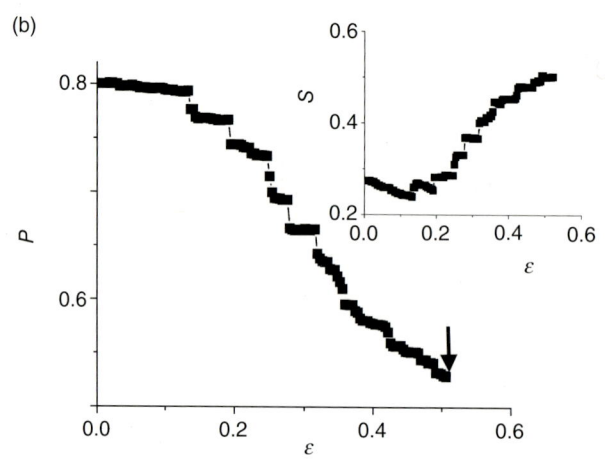

Figure 1.5 (a) Changes in the total number of bonds in the sample at the initial crosslinking ratio $P(0) = 0.8$ (circles) and at $P = 1.0$ (triangles) during the deformation shown in Figure 1.4; (b) Dependence of P on strain ε, calculated for the same simulation. Inset: Saturation s in the labile bond network as a function of strain. The vertical arrows in (a) and (b) mark the strain at which the sample fractured.

1.3.3
Tensile Strength of Nanogel Samples

Finally, an investigation was made into how the stability of the samples depends on V_t, the rate of the tensile deformation. Calculations were performed for rates V_t in the range from 10^{-4} through $10^{-2} \times d/\tau$. At any given rate, a stress–strain curve similar to that shown in Figure 1.4 was calculated for the largest sample with

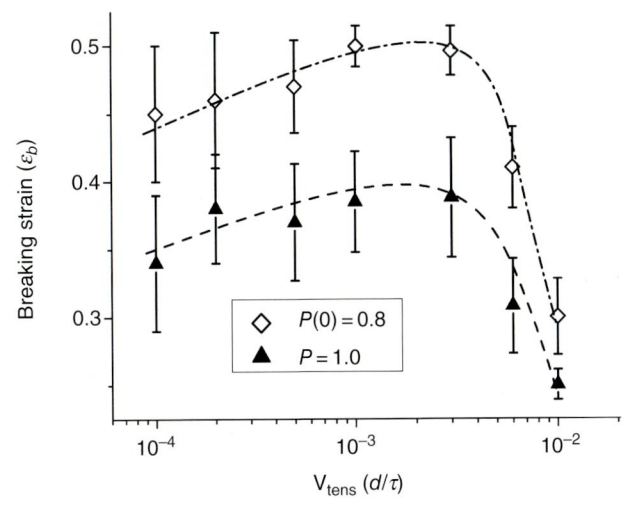

Figure 1.6 Dependence of σ_b on tensile rate, shown on a logarithmic scale for the largest sample at $P(0) = 0.8$ (diamonds) and $P = 1.0$ (triangles).

a randomly placed crack on its surface. The strain at which the sample fractured into two pieces, ε_b, was determined in each simulation as the strain at which the stress–strain curve dropped sharply to zero. The results obtained for ε_b were averaged over eight independent simulations made with different positions of the crack. The results for the $P = 0.8$ and $P = 1.0$ samples are summarized in Figure 1.6, where the points show the averaged values for ε_b and the error bars mark the standard deviations. The plots reveal a maximum at a tensile rate of $V_t \sim 2 \times 10^{-3} \times d/\tau$ for both dual and permanently crosslinked samples. Note that a maximum in the dependence of ε_b on the tensile rate is known to occur for permanently crosslinked elastomers in the regime where viscoelastic effects are important [33]. It is also evident from Figure 1.6 that, at any tensile rate, the ε_b calculated for the $P = 0.8$ sample was from ~20% (at the fastest tensile rates $V_t \geq 6 \times 10^{-3} \times d/\tau$) to 30% (at slower tensile rates, $V_t < 6 \times 10^{-3} \times d/\tau$) higher than that for the $P = 1.0$ sample. This was in accordance with the results described above for the calculations at constant velocity. Of note, it follows from Figure 1.6 that the stress–strain curves shown in Figure 1.4 were computed for conditions near the maximum of the $\varepsilon_b(V_t)$ dependence.

1.3.4 Modeling Viscoelastic Nanogel Particles

In Section 1.3.3, the individual nanogel particles were modeled as purely elastic objects, and the relaxation processes within the system were due solely to the rearrangement of bonds interconnecting the nanogels. In order to capture viscoelastic behavior, the gel lattice spring model (gLSM) [34] was introduced into the computational framework, so as to allow a generalization of the methodology to a broader range of materials. To formulate this gLSM, we first determined the energy density within the deformed material through the use of a phenomenological model of viscoelasticity. A finite element approximation was then employed to describe the deformation field in terms of a set of nodal coordinates. Finally, Newton's second law was used to derive the equations of motion for the nodal points. Each of these steps is described below.

First, an isotropic solid body that undergoes a time-dependent deformation $\mathbf{X} \to \mathbf{x}(\mathbf{X},t)$ is considered. Here, $\mathbf{X}$ represents the coordinates of a point within the material in the initial, undeformed state, whereas $\mathbf{x}$ is the position of the same point upon deformation at time t. The local strain is characterized by the Finger tensor [35] $\hat{\mathbf{B}}(t) = \hat{\mathbf{F}}(t) \cdot \hat{\mathbf{F}}^T(t)$, where

$$[\hat{\mathbf{F}}(t)]_{ij} = \frac{\partial x_i(\mathbf{X},t)}{\partial X_j}, \quad i,j = 1,2,3, \tag{1.6}$$

is the deformation-gradient tensor, and the superscript "T" represents the transposition operation. In purely elastic solids, the local stresses at a time t depend on the local strains at the same time t; that is, the elastic stress tensor $\hat{\boldsymbol{\sigma}}_{el}(t)$ is a function of the strain tensor $\hat{\mathbf{B}}(t)$. If, however, the solid is viscoelastic, then the local stresses depend on the deformation history. Hence, the stress tensor $\hat{\boldsymbol{\sigma}}^*(t)$ is a *functional* of the relative strain tensor $\hat{\mathbf{b}}(t,t')$, which characterizes deformations in the body at time t relative to the state of the body at $t' \leq t$ [36]. The relative strain tensor can be determined through the following decomposition:

$$\hat{\mathbf{b}}(t,t') = \hat{\mathbf{F}}(t) \cdot \hat{\mathbf{C}}^{-1}(t') \cdot \hat{\mathbf{F}}^T(t), \tag{1.7}$$

where $\hat{\mathbf{C}}(t) = \hat{\mathbf{F}}^T(t) \cdot \hat{\mathbf{F}}(t)$ is the left Cauchy–Green strain tensor; $\hat{\mathbf{b}}(t,0) = \hat{\mathbf{B}}(t)$, since the body is assumed to be undeformed at $t' = 0$.

The constitutive equation (i.e., the stress–strain relationship) can be determined if the energy dependence on the strain is known. It is assumed that the strain energy density, U, which is defined per unit volume of unstrained material, is represented as a sum of two contributions: $U = U_{el} + U^*$, where U_{el} describes the purely elastic deformations and U^* is the viscoelastic contribution to the strain energy. Correspondingly, the stress tensor also consists of the two contributions: $\hat{\boldsymbol{\sigma}}(t) = \hat{\boldsymbol{\sigma}}_{el}(t) + \hat{\boldsymbol{\sigma}}^*(t)$. The dependence of the stress tensors $\hat{\boldsymbol{\sigma}}_{el}(t)$ and $\hat{\boldsymbol{\sigma}}^*(t)$ on the respective strain tensors $\hat{\mathbf{B}}(t)$ and $\hat{\mathbf{b}}(t,t')$ is determined by the choice of U_{el} and U^*.

The elastic energy density U_{el} depends only on $\hat{\mathbf{B}}$ through the invariants of this tensor I_i, $i = 1,2,3$, that is, $U_{el} = U_{el}(I_1, I_2, I_3)$ [35]. The invariants are calculated as follows:

$$I_1 = \mathrm{tr}\,\hat{\mathbf{B}}, \quad I_2 = \frac{1}{2}[(\mathrm{tr}\,\hat{\mathbf{B}})^2 - \mathrm{tr}(\hat{\mathbf{B}}^2)], \quad I_3 = \det \hat{\mathbf{B}}. \tag{1.8}$$

It is worth noting that $I_3^{1/2} = dV/dV_0$ is the relative volumetric change in a material element due to the deformation, where dV and dV_0 are the element volumes in the deformed and undeformed states, respectively. Below, the notation $J = I_3^{1/2}$ will also be used. The constitutive equation for the purely elastic stress can written in the following general form: [35]

$$\hat{\sigma}_{el} = 2I_3^{-1/2}(w_2 I_2 + w_3 I_3)\hat{I} + 2I_3^{-1/2} w_1 \hat{B} - 2I_3^{1/2} w_2 \hat{B}^{-1}, \qquad (1.9)$$

where $\hat{I}$ is a unit tensor and

$$w_i = \frac{\partial}{\partial I_i} U_{el}(I_1, I_2, I_3), \quad i = 1, 2, 3. \qquad (1.10)$$

To specify the elastic contribution to the strain energy of the nanogels, the so-called "neo-Hookean compressible material" model is employed [36, 37]. Within this model, the elastic strain energy depends only on I_1 and I_3, and has the following form:

$$U_{el} = 1/2 c_0 (I_1 - 3) + 1/2 K \log^2(J). \qquad (1.11)$$

The first term on the right-hand-side of Equation 1.11 depends only on I_1, and describes the contribution from the shear deformations. The second term on the right-hand-side of Equation 1.11 is the energy of bulk deformations, that is a function of the volumetric change J. The c_0 and K are the model parameters, which are proportional to the shear and bulk moduli, respectively. Note that $K \gg c_0$, since the bulk modulus is usually much greater that the shear modulus. The substitution of Equation 1.11 into Equation 1.9 results in the following equation for the purely elastic stress contribution:

$$\hat{\sigma}_{el}(t) = c_0 J^{-1}(t)\hat{B}(t) + K J^{-1}(t)\log[J(t)]\hat{I} \qquad (1.12)$$

The general form of the viscoelastic contribution to the energy density U^* also depends only on the invariants $I_i(t,t')$, $i = 1,2,3$, of the relative stress tensor $\hat{b}(t,t')$. In polymeric materials, the shear deformations exhibit strong relaxation effects, whereas the bulk deformations are essentially purely elastic. Therefore, it can be assumed that U^* only includes the contribution from the relaxing shear stresses. The latter contribution can be generalized from the elastic neo-Hookean term in Equation 1.11 (the first term that depends on $I_1(t)$) to the case of viscoelastic behavior, so that U^* depends on the first invariant $I_1(t,t')$ of the relative strain tensor. The following simple generalization of the neo-Hookean term is utilized [36]:

$$U^* = 1/2 \chi(t,0)[I_1(t,0) - 3] + 1/2 \int_0^t \frac{\partial \chi}{\partial t'}(t,t')[I_1(t,t') - 3] dt'. \qquad (1.13)$$

Here, $\chi(t,t')$ gives the viscoelastic strain energy generated at time t' that remains unrelaxed at time $t \geq t'$. The corresponding constitutive equation for the viscoelastic stress is [36]:

$$\hat{\boldsymbol{\sigma}}^*(t) = \chi(t,0) J^{-1}(t) \hat{\mathbf{b}}(t,0) + J^{-1}(t) \int_0^t \frac{\partial \chi}{\partial t'}(t,t') \hat{\mathbf{b}}(t,t') dt'. \tag{1.14}$$

The viscoelastic behavior of the nanogels is modeled by assuming a simple exponential relaxation:

$$\chi(t,t') = c_0^* \exp[-(t-t')/\tau_R], \tag{1.15}$$

where c_0^* contributes to the unrelaxed shear modulus and τ_R is the relaxation time.

After the stress–strain relationships (Equations 1.12 and 1.14) have been specified, the dynamics of each nanogel particle is described by the following continuum equation:

$$\rho[\partial_t \mathbf{v} + (\mathbf{v} \cdot \nabla)\mathbf{v}] = \nabla \cdot (\boldsymbol{\sigma}_{el} + \boldsymbol{\sigma}^*). \tag{1.16}$$

Here, ρ is the mass density, which depends on the volumetric changes J, and $\rho = \rho_0 J^{-1}$, where ρ_0 is the nanogel density in the undeformed state. For each nanogel particle, Equation 1.16 is subject to boundary conditions due to the interparticle interactions.

The numerical integration of Equation 1.16 can be readily performed using the gLSM; this entails approximating a nanogel particle by a number of finite elements and solving the equations of motion for the nodal points of the elements. The gLSM approach is illustrated by considering the 2-D nanogel particle shown in Figure 1.7a. The nanogel is modeled as a three-dimensional (3-D) particle confined in a slit of thickness $H = \lambda_\perp H_0$, where H_0 is the particle height in the undeformed state, and $\lambda_\perp$ is the uniform compressive strain imposed on the particle in the direction perpendicular to the slit surface. It is assumed that motion of the particle along the slit surface is frictionless, so that the particle dynamics can be considered as purely 2-D.

The shape of the hexagonal nanogel particle shown in Figure 1.7a is best captured by six equal triangular finite elements, each of which is labeled by the integer number $m = 1,2,\ldots,6$ (see Figure 1.7a). The nodes within the element m are labeled by $n = 1,2,3$ in the counter-clockwise direction (see Figure 1.7b). The position of a node is given by $\mathbf{x}_n(m)$. In the undeformed state, the elements are assumed to have a uniform density ρ_0, and the elemental area is A_0. In the finite element approximation, the total energy of a system is equal to the sum of the energies of the elements; that is, $W = \int U dV_0 \approx H_0 A_0 \sum_m U(m)$, and each elemental energy $U(m)$ is expressed in terms of the nodal coordinates. The force acting on the node n belonging to the element m is obtained by differentiating W with respect to $\mathbf{x}_n(m)$. The elemental contribution to the equation of motion is then written as:

$$1/3 \rho_0 \frac{d^2}{dt^2} \mathbf{x}_n(m) = -\frac{\partial U(m)}{\partial \mathbf{x}_n(m)}. \tag{1.17}$$

If a node is common to several adjacent elements, then the equation of motion for the node is obtained by summation of the elemental contributions given by Equation 1.17.

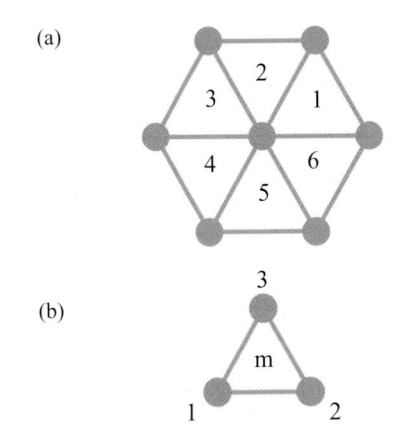

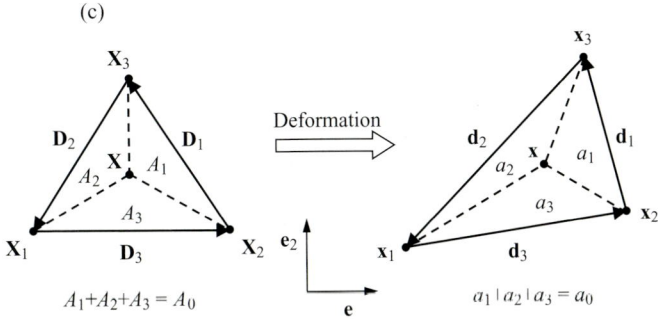

Figure 1.7 (a) Schematic of a viscoelastic nanogel cluster approximated by six triangular elements. Each element is labeled by an integer number m = 1, 2, ... , 6; (b) Notation for the node labeling in an *m*th viscoelastic triangular element. The nodes within the element are labeled as 1, 2, or 3, where the node shared by all elements in the center of the cluster is marked as 1, and the rest of the nodes are marked sequentially in a counterclockwise direction; (c) Triangular element in undeformed (left) and deformed (right) states. **X** is an arbitrary point within the unit. The vectors $\mathbf{e}_i$ define the 2-D reference frame used in the simulations. A_i and a_i are the areas of the triangles (as shown in the figure), and A_0 and a_0 are the total area of the unit before and after deformation.

First, let us consider the triangular element m in the hexagonal gel (below, the element label is omitted for brevity). As indicated in Figure 1.7c, the element nodes have the coordinates $\mathbf{X}_n$, $n = 1,2,3$ in the undeformed state. The edge vectors $\mathbf{D}_n$, $n = 1,2,3$, are also introduced, where

$$\mathbf{D}_1 = \mathbf{X}_3 - \mathbf{X}_2, \mathbf{D}_2 = \mathbf{X}_1 - \mathbf{X}_3, \mathbf{D}_3 = \mathbf{X}_2 - \mathbf{X}_1, \tag{1.18}$$

so that the edge n is located opposite to the node n (see Figure 1.7c). Note that $\sum_n \mathbf{D}_n = 0$. Any point **X** within the triangle can be uniquely parameterized using the local triangular coordinates L_n [38]:

$$\mathbf{X} = \sum_{n=1}^{3} L_n \mathbf{X}_n, \tag{1.19}$$

The values of L_n, which are also known as the shape functions, are defined as $L_n = A_n / A_0$, where A_n is the area of the triangle formed by the point $\mathbf{X}$ and the end points of the edge n, and $\sum_n A_n = A_0$ is the total area of the undeformed triangle. The value of A_n can be determined through the coordinates of the point $\mathbf{X}$ and the nodal coordinates as:

$$A_n = 1/2 \|\mathbf{D}_n \times (\mathbf{X} - \mathbf{X}_{n-1})\|. \tag{1.20}$$

The above definitions are illustrated in Figure 1.7c.

Displacement of the nodal points $\mathbf{X}_n \to \mathbf{x}_n(t)$ results in deformation of the interior of the element, and the areas A_n also change, $A_n \to a_n$ (as illustrated in Figure 1.7c). The total area of the deformed triangle is $a_0 = \sum_n a_n$. The deformation $\mathbf{X} \to \mathbf{x}(\mathbf{X})$ is then approximated by linear functions of $\mathbf{X}$. In this case, $a_n / a_0 = A_n / A_0 = L_n$, so the position of the point $\mathbf{x}(\mathbf{X})$ in the deformed triangle is characterized by the same triangular coordinates L_n as for the point $\mathbf{X}$ of the undeformed element (Figure 1.7c):

$$\mathbf{x}(\mathbf{X}, t) = \sum_{n=1}^{3} L_n(\mathbf{X}) \mathbf{x}_n(t). \tag{1.21}$$

The above equation indicates explicitly that $L_n = L_n(\mathbf{X})$, according to Equation 1.20.

By using Equation 1.21 to approximate the deformation field within a triangular element, it is possible to determine the element strain–energy density U as a function of the nodal points. To facilitate this computation, the base vectors $\mathbf{g}_i$ are introduced as

$$\mathbf{g}_i = \frac{\partial \mathbf{x}}{\partial X_i}, \; i = 1, 2, 3, \tag{1.22}$$

and the matrix element for the left Cauchy–Green tensor is given in terms of $\mathbf{g}_i$ as

$$[\mathbf{C}(t)]_{ij} = \mathbf{g}_i(t) \cdot \mathbf{g}_j(t) \text{ and } [\mathbf{C}^{-1}(t)]_{ij} = \varepsilon_{ikl} \varepsilon_{jmn} \frac{(\mathbf{g}_k \times \mathbf{g}_l) \cdot (\mathbf{g}_m \times \mathbf{g}_n)}{[\mathbf{g}_1 \cdot (\mathbf{g}_2 \times \mathbf{g}_3)]^2}, \tag{1.23}$$

where ε_{ijk} is the Levi–Civita tensor. According to Equations 1.21 and 1.22, the base vectors depend on the nodal coordinates as

$$\mathbf{g}_i(t) = \sum_{n=1}^{3} \frac{\partial L_n}{\partial X_i} \mathbf{x}_n(t). \tag{1.24}$$

The time dependence of the base vectors can be expressed in terms of the instantaneous values of the edge vectors $\mathbf{d}_n$ as

$$\mathbf{g}_1(t) = \frac{1}{2A_0} [(\mathbf{D}_2 \cdot \mathbf{e}_2) \mathbf{d}_1(t) - (\mathbf{D}_1 \cdot \mathbf{e}_2) \mathbf{d}_2(t)],$$

$$\mathbf{g}_2(t) = -\frac{1}{2A_0} [(\mathbf{D}_2 \cdot \mathbf{e}_1) \mathbf{d}_1(t) - (\mathbf{D}_1 \cdot \mathbf{e}_1) \mathbf{d}_2(t)], \tag{1.25}$$

$$\mathbf{g}_3(t) = \lambda_\perp \mathbf{e}_3,$$

where $\mathbf{d}_n(t) = \mathbf{x}_{n+2}(t) - \mathbf{x}_{n+1}(t)$ and $\mathbf{D}_n$ are defined by Equation 1.18. The invariants of the strain tensors are calculated by substituting Equation 1.25 into Equation 1.23 to obtain

$$I_1(t) \equiv \mathbf{g}_1^2 + \mathbf{g}_2^2 + \mathbf{g}_3^2 = \sum_{n,m} \Gamma_{nm}^{(0)} \mathbf{x}_n(t) \cdot \mathbf{x}_m(t), \tag{1.26}$$

$$I_1(t,t') \equiv \mathrm{tr}[\mathbf{C}(t) \cdot \mathbf{C}^{-1}(t')] = \sum_{n,m} \Gamma_{nm}(t') \mathbf{x}_n(t) \cdot \mathbf{x}_m(t). \tag{1.27}$$

The matrix elements $\Gamma_{nm}^{(0)}$ depend on the element shape in the equilibrium state, whereas the values $\Gamma_{nm}(t')$ retain information concerning the deformation history, namely:

$$\Gamma_{nm}^{(0)} = \frac{\mathbf{D}_n \cdot \mathbf{D}_m}{(2A_0)^2}, \quad \Gamma_{nm}(t) = \frac{\mathbf{d}_n(t) \cdot \mathbf{d}_m(t)}{[\mathbf{d}_n(t) \times \mathbf{d}_m(t)]^2}. \tag{1.28}$$

Similarly, the relative volume ratio between the deformed and undeformed gel, J, can be calculated as

$$J(t) \equiv \mathbf{g}_1(t) \cdot [\mathbf{g}_2(t) \times \mathbf{g}_3(t)] = \lambda_\perp (2A_0)^{-1} \mathbf{e}_3 \cdot [\mathbf{d}_1(t) \times \mathbf{d}_2(t)]. \tag{1.29}$$

The substitution of Equations 1.26, 1.27 and 1.29 into Equations 1.11, 1.13 and 1.16 yields the elemental energy density as a function of the nodal coordinates:

$$U(t) = c_0 \sum_{nm} \Gamma_{nm}^{(0)} (\mathbf{x}_n(t) \cdot \mathbf{x}_m(t)) + 1/2 K \log^2[\lambda_\perp (2A_0)^{-1} \mathbf{e}_3 \cdot (\mathbf{d}_1(t) \times \mathbf{d}_2(t))]$$
$$+ c^* \sum_{nm} \theta_{nm}(t)(\mathbf{x}_n(t) \cdot \mathbf{x}_m(t)) \tag{1.30}$$

Here, the dimensionless functions $\theta_{nm}(t)$ describe the relaxation processes and are obtained by solving the following rate equations:

$$\tau_R \frac{d\theta_{mn}(t)}{dt} = -\theta_{mn}(t) + \Gamma_{mn}(t), \tag{1.31}$$

where τ_R is the relaxation time, and $\Gamma_{mn}(t)$ is determined by Equation 1.28. Equation 1.31 was obtained by differentiating Equation 1.13, with Equation 1.15 taken into account. Finally, the force on the node n of the element m is calculated according to Equation 1.17.

It can now be demonstrated that the formulation developed above captures the creep and stress relaxation behavior that is characteristic of viscoelastic solids. Below are presented the results of computer simulations performed for the triangular finite element and the entire hexagonal nanogel particle. The dynamic equations were transformed to the dimensionless form using the length and time scales of $L_0 = \Delta$ and $T_0 = (\rho_0 \Delta^2 / c_0)^{1/2}$, respectively, where Δ is the lateral size of the

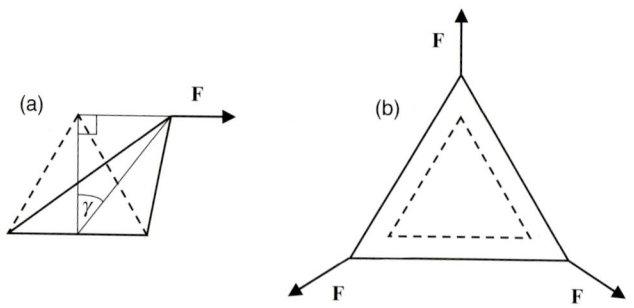

Figure 1.8 Scheme depicting (a) shear and (b) bulk modes of deformation of a triangular element shown in Figure 1.7b. The equilibrium shape of the element is shown by the dashed lines.

undeformed triangular element. The model parameters were chosen to be $\lambda_\perp = 1$, $K/c_0 = 4$, $c_0^*/c_0 = 1$, and $\tau_R = 0.1$. The undeformed, relaxed configuration was used for the initial condition. The material's behavior was tested for the bulk and shear deformation modes. For the triangular element, the deformation modes are shown schematically in Figure 1.8. The bulk and shear deformations of the hexagonal particle were introduced in a similar manner.

The creep behavior of the triangular element was simulated by applying forces of $\|\mathbf{F}\| = 10$ to each of the element nodes, as shown in Figure 1.8. The dynamic equations were then solved to determine the time-dependent strain $\varepsilon(t)$; the results are shown in Figure 1.9a. The shear strain is indicated by the solid line and was defined as $\varepsilon_{sh}(t) = \tan\gamma(t)$, where γ is the shear angle (see Figure 1.8a). The bulk strain is shown by the dashed line, and was calculated as $\varepsilon_b(t) = A(t)/A_0 - 1$, where $A(t)$ and A_0 are the element areas in the deformed and undeformed states, respectively. Figure 1.9a demonstrates that, due to the viscoelasticity of the material, the application of the external force leads to a gradual build-up of the strain.

The stress relaxation behavior of the triangular element was obtained by determining the forces that develop in the material after an instantaneous deformation, which then is kept constant. Figure 1.9b shows the results of simulations at the instantaneous shear and bulk strains of $\varepsilon_{sh} = 0.207$ and $\varepsilon_b = 0.718$, respectively. Figure 1.9b shows that the nodal force, $F(t)$, acquires its maximum value $F(0)$ at an initial moment of deformation, and then decreases as the viscoelastic relaxation takes place. It can be seen in Figure 1.9b that the shear stress relaxation is noticeably stronger than the bulk stress relaxation.

When six triangular elements are put together to form a hexagonal particle, the resulting particle "inherits" the viscoelastic properties of the constituent elements. Figure 1.10 shows the creep behavior of the hexagonal particle after the shear and bulk stresses are applied. To model the creep behavior under shear, the two bottom nodes were pinned to their equilibrium positions, the shear forces of $\|\mathbf{F}\| = 10$ were applied to the two top nodes, and the particle height and distance between the

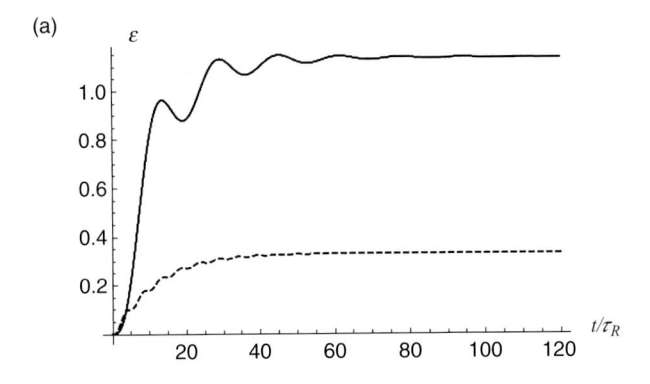

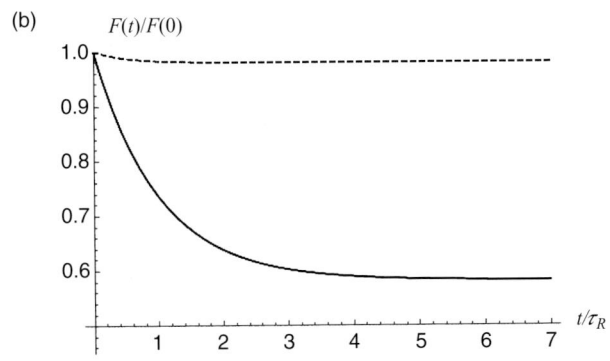

Figure 1.9 (a) Creep and (b) stress relaxation of a triangular element under the shear (solid curves) and bulk (dashed curves) deformations (see text for notations); τ_R is the relaxation time, as defined in Equation 1.15.

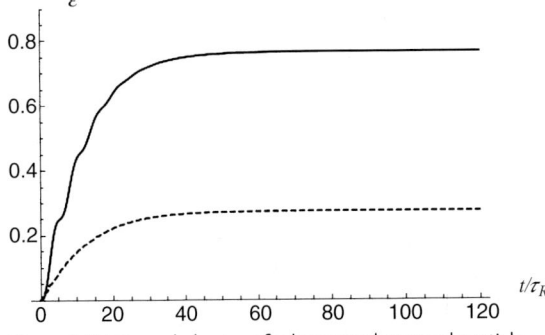

Figure 1.10 Creep behavior of a hexagonal nanogel particle (Figure 1.7a) under the shear (solid curve) and bulk (dashed curve) modes of deformation.

top nodes were kept constant. The creep behavior under dilatation was modeled by applying the pressure of $p = 10$ that pulls the surface nodes outwards. (The pressure p was computed as the value of the applied force divided by the current length of an edge.) By comparing Figures 1.9a and 1.10, it can be seen that the individual triangular elements and the hexagonal particle exhibit similar viscoelastic behaviors.

The approach outlined above provides a powerful method for modeling the behavior of deformable materials that encompass viscoelastic behavior. In future studies, this model will be built on to determine how the viscoelasticity of the nanogels affects the macroscopic response of the dual crosslinked material described in this chapter.

1.4
Conclusions

To summarize, the aim of the present studies was to demonstrate how computational modeling can be used to design self-healing materials and coatings. To that end, previous studies on nanoscopic gel particles that are interconnected into a microscopic network by both stable and labile bonds were reviewed. New calculations for modeling the viscoelastic behavior of the individual nanogel particles were also described.

To demonstrate the self-healing behavior of dual crosslinked polymeric materials, the response of a network of deformable nanogels to a tensile stress was modeled; this showed that the introduction of a small fraction of labile crosslinks can lead to dramatic improvements in the strength of a material. The rapid reforming of these labile bonds provides the structural rearrangement that preserves the mechanical integrity of the sample.

Analogies can drawn with other experimental systems that indicate the validity of these predictions. For example, it is useful to recall the polydisulfide chains that contribute to the unique properties of rubber [39]. In particular, the reshuffling of the labile S–S bonds in the polysulfide crosslinks as the rubber is deformed is what contributes to the toughness of this material [40]. Recently, investigations have shown that polymer chains which encompass a significant fraction of hydrogen bonds can also undergo a rapid structural rearrangement due to bond breaking and remaking that imparts self-healing properties to the bulk material [3].

Acknowledgments

The authors gratefully acknowledge financial support from the DOE (for partial support of G.V.K. and S.F.D.), and ONR (for the partial support of V.V.Y.). G.V.K. also acknowledges partial support from NSF through TeraGrid resources provided by NCSA.

References

1 Caruso, M.M., Davis, D.A., Shen, Q., Odom, S.A., Sottos, N.R., White, S.R. and Moore, J.S. (2009) Mechanically-induced chemical changes in polymeric materials. *Chemical Reviews*, **109**, 5755–98.

2 Chen, X., Dam, M.A., Ono, K., Mal, A., Shen, H., Nutt, S.R., Sheran, K. and Wudl, F.A. (2002) Thermally re-mendable cross-linked polymeric material. *Science*, **295**, 1698–702.

3 Cordier, P., Tournilhac, F., Soulie-Ziakovic, C. and Leibler, L. (2008) Self-healing and thermoreversible rubber from supramolecular assembly. *Nature*, **451**, 977–80.

4 Amendola, V. and Meneghetti, M. (2009) Self-healing at the nanoscale. *Nanoscale*, **1**, 74–88.

5 Trask, R.S., Williams, H.R. and Bond, I.P. (2007) Self-healing polymer composites: mimicking nature to enhance performance. *Bioinspiration and Biomimetics*, **2**, P1–9.

6 Balazs, A.C. (2007) Modeling self-healing materials. *Materials Today*, **10**, 18–23.

7 Wool, R.P. (2008) Self-healing materials: a review. *Soft Matter*, **4**, 400–18.

8 Wu, D.Y., Meure, S. and Solomon, D. (2008) Self-healing polymeric materials: a review of recent developments. *Progress in Polymer Science*, **33**, 479–522.

9 Hickenboth, C.R. *et al.* (2007) Biasing reaction pathways with mechanical force. *Nature*, **446**, 423–7.

10 Kolmakov, G.V., Matyjaszewski, K. and Balazs, A.C. (2009) Harnessing labile bonds between nanogel particles to create self-healing materials. *ACS Nano*, **3**, 885–92.

11 Min, K. and Matyjaszewski, K. (2005) Atom-transfer radical atom polymerization in microemulsion. *Macromolecules*, **38**, 8131–4.

12 Min, K., Gao, H. and Matyjaszewski, K. (2006) Development of an ab initio emulsion transfer radical polymerization: from microemulsion to emulsion. *Journal of the American Chemical Society*, **128**, 10521–6.

13 Smith, B.L., Schaffer, T.E., Viani, M., Thompson, J.B., Frederick, N.A., Kindt, J., Belcher, A., Stucky, G.D., Morse, D.E. and Hansma, P.K. (1999) Molecular mechanistic origin of the toughness of natural adhesive, fibres and composites. *Nature*, **399**, 761–3.

14 Kolmakov, G.V., Revanur, R., Tangirala, R., Emrick, T., Russell, T.P., Crosby, A.J. and Balazs, A.B. (2010) Using nanoparticle-filled microcapsules for site-specific healing of damaged substrates: creating a "repair-and-go" system. *ACS Nano*, **4**, 1115–23.

15 Therriault, D., White, S.R. and Lewis, J.A. (2003) Chaotic mixing in three-dimensional microvascular networks fabricated by direct-write assembly. *Nature Materials*, **2**, 265–71.

16 Ashurst, W.T. and Hoover, W.G. (1976) Microscopic fracture studies in the two-dimensional triangular lattice. *Physical Review B*, **14**, 1465–73.

17 Buxton, G.A., Care, C.M. and Cleaver, D.J. (2001) A lattice spring model of heterogeneous materials with plasticity. *Modelling and Simulation in Materials Science and Engineering*, **9**, 485–97.

18 Buxton, G.A. and Balazs, A.C. (2004) Modeling the dynamic fracture of polymer blends processed under shear. *Physical Review B*, **69**, 054101.

19 Yashin, V.V. and Balazs, A.C. (2006) Pattern formation and shape changes in self-oscillating polymer gels. *Science*, **314**, 798–801.

20 Kuksenok, O., Yashin, V.V. and Balazs, A.C. (2007) Mechanically induced chemical oscillations and motion in responsive gels. *Soft Matter*, **3**, 1138–44.

21 Bell, G.I. (1978) Models for the specific adhesion of cells to cells. *Science*, **200**, 618–27.

22 Chang, K.C., Tees, D.F.J. and Hammer, D.A. (2000) The state diagram for cell adhesion under flow: leukocyte rolling and firm adhesion. *Proceedings of the National Academy of Sciences of the United States of America*, **97**, 11262–7.

23 Wiita, A.P., Ainavarapu, S.R.K., Huang, H.H. and Fernandez, J.M. (2006) Force-dependent chemical kinetics of disulfide bond reduction observed with

single-molecule techniques. *Proceedings of the National Academy of Sciences of the United States of America*, **103**, 7222–7.
24 Bell, G.I., Dembo, M. and Bongrand, P. (1984) Cell adhesion. Competition between nonspecific repulsion and specific bonding. *Biophysical Journal*, **45**, 1051–64.
25 King, M.R. and Hammer, D.A. (2001) Multiparticle adhesive dynamics: hydrodynamic recruitment of rolling leukocytes. *Proceedings of the National Academy of Sciences of the United States of America*, **98**, 14919–24.
26 Bhatia, S.K., King, M.R. and Hammer, D.A. (2003) The state diagram for cell adhesion mediated by two receptors. *Biophysical Journal*, **84**, 2671–90.
27 Sanderson, R.T. (1976) *Chemical Bonds and Bond Energy*, 2nd edn, Academic Press, New York.
28 Lawn, B.R. (1993) *Fracture of Brittle Solids*, 2nd edn, Cambridge University Press, New York.
29 Dirama, T.E., Varshney, V., Anderson, K.L., Shumaker, J.A. and Johnson, J.A. (2008) Coarse-grained molecular dynamics simulations of ionic polymer networks. *Mechanics of Time-Dependent Materials*, **12**, 205–20.
30 Tsige, M., Lorentz, C.D. and Stevens, M.J. (2004) Role of network connectivity on the mechanical properties of highly cross-linked polymers. *Macromolecules*, **37**, 8466–72.
31 Budinski, K.G. and Budinski, M.K. (2004) *Engineering Materials: Properties and Selection*, 8th edn, Prentice-Hall, Columbus.
32 Mukherji, D. and Abrams, C.F. (2008) Microvoid formation and strain hardening in highly cross-linked polymer networks. *Physical Review E*, **78**, 050801(R).
33 Gents, A.N. (1994) Strength of elastomers, *Science and Technology of Rubber*, 2nd edn (eds J.E. Mark, B. Erman and F.R. Eirich), Academic Press, New York, pp. 471–512.
34 Yashin, V.V. and Balazs, A.C. (2007) Theoretical and computational modeling of self-oscillating polymer gels. *Journal of Chemical Physics*, **126**, 124707.
35 Atkin, R.J. and Fox, N. (1980) *An Introduction to the Theory of Elasticity*, Longman, New York.
36 Drozdov, A.D. (1996) *Finite Elasticity and Viscoelasticity: A Course in the Nonlinear Mechanics of Solids*, World Scientific, Singapore.
37 Bonet, J. and Wood, R.D. (1997) *Nonlinear Continuum Mechanics for Finite Element Analysis*, Cambridge University Press, New York.
38 Zienkiewicz, O.C. (1977) *The Finite Element Method*, 3rd edn, McGraw-Hill, London.
39 Tobolsky, A.V. and MacKnight, W.J. (1965) *Polymeric Sulfur and Related Polymers*, Wiley-Interscience, New York.
40 Aklonis, J.J. and McKnight, W.J. (1983) *Introduction to Polymer Viscoelasticity*, 2nd edn, Wiley-Interscience, New York.

2
Synthesis and Characterization of Polymeric Nanogels

Yoshifumi Amamoto, Hideyuki Otsuka and Atsushi Takahara

2.1
Introduction

Polymeric nanogels have been widely studied from both scientific and engineering perspectives, because their unique structures produce some interesting properties. Their crosslinked nanometer-scale structures enable them to capture some substrates such as biomolecules into the nanosize gels, and then to release them by external stimulation. Moreover, the properties of nanogels are different from those of macroscopic gels; typically, they have faster swelling and shrinking rates. Consequently, many synthetic methods have been developed to prepare various types of nanogel, and a wide range of applications have been investigated, including drug-delivery systems [1–3], microreactors [4], oil removal [5], sensors [6, 7], and glass-reinforcing agents [8]. In order to design the structures of nanogels, and to identify satisfied functionalities, it is important to acquire an understanding of the methods used in their synthesis. In addition, their characterization provides important information about their structures. In this chapter, recent progress in polymeric nanogels is reviewed from the perspectives of their synthesis and characterization. Although, the applications of these materials are not described in detail at this point, a plethora of information on this subject is available elsewhere [1–8].

To date, a large number of methods have been reported on the synthesis and characterization of nanogels, and it is difficult to review them all; nonetheless, some important techniques are discussed in this chapter. Initially, the synthetic methods are categorized from two aspects, namely chemical and physical nanogel synthesis (Figure 2.1). Chemical nanogels with degradable natures or decrosslinking abilities are also introduced as a recent development. Because it is more meaningful to appreciate the synthetic methods used to obtain target structures and sizes, they are introduced systematically. Methods used for the characterization of nanogels are introduced briefly later in the chapter.

Recent reports have employed a variety of terms when referring to a nanometer-sized gel, including microgel, nanoparticle, core crosslinked star polymer, and

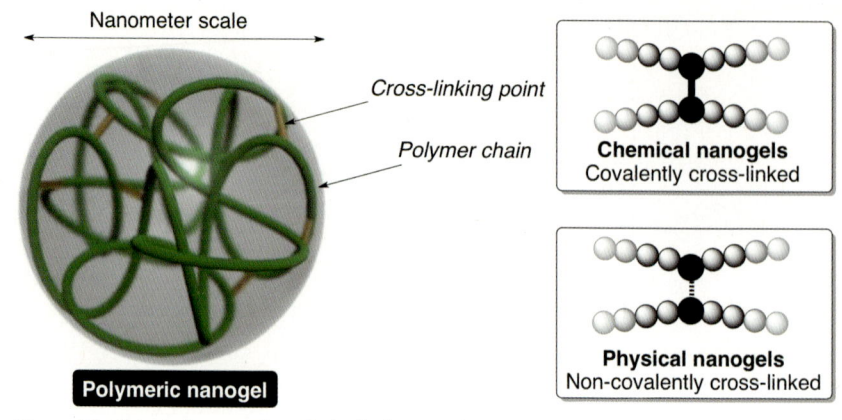

Figure 2.1 Chemical nanogels and physical nanogels.

crosslinked micelle. According to an IUPAC report, a "nanogel" is defined as a particle of gel of any shape with an equivalent diameter of between approximately 1 and 100 nm [9]. Although gel particles ranging from 100 nm to 100 μm in size were defined as microgels [9], those with diameters up to several hundred nanometers are also introduced in this chapter.

2.2
Synthesis of Polymeric Nanogels

The methods used to synthesize nanogels can be categorized according to features such as size, polymerization method, and nanometer scale-making. In this section, the methods used to prepare gels and control their nanostructures are explained in systematic fashion.

The methods usually are of two types, based on their crosslinking structures. The first crosslinking method employs covalent bonds, while the second method uses noncovalent bonds such as ionic bonds, hydrogen bonds, and coordination bonds. The gels obtained by these methods are defined as *chemical gels* and *physical gels*, respectively. In the case of chemical gels, the gel structures are usually stable and rigid, and difficult to exchange. However, in the case of physical gels, sol–gel transitions can proceed easily by external stimulation; in other words, physical gels can collapse easily as a result of changes in their environment. Recently, chemical gels with degradable abilities have been developed by using reversible covalent bonds; these possess not only stable structures but also de-crosslinking abilities that respond to specific stimuli. The synthetic methods used to create polymeric nanogels are described in this section, and the various details of chemical nanogels, physical nanogels, and degradable chemical nanogels are discussed.

For the synthesis of nanoscale structures, it is important to control not only their size but also their shape. There are two major approaches for creating nanometer-scale particles: (i) the "top-down" approach, in which mechanical methods such as lithography are used; and (ii) the "bottom-up" approach, in which the nanometer scale is controlled by designing molecular structures and assemblies. However, because the top-down approach might be more suitable for synthesizing micron-sized (µm) particles [10–12], very few reports have been made on the formation of nanogels. Hence, attention in this chapter is focused mainly on bottom-up approaches and, in particular, for chemical gels with nanosize control. For example, polymerizations in templates and in dilute solutions, and the crosslinking reactions of linear polymer chain ends and of linear polymer side chains, represent bottom-up approaches to the synthesis of nanogels.

2.2.1
Chemical Nanogels

The crosslinking points of chemical nanogels consist of covalent bonds, which are formed during and after polymerization (Figure 2.2). In the case of the "during-polymerization method," a bifunctional crosslinking agent such as methylenebis(acrylamide) (BIS) or divinyl benzene (DVB) is used. In contrast, monomers with reactive groups are polymerized via the "after-polymerization method," with either chemical or physical stimulation being provided to create the crosslinking points.

2.2.1.1 Polymerization in Templates
This method requires some nano-spaces to polymerize a monomer. Typically, surfactant polymerizations – so-called "miniemulsion polymerization" and "inverse miniemulsion polymerization" – are used, which require the surfactants to be dispersed oil-in-water and water-in-oil emulsions, respectively. Unfortunately, however, these polymerizations require the surfactants to be removed after polymerization, and this is not possible in some cases. An easier-to-use approach is that of surfactant-free emulsion polymerization (SFEP), which includes "suspension polymerization" or "precipitation polymerization," although in these cases it is

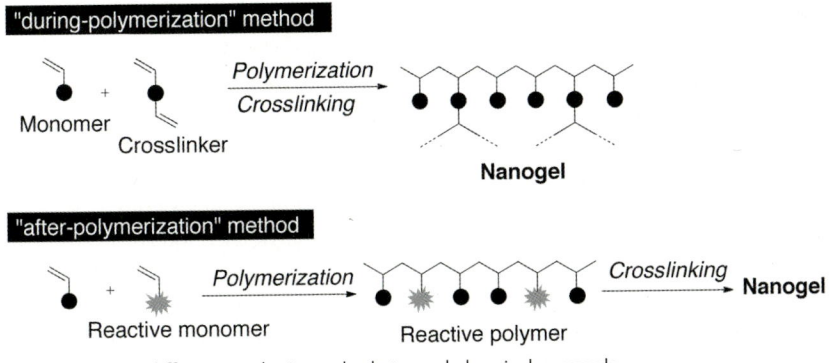

Figure 2.2 Two different synthetic methods toward chemical nanogels.

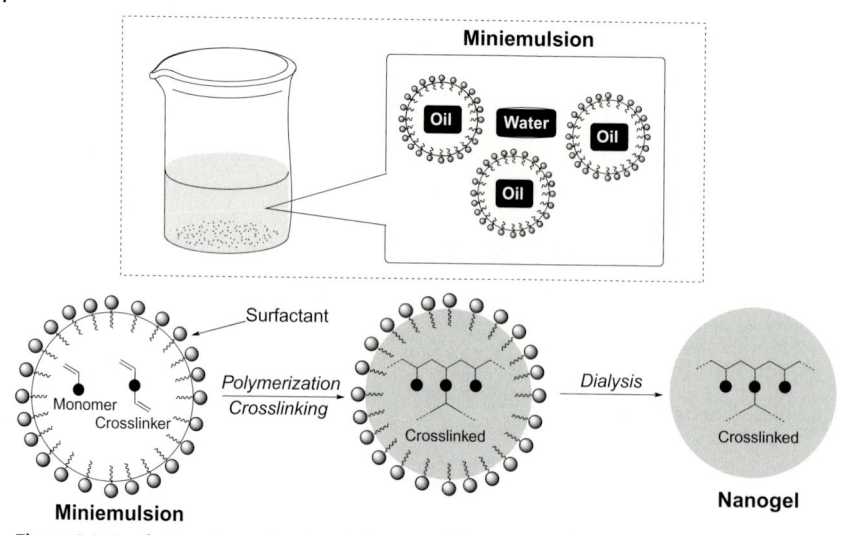

Figure 2.3 Synthetic scheme for chemical nanogels by miniemulsion polymerization.

difficult to control the polymer size and to create nanometer-sized gels. The process of suspension polymerization is introduced here, in addition to templates that employ inorganic nanoparticles.

2.2.1.1.1 Miniemulsion Polymerization Miniemulsion polymerization is conducted using surfactants such as sodium dodecyl sulfate (SDS) and sodium dodecylbenzene sulfonate (SDBS), whereby the monomers and crosslinkers (such as bifunctional compounds) are polymerized in oil dispersed in excess water. Typically, hydrophobic monomers are normally used for this method (Figure 2.3), with polymerization carried out as follows. First, the monomer and divinyl monomer, with or without an organic solvent, are added to a surfactant aqueous solution. The mixed solution is then stirred to create a miniemulsion, after which a hydrophobic initiator is added; following polymerization, the surfactant is removed by dialysis. The most important point here is that the sizes of the nanogels can be controlled via the amount of surfactant present. Surfactant-initiated and surfactant-through polymerizations are described in greater detail in Section 2.2.1.2.3.

Although both miniemulsion polymerization and surfactant-free emulsion polymerization (SFRP) including precipitation polymerization are used to produce poly(N-isopropylacrylamide) (PNIPAM) nanogels, the SFEP method permits a much easier control of the polymer size, allowing small-sized nanogels to be prepared [13]. Precipitation polymerization is described in Section 2.2.1.1.7. Following pioneering studies conducted by Pelton and coworkers [13, 14], the most frequently used preparations of PNIPAM nanogels have employed potassium persulfate as an initiator, BIS as a crosslinker, and SDS as a surfactant [15, 16]. The main reason for producing PNIPAM nanogels is to investigate their thermosensitivity, and to compare their properties with those of PNIPAM macrogels. Notably, as a

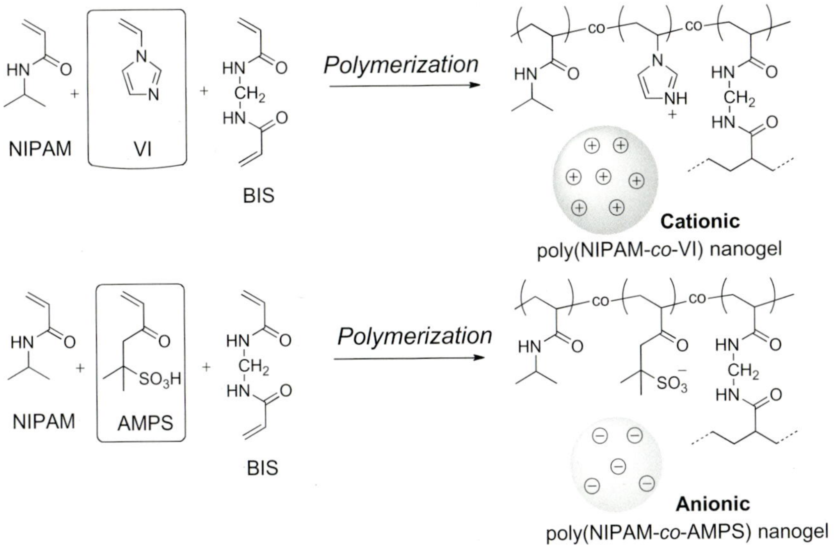

Figure 2.4 Synthetic scheme for poly(NIPAM-co-VI) and poly(NIPAM-co-AMPS) nanogels.

polymer PNIPAM is hydrophilic at room temperature, but hydrophobic at higher temperatures.

The copolymerization of N-isopropylacrylamide (NIPAM) with other monomers, in the presence of a bifunctional monomer, was also carried out (Figure 2.4). An example of this was the copolymerization of NIPAM with electrolyte monomers in order to prepare cationic nanogels. In this case, 1-vinylimidazole (VI) was copolymerized with NIPAM in the presence of BIS to prepare cationically charged poly(NIPAM-co-VI) nanogels [17]. Anionically charged nanogels were also prepared by using 2-acrylamido-2-methylpropane sulfonic acid (AMPS), so as to achieve the formation of a polyelectrolyte complex between the poly(NIPAM-co-VI) and poly(NIPAM-co-AMPS) nanogels [18]. Polyampholyte nanogels were also synthesized via the copolymerization of acrylic acid (AA), VI, BIS, and NIPAM [19]. By employing the thermosensitive abilities of PNIPAM, it was possible to prepare core–shell nanogels consisting of a poly(acrylonitrile) core and a PNIPAM shell. To form the core, acrylonitrile was polymerized in the presence of a bifunctional crosslinker under low-temperature conditions in a miniemulsion, while the NIPAM monomer was retained outside the emulsion. On subsequent heating, the NIPAM monomer became sufficiently hydrophobic to undergo polymerization in the emulsion [20].

More recently, controlled polymerization methods capable of synthesizing well-defined polymers with specific molecular weights, molecular weight distributions, and compositions, have been used in miniemulsions to prepare nanogels. In particular, a controlled/living radical polymerization is suitable for these systems because the radicals are stable in water (for a review, see Ref. [21]). The method of atom transfer radical polymerization (ATRP) is widely used to prepare polymers

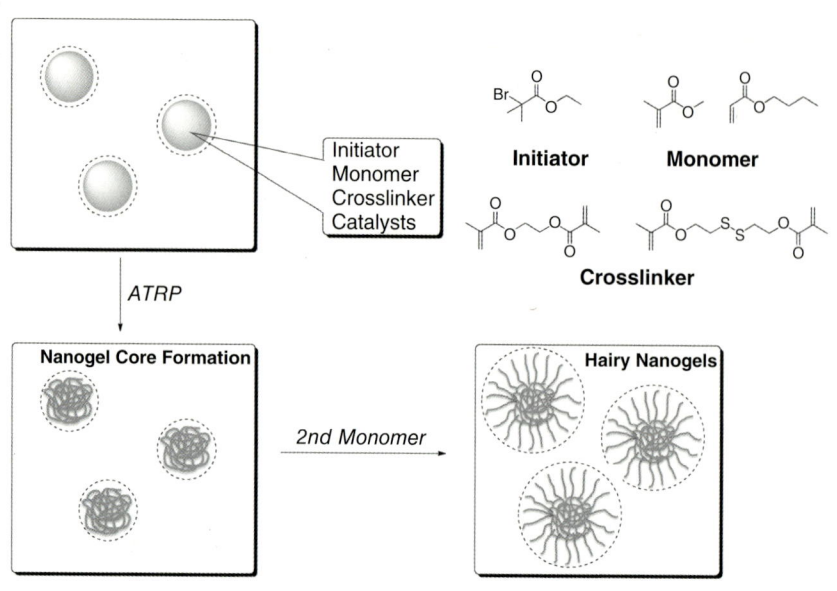

Figure 2.5 One-pot synthesis of hairy nanogels by ATRP in miniemulsion.

under controlled conditions [22]. For example, by using ATRP, di(ethylen glycol) methyl ether methacrylate was successfully polymerized in miniemulsion by using polyoxyethylene oleyl ether (Brij 98) as a surfactant. Use of this method in the presence of oleic acid-coated Fe_3O_4 nanoparticles led to the formation of thermally responsive magnetic nanogels [23]. Likewise, "hairy nanogels" – the structure of which resembles a core-crosslinked star polymer – were also prepared via a two-step polymerization in one-pot. In this case, a methacrylic ester crosslinker was polymerized in an emulsion using Brij 98 to form nanogels and, subsequently, a second monomer such as methyl methacrylate (MMA) was polymerized from the nanogels (Figure 2.5) [24]. Hence, by combining the ATRP method and "click chemistry" it was possible to synthesize well-defined nanogels [25].

Nitroxide-mediated radical polymerization (NMRP), a controlled radical polymerization process, has also been used to synthesize nanogels. In a miniemulsion of SDBS, styrene and DVB were polymerized via an NMRP method from polystyrene (PSt), with 2,2,6,6-tetramethylpiperidinyl-1-oxy (TEMPO) at the chain end to form nanogels [26]. Although other controlled miniemulsion polymerization procedures have been used to create nanogels, the reports are limited in number [27].

Finally, in a unique approach to surfactant-free miniemulsion polymerization, a macromer rather than a conventional low-molar mass surfactant was used, whereby dextran-lactate-2-hydroxyethyl methacrylate was polymerized with PNIPAM in the absence of a low-molar mass surfactant [28]. In this case, the lactate-2-hydroxyethyl methacrylate grafts and a dextran backbone were used as the hydrophobic and hydrophilic units, respectively, to construct an amphiphilic macromer that functioned as a surfactant to facilitate micelle formation during the miniemulsion polymerization. Thus, no surfactant was needed to form the nanogels [28].

2.2.1.1.2 Inverse Miniemulsion Polymerization

Inverse miniemulsion methods are used to polymerize hydrophilic monomers such as electrolyte monomers and poly(ethylene glycol) (PEG)-containing monomers. The two main differences between these miniemulsion methods are that: (i) water-soluble monomers are polymerized in aqueous solutions dispersed in excess oil; and (ii) oil-soluble surfactants, such as sorbitan monooleate (Span-80) and sodium bis(2-ethylhexyl) sulfosuccinate (AOT), are used. The reaction procedures are similar to those used for the miniemulsion methods.

The signature use of the inverse miniemulsion method is to prepare polyelectrolyte nanogels. Cationic nanogels, which can interact with anionic DNA, have been prepared via the inverse miniemulsion method, whereby a cationic monomer, (3-acrylamidopropyl)-trimethylammonium chloride, was polymerized with BIS in inverse miniemulsion using both L-α-phosphatidylcholine (lecithin) and AOT surfactants [29]. In another report, the copolymerization of 2-acryloxyethyltrimethylammonium chloride and 2-hydroxyethylacrylate to afford cationic nanogels was indicated. In this case, PEG diacrylate was used as a crosslinker and laureth-3 as a surfactant, such that an ion complex of the nanogels and DNA was formed [30]. [2-(Methacryloyloxy)ethyl]-trimethyl-ammonium chloride was also polymerized for use as a carrier of small interfering RNA (siRNA) [31]. The main point of this report was that dextran hydroxyethyl methacrylate (dex-HEMA) and dextran methacrylate, the structures of which include dextran with vinyl groups in the side chains, were used instead of divinyl-type crosslinkers [31]. By contrast, anionic nanogels were prepared by using an AMPS monomer [32], whereby AMPS was copolymerized with acrylamide (AM) and BIS, and AOT was used as the surfactant.

Neutral monomers such as N-vinylformamide (NVF) were also polymerized via an inverse miniemulsion [33]. More recently, a controlled polymerization, particularly with the ATRP method, in inverse miniemulsion has been developed [34]. For example, an inverse miniemulsion was formed by Span 80, and oligo(ethylene glycol) monomethyl ether methacrylate with disulfide-functionalized dimethacrylate as a crosslinker was polymerized to form nanogels [35, 36]. In this system, styrene could be polymerized after the formation of nanogels, and chain extension from the core was enabled to form hairy nanogels [35].

2.2.1.1.3 Suspension Polymerization

The SFEP process is divisible into suspension and precipitation polymerizations, where the former method uses an oil-in-water emulsion (achieved by stirring), and the latter is based on differences in the solubilities of the monomer and polymer in a solvent (see Section 2.2.1.1.7). Suspension polymerization represents a simple approach to the polymerization of hydrophobic monomers, because it is not require that the surfactant is removed after polymerization. However, when using this approach it is difficult to control the polymer size, and to create smaller nanogels. Previously, poly(2-vinylpyridine)

(P2VP) nanogels were prepared by the copolymerization of VP and DVB in water (with stirring) [37], while PNIPAM nanogels were also prepared via suspension polymerization [38]. In both cases, the use of a surfactant in the polymerization was avoided because of any potential interaction of the nanogels with the surfactants.

2.2.1.1.4 Polymerization in Liposomes

In a unique method used to prepare hydrophilic nanogels, liposomes are used as nanometer-scale reactors. Liposomes are formed from spherical lipid bilayer membranes consisting of phospholipid assemblies, the hydrophilic groups of which face both the inside and outside of the spheres, while the inside spaces are used as nanometer-scale reactors to polymerize hydrophilic monomers (for a review, see Ref. [39]). Previously [40, 41], the procedures for photo-polymerization in liposome reactors have been identified as:

- The encapsulation of monomers into the liposomes.
- Dilution of the suspension to prevent polymerization outside the liposomes.
- Photopolymerization inside the liposomes.
- Solubilization of the lipid bilayer, using detergents.
- Removal of the mixed phospholipid-detergent micelles by dialysis.
- Drying the hydrophilic nanogels by gentle evaporation in a temperature gradient [40].

In this system, polymerization in the outer liposomes represents a major challenge although, by using ascorbic acid as a radical scavenger, the side reaction could be prevented and monodisperse, hydrophilic nanogels obtained [42]. By using this method, PNIPAM, polyacrylamide (PAM) and poly(vinylimidazole) (PVI) nanogels were synthesized, using BIS as a crosslinker [40–42]. Nanogels were also obtained by the polymerization of dex-HEMA containing 1-stearoyl-2-oleoyl-sn-glycero-3-phosphocholine liposomes; the subsequent release of both bovine serum albumin (BSA) and lysozyme from the dextran nanogels was investigated [43].

2.2.1.1.5 Polymerization on Inorganic Nanoparticles

In this approach, inorganic nanoparticles were used as templates to form hollow nanogels (nanogel capsules). Initially, the crosslinking reactions were carried out on the surface of nanoparticles, and the nanoparticles then removed. The first such example was a silica nanoparticle on which NVF and 2-bis[2,2'-di(N-vinylformamido)-ethoxy] propane as a crosslinker were polymerized under a P2VP stabilizer. Subsequently, the silica particles were removed by hydrolysis in aqueous NaOH to form the hollow nanogels [44]. When gold nanoparticles (AuNPs) were used in this method, the NIPAM and BIS were polymerized on AuNPs that had been immobilized with NH_2-terminated PNIPAM, after which the NPs were removed (using KCN) to afford the hollow nanogels [45]. Another example in which Au templates were used involved a crosslinking reaction of diblock copolymers with an active ester for amine in their side chain on AuNPs, with the Au being removed later by oxidative etching [46].

2.2 Synthesis of Polymeric Nanogels

2.2.1.1.6 Polymerization in Polymer Solutions Recently, the synthesis of small polymeric nanogels of DVB less than 10 nm in size has been reported [47]. The polymerization of DVB was conducted in a poly(4-vinylpyridine) (P4VP) solution, that prohibited the aggregation of poly-DVB. The latter was purified by serial centrifugation in water under acidic conditions, which rendered the P4VP water-soluble.

2.2.1.1.7 Dispersion Polymerization and Precipitation Polymerization Both, dispersion polymerization and precipitation polymerization are performed in cases where the monomers and polymers are soluble and insoluble in a specific solvent, respectively. During polymerization, precipitation may occur because the polymer aggregates are insoluble in the solvent. In the presence or absence of dispersion stabilizers, these procedures can be defined as dispersion polymerization or precipitation polymerization, respectively, although in some cases the difference between these terms is not distinguished, and precipitation polymerization is used as an umbrella term. One advantage is that no surfactant is necessary in this type of polymerization, and purification is very easily carried out.

Dispersion Polymerization Dispersion polymerization has been widely used to prepare PNIPAM nanogels, because the monomer is hydrophilic and the polymer is hydrophobic in water under heating conditions. Stable nanogels can be obtained in the presence of a dispersion stabilizer, and this simple polymerization method can provide remarkably uniform-sized PNIPAM nanogels [13]. Furthermore, core–shell structures can be obtained using this method.

The copolymerization of NIPAM with some monomers has also been reported; for example, poly(NIPAM-co-allylamine) nanogels [48, 49], poly(NIPAM-co-AA-co-rhodamine) nanogels [50], and PNIPAM–poly(N-isopropylmethacrylamide) (PNIPMAM) [51] nanogels with nanophase-separated structures were synthesized by using SDS as a stabilizer. Allylamine was used for the immobilization of spiropyran [48], and the rhodamine acted as a fluorescent moiety [50]. Core–shell nanogels were also obtained by dispersion polymerization in order to investigate their thermosensitivity. For this, nanogels with PNIPAM in the shell and PNIPMAM in the core were prepared, with lower critical solution temperatures (LCSTs) in aqueous solution of 34 °C and 44 °C, respectively [52]. Thus, core–shell nanogels with a PNIPAM hydrogel core and a PEG–monomethacrylate (PEGMA) shell [53], and/or with a PNIPAM core and a P4VP shell, were synthesized [54].

Precipitation Polymerization The use of precipitation polymerization to prepare nanogels is limited because polymer aggregates with larger sizes segregate very easily. Previously, PNIPAM nanogels were synthesized by precipitation polymerization [13], and the copolymerization of NIPAM and AA was also conducted in the presence of BIS to form Janus nanogels [55]. Poly(NIPAM-co-AA) nanogels were also prepared via a semibatch process, which provided a fine thermosensitivity under a lower pH condition [56]. The reversible addition-fragmentation chain transfer (RAFT) precipitation polymerization of NIPAM was performed by using water-soluble macromolecular chain-transfer agents [57]. In addition, when chitosan with amino groups (which can dissolve in water) was reacted with ethylenediaminetetra-acetic dianhydride, the formation of an amide linkage permitted the formation of a precipitate [58].

2.2.1.2 Crosslinked Micelles and Star Polymers

A crosslinked micelle consists of a nanogel core and shell arms, where the shell arms prevent aggregation. Although various terms have been used to refer to these structures – such as "star polymers," "star-like microgels," "star-like nanogels," and "core-crosslinked star polymers" – the most suitable expression for each synthetic method is used in this chapter. The various synthetic methods which have been proposed for their formation generally involve living polymerization techniques, and include the "arm-first," "core-first", "micelle crosslinking," and "block copolymer crosslinking" methods. The main advantage of these methods is that they permit the easy preparation of nanogels with sizes smaller than 10 nm.

2.2.1.2.1 The Arm-First Method
The arm-first method (which is generally used in star polymer chemistry) is conducted via a living polymerization technique, whereby the monomers are first polymerized to form a linear chain, after which the chain end is crosslinked by a bifunctional monomer (Figure 2.6). The steric repulsion of the linear chains causes the macroscopic gelation to be depressed. A controlled/living polymerization procedure enables the formation of nanogels with smaller sizes and objective structures; indeed, certain functional groups can be introduced in specific regions, such as the cores and the chain ends. Examples of living polymerization include radical, cationic, and ring-opening polymerizations; the crosslinking reaction of linear polymers following micelle formation is described in Section 2.2.1.2.3.

Living radical polymerization has been developed as a useful technique to form well-defined polymers, not only because it is unnecessary to remove the water, but

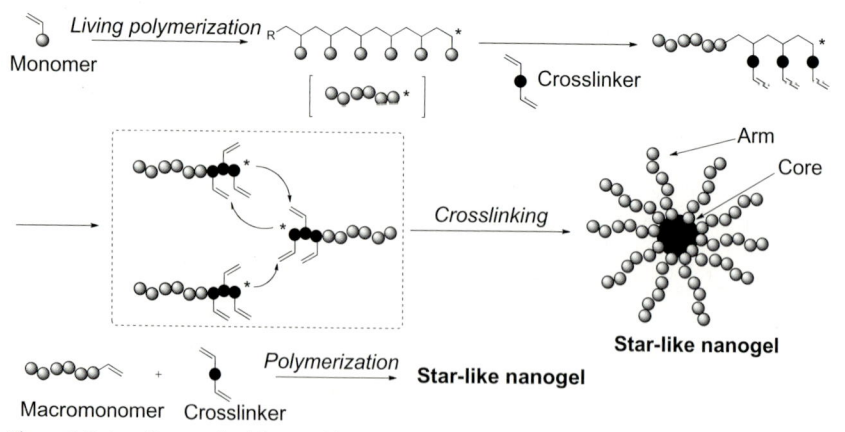

Figure 2.6 Arm-first method for star-like nanogel synthesis.

also because radical polymerization is tolerant of many functional groups. The ATRP, NMRP, and RAFT polymerization techniques have mainly been used to create star polymers (for reviews, see Refs [59–60]). For example, amphiphilic and thermosensitive star polymers were prepared by the copolymerization of ethylene glycol dimethacrylate (EGDMA) and a phosphine-ligand monomer [CH_2=$CH(C_6H_4)$ PPh_2] from PEGMA-b-PMMA, via a $RuCl_2(PPh_3)_3$-catalyzed living radical polymerization [61]. During the crosslinking reaction, ligand exchanges occurred between the catalyst and the phosphine-ligand in the polymers, and ruthenium(II) was immobilized in the cores [61]. By using bifunctional and trifunctional initiators in the arm synthesis, both dumbbell and tripartite nanogels were also synthesized by crosslinking [62]. The arm and core consisted of PEGMA, and poly(butyl acrylate) and EGDMA, respectively.

Living cationic polymerization has also been used in the preparation of monodispersed star polymers. In one example, isobutyl vinyl ether arms, prepared with cationogen/$EtAlCl_2$ at 0°C in hexane in the presence of ethyl acetate, were crosslinked by a small amount of 1,4-cyclohexanedimethanol divinyl ether to form nanogel cores. It should be noted that star polymers with a narrow molecular weight distribution (M_w/M_n = 1.1–1.2) were obtained in a quantitative yield [63]. Thus, a gold nanocluster was formed in the core, which could be used as a reusable and durable catalyst for an aerobic alcohol oxidation [64].

Ring-opening polymerization (ROP), by which various functional groups can be introduced in the main chain, was also used to form star polymers. For example, ε-caprolactone (CL) was polymerized in the first step, after which bislactone (4,4′-bioxepanyl-7,7′-dione, BOD) was added and polymerized, to yield a narrow molecular distribution (M_w/M_n = 1.12–1.22) [65]. Because the main chain of the core consisted of ester groups, the nanogels proved to be biodegradable [65]. Other examples included the core-crosslinked PEG-armed polyphosphoester nanogels, which could be prepared via a one-step ROP [66]. A review of the general synthesis of PEG-armed star polymers is provided in Ref. [67].

In a combination of these two methods, both ATRP and the ROP were carried out to form selectively degradable star-like nanogels. In this case, a multifunctional initiator was used for the ATRP and ROP, and the polymerization was carried out using a selective method. Initially, a nondegradable monomer (styrene or MMA) and a degradable monomer (CL) were polymerized by ATRP and ROP, respectively. Next, a nondegradable divinyl monomer (DVB or EGDMA), or a degradable bifunctional monomer (BOD, etc.), was polymerized to form star-like nanogels. In this way it was possible to produce four types of star polymer, all of which were segment-selectively degradable [68].

2.2.1.2.2 **The Core-First Method** The core-first method (which also is used in star polymer chemistry) comprises two steps: the nanogel core is formed in the first step, after which the arms are synthesized by polymerization from the nanogels, to form hairy nanogels (Figure 2.7). Although, in star polymer chemistry, this term is used to refer to polymerization from multifunctional low-molecular-weight initiators to obtain defined arm numbers, it is used in a broad sense here. The

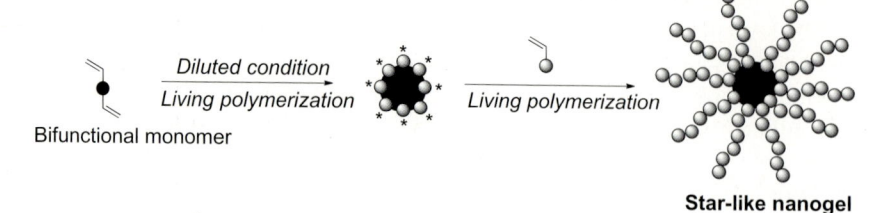

Figure 2.7 Core-first method for star-like nanogel synthesis.

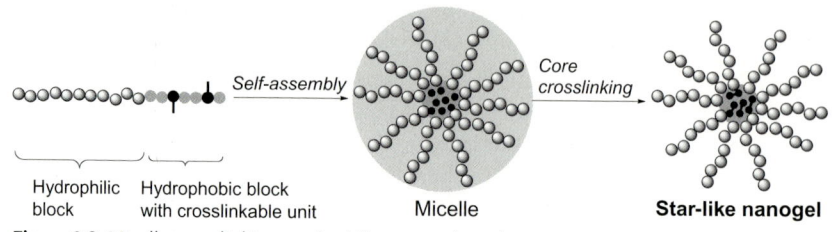

Figure 2.8 Micelle crosslinking method for nanogel synthesis.

method is easy to use, but the structures of the obtained nanogels are difficult to control. It should be noted that the core-first methods have been conducted with or without using surfactants. For a discussion of the use of surfactants, see Sections 2.2.1.1.1 and 2.2.1.1.2.

The core-first methods were also performed by using the ATRP method; in this case, a bifunctional monomer ethylene glycol diacrylate (EGDA) was polymerized in a dilute condition, after which the second monomer, methyl acrylate (MA), was added to form star-like nanogels [69]. Furthermore, a third monomer was polymerized from the star-like nanogels to form diblock copolymer arms [69]. By using the free radical polymerization method, core-crosslinked star clusters were reported; DVB was polymerized under AIBN as a radical initiator in the first step, after which the MMA was polymerized through the poly-PDVB cores [70]. The clusters consisted of some nanogels and numerous connected chains between the nanogels, and their size could be controlled by the solvents [70].

2.2.1.2.3 Micelle Crosslinking Method This method provides crosslinked micelles, whereby surfactants or diblock copolymers are self-assembled in any solvent, and the micelles then crosslinked using any method. Herein, two types of crosslinking methods are introduced – a side chain crosslinking type, and a chain end crosslinking type (Figure 2.8). The side chain crosslinking type is carried out using any functional groups in the side chain, with the crosslinking reaction proceeding by reaction of the functional groups. In contrast, the chain end crosslinking type has an initiating group or a polymerizable group at the chain end, and the crosslinking reaction proceeds by polymerization or copolymerization, using multifunctional monomers.

The reactions of the side chain crosslinking type are usually performed by photo-, thermal- or chemical-crosslinking reactions. For example, cinnamoyl groups were introduced into PEG-b-polyglycidol block copolymers by living anionic polymerization, and ultraviolet (UV) light was used to irradiate aqueous media to form a crosslinked micelle [71]. The RAFT polymerization method was also used to introduce cinnamoyl groups into PEG-b-poly(2-(diethylamino)ethyl methacrylate (PDEAEMA) [72]. In this system, the formation and deformation of the micelles in water could be controlled by the pH based on the reversible protonation of amino groups, and the crosslinked micelles were formed by the irradiation of UV light under basic conditions [72]. Coumarin groups were also used as photocrosslinkable groups, where photo-dimerization occurred under irradiation at $\lambda > 310$ nm, and the photo-cleavage of cyclobutane bridges occurred under irradiation at $\lambda < 260$ nm. The PEG-b-poly[2-(2-methoxyethoxy)ethyl methacrylate] with the coumarin groups in the side chain formed nanogels under UV irradiation, and their crosslinking densities could be controlled reversibly by the UV wavelength [73]. In contrast, when the thermal crosslinking reaction of the side chain was attempted, the poly(4-vinylbenzocyclobutene-b-poly(butadiene) (PVBCB-b-PB) diblock copolymers were self-assembled to form spherical micelles, and the PVBCB block was crosslinked through thermal dimerization at 200–240 °C [74].

Some examples of the crosslinking reactions of micelles in the side chains by chemical reactions have also been described. For example, PEG-b-poly(glycerol monomethacrylate) (PGMA)-b-PDEAEMA triblock copolymers formed micelles in an aqueous solution below pH 9, while reaction of the carboxylic acid in the side chain and the addition of diamine under an amidation activator, led to the formation of shell crosslinked micelles [75]. The polyferrocenylsilane-b-polyisoprene (PI) block copolymers formed cylindrical micelles in a PI-selective solvent such as hexane, and the cylindrical micelles with vinyl groups in the PI corona were crosslinked through a Pt(0)-catalyzed hydrosilylation reaction using 1,1,3,3-tetramethyl disiloxane as a crosslinker at room temperature [76]. The length of the cylinders could be controlled by changing the solvent.

The chain end crosslinking types can be divided into a macroinitiator type, with an initiating group, and a macromer type, with a polymerizable group at the polymer chain end. Although some examples of polymer reactions between surfactant chain ends and added polymers have been reported [77, 78], the details are not included at this point. As with the macroinitiator type, controlled polymerization techniques have been used, and the initiators (which were directed to the core) polymerized the bifunctional monomers to form nanogels. For example, poly(ethylene oxide) (PEO)-b-PSt diblock copolymers with chloride atoms in the PSt chain end formed micelles in aqueous media, and EGDA was polymerized by the ATRP method [79]. Furthermore, star-like hollow nanogels were also synthesized via the ATRP method, using the PEO-b-PBMA-Cl diblock copolymer [80]. A macroRAFT agent, consisting of a hydrophilic PEG block and a hydrophobic dodecyl chain, formed micelles in aqueous media, after which the copolymerization of N,N-diethylacrylamide and BIS was carried out to form nanogels [81].

In the case of the macromer type, PEG with styrene and carboxyl groups at both ends was synthesized, and a radical copolymerization of DEAEMA and the vinyl group in aqueous media was then carried out to afford nanogels [82]. In another example, a triblock copolymer consisting of PEG-*b*-poly(propylene glycol)-*b*-PEG with poly(lactic acid) (PLA) oligomer and a vinyl compound in its chain ends was converted into flower-like micelles in an aqueous medium, being directly polymerized to form nanogels [83].

2.2.1.2.4 Block Copolymer Crosslinking Method This method is performed by the crosslinking reaction of diblock copolymers, without a surfactant agent. Although an amphiphilic structure is not needed, the crosslinking reaction must be performed under high-concentration conditions in order to prevent intramolecular crosslinking. For example, PMMA diblock copolymers with alkoxyamine units as thermally crosslinkable units in one segment are synthesized by the ATRP method. By heating the diblock copolymers in an anisole solution, star-like nanogels were formed as a result of the crosslinking reaction of the diblock copolymers [84]. Furthermore, two types of complementary reactive diblock copolymer with different molecular weights were synthesized, and hetero-arm star-like nanogels with different arm lengths were formed (Figure 2.9) [85–87].

2.2.1.3 Polymer Reaction of Linear Chain, Nanoparticles, and Nanocomplexes

Although previous sections have mainly focused on polymerization techniques, where the crosslinking points are formed by *in situ* polymerizations, this section refers to the crosslinking reaction of linear chains, nanoparticles, and nanocomplexes, the crosslinking points of which are formed *after* polymerization. Details of the polymer reactions of micelles and block copolymers have been described previously, in Sections 2.2.1.2.3 and 2.2.1.2.4.

2.2.1.3.1 Crosslinking of Linear Chains This method normally involves either of two approaches: (i) the irradiation of a beam to generate a radical; and (ii) the

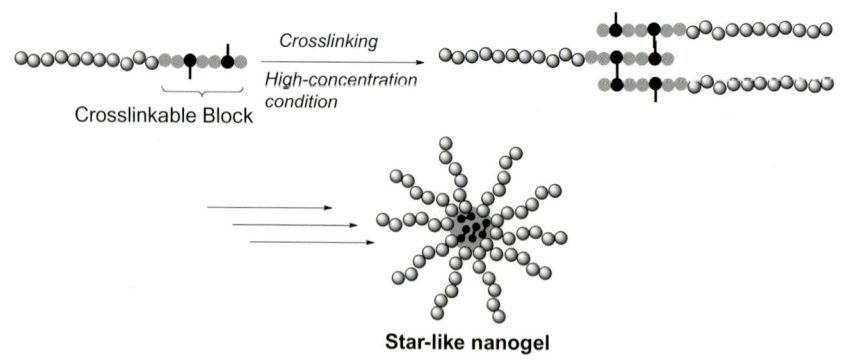

Figure 2.9 Block copolymer crosslinking method for star-like nanogel synthesis.

chemical reaction of crosslinkable groups in the polymer side chains. Beam irradiation provides an easy means of preparing nanogels without any added compound. In the first example, which is a crosslinking reaction of poly(acrylic acid) (PAA), linear PAA (in dilute aqueous solution) was irradiated by short pulses of fast electrons, which led to the generation of C–C double bonds and radicals in the main chain [88]. The addition of radicals to the C–C double bonds then led to the formation of a crosslinking structure [88]. The pulse electron beam was also used to synthesize poly(vinyl methyl ether) nanogels in an aqueous solution [89]; a photo-Fenton reaction [90] and quantum-ray irradiation [91] were also used to form nanogels in dilute solution.

In the case of a chemical reaction, isocyanate-functionalized linear polymers were prepared via a RAFT copolymerization of MMA and isocyanatoethyl methacrylate. The addition of 2,2′-(ethylenedioxy)diethylamine to the linear polymers in dilute solution led to the efficient formation of crosslinking groups to form nanogels [92].

2.2.1.3.2 **Crosslinking of Nanoparticles** The crosslinking reactions of nanoparticles are divisible into three types: a side chain type; a main chain type; and a surface crosslinking type. As an example of the side chain-type reaction, PNIPAM with a maleinimido group was prepared by the radical copolymerization of NIPAM and 2-dimethylmaleinimido ethylacrylamide. Following the formation of nanoparticles over the LCST using a surfactant, a photo-crosslinking reaction of the maleinimido group was carried out to form nanogels [93].

In the case of the main chain-type reaction, the polymerization was performed by polycondensation under suspension to obtain polyester nanoparticles with the cinnamic group in the main chain [94]. Following UV irradiation, the crosslinking reaction proceeded to form nanogels, the sixes of which could be changed reversibly, according to the wavelength of the UV light employed [94].

The surface cross-linking of nanoparticles was performed in order to coat biomolecules with thin gel films. The first example involved horseradish peroxidase, the NH_2 groups of which were displaced by polymerizable units by using *N*-acryloxysuccinimide under mild conditions [95]. The polymerizable units were copolymerized with NIPAM and BIS to form surface-crosslinked nanogels [95]. Likewise, bovine carbonic anhydrase was also shown to be encapsulated into surface-crosslinked nanogels [96].

2.2.1.3.3 **Crosslinking of Nanocomplexes** This procedure has been performed by crosslinking *in situ* polymer complexes to conjugate and immobilize biomolecules, into nanogels. For example, a complex of plasmid DNA and thiol-functionalized six-arm PEG was formed in dimethylsulfoxide (DMSO), whereby the crosslinking reaction of the SH group and added dithio-bis-maleimidoethane provided the formation of nanogels [97]. In addition, a complex of thiolated heparin and PEG in DMSO was also used to form heparin nanogels by sonication of the complex, followed by extensive dialysis to remove the PEG chains [98].

2.2.2
Physical Nanogels

Physical nanogels are crosslinked by noncovalent bonds, such as hydrogen bonds, coordination bonds, and ionic bonds. Such nanogels have the specific characteristic of sol–gel reversibility by an external stimulus, which leads to their application in areas such as drug-delivery systems. In contrast to chemical nanogel syntheses, the formation of crosslinking points must be carried out *after* polymerization; consequently, during the design of their structure it is very important to form objective structures. In this section, the preparation of physical nanogels from polymer precursors is introduced according to the type of crosslinking point, whether hydrogen bonding, ionic bonding, coordination bonding, and/or hydrophobic interaction. The formation of nanogels from low-molecular-weight compounds, using noncovalent bonds, is not described.

2.2.2.1 Hydrogen Bonding

Hydrogen bonding represents one of the most standard methods to form physical gels. In general, linear polymers with hydrogen-bonding units are designed, and the objective nanogels formed, by a stimulus or change in the environment (Figure 2.10). For example, benzamide groups were used as self-complementary hydrogen-bonding units. When PMMA with dendritic benzamide groups was synthesized by RAFT polymerization, it was confirmed that the nanogel had been formed by

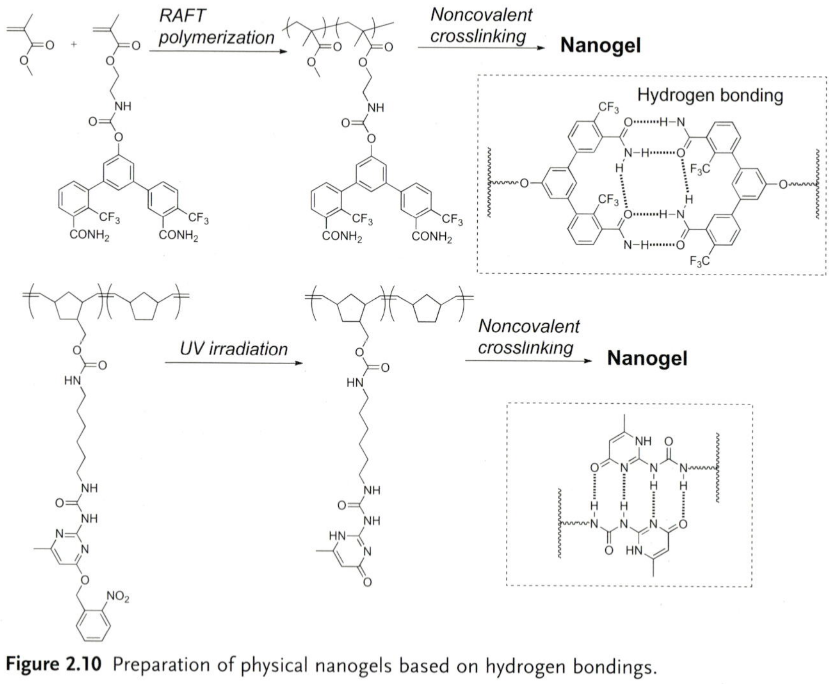

Figure 2.10 Preparation of physical nanogels based on hydrogen bondings.

hydrogen bonds, not in tetrahydrofuran, but rather in toluene [99]. The interesting point here is that the intermolecular hydrogen bonds were formed easily to create large aggregates in the case of using a lower content of hydrogen-bonding units [99]. 2-Ureido-pyrimidinone groups were also used as hydrogen-bonding units. In this case, polymers with 2-ureido-pyrimidinone groups in the side chain were synthesized, using UV irradiation, to create the corresponding precursor polymers, and the resultant polymers were able to form metastable nanogels. In fact, these nanogels were so stable that they could be estimated using gel-permeation chromatography (GPC), while the hydrogen bonding could be removed by the addition of acid [100]. Another example involved the use of biopolymers, based on hydrogen-bonded nucleobase pairing. In this case, a tetrafunctional avidin–peptide nucleic acid and trifunctional DNA were used to form nanogels, with assembly and disassembly being controlled by the temperature [101].

2.2.2.2 Ionic Bonding and Coordination Bonding

In the case of ionic bonding, the electrostatic attractive force is the driving force in creating the crosslinking points. The combination of ion bonds is divisible into two types: (i) anionic polymer–cationic polymer complexes; and (ii) anionic polymer–metal ion complexes. Macroscopic gelation is prevented by using either diblock copolymers or nano templates.

Anionic polymer–cationic polymer complexes are useful for the synthesis of nanogels; the simple mixing of oppositely charged block copolymers, such as PEG-block-poly(L-lysine) and PEG-block-poly(α,β-aspartic acid), would afford nanogels (Figure 2.11) [102]. More recently, PEG-block-polyions containing a disulfide linkage between the PEG chains and the charged segments were synthesized, and diblock copolymer polyion-complexes also developed. For example, PEG-SS-P(Asp) with carboxyl groups in the side chain, and PEG-SS-P[Asp(DET)] with amine groups in the side chain, formed polyion-complex micelles. It should be noted that the disulfide bonds were degraded by reduction to produce hollow nanocapsules [103]. In another example, the PEG-b-P[Asp(DET)] diblock copolymer with no disulfide bonds formed ion complexes with anion-modified cytochrome c (CytC). Here, the ion complexes were dissociated at pH 5.5, and the

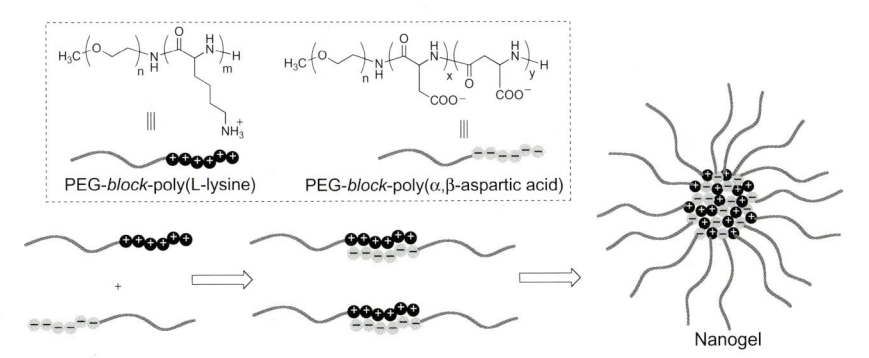

Figure 2.11 Preparation of physical nanogels based on polyion complexes.

controlled release of CytC was investigated [104]. Anionic polymer–cationic polymer complexation is also useful for capturing DNA, which is one of the anionic polymers, in cationic diblock copolymer micelles. For example, polymer nanogels can be formed between poly(ethylene imine) (PEI) as a cationic polymer and DNA, and can be stabilized by amine-reactive Pluronic, which is a triblock copolymer consisting of PEG and poly(propylene oxide) (PPO) [105].

The anionic polymer–metal ion complexes represent an easy-to-use approach for forming ion-complex nanogels. For this, sodium alginate was encapsulated into a liposome template, while a subsequent by ion exchange to calcium ions under heating produced ionic nanogels [106]. Pluronic was also used to form anionic polymer–metal ion complexes; by adding calcium ions, Pluronic formed polyion complexes with PAA, while the PEG chains prevented the formation of larger aggregates [107]. PEG–SS–siRNA and calcium ions formed ion-complex nanogels under phosphate conditions [108]; in this case, the PEG was removed and the nanogel applied as an siRNA delivery carrier [108].

In the case of coordination bonding, the crosslinking points consist of coordination bonds, such as a metal complex. For example, carboxylic-functionalized poly(β-aminoester)s with PEG side chains were treated with cisplatin to afford nanogels [109]. Frequently used methods to make crosslinking points were connected by ruthenium–terpyridine complexes. As many examples of metallopolymers containing ruthenium–terpyridine complexes have been reported to date [110, 111], these connections may serve as good candidates for the preparation of nanogels.

2.2.2.3 Hydrophobic Interaction

Hydrophobic interaction is not a chemical bond, but rather a intermolecular interaction, by which hydrophobic molecules in aqueous solution form assemblies to prevent the entropy loss of water. Because many studies on nanogel formation via hydrophobic interaction have been reported, this is regarded as one category of the crosslinking methods described in this chapter. For example, a hydrophilic polymer with hydrophobic low molecules in the side chains will aggregate in water because the hydrophobic parts assemble and behave like crosslinking points. By modifying the alkyl chains into PEG in the side chain, nanogels have been formed [112]. However, simple diblock copolymer micelles are not regarded as nanogels because the assembled hydrophobic parts are large, and it is unreasonable to assume that they are crosslinking points. In order to overcome this problem, cholesterol-bearing pullulan (CHP) nanogels, which are very stable in aqueous media, have been investigated; details of CHP and amphiphilic graft copolymers are presented in the following subsections.

2.2.2.3.1 **Cholesterol-Bearing Pullulan** Cholesterol-bearing pullulan (CHP), which consists of hydrophilic pullulan in the main chain and hydrophobic cholesterol in the side chain, can form nanogels in aqueous media (Figure 2.12) [113]. The hydrophobic interaction among cholesterol units leads to the formation of quasi-crosslinking points, and the nanogels are so stable that they can be measured

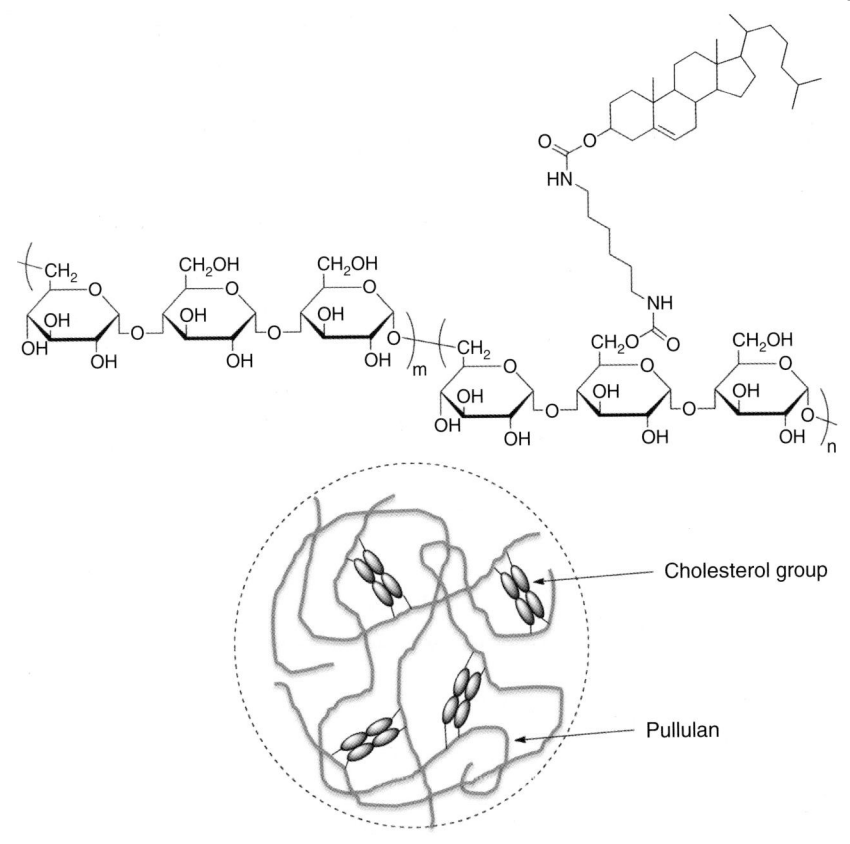

Figure 2.12 Chemical structure of cholesterol-bearing pullulan (CHP), and structure model for a CHP nanogel.

using GPC [113]. Furthermore, the CHP can form complexes with proteins, and the crosslinking points can be destroyed by the addition of β-cyclodextrin (β-CD) [114, 115]. These materials have been widely applied to both protein- and drug-delivery systems [116].

Based on recently reported synthetic methods, major developments have been made in the creation of CHP nanogels. In fact, CHP nanogels have been reported with two types of crosslinking point, that involve the hydrophobic interaction of cholesterol, and also siloxane covalent bonds [117]. The chemical structure of CHP has also been updated such that, instead of pullulan, cholesteryl-bearing highly branched polysaccharide mannan was used, with the nanogels being formed in dilute aqueous solution [118]. A highly branched cyclic dextrin derivative that is partly substituted with cholesterol groups also has the ability to form nanogels [119].

When, recently, the higher order structures of CHP nanogels were investigated, those with acrylate groups were shown to react with thiol group-modified tetra-arm

PEG. As a result of the intermolecular crosslinking of the CHP nanogels, raspberry-like assemblies of crosslinked nanogels were formed, and these were applied to protein delivery systems [120]. The CHP nanogels were also incorporated into the side chain in PNIPAM, which was prepared by free-radical copolymerization. In an aqueous solution, grape-like (botryoidal) clusters were formed, the size of which was controlled by the temperature [121].

2.2.2.3.2 **Amphiphilic Graft Copolymer** Amphiphilic graft copolymers can form polymer nanogels as well as CHPs. In order to synthesize graft copolymers with desired structures, it is necessary to employ controlled polymerization techniques, because the molecular weight of the hydrophobic segment should be low. For example, NIPAM was only slightly polymerized from polysaccharides by RAFT polymerization, while polymeric nanogels were formed under LCST conditions [122]. PLA, as a hydrophobic group, was also introduced into pullulan [123] and dextran [124] via the triethyl amine catalyst polymerization of lactic acid and the addiction reaction of PLA, respectively. In both cases, the graft copolymers proved to be biodegradable, and the nanogels were formed in aqueous media. In contrast, a hydrophilic monomer was polymerized from a hydrophobic polymer; 2-methacryloyloxyethyl phosphorylcholine (MPC), the polymer of which demonstrated biocompatibility, was polymerized from hydrophobic polyphosphate using the ATRP method. This polymer assembly, the formation of which was confirmed in aqueous solution, was also shown to be biodegradable, especially under basic conditions [125].

As another example, a complex of a dextran with an alkyl chain (C_{12}) and poly(β-CD) formed nanogels, because inclusion of the alkyl chain into β-CD led it form crosslinking points via a host–guest-type hydrophobic interaction [126]. In addition to graft polymers, branched polymers may also serve as precursors for nanogels. A fifth-generation poly(amidoamine) dendrimer modified with cholesteryl groups was used to form nanogels, and the reversible transformation between nanogels and microgels was accomplished by temperature control [127].

2.2.2.4 **Protein Denaturation**
The ability to obtain nanogels via protein denaturation, simply by heating the proteins, means that nanogels should be very useful when developing "green" delivery systems. The heating of oppositely charged proteins led to the production of narrowly distributed nanogels. For example, when ovalbumin and lysozyme – the two main proteins in hen egg white – were mixed at pH 5.5, and subsequently heated at 80 °C for 90 min, nanogels with stable structures and narrow distributions were obtained [128]. The proteins in the nanogels were shown to be in the denatured state, and the crosslinking points of the nanogels consisted of intermolecular interactions, hydrogen bonds, and disulfide bonds [128]. Another such combination was also examined by mixing and heating ovalbumin and ovotransferrin, to prepare nanogels [129]. By heating the protein with sugar, core–shell-type nanogels were obtained, whereas heating a mixture of chitosan and ovalbumin at 80 °C for 20 min led to the formation of core–shell nanogels with a chitosan shell

and ovalbumin core [130]. Core–shell nanogels with a lysozyme core and a dextran shell were also obtained in similar fashion [131]. It should be noted that, although protein denaturation is not involved, chitosan is able to form nanogels when ammonia is added, in an inverse miniemulsion [132].

2.2.3
Degradable Chemical Nanogels

Although chemical nanogels are stable under environmental changes, the sol–gel transition is difficult to achieve. Recently, chemical nanogels with degradable abilities have been developed by using reversible covalent bonds such as disulfide bonds [23, 35, 36, 122, 133], ester bonds [28, 65, 68], acetal bonds [33, 44], and alkoxyamine units [85–87]. Most of these nanogels are expected to be applied as biodegradable nanogels or for drug release by a specific stimulus. In this section, degradable nanogels are described on the basis of their bond category, while those with the ability to transform their structure between linear polymers and nanogels, in addition to being degradable, are described below. Details regarding the synthetic methods employed were provided in Sections 2.2.1 and 2.2.2.

Disulfide bonds are widely used as reversible covalent bonds because they can be cleaved into thiol groups by reduction, and the thiol groups then oxidized to disulfide bonds. Based on this system, nanogels with disulfide groups were prepared by the polymerization of divinyl compounds with disulfide bonds. For example, nanogels were synthesized in an inverse miniemulsion using either di(methacryloyl polyethyleneglycol) disulfide [35] via ATRP, or 5,5′-dithio-bis(2-nitrobenzoic acid) via a condensation reaction [133]; these compounds were then reduced in the presence of tributyl phosphine and tris(2-carboxyethyl) phosphine, respectively. Furthermore, biomolecules were used for the reduction; the de-crosslinking reaction was accomplished in the presence of glutathione, one of the oligopeptides [23, 36], after which biodegradation, *in vitro* release and bioconjugation were each investigated [36].

Ester bonds are also known as biodegradable units. In particular, PLA has been studied as a biodegradable polymer. Dextran-lactate-2-hydroxyethyl methacrylate was copolymerized with NIPAM in miniemulsion, and PLA introduced in their crosslinking points. The de-crosslinking reaction was performed by hydrolysis using phosphate-buffered saline [28]. Furthermore, core-crosslinked star polymers were prepared as biodegradable polymers. A bislactone, BOD, was polymerized via a ring-opening reaction, and ester bonds were introduced into the core [65, 68]; the hydrolysis was conducted under acidic conditions. In the case of a combination of the ATRP method and the ROP, a part-selective biodegradable reaction can be accomplished [68].

Acetal bonds are easily degradable under acidic conditions. Nanogels with acetal bonds in the crosslinking points were synthesized by the inverse emulsion polymerization of NVF with 2-bis[2,2′-di(N-vinylformamido) ethoxy]propane, and their degradable reaction was carried out under several acidic conditions. As the pH was decreased, the rate of degradation was increased [33]. Furthermore, the hollow

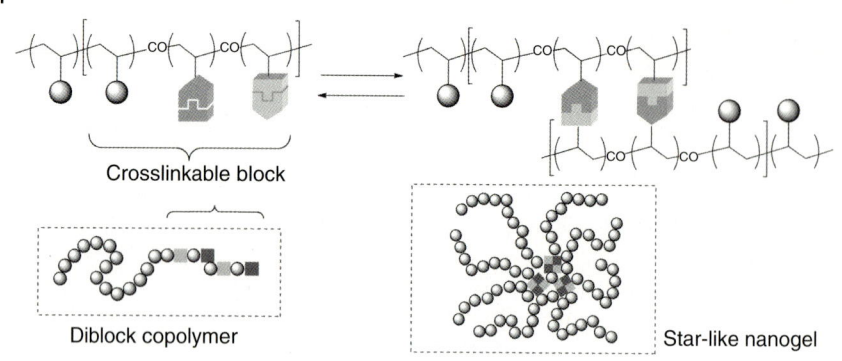

Figure 2.13 Structural transformation between linear diblock copolymers with complementarily reactive alkoxyamine units in the side chains and star-like nanogel.

nanogels prepared using a nanoparticle template method also demonstrated degradable properties [44].

Structural transformations between linear polymers and nanogels have been attempted by designing linear polymers with reversible covalent bonds. For example, diblock copolymers with alkoxyamine units in their side chains were designed. The alkoxyamine employed is an adduct of the styryl radical and nitroxide radical, and their units can crossover under heating conditions. In the first step, the crosslinking reaction was carried out by heating diblock copolymers with alkoxyamine units in the side chains, and star-like nanogels were formed as a result of the radical crossover reaction. Furthermore, the structural transformation from nanogels to diblock copolymers was accomplished by the crossover reaction of the alkoxyamine at the crosslinking points and the addition of low-mass alkoxyamine (Figure 2.13) [85–87]. On the other hand, a structural transformation was attempted using reversible disulfide bonds; an amphiphilic graft copolymer consisting of polysaccharides and NIPAM with SH groups in their grafting chain ends was synthesized by RAFT polymerization; subsequently, at a temperature over the LCST nanogels crosslinked by disulfide bonds were formed by air oxidation. Furthermore, the nanogels were treated by tris(2-carboxyethyl) phosphine, and linear polymers obtained below the LCST. Linear PEG with the thiol group was also synthesized, and its structural transformation accomplished [133].

2.3
Characterization of Polymeric Nanogels

Following their synthesis, the polymeric nanogels must be characterized to confirm the formation of objective nanogels. In particular, the size, size distribution, and shape (chain dimensions) of the nanogels are important factors for understanding their properties and considering their applications. Although many types of characterization method have been developed, there is no simple approach to characterizing the nanogel structures. Consequently, several methods are used, with the combined results being used to confirm the structures. Clearly, in this situation an appropriate characterization method must be selected, and a precise analysis performed.

The two main methods used to characterize polymer nanogels are *scattering* (e.g., dynamic scattering and static scattering) and *microscopy* (e.g., electron microscopy and atomic force microscopy, AFM). The data obtained with these methods are complementary when confirming the nanogel structures, and both types of investigation are generally conducted for synthesized polymeric nanogels. Whilst scattering measurements present average values (as indirect information in reciprocal space), the shapes and sizes of the nanogels can be observed directly by using microscopy; the determination of an average value is difficult, however.

Scattering measurements and microscopic observations are generally performed in solution and bulk states, respectively. Whilst dynamic scattering measurements provide information regarding the hydrodynamic radius (R_h) and size distribution, static scattering measurements provide the weight-average molecular weight (M_w), the radius of gyration (R_g), the second virial coefficient (A_2), and the chain dimensions. Microscopic observations reveal the chain dimension, size, and size distribution.

As an example, Figure 2.14 shows the 3-D AFM images of star-like nanogels with different arm lengths, and the molecular weights of their precursor diblock

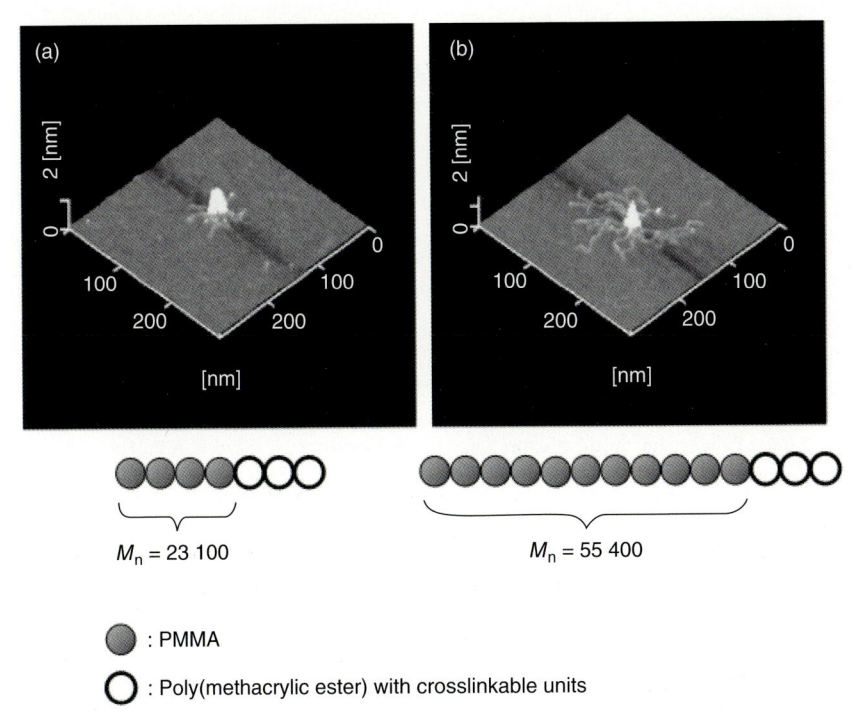

Figure 2.14 Three-dimensional AFM images of star-like nanogels with different arm lengths, and their molecular weights of the corresponding precursor diblock copolymers.

copolymers. In this case, images consisting of gel parts and several connecting chains can be confirmed. The lengths of the connecting chains are seen to depend on the lengths of the corresponding diblock copolymer precursors, indicating that the polymer chains can be observed as single chains [87]. Furthermore, hetero-arm star-like nanogels were synthesized from mutually reactive diblock copolymers with different molecular weights, and the connecting chains of two different lengths also confirmed via AFM images [87]. When hairy nanoparticles were investigated by AFM observations, the hairy chains were seen to become longer with increasing reaction times [24].

Other methods, including viscosity, GPC, and thermal analysis are also used when characterizing the nanogels:

- *Viscosity* measurements enable the analysis of the chain dimensions.
- *GPC* provides the relative number average molecular weight (M_n), the molecular weight distribution (M_w/M_n), and the chain dimension.
- *Thermal analysis* can be used to reveal two different components, such as in core–shell structures.

For reviews and additional information on this topic, see Refs [134, 135].

References

1 Oh, J.K., Drumright, R., Siegwart, D.J. and Matyjaszewski, K. (2008) The development of microgels/nanogels for drug delivery applications. *Progress in Polymer Science*, **33**, 448–77.

2 Hamidi, M., Azadi, A. and Rafiei, P. (2008) Hydrogel nanoparticles in drug delivery. *Advanced Drug Delivery Reviews*, **60**, 1638–49.

3 Kabanov, A.V. and Vinogradov, S.V. (2009) Nanogels as pharmaceutical carriers: finite networks of infinite capabilities. *Angewandte Chemie International Edition*, **48**, 5418–29.

4 Zhang, J., Xu, S. and Kumacheva, E. (2004) Polymer microgels: reactors for semiconductor, metal, and magnetic nanoparticles. *Journal of the American Chemical Society*, **126**, 7908–14.

5 Quevedo, J.A., Patel, G. and Pfeffer, R. (2009) Removal of oil from water by inverse fluidization of aerogels. *Industrial and Engineering Chemistry Research*, **48**, 191–201.

6 Oishi, M., Tamura, A., Nakamura, T. and Nagasaki, Y. (2009) A smart nanoprobe based on fluorescence-quenching PEGylated nanogels containing gold nanoparticles for monitoring the response to cancer therapy. *Advanced Functional Materials*, **19**, 827–34.

7 Gota, C., Okabe, K., Funatsu, T., Harada, Y. and Uchiyama, S. (2009) Hydrophilic fluorescent nanogel thermometer for intracellular thermometry. *Journal of the American Chemical Society*, **131**, 2766–7.

8 Mattsson, J., Wyss, H.M., Fernández-Nieves, A., Miyazaki, K., Hu, Z., Reichman, D.R. and Weitz, D.A. (2009) Soft colloids make strong glasses. *Nature*, **462**, 83–6.

9 Alemán, J., Chadwick, A.V., He, J., Hess, M., Horie, K., Jones, R.G., Kratochvíl, P., Meisel, I., Mita, I., Moad, G., Penczek, S., Stepto, R.F.T. and Jones, R.G. (2007) Definitions of terms relating to the structure and processing of sols, gels, networks, and inorganic–organic hybrid materials. *Pure and Applied Chemistry*, **79**, 1801–29.

10 Kim, J.-W., Utada, A.S., Fernández-Nieves, A., Hu, Z. and Weitz, D.A. (2007) Fabrication of monodisperse gel

shells and functional microgels in microfluidic devices. *Angewandte Chemie International Edition*, **46**, 1819–22.
11. Hong, Y., Krsko, P. and Libera, M. (2004) Protein surface patterning using nanoscale PEG hydrogels. *Langmuir*, **20**, 11123–6.
12. Nie, Z., Xu, S., Seo, M., Lewis, P.C. and Kumacheva, E. (2005) Polymer particles with various shapes and morphologies produced in continuous microfluidic reactors. *Journal of the American Chemical Society*, **127**, 8058–63.
13. Pelton, R. (2000) Temperature-sensitive aqueous microgels. *Advances in Colloid and Interface Science*, **85**, 1–33.
14. Pelton, R. and Chibante, P. (1986) Preparation of aqueous lattices with N-isopropylacrylamide. *Colloids and Surfaces A: Physicochemical and Engineering Aspects*, **20**, 247–56.
15. Gao, J. and Hu, Z. (2002) Optical properties of *N*-isopropylacrylamide microgel spheres in water. *Langmuir*, **18**, 1360–7.
16. Wu, J., Huang, G. and Hu, Z. (2003) Interparticle potential and the phase behavior of temperature-sensitive microgel dispersions. *Macromolecules*, **36**, 440–8.
17. Ikkai, F. and Shibayama, M. (2007) Gel-size dependence of temperature-induced microphase separation in weakly-charged polymer gels. *Polymer*, **48**, 2387–94.
18. Miyake, M., Ogawa, K. and Kokufuta, E. (2006) Light-Scattering study of polyelectrolyte complex formation between anionic and cationic nanogels in an aqueous salt-free system. *Langmuir*, **22**, 7335–41.
19. Ogawa, K., Nakayama, A. and Kokufuta, E. (2003) Preparation and characterization of thermosensitive polyampholyte nanogels. *Langmuir*, **19**, 3178–84.
20. Sahiner, N., Alb, A.M., Graves, R., Mandal, T., McPherson, G.L., Reed, W.F. and John, V.T. (2007) Core-shell nanohydrogel structures as tunable delivery systems. *Polymer*, **48**, 704–11.
21. Cunningham, M.F. (2008) Controlled/living radical polymerization in aqueous dispersed systems. *Progress in Polymer Science*, **33**, 365–98.
22. Matyjaszewski, K. and Tsarevsky, N.V. (2009) Nanostructured functional materials prepared by atom transfer radical polymerization. *Nature Chemistry*, **1**, 276–88.
23. Dong, H., Mantha, V. and Matyjaszewski, K. (2009) Thermally responsive PM(EO)$_2$MA magnetic microgels via activators generated by electron transfer atom transfer radical polymerization in miniemulsion. *Chemistry of Materials*, **21**, 3965–72.
24. Min, K., Gao, H., Yoon, J.A., Wu, W., Kowalewski, T. and Matyjaszewski, K. (2009) One-pot synthesis of hairy nanoparticles by emulsion ATRP. *Macromolecules*, **42**, 1597–603.
25. Xu, L.Q., Yao, F., Fu, G.-D. and Shen, L. (2009) Simultaneous "click chemistry" and atom transfer radical emulsion polymerization and prepared well-defined cross-linked nanoparticles. *Macromolecules*, **42**, 6385–92.
26. Zetterlund, P.B., Alam, N. and Okubo, M. (2009) Effects of the oil–water interface on network formation in nanogel synthesis using nitroxide-mediated radical copolymerization of styrene/divinylbenzene in miniemulsion. *Polymer*, **50**, 5661–7.
27. Sisson, A.L., Steinhilber, D., Rossow, T., Welker, P., Licha, K. and Haag, R. (2009) Biocompatible functionalized polyglycerol microgels with cell penetrating properties. *Angewandte Chemie International Edition*, **48**, 7540–5.
28. Huang, X., Misra, G.P., Vaish, A., Flanagan, J.M., Sutermaster, B. and Lowe, T.L. (2008) Novel nanogels with both thermoresponsive and hydrolytically degradable properties. *Macromolecules*, **41**, 8339–45.
29. Sahiner, N., Godbey, W.T., McPherson, G.L. and John, V.T. (2006) Microgel, nanogel and hydrogel–hydrogel semi-IPN composites for biomedical applications: synthesis and characterization. *Colloid and Polymer Science*, **284**, 1121–9.
30. McAllister, K., Sazani, P., Adam, M., Cho, M.J., Rubinstein, M., Samulski, R.J. and DeSimone, J.M. (2002)

Polymeric nanogels produced via inverse microemulsion polymerization as potential gene and antisense delivery agents. *Journal of the American Chemical Society*, **124**, 15198–207.

31 Raemdonck, K., Naeye, B., Buyens, K., Vandenbroucke, R.E., Høgset, A., Demeester, J. and De Smedt, S.C. (2009) Biodegradable dextran nanogels for RNA interference: focusing on endosomal escape and intracellular siRNA delivery. *Advanced Functional Materials*, **19**, 1406–15.

32 Bhardwaj, P., Singh, V., Aggarwal, S. and Mandal, U.K. (2009) Poly(acrylamide-*co*-2-acrylamido-2-methyl-1-propanesulfonic acid) nanogels made by inverse microemulsion polymerization. *Journal of Macromolecular Science*, Part A, **46**, 1083–94.

33 Shi, L., Khondee, S., Linz, T.H. and Berkland, C. (2008) Poly(*N*-vinylformamide) nanogels capable of pH-sensitive protein release. *Macromolecules*, **41**, 6546–54.

34 Oh, J.K., Bencherif, S.A. and Matyjaszewski, K. (2009) Atom transfer radical polymerization in inverse miniemulsion: a versatile route toward preparation and functionalization of microgels/nanogels for targeted drug delivery applications. *Polymer*, **50**, 4407–23.

35 Oh, J.K., Tang, C., Gao, H., Tsarevsky, N.V. and Matyjaszewski, K. (2006) Inverse miniemulsion ATRP: a new method for synthesis and functionalization of well-defined water-soluble/cross-linked polymeric particles. *Journal of the American Chemical Society*, **128**, 5578–84.

36 Oh, J.K., Siegwart, D.J., Lee, H., Sherwood, G., Peteanu, L., Hollinger, J.O., Kataoka, K. and Matyjaszewski, K. (2007) Biodegradable nanogels prepared by atom transfer radical polymerization as potential drug delivery carriers: synthesis, biodegradation, in vitro release, and bioconjugation. *Journal of the American Chemical Society*, **129**, 5939–45.

37 Bradley, M., Vincent, B., Warren, N., Eastoe, J. and Vesperinas, A. (2006) Photoresponsive surfactants in microgel dispersions. *Langmuir*, **22**, 101–5.

38 Mears, S.J., Deng, Y., Cosgrove, T. and Pelton, R. (1997) Structure of sodium dodecyl sulfate bound to a poly(NIPAM) microgel particle. *Langmuir*, **13**, 1901–6.

39 Kazakov, S. and Levon, K. (2006) Liposome-nanogel structures for future pharmaceutical applications. *Current Pharmaceutical Design*, **12**, 4713–28.

40 Kazakov, S., Kaholek, M., Teraoka, I. and Levon, K. (2002) UV-induced gelation on nanometer scale using liposome reactor. *Macromolecules*, **35**, 1911–20.

41 Kazakov, S., Kaholek, M., Kudasheva, D., Teraoka, I., Cowman, M.K. and Levon, K. (2003) Poly(*N*-isopropylacrylamide-*co*-1-vinylimidazole) hydrogel nanoparticles prepared and hydrophobically modified in liposome reactors: atomic force microscopy and dynamic light scattering study. *Langmuir*, **19**, 8086–93.

42 Schillemans, J.P., Flesch, F.M., Hennink, W.E. and van Nostrum, C.F. (2006) Synthesis of bilayer-coated nanogels by selective cross-linking of monomers inside liposomes. *Macromolecules*, **39**, 5885–90.

43 Van Thienen, T.G., Raemdonck, K., Demeester, J. and De Smedt, S.C. (2007) Protein release from biodegradable dextran nanogels. *Langmuir*, **23**, 9794–801.

44 Shi, L. and Berkland, C. (2007) Acid-labile polyvinylamine micro- and nanogel capsules. *Macromolecules*, **40**, 4635–43.

45 Singh, N. and Lyon, L.A. (2007) Au nanoparticle templated synthesis of pNIPAm nanogels. *Chemistry of Materials*, **19**, 719–26.

46 Liu, X. and Basu, A. (2009) Core functionalization of hollow polymer nanocapsules. *Journal of the American Chemical Society*, **131**, 5718–19.

47 Zhang, K., Gui, Z., Chen, D. and Jiang, M. (2009) Synthesis of small polymeric nanoparticles sized below 10 nm via polymerization of a cross-linker in a glassy polymer matrix. *Chemical Communications*, 6234–6.

48 Garcia, A., Marquez, M., Cai, T., Rosario, R., Hu, Z., Gust, D., Hayes, M., Vail, S.A. and Park, C.-D. (2007) Photo-,

thermally, and pH-responsive microgels. *Langmuir*, **23**, 224–9.

49 Huang, G. and Hu, Z. (2007) Phase behavior and stabilization of microgel arrays. *Macromolecules*, **40**, 3749–56.

50 Wong, J.E., Müller, C.B., Díez-Pascual, A.M. and Richtering, W. (2009) Study of layer-by-layer films on thermoresponsive nanogels using temperature-controlled dual-focus fluorescence correlation spectroscopy. *Journal of Physical Chemistry B*, **113**, 15907–13.

51 Keerl, M., Pedersen, J.S. and Richtering, W. (2009) Temperature sensitive copolymer microgels with nanophase separated structure. *Journal of the American Chemical Society*, **131**, 3093–7.

52 Berndt, I., Pedersen, J.S. and Richtering, W. (2006) Temperature-sensitive core–shell microgel particles with dense shell. *Angewandte Chemie International Edition*, **45**, 1737–41.

53 Iijima, M., Yoshimura, M., Tsuchiya, T., Tsukada, M., Ichikawa, H., Fukumori, Y. and Kamiya, H. (2008) Direct measurement of interactions between stimulation-responsive drug delivery vehicles and artificial mucin layers by colloid probe atomic force microscopy. *Langmuir*, **24**, 3987–92.

54 Li, X., Zuo, J., Guo, Y. and Yuan, X. (2004) Preparation and characterization of narrowly distributed nanogels with temperature-responsive core and pH-responsive shell. *Macromolecules*, **37**, 10042–6.

55 Suzuki, D., Tsuji, S. and Kawaguchi, H. (2007) Janus microgels prepared by surfactant-free Pickering emulsion-based modification and their self-assembly. *Journal of the American Chemical Society*, **129**, 8088–9.

56 Zhang, Q., Zha, L., Ma, J. and Liang, B. (2009) A novel route to prepare pH- and temperature-sensitive nanogels via a semibatch process. *Journal of Colloid and Interface Science*, **330**, 330–6.

57 An, Z., Shi, Q., Tang, W., Tsung, C.-K., Hawker, C.J. and Stucky, G.D. (2007) Facile RAFT Precipitation polymerization for the microwave-assisted synthesis of well-defined, double hydrophilic block copolymers and nanostructured hydrogels. *Journal of the American Chemical Society*, **129**, 14493–9.

58 Shen, X., Zhang, L., Jiang, X., Hu, Y. and Guo, J. (2007) Reversible surface switching of nanogel triggered by external stimuli. *Angewandte Chemie International Edition*, **46**, 7104–7.

59 Blencowe, A., Tan, J.F., Goh, T.K. and Qiao, G.G. (2009) Core cross-linked star polymers via controlled radical polymerization. *Polymer*, **50**, 5–32.

60 Gao, H. and Matyjaszewski, K. (2009) Synthesis of functional polymers with controlled architecture by CRP of monomers in the presence of cross-linkers: from stars to gels. *Progress in Polymer Science*, **34**, 317–50.

61 Terashima, T., Ouchi, M., Ando, T., Kamigaito, M. and Sawamoto, M. (2007) Amphiphilic, thermosensitive ruthenium(II)-bearing star polymer catalysts: one-pot synthesis of PEG armed star polymers with ruthenium(II)-enclosed microgel cores via metal-catalyzed living radical polymerization. *Macromolecules*, **40**, 3581–8.

62 He, T., Adams, D.J., Butler, M.F., Cooper, A.I. and Rannard, S.P. (2009) Polymer nanoparticles: shape-directed monomer-to-particle synthesis. *Journal of the American Chemical Society*, **131**, 1495–501.

63 Shibata, T., Kanaoka, S. and Aoshima, S. (2006) Quantitative synthesis of star-shaped poly(vinyl ether)s with a narrow molecular weight distribution by living cationic polymerization. *Journal of the American Chemical Society*, **128**, 7497–504.

64 Kanaoka, S., Yagi, N., Fukuyama, Y., Aoshima, S., Tsunoyama, H., Tsukuda, T. and Sakurai, H. (2007) Thermosensitive gold nanoclusters stabilized by well-defined vinyl ether star polymers: reusable and durable catalysts for aerobic alcohol oxidation. *Journal of the American Chemical Society*, **129**, 12060–1.

65 Wiltshire, J.T. and Qiao, G.G. (2006) Degradable core cross-linked star polymers via ring-opening polymerization. *Macromolecules*, **39**, 4282–5.

66 Xiong, M.-H., Wu, J., Wang, Y.-C., Li, L.-S., Liu, X.-B., Zhang, G.-Z., Yan, L.-F. and Wang, J. (2009) Synthesis of PEG-armed and polyphosphoester core-cross-linked nanogel by one-step ring-opening polymerization. *Macromolecules*, **42**, 893–6.

67 Lapienis, G. (2009) Star-shaped polymers having PEO arms. *Progress in Polymer Science*, **34**, 852–92.

68 Wiltshire, J.T. and Qiao, G.G. (2006) Selectively degradable core cross-linked star polymers. *Macromolecules*, **39**, 9018–27.

69 Gao, H. and Matyjaszewski, K. (2008) Synthesis of star polymers by a new "core-first" method: sequential polymerization of cross-linker and monomer. *Macromolecules*, **41**, 1118–25.

70 Goh, T.K., Sulistio, A.P., Blencowe, A., Johnson, J.W. and Qiao, G.G. (2007) Synthesis and characterization of core cross-linked star clusters by conventional free-radical polymerization. *Macromolecules*, **40**, 7819–26.

71 Jamróz-Piegza, M., Wałach, W., Dworak, A. and Trzebicka, B. (2008) Polyether nanoparticles from covalently crosslinked copolymer micelles. *Journal of Colloid and Interface Science*, **325**, 141–8.

72 Yusa, S., Sugahara, M., Endo, T. and Morishima, Y. (2009) Preparation and characterization of a pH-responsive nanogel based on a photo-cross-linked micelle formed from block copolymers with controlled structure. *Langmuir*, **25**, 5258–65.

73 He, J., Tong, X. and Zhao, Y. (2009) Photoresponsive nanogels based on photocontrollable cross-links. *Macromolecules*, **42**, 4845–52.

74 Sakellariou, G., Avgeropoulos, A., Hadjichristidis, N., Mays, J.W. and Baskaran, D. (2009) Functionalized organic nanoparticles from core-crosslinked poly(4-vinylbenzocyclobutene-b-butadiene) diblock copolymer micelles. *Polymer*, **50**, 6202–11.

75 Fujii, S., Cai, Y., Weaver, J.V.M. and Armes, S.P. (2005) Syntheses of shell cross-linked micelles using acidic ABC triblock copolymers and their application as pH-responsive particulate emulsifiers. *Journal of the American Chemical Society*, **127**, 7304–5.

76 Wang, X., Liu, K., Arsenault, A.C., Rider, D.A., Ozin, G.A., Winnik, M.A. and Manners, I. (2007) Shell-cross-linked cylindrical polyisoprene-b-polyferrocenylsilane (PI-*b*-PFS) block copolymer micelles: one-dimensional (1D) organometallic nanocylinders. *Journal of the American Chemical Society*, **129**, 5630–9.

77 Vinogradov, S.V. (2006) Colloidal microgels in drug delivery applications. *Current Pharmaceutical Design*, **12**, 4703–12.

78 Vinogradov, S.V., Kohli, E. and Zeman, A.D. (2006) Comparison of nanogel drug carriers and their formulations with nucleoside 50-triphosphates. *Pharmaceutical Research*, **23**, 920–30.

79 Li, W. and Matyjaszewski, K. (2009) Star polymers via cross-linking amphiphilic macroinitiators by AGET ATRP in aqueous media. *Journal of the American Chemical Society*, **131**, 10378–9.

80 Li, W., Matyjaszewski, K., Albrecht, K. and Müller, M. (2009) Reactive surfactants for polymeric nanocapsules via interfacially confined miniemulsion ATRP. *Macromolecules*, **42**, 8228–33.

81 Rieger, J., Grazon, C., Charleux, B., Alaimo, D. and Jérôme, C. (2009) PEGylated Thermally responsive block copolymer micelles and nanogels via in situ RAFT aqueous dispersion polymerization. *Journal of Polymer Science Part A–Polymer Chemistry*, **47**, 2373–90.

82 Hayashi, H., Iijima, M., Kataoka, K. and Nagasaki, Y. (2004) pH-Sensitive nanogel possessing reactive PEG tethered chains on the surface. *Macromolecules*, **37**, 5389–96.

83 Yang, Z. and Ding, J. (2008) A thermosensitive and biodegradable physical gel with chemically crosslinked nanogels as the building block. *Macromolecular Rapid Communications*, **29**, 751–6.

84 Amamoto, Y., Higaki, Y., Matsuda, Y., Otsuka, H. and Takahara, A. (2007) Programmed thermodynamic formation and structure analysis of star-like

nanogels with core cross-linked by thermally exchangeable dynamic covalent bonds. *Journal of the American Chemical Society*, **129**, 13298–304.

85 Amamoto, Y., Higaki, Y., Matsuda, Y., Otsuka, H. and Takahara, A. (2007) Programmed formation of nanogels via a radical crossover reaction of complementarily reactive diblock copolymers. *Chemistry Letters*, **35**, 774–5.

86 Amamoto, Y., Maeda, T., Kikuchi, M., Otsuka, H. and Takahara, A. (2009) Rational approach to star-like nanogels with different arm lengths: formation by dynamic covalent exchange and their imaging. *Chemical Communications*, 689–91.

87 Amamoto, Y., Kikuchi, M., Masunaga, H., Sasaki, S., Otsuka, H. and Takahara, A. (2010) Intelligent build-up of complementarily reactive diblock copolymers via dynamic covalent exchange toward symmetrical and miktoarm star-like nanogels. *Macromolecules*, **43**, 1785–91.

88 Kadlubowski, S., Grobelny, J., Olejniczak, W., Cichomski, M. and Ulanski, P. (2003) Pulses of fast electrons as a tool to synthesize poly(acrylic acid) nanogels. Intramolecular cross-linking of linear polymer chains in additive-free aqueous solution. *Macromolecules*, **36**, 2484–92.

89 Schmidt, T., Janik, I., Kadłubowski, S., Ulański, P., Rosiak, J.M., Reichelt, R. and Arndt, K. (2005) Pulsed electron beam irradiation of dilute aqueous poly(vinyl methyl ether) solutions. *Polymer*, **46**, 9908–18.

90 Xu, D., Hong, J., Sheng, K., Dong, L. and Yao, S. (2007) Preparation of polyethyleneimine nanogels via photo-Fenton reaction. *Radiation Physics and Chemistry*, **76**, 1606–11.

91 Akiyama, Y., Fujiwara, T., Takeda, S., Izumi, Y. and Nishijima, S. (2007) Preparation of stimuli-responsive protein nanogel by quantum-ray irradiation. *Colloid and Polymer Science*, **285**, 801–7.

92 Beck, J.B., Killops, K.L., Kang, T., Sivanandan, K., Bayles, A., Mackay, M.E., Wooley, K.L. and Hawker, C.J. (2009) Facile preparation of nanoparticles by intramolecular cross-linking of isocyanate functionalized copolymers. *Macromolecules*, **42**, 5629–35.

93 Kuckling, D., Vo, C.D., Adler, H.-J.P., Völkel, A. and Cölfen, H. (2006) Preparation and characterization of photo-cross-linked thermosensitive PNIPAAm nanogels. *Macromolecules*, **39**, 1585–91.

94 Shi, D., Matsusaki, M., Kaneko, T. and Akashi, M. (2008) Photo-cross-linking and cleavage induced reversible size change of bio-based nanoparticles. *Macromolecules*, **41**, 8167–72.

95 Yan, M., Ge, J., Liu, Z. and Ouyang, P. (2006) Encapsulation of single enzyme in nanogel with enhanced biocatalytic activity and stability. *Journal of the American Chemical Society*, **128**, 11008–9.

96 Yan, M., Liu, Z., Lu, D. and Liu, Z. (2007) Fabrication of single carbonic anhydrase nanogel against denaturation and aggregation at high temperature. *Biomacromolecules*, **8**, 560–5.

97 Mok, H. and Park, T.G. (2006) PEG-Assisted DNA solubilization in organic solvents for preparing cytosol specifically degradable PEG/DNA nanogels. *Bioconjugate Chemistry*, **17**, 1369–72.

98 Bae, K.H., Mok, H. and Park, T.G. (2008) Synthesis, characterization, and intracellular delivery of reducible heparin nanogels for apoptotic cell death. *Biomaterials*, **29**, 3376–83.

99 Seo, M., Beck, B.J., Paulusse, J.M.J., Hawker, C.J. and Kim, S.Y. (2008) Polymeric nanoparticles via noncovalent cross-linking of linear chains. *Macromolecules*, **41**, 6413–18.

100 Foster, E.J., Berda, E.B. and Meijer, E.W. (2009) Metastable supramolecular polymer nanoparticles via intramolecular collapse of single polymer chains. *Journal of the American Chemical Society*, **131**, 6964–6.

101 Cao, R., Gu, Z., Hsu, L., Patterson, G.D. and Armitage, B.A. (2003) Synthesis and characterization of thermoreversible biopolymer microgels based on hydrogen bonded nucleobase pairing. *Journal of the American Chemical Society*, **125**, 10250–6.

102 Harada, A. and Kataoka, K. (1999) Chain length recognition: core-shell supramolecular assembly from oppositely charged block copolymers. *Science*, **283**, 65–7.

103 Dong, W.-F., Kishimura, A., Anraku, Y., Chuanoi, S. and Kataoka, K. (2009) Monodispersed polymeric nanocapsules: spontaneous evolution and morphology transition from reducible hetero-PEG PIC micelles by controlled degradation, *Journal of the American Chemical Society* **131**, 3804–5.

104 Lee, Y., Ishii, T., Cabral, H., Kim, H.J., Seo, J.-H., Nishiyama, N., Oshima, H., Osada, K. and Kataoka, K. (2009) Charge-conversional polyionic complex micelles–efficient nanocarriers for protein delivery into cytoplasm. *Angewandte Chemie International Edition*, **48**, 5309–12.

105 Lee, J.I. and Yoo, H.S. (2008) Pluronic decorated-nanogels with temperature-responsive volume transitions, cytotoxicities, and transfection efficiencies. *European Journal of Pharmaceutics and Biopharmaceutics*, **70**, 506–13.

106 Hong, J.S., Vreeland, W.N., DePaoli Lacerda, S.H., Locascio, L.E., Gaitan, M. and Raghavan, S.R. (2008) Liposome-templated supramolecular assembly of responsive alginate nanogels. *Langmuir*, **24**, 4092–6.

107 Bronich, T.K., Bontha, S., Shlyakhtenko, L.S., Bromberg, L., Hatton, T.A. and Kabanov, A.V. (2006) Template-assisted synthesis of nanogels from Pluronic-modified poly(acrylic acid). *Journal of Drug Targeting*, **14**, 357–66.

108 Zhang, M., Ishii, A., Nishiyama, N., Matsumoto, S., Ishii, T., Yamasaki, Y. and Kataoka, K. (2009) PEGylated calcium phosphate nanocomposites as smart environment-sensitive carriers for siRNA delivery. *Advanced Materials*, **21**, 3520–5.

109 Jin, W., Xu, P., Zhan, Y., Shen, Y., Van Kirk, E.A., Alexander, B., Murdoch, W.J., Liu, L. and Isaak, D.D. (2007) Degradable cisplatin-releasing core-shell nanogels from zwitterionic poly(β-aminoester)-graft-PEG for cancer chemotherapy. *Drug Delivery*, **14**, 279–86.

110 Andres, P.R. and Schubert, U.S. (2004) New functional polymers and materials based on 2,2′:6′,2″-terpyridine metal complexes. *Advanced Materials*, **16**, 1043–68.

111 Ievins, A.D., Wang, X., Moughton, A.O., Skey, J. and O'Reilly, R.K. (2008) Synthesis of core functionalized micelles and shell cross-liked nanoparticles. *Macromolecules*, **41**, 2998–3006.

112 Nagahama, K., Hashizume, M., Yamamoto, H., Ouchi, T. and Ohya, Y. (2009) Hydrophobically modified biodegradable poly(ethylene glycol) copolymers that form temperature-responsive nanogels. *Langmuir*, **25**, 9734–40.

113 Akiyoshi, K., Deguchi, S., Moriguchi, N., Yamaguchi, S. and Sunamoto, J. (1993) Self-aggregates of hydrophobized polysaccharides in water. Formation and characteristics of nanoparticles. *Macromolecules*, **26**, 3062–8.

114 Nishikawa, T., Akiyoshi, K. and Sunamoto, J. (1996) Macromolecular complexation between bovine serum albumin and the self-assembled hydrogel nanoparticle of hydrophobized polysaccharides. *Journal of the American Chemical Society*, **118**, 6110–15.

115 Inomoto, N., Osaka, N., Suzuki, T., Hasegawa, U., Ozawa, Y., Endo, H., Akiyoshi, K. and Shibayama, M. (2009) Interaction of nanogel with cyclodextrin or protein: study by dynamic light scattering and small-angle neutron scattering. *Polymer*, **50**, 541–6.

116 Kobayashi, H., Katakura, O., Morimoto, N., Akiyoshi, K. and Kasugai, S. (2009) Effects of Cholesterol-bearing pullulan (CHP)-nanogels in combination with prostaglandin E1 on wound healing. *Journal of Biomedical Materials Research Part B: Applied Biomaterials*, **91**, 55–61.

117 Yamane, S., Sasaki, Y. and Akiyoshi, K. (2008) Siloxane-crosslinked polysaccharide nanogels for potential biomedical applications. *Chemistry Letters*, **37**, 1282–3.

118 Akiyama, E., Morimoto, N., Kujawa, P., Ozawa, Y., Winnik, F.M. and Akiyoshi, K. (2007) Self-assembled nanogels of cholesteryl-modified polysaccharides:

effect of the polysaccharide structure on their association characteristics in the dilute and semidilute regimes. *Biomacromolecules*, **8**, 2366–73.
119 Ozawa, Y., Sawada, S., Morimoto, N. and Akiyoshi, K. (2009) Self-assembled nanogel of hydrophobized dendritic dextrin for protein delivery. *Macromolecular Bioscience*, **9**, 694–701.
120 Hasegawa, U., Sawada, S., Shimizu, T., Kishida, T., Otsuji, E., Mazda, O. and Akiyoshi, K. (2009) Raspberry-like assembly of cross-linked nanogels for protein delivery. *Journal of Controlled Release*, **140**, 312–17.
121 Morimoto, N., Winnik, F.M. and Akiyoshi, K. (2007) Botryoidal Assembly of cholesteryl-pullulan/poly(*N*-isopropylacrylamide) nanogels. *Langmuir*, **23**, 217–23.
122 Morimoto, N., Qiu, X., Winnik, F.M. and Akiyoshi, K. (2008) Dual Stimuli-Responsive Nanogels by self-assembly of polysaccharides lightly grafted with thiol-terminated poly(*N*-isopropylacrylamide) chains. *Macromolecules*, **41**, 5985–7.
123 Cho, J., Park, W. and Na, K. (2009) Self-organized nanogels from pullulan-g-poly(L-lactide) synthesized by one-pot method: physicochemical characterization and *in vitro* doxorubicin release. *Journal of Applied Polymer Science*, **113**, 2209–16.
124 Nagahama, K., Mori, Y., Ohya, Y. and Ouchi, T. (2007) Biodegradable nanogel formation of polylactide-grafted dextran copolymer in dilute aqueous solution and enhancement of its stability by stereocomplexation. *Biomacromolecules*, **8**, 2135–41.
125 Iwasaki, Y. and Akiyoshi, K. (2004) Design of biodegradable amphiphilic polymers: well-defined amphiphilic polyphosphates with hydrophilic graft chains via ATRP. *Macromolecules*, **37**, 7637–42.
126 Daoud-Mahammed, S., Couvreur, P. and Gref, R. (2007) Novel self-assembling nanogels: stability and lyophilisation studies. *International Journal of Pharmaceutics*, **332**, 185–91.
127 Zhang, D., Hamilton, P.D., Kao, J.L.-F., Venkataraman, S., Wooley, K.L. and Ravi, N. (2007) Formation of nanogel aggregates by an amphiphilic cholesteryl-poly(amidoamine) dendrimer in aqueous media. *Journal of Polymer Science Part A – Polymer Chemistry*, **45**, 2569–75.
128 Yu, S., Yao, P., Jiang, M. and Zhang, G. (2006) Nanogels prepared by self-assembly of oppositely charged globular proteins. *Biopolymers*, **83**, 148–58.
129 Hu, J., Yu, S. and Yao, P. (2007) Stable amphoteric nanogels made of ovalbumin and ovotransferrin via self-assembly. *Langmuir*, **23**, 6358–64.
130 Yu, S., Hu, J., Pan, X., Yao, P. and Jiang, M. (2006) Stable and pH-sensitive nanogels prepared by self-assembly of chitosan and ovalbumin. *Langmuir*, **22**, 2754–9.
131 Li, J., Yu, S., Yao, P. and Jiang, M. (2008) Lysozyme-dextran core-shell nanogels prepared via a green process. *Langmuir*, **24**, 3486–92.
132 Brunel, F., Véron, L., Ladavière, C., David, L., Domard, A. and Delair, T. (2009) Synthesis and structural characterization of chitosan nanogels. *Langmuir*, **25**, 8935–43.
133 Groll, J., Singh, S., Albrecht, K. and Moeller, M. (2009) Biocompatible and degradable nanogels via oxidation reactions of synthetic thiomers in inverse miniemulsion. *Journal of Polymer Science Part A – Polymer Chemistry*, **47**, 5543–9.
134 Borsali, R. and Pecora, R. (eds) (2008) *Soft Matter Characterization*, Springer, New York, USA.
135 Tanaka, T. (ed.) (2000) *Experimental Methods in Polymer Science*, Academic Press, San Diego, USA.

3
Stimulus-Responsive Polymers at Nanointerfaces
Roshan Vasani, Martin Cole, Amanda V. Ellis and Nicolas H. Voelcker

3.1
Introduction

Stimulus-responsive polymers (SRPs) are synthetic macromolecules that undergo changes in their properties in response to a defined external stimulus. A range of responsive polymers exist that are sensitive to different stimuli (triggers) such as light, temperature, electric field, magnetic field, pH, and chemicals. The types of response associated with the stimuli also vary greatly, including structural changes, sol–gel transitions, changes to optical properties, solubility changes, and micelle formation. The most widely studied stimuli are changes in pH, temperature, light intensity or wavelength, ionic strength and electric field strength [1, 2]. Additionally, SRPs can be designed that exhibit responses to more than one type of stimulus, such as temperature- and pH-responsive systems [3] or light- and temperature-responsive polymers [4]. The responsiveness of SRPs is somewhat comparable to the behavior of certain biomacromolecules (proteins, DNA, RNA) in natural systems. This biomimicry, combined with the additional properties of synthetic (abiotic) macromolecules, is being exploited for manifold applications in the life sciences, such as controlled drug release [5–7], cell and tissue culture [1, 8], actuators [9], and sensing [10]. With regards to biological applications, it is important to be aware that the physiologically relevant window of pH, temperature or ionic strength is limited [11], and that any polymer responsive to these environmental factors must be carefully matched to the biological application in mind.

Interfacing SRPs with other, non-stimulus-responsive materials conveys extra advantages, and expands the repertoire of the SRPs. The combination with nanostructured materials such as nanoparticles, nanotubes or nanoporous materials, allows the construction of novel functional materials. For example, liposomes functionalized with poly(*N*-isopropylacrylamide) (PNIPAAm) chains combine the thermoresponsive nature of the polymer with the drug-carrying capacity of the liposome [12]. Likewise, iron oxide nanoparticles combined with PNIPAAm allow for the *in vivo* targeting of the coated nanoparticles by using an externally applied magnetic field. Such targeting can employed for targeted drug delivery [13].

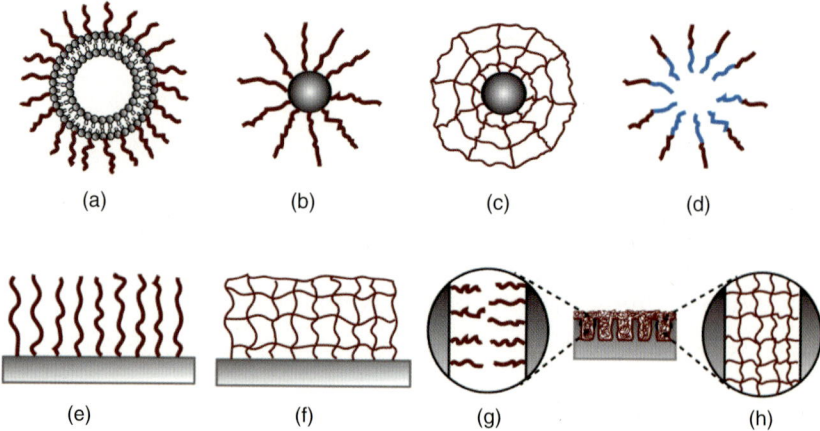

Figure 3.1 Schematic illustration of some key types of stimulus-responsive polymer nanointerfaces. (a) Polymer-decorated liposomes; (b) Polymer chains coupled to nanoparticles; (c) Nanoparticle encapsulated in polymer gels; (d) Diblock copolymer micelles; (e) Polymer brushes grafted to or from a flat surface; (f) Hydrogels grafted onto a flat surface; (g) Polymer chains grafted to or from porous membranes; (h) Polymer hydrogels within porous membranes.

Coating materials with SRPs, for example in the form of polymer brushes, sets the stage for the development of advanced biosensors, scaffolds for tissue engineering, cell culture materials and textiles, as well as microfluidic and microelectromechanical devices [10, 14–17]. SRPs can form nanointerfaces with flat and porous surfaces, and also with different types of inorganic and organic nanoparticles. Some of the key types of SRP nanointerfaces are depicted in Figure 3.1, and these will be the subject of this chapter. In addition to the generation of stimulus-responsive nanointerfaces, recent developments in terms of applying these nanointerfaces to drug delivery, microfluidics, molecular separation, and tissue culture will be highlighted.

3.2
Types of Stimulus-Responsive Polymer

3.2.1
Thermoresponsive Polymers

The class of SRPs that respond to changes in the temperature of their immediate environment are known as *thermoresponsive polymers*. The response is usually manifested as a change in the molecular conformation of the polymer in solution, which leads to a change in a macroscopic property, for example, density or sol–gel state (volume phase transition). At the transition temperature, entropic or enthalpic factors reduce the affinity of the polymer chains to the solvent molecule, while at the same time increasing the affinity of the monomer units to themselves [18].

This results in a breaking of the hydrogen bonds between the polymer and the solvent, and the subsequent formation of intra- and inter-molecular hydrogen bonds within the polymer chains. As a result, the chain expels the solvent molecules and collapses to form a globule, with the hydrophilic domains tucked inwards and the hydrophobic domains exposed at the surface [19]. This transition between an extended hydrophilic coil and a compact hydrophobic globule is typically reversible. *Hydrogels* (crosslinked polymer networks) formed from thermoresponsive polymers have been found to undergo similar volume–phase transitions. However, as opposed to free polymer chains, the hydrogels generally exhibit bulk swelling–shrinking at the transition temperature.

Some polymers become less soluble and collapse when the temperature of their environment is increased to or beyond the transition temperature. In this case, the transition temperature is called the lower critical solution temperature (LCST) of the polymer. In the case of hydrogels, this transition temperature is known as the lower gel transition temperature (LGTT). In contrast to this, some polymers become more soluble on increasing the temperature above the upper critical solution temperature (UCST), or the upper gel transition temperature (UGTT) in the case of hydrogels. These critical temperatures can generally be tuned over a certain range to suit the applications by changing the molecular weight of the polymer, by adding comonomers, or by the addition of additives such as salts or surfactants [20].

3.2.1.1 LCST Polymers

PNIPAAm is arguably the most extensively studied polymer exhibiting LCST behavior [1, 8, 18]. The LCST of this polymer has been shown to be in the range of 31–34 °C [18], which falls into a physiologically relevant temperature regime. The polymer shows good biocompatibility [21], which predestines it for use in biotechnological and biomedical fields, for example as a vehicle for the controlled release of drugs [2, 5, 15], for the capture and release of cells and proteins [1], and for cell sheet engineering [8]. Despite these favorable properties, United States Food and Drug Administration (FDA) approval has not been forthcoming, due to the nondegradability of this polymer *in vivo*, and also to the fact that the acrylamide-based polymers activate platelets on contact with blood [7]. Another thermoresponsive polymer known for its biocompatibility is poly(N-vinyl caprolactam) (PVCL). Similar to PNIPAAm, PVCL exhibits a LCST at around 32 °C, but has an additional interesting property, namely that it is able to form complexes with organic compounds – an ability that has been exploited for the purification of enzymes [22]. One other thermoresponsive polymer that has recently come under scrutiny is poly(methyl vinyl ether) (PMVE) [23], mainly because of its physiologically relevant LCST of 37 °C. Finally, poly(2-isopropyl-2-oxazoline) (PiPrOx) has LCST behavior, and can be regarded as a pseudo-peptide and isomer of PNIPAAm and polyleucine [24]. The LCST of this polymer in aqueous solutions is in the range of 36–70 °C. One interesting property of PiPrOx is that its LCST behavior is only reversible if it is immediately cooled back past its LCST; prolonged exposure of the polymer to elevated temperatures (~65 °C) causes crystallization of the precipitated polymer

[25]. PiPrOx has shown potential for drug-delivery and biomaterials application, because its LCST is close to body temperature and can be easily tuned by altering the molecular weight and concentration [25]; moreover, the polymer is considered to be nontoxic [26].

3.2.1.2 UCST Polymers

As mentioned earlier, there exists another group of thermoresponsive polymers that exhibit a UCST. Copolymers of acrylamide (AAm) and acrylic acid (AAc) [27], triblock copolymers of poly(ethylene oxide) (PEO) and poly(lactic-co-glycolic acid) (PLGA) [28, 29], and also certain copolymers of PNIPAAm [30, 31], have been reported as UCST polymers or polymers showing both UCST and LCST behaviors under certain conditions. Triblock copolymers of PEO and poly(L-lactic acid) (PLLA), which exist as sols above the UCST of about 45 °C and form gels at body temperature, have been studied as injectable drug-delivery materials [32]. Gels of this copolymer can be mixed, loaded with drugs and heated to the UCST (sol phase) for injecting into the body; on cooling, a gel is formed that demonstrates a sustained release of the drug. Block copolymers of poly[2-(3-ethoxy)ethoxyethyl vinyl ether] (PEOEOVE) and poly(2-methoxyethyl vinyl ether) (PMOVE) are also examples of polymers that show both LCST and UCST behaviors [33]. At room temperature, PEOEOVE-PMOVE are completely soluble in water, but on heating the solution to 42 °C a clear gel is formed; subsequent heating of the diblock copolymer to 57 °C results in the formation of a sol once again. The thermoresponsive polymers discussed here, along with their properties and applications, are listed in Table 3.1.

3.2.2
pH-Responsive Polymers

pH-responsive polymers respond to small changes in the pH of their local environment. This is of interest to biomedical applications, since the pH varies in different parts of the body, and also among different types of tissues and cell components (Table 3.2). Additionally, infected, inflamed and cancerous tissues generally have a different pH profile to normal and healthy tissues [2]. Hence, polymers and polymeric materials that can respond to changes in pH can, potentially, be used to target specific areas of the body, tissues, or cellular components for the delivery of therapeutics. In addition to this, pH-responsive polymers have also been used for enzyme immobilization, molecular separation, and actuators. pH-responsive polymers carry ionizable groups (e.g., amino, carboxylic or phosphoric groups) in their side or main chains, the pK_a values of which range from 3 to 10 [2]. They are also classified as weak polyelectrolytes and, depending on the type of ionizable group, they can be either polycations or polyanions.

3.2.2.1 Polycations
Polycations are weak polybases having weakly basic groups in their structure. As a result, these polymers become protonated at low pH, leading to conformational

Table 3.1 A summary of the thermoresponsive polymers discussed, and their properties and applications.

Polymer	Abbreviation	LCST/UCST	Biodegradable	Biocompatible	Applications	Reference
Poly(N-isopropylacrylamide)	PNIPAAm	LCST	No	+++	Drug-delivery systems, tissue culture surfaces	[8]
Poly(N-vinyl caprolactam)	PVCL	LCST	No	+++	Purification of enzymes	[34]
Poly(methyl vinyl ether)	PMVE	LCST	No	Not tested	–	[23]
Poly(2-isopropyl-2-oxazoline)	PiPrOx	LCST	Yes	+++	Stable crystalline microspheres for protein recognition	[25]
Poly(acrylamide-co-acrylic acid)	P(AAm-co-AAc)	UCST	No	++	UCST-driven functional gates for flow control	[27]
Poly(ethylene oxide-b-(lactic-co-glycolic acid)-b-ethylene oxide)	PEO-PLGA-PEO	UCST	Yes	+++	Sustained drug release	[29]
Poly(ethylene oxide-b-(L-lactic acid)-b-ethylene oxide)	PEO-PLLA-PEO	UCST	Yes	+++	Sustained drug release *in vivo*	[32]
Poly[(2-(3-ethoxy)ethoxyethyl vinyl ether)-b-(2-methoxyethyl vinyl ether)]	PEOEOVE-PMOVE	Both	No	Not tested	–	[33]
Poly(N-isopropylacrylamide)-gelatin conjugates	PNIPAAm-Gelatin	Both	Yes	++	Injectable cell scaffolds	[31]

3.2 *Types of Stimulus-Responsive Polymer* | 63

Table 3.2 pH values in various tissues and cellular compartments. Adapted from Ref. [2].

Tissue/cellular compartment	pH
Blood	7.4–7.5
Stomach	1.3–3.0
Duodenum	4.8–8.2
Colon	7.0–7.5
Early endosome	6.0–6.5
Late endosome	5.0–6.0
Lysosome	4.5–5.0
Golgi	6.4
Tumor, extracellular	7.2–6.5

changes of the polymer chains. The presence of the positive charges on adjacent monomers increases the repulsive forces between them, and causes the polymer chain to extend. However, as the pH of the environment is increased, the polymer deprotonates, and this leads to increased intramolecular interactions, driven by electrostatic forces such as hydrogen bonding or by hydrophobic effects, that in turn cause a collapse of the polymer chains [35]. In the case of crosslinked polycationic hydrogels, a swelling–deswelling behavior is seen on changing from low to high pH. Poly(N,N-dimethylaminoethylmethacrylate) (PDMAEMA) is a good example of this effect. Surface-grafted PDMAEMA expands and increases surface wettability at low pH (pH = 2), but collapses at high pH (pH = 9), rendering the surface more hydrophobic (Figure 3.2a) [35]. This change in wettability is completely reversible (Figure 3.2b–c). Another example of a polycationic polymer is chitosan, a copolymer of β-(1,4)-linked-2-acetamido-2-deoxy-D-glucopyranose and 2-amino-2-deoxy-D-glucopyranose that is obtained from naturally occurring chitin by deactylation. This polymer changes from a swollen to a collapsed state at a physiologically relevant pH of about 7. A chitosan hydrogel crosslinked with glycidoxypropyltrimethoxysilane (GPTMS), referred to as chitosan/GPTMS, has been grafted onto a porous silicon film and used as a nanovalve to control the release of insulin molecules loaded into the pores [36].

Polycationic SRP brushes made from poly([2-(methacryloyloxy)ethyl]trimethylammoniumchloride) (PMETAC) can be switched between the collapsed or extended states [37] via immersion in 1 M NaCl or pure water, respectively. Interestingly, the polymer can be locked in the collapsed state by the addition of 0.5 M $MgSO_4$ (subsequent water immersion was unable to reswell the chains). Washing with 1 M NaCl then "unlocked" the $MgSO_4$, inducing conformational collapse [37].

3.2.2.2 Polyanions

Polyanions are weak polyacids with acidic groups on the polymer main chain or side chains. Those groups are ionized when exposed to solutions with high pH. The ionization (deprotonation) increases the electrostatic repulsion between adjacent monomers in the polymer chain, thus causing the polymer to adopt an

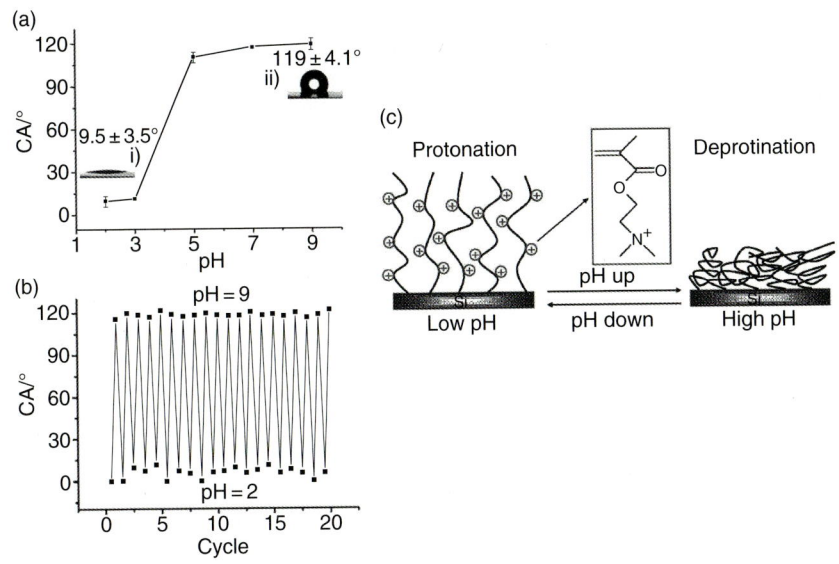

Figure 3.2 (a) Change in contact angle (CA) versus pH of a silicon surface grafted with pH-sensitive polymer PDMAEMA. The images i) and ii) show photographs of the contact angle of the surface pretreated with an aqueous solution at pH 2 and pH 9, respectively; (b) Water CA at two different pH values for a PDMAEMA-modified substrate. Half cycle: pH = 2; integral cycles: pH = 9; (c) Schematic showing the reversible conformations of PDMAEMA at different pH-values [35].

expanded coil conformation. On exposing the polymer to solutions of low pH, the polyanions become protonated and subsequently collapse. Poly(acrylic acid) (PAAc) is a polyacid (pK_a ~ 6) that shows this type of reversible behavior. Surface-grafted brushes of PAAc on a gold substrate collapse at pH 5 and assume an expanded confirmation at pH 9 [38]. Others [39] found that star polymers of PAAc which had hydrodynamic diameters of about 30 nm at pH 9 showed a gradual decrease in diameter to 23 nm upon acidifying the solution to pH 2. However, the collapsed star polymers swelled back to diameters of about 27 nm when the pH was raised back to 9. Another well-studied polyacid is poly(propylacrylic acid) (PPAc) [40], which transitions from a hydrophilic entity at physiological pH to a hydrophobic polymer at pH 6. The hydrophobic form of this polymer was found to have strong cell membrane-disrupting properties, and consequently has been studied extensively for applications in liposome-mediated drug delivery. Other alkylacrylic acid-based polymers, such as poly(ethylacrylic acid) (PEAc) and poly(butylacrylic acid) (PBAc) have also been studied as pH-responsive polymers [41].

3.2.2.3 Polyzwitterions

Zwitterionic polymers have also been employed as pH-responsive materials. For example, Liu *et al.* [42] reported a diblock copolymer, poly(4-vinyl benzoic acid-block-2-(diethylamino)ethyl methacrylate) (PVBA-*b*-PDEA), where the PDEA block is a polycation that can dissolve in acidic pH but becomes insoluble above pH 7.1. In contrast, the PVBA is a polyanion, which is insoluble at acidic pH but becomes soluble above pH 6.2. As a result, the copolymer exists as a PVBA-core micelle at pH 2 and a PDEA-core micelle at pH 10. Additionally, the polymer was found to precipitate in pH 6.6–8.3 solutions because the isoelectric point (IEP) of the polyzwitterion was found to be around neutral pH. The synthesis of a pH-responsive polyzwitterion based on block copolymers PDMAEMA and poly(potassium acrylate) (PPA) has also been reported. Here, the PDMAEMA block of the polymer provides the $-NH_3^+$ group, while the PPA block provides the $-COO^-$ group. This block copolymer has a net positive charge in acidic pH, an IEP of pH 8.9, and a net negative charge above the IEP [43]. Mi et al. prepared a copolymer film composed of positively charged quaternary amine [2-(acryloyloxy)ethyl] trimethyl ammonium chloride (TMA) and negatively charged 2-carboxy ethyl acrylate, (CAA) monomers [44]. This polymer was positively charged in acidic environments (pH 4.5), and had no net charge at neutral and basic pH. Bacterial cells adhered to these films at low pH values, and were released on increasing the pH.

3.2.3
Photoresponsive Polymers

The use of light to control SRPs is particularly intriguing because, unlike temperature and pH, light can be focused onto a region of interest or, by using a mask, applied in a pattern. This sets the stage for more sophisticated, and temporally as well as spatially controlled switching. Photoresponsive polymers contain chromophores, which undergo structural isomerization or cyclization upon irradiation with light of a certain wavelength. This structural change affects the polarity of the polymer, leading to a transition between an extended hydrophilic conformation and a compact hydrophobic globule in the polymer (similar to the situation observed for thermoresponsive polymers), or to a change in the color of the polymer. Photoisomerization is reversible (typically by exposure to another wavelength of light), and so are the macroscopic changes occurring for photoresponsive polymers in solution, in hydrogels, or mounted on surfaces. Unfortunately, unlike thermoresponsive and pH-responsive polymers, photoresponsive polymers are prone to degradation, with the number of cycles achieved before stimulus responsiveness is lost depending on the stability of the molecules towards irreversible photoinduced degradation.

The photoisomerization of azobenzene groups between *cis* and *trans* conformations is an example of photoresponsive switching, which has been employed for applications such as control of wettability at polymer interfaces [45, 46]. Jiang et al. prepared poly{2-[4-phenylazophenoxy] ethyl acrylate-*co*-acrylic acid} by functionalization of poly(acryloyl chloride) with 2-(4-phenylazophenoxy) ethanol to introduce azobenzene units in the polymer's side chains [46]. When the polymer layers were adsorbed onto microstructured silicon pillars, the water contact angle switched repeatedly from 153° (under visible light) to 87° [under ultraviolet (UV) light[[46]. Azobenzene side chains have also been integrated into poly(malonic esters) and poly(silicate esters), where they were investigated for the application of optical memory storage [47]. Other photoisomerizable groups such as spiropyran [1′,3′-dihydro-1′,3′,3′-trimethyl-6-nitrospiro(2H-1-benzopyran-2,2′-2H-indole)] have been used as a dopant in poly(ethyl methacrylate)-co-poly(methyl acrylate) random copolymer films for studies of controlled wettability [48]. Similarly, Cheng et al. used spirooxazine {5-chloro-1,3-dyhydro-1,3,3-trimethylspiro[2H-indole-2,3′-[3H]-naphth-1,4-oxazine]} as a dopant with random copolymers of 3-glycidoxypropyltrimethoxysilane and tetracetoxysilane for studies of photochromic fabrics [17]. The incorporation of spiropyran groups into electrospun poly(vinylidene fluoride-co-hexafluoropropylene) fibers allowed a light-sensitive control over color, without altering the surface properties [49]. Another example of a light-responsive polymer is poly(N-benzyl-N′-(4-vinylbenzyl)-4,4′-bipyridium dichloride) or P(BpyClCl) [50]. The polyelectrolyte BpyClCl monomer contains the light-sensitive viologen group, with a viologen dication (BV^{2+}). Upon UV-irradiation, the transfer of an electron from the counteranion to the BV^{2+} results in the formation of a BV^+ viologen radical cation that readily oxidizes in air to the dication. This switching between monocationic and dicationic states can be used for the reduction of metal ions.

Finally it is also worth mentioning that, in some cases, the coupling of SRPs to certain plasmonic nanomaterials enables the fabrication of indirectly photoresponsive polymers. For example, gold nanoparticles embedded into PNIPAAm gels can be heated by using light of different wavelengths (for different types of particles), causing the subsequent LCST response of PNIPAAm gel [51].

3.2.4
Dual and Multi-Stimulus-Responsive Polymers

SRPs which can respond to multiple environmental triggers are expected to play key roles in the design of materials with incorporated multifunctionality, adapted to meeting the demands of advanced applications in the life sciences, molecular separation, and microfluidics. The extra level of sophistication in dual SRPs may assist in enhancing reversibility by increasing the switching speed and working life of the polymers [1]. SRPs based on homopolymers commonly respond strongly to a particular stimulus, but may also be somewhat responsive to secondary stimuli. Although secondary triggers allow fine tuning of the SRPs' properties, they often do not exhibit a critical transition point. An example of this behavior is the known sensitivity of PNIPAAm [52–55], which is primarily a thermally responsive polymer, towards changes in ionic strength. However, copolymers combining monomer constituents of two or more different SRPs engender true dual or even multi-stimulus-responsiveness. Copolymerization of course requires monomers to be compatible with respect to solvent and reactivity [56–59]. While this area of SRP research is in its infancy, a number of dual and multi-responsive SRPs have been designed [60–64]. Combinations of thermo- and pH-responsive or thermo- and photo-responsive polymers have attracted most attention [63–66]. Xia et al. prepared dual-responsive PNIPAAm-co-PAAc coatings on substrates with a nanostructured topography [63]. The polymer-modified structures featured tunable wettability, such that the water contact angle could be controlled between 8° and 149° by either changing the temperature from 21 to 45 °C or the pH from 2 to 11 [63]. The preparation of SRPs that include photoisomerizable components tends to require additional efforts in organic synthesis [65, 67]. A common approach is to modify a photochromic compound to incorporate a vinyl moiety (typically acrylate or acrylamide), so as to allow copolymerization along with another monomer [64, 65]. An alternative approach includes the addition of a reactive functional group to the photochromic molecule, so that it may be coupled to a copolymer that includes reactive side groups after polymerization [60]. Edahiro et al. fabricated copolymer coatings of temperature-responsive NIPAAm and the photoresponsive nitrospiropyran (NSp) -derivatized acrylamide [66]. This dual stimulus-responsiveness allowed both the temporal and spatial control of cell attachment [66]. Further studies conducted by Edahiro and coworkers have included characterization of the surface hydration of PNSp-NIPAAm coatings by using a quartz crystal microbalance (QCM) [67]. The results showed a rapid hydration of the coating at 19 °C under UV irradiation, but at temperatures between 25 and 35 °C the collapsed property of the PNIPAAm component dominated and UV irradiation did not result in hydration [67]. In an investigation conducted by Garcia et al. [60], a crosslinked copolymer of NIPAAm, allylamine and N,N'-methylenebisacrylamide was prepared in the form of gel particles. Synthesized photoisomerizable NSp-COOH molecules were then coupled to the free amine groups within the copolymer particles, using 1-ethyl-3-(3-dimethylaminopropyl)carbodiimide (EDC) as a coupling reagent. The diameter of the gel particles produced was found to be sensitive to three types of changes, namely pH, light, and temperature. The temperature of the environment was found to provide the greatest control over the size of nano/microgel particles of 65–140 nm [60]. Another triple-responsive copolymer has been reported by Yu et al. [68], who also combined temperature, light and pH sensitivity in a single copolymer. In this study, hydroxyl groups in poly(NIPAAm-co-N-hydroxymethylacrylamide) (PNIPAAm-co-PNHMA) were used to couple a photoresponsive molecule 2-diazo-1,2-naphthoquinone-5-sulfonyl chloride (DNQsc), which undergoes a Wolff rearrangement upon illumination with UV light (405 nm). The structural change in the DNQ to a more polar (hydrophilic) substituent induced an increase in the copolymer's LCST, the magnitude of which was dependent on the duration of exposure to light. The LCST could also be shifted to a higher or lower temperature via pH changes [68]. The incorporation of crown ether molecules into a PNIPAAm coating via copolymerization of NIPAAm and benzo-18-crown-6-acrylamide (BCAm) allowed the coating properties (hydration) to be controlled not only via temperature but also by the potassium ion concentration [69]. The binding of K^+ ions to their crown ether hosts significantly increased the hydration of the coating [70]. The mechanism is based on an increase in the hydrophilic nature when K^+ ions are bound, causing the position of the LCST of the copolymer to shift to higher temperatures, for example, above 37 °C. This effect allows for the manipulation of cell attachment; an example is the selective detachment of dead cells, which release their intracellular K^+ while neighboring live cells remained unaffected [69]. This study exemplifies the potential of dual-stimulus-responsive polymers for the design of cell microreactors. For the sake of completeness, it should be noted that another strategy to produce multi-stimulus-responsive systems consists of attaching homopolymeric SRPs to nanoparticles of gold or iron which are optically and magnetically active materials, respectively (*vide infra*) [71].

3.3 Generating Stimulus-Responsive Interfaces

Although SRPs have applications in bulk and solution form, the potential of these polymers is best harnessed when generating SRPs at interfaces, either in the form of nanoparticles or nanotubes, or in the form of flat or porous surfaces. Stimulus-responsive interfaces can be generated either via covalent or noncovalent routes.

3.3.1 Covalent Routes

The covalent routes, which usually follow one of two approaches – "grafting-from" or "grafting-to" – are employed to permanently immobilize SRPs to surfaces [72, 73].

3.3.1.1 "Grafting-From" Techniques

In the "grafting-from" approach, the surfaces are first functionalized with initiator molecules, such that polymerization occurs directly from the surface. Controlled radical polymerization (CRP) techniques, which include atom transfer radical polymerization (ATRP) [74] and reversible addition–fragmentation chain transfer (RAFT) [75] polymerization, have proven particularly popular for the preparation of graft polymers [72, 76–78], due mainly to the highly controlled nature of chain growth during polymerization. Although a narrow polydispersity is not a critical criterion for responsiveness, controlled growth methods are favorable for the preparation of SRPs, since the molecular weight (MW) can have a significant effect on the responsive nature of the polymers [63, 79]. Other less frequently used CRP techniques include nitroxide-mediated polymerization and photoiniferter-mediated polymerization. Free-radical polymerization and ring-opening polymerization (ROP) have also been employed [80]; the former of these processes constitutes a rapid and simple technique for SRP grafting from surfaces, that can be applied in conjunction with surface-polymerizable groups [81–83]. In all of these techniques, the grafting density can be controlled by the surface density of initiator groups, and also by the graft conditions [72].

3.3.1.2 "Grafting-To" Techniques

In contrast, the "grafting-to" approach involves the anchoring of preformed polymer chains with reactive functionalities at the chain termini onto a surface, by using suitable linker molecules [84–86]. The maximum chain density in this approach is somewhat limited, since the attached polymer chains sterically hinder the proximal attachment of further polymer chains. In an ideal solvent, a surface with a so-called "mushroom" conformation results, where the polymer chains are separated by greater than or equal to twice the radius of gyration (R_g). This can be somewhat rectified by a disruption of chain solvation, allowing the R_g of the polymer chains to be decreased so that higher graft densities may be achieved. A reduction of R_g can be realized by increasing the salt concentration, as is the case when grafting poly(ethylene glycol) (PEG) to surfaces ("cloud point grafting") [87], or by exploiting the stimulus-responsive nature of a given polymer such as PNIPAAm. Thus, when transferred to an ideal solvent, the polymer chains adopt a space-filling or "brush" conformation.

Both, plasma polymerization and electron-beam polymerization are also used to covalently graft SRPs from or to surfaces. Whilst these methods afford a good control over the polymer growth kinetics, they provide very little control over the polymer structure and functionality in comparison to the above-listed wet chemical approaches. The preparation of surface coatings via electron beam polymerization (or radiation grafting) can be achieved via a number of different avenues [88–90]. Typically, the substrates are exposed to the electron beam either upon immersion in monomer solution (wet) [90, 91], or after pre-casting a polymer onto the substrate (dry) [89]. The process involves the generation of radicals both in the monomer/polymer and at the substrate surface, so as to allow covalent attachment and crosslinking of the resultant polymeric coating. Although the electron-beam grafting of SRPs has been quite successful [92–95], it has attracted very little attention when compared to purely wet chemical polymerization approaches, most likely due to the expensive instrumentation required.

Plasma polymerization, on the other hand, involves exposure of the substrate to a plasma gas containing the desired monomer. In this case, deposition is generally independent of the substrate type, and can be controlled with nanometric precision. The variation in deposition power allows control over the extent of fragmentation, crosslinking, and retention of monomer structure. The method is scalable, solvent-free, does not adversely affect the bulk material, and generally affords pinhole-free coatings [96]. The limitations of plasma polymerization include a lack of SRP-forming monomers that have suitable vapor pressure, and a lack of fine control over the polymer structure and functionality [96–98].

3.3.2
Noncovalent Routes

Noncovalent routes to generate stimulus-responsive interfaces generally rely on intermolecular forces, including hydrogen bonding, π–π stacking, van der Waals forces and hydrophobic effects, to self-assemble polymer chains into aggregate structures or to attach them to surfaces. In terms of surface attachment, these methods face issues of steric hindrance similar to those of the grafting-to approaches. Further drawbacks include the nonhomogeneous and unstable attachment of polymers, and the so-called "pancake effect," when the polymers are adsorbed flat onto a substrate and allow minimal rearrangement when compared to end-point-tethered polymers. Nevertheless, noncovalent routes are well suited to the preparation of micelles of amphiphilic block copolymers [99], and also for the coupling of polymers onto metallic or magnetic nanoparticles [100].

3.3.3
Interfacing SRPs with Flat and Porous Substrates

SRP-functionalized flat and porous substrates have found applications in sensing [10], drug delivery [15], cell sheet engineering [8, 16], separation [101], microfluidics [102], and smart optical systems [103]. Flat silicon is a more commonly studied substrate, owing to its biological inertness, its crystallinity, and low roughness [104]. Silicon wafers have been grafted with the thermoresponsive polymer PNIPAAm using surface-initiated ATRP [14, 105, 106], and such surfaces have been tested for the thermoresponsive capture and release of cells. Other polymers, such as pH-sensitive PDMAEMA, have also been grafted onto silicon surfaces using the same technique. Additionally, SRP-modified glass, as well as other transparent substrates such as tissue culture polystyrene (TCPS) [95, 107], are also commonly used as they are optically transparent. Cunliffe et al. prepared PNIPAAm-covered glass surfaces by using a "grafting-to" technique in order to immobilize carboxyl-terminated polymers onto a (3-aminopropyl)triethoxysilane (APTES) -functionalized glass surface [108]. Cheng et al. grafted PNIPAAm from the surface of 4-(chloromethyl)phenyltrichlorosilane-modified glass surfaces using surface-initiated ATRP [109], while Akiyama et al. used electron-beam polymerization to fabricate PNIPAAm hydrogel layers on TCPS dishes [94]. In addition to the above methods, plasma polymerization has been used for the generation PNIPAAm [97, 110, 111] and other SRP (from N,N-diethylacrylamide and 1-amino-2-propanol monomers) [112, 113] -modified surfaces. Surfaces such as gold and mica have also been modified by SPRs [114–116].

Porous materials such as porous silicon and porous alumina have also been investigated as substrates for stimulus-responsive polymers, since control over pore size at the nanoscale is important in molecular separation and drug delivery. Fu et al. employed surface-initiated ATRP to generate PNIPAAm coatings on the surface of porous alumina templates, and examined the changes in surface topology and also the wettability of the surfaces on temperature modulation [117]. Li et al. [118] also used surface-initiated ATRP to graft PNIPAAm onto the inner pore walls of a porous alumina membrane, and demonstrated a thermoresponsive swelling of the polymer and control over the flux of molecules through the membrane. Lokuge et al. [119] prepared a stimulus-responsive porous membrane to control molecular flux via the ATRP-grafting of PNIPAAm onto a gold layer deposited on the pores of track-etched polycarbonate membranes. Chu et al. prepared two SRP coatings on porous membranes, one based on PNIPAAm showing LCST behavior [120], and the other based on an interpenetrating network of (poly(acrylamide)/PAAc (PAAm/PAAc)) showing UCST behavior [121]. In both cases, thermoresponsive control of the transport of molecules through the porous membrane was demonstrated. Recently, Segal *et al.* grafted amine-terminated PNIPAAm chains onto a functionalized pSi surface [85], and also prepared porous silicon substrates containing crosslinked PNIPAAm hydrogel within the pores by first impregnating the pores with a solution containing the monomer, initiator and crosslinking agent, and then allowing the polymerization reaction to take place within the pores [103]. This hybrid structure was then studied, using interferometric reflectance spectroscopy, to monitor changes in the optical thickness of the porous layer below and above the LCST of the polymer component. Porous surfaces coated with SRPs have also been studied for the culturing of cells. For example, Hernández-Guerrero [78] reported the formation of honeycomb porous polystyrene (PS) films grafted with PNIPAAm, using RAFT polymerization. In this case, the PNIPAAm-grafted honeycomb surface was found to allow for a better cell attachment as compared to the flat PS–PNIPAAm films.

Phase separation has been used to prepare dual stimulus-responsive porous membranes from polystyrene-block-poly(*N*,*N*-dimethylaminoethylmethacrylate) (PS-*b*-PDMAEMA) in a one-step approach [122], where the PDMAEMA component allowed a stimulus-responsive control over the pore size. In fact, the membrane prepared allowed the separation of mixtures of different-sized silica nanoparticles in solution, in addition to providing pH- and thermoresponsive control over the swelling and deswelling [122].

3.3.4
Nanoparticles and Nanotubes

Recently, SRPs have received particular attention for applications incorporating nanoparticles, which represent key research areas in areas of drug delivery and medical imaging. Here, a distinction is made between the most common types of nanoparticle, whether hard nanoparticles (e.g., silica nanoparticles), metallic nanoparticles and quantum dots (nanocrystals), or soft nanoparticles such as micelles and liposomes.

3.3.4.1 Silica, Metal, and Metal Oxide Nanoparticles

Zhang *et al.* used ATRP to graft PNIPAAm from the surface of silica nanoparticles using surface-bound silane-based initiators [123]. In a recent report, Xu *et al.* described the ATRP grafting of a novel UV-responsive polyelectrolyte, P(BpyClCl), also from the surface of silica nanoparticles [50].

Several different types of metal nanoparticle and nanorod modified with SRPs have been described [80, 124–126]. The addition of a SRP shell to the nanoparticles provides a means of controlling the surface interactions of nanoparticles, such that they can be switched between a colloidally stable aqueous solution or an insoluble precipitate. In particular, the use of gold nanoparticles (AuNPs) can afford control over the thermal collapse of an attached SRP by the application of light with a wavelength at or near the plasmon resonance absorption maximum of the AuNPs [126]. Shen *et al.* reported the functionalization of AuNPs with a hyperbranched polyglycerol modified with NIPAAm monomers (HPG-NIPAAm) [100]. The HPG-NIPAAm-coated AuNPs exhibited both thermoresponsive and pH-responsive behaviors; that is, the LCST of the polymer could be increased from 24 °C to 55 °C by changing the pH of the solution. A drastic red-shift (~50 nm) in the surface plasmon band of the coated AuNPs, with a corresponding change in solution color from clear red to opaque purple, was observed on switching the temperature of the solution across the LCST.

Additionally, SRP-coated magnetic nanoparticles have been the subject of recent interest, owing to the ability to remotely control (magnetic targeting) the distribution of such particles within the body, while simultaneously effecting a responsive release of drugs encapsulated in the SRP shells. This was demonstrated by Zhang et al., who encapsulated Fe_3O_4 within a thermoresponsive shell of dextran-g-PNIPAAm-co-poly(N,N-dimethylacrylamide) (PDMA) gel by crosslinking the polymer chains using 1,6-diaminohexane in the presence of magnetic nanoparticles [13]. Likewise, Purushotham et al. employed dispersion free-radical polymerization to encapsulate the magnetic nanoparticles in PNIPAAm hydrogel spheres [127], and induced a reversible collapse of the hydrogel by using an alternating magnetic field (AMF) to heat the magnetic nanoparticles and, correspondingly, the PNIPAAm gel.

3.3.4.2 Quantum Dots

Quantum dots (QDs) are fluorescent particles used to label cells in biomedical research. Their most attractive features include a high stability compared to other fluorescent probes, a broad excitation range with a narrow emission spectra, and the control of emission color based on size. Unfortunately, QDs – especially those based on $Cd[2+]$ – are highly toxic to cells, and this limits their application *in vivo*. Consequently, methods to reduce such toxicity are currently being sought, with one proposal being the encapsulation of QDs within polymer coatings. As a result, several different types of polymer nanocomposite based on QD–SRP conjugates have been produced [128–134], not only to reduce toxicity of the QDs but also to increase their photoluminescence or cellular uptake [86]. For example, PNIPAAm hydrogel nanospheres containing reactive thiol groups have been prepared, using precipitation polymerization with methylene bisacrylamide and N,N-cysteine-bis-acrylamide, followed by treatment with 1,4-dithiothreitol to break the S–S bonds and to form reactive thiol groups on the hydrogel particles. These nanospheres were then reacted with CdTe QDs which also contained free thiol groups, obtained by reacting sodium hydrogen telluride with a mixture of cadmium perchlorate and thioglycolic acid [86]. The hydrogel spheres adopted a donut shape (Figure 3.3f) around each QD, suggesting that the QDs were encapsulated in the center of the hydrogel nanospheres. Encapsulation of the QDs in PNIPAAm hydrogels enhanced their photoluminescence on increasing the temperature beyond the LCST of PNIPAAm [86, 134]. Additionally, the hydrogel-encapsulated QDs demonstrated an enhanced intra-tumoral uptake when injected into mice bearing JHU-31 tumors, as compared to nonencapsulated QDs (Figure 3.3e), with an enhanced fluorescent labeling of the tumor (Figure 3.3a–d) that would be valuable for

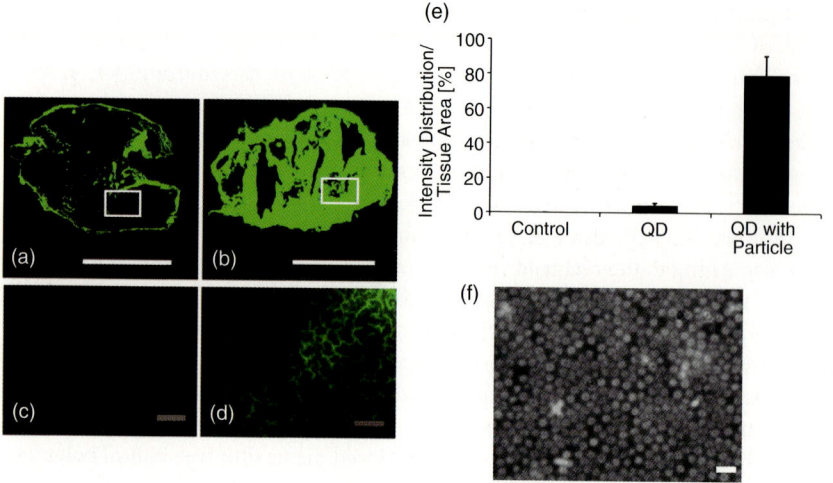

Figure 3.3 (a,b) Distribution analysis of both QD (a) and QD–PNIPAAm (b) in tumor tissue, using a microarray scanner. Scale bars = 500 μm; (c,d) Fluorescence microscopy images of the marked areas in panels (a) and (b). Scale bars = 200 μm; (e) Chart showing the difference in accumulation of QD and QD–PNIPAAm in tumor tissue by comparing their respective intensity distribution per tissue area; (f) Donut-shaped QD–PNIPAMs (200 nm diameter). Note the light external PNIPAAm ring and the dark QD central core. Scale bar = 500 nm. Adapted from Ref. [86].

diagnostic imaging. Likewise, Hou *et al.* synthesized QDs that were grafted with PNIPAAm chains by first generating PNIPAAm chains with thiol end groups by RAFT polymerization, and then reacting these with freshly prepared CdS QDs [135]. This system was used to determine the dependence of QD luminescence on the interdot distance. The distance between particles was controlled using the LCST of the grafted PNIPAAm, so as to alter the aggregation behavior of the composite particles.

3.3.4.3 Micelles

Soft nanoparticles comprised of amphiphilic block-copolymer micelles show considerable potential for drug-delivery applications, and have been extensively reviewed [136–138]. The addition of SRPs into these micellar structures can afford a stimulus-responsive control over drug release [139]. For example, a biodegradable thermosensitive micelle forming a triblock copolymer of PNIPAAm with poly[(R)-3-hydroxybutyrate] (PHB) was synthesized using ATRP [99]. First, the PHB-diol was reacted with ATRP initiator 2-bromoisobutyryl bromide to obtain the Br–PHB–Br macroinitiator, which was then subjected to ATRP with NIPAAm; this resulted in the triblock polymer, PNIPAAm–PHB–PNIPAAm. The micelles formed from these polymers possessed a hydrophobic PHB core and a PNIPAAm corona. The low critical micellar concentrations (CMCs) required for micelle formation, combined with the highly hydrophobic core (which can be used to load hydrophobic drugs), make this an ideal material for the delivery of drugs. Additionally, the micelle was found to be noncytotoxic and also biodegradable.

RAFT polymerization has also been used successfully to prepare thermoresponsive micelles. For example, Akimoto et al. prepared thermoresponsive hydroxyl-terminated poly(N-isopropylacrylamide-co-N,N-dimethylmethacrylate) (PNIPAAm-co-PDMA) using RAFT polymerization, followed by ROP, to synthesize a PDLLA block at the –OH terminal [140]. The resulting diblock polymer, (PNIPAAm-co-PDMA)-b-PDLLA, had an LCST of about 40 °C. The micelles formed from this polymer showed an enhanced cellular uptake above the LCST; this phenomenon was most likely due to hydrophobic interactions between the PNIPAAm corona above the LCST and the cell surface.

Several other methods have also been reported for the synthesis of SRP-based micelles [3, 141–143]. For example, by using thermosensitive and pH-sensitive polymers based on poly(N-isopropylacrylamide)-b-[poly(ethyl acrylate)-g-poly(2-vinylpyridine)] (PNIPAAm-b-(PEA-g-P2VP)), Feng et al. [4] demonstrated a switching between random unimers (single macromolecules without ordered conformation), unimeric micelles (single macromolecules in micelle-like conformation), and core–shell micelles (micellar aggregates of multiple macromolecules). These dual stimulus-responsive copolymers showed a so-called "schizophrenic" behavior, where changes to the pH or temperature can result in an inversion of the core and the shell [4]. Smith et al. [144] have also reported such behavior in core–shell micelles prepared from PDEAEMA-b-PNIPAAm. Efforts to stabilize stimulus-responsive micelles via covalent crosslinking have also been explored [145]. Finally, micelles of poly(acryloyl glucosamine)-b-poly(N-isopropylacrylamide), prepared by RAFT polymerization, could also be crosslinked by a second RAFT polymerization with a linker, 3,9-divinyl-2,4,8,10-tetraoxaspiro[5.5]undecane, to provide acid-degradable acetal linkages [146].

3.3.4.4 Liposomes

In contrast to micelles, liposomes (i.e., vesicles with a lipid bilayer membrane) and polymersomes (i.e., vesicles with a membrane made from amphiphilic block copolymers) contain water-filled interiors – an architecture which allows the encapsulation of hydrophilic drugs [147, 148]. Stimulus-responsive properties can be imparted to the liposomes by decorating their surfaces with SRPs. As an example, Kono et al. [149] reported the synthesis of a thermoresponsive liposome prepared by tethering copolymers of PNIPAAm and octadecyl acrylate (ODA) onto the surface of the liposome via the adsorption of octadecyl chains onto the lipid bilayers [149]. Another group, recently reported the synthesis of the thermoresponsive polymer poly(N-(2-hydroxypropyl)methacrylamide mono/dilactate) (PHPMA mono/dilactate) from mono- and dilactate-modified HPMA monomers and thiocholesterol as the chain transfer agent. By using the thiocholesterol as an anchor, the liposomes could be coated with the polymer, rendering them responsive to temperature changes, with LCSTs between 13 and 65 °C [12]. Lee and coworkers synthesized cholesterol-terminated PAAc chains using a nitroxide-mediated controlled radical polymerization, and then proceeded to crosslink the PAAc chains using 2,2′-(ethylenedioxy)-bis(ethylamine) [150]. This resulted in the formation of a cage-like structure around the liposome membrane. These caged liposomes were found to be more stable than liposomes with anchored polymer chains, and also showed the pH-responsive leakage of a fluorescent dye (further information on these liposomes is provided in Section 4.1).

Polymersomes (also referred to as polymeric vesicles) are polymer equivalents of liposomes. Like liposomes, polymersomes are often used for drug-delivery applications [137, 151], as they have an increased stability over liposomes and can also transport both hydrophobic and hydrophilic molecules, enabling the delivery of genes, proteins, and anticancer agents. Polymersomes are commonly prepared using polar/nonpolar block copolymers and a mix of polar/nonpolar solvents that, when phase-separated, act as a template for assembly of the hydrophobic and hydrophilic blocks [138, 152]. Recently, much effort has been directed towards the development of stimulus-responsive polymersomes [147].

3.3.4.5 Nanotubes

Carbon nanotubes (CNTs) possess a wide range of salient and unique properties, such as remarkable mechanical strength and stiffness, high surface areas, and outstanding charge transport characteristics [153]. However, CNTs are not biocompatible, and are also insoluble in aqueous solvents, which limits their use for biological applications [154]. In order to improve both biocompatibility and water solubility, a popular approach is to modify the CNT surfaces with biocompatible and soluble polymers. In expanding the repertoire of CNTs even further, recent investigations have demonstrated the surface attachment of SRPs to CNTs in order to make the latter sensitive to their environment and to gain control over CNT solubility [153, 155, 156]. Surface-initiated ATRP has been used to graft thermoresponsive PNIPAAm from the surface of CNTs [156–158]. For example, Kong *et al.* demonstrated the grafting of PNIPAAm from the outer surface of initiator-functionalized multi-walled carbon nanotubes (MWNTs), and showed their solubilization in water at room temperature and reversible aggregation at temperatures above the LCST [157]. Instead of ATRP, recent studies have focused on covalently attaching RAFT chain transfer agents to CNTs [159], with a view to producing soft polymeric shells around CNTs via RAFT polymerization. For example, You *et al.* used the RAFT technique to prepare CNT–PNIPAAm conjugates, where the PNIPAAm was attached via cleavable disulfide tethers [154]. Furthermore, the covalent grafting of hyperbranched poly(amido amine) polymers onto CNTs resulted in composites that underwent sol–gel switching in response to sonication and temperature changes [160]. However, while the above-described methods have been shown to produce a high grafting density (ca.50 wt%) on the CNTs, there is a synthetic downside. In order to immobilize the ATRP initiator onto the surface of the CNTs, strong oxidative acids must first be used to generate the desired carboxyl functionality on the CNT surface, which can lead to their undesirable shortening [155]. In pursuing an alternative strategy, Yang *et al.* first used the surfactant sodium dodecylbenzene sulfonate (SDBS) and sonication, to obtain a stable suspension of individual CNTs through micellization [155], and then performed graft polymerization in the micelles, using acrylonitrile as the monomer and azo-bis-*iso*-butyronitrile (AIBN) as the initiator to obtain poly(acrylonitrile)-grafted CNTs (CNT–PANs). The authors then proceeded to hydrolyze the PAN to graft the pH-responsive polymer PAAc onto the CNTs, as CNT–PAAc. The latter was readily dispersed in aqueous solutions at pH 5 or above; however, on reducing the pH to below 5, a precipitation of the CNT–PAAc was observed. At a higher pH, the repulsive force between the negative charges on the PAAc prevented agglomeration of the CNT–PAAc, and in turn prevented precipitation. However, at pH < 5.0, protonation of the PAAc units would lead to a reduction of the repulsive forces, with a subsequent increase in hydrogen-bonding interactions between the –COOH groups on adjacent CNT–PAAc, causing agglomeration. In another report, micelles with a thermoresponsive azide-terminated PDMA in the corona, and PNIPAAm in the core, were grafted onto the surface of alkyne-functionalized MWNTs, using a Cu(I)-catalyzed [3+2]-Huisgen cycloaddition or "click" coupling [161]. The micelles were prepared by heating the copolymer solution above the critical micelle temperature (CMT), which varied between 30 °C and 50 °C, depending on the polymer concentration and the PNIPAAm chain length. In another study, poly(pentafluorophenylacrylate) derivatized with *N*-(6-aminohexyl)-4-(pyren-4-yl) butanamide was used to prepare a thermoresponsive SRP of poly(*N*-cyclopropylacrylamide) (PNCPA) with pyrene side groups. In exploiting the strong affinity of pyrene for the sidewalls of CNTs via π–π stacking interactions, adsorption of the polymer onto the CNTs permitted a thermal control over the dispersion and precipitation of CNTs [162].

Other types of nanotube have also been interfaced with SRPs. For example, titanate nanotubes with exposed surface Ti–OH groups provided "chemical handles" for the covalent attachment of organosilane initiators. Subsequent surface-initiated ATRP with the NIPAAm monomer resulted in thermoresponsive core nanotube–shell PNIPAAm conjugates [163]. Finally, organic peptide nanotubes have been grafted with PNIPAAm via surface initiator groups [76]. Peptide-based nanotubes offer an alternative to inorganic nanotubes and are, potentially, more suitable for drug-delivery and biomaterial applications.

3.4
Applications of Stimulus-Responsive Polymers at Interfaces

3.4.1
Controlled Drug Delivery

Unfortunately, many of the therapeutic agents currently in use have undesirable side effects that severely limit their clinical application. For example, most drugs used to treat cancer are also highly toxic to the normal cells of the body [15]. This problem can be resolved by increasing the selectivity of a drug, by targeting it to a particular tissue, and/or by increasing the drug's potency – all of which will lead to a reduction in the required dosage. Sadly, the scope for further improvements with many current drugs is limited, and consequently the controlled and site-specific delivery of drugs within the body becomes very important when treating diseases. The major goal of a controlled drug delivery system (CDDS) is to increase the therapeutic efficacy of a drug while reducing its side effects [148]. Stimulus-responsive materials represent particularly attractive options for the development of a CDDS, as they can be designed to take advantage of very small physiological changes within the body to release their drug payload [2]. The subject of stimulus-responsive drug delivery has been widely reviewed [104, 148, 164, 165]; consequently, only some of the recent developments in the use of SRPs at interfaces for drug delivery will be described at this point.

Tsukagoshi *et al.* reported the use of a thin film of PNIPAAm grafted onto flat silicon surfaces for the loading and controlled release of aspirin [15]. The collapse of the PNIPAAm brush above the LCST was used to trap and hold the drug molecules, which could then be released simply by lowering the temperature of the system below the LCST. Wu *et al.* demonstrated a pH-controlled release of insulin from porous silicon oxide substrates, using a pH-responsive polycationic chitosan hydrogel as a "gatekeeper" film [36]. In this case, the porous silicon scaffolds loaded with insulin were capped using a chitosan hydrogel; subsequently, in solution at pH 7.4 the hydrogel would collapse and become impermeable to the drug. However, at a lower pH (ca. 6) the hydrogel would expand and facilitate insulin release. Other studies on the modification of mesoporous silica particles with pNIPAAm, either by the surface-initiated ATRP of NIPAAm [77] or the free-radical copolymerization of NIPAAm and MBAAm [166] within the pores, have demonstrated high storage capacities for ibuprofen, and its controlled release.

In addition to flat and porous surfaces, responsive nanoparticles have also been studied for the development of CDDS [80]. For example, thermoresponsive magnetic nanoparticles prepared by coating magnetic iron oxide nanoparticles with thermosensitive polymers have been reported [13]. These composite nanoparticles can ideally be targeted to tumor sites by applying an external magnetic field. *In vitro* release experiments have shown that heating the drug-loaded SRP-coated magnetic nanoparticles to temperatures above the LCST, led to an increase in the rate of release of the antineoplastic agent doxorubicin (DOX). Recently, Purushotham *et al.* showed that PNIPAAm-coated iron oxide nanoparticles could be remotely targeted to a desired body site by using a magnetic field, and then also heated to temperatures above the LCST by using an AMF [127]. This constitutes a powerful CDDS design, which can be used to induce cancer cell death *in vivo* by hyperthermia, while at the same time increasing the release rate of drugs loaded into the thermoresponsive polymeric shell.

Another class of stimulus-responsive nanoparticles commonly studied for drug-delivery roles is that of liposomes [164]. Conventionally, responsive polymers have been anchored on the surface of liposomes via hydrophobic alkyl or cholesterol tethers [12, 149]. Such liposomes, when loaded with a dye, were found to leak and release the dye at temperatures above the polymer's LCST. At this raised temperature, the hydrophobic polymer interacts strongly with the lipid bilayer vesicle membrane, thereby increasing its permeability. By using a similar synthesis and crosslinking the liposome-anchored PAAc chains, Lee and coworkers formed a pH-responsive polymeric cage-like structure encapsulating the (polymer-caged) liposome (PCL) [150]. The caged vesicles were found to be much more stable than bare liposomes, and to retain their spherical shape after freeze-drying and rehydration. Additionally, only about 5% leakage of the calcein dye loaded into the PCLs was found to occur after 500 h of incubation in 90% fetal bovine serum; this was in contrast to the 50% leakage seen in the case of bare liposomes. PCLs loaded with calcein as a model drug showed a higher release at a lower pH (~80% at pH 4.0) than at a higher pH value (pH 5.5, ~45% release). The authors attributed this greater drug release at lower pH values to changes in the hydrodynamic diameter (~69% reduction) of the PCLs, which caused a destabilization of the bilayer membrane of the liposome and resulted in leakage of the drug.

SRPs such as poly(*tert*-butyl acrylate)-*b*-poly(*N*-isopropylacrylamide) (PtBA-*b*-PNIPAAm) have been used to form micellar structures, and have shown potential for use as vessels to contain and deliver hydrophobic drugs [138]. The preparation of PtBA-*b*-PNIPAAm by ATRP, with subsequent micellization in dimethylformamide (DMF) and water in the presence of the drug naproxen, allowed drug encapsulation and circumvented the drug's poor water-solubility. Furthermore, by switching the temperature to 40 °C (above the LCST) the amount of naproxen released was substantially greater than at 25 °C (82% versus 22%) [138].

3.4.2
Control of Biointerfacial Interactions

The best-known application of SRPs at interfaces is the ability to switch a surface from a cell- or protein-adhesive to a nonadhesive state, an effect which has been achieved almost routinely with thermoresponsive polymers and copolymers of PNIPAAm [16, 62, 88, 91, 167]. Particular attention has been paid to the reversible attachment of cells to and from SRP coatings for cell sheet and tissue engineering, and this field has been extensively reviewed [8, 168–170]. In general, the transition of surface-attached SRPs with LCST behavior from a highly hydrated and extended chain conformation to a more rigid, dehydrated and collapsed conformation with more hydrophobic character, has been shown to transform a surface from repellent to adhesive for both proteins and cells [94, 171]. Coatings of PNIPAAm are popular examples; the reversibility of the chain collapse allows the release of adhering proteins and cells on demand, simply by cooling the system to below the LCST [107, 167]. However, some reports exist of PNIPAAm coatings where the change from a swollen to a collapsed state (or *vice versa*) was insufficient to either induce or reverse protein or cell adhesion [94, 172–174]. SRP-assisted gentle cell detachment has benefits over harsher methods of cell removal, such as mechanical scraping or the enzymatic degradation of cell–protein junctions [94, 97, 175], mainly because it improves cell viability and allows the release of cell sheets with their associated extracellular matrix (ECM) proteins intact. A number of cell types have been investigated for this role, including fibroblasts, epithelial cells, endothelial cells, hepatocytes, macrophages, microglial cells and stem cells. Currently, tissue culture ware for thermoreversible cell lift-off is commercially available [1]. Cell sheets collected using thermoresponsive surfaces have been implanted in studies aimed at the repair of corneal epithelial dysfunction, the repair of damaged myocardial tissue, and regeneration of the urothelium and the periodontal ligament [176–178]. Today, the patterned 2-D coculture of cells and the engineering of 3-D cell-sheet stacks is broadening the scope of regenerative medicine applications made possible by SRPs at interfaces [169, 179]. For example, two different cell types – hepatocytes and endothelial cells – have been cocultured on surfaces patterned with PNIPAAm and PNIPAAm-poly(n-butyl methacrylate) (PNIPAAm-PBMA), prepared using electron-beam irradiation through a mask [179]. Hepatocytes in culture at 27 °C adhered to the PNIPAAm-PBMA regions, which were collapsed and cell-adhesive. The removal of unbound hepatocytes, and subsequent culture of endothelial cells at 37 °C (above the LCST of PNIPAAm homopolymer), allowed the attachment of endothelial cells to the PNIPAAm spots, while cooling of the coculture below the LCST of both components allowed complete release of the cell sheet [179]. Although the mechanism of cell detachment remains poorly understood, changes in cell morphology and metabolic activity appear to play a role in this situation [95, 180].

In addition to providing control over mammalian cell attachment and detachment, SRP coatings have been employed to control bacteria, lipid vesicles, DNA, drug molecules, and proteins at interfaces [181–185]. Protein adsorption is a key descriptor of the performance of materials in life sciences applications, and there is a significant demand for materials providing an increased control over protein adsorption. Recently, SRPs have shown promise in providing a means for controlling *when* adsorption occurs (temporal control), and also *where* it occurs onto a surface (spatial control). For example, by using a microfluidic device containing a microheater, Huber *et al.* were able to thermally modulate the conformation of a grafted PNIPAAm coating, with millisecond response times [182]. In this case, by flowing a solution of fluorescently labeled myoglobin across the microheater, proteins were adsorbed onto the PNIPAAm surface when the heater was switched on ($T >$ LCST), and desorbed when the heater was switched off ($T <$ LCST). Heinz *et al.* prepared PNIPAAm coatings via plasma-activated radical grafting, whereby a semi-quantitative time-of-flight secondary ion mass spectrometry (ToF-SIMS) study on the adsorption of bovine serum albumin (BSA) at different temperatures showed that the PNIPAAm coatings were protein-resistant at 20 °C, but adhesive at 37 °C [171]. Likewise, when Cole *et al.* applied ultra-surface-sensitive ToF-SIMS and principle component analysis (PCA) to study the temperature-mediated adsorption of BSA and lysozyme (Lys) [81], they showed that adsorption occurred only at 37 °C, and not at 20 °C. Likewise, atomic force spectroscopy, using a protein-coated colloid probe, showed that adhesive interactions existed between the protein-coated probe and PNIPAAm coatings above 32.5 °C [186], whereas repulsive forces were observed for a protein-coated probe at or below 27.5 °C. Cheng *et al.* examined the temperature-dependent adsorption of BSA, anti ferritin antibody (anti-Fe) and fibrinogen (Fg) on plasma-polymerized NIPAAm coatings using radiolabeling and ToF-SIMS [172]. The anti-Fe and Fg adsorbed to the coating above its LCST was found to remain irreversibly bound to the surface when the temperature was reduced below the LCST. The BSA adsorption was only partially reversible upon cooling, which indicated that in this case the coating was suitable only for one-off switching [172]. These results were consistent with those of Yamato et al., who found that Fn adsorbed to PNIPAAm coatings above the LCST remained on the surface after cooling to a temperature below the LCST [107].

Although, recently, cell and protein studies using thermoresponsive SRP coatings have been dominant, several other reports have focused on different means to control protein adsorption and cell attachment, including photoisomerization [84] and the switching of pH and ionic strength [187, 188]. In PNSp-co-PMMA, UV light induces an isomerization from the nonpolar spiropyran to the zwitterionic merocyanine forms [84]. Following the exposure of PNSp-co-PMMA-coated glass to 50% platelet-poor plasma (PPP) solution, with and without in-situ UV illumination of the coatings, Higuchi et al. conducted a series of enzyme-linked immunosorbent assays (ELISAs) to measure the amount of Fg present. When the photoresponsive PNSp-co-PMMA surface was seen to show a decrease in Fg after illumination, the initial interpretation was that switching of the NSp molecules had induced the release of Fg. However, subsequent experiments involving exposure of the pre-irradiated PNSp-co-PMMA (PNsp in the zwitterionic merocyanine form) to PPP solution actually showed a 1.2-fold increase in adsorbed Fg, as compared to the nonirradiated coatings [84]. When Kikuchi et al. spin-coated blended PNSp-co-PMMA and PEG onto TCPS, the PNSp-co-PMMA/PEG-coated surface resisted cell adhesion. However, illumination of the coating with UV light induced an isomerization of the spiropyran groups within the PNSp-co-PMMA to a polar state; this resulted in a release of the entangled PEG molecules, and allowed cells to adhere to the PNSp-co-PMMA. By using a photo mask and multiple illumination/culture cycles, multiple cell types could be deposited sequentially, and cultured in close proximity to each other in complex 2-D patterns (Figure 3.4) [189]. The coupling of cell-adhesive RGD peptide sequences to azobenzene units attached to PMMA surfaces via alkyl spacers allowed the availability of the peptide sequence for cell binding to be controlled by illumination with UV light, which altered the azobenze conformation [190]. Nakayama et al. generated photoresponsive polymers by functionalizing polystyrene with malachite green groups as the photoresponsive component [191]. The resultant polymer, poly(styrene-co-diphenyl(4-vinylphenyl)methaneleucohydroxide) could be switched by UV light to a more polar state, which in turn enhanced cell attachment and proliferation. Unfortunately, the number of studies of light-responsive surfaces for control over biointerfacial interactions is limited, most likely due to the relatively long duration of cell culture, coupled with the tendency for photoisomerized molecules to relax or oxidize irreversibly. The inherently mixed hydrophobic/hydrophilic and neutral/charged character of the photoresponsive molecules in both states arguably prevents the proper design of adhesive and repellent states. In a demonstration of a pH-responsive coating, the adsorption of alpha-chymotrypsin on binary brush coatings of poly-(2-vinylpyridine) (P2VP) and PAAc could be reversibly controlled

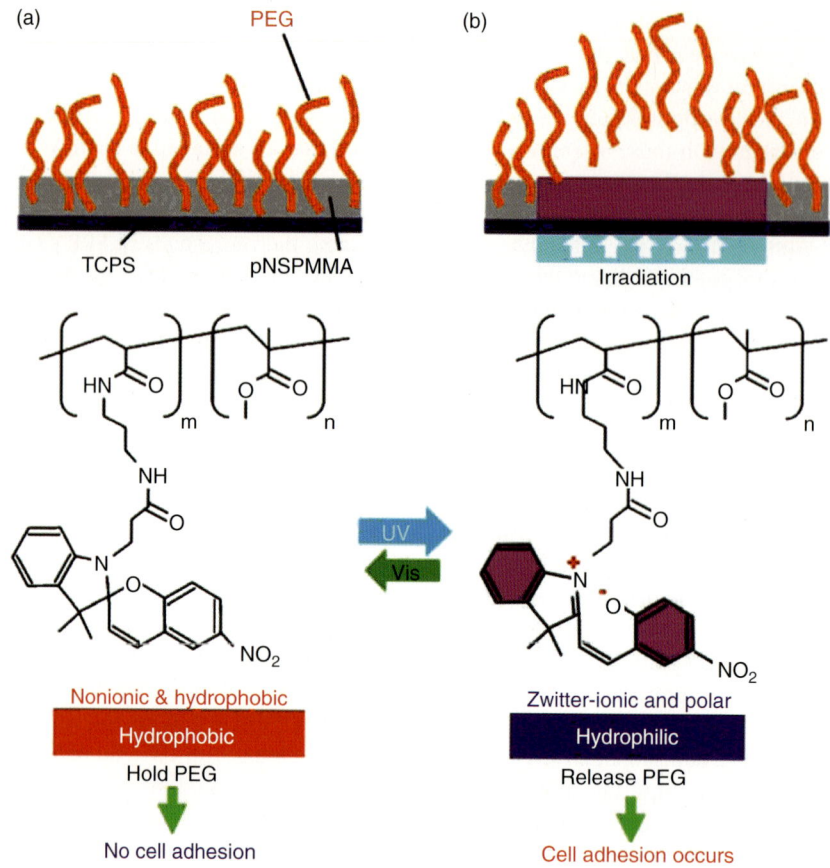

Figure 3.4 (a) PEG is used as a cell-adhesion inhibitor on the PNSp-co-PMMA surface. Before UV irradiation, PNSp-co-PMMA is hydrophobic, so that the PEG molecules are retained on the surface; (b) The PNSp-co-PMMA/PEG surface is UV-irradiated in the presence of aqueous buffer to isomerize the spyropyran within PNSp-co-PMMA to its more polar form, and release PEG from the surface. The spyropyran returns to its hydrophobic state in visible light, after which the surface can promote cell adhesion [189].

through the solution pH and either a low or high ionic strength [188]. Reversible protein adsorption was possible from a pH 9 buffer, but not from pH 4 buffer.

3.4.3
Microfluidic Valves

Microfluidic devices have channels within the device where one or more dimensions are below 1 mm. The consequence of manipulating fluids at this scale are that the surface area-to-volume ratio of such devices are high, that effects such as surface tension become very important, and that the fluid flow is laminar due to a low Reynolds numbers. Yet, at the same time, these devices have been miniaturized such that they require only minute volumes of fluid and demonstrate a rapid thermal relaxation. Taken together, these properties are of benefit to a raft of analytical "lab-on-a-chip" applications, including the determination of chemical and enzyme kinetics [192, 193, 230], temperature [194], fluid viscosity [195, 196], chemical binding coefficients [197], and diffusion coefficients [198]. Other common applications for microfluidic devices include capillary electrophoresis [199, 200], enzyme assays [201], immunoassays [202–205], DNA analyses [204], isoelectric focusing [206], mass spectrometry [207], polymerase chain reaction [208], chemical gradient formation [209, 210], cell sorting [211, 212], and cell templating [213].

One fundamental challenge in the design of microfluidic devices lies in the need to accurately control the transport of fluid at any given time, in order to supply or remove the reagents, analytes, or reaction products. While externally actuated valves have progressed to lab-on-a-chip applications, these rely on significant off-chip hardware, such as vacuum pumps and switching solenoids, and this limits their use as point-of-care devices due to the sophisticated instrumentation required. Lab-on-a-chip devices have, therefore, provided the impetus for the development of a variety of new actuating methods such as internally actuated microvalves on the basis of SRPs, which can be triggered thermally, magnetically, optically, or by a change in pH [214]. Examples of such systems include thermoresponsive PNIPAAm brushes within microchannels, which undergo swelling and collapse in response to a stimulus to open or close fluidic pathways. Microheaters have had an obvious use in the control over swelling and collapse of PNIPAAm brushes acting as actuator valves [215]. By hydrating the grafted PNIPAAm below its LCST, Idota et al. were able to completely block a microchannel, with a slight temperature increase between 30–35 °C resulting in collapse of the polymer chains and the resurrection of fluid flow [216]. Limiting the heating to a confined area may be difficult, however, and in a situation where cooling is passive the ability to open and close such a valve quickly may be problematic. Satarkar et al. used a nanocomposite hydrogel consisting of temperature-responsive PNIPAAm impregnated with magnetic Fe_3O_4 nanoparticles to demonstrate a remote-controlled actuation [217]. In this case, the collapse and swelling of the polymer was controlled by the remote application of an alternating magnetic field (AMF) to the nanocomposite within a Y-junction microchannel. When the AMF was turned OFF, the cooling of the microvalve led to a closing of the valve, whereas when the AMF was ON, the temperature increased above the LCST, which led to an opening of the valve. Notably, the system proved to be highly reproducible over extended periods. In a similar system, when Ghosh et al. used ferromagnetic nanoparticles (Fe_3O_4) encapsulated within PNIPAAm inside a 300 μm microchannel, they observed up to 80% volume shrinkage under the application of an AMF, with a response time of ~3 s that was significantly faster than thermal actuation [218]. The microvalve also exhibited a faster response time compared to a macrovalve (PNIPAAm monolith inside a 1500 μm-diameter channel).

In *optically actuated valves*, the use of light is particularly attractive, as it can be easily directed to specific parts of a microchannel and allow for a more precise control of the microvalve. For example, Sershen et al. embedded metallic nanoparticles with a distinct and strong optical absorption profile, such as gold colloids (absorbance maximum at 532 nm) and nanoshells (absorbance maximum 832 nm) into a thermally responsive PNIPAAm [51]. Subsequently, the gold colloid nanocomposite material was shown to collapse under green light, whereas the gold–nanoshell nanocomposite collapsed under near-infrared (NIR) irradiation. Although the response time was very slow at the macroscale, the authors suggested that scaling in size to a microvalve would give rise to faster response times. Finally, Sugiura et al. employed light-responsive copolymers to actuate microvalves inside microfluidic channels [219, 220].

Both, thermoresponsive and light-responsive microvalves were prepared by photoinitiated radical polymerization with N,N'-methylene bis(acrylamide) (MBAAm), NIPAAm, and an acrylated spirobenzopyran molecule as monomer [219]. Irradiation of the gel valves at 420–440 nm wavelength light resulted in valve deformation, allowing the flow of blue dye though the channels in a polydimethylsiloxane (PDMS) chip. A temperature increase to 46 °C induced the collapse of all gel valves within 5 s, while directed light irradiation could be used to selectively open specific valves for flow within 30 s [220]. An alternative approach employed the preparation of a gel layer, also from MBAAm, NIPAAm and an acrylated spirobenzopyran, that was sandwiched between a glass slide and a PDMS upper sheet containing only inlets and outlets (Figure 3.5) [220]. Irradiation of the area between two isolated ports caused a localized collapse of the responsive sheet sandwiched within the chip, so as to form a channel connecting the ports that could then be used to transport a suspension of fluorescent beads [220]. Another type of internally actuated microvalves is based on SRPs that respond to changes in local pH by a drastic change in volume, thus eliminating the need for an external energy source. Ayala et al. created a PDMS-valve using soft-lithography, where the transparent channels were filled with a hydrogel of 2-hydroxyethyl methacrylate and AAc crosslinked with ethylene glycol dimethacrylate (EGDMA) [221]. The hydrogel was found to expand at high pH and to contract at low pH, with an average response time of 2.5 s. In taking this concept one step further, Liu et al. used a new multiplexing technology to construct arrays of PDMS microvalves containing pH-responsive hydrogel microspheres of copolymers from 4-hydroxybutyl acrylate (4-HBA), AAc and EGDMA. Multiple laminar streams of buffers at different pH allowed the independent opening or closing of the microvalves [222]. Beebe et al. employed a UV-initiated polymerization through a photomask to construct pH-responsive microvalves in desired locations within microfluidic channels [102]. This approach also permitted the polymerization of SRPs with different responsive components, or different transition points within a microfluidic device. In this way, the direction of fluid flow could be controlled by using inversely swelling–collapsing pH-responsive valves at a microfluidic T-junction. At low pH, one valve was collapsed while the other was swollen, so that flow was directed past the collapsed valve; however, at high pH the opposite effect would occur, and the flow was redirected. Under intermediate pH conditions, both valves were swollen such that the channel was sealed [102].

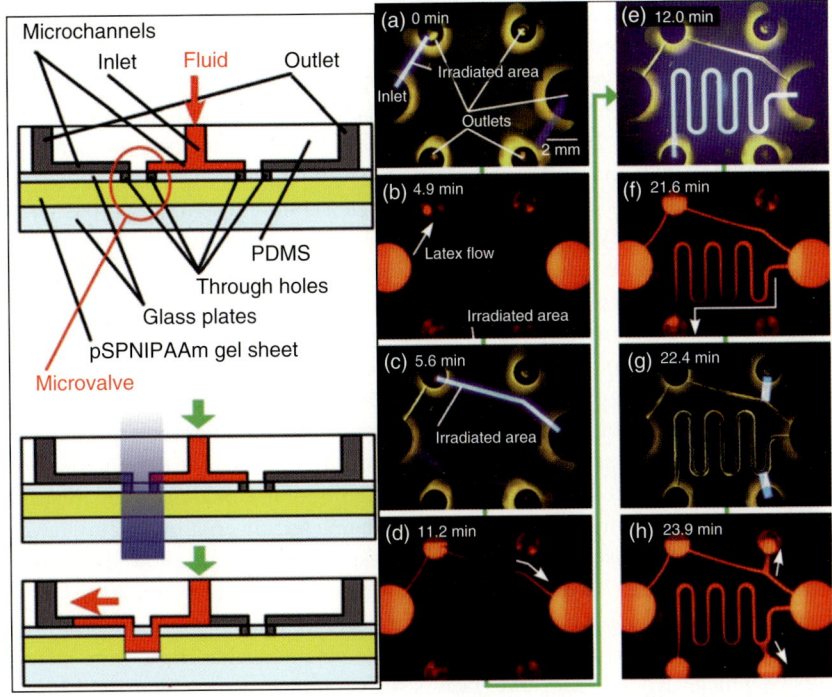

Figure 3.5 Microfluidic system for on-demand formation of arbitrary microchannels by micropatterned light irradiation. A PDMS microchannel network equipped with inlets/outlets above a responsive gel sheet can be used for independent and parallel flow control in microchannels that are prepared on demand by the application of light through a mask. The images (a–h) show four consecutive micropatterned light irradiations for channel formation and the subsequent flow of fluorescently labeled latex particles (direction indicated by white arrows) through the newly formed microchannels [220].

Clearly, these examples of thermally, magnetically, optically and pH-regulated actuators afford simplicity and reliability of the functional components of microfluidic devices, and also remove the need for complicated peripheral lines or solenoid-based valves.

3.4.4
Molecular Separation

A key aspect of SRPs that has attracted profound interest is the possibility of dynamic control over the permeation of chemicals through mesoporous [223] and nanoporous [224] membranes for bioseparations [225, 226], or the interaction of ions [227] with stimulus-responsive surfaces. By grafting stimulus-responsive polymers to or from porous membranes, it becomes possible to control temporally the permeation of small molecules or nanoparticles [27, 119, 120, 122, 228]. Fundamentally, molecular separations rely on a complex interrelationship between pore size, polymer molecular weight, and grafting density on the one hand, and permeation flux on the other hand. Fu et al. achieved a reversible control of the size and surface energy of the pores in mesoporous silica by grafting PNIPAAm [223]. The resultant polymer-grafted particles allowed the adsorption and release of rhodamine 6G to be dynamically controlled by switching the temperature around the LCST. In another example, environment-responsive ultrafiltration membranes were created by the modification of a commercial microporous polyvinylidene fluoride (PVDF) membrane with a salt-responsive hydrogel composed of poly(N-vinyllactams) crosslinked with bisacrylamide [226]. The LCST of the hydrogel could be decreased by the addition of salts; this resulted in a shift from a hydrophilic state at ambient temperature (open polymer network structure) in the absence of salt to a hydrophobic state (collapsed network structure) in the presence of salt. The collapse of the network onto the supporting microporous membrane, accompanied by the expulsion of surrounding water molecules, led to a significant increase in membrane permeability. By changing the membrane pore and surface properties dynamically during ultrafiltration, a model protein mixture of BSA, human immunoglobulin G (HIgG), equine ferritin and thyroglobulin was investigated. In a closed valve position, a selected membrane allowed the total transmission of BSA, a moderate transmission of HIgG, and low transmission of ferritin and thyroglobulin. In the open valve position, the transmission of ferritin and thyroglobulin was increased, while results obtained with HIgG were inconclusive, which suggested that other factors such as protein aggregation and adsorption might be involved. The results obtained demonstrated the potential for the use of these membranes for multicomponent protein separation [226]. Finally, the polymer PVCL was used for the thermoresponsive separation of lactate dehydrogenase (LDH), using dye affinity chromatography [34, 229].

3.5
Summary and Future Perspectives

Today, the field of SRPs has advanced far beyond the well-known thermoresponsive PNIPAAm polymers with LCST behavior, with a range of polymers having been synthesized that are sensitive to various environmental stimuli (triggers) such as light, temperature, electric field, magnetic field, pH, and chemicals. The types of response associated with the stimuli include structural changes, sol–gel transitions, changes to optical properties, solubility changes, and micelle formation. Thermoresponsive polymers have been designed with both LCST and UCST behavior characterized by physiologically relevant critical temperatures, such as PiPrOx and triblock copolymers of PEO and PLLA, respectively. pH-responsive polymers that react to small changes in the pH of their local environment are either polycationic (such as PDMAEMA), polyanionic (such as PAAc), or zwitterionic (such as PVBA-b-PDEA). There is indeed a broad palette of polyelectrolytes with pH-responsive behavior that depends on the desired critical pH. Within the realm of photoresponsive polymers fall those with chromophores in the side chain, that undergo structural isomerization or cyclization reactions upon irradiation, and which include azobenzene, spiropyrans, or viologen. The irradiation of such a polymer leads to significant changes in its properties, notably in terms of wettability. These polymers are particularly intriguing because of the potential for patterning, as well as for temporally and spatially controlled switching. Currently, an emerging field of research involves stimulus-responsive copolymers that exhibit responses to more than one type of stimulus, such as thermoresponsive and pH-responsive systems, or light-responsive and thermoresponsive systems. Recently, the fabrication of these SRPs has benefitted significantly from the range of experimentally amenable CRP techniques, such as ATRP and RAFT, that are now available even to nonspecialist laboratories. These CRP techniques have also had an impact on the tuning of the SRPs' switching behavior, since the critical environmental conditions at which switching occurs depend on the molecular weight of the polymer, while the sharpness of the response decreases with increasing polydispersity. In many cases, knowledge of the underlying mechanisms of stimulus-responsiveness in SRPs is still limited; consequently, there is a conspicuous need for a better fundamental understanding of the effects at play, in order to harness the full potential of these materials.

Clearly, the combination of SRPs with other, non-stimulus-responsive materials, such as hard or soft nanoparticles, nanotubes or nanoporous materials, and the attachment of SRPs onto or from flat and porous surfaces, allows the fabrication of advanced materials presenting smart nanointerfaces. The fabrication of these nanointerfaces proceeds via covalent grafting approaches, plasma polymerization, and electron-beam polymerization, as well as noncovalent routes, with the latter routes being particular suitable for the preparation of soft nanointerfaces such as micelles, liposomes, and polymersomes. The attachment of SRPs to metallic or magnetic nanoparticles allows the preparation of dual-stimulus-responsive materials that are capable of exploiting the plasmonic and magnetic effects of the nanoparticles, respectively. It has been shown in several studies that the functionalization of QDs with SRPs can improve their biocompatibility and optical properties; likewise, the biocompatibility and water-solubility of CNTs is improved remarkably when SRPs are coupled to the CNT surface.

The generation of such smart nanointerfaces continues to open new vistas for a whole raft of applications, including drug delivery, tissue engineering, cell culture, molecular separation, microfluidics, and biosensors. With regards to drug delivery, the advent of SRPs has led to a paradigm shift in controlled drug release, allowing the preparation of CDDS with excellent control over release kinetics and release location. Such SRP-containing drug-delivery vehicles can exploit minute physiological changes in the body to trigger release of the drug. SRPs at nanointerfaces have also played a significant role in controlling biomolecule–surface and cell–surface interactions, which is highly relevant to a variety of biomedical applications; this is best exemplified by the successful use of PNIPAAm and PNSp-*co*-PMMA coatings for cell sheet engineering. The need for actuators in microfluidic and lab-on-a-chip devices has spawned research into the development of a variety of new actuating methods (such as microvalves) on the basis of SRPs, which can be triggered thermally, magnetically, optically, or by a change in pH. Finally, the grafting of SRPs to or from porous membranes permits temporal control of the permeation of small molecules or nanoparticles, which is a powerful feature of molecular separation devices.

The future will, no doubt, witness the development of even smarter nanointerfaces capable of exploiting the disruptive properties of nanostructured materials and of the fine-tuned (multi) stimulus-responsiveness of polymers synthesized via increasingly sophisticated polymerization procedures. Such developments will surely pave the road to a plethora of industrial applications.

List of Abbreviations

4-HBA	hydroxybutyl acrylate
AAc	acrylic acid
AAm	acrylamide
AIBN	azo-bis-*iso*-butyronitrile
AMF	alternating magnetic field
Anti-Fe	anti ferritin antibody
APTES	(3-aminopropyl)triethoxysilane
ATRP	atom transfer radical polymerization
BCAm	benzo-18-crown-6-acrylamide
BSA	bovine serum albumin
CAA	2-carboxy ethyl acrylate
CDDS	controlled drug-delivery system
CMC	critical micellar concentration
CMT	critical micelle temperature
CNT	carbon nanotube
CNT-PAAc	carbon nanotubes grafted with PAAc
CNT-PAN	poly(acrylonitrile)-grafted CNTs
CRP	controlled radical polymerization
DEAM	N,N-diethylacrylamide
DNQsc	2-diazo-1,2-naphthoquinone-5-sulfonyl chloride
DOX	doxorubicin
EDC	1-ethyl-3-(3-dimethylaminopropyl)carbodiimide
EGDMA	ethylene glycol dimethacrylate
FBS	fetal bovine serum
FDA	Food and Drug Administration
Fg	fibrinogen
GPTMS	glycidoxypropyltrimethoxysilane
HIgG	human immunoglobulin G
HPG-NIPAAm	hyperbranched polyglycerol modified with NIPAAm
IEP	isoelectric point
LCST	lower critical solution temperature

LGTT	lower gel transition temperature
Lys	lysozyme
MALDI	matrix-assisted laser desorption/ionization
MBAAm	N,N'-methylene bis(acrylamide)
MWNT	multi-walled carbon nanotube
NSp	nitrospiropyran
NIPAAm	N-isopropylacrylamide
ODA	octadecyl acrylate
PAAc	poly(acrylic acid)
PAAm	poly(acrylamide)
PBAc	poly(butylacrylic acid)
P(BpyClCl)	poly(N-benzyl-N'-(4-vinylbenzyl)-4,4'-bipyridium dichloride)
PCA	principal component analysis
PCL	polymer-caged liposome
PCR	polymerase chain reaction
PDMA	poly(N,N-dimethylacrylamide)
PDMAEMA	poly(N,N-dimethylaminoethylmethacrylate)
PDMS	polydimethylsiloxane
PEAc	poly(ethylacrylic acid)
PEG	polyethylene glycol
PEO	poly(ethylene oxide)
PEOEOVE	poly[2-(3-ethoxy)ethoxyethyl vinyl ether]
PHB	poly[(R)-3-hydroxybutyrate]
PHPMA mono/dilactate	poly(N-(2-hydroxypropyl)methacrylamide mono/dilactate)
PiPrOx	poly(2-isopropyl-2-oxazoline)
PLGA	poly(lactic-co-glycolic acid)
PLLA	poly(L-lactic acid)
PMETAC	poly([2-(methacryloyloxy)ethyl]trimethylammonium chloride)
PMOVE	poly(2-methoxyethyl vinyl ether)
PMVE	poly(methyl vinyl ether)
PNCPA	poly(N-cyclopropylacrylamide)
PNIPAAm	poly(N-isopropylacrylamide)
PNIPAAm-b-(PEA-g-P2VP)	poly(N-isopropylacrylamide)-b-[poly(ethyl acrylate)-g-poly(2-vinylpyridine)]
PNIPAAm-PBMA	PNIPAAm-poly(n-butyl methacrylate)
PNIPAAm-co-PAAc)	poly(N-isopropylacrylamide-co-acrylic acid)
PNIPAAm-co-PDMA)	poly(N-isopropylacrylamide-co-N,N-dimethylmethacrylate)
PNIPAAm-co-PNHMA	poly(NIPAAm-co-N-hydroxymethylacrylamide)
PNSp-co-PMMA	poly(nitrospiropyran-co-methyl methacrylate)
PNSp	poly(nitrospiropyran)
PPA	poly(potassium acrylate)

PPAc	poly(propylacrylic acid)
pPNIPAAm	plasma polymerized NIPAAm
PS	polystyrene
PS-b-PDMAEMA	polystyrene-block-poly(N,N-dimethylaminoethylmethacrylate)
PVCL	poly(N-vinyl caprolactam)
P2VP	poly-(2-vinylpyridine)
PVBA-b-PDEA	poly(4-vinyl benzoic acid-block-2-(diethylamino) ethyl methacrylate)
PVDF	polyvinylidene fluoride
QCM	quartz crystal microbalance
QD	quantum dot
RAFT	reversible addition–fragmentation chain transfer
R_g	radius of gyration
SDBS	sodium dodecylbenzene sulfonate
SRP	stimulus-responsive polymer
TCPS	tissue culture polystyrene
ToF-SIMS	time-of-flight secondary ion mass spectrometry
UCST	upper critical solution temperature
UGTT	upper gel transition temperature

References

1 Cole, M.A., Voelcker, N.H., Thissen, H. and Griesser, H.J. (2009) Stimuli-responsive interfaces and systems for the control of protein-surface and cell-surface interactions. *Biomaterials*, **30** (9), 1827–50.

2 Schmaljohann, D. (2006) Thermo- and pH-responsive polymers in drug delivery. *Advanced Drug Delivery Reviews*, **58** (15), 1655–70.

3 Wei, H., Zhang, X.-Z., Cheng, H., Chen, W.-Q., Cheng, S.-X. and Zhuo, R.-X. (2006) Self-assembled thermo- and pH responsive micelles of poly(10-undecenoic acid-b-N-isopropylacrylamide) for drug delivery. *Journal of Controlled Release*, **116** (3), 266–74.

4 Feng, H., Zhao, Y., Pelletier, M., Dan, Y. and Zhao, Y. (2009) Synthesis of photo- and pH-responsive composite nanoparticles using a two-step controlled radical polymerization method. *Polymer*, **50** (15), 3470–7.

5 Kikuchi, A. and Okano, T. (2002) Pulsatile drug release control using hydrogels. *Advanced Drug Delivery Reviews*, **54** (1), 53–77.

6 Henna, V., Antti, L., Heikki, T. and Jouni, H. (2008) Drug release characteristics of physically cross-linked thermosensitive poly(N-vinylcaprolactam) hydrogel particles. *Journal of Pharmaceutical Sciences*, **97** (11), 4783–93.

7 Hatefi, A. and Amsden, B. (2002) Biodegradable injectable in situ forming drug delivery systems. *Journal of Controlled Release*, **80** (1–3), 9–28.

8 Da Silva, R.M.P., Mano, J.F. and Reis, R.L. (2007) Smart thermoresponsive coatings and surfaces for tissue engineering: switching cell-material boundaries. *Trends in Biotechnology*, **25** (12), 577–83.

9 Harmon, M.E., Tang, M. and Frank, C.W. (2003) A microfluidic actuator based on thermoresponsive hydrogels. *Polymer*, **44** (16), 4547–56.

10 Valiaev, A., Abu-Lail, N.I., Lim, D.W., Chilkoti, A. and Zauscher, S. (2006) Microcantilever sensing and actuation

with end-grafted stimulus-responsive elastin-like polypeptides. *Langmuir*, **23** (1), 339–44.
11. Liua, Y.-Y., Shaob, Y.-H. and Lu, J. (2006) Preparation, properties and controlled release behaviors of pH-induced thermosensitive amphiphilic gels. *Biomaterials*, **27**, 4016–24.
12. Paasonen, L., Romberg, B., Storm, G., Yliperttula, M., Urtti, A. and Hennink, W.E. (2007) Temperature-sensitive poly(N-(2-hydroxypropyl)methacrylamide mono/dilactate)-coated liposomes for triggered contents release. *Bioconjugate Chemistry*, **18** (6), 2131–6.
13. Zhang, J. and Misra, R.D.K. (2007) Magnetic drug-targeting carrier encapsulated with thermosensitive smart polymer: core-shell nanoparticle carrier and drug release response. *Acta Biomaterialia*, **3** (6), 838–50.
14. Li, L., Zhu, Y., Li, B. and Gao, C. (2008) Fabrication of thermoresponsive polymer gradients for study of cell adhesion and detachment. *Langmuir*, **24** (23), 13632–9.
15. Tsukagoshi, T., Kondo, Y. and Yoshino, N. (2007) Preparation of thin polymer films with controlled drug release. *Colloids and Surfaces B: Biointerfaces*, **57** (2), 219–25.
16. Hirose, M., Kwon, O.H., Yamato, M., Kikuchi, A. and Okano, T. (2000) Creation of designed shape cell sheets that are noninvasively harvested and moved onto another surface. *Biomacromolecules*, **1** (3), 377–81.
17. Cheng, T., Lin, T., Brady, R. and Wang, X. (2008) Photochromic fabrics with improved durability and photochromic performance. *Fibers and Polymers*, **9** (5), 521–6.
18. Schild, H.G. (1992) Poly (N-isopropylacrylamide): experiment, theory and application. *Progress in Polymer Science*, **17** (2), 163–249.
19. Graziano, G. (2000) On the temperature-induced coil to globule transition of poly-N-isopropylacrylamide in dilute aqueous solutions. *International Journal of Biological Macromolecules*, **27** (1), 89–97.
20. Kumar, A., Srivastava, A., Galaev, I.Y. and Mattiasson, B. (2007) Smart polymers: physical forms and bioengineering applications. *Progress in Polymer Science*, **32** (10), 1205–37.
21. Vihola, H., Laukkanen, A., Valtola, L., Tenhu, H. and Hirvonen, J. (2005) Cytotoxicity of thermosensitive polymers poly(N-isopropylacrylamide), poly(N-vinylcaprolactam) and amphiphilically modified poly(N-vinylcaprolactam). *Biomaterials*, **26** (16), 3055–64.
22. Prabaharan, M., Grailer, J.J., Steeber, D.A. and Gong, S. (2008) Stimuli-responsive chitosan-graft-poly(N-vinylcaprolactam) as a promising material for controlled hydrophobic drug delivery. *Macromolecular Bioscience*, **8** (9), 843–51.
23. Arndt, K.-F., Schmidt, T. and Reichelt, R. (2001) Thermo-sensitive poly(methyl vinyl ether) micro-gel formed by high energy radiation. *Polymer*, **42** (16), 6785–91.
24. Schlaad, H., Diehl, C., Gress, A., Meyer, M., Demirel, A.L., Nur, Y. and Bertin, A. (2010) Poly(2-oxazoline)s as smart bioinspired polymers. *Macromolecular Rapid Communications*, **31** (6), 511–25.
25. Diehl, C. and Schlaad, H. (2009) Polyoxazoline-based crystalline microspheres for carbohydrate-protein recognition. *Chemistry – A European Journal*, **15** (43), 11469–72.
26. Gaertner, F.C., Luxenhofer, R., Blechert, B., Jordan, R. and Essler, M. (2007) Synthesis, biodistribution and excretion of radiolabeled poly(2-alkyl-2-oxazoline)s. *Journal of Controlled Release*, **119** (3), 291–300.
27. Chu, L.-Y., Li, Y., Zhu, J.-H. and Chen, W.-M. (2005) Negatively thermoresponsive membranes with functional gates driven by zipper-type hydrogen-bonding interactions. *Angewandte Chemie International Edition*, **44** (14), 2124–7.
28. Jeong, B., Bae, Y.H. and Kim, S.W. (1999) Thermoreversible gelation of PEG–PLGA–PEG triblock copolymer aqueous solutions. *Macromolecules*, **32** (21), 7064–9.
29. Jeong, B., Bae, Y.H. and Kim, S.W. (2000) Drug release from biodegradable injectable thermosensitive hydrogel of PEG-PLGA-PEG triblock copolymers. *Journal of Controlled Release*, **63** (1–2), 155–63.

30 Yoshioka, H., Mori, Y., Tsukikawa, S. and Kubota, S. (1998) Thermoreversible gelation on cooling and on heating of an aqueous gelatin-poly(*N*-isopropylacrylamide) conjugate. *Polymers for Advanced Technologies*, **9** (2), 155–8.

31 Ohya, S., Nakayama, Y. and Matsuda, T. (2004) In vivo evaluation of poly(*N*-isopropylacrylamide) (PNIPAM)-grafted gelatin as an in situ-formable scaffold. *Journal of Artificial Organs*, **7** (4), 181–6.

32 Jeong, B., Bae, Y.H., Lee, D.S. and Kim, S.W. (1997) Biodegradable block copolymers as injectable drug-delivery systems. *Nature*, **388** (6645), 860–2.

33 Sugihara, S., Kanaoka, S. and Aoshima, S. (2005) Double thermosensitive diblock copolymers of vinyl ethers with pendant oxyethylene groups: unique physical gelation. *Macromolecules*, **38** (5), 1919–27.

34 Galaev, I.Y., Warrol, C. and Mattiasson, B. (1994) Temperature-induced displacement of proteins from dye-affinity columns using an immobilized polymeric displacer. *Journal of Chromatography A*, **684** (1), 37–43.

35 Zhang, Q., Xia, F., Sun, T., Song, W., Zhao, T., Liu, M. and Jiang, L. (2008) Wettability switching between high hydrophilicity at low pH and high hydrophobicity at high pH on surface based on pH-responsive polymer. *Chemical Communications*, **10**, 1199–201.

36 Wu, J. and Sailor, M.J. (2009) Chitosan hydrogel-capped porous SiO2 as a pH responsive nano-valve for triggered release of insulin. *Advanced Functional Materials*, **19** (5), 733–41.

37 Moya, S., Azzaroni, O., Farhan, T., Osborne, V.L. and Huck, W.T.S. (2005) Locking and unlocking of polyelectrolyte brushes: toward the fabrication of chemically controlled nanoactuators. *Angewandte Chemistry International Edition*, **44** (29), 4578–81.

38 Van Camp, W., Du Prez, F.E., Alem, H., Demoustier-Champagne, S., Willet, N., Grancharov, G. and Duwez, A.-S. (2010) Poly(acrylic acid) with disulfide bond for the elaboration of pH-responsive brush surfaces. *European Polymer Journal*, **46** (2), 195–201.

39 Connal, L.A., Li, Q., Quinn, J.F., Tjipto, E., Caruso, F. and Qiao, G.G. (2008) pH-responsive poly(acrylic acid) core cross-linked star polymers: morphology transitions in solution and multilayer thin films. *Macromolecules*, **41** (7), 2620–6.

40 Kyriakides, T.R., Cheung, C.Y., Murthy, N., Bornstein, P., Stayton, P.S. and Hoffman, A.S. (2002) pH-sensitive polymers that enhance intracellular drug delivery in vivo. *Journal of Controlled Release*, **78** (1–3), 295–303.

41 Stayton, P., El-Sayed, M., Murthy, N., Bulmus, V., Lackey, C., Cheung, C. and Hoffman, A. (2005) "Smart" delivery systems for biomolecular therapeutics. *Orthodontics and Craniofacial Research*, **8** (3), 219–25.

42 Liu, S. and Armes, S.P. (2002) Polymeric surfactants for the new millennium: a pH-responsive, zwitterionic, schizophrenic diblock copolymer13. *Angewandte Chemie International Edition*, **41** (8), 1413–16.

43 Sun, Y., Liu, Y., Zhao, G., Zhou, X., Gao, J. and Zhang, Q. (2008) Preparation of pH-responsive silver nanoparticles by RAFT polymerization. *Journal of Materials Science*, **43** (13), 4625–30.

44 Mi, L., Bernards, M.T., Cheng, G., Yu, Q. and Jiang, S. (2010) pH responsive properties of non-fouling mixed-charge polymer brushes based on quaternary amine and carboxylic acid monomers. *Biomaterials*, **31** (10), 2919–25.

45 Xin, B.W. and Hao, J.C. (2010) Reversibly switchable wettability. *Chemical Society Reviews*, **39** (2), 769–82.

46 Jiang, W.H., Wang, G.J., He, Y.N., Wang, X.G., An, Y.L., Song, Y.L. and Jiang, L. (2005) Photo-switched wettability on an electrostatic self-assembly azobenzene monolayer. *Chemical Communications*, **28**, 3550–2.

47 Lee, M.J., Jung, D.H. and Han, Y.K. (2006) Photo-responsive polymers and their applications to optical memory. *Molecular Crystals and Liquid Crystals*, **444**, 41–50.

48 Anastasiadis, S.H., Lygeraki, M.I., Athanassiou, A., Farsari, M. and Pisignano, D. (2008) Reversibly photo-responsive polymer surfaces for

controlled wettability. *Journal of Adhesion Science and Technology*, **22** (15), 1853–68.
49. Wang, M., Vail, S.A., Keirstead, A.E., Marquez, M., Gust, D. and Garcia, A.A. (2009) Preparation of photochromic poly(vinylidene fluoride-*co*-hexafluoropropylene) fibers by electrospinning. *Polymer*, **50** (16), 3974–80.
50. Xu, F.J., Su, F.B., Deng, S.B. and Yang, W.T. (2010) Novel stimuli-responsive polyelectrolyte brushes. *Macromolecules*, **43** (5), 2630–3.
51. Sershen, S.R., Mensing, G.A., Ng, M., Halas, N.J., Beebe, D.J. and West, J.L. (2005) Independent optical control of microfluidic valves formed from optomechanically responsive nanocomposite hydrogels. *Advanced Materials*, **17** (11), 1366–8.
52. Routh, A.F. and Vincent, B. (2002) Salt-induced homoaggregation of poly(*N*-isopropylacrylamide) microgels. *Langmuir*, **18** (14), 5366–9.
53. Hoffmann, J., Plotner, M., Kuckling, D. and Fischer, W.J. (1999) Photopatterning of thermally sensitive hydrogels useful for microactuators. *Sensors and Actuators A, Physical*, **77** (2), 139–44.
54. Park, T.G. and Hoffman, A.S. (1993) Sodium chloride-induced phase-transition in nonionic poly(*N*-isopropylacrylamide) gel. *Macromolecules*, **26** (19), 5045–8.
55. Jhon, Y.K., Bhat, R.R., Jeong, C., Rojas, O.J., Szleifer, I. and Genzer, J. (2006) Salt-induced depression of lower critical solution temperature in a surface-grafted neutral thermoresponsive polymer. *Macromolecular Rapid Communications*, **27** (9), 697–701.
56. Erbil, C., Aras, S. and Uyanik, N. (1999) Investigation of the effect of type and concentration of ionizable comonomer on the collapse behavior of *N*-isopropylacrylamide copolymer gels in water. *Journal of Polymer Science Part A: Polymer Chemistry*, **37** (12), 1847–55.
57. Christodoulakis, K.E. and Vamvakaki, M. (2010) Amphoteric core-shell microgels: contraphilic two-compartment colloidal particles. *Langmuir*, **26** (2), 639–47.
58. Wang, J.Y., Chen, L., Zhao, Y.P., Guo, G. and Zhang, R. (2009) Cell adhesion and accelerated detachment on the surface of temperature-sensitive chitosan and poly(*N*-isopropylacrylamide) hydrogels. *Journal of Materials Science: Materials in Medicine*, **20** (2), 583–90.
59. Ebara, M., Yamato, M., Hirose, M., Aoyagi, T., Kikuchi, A., Sakai, K. and Okano, T. (2003) Copolymerization of 2-carboxyisopropylacrylamide with *N*-isopropylacrylamide accelerates cell detachment from grafted surfaces by reducing temperature. *Biomacromolecules*, **4** (2), 344–9.
60. Garcia, A., Marquez, M., Cai, T., Rosario, R., Hu, Z.B., Gust, D., Hayes, M., Vail, S.A. and Park, C.D. (2007) Photo-, thermally, and pH-responsive microgels. *Langmuir*, **23** (1), 224–9.
61. Zareie, H.M., Bulmus, E.V., Gunning, A.P., Hoffman, A.S., Piskin, E. and Morris, V.J. (2000) Investigation of a stimuli-responsive copolymer by atomic force microscopy. *Polymer*, **41** (18), 6723–7.
62. Ito, T. and Yamaguchi, T. (2006) Controlled release of model drugs through a molecular recognition ion gating membrane in response to a specific ion signal. *Langmuir*, **22** (8), 3945–9.
63. Xia, F., Feng, L., Wang, S.T., Sun, T.L., Song, W.L., Jiang, W.H. and Jiang, L. (2006) Dual-responsive surfaces that switch superhydrophilicity and superhydrophobicity. *Advanced Materials*, **18** (4), 432–6.
64. Sumaru, K., Kameda, M., Kanamori, T. and Shinbo, T. (2004) Characteristic phase transition of aqueous solution of poly(*N*-isopropylacrylamide) functionalized with spirobenzopyran. *Macromolecules*, **37** (13), 4949–55.
65. Ivanov, A.E., Eremeev, N.L., Wahlund, P.O., Galaev, I.Y. and Mattiasson, B. (2002) Photosensitive copolymer of *N*-isopropylacrylamide and methacryloyl derivative of spyrobenzopyran. *Polymer*, **43** (13), 3819–23.
66. Edahiro, J., Sumaru, K., Tada, Y., Ohi, K., Takagi, T., Kameda, M., Shinbo, T., Kanamori, T. and Yoshimi, Y. (2005) In situ control of cell adhesion using photoresponsive culture surface. *Biomacromolecules*, **6** (2), 970–4.

67 Edahiro, J.I., Sumaru, K., Takagi, T., Shinbo, T., Kanamori, T. and Sudoh, M. (2008) Analysis of photo-induced hydration of a photochromic poly(N-isopropylacrylamide)–spiropyran copolymer thin layer by quartz crystal microbalance. *European Polymer Journal*, **44** (2), 300–7.

68 Yu, Y.-Y., Tian, F., Wei, C. and Wang, C.-C. (2009) Facile synthesis of triple-stimuli (photo/pH/thermo) responsive copolymers of 2-diazo-1,2-naphthoquinone-mediated poly(N-isopropylacrylamide-co-N-hydroxymethylacrylamide). *Journal of Polymer Science Part A – Polymer Chemistry*, **47** (11), 2763–73.

69 Okajima, S., Sakai, Y. and Yamaguchi, T. (2005) Development of a regenerable cell culture system that senses and releases dead cells. *Langmuir*, **21** (9), 4043–9.

70 Ito, T., Sato, Y., Yamaguchi, T. and Nakao, S. (2004) Response mechanism of a molecular recognition ion gating membrane. *Macromolecules*, **37** (9), 3407–14.

71 Budhlall, B.M., Marquez, M. and Velev, O.D. (2008) Microwave, photo- and thermally responsive PNIPAm-gold nanoparticle microgels. *Langmuir*, **24** (20), 11959–66.

72 Barbey, R., Lavanant, L., Paripovic, D., Schuwer, N., Sugnaux, C., Tugulu, S. and Klok, H.A. (2009) Polymer brushes via surface-initiated controlled radical polymerization: synthesis, characterization, properties, and applications. *Chemical Reviews*, **109** (11), 5437–527.

73 Nebhani, L. and Barner-Kowollik, C. (2009) Orthogonal transformations on solid substrates: efficient avenues to surface modification. *Advanced Materials*, **21** (34), 3442–68.

74 Matyjaszewski, K. (1997) Mechanistic and synthetic aspects of atom transfer radical polymerization. *Journal of Macromolecular Science, Part A: Pure and Applied Chemistry*, **A34** (10), 1785–801.

75 Chiefari, J., Chong, Y.K., Ercole, F., Krstina, J., Jeffery, J., Le, T.P.T., Mayadunne, R.T.A., Meijs, G.F., Moad, C.L., Moad, G., Rizzardo, E. and Thang, S.H. (1998) Living free-radical polymerization by reversible addition-fragmentation chain transfer: the RAFT process. *Macromolecules*, **31** (16), 5559–62.

76 Couet, J. and Biesalski, M. (2006) Surface-initiated ATRP of N-isopropylacrylamide from initiator-modified self-assembled peptide nanotubes. *Macromolecules*, **39** (21), 7258–68.

77 Zhou, Z.Y., Zhu, S.M. and Zhang, D. (2007) Grafting of thermoresponsive polymer inside mesoporous silica with large pore size using ATRP and investigation of its use in drug release. *Journal of Materials Chemistry*, **17** (23), 2428–33.

78 Hernandez-Guerrero, M., Min, E., Barner-Kowollik, C., Muller, A.H.E. and Stenzel, M.H. (2008) Grafting thermoresponsive polymers onto honeycomb structured porous films using the RAFT process. *Journal of Materials Chemistry*, **18** (39), 4718–30.

79 Xia, Y., Yin, X.C., Burke, N.A.D. and Stover, H.D.H. (2005) Thermal response of narrow-disperse poly(N-isopropylacrylamide) prepared by atom transfer radical polymerization. *Macromolecules*, **38** (14), 5937–43.

80 Motornov, M., Roiter, Y., Tokarev, I. and Minko, S. (2010) Stimuli-responsive nanoparticles, nanogels and capsules for integrated multifunctional intelligent systems. *Progress in Polymer Science*, **35** (1–2), 174–211.

81 Cole, M.A., Jasieniak, M., Thissen, H., Voelcker, N.H. and Griesser, H.J. (2009) Time-of-flight-secondary ion mass spectrometry study of the temperature dependence of protein adsorption onto poly(N-isopropylacrylamide) graft coatings. *Analytical Chemistry*, **81** (16), 6905–12.

82 Kurkuri, M.D., Nussio, M.R., Deslandes, A. and Voelcker, N.H. (2008) Thermosensitive copolymer coatings with enhanced wettability switching. *Langmuir*, **24** (8), 4238–44.

83 Ishida, N. and Biggs, S. (2007) Direct observation of the phase transition for a poly(N-isopropylacrylamide) layer grafted onto a solid surface by AFM and

QCM-D. *Langmuir*, **23** (22), 11083–8.

84 Higuchi, A., Hamamura, A., Shindo, Y., Kitamura, H., Yoon, B.O., Mori, T., Uyama, T. and Umezawa, A. (2004) Photon-modulated changes of cell attachments on poly(spiropyran-*co*-methyl methacrylate) membranes. *Biomacromolecules*, **5** (5), 1770–4.

85 Segal, E., Perelman, L.A., Moore, T., Kesselman, E. and Sailor, M.J. (2009) Grafting stimuli-responsive polymer brushes to freshly-etched porous silicon. *Physical Review E*, **6** (7), 1717–20.

86 Nair, A., Shen, J., Thevenot, P., Zou, L., Cai, T., Hu, Z. and Tang, L. (2008) Enhanced intratumoral uptake of quantum dots concealed within hydrogel nanoparticles. *Nanotechnology*, **19** (48), 485102.

87 Kingshott, P., Thissen, H. and Griesser, H.J. (2002) Effects of cloud-point grafting, chain length, and density of PEG layers on competitive adsorption of ocular proteins. *Biomaterials*, **23** (9), 2043–56.

88 Kushida, A., Yamato, M., Konno, C., Kikuchi, A., Sakurai, Y. and Okano, T. (1999) Decrease in culture temperature releases monolayer endothelial cell sheets together with deposited fibronectin matrix from temperature-responsive culture surfaces. *Journal of Biomedical Materials Research*, **45** (4), 355–62.

89 Christman, K.L., Schopf, E., Broyer, R.M., Li, R.C., Chen, Y. and Maynard, H.D. (2009) Positioning multiple proteins at the nanoscale with electron beam cross-linked functional polymers. *Journal of the American Chemical Society*, **131** (2), 521–7.

90 Sofia, S.J. and Merrill, E.W. (1998) Grafting of PEO to polymer surfaces using electron beam irradiation. *Journal of Biomedical Materials Research*, **40** (1), 153–63.

91 Yamada, N., Okano, T., Sakai, H., Karikusa, F., Sawasaki, Y. and Sakurai, Y. (1990) Thermo-responsive polymeric surfaces; control of attachment and detachment of cultured cells. *Makromolekulare Chemie – Rapid Communications*, **11** (11), 571–6.

92 Wu, M.H., Chen, J., Qian, Q. and Bao, B.R. (2002) Thermo-responsive interpenetrating polymer networks composed of AAc/AAm/NMA by radiation grafting. *Journal of Radioanalytical and Nuclear Chemistry*, **252** (3), 531–5.

93 Idota, N., Tsukahara, T., Sato, K., Okano, T. and Kitamori, T. (2009) The use of electron beam lithographic graft-polymerization on thermoresponsive polymers for regulating the directionality of cell attachment and detachment. *Biomaterials*, **30** (11), 2095–101.

94 Akiyama, Y., Kikuchi, A., Yamato, M. and Okano, T. (2004) Ultrathin poly(N-isopropylacrylamide) grafted layer on polystyrene surfaces for cell adhesion/detachment control. *Langmuir*, **20** (13), 5506–11.

95 Yamato, M., Okuhara, M., Karikusa, F., Kikuchi, A., Sakurai, Y. and Okano, T. (1999) Signal transduction and cytoskeletal reorganization are required for cell detachment from cell culture surfaces grafted with a temperature-responsive polymer. *Journal of Biomedical Materials Research*, **44** (1), 44–52.

96 Siow, K.S., Britcher, L., Kumar, S. and Griesser, H.J. (2006) Plasma methods for the generation of chemically reactive surfaces for biomolecule immobilization and cell colonization – a review. *Plasma Processes and Polymers*, **3** (6/7), 392–418.

97 Canavan, H.E., Cheng, X.H., Graham, D.J., Ratner, B.D. and Castner, D.G. (2006) A plasma-deposited surface for cell sheet engineering: advantages over mechanical dissociation of cells. *Plasma Processes and Polymers*, **3** (6–7), 516–23.

98 Bullett, N.A., Talib, R.A., Short, R.D., McArthur, S.L. and Shard, A.G. (2006) Chemical and thermoresponsive characterization of surfaces formed by plasma polymerization of N-isopropyl acrylamide. *Surface and Interface Analysis*, **38** (7), 1109–16.

99 Loh, X.J., Zhang, Z.-X., Wu, Y.-L., Lee, T.S. and Li, J. (2008) Synthesis of novel biodegradable thermoresponsive triblock copolymers based on poly[(*R*)-3-hydroxybutyrate] and poly(*N*-isopropylacrylamide) and their formation

of thermoresponsive micelles. *Macromolecules*, **42** (1), 194–202.

100 Shen, Y., Kuang, M., Shen, Z., Nieberle, J., Duan, H. and Frey, H. (2008) Gold nanoparticles coated with a thermosensitive hyperbranched polyelectrolyte: towards smart temperature and pH nanosensors. *Angewandte Chemie International Edition*, **47** (12), 2227–30.

101 Castellanos, A., DuPont, S.J., Heim, A.J., Matthews, G., Stroot, P.G., Moreno, W. and Toomey, R.G. (2007) Size-exclusion capture and release separations using surface-patterned poly(N-isopropylacrylamide) hydrogels. *Langmuir*, **23** (11), 6391–5.

102 Beebe, D.J., Moore, J.S., Bauer, J.M., Yu, Q., Liu, R.H., Devadoss, C. and Jo, B.H. (2000) Functional hydrogel structures for autonomous flow control inside microfluidic channels. *Nature*, **404** (6778), 588.

103 Segal, E., Perelman, L.A., Cunin, F., Di Renzo, F., Devoisselle, J.M., Li, Y.Y. and Sailor, M.J. (2007) Confinement of thermoresponsive hydrogels in nanostructured porous silicon dioxide templates. *Advanced Functional Materials*, **17** (7), 1153–62.

104 McInnes, S.J.P., Thissen, H., Choudhury, N.R. and Voelcker, N.H. (2009) New biodegradable materials produced by ring opening polymerization of poly(L-lactide) on porous silicon substrates. *Journal of Colloid and Interface Science*, **332** (2), 336–44.

105 Yu, W.H., Kang, E.T., Neoh, K.G. and Zhu, S. (2003) Controlled grafting of well-defined polymers on hydrogen-terminated silicon substrates by surface-initiated atom transfer radical polymerization. *Journal of Physical Chemistry B*, **107** (37), 10198–205.

106 Xu, F.J., Zhong, S.P., Yung, L.Y.L., Kang, E.T. and Neoh, K.G. (2004) Surface-active and stimuli-responsive polymer-Si(100) hybrids from surface-initiated atom transfer radical polymerization for control of cell adhesion. *Biomacromolecules*, **5** (6), 2392–403.

107 Yamato, M., Konno, C., Kushida, A., Hirose, M., Utsumi, M., Kikuchi, A. and Okano, T. (2000) Release of adsorbed fibronectin from temperature-responsive culture surfaces requires cellular activity. *Biomaterials*, **21** (10), 981–6.

108 Cunliffe, D., de las Heras Alarcon, C., Peters, V., Smith, J.R. and Alexander, C. (2003) Thermoresponsive surface-grafted poly(N-isopropylacrylamide) copolymers: effect of phase transitions on protein and bacterial attachment. *Langmuir*, **19** (7), 2888–99.

109 Beiyi, C., Xu, F.J., Ning, F., Neoh, K.G., Kang, E.T., Wei Ning, C. and Vincent, C. (2008) Engineering cell de-adhesion dynamics on thermoresponsive poly(N-isopropylacrylamide). *Acta Biomaterialia*, **4** (2), 218–29.

110 Cheng, X.H., Wang, Y.B., Hanein, Y., Bohringer, K.F. and Ratner, B.D. (2004) Novel cell patterning using microheater-controlled thermoresponsive plasma films. *Journal of Biomedical Materials Research Part A: Early View*, **70A** (2), 159–68.

111 Pan, Y.V., Wesley, R.A., Luginbuhl, R., Denton, D.D. and Ratner, B.D. (2001) Plasma polymerized N-isopropylacrylamide: synthesis and characterization of a smart thermally responsive coating. *Biomacromolecules*, **2** (1), 32–6.

112 Bhattacharyya, D., Pillai, K., Chyan, O.M.R., Tang, L.P. and Timmons, R.B. (2007) A new class of thin film hydrogels produced by plasma polymerization. *Chemistry of Materials*, **19** (9), 2222–8.

113 Chu, L.Q., Zou, X.N., Knoll, W. and Forch, R. (2008) Thermosensitive surfaces fabricated by plasma polymerization of N,N-diethylacrylamide. *Surface and Coatings Technology*, **202** (10), 2047–51.

114 Ernst, O., Lieske, A., Jager, M., Lankenau, A. and Duschl, C. (2007) Control of cell detachment in a microfluidic device using a thermoresponsive copolymer on a gold substrate. *Lab on a Chip*, **7** (10), 1322–9.

115 Ernst, O., Lieske, A., Holländer, A., Lankenau, A. and Duschl, C. (2008) Tuning of thermo-responsive self-assembly monolayers on gold for cell-type-specific control of adhesion. *Langmuir*, **24** (18), 10259–64.

116. Malham, I.B. and Bureau, L. (2009) Density effects on collapse, compression, and adhesion of thermoresponsive polymer brushes. *Langmuir*, **26** (7), 4762–8.
117. Fu, Q., Rao, G.V.R., Basame, S.B., Keller, D.J., Artyushkova, K., Fulghum, J.E. and Lopez, G.P. (2004) Reversible control of free energy and topography of nanostructured surfaces. *Journal of the American Chemical Society*, **126** (29), 8904–5.
118. Li, P.-F., Xie, R., Jiang, J.-C., Meng, T., Yang, M., Ju, X.-J., Yang, L. and Chu, L.-Y. (2009) Thermo-responsive gating membranes with controllable length and density of poly(N-isopropylacrylamide) chains grafted by ATRP method. *Journal of Membrane Science*, **337** (1–2), 310–17.
119. Lokuge, I., Wang, X. and Bohn, P.W. (2007) Temperature-controlled flow switching in nanocapillary array membranes mediated by poly(N-isopropylacrylamide) polymer brushes grafted by atom transfer radical polymerization. *Langmuir*, **23** (1), 305–11.
120. Chu, L.Y., Niitsuma, T., Yamaguchi, T. and Nakao, S. (2003) Thermoresponsive transport through porous membranes with grafted PNIPAM gates. *AichE Journal*, **49** (4), 896–909.
121. Liang-Yin, C., Yan, L., Jia-Hua, Z. and Wen-Mei, C. (2005) Negatively thermoresponsive membranes with functional gates driven by zipper-type hydrogen-bonding interactions13. *Angewandte Chemie International Edition*, **44** (14), 2124–7.
122. Schacher, F., Ulbricht, M. and Muller, A.H.E. (2009) Self-supporting, double stimuli-responsive porous membranes from polystyrene-block-poly(N,N-dimethylaminoethyl methacrylate) diblock copolymers. *Advanced Functional Materials*, **19** (7), 1040–5.
123. Zhang, K., Ma, J., Zhang, B., Zhao, S., Li, Y., Xu, Y., Yu, W. and Wang, J. (2007) Synthesis of thermoresponsive silica nanoparticle/PNIPAM hybrids by aqueous surface-initiated atom transfer radical polymerization. *Materials Letters*, **61** (4–5), 949–52.
124. Kim, J.H., Na, K.H., Kang, C.J. and Kim, Y.S. (2005) A disposable thermopneumatic-actuated micropump stacked with PDMS layers and ITO-coated glass. *Sensors and Actuators A, Physical*, **120** (2), 365–9.
125. Elaissari, A. and Bourrel, V. (2001) Thermosensitive magnetic latex particles for controlling protein adsorption and desorption. *Journal of Magnetism and Magnetic Materials*, **225** (1-2), 151–5.
126. Wei, Q.S., Ji, J. and Shen, J.C. (2008) Synthesis of near-infrared responsive gold nanorod/PNIPAAm core/shell nanohybrids via surface initiated ATRP for smart drug delivery. *Macromolecular Rapid Communications*, **29** (8), 645–50.
127. Purushotham, S. *et al* (2009) Thermoresponsive core–shell magnetic nanoparticles for combined modalities of cancer therapy. *Nanotechnology*, **20** (30), 305101.
128. Zhou, L., Gao, C. and Xu, W.J. (2009) Amphibious polymer-functionalized CdTe quantum dots: synthesis, thermoresponsive self-assembly, and photoluminescent properties. *Journal of Materials Chemistry*, **19** (31), 5655–64.
129. Tagit, O., Tomczak, N., Benetti, E.M., Cesa, Y., Blum, C., Subramaniam, V., Herek, J.L. and Vancso, G.J. (2009) Temperature-modulated quenching of quantum dots covalently coupled to chain ends of poly(N-isopropyl acrylamide) brushes on gold. *Nanotechnology*, **20** (18), 6.
130. Shen, L., Pich, A., Fava, D., Wang, M.F., Kumar, S., Wu, C., Scholes, G.D. and Winnik, M.A. (2008) Loading quantum dots into thermoresponsive microgels by reversible transfer from organic solvents to water. *Journal of Materials Chemistry*, **18** (7), 763–70.
131. Janczewski, D., Tomczak, N., Han, M.Y. and Vancso, G.J. (2009) Stimulus responsive PNIPAM/QD hybrid microspheres by copolymerization with surface engineered QDs. *Macromolecules*, **42** (6), 1801–4.
132. Fu, H.K., Kuo, S.W., Huang, C.F., Chang, F.C. and Lin, H.C. (2009) Preparation of the stimuli-responsive ZnS/PNIPAM hollow spheres. *Polymer*, **50** (5), 1246–50.
133. GhoshMitra, S., Cai, T., Ghosh, S., Neogi, A., Hu, Z. and Mills, N. (2008)

Microbial growth response to hydrogel encapsulated quantum dot nanospheres. *Materials Research Society Symposium Proceedings*, **1064**, 6.

134 Larson, D.R., Zipfel, W.R., Williams, R.M., Clark, S.W., Bruchez, M.P., Wise, F.W. and Webb, W.W. (2003) Water-soluble quantum dots for multiphoton fluorescence imaging in vivo. *Science*, **300** (5624), 1434–6.

135 Hou, Y., Ye, J., Gui, Z. and Zhang, G. (2008) Temperature-modulated photoluminescence of quantum dots. *Langmuir*, **24** (17), 9682–5.

136 Muthu, M.S., Rajesh, C.V., Mishra, A. and Singh, S. (2009) Stimulus-responsive targeted nanomicelles for effective cancer therapy. *Nanomedicine*, **4** (6), 657–67.

137 Onaca, O., Enea, R., Hughes, D.W. and Meier, W. (2009) Stimuli-responsive polymersomes as nanocarriers for drug and gene delivery. *Macromolecular Bioscience*, **9** (2), 129–39.

138 Li, G.Y., Guo, L. and Ma, S.M. (2009) Self-assembly and drug delivery behaviors of thermo-sensitive poly(t-butyl acrylate)-b-poly (N-isopropylacrylamide) micelles. *Journal of Applied Polymer Science*, **113** (2), 1364–8.

139 Chang, C., Wei, H., Quan, C.Y., Li, Y.Y., Liu, J., Wang, Z.C., Cheng, S.X., Zhang, X.Z. and Zhuo, R.X. (2008) Fabrication of thermosensitive PCL-PNIPAAm-PCL triblock copolymeric micelles for drug delivery. *Journal of Polymer Science Part A: Polymer Chemistry*, **46** (9), 3048–57.

140 Akimoto, J., Nakayama, M., Sakai, K. and Okano, T. (2009) Temperature-induced intracellular uptake of thermoresponsive polymeric micelles. *Biomacromolecules*, **10** (6), 1331–6.

141 Determan, M.D., Cox, J.P., Seifert, S., Thiyagarajan, P. and Mallapragada, S.K. (2005) Synthesis and characterization of temperature and pH-responsive pentablock copolymers. *Polymer*, **46** (18), 6933–46.

142 Alvarez-Lorenzo, C., Deshmukh, S., Bromberg, L., Hatton, T.A., Sández-Macho, I. and Concheiro, A. (2007) Temperature- and light-responsive blends of pluronic F127 and poly(N,N-dimethylacrylamide-co-methacryloyloxyazobenzene). *Langmuir*, **23** (23), 11475–81.

143 Wu, X.L., Kim, J.H., Koo, H., Bae, S.M., Shin, H., Kim, M.S., Lee, B.-H., Park, R.-W., Kim, I.-S., Choi, K., Kwon, I.C., Kim, K. and Lee, D.S. (2010) Tumor-targeting peptide conjugated pH-responsive micelles as a potential drug carrier for cancer therapy. *Bioconjugate Chemistry*, **21** (2), 208–13.

144 Smith, A.E., Xu, X., Kirkland-York, S.E., Savin, D.A. and McCormick, C.L. (2010) "Schizophrenic" self-assembly of block copolymers synthesized via aqueous RAFT polymerization: from micelles to vesicles. *Macromolecules*, **43** (3), 1210–17.

145 Pascual, S. and Monteiro, M.J. (2009) Shell-crosslinked nanoparticles through self-assembly of thermoresponsive block copolymers by RAFT polymerization. *European Polymer Journal*, **45** (9), 2513–19.

146 Zhang, L., Bernard, J., Davis, T.P., Barner-Kowollik, C. and Stenzel, M.H. (2008) Acid-degradable core-crosslinked micelles prepared from thermosensitive glycopolymers synthesized via RAFT polymerization. *Macromolecular Rapid Communications* **29** (2), 123–9.

147 Meng, F.H., Zhong, Z.Y. and Feijen, J. (2009) Stimuli-responsive polymersomes for programmed drug delivery. *Biomacromolecules*, **10** (2), 197–209.

148 Ganta, S., Devalapally, H., Shahiwala, A. and Amiji, M. (2008) A review of stimuli-responsive nanocarriers for drug and gene delivery. *Journal of Controlled Release*, **126** (3), 187–204.

149 Kono, K., Hayashi, H. and Takagishi, T. (1994) Temperature-sensitive liposomes: liposomes bearing poly (N-isopropylacrylamide). *Journal of Controlled Release*, **30** (1), 69–75.

150 Lee, S.-M., Chen, H., Dettmer, C.M., O'Halloran, T.V. and Nguyen, S.T. (2007) Polymer-caged liposomes? A pH-responsive delivery system with high stability. *Journal of the American Chemical Society*, **129** (49), 15096–7.

151 Lopresti, C., Lomas, H., Massignani, M., Smart, T. and Battaglia, G. (2009) Polymersomes: nature inspired nanometer sized compartments. *Journal of Materials Chemistry*, **19** (22), 3576–90.

152. Meng, F.H., Hiemstra, C., Engbers, G.H.M. and Feijen, J. (2003) Biodegradable polymersomes. *Macromolecules*, **36** (9), 3004–6.
153. Hong, C.-Y. and Pan, C.-Y. (2008) Functionalized carbon nanotubes responsive to environmental stimuli. *Journal of Materials Chemistry*, **18** (16), 1831–6.
154. You, Y.-Z., Hong, C.-Y. and Pan, C.-Y. (2007) Preparation of smart polymer/carbon nanotube conjugates via stimuli-responsive linkages. *Advanced Functional Materials*, **17** (14), 2470–7.
155. Yang, D., Hu, J. and Wang, C. (2006) Synthesis and characterization of pH-responsive single-walled carbon nanotubes with a large number of carboxy groups. *Carbon*, **44** (15), 3161–7.
156. Xu, G., Xia, R., Wang, H., Meng, X. and Zhu, Q. (2008) Grafting of thermoresponsive polymer from the surface of functionalized multiwalled carbon nanotubes via atom transfer radical polymerization. *Chinese Science Bulletin*, **53** (15), 2297–306.
157. Kong, H., Li, W., Gao, C., Yan, D., Jin, Y., Walton, D.R.M. and Kroto, H.W. (2004) Poly(N-isopropylacrylamide)-coated carbon nanotubes: temperature-sensitive molecular nanohybrids in water. *Macromolecules*, **37** (18), 6683–6.
158. Kong, H., Gao, C. and Yan, D. (2004) Constructing amphiphilic polymer brushes on the convex surfaces of multi-walled carbon nanotubes by in situ atom transfer radical polymerization. *Journal of Materials Chemistry*, **14** (9), 1401–5.
159. Ellis, A.V., Waterland, M.R. and Quinton, J. (2007) Water-soluble carbon nanotube chain-transfer agents (CNT-CTAs). *Chemistry Letters*, **36** (9), 1172–3.
160. You, Y.Z., Yan, J.J., Yu, Z.Q., Cui, M.M., Hong, C.Y. and Qu, B.J. (2009) Multi-responsive carbon nanotube gel prepared via ultrasound-induced assembly. *Journal of Materials Chemistry*, **19** (41), 7656–60.
161. Liu, J., Nie, Z., Gao, Y., Adronov, A. and Li, H. (2008) "Click" coupling between alkyne-decorated multiwalled carbon nanotubes and reactive PDMA-PNIPAM micelles. *Journal of Polymer Science Part A: Polymer Chemistry*, **46** (21), 7187–99.
162. Etika, K.C., Jochum, F.D., Theato, P. and Grunlan, J.C. (2009) Temperature controlled dispersion of carbon nanotubes in water with pyrene-functionalized poly(N-cyclopropylacrylamide). *Journal of the American Chemical Society*, **131** (38), 13598–9.
163. Gao, Y., Zhou, Y.F. and Yan, D.Y. (2009) Temperature-sensitive and highly water-soluble titanate nanotubes. *Polymer*, **50** (12), 2572–7.
164. Drummond, D.C., Zignani, M. and Leroux, J.-C. (2000) Current status of pH-sensitive liposomes in drug delivery. *Progress in Lipid Research*, **39** (5), 409–60.
165. Alexander, C. and Shakesheff, K.M. (2006) Responsive polymers at the biology/materials science interface. *Advanced Materials*, **18** (24), 3321–8.
166. Zhu, S.M., Zhou, Z.Y. and Zhang, D. (2007) Control of drug release through the in situ assembly of stimuli-responsive ordered mesoporous silica with magnetic particles. *ChemPhysChem*, **8** (17), 2478–83.
167. Huber, D.L., Manginell, R.P., Samara, M.A., Kim, B.-I. and Bunker, B.C. (2003) Programmed adsorption and release of proteins in a microfluidic device. *Science*, **301** (5631), 352–4.
168. Kikuchi, A. and Okano, T. (2005) Nanostructured designs of biomedical materials: applications of cell sheet engineering to functional regenerative tissues and organs. *Journal of Controlled Release*, **101** (1–3), 69–84.
169. Yang, J., Yamato, M., Shimizu, T., Sekine, H., Ohashi, K., Kanzaki, M., Ohki, T., Nishida, K. and Okano, T. (2007) Reconstruction of functional tissues with cell sheet engineering. *Biomaterials*, **28**, 5033–43.
170. Nagase, K., Kobayashi, J. and Okano, T. (2009) Temperature-responsive intelligent interfaces for biomolecular separation and cell sheet engineering. *Journal of the Royal Society, Interface*, **6**, S293–309.
171. Heinz, P., Bretagnol, F., Mannelli, I., Sirghi, L., Valsesia, A., Ceccone, G., Gilliland, D., Landfester, K., Rauscher,

H. and Rossi, F. (2008) Poly(N-isopropylacrylamide) grafted on plasma-activated poly(ethylene oxide): thermal response and interaction with proteins. *Langmuir*, **24** (12), 6166–75.

172 Cheng, X., Canavan, H.E., Graham, D.J., Castner, D.G. and Ratner, B.D. (2006) Temperature-dependent activity and structure of adsorbed proteins on plasma polymerized N-isopropyl acrylamide. *Biointerphases*, **1** (1), 61–72.

173 Hirata, I., Okazaki, M. and Iwata, H. (2004) Simple method for preparation of ultra-thin poly(N-isopropylacrylamide) hydrogel layers and characterization of their thermoresponsive properties. *Polymer*, **45** (16), 5569–78.

174 Mizutani, A., Kikuchi, A., Yamato, M., Kanazawa, H. and Okano, T. (2008) Preparation of thermoresponsive polymer brush surfaces and their interaction with cells. *Biomaterials*, **29** (13), 2073–81.

175 Canavan, H.E., Cheng, X.H., Graham, D.J., Ratner, B.D. and Castner, D.G. (2005) Cell sheet detachment affects the extracellular matrix: a surface science study comparing thermal liftoff, enzymatic, and mechanical methods. *Journal of Biomedical Materials Research Part A*, **75A** (1), 1–13.

176 Miyahara, Y., Nagaya, N., Kataoka, M., Yanagawa, B., Tanaka, K., Hao, H., Ishino, K., Ishida, H., Shimizu, T., Kangawa, K., Sano, S., Okano, T., Kitamura, S. and Mori, H. (2006) Monolayered mesenchymal stem cells repair scarred myocardium after myocardial infarction. *Nature Medicine*, **12** (4), 459–65.

177 Nishida, K., Yamato, M., Hayashida, Y., Watanabe, K., Yamamoto, K., Adachi, E., Nagai, S., Kikuchi, A., Maeda, N., Watanabe, H., Okano, T. and Tano, Y. (2004) Corneal reconstruction with tissue-engineered cell sheets composed of autologous oral mucosal epithelium. *New England Journal of Medicine*, **351** (12), 1187–96.

178 Yang, J., Yamato, M., Kohno, C., Nishimoto, A., Sekine, H., Fukai, F. and Okano, T. (2005) Cell sheet engineering: recreating tissues without biodegradable scaffolds. *Biomaterials*, **26** (33), 6415–22.

179 Tsuda, Y., Kikuchi, A., Yamato, M., Nakao, A., Sakurai, Y., Umezu, M. and Okano, T. (2005) The use of patterned dual thermoresponsive surfaces for the collective recovery as co-cultured cell sheets. *Biomaterials*, **26** (14), 1885–93.

180 Okano, T., Yamada, N., Okuhara, M., Sakai, H. and Sakurai, Y. (1995) Mechanism of cell detachment from temperature-modulated, hydrophilic-hydrophobic polymer surfaces. *Biomaterials*, **16** (4), 297–303.

181 Cunliffe, D., Alarcon, C.D., Peters, V., Smith, J.R. and Alexander, C. (2003) Thermoresponsive surface-grafted poly(N-isopropylacrylamide) copolymers: effect of phase transitions on protein and bacterial attachment. *Langmuir*, **19** (7), 2888–99.

182 Huber, D.L., Manginell, R.P., Samara, M.A., Kim, B.I. and Bunker, B.C. (2003) Programmed adsorption and release of proteins in a microfluidic device. *Science*, **301** (5631), 352–4.

183 Kurisawa, M., Yokoyama, M. and Okano, T. (2000) Gene expression control by temperature with thermoresponsive polymeric gene carriers. *Journal of Controlled Release*, **69** (1), 127–37.

184 Theato, P. and Zentel, R. (2003) Alpha, omega-functionalized poly-N-isopropylacrylamides: controlling the surface activity for vesicle adsorption by temperature. *Journal of Colloid and Interface Science*, **268** (1), 258–62.

185 Zintchenko, A., Ogris, M. and Wagner, E. (2006) Temperature dependent gene expression induced by PNIPAM-based copolymers: potential of hyperthermia in gene transfer. *Bioconjugate Chemistry*, **17** (3), 766–72.

186 Cole, M.A., Voelcker, N.H., Thissen, H., Horn, R.G. and Griesser, H.J. (2010) Colloid probe AFM study of thermal collapse and protein interactions of poly(N-isopropylacrylamide) coatings. *Soft Matter*, **6**, 2657–67.

187 Hyun, J., Lee, W.K., Nath, N., Chilkoti, A. and Zauscher, S. (2004) Capture and release of proteins on the nanoscale by stimuli-responsive elastin-like polypeptide "switches". *Journal of the American Chemical Society*, **126** (23), 7330–5.

188 Uhlmann, P., Houbenov, N., Brenner, N., Grundke, K., Burkert, S. and Stamm, M. (2007) In-situ investigation of the adsorption of globular model proteins on stimuli-responsive binary polyelectrolyte brushes. *Langmuir*, **23** (1), 57–64.

189 Kikuchi, K., Sumaru, K., Edahiro, J.I., Ooshima, Y., Sugiura, S., Takagi, T. and Kanamori, T. (2009) Stepwise assembly of micropatterned co-cultures using photoresponsive culture surfaces and its application to hepatic tissue arrays. *Biotechnology and Bioengineering*, **103** (3), 552–61.

190 Auernheimer, J., Dahmen, C., Hersel, U., Bausch, A. and Kessler, H. (2005) Photoswitched cell adhesion on surfaces with RGD peptides. *Journal of the American Chemical Society*, **127** (46), 16107–10.

191 Nakayama, Y., Furumoto, A., Kidoaki, S. and Matsuda, T. (2003) Photocontrol of cell adhesion and proliferation by a photoinduced cationic polymer surface. *Photochemistry and Photobiology*, **77** (5), 480–6.

192 Bringer, M.R., Gerdts, C.J., Song, H., Tice, J.D. and Ismagilov, R.F. (2004) Microfluidic systems for chemical kinetics that rely on chaotic mixing in droplets. *Philosophical Transactions of the Royal Society of London Series A: Mathematical, Physical and Engineering Sciences*, **362** (1818), 1087–104.

193 Song, H. and Ismagilov, R.F. (2003) Millisecond kinetics on a microfluidic chip using nanoliters of reagents. *Journal of the American Chemical Society*, **125** (47), 14613–19.

194 Seger-Sauli, U., Panayiotou, M., Schnydrig, S., Jordan, M. and Renaud, P. (2005) Temperature measurements in microfluidic systems: heat dissipation of negative dielectrophoresis barriers. *Electrophoresis*, **26** (11), 2239–46.

195 Cubaud, T. and Mason, T.G. (2009) High-viscosity fluid threads in weakly diffusive microfluidic systems. *New Journal of Physics*, **11** (7), 075029.

196 Keen, S., Yao, A., Leach, J., Di Leonardo, R., Saunter, C., Love, G., Cooper, J. and Padgett, M. (2009) Multipoint viscosity measurements in microfluidic channels using optical tweezers. *Lab on a Chip*, **9**, 2059–62.

197 Ridgeway, W.K., Seitaridou, E., Phillips, R. and Williamson, J.R. (2009) RNA-protein binding kinetics in an automated microfluidic reactor. *Nucleic Acids Research*, **37** (21), e142.

198 Kamholz, A.E., Schilling, E.A. and Yager, P. (2001) Optical measurement of transverse molecular diffusion in a microchannel. *Biophysical Journal*, **80** (4), 1967–72.

199 Perrin, D., Frémaux, C. and Shutes, A. (2010) Capillary microfluidic electrophoretic mobility shift assays: application to enzymatic assays in drug discovery. *Expert Opinion on Drug Discovery*, **5** (1), 51–63.

200 Wallingford, R.A. and Ewing, A.G. (1988) Capillary zone electrophoresis with electrochemical detection in 12.7 microns diameter columns. *Analytical Chemistry*, **60** (18), 1972–5.

201 Atalay, Y., Verboven, P., Vermeir, S., Vergauwe, N., Delport, F., Nicolaï, B. and Lammertyn, J. (2008) Design optimization of an enzymatic assay in an electrokinetically-driven microfluidic device. *Microfluidics and Nanofluidics*, **5** (6), 837–49.

202 Hatch, A., Kamholz, A.E., Hawkins, K.R., Munson, M.S., Schilling, E.A., Weigl, B.H. and Yager, P. (2001) A rapid diffusion immunoassay in a T-sensor. *Nature Biotechnology*, **19** (5), 461–5.

203 Lin, F.Y., Gao, Y., Li, D. and Sherman, P.M. (2010) Development of microfluidic-based heterogeneous immunoassays. *Frontiers in Bioscience*, **2**, 73–84.

204 Liu, P. and Mathies, R.A. (2009) Integrated microfluidic systems for high-performance genetic analysis. *Trends in Biotechnology*, **27** (10), 572–81.

205 Reichmuth, D.S., Wang, S.K., Barrett, L.M., Throckmorton, D.J., Einfeld, W. and Singh, A.K. (2008) Rapid microchip-based electrophoretic immunoassays for the detection of swine influenza virus. *Lab on a Chip*, **8** (8), 1319–24.

206 Macounova, K., Cabrera, C.R., Holl, M.R. and Yager, P. (2000) Generation of natural pH gradients in microfluidic

channels for use in isoelectric focusing. *Analytical Chemistry*, **72** (16), 3745–51.

207 Winkle, R.F., Nagy, J.M., Cass, A.E. and Sharma, S. (2008) Towards microfluidic technology-based MALDI-MS platforms for drug discovery: a review. *Expert Opinion on Drug Discovery*, **3** (11), 1281–92.

208 Zhang, C. and Xing, D. (2007) Miniaturized PCR chips for nucleic acid amplification and analysis: latest advances and future trends. *Nucleic Acids Research*, **35** (13), 4223–37.

209 Cabrera, C.R., Finlayson, B. and Yager, P. (2000) Formation of natural pH gradients in a microfluidic device under flow conditions: model and experimental validation. *Analytical Chemistry*, **73** (3), 658–66.

210 Liu, X. and Abbott, N.L. (2009) Electrochemical generation of gradients in surfactant concentration across microfluidic channels. *Analytical Chemistry*, **81** (2), 772–81.

211 Studer, V., Jameson, R., Pellereau, E., Pépin, A. and Chen, Y. (2004) A microfluidic mammalian cell sorter based on fluorescence detection. *Microelectronic Engineering*, **73–74**, 852–7.

212 Chao, T.C. and Ros, A. (2008) Microfluidic single-cell analysis of intracellular compounds. *Journal of the Royal Society, Interface*, **5** (Suppl. 2), S139–50.

213 Marcus, J.S., Anderson, W.F. and Quake, S.R. (2006) Microfluidic single-cell mRNA isolation and analysis. *Analytical Chemistry*, **78** (9), 3084–9.

214 Dong, L. and Jiang, H. (2007) Autonomous microfluidics with stimuli-responsive hydrogels. *Soft Matter*, **3** (10), 1223–30.

215 Yu, C., Mutlu, S., Selvaganapathy, P., Mastrangelo, C.H., Svec, F. and Frechett, J.M.J. (2003) Flow control valves for analytical microfluidic chips without mechanical parts based on thermally responsive monolithic polymers. *Analytical Chemistry*, **75** (8), 1958–61.

216 Idota, N., Kikuchi, A., Kobayashi, J., Sakai, K. and Okano, T. (2005) Microfluidic valves comprising nanolayered thermoresponsive polymer-grafted capillaries. *Advanced Materials*, **17** (22), 2723–7.

217 Satarkar, N.S., Zhang, W., Eitel, R.E. and Hilt, J.Z. (2009) Magnetic hydrogel nanocomposites as remote controlled microfluidic valves. *Lab on a Chip*, **9**, 1773–9.

218 Ghosh, S. et al (2009) Oscillating magnetic field-actuated microvalves for micro- and nanofluidics. *Journal of Physics D: Applied Physics*, **42** (13), 135501.

219 Sugiura, S., Sumaru, K., Ohi, K., Hiroki, K., Takagi, T. and Kanamori, T. (2007) Photoresponsive polymer gel microvalves controlled by local light irradiation. *Sensors and Actuators A, Physical*, **140** (2), 176–84.

220 Sugiura, S., Szilagyi, A., Sumaru, K., Hattori, K., Takagi, T., Filipcsei, G., Zrinyi, M. and Kanamori, T. (2009) On-demand microfluidic control by micropatterned light irradiation of a photoresponsive hydrogel sheet. *Lab on a Chip*, **9** (2), 196–8.

221 Ayala, V.C. et al (2007) Design, construction and testing of a monolithic pH-sensitive hydrogel-valve for biochemical and medical application. *Journal of Physics: Conference Series*, **90** (1), 012025.

222 Liu, C. et al (2007) Arrayed pH-responsive microvalves controlled by multiphase laminar flow. *Journal of Micromechanics and Microengineering*, **17** (10), 1985.

223 Fu, Q., Rao, G.V.R., Ista, L.K., Wu, Y., Andrzejewski, B.P., Sklar, L.A., Ward, T.L. and López, G.P. (2003) Control of molecular transport through stimuli-responsive ordered mesoporous materials. *Advanced Materials*, **15** (15), 1262–6.

224 Bohn, P.W. (2009) Nanoscale control and manipulation of molecular transport in chemical analysis. *Annual Review of Analytical Chemistry*, **2** (1), 279–96.

225 Jeong, B. and Gutowska, A. (2002) Lessons from nature: stimuli-responsive polymers and their biomedical applications. *Trends in Biotechnology*, **20** (7), 305–11.

226 Huang, R., Kostanski, L.K., Filipe, C.D.M. and Ghosh, R. (2009) Environment-responsive hydrogel-based ultrafiltration membranes for protein bioseparation. *Journal of Membrane Science*, **336** (1–2), 42–9.

227 Yameen, B., Ali, M., Neumann, R., Ensinger, W., Knoll, W. and Azzaroni, O. (2009) Ionic transport through single solid-state nanopores controlled with thermally nanoactuated macromolecular gates. *Small*, **5** (11), 1287–91.

228 Rao, G.V. and Lopez, G.P. (2000) Encapsulation of poly(N-isopropyl acrylamide) in silica: a stimuli-responsive porous hybrid material that incorporates molecular nano-valves. *Advanced Materials*, **12** (22), 1692.

229 Galaev, I.Y. and Mattiasson, B. (1999) Smart polymers and what they could do in biotechnology and medicine. *Trends in Biotechnology*, **17** (8), 335–40.

230 Han, Z., Li, W., Huang, Y. and Zheng, B. (2009) Measuring rapid enzymatic kinetics by electrochemical method in droplet-based microfluidic devices with pneumatic valves. *Analytical Chemistry*, **81** (14), 5840–5.

4
Self-Assembled Peptide Nanostructures and Their Controlled Positioning on Surfaces

Maria Farsari and Anna Mitraki

4.1
Introduction

During the past decade, self-assembling peptides have emerged as a novel generation of materials that can be designed to attain one-dimensional (1-D), two-dimensional (2-D), and three-dimensional (3-D) nanostructures with open-ended potential applications [1–3]. These peptides form supramolecular structures such as ribbons, nanotubes, and fibers which, in turn, often form gels [4]. The most compelling advantage of peptide-based materials is the possibility of conferring tailor-made assembly and materials properties by effecting simple amino acid changes at the sequence level. Moreover, they can assemble under mild aqueous conditions at ambient temperatures, thus avoiding the need for harsh solvents and elevated temperatures. Despite the "very soft" preparation and manufacturing conditions that are required, many peptide nanostructures – once assembled – can be remarkably strong and rigid, attaining Young's moduli of up to 19 GPa [5]. They can also withstand solvents and temperatures up to approximately 120 °C, which makes them good candidates for interfacing with the world of "hard materials" [6].

Of particular interest is the possibility to use these self-assembling peptides as templates for the growth of inorganic materials, such as metals (silver, gold, and platinum), ferromagnetic metals (cobalt and nickel), silica, and calcium phosphates. Another possibility involves the introduction of cell-adhesion motifs to the peptide building blocks, so as to enable tissue growth and regeneration on peptide scaffolds. The templating of metals was one of the first proof-of concept applications to have been demonstrated for self-assembling peptides, when Ehud Gazit and colleagues, while studying the basic recognition elements involved in the self-assembly of the Alzheimer disease-related beta-peptide, discovered that

Nanomaterials for the Life Sciences Vol.10: Polymeric Nanomaterials. Edited by Challa S. S. R. Kumar
Copyright © 2011 WILEY-VCH Verlag GmbH & Co. KGaA, Weinheim
ISBN: 978-3-527-32170-4

the diphenylalanine (Phe-Phe) peptide was in fact self-assembling into hollow nanotubes. The straightforward introduction of silver nitrate into these hollow nanotubes, followed by reduction with citric acid and subsequent digestion of the peptide "mold" with proteinase K, led to the fabrication of silver nanowires with a diameter of 20 nm [7]. The peptide nanotubes were subsequently shown to act as recognition and surface-enhancing elements in bioelectrochemical sensor applications [8, 9]. Concurrently, Scheibel and Lindquist introduced cysteine residues at the sequence of a self-assembling protein fragment from yeast, whereby the exposed cysteines acted as nucleation sites for gold nanoparticles [10]; following silver enhancement, the formation of conducting nanowires was demonstrated. Clearly, cysteines introduced through rational design at the sequence of short self-assembling peptide fibrils might, in time, become regarded as a general strategy for the templating of metal nanoparticles, and the formation of conductive nanowires [11].

In recent years, there has been an exponential increase in the number of reports concerning the application of peptide-based biomaterials; these have ranged from the materials sciences and bionanotechnology to biomedicine and biotechnology [12–15]. As the precise positioning of peptide-based nanostructures is an essential part of their use in technological applications, there is a compelling need to develop methods for their controlled assembly, positioning and integration into microsystems, devices, and scaffolds. The controlled positioning methods developed for carbon and inorganic nanostructures often employ harsh temperatures and solvent conditions that are incompatible with protein materials. Consequently, it will be necessary to develop novel methods that are appropriate for protein and peptide positioning. The "soft" methods targeted at peptide positioning, and reported during the past three years, form the basis of this chapter.

4.2
Vertical and Horizontal Alignment on Surfaces

In 2006, Reches and Gazit reported that aromatic dipeptide nanotubes were able to form vertical "nanoforests" on siliconized glass surfaces, upon evaporation from hexafluoroisopropanol (HFIP) [16]. HFIP is a highly volatile, chaotropic solvent that retains the dipeptide in a dissolved, monomeric form, and is widely used to dissolve commercial textile fibers in preparation for spinning. Upon rapid evaporation of the HFIP, a rapid preferential growth in the vertical axis results in self-assembly and stacking along this direction (Figure 4.1), although most nanotechnological applications require a horizontal positioning and alignment of the nanostructures. Hence, a horizontal alignment was achieved by attaching magnetite nanoparticles to the peptide nanotubes, with the attachment most likely effected through hydrophobic interactions. The decorated tubes could then be aligned by the external application of a magnetic field (Figure 4.2). Such vertical and horizontal alignments of peptide nanotubes represented the first proof-of-principle of the controlled deposition of self-assembled peptide nanostructures.

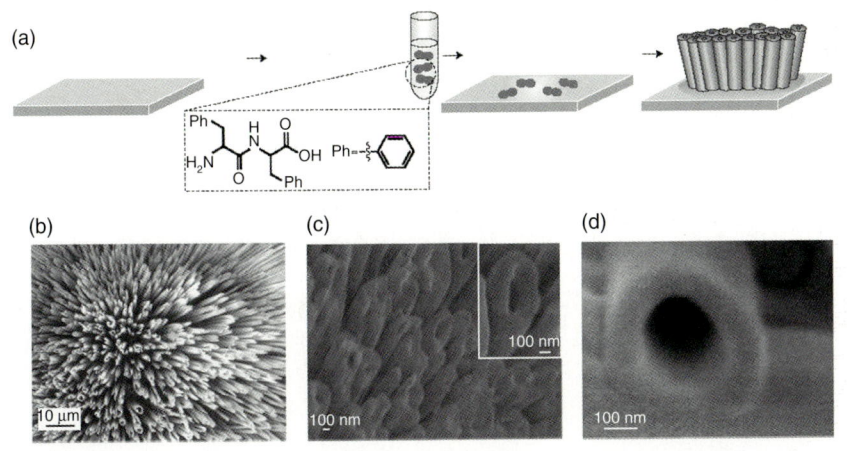

Figure 4.1 (a) Vertically aligned nanotubes formed following the evaporation of a solution of diphenylalanine peptide in hexafluoroisopropanol (HFIP) on siliconized glass; (b) Scanning electron microscopy image of the nanotube forest; (c, d) Cold-field emission gun high-resolution electron microscopy (CFEG-HRSEM) analysis, showing the hollow nature of the nanotubes. From ref. 16, reprinted with permission from Macmillan Publishers LTD.

4.3
Printing Using Inkjet Technology

Today, inkjet printing technology is used widely for the controlled deposition of polymers, resins, and even conductive multi-walled carbon nanotubes (MWNTs) [17]. When Adler-Abramovich and Gazit dissolved peptide nanotubes and spheres in 50% ethanol:water mixtures, and filled this solution into the cartridge of a commercially available inkjet printer [18], they were able to print patterns on commercial transparency foil, or on the conductive site of an indium tin oxide (ITO) plastic. A subsequent examination, using scanning electron microscopy (SEM), confirmed the deposition of the nanostructures within the desired area, one example being a letter character (Figure 4.3). Both, single and multiple printings could be afforded, and the patterned structures also demonstrated a durable nature. Indeed, at eight months after printing the printed areas remained clearly visible and of similar appearance to the freshly printed version.

4.4
Vapor Deposition Methods

The use of vapor deposition methods for peptide nanostructures has been attempted very recently [19]. When the diphenylalanine nanotube powders (as described above) were heated at 200 °C, the peptide molecules rapidly evaporated due to the inherent volatility of the aromatic moieties, and subsequently

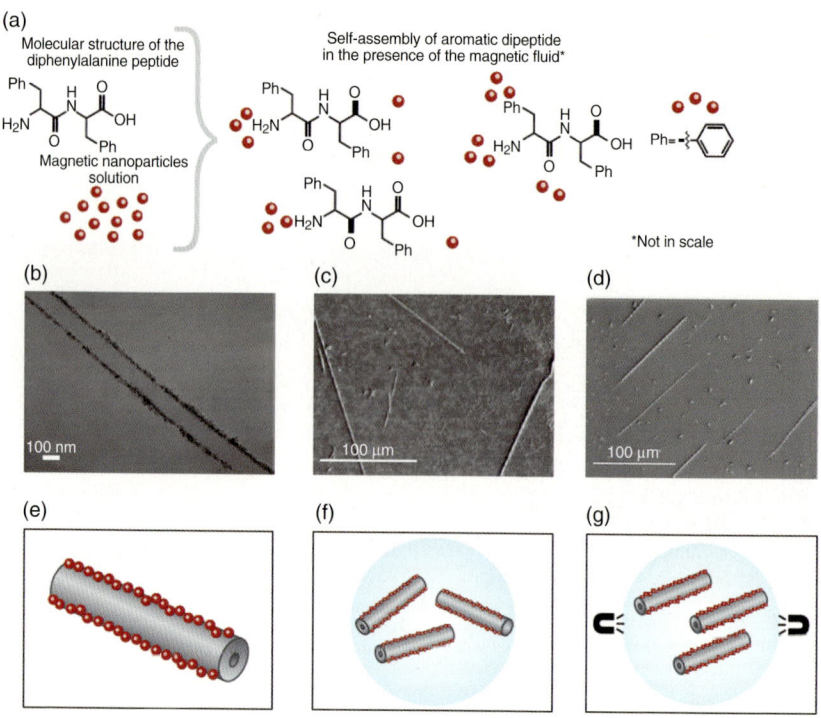

Figure 4.2 (a) Schematic representation of assembly of nanotubes in the presence of magnetite nanoparticles; (b) Transmission electron microscopy image of a nanotube with attached nanoparticles; (c) Scanning electron microscopy images showing the random orientation of nanoparticle-coated nanotubes; (d) Horizontally aligned nanotubes after exposure to an external magnetic field; (e–g) Schematic representations of a nanotube coated with magnetic nanoparticles (e), the random orientation of nanotubes (f), and their horizontal alignment after exposure to the magnetic field (g). From ref. 16, reprinted with permission from Macmillan Publishers LTD.

condensed onto a substrate surface placed approximately 2 cm above the heating source (Figure 4.4). After condensing, they formed hollow nanotubes that were oriented perpendicular to the substrate surface, and were coated homogeneously over an area of approximately 10 cm^2 (Figure 4.4). The surface density and average length of the deposited nanotubes could be controlled by adjusting the deposition parameters. This technique is applicable to a wide variety of substrates, including glass, carbon, gold, and silicon dioxide. The orderly deposited nanoforests of aromatic nanotubes were reminiscent of the "lotus-type" surfaces – that is, surfaces that combine the microstructuring and nanostructuring of a low-energy surface. In fact, nanotube-coated glass was shown to have an eightfold increase in hydrophobicity compared to bare glass surfaces (contact angle of 125°, compared to 15° for bare glass). These biocompatible, highly hydrophobic surfaces may have numerous applications for self-cleaning surfaces and solar cells. Moreover, the combination of this "bottom-up" deposition technology with the classic "top-down" patterning technology may provide a route for large-scale technological

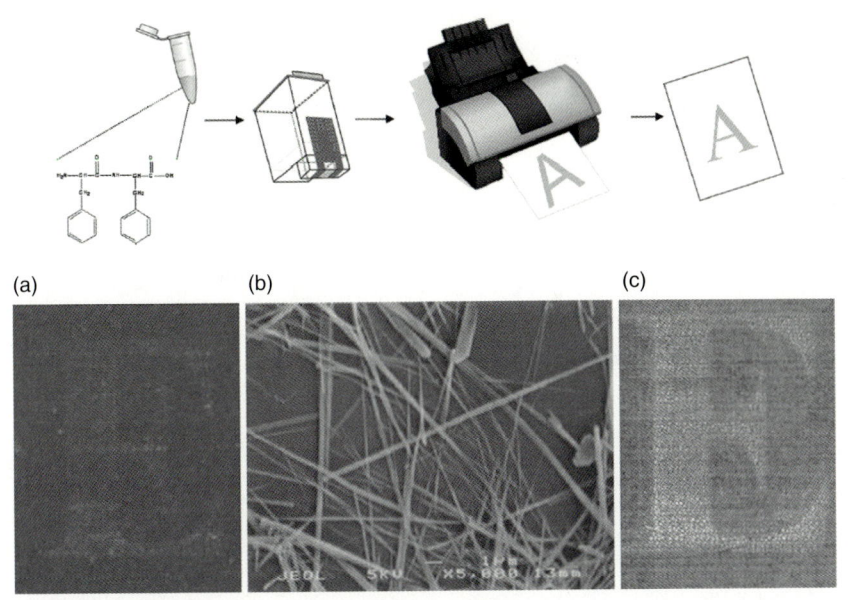

Figure 4.3 Top row: A schematic representation of the patterning procedure of peptide nanotubes using an inkjet printer. Bottom row: (a) A single printing of the letter "E"; (b) SEM image of the printed area in panel (a); (c) The same letter after multiple printing (10 times). From ref. 18.

applications. For example, when a microfluidic pattern was created on a silicon dioxide chip, the area outside the channels was covered with peptide nanotubes. Following exposure to water vapor and cooling to 5 °C, the water was able to condense on the channels where the surface remained hydrophilic, forming liquid micro-channels [19]. Such a vapor deposition method can also be combined with standard photolithographic methods to fabricate regularly patterned, peptide nanotube-coated surfaces. This was very recently demonstrated by using reactive ion etching (RIE) to pattern regularly spaced holes on a silicon wafer covered with a thermally grown, 5 μm-thick silicon dioxide layer [20]. Subsequently, a nanoforest of vertically aligned nanotubes of height approximately 5 μm was deposited by vapor deposition. A critical period of etching (60 s) with 49% hydrofluoric acid caused a partial etching of the silicon dioxide layer, and resulted in the selective removal of the peptide nanotubes above the layer. The nanotubes patterned inside the holes were not damaged, and retained their position and structure; as a result, a regularly patterned array of holes covered with peptide nanotubes was fabricated. It should be noted that chemical modifications of the peptide material might be unavoidable when using these vapor deposition methods. For example, diphenylalanine nanotubes undergo cyclization during the evaporation process, as verified using mass spectrometry (Figure 4.4). Moreover, these methods rely heavily on properties that are inherent to the aromatic nanotubes, such as volatility.

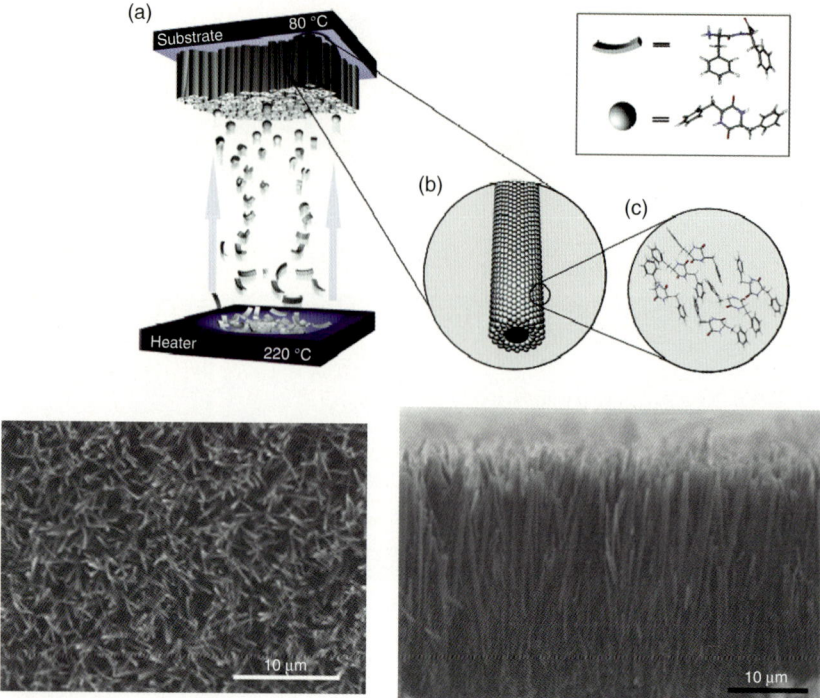

Figure 4.4 Top row: Schematic representation of the vapor deposition process. (a) The diphenylalanine peptide powder is heated at 220 °C, causing the peptide to attain a cyclic structure upon evaporation and to self-assemble into nanotubes on a surface placed above the heater; (b) A single peptide nanotube composed of the cyclic dipeptide molecules; (c) The conformation of six dipeptide molecules within the nanotube moiety, after energy minimization. The inset shows schematic representations of the linear conformation of the dipeptide (rectangular shape) and the cyclic conformation (round shape). Bottom row: SEM images of the top view (left) and side view (right) of the vertically aligned nanotubes. From ref. 19, reprinted with permission from Macmillan Publishers LTD.

4.5
Positioning Using Dielectrophoresis

Dielectrophoresis was used previously for the positioning of cells and DNA, and more recently for peptide nanotubes [21]. The suspension of polarizable particles in an inhomogeneous electric field induces different electric forces on the charges on each half of the dipole, and can be used to orient and connect peptide nanofibers and nanotubes to electrodes. A microchip was fabricated as follows: a layer of SiO_2 was spun on a silicon wafer, after which a 1.5 μm resist layer was spun on top of the oxide. The electrodes were subsequently patterned using a positive photolithography process. Following development of the resist, a titanium and, subsequently, a gold layer were deposited on the wafer and the electrodes defined

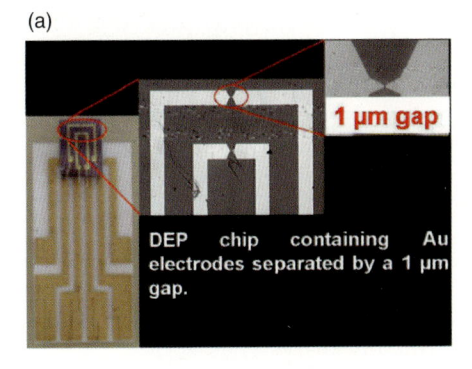

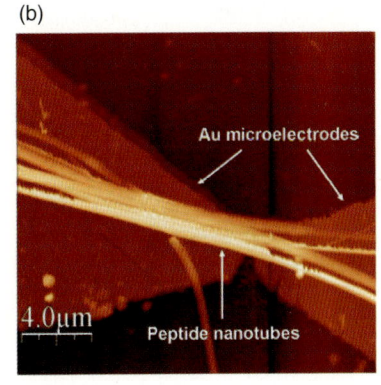

Figure 4.5 (a) A dielectrophoresis microchip fabricated by optical lithography, showing the gap between the electrodes; (b) Atomic force microscopy image of a bundle of peptide nanotubes aligned with dielectrophoresis between the electrodes. From ref. 21.

by a lift-off process using acetone. As a result, gold microelectrodes with a 1 μm gap between them were fabricated (Figure 4.5a). An aqueous solution of peptide nanotubes was deposited and subjected to dielectrophoresis. Under appropriate conditions of voltage, frequency and time applied, bundles of nanotubes and even some single nanotubes were successfully aligned between the electrodes (Figure 4.5b). The main advantage of this method relates to its compatibility with aqueous solutions; moreover, the peptide nanostructures can be manipulated and positioned in a controlled manner, opening the route to applications such as field-effect transistors (FETs), microelectrode array (MEA) biochips, and biosensing.

4.6
Laser Patterning

4.6.1
Laser-Induced Forward Transfer

Over the past few years, much interest has been expressed, and progress made, in the direct laser writing of biological molecules [22–27]. One such process, referred to as laser-induced forward transfer (LIFT), is a direct-write, noncontact method which offers an interesting and versatile alternative to conventional arraying techniques. The main benefits are that LIFT does not require the use of expensive photolithographic equipment, nor does it suffer from clogging and contamination problems, as are encountered with deposition and ink-jet printing [28]. In the case of LIFT, a single pulse from a focused laser beam is used to transfer a solution from a donor-coated surface to an acceptor surface. The laser

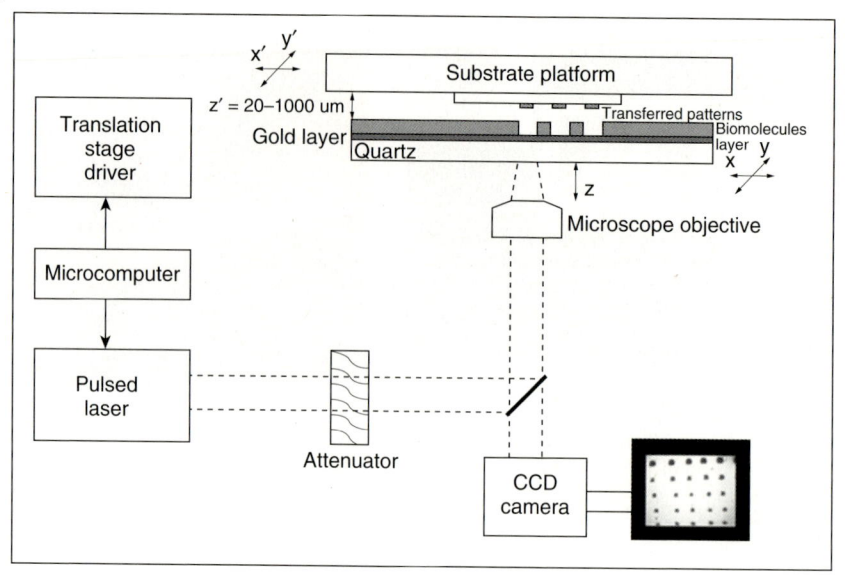

Figure 4.6 Schematic diagram of the LIFT experimental set-up.

beam is first focused onto a metallic layer which coats the transparent donor surface; this so-called "target" acts as an energy-conversion material, converting the laser energy via absorption and conduction into the heat that is used to expel the target material. The transferred material is collected on a substrate, placed in close proximity and parallel to the donor surface. The advantage of LIFT over other techniques is its ability to provide a high-resolution, noncontact, direct, flexible and parallel transfer of more than one type of material, but not to mix them, as different targets can be used. The experimental set-up is shown in Figure 4.6.

For peptides, two different LIFT methodologies have been demonstrated: (i) the direct transfer of peptides [29]; and (ii) the self-assembly of peptides on LIFT-patterned biotin, using thiol chemistry and an avidin–biotin-mediated assembly [27].

4.6.1.1 Direct Transfer

In the direct transfer approach, the peptides used were self-assembled peptides derived from sequences of a natural fibrous protein, the adenovirus fiber (see Ref. [30]). First, a solution consisting of peptides dissolved in water was spread onto quartz coated with a gold thin-layer film, for use as a target. With one laser pulse, a small quantity (which was related to the laser beam dimension and the thickness of the film) was transferred from the liquid solution to the receiving glass substrate as droplets (Figure 4.7).

The mechanism of peptide transfer with respect to the laser–target interaction can be described as follows. First, the incident laser energy interacts with the absorption gold layer on the target. At low intensities, the processes relate to a

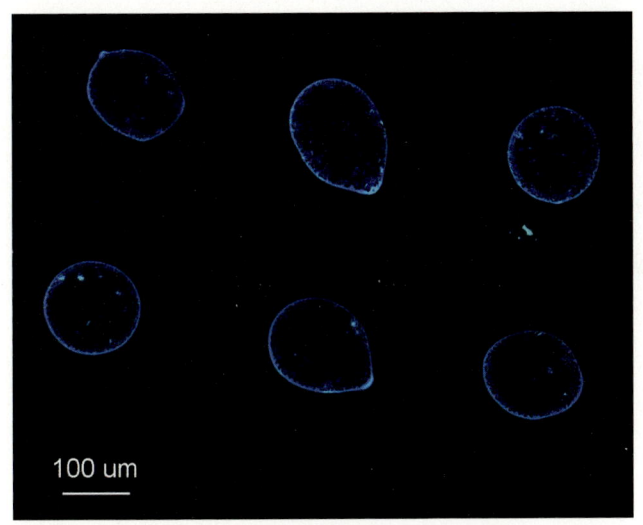

Figure 4.7 Fluorescence microscopy image of six thioflavin-stained peptide droplets on a gold surface. The presence of fibrils inside the spot can be seen through the blue color given by thioflavin (emission wavelengths at 485 nm). From ref. 29.

simple excitation in the layer, with low thermal effects causing only a minor perturbation of the target [26, 31]. For sufficient laser energies, however, the absorption layer acts as an energy-conversion material, whereby the energy absorbed is transformed into heat, while the liquid in contact with the metallic layer is "forwarded" towards the substrate. A small quantity of material, which is related to the laser beam dimension, is transferred from the viscous solution to the receiving glass substrate in droplet form, and in rows.

4.6.1.2 Self-Assembly of Peptides on LIFT-Patterned Biotin

In the second approach, LIFT was used to generate high-density photosensitive biotin (photobiotin) arrays on ORMOCER®-coated surfaces. The peptide microarrays were then generated and assembled by utilizing avidin–biotin and thiol chemistry. ORMOCER® is an organically modified ceramic to which photobiotin can be attached irreversibly when exposed to ultraviolet (UV) light from a laser or a lamp [32]. However, the technique is not limited to ORMOCER®; in fact, photobiotin can be attached photolytically to a variety of organic and inorganic materials, thus providing the methodology with a wide flexibility and applicability. When the biotin had been immobilized on ORMOCER®, it was first incubated with avidin and, subsequently, with an iodoacetamide-functionalized biotin, N-(biotinoyl)-N'-(iodoacetyl) ethylenediamine. Finally, the samples were immersed in an aqueous solution of peptides that had been functionalized with cysteine (a thiol-containing amino-acid). Assembly of the peptide fibrils on the photobiotin spots was initiated through the controlled evaporation of the water, although in order for this to occur

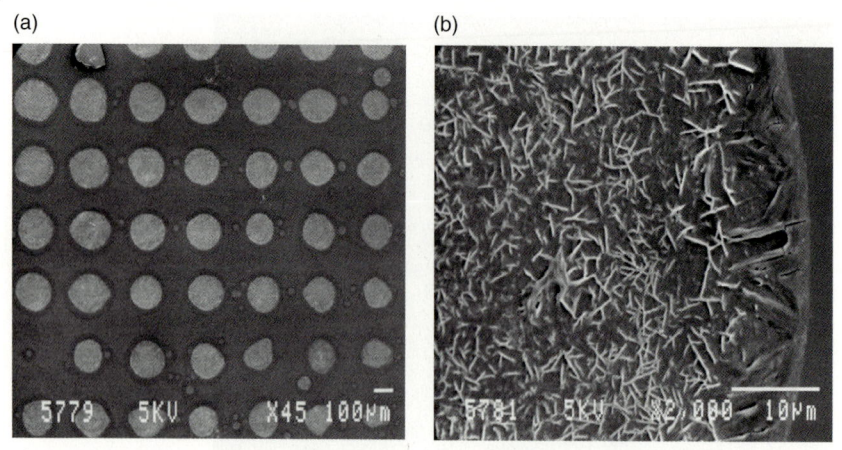

Figure 4.8 SEM images of a peptide micro-array (a) and detail of a spot (b). From ref. 27.

a seed or "anchoring point" was first required. Such a point was provided by the bonding between the iodoacetamide group in the biotin derivative and the thiol group in the cysteine.

The SEM images of some spots obtained using LIFT are shown in Figure 4.8. As noted above, arrays of photobiotin were first printed on ORMOCER®, and subsequently functionalized with the appropriate biomolecules. The fibrils have assembled only on those locations where photobiotin was immobilized with LIFT. The magnified detail of one spot is shown in Figure 4.8b, where the first layer of immobilized peptides, onto which the fibrils grow further, is clearly visible.

By combining the directional deposition of the peptides on specific places and different surfaces by LIFT, and by employing sequence manipulations, it is possible to enable the specific fabrication of large number of different structures, each of which can be used in diverse applications such tissue repair, patterning, miniaturized solar cells, and optical and electronic devices.

4.6.2
Nonlinear Lithography

The precise 3-D patterning of amyloid fibrils has been demonstrated on 3-D microstructures fabricated using femtosecond laser nonlinear lithography [33], which is based on the phenomenon of two-photon polymerization. When the beam of an ultrafast infrared laser is tightly focused into the volume of a photosensitive material, the polymerization process can be initiated by two-photon absorption within the focal region. However, by moving the laser focus in 3-D fashion through the photosensitive material, arbitrary 3-D structures can be fabricated [34].

In these studies, as previously with LIFT, the 3-D structures were first functionalized with photobiotin, after which the thiol chemistry and self-assembly of

4.6 Laser Patterning

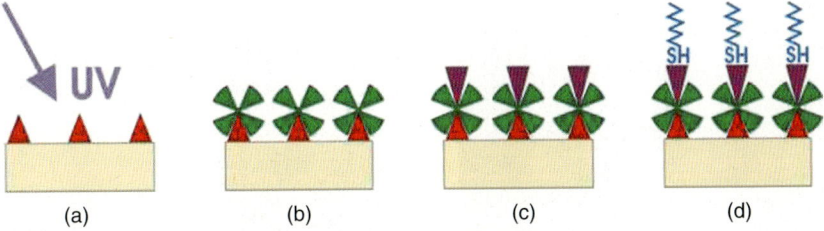

Figure 4.9 Functionalization of ORMOCER® for peptide fibril growth. A thin layer of photobiotin (red triangles) is deposited on the ORMOCER® surface and exposed to UV light (a) before being further functionalized with avidin (green crosses) (b) and iodoacetamide-functionalized biotin (purple triangles) (c). The final step is the attachment of the cysteine-containing peptide through the SH–iodoacetamide reaction (d). From ref. 33.

peptide fibrils were exploited. The 3-D structures consisted of ORMOCER®, onto which biotin was first immobilized (Figure 4.9a), followed by an incubation with avidin (Figure 4.9b) and, subsequently, with iodoacetamide-functionalized biotin, N-(biotinoyl)-N'-(iodoacetyl) ethylenediamine (Figure 4.9c). Finally, the functionalized 3-D structures were immersed in an aqueous solution of peptides that contained a cysteine residue. The peptide solutions were "aged" so that self-assembled fibrils were already formed in solution. The self-assembly of the peptide fibrils into bridges on the structures was initiated through the controlled evaporation of water (Figure 4.9d), although in order for this to occur between two specific positions a seed or "anchoring point" was required. This was provided by the covalent bond formation between the iodoacetamide group in the biotin derivative, and the thiol group of the cysteine. The requirement for "seeding" – and therefore the selectivity of the technique – was demonstrated by using of peptides that did not contain cysteine. In this case, self-assembled peptide fibrils were formed in solution, but there was no peptide fibril attachment on the functionalized structures.

Peptide fibrils were visualized using either SEM or the fluorescence emission of the dye thioflavin. A thioflavin fluorescence image and a SEM image of the directed 3-D assembly of the peptide fibrils is shown in Figure 4.10a and b, respectively. In this case, a series of 3-D columns was fabricated and subsequently functionalized. The peptide bridges appeared to form over the shortest distance, as can be seen in the SEM image of Figure 4.11a. The length and the diameter of the peptide fibrils were dependent on the design of the 3-D structures – that is, the distance between them and the diameter of the fibril support (Figures 4.11b and c). Once the fibril bridges had been formed they remained at their position, even when the sample was immersed in water for up to 24 h.

The major advantage of the laser techniques described above is their compatibility with aqueous solutions, and the possibility of direct writing without the use of a mask, mold, or stamp. Furthermore, they exploit biotin–avidin and thiol

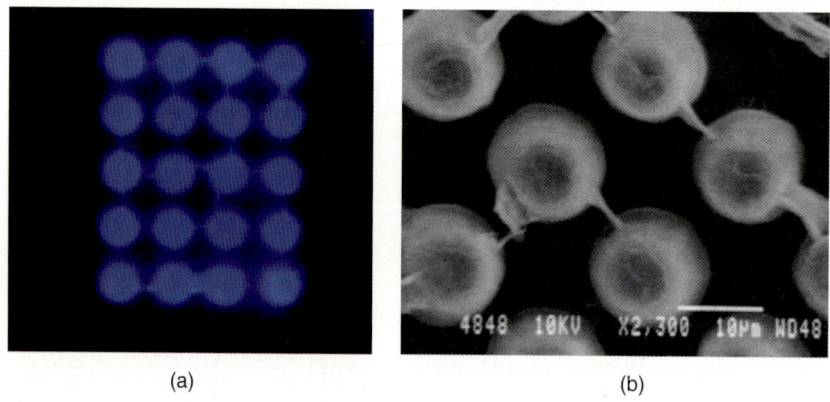

116 | *4 Self-Assembled Peptide Nanostructures and Their Controlled Positioning on Surfaces*

Figure 4.10 (a) Thioflavin fluorescence image and (b) SEM image of a 3-D column array, with peptide bridges self-assembled between the columns. From ref. 33.

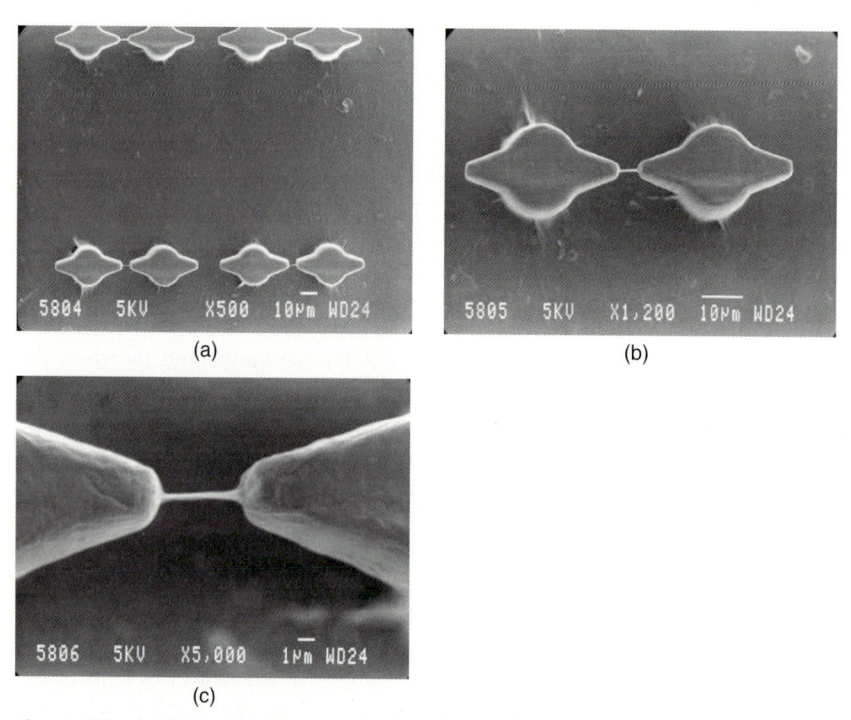

Figure 4.11 (a) SEM image of a series of 3-D ORMOCER® columns, with peptide fibril bridges self-assembled between them; (b) One pair of 3D ORMOCER® columns, with peptide fibril bridges self-assembled between them; (c) Detail of the self-assembled fiber bridge. From ref. 33.

chemistry without the need for covalent modification of the peptide moieties. Thiol-functionalized biomolecules can be easily produced on the laboratory scale, and may even be available commercially; hence, this method should be applicable not only to peptides but also to other self-assembling biomolecules. Amine-reactive biotinylation reagents are also commercially available, and can be used as an alternative to the thiol-reactive biotinylation reagent used in these studies. Furthermore, as photobiotin can be easily attached to many different materials, this technique is suitable for a wide variety of applications and materials.

4.7
Summary and Perspectives

Self-assembling peptide nanowires and nanotubes represent a relatively new class of materials, with open-ended possibilities for their application. With this in mind, reproducible and well-controlled methods must be developed for the positioning and integration of these bionanoassemblies with microsystems. Moreover, these methods must be compatible with aqueous solutions and moderate temperatures when compared to well-established methods for the positioning and integration of inorganic nanostructures. In this chapter, the relatively few existing methods used to control the positioning of peptide nanostructures have been reviewed. Nonetheless, in the years to come it should be expected that there will be a considerable increase in the number of reports relating to controlled positioning strategies. For this, a true interdisciplinary approach will be needed in order to combine tools from diverse fields such as peptide chemistry, physics, bioanalysis, and engineering.

Acknowledgments

A. M. gratefully acknowledges funding from the European Union (STREP NMP-CT-2006-033256, "BeNatural."

References

1 Gilead, S. and Gazit, E. (2005) Self-organization of short peptide fragments: from amyloid fibrils to nanoscale supramolecular assemblies. *Supramolecular Chemistry*, **17** (1–2), 87–92.

2 Rajagopal, K. and Schneider, J.P. (2004) Self-assembling peptides and proteins for nanotechnological applications. *Current Opinion in Structural Biology*, **14** (4), 480–6.

3 Zhang, S.G. (2003) Fabrication of novel biomaterials through molecular self-assembly. *Nature Biotechnology*, **21** (10), 1171–8.

4 Aggeli, A., Bell, M., Boden, N., Keen, J.N., Knowles, P.F., McLeish, T.C., Pitkeathly, M. and Radford, S.E. (1997) Responsive gels formed by the spontaneous self-assembly of peptides into polymeric beta-sheet tapes. *Nature*, **386** (6622), 259–62.

5. Kol, N., Adler-Abramovich, L., Barlam, D., Shneck, R.Z., Gazit, E. and Rousso, I. (2005) Self-assembled peptide nanotubes are uniquely rigid bioinspired supramolecular structures. *Nano Letters*, **5** (7), 1343–6.
6. Adler-Abramovich, L., Reches, M., Sedman, V.L., Allen, S., Tendler, S.J.B. and Gazit, E. (2006) Thermal and chemical stability of diphenylalanine peptide nanotubes: implications for nanotechnological applications. *Langmuir*, **22** (3), 1313–20.
7. Reches, M. and Gazit, E. (2003) Casting metal nanowires within discrete self-assembled peptide nanotubes. *Science*, **300** (5619), 625–7.
8. Yemini, M., Reches, M., Gazit, E. and Rishpon, J. (2005) Peptide nanotube-modified electrodes for enzyme-biosensor applications. *Analytical Chemistry*, **77** (16), 5155–9.
9. Yemini, M., Reches, M., Rishpon, J. and Gazit, E. (2005) Novel electrochemical biosensing platform using self-assembled peptide nanotubes. *Nano Letters*, **5** (1), 183–6.
10. Scheibel, T., Parthasarathy, R., Sawicki, G., Lin, X.M., Jaeger, H. and Lindquist, S.L. (2003) Conducting nanowires built by controlled self-assembly of amyloid fibers and selective metal deposition. *Proceedings of the National Academy of Sciences of the United States of America*, **100** (8), 4527–32.
11. Kasotakis, E., Mossou, E., Adler-Abramovich, L., Mitchell, E.P., Forsyth, V.T., Gazit, E. and Mitraki, A. (2009) Design of metal-binding sites onto self-assembled peptide fibrils. *Biopolymers*, **92** (3), 164–72.
12. Fairman, R. and Akerfeldt, K.S. (2005) Peptides as novel smart materials. *Current Opinion in Structural Biology*, **15** (4), 453–63.
13. Gazit, E. (2007) Use of biomolecular templates for the fabrication of metal nanowires. *FEBS Journal*, **274** (2), 317–22.
14. Gelain, F., Horii, A. and Zhang, S.G. (2007) Designer self-assembling peptide scaffolds for 3-D tissue cell cultures and regenerative medicine. *Macromolecular Bioscience*, **7** (5), 544–51.
15. Pepe-Mooney, B.J. and Fairman, R. (2009) Peptides as materials. *Current Opinion in Structural Biology*, **19** (4), 483–94.
16. Reches, M. and Gazit, E. (2006) Controlled patterning of aligned self-assembled peptide nanotubes. *Nature Nanotechnology*, **1**, 195–200.
17. Calvert, P. (2001) Inkjet printing for materials and devices. *Chemistry of Materials*, **13** (10), 3299–305.
18. Adler-Abramovich, L. and Gazit, E. (2008) Controlled patterning of peptide nanotubes and nanospheres using inkjet printing technology. *Journal of Peptide Science*, **14** (2), 217–23.
19. Adler-Abramovich, L., Aronov, D., Beker, P., Yevnin, M., Stempler, S., Buzhansky, L., Rosenman, G. and Gazit, E. (2009) Self-assembled arrays of peptide nanotubes by vapour deposition. *Nature Nanotechnology*, **4** (12), 849–54.
20. Shklovsky, J., Beker, P., Amdursky, N., Gazit, E. and Rosenman, G. (2009) Bioinspired peptide nanotubes: deposition technology and physical properties. *Materials Science and Engineering B*, **169** (1–3), 62–6.
21. Castillo, J., Tanzi, S., Dimaki, M. and Svendsen, W. (2008) Manipulation of self-assembly amyloid peptide nanotubes by dielectrophoresis. *Electrophoresis*, **29** (24), 5026–32.
22. Zergioti, I., Karaiskou, A., Papazoglou, D.G., Fotakis, C., Kapsetaki, M. and Kafetzopoulos, D. (2005) Femtosecond laser microprinting of biomaterials. *Applied Physics Letters*, **86** (16), 163902.
23. Chrisey, D.B. (2000) Materials processing – the power of direct writing. *Science*, **289** (5481), 879–81.
24. Ringeisen, B.R., Callahan, J., Wu, P.K., Pique, A., Spargo, B., McGill, R.A., Bucaro, M., Kim, H., Bubb, D.M. and Chrisey, D.B. (2001) Novel laser-based deposition of active protein thin films. *Langmuir*, **17** (11), 3472–9.
25. Dinca, V., Catherine, J., Mourka, A., Georgiou, S., Farsari, M. and Fotakis, C. (2009) 2D and 3D biotin patterning by ultrafast lasers. *International Journal of Nanotechnology*, **6** (1–2), 88–98.
26. Dinca, V., Ranella, A., Farsari, M., Kafetzopoulos, D., Dinescu, M., Popescu, A. and Fotakis, C. (2008) Quantification

27. Dinca, V., Kasotakis, E., Mourka, A., Ranella, A., Farsari, M., Mitraki, A. and Fotakis, C. (2008) Fabrication of amyloid peptide micro-arrays using laser-induced forward transfer and avidin-biotin mediated assembly. *Physica Status Solidi C: Current Topics in Solid State Physics*, **2008**, 3576–9.
28. Ekins, R. and Chu, F.W. (1999) Microarrays: their origins and applications. *Trends in Biotechnology*, **17** (6), 217–18.
29. Dinca, V., Kasotakis, E., Catherine, J., Mourka, A., Mitraki, A., Popescu, A., Dinescu, M., Farsari, M. and Fotakis, C. (2007) Development of peptide-based patterns by laser transfer. *Applied Surface Science*, **254** (4), 1160–3.
30. Papanikolopoulou, K., Schoehn, G., Forge, V., Forsyth, V.T., Riekel, C., Hernandez, J.F., Ruigrok, R.W.H. and Mitraki, A. (2005) Amyloid fibril formation from sequences of a natural beta-structured fibrous protein, the adenovirus fiber. *Journal of Biological Chemistry*, **280** (4), 2481–90.
31. Serra, P., Fernandez-Pradas, J.M., Berthet, F.X., Colina, M., Elvira, J. and Morenza, J.L. (2004) Laser direct writing of biomolecule microarrays. *Applied Physics A: Materials Science and Processing*, **79** (4–6), 949–52.
32. Drakakis, T.S., Papadakis, G., Sambani, K., Filippidis, G., Georgiou, S., Gizeli, E., Fotakis, C. and Farsari, M. (2006) Construction of three-dimensional biomolecule structures employing femtosecond lasers. *Applied Physics Letters*, **89** (14), article no. 144108.
33. Dinca, V., Kasotakis, E., Catherine, J., Mourka, A., Ranella, A., Ovsianikov, A., Chichkov, B.N., Farsari, M., Mitraki, A. and Fotakis, C. (2008) Directed three-dimensional patterning of self-assembled peptide fibrils. *Nano Letters*, **8** (2), 538–43.
34. Sun, H.-B. and Kawata, S. (2004) Two-photon photopolymerization and 3D lithographic microfabrication, in *NMR. 3D Analysis. Photopolymerization*, vol. 170 (ed. N. Fatkullin), Springer, Berlin, Heidelberg, pp. 169–273.

5
Multifunctional Pharmaceutical Nanocarriers: Promises and Problems
Vladimir P. Torchilin

5.1
Introduction

The use of nanoparticulate pharmaceutical carriers to enhance the *in vivo* efficiency of many drugs has become well established over the past decade, both in pharmaceutical research and clinical settings. Various pharmaceutical nanocarriers, such as polymeric nanoparticles, nanocapsules, liposomes, micelles, solid lipid nanoparticles, niosomes, carbon nanotubes (CNTs), dendrimers, and many others, are widely used for the experimental and clinical delivery of therapeutic and diagnostic agents [1, 2]. As nanoparticulate pharmaceutical carriers are often cleared rapidly from the body, are unstable under physiological conditions, are taken up by the mononuclear phagocytic system (MPS), and do not specifically target the site of pathology, the surface modification of these carriers is often used to adjust their properties in a desirable fashion. In the past, many reports on drug-delivery systems (DDSs) describe DDSs that combine several properties/functions, such as a prolonged blood circulation, the ability to accumulate specifically in target areas, to contrast properties, and sensitivity to local conditions inside the target [3]. Current efforts are directed towards preparing reasonably simple, cheap, and easy to use multifunctional DDSs, which are expected to be able to:

- bear a sufficient load of a drug or DNA-related material;
- to circulate long for long time periods;
- to accumulate specifically in the target area;
- to provide real-time information concerning their biodistribution and accumulation;
- to penetrate cells and even target individual intracellular compartments; and
- to release their load after having responded to local stimuli.

Although the amount of research conducted in this area continues to increase (for reviews, see Refs [3–5]), the development of DDSs that meet all listed requirements remains a very "hot" topic. Ultimately, the desired multifunctionality of pharmaceutical nanocarriers is normally achieved by the correct combination of various

Nanomaterials for the Life Sciences Vol.10: Polymeric Nanomaterials. Edited by Challa S. S. R. Kumar
Copyright © 2011 WILEY-VCH Verlag GmbH & Co. KGaA, Weinheim
ISBN: 978-3-527-32170-4

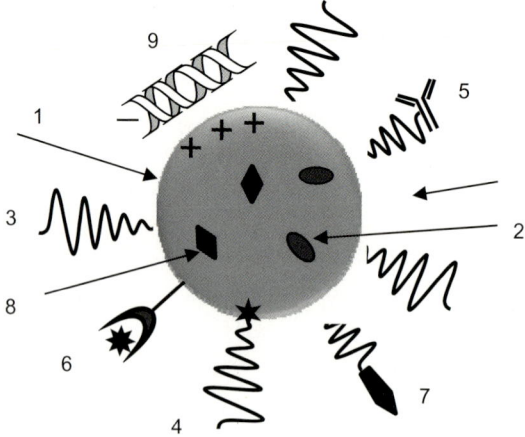

Figure 5.1 Schematic structure of multifunctional "super" nanocarrier. The nanocarrier (**1**) is loaded with a drug (**2**), and can be additionally modified (separately, simultaneously, or in any combination) with a nondetachable (**3**) or detachable (**4**) protective polymeric coat for longevity, targeting ligand attached to the distal tips of protective polymer chains (**5**), contrast moiety (**6**) for visualization, and cell-penetrating peptide (**7**) for intracellular delivery. The system can also carry superparamagnetic particles (**8**) for visualization or magnetic targeting, and positively charged sites for complexing DNA/ODN/siRNA (**9**).

functional moieties/modifiers on their surface. Such moieties/modifiers, which are added on top of the nanoparticle drug/DNA load, frequently include soluble synthetic polymers (to impart longevity), specific ligands (to achieve a targeting effect), pH- or temperature-sensitive components (to bring about stimuli-sensitivity), and contrast moieties (to introduce the imaging component) (see Figure 5.1). At this point, some paradigms already established in the area of multifunctional pharmaceutical nanocarriers will be briefly discuss, together with details of the most recent developments.

5.2
Established Paradigms: Longevity and Targetability

The creation of long-circulating and targeted DDSs is clearly the most developed approach, since long-circulating pharmaceuticals and pharmaceutical nanocarriers today represent an important and area of biomedical research [6–8]. The longevity of drug carriers allows for the maintenance of a required level of a pharmaceutical agent in the blood, for extended time intervals. In addition, long-circulating drug-containing microparticulates or large macromolecular aggregates can accumulate slowly, via either "passive" targeting or an impaired filtration mechanism, or via an enhanced permeability and retention (EPR) effect [9] at pathological sites with compromised and leaky vasculature (such as tumors,

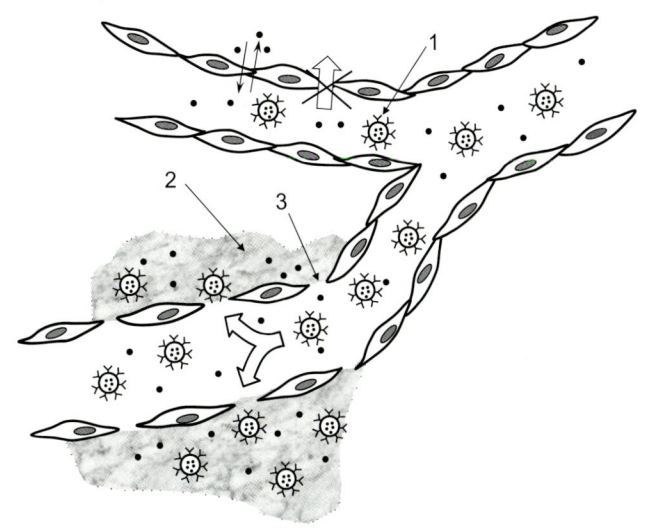

Figure 5.2 Schematic of the enhanced permeability and retention (EPR) effect. The normal continuous vasculature does not allow for the extravasation of large molecules and small particles (**1**) in normal tissues. However, extravasation and accumulation in the interstitium become possible in pathological areas (tumors, infarcts, inflammations (**2**), with the compromised (leaky) vasculature (**3**).

inflammations, and infarcted areas), and subsequently enhance the delivery of pharmaceutical agents into such areas [9, 10] (see Figure 5.2). The prolonged circulation can also help to achieve a better targeting effect for targeted (specific ligand-modified) drugs and drug carriers, allowing additional time for their interaction with the target. An excellent review on gene delivery by long-circulating nanocarriers is provided in Ref. [11].

5.2.1
Polymers for Longevity

The most usual way to obtain long-circulating nanoparticles is to coat them with certain hydrophilic and flexible polymers, such as poly(ethylene glycol) (PEG). In fact, the use of PEG was first suggested with liposomes to sterically hinder the interactions of blood components with the liposomes' surfaces, and to reduce the binding of plasma proteins with PEG-particles [12–14]. The molecular mechanism of PEG's protective effect is based on its participation in the repulsive interactions between PEG-grafted membranes and other particles [15], and on the properties of a flexible polymer molecule in solution. This includes the formation of a polymeric hydrogel layer over the particle surface, even at low polymer concentrations (when it closely resembles water) [16, 17]. This not only prevents drug–carrier interaction with opsonins but also slows down the rapid capture of the liposomes by the reticuloendothelial system (RES) [18]. To summarize, the mechanisms of preventing opsonization by PEG include the shielding of the surface charges, an increased surface hydrophilicity [19], an enhanced repulsive interaction between polymer-coated nanocarriers and blood components [15], and formation of the polymeric layer over the particle surface, which then becomes impermeable to other solutes, even at relatively low surface polymer concentrations [17, 19].

Although many polymers have been investigated as possible steric protectors for nanoparticulate drug carriers [7], the majority of studies into long-circulating drugs and drug carriers have been conducted by using PEG as a sterically protecting polymer, mainly because of an attractive combination of PEG's properties which include:

- Excellent solubility in aqueous solutions and an ability to bind many water molecules
- High flexibility of the polymer chain
- Very low toxicity, immunogenicity, and antigenicity
- A lack of accumulation in the RES cells
- Minimum influence on the specific biological properties of modified pharmaceuticals [20, 21].

It is also important that PEG is not biodegradable, and subsequently does not form any toxic metabolites. In contrast, PEG molecules with a molecular weight below 40 kDa are easily excreted from the body, via the kidneys. From a practical point of view, PEG is readily available on a commercial basis, with a variety of molecular weights. Those PEGs normally used to modify drugs and drug carriers typically have molecular weights ranging 1 to 20 kDa. Single-terminus reactive (semi-telehelic) PEG derivatives are often used for the modification of pharmacologically important substances, without the formation of crosslinked aggregates and heterogenic products. Currently, many chemical approaches are available for the synthesis of activated derivatives of PEG, and to couple these derivatives with a variety of drugs and drug carriers [20, 22].

Other amphiphilic polymers with highly soluble and flexible hydrophilic moieties, such as amphiphilic poly(acryl amide) (PAA), poly(vinyl pyrrolidone) (PVP), poly(acryloyl morpholine) (PAcM), poly(oxazolines), and poly(vinyl alcohol) (PVA) have been also used successfully as liposome steric protectors [7, 17, 23–26].

5.2.2
Long-Circulating Liposomes

In order that PEG becomes capable of incorporation into the liposomal membrane, the reactive derivative of hydrophilic PEG must first be single terminus-modified with a hydrophobic moiety [normally the residue of phosphatidyl ethanolamine (PE) or long-chain fatty acid is attached to a PEG-hydroxysuccinimide ester] [12, 27]. In the majority of protocols, PEG–PE is used, but this must be added to the lipid mixture prior to liposome formation. Alternatively, it was suggested that, in order to synthesize single end-reactive derivatives, the PEG should be coupled with certain reactive groups (e.g., maleimide) on the surface of pre-prepared liposomes; this is referred to as the "post-coating method" [28]. The spontaneous incorporation of PEG–lipid conjugates into the liposome membrane from PEG–lipid micelles was also shown to be very effective, and did not disturb the vesicles [29].

From a clinical point of view, it is extremely important that various long-circulating liposomes of a relatively small size (100–200 nm) can accumulate effectively in many tumors, via the "impaired filtration" mechanism [9, 10, 30, 31]. As a result, PEG-coated and other long-circulating liposomes were prepared that contained a variety of anticancer agents, including doxorubicin, arabinofuranosyl-cytosine, adriamycin, and vincristine [32, 33]. The greatest success was achieved with PEG–liposome-incorporated doxorubicin, which demonstrated good clinical results [10, 34]. An analysis of the pharmacokinetics of long-circulating nanocarriers (using PEG–liposomes) has been conducted [35].

The stability of PEGylated nanocarriers may not always be favorable for drug delivery. For example, drug-containing nanocarriers may become less able to easily release the drug; alternatively, the presence of a PEG coat on the nanocarrier's surface may preclude the contents from escaping the endosome. Consequently, an additional function must be added to long-circulating PEGylated pharmaceutical carriers, which allows for the detachment of PEG chains under the action of certain local stimuli characteristic of pathological areas. These might include a decreased pH value or an increased temperature, as typically occurs in inflamed and neoplastic regions. Thus, the chemistry was developed to detach PEG from the lipid anchor under the desired conditions. For this, polymeric components with pH-sensitive (pH-cleavable) bonds are often used to produce stimuli-responsive DDSs that are stable in the circulation, or in normal tissues, but which acquire the ability to degrade and release the entrapped drugs in body areas or cell compartments with a lowered pH, such as tumors, infarcts, zones of inflammation, or cell cytoplasm or endosomes [36, 37]. Since at "acidic" sites the pH falls from the normal physiological value of 7.4 to pH 6 and below, the chemical bonds used to date for preparing acid pH-sensitive carriers have included vinyl esters, double esters, and hydrazones; all of these are quite stable at pH ca. 7.5, but are hydrolyzed relatively rapidly at pH values of 6 and below [38, 39]. Currently, a variety of liposomes [40, 41] and micelles [42, 43] have been described that include components with the above-mentioned bonds, as well as a variety of drug conjugates capable of releasing drugs such as adriamycin [44], paclitaxel [45], doxorubicin [46], and DNA [47–49] in acidic cell compartments (endosomes) and pathological body areas under acidosis. PEGylated liposomes with a pH-sensitive detachable PEG coat attached to the surfaces of liposomes and micelles via the hydrazone bond, were used for an enhanced transfection of tumor cells both *in vitro* and *in vivo* (in this case, the plasmid encoding for the green fluorescent protein was used) [50, 51]. Detachment of the PEG coat was also achieved by using ester bonds for the PEG attachment, that could be cleaved by the action of esterases [52].

5.2.3
Nonliposomal Long-Circulating DDSs

Synthetic amphiphilic polymers also have been used for the steric stabilization of polymeric particles with the surface of a hydrophobic nature, in order to prolong their circulation in the blood and to alter their biodistribution. These polymers demonstrate the ability to be easily adsorbed on the surface of the particulate carrier, due to hydrophobic interactions. On the other hand, their hydrophilic portion is exposed in solution, and effectively protects those particulates against interactions with the plasma proteins and with cells. Surface modification of the particle can be achieved by one of two methods: (i) the absorption of a polymer onto a particle surface; or (ii) the chemical grafting of polymer chains onto a particle. Possible examples of the first case include the absorption of a series of polyethylene oxide (PEO) and polypropylene oxide (PPO) copolymers onto the surface of polystyrene latex particles, via the hydrophobic interaction mechanism.

Another important type of amphipathic polymer includes copolymers where, in an aqueous medium, the hydrophobic block is able to form a solid phase (particle), while the hydrophilic part remains as a surface-exposed protective "cloud." In order to provide a brief example of the chemically grafted onto a solid core protecting polymer, mention should be made that the use of a poly(lactic-co-glycolic acid) (PLGA)–PEG copolymer enables the preparation of long-circulating particles with an insoluble (solid) PLAGA core and a water-soluble PEG shell that is linked covalently to the core [53]. Similar effects on the longevity and biodistribution of microparticulate drug carriers might be achieved by the direct chemical attachment of protective PEO chains onto the surface of preformed particles [54].

When PEGylation was performed with poly(ε-caprolactone) (PCL) nanoparticles, their opsonization and phagocytosis was significantly suppressed [55]. For a variety of polymeric nanoparticles, the increase in PEG content resulted in an increased blood circulation time and a reduced liver uptake [56, 57]. The use of terminus-activated PEG to coat PLGA nanoparticles provided a convenient platform for the further derivatization of nanoparticles, for example, with targeting ligands [58]. Additionally, PEGylation was shown to enhance the diffusion of PLGA nanoparticles into the human cervical mucus [59]. Triple-layered nanoparticles consisting of PLGA, lecithin, and PEG were also described as a promising controlled release system [60].

Recently, many studies have been conducted to investigate the surface modification of superparamagnetic nanoparticles. Similarly to other nanocarriers, PEG-modified magnetite nanospheres demonstrated an increased colloidal stability and an improved localization in lymph nodes [61]. Subsequently, cell culture experiments have confirmed that surface PEGylation alters the interaction of modified iron oxide particles with fibroblasts [62]. Nanomagnetite particles have also been prepared by the deposition of a PCL–PEG copolymer on their surface [63].

The grafting of PEG onto the surface of gold particles via mercaptosilanes led, not unexpectedly, to a decrease in protein adsorption onto modified particles, and also to a reduced platelet adhesion [64]. In general, the PEGylation of gold nanoparticles has proven to be an effective means of reducing their MPS uptake and clearance from the body, with the efficiency of protection being dependent on the PEG density [65, 66]. An effective "click" chemistry was developed for the PEGylation of gold nanoparticles [67], while a fluorescence-based assay was developed to control the quantity of the grafted PEG [68]. The use of PEGylated gold nanoparticles has been suggested, among other applications, for photothermal tumor therapy (ablation), as they are known to accumulate in tumor tissues via the EPR effect, mainly because of their long-circulation properties [69]. Likewise, PEGylated gold nanoparticles of 3.7 nm were shown to penetrate into nuclei of HeLa cells [70]. The use of PEG-coated gold nanoparticles has also been suggested as a contrast medium for computed tomography (CT) investigations [71].

Dendrimers, which are currently considered as promising drug-delivery carriers, have also been modified with PEG to increase their longevity and accumulation; notably, studies have been conducted with PEGylated polyamidoamine (PAMAM) [72]. PEGylated methotrexate-conjugated poly-L-lysine dendrimers were shown to accumulate efficiently in solid tumors via the EPR effect [73]. The pharmacokinetics of PEGylated dendrimers was found to depend on the structure of the dendrimer, and also the quantity and length of the grafted PEG; consequently, the higher-generation dendrimers with longer PEG chains attached and a higher density of attachment demonstrated a greater blood retention [74, 75]. Importantly, PEGylation was shown to reduce the toxicity of positive charge-bearing dendrimers, such as PAMAM [76, 77].

The biomedical application of CNTs also depends on their PEGylation, and this can be achieved by a simple sonication of single-wall nanotubes in the presence of PEG–PE [78]. Upon intravenous administration to mice, PEGylated nanotubes demonstrate the longer circulation times that are essential for their future application [79]. Fluorescently labeled PEGylated CNTs have been shown to penetrate cells and even cell nuclei, and can be used for intracellular drug delivery [80]. When a PEG–doxorubicin conjugate was adsorbed onto CNTs, the modified nanotubes were suggested as being effective as local cancer chemotherapy, as they accumulate well in the lymph nodes [81]. The use of PEGylated CNTS has also suggested as vehicles for the delivery of antisense oligonucleotides [82].

Quantum dots (QDs) modified via PEGylation have also been shown, in mice, to have a lesser uptake in the MPS organs (e.g., spleen, liver) and a more prolonged blood circulation time compared to other nanoparticulates [83]. In general, PEGylation causes an increase in the biocompatibility of QDs, but reduces their toxicity [84, 85], the extent of such an effect being dependent on the PEG length and density [86].

5.2.4
Combination of Longevity and Targeting

The combination, on the same nanocarrier, of a protecting polymer for longevity and of a targeting moiety for directed transport, can clearly provide a multifunctional system. In order to achieve a more selective targeting of the PEG-coated

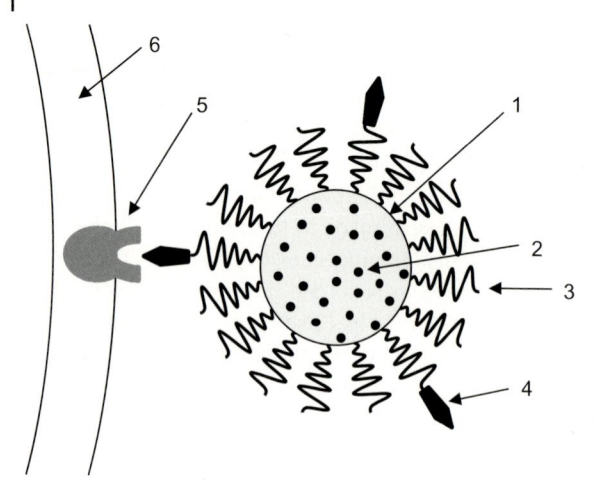

Figure 5.3 Schematic structure of long-circulating targeted nanocarrier. (**1**) Nanocarrier; (**2**) Drug; (**3**) Sterically protecting polymer (usually, PEG) grafted onto the surface of the nanocarrier; (**4**) Targeting ligand (antibody, folate, transferrin) chemically coupled with distant tips of some of the protecting polymer grafted chains; (**5**) Specific receptor targeted with the attached ligand; (**6**) Cell membrane.

nanocarriers, it is beneficial not to attach the targeting ligand to the particles directly, but rather via a PEG spacer arm. In this way, the ligand will be extended outside the dense PEG brush, thus excluding any steric effects that might hinder binding to the target receptors (Figure 5.3). Currently, a variety of advanced technologies have been used, where the targeting moiety is usually attached above the protecting polymer layer by coupling it with the distal water-exposed terminus of the activated, liposome-grafted polymer molecule [87, 88]. For this purpose, a number of derivatives of end-group functionalized lipopolymers of the general formula X–PEG–PE were introduced [20, 89], where X represents a reactive functional group-containing moiety that might include *p*-nitrophenylcarbonyl–PEG–PE (pNP–PEG–PE) [88, 90].

There are, however, certain issues to be considered when designing such systems:

- The ligand (whether an antibody, another protein, peptide, or carbohydrate) attached to the carrier surface may increase the rate of its uptake by the RES, despite the presence of a sterically protecting graft (see Ref. [91]).

- Ligand-bearing, long-circulating-time nanocarriers could facilitate the development of an unwanted immune response (as shown with the raising of antiliposome antibodies), the extent of which depends on the type of the ligand (small peptides or Fv fragments are less immunogenic than a complete IgG molecule) and the liposome composition [92, 93].

- The amount of ligand attached to the carrier may be critical to ensure successful binding with the target while maintaining the extended circulation of the carrier.

Several general strategies have been employed to assemble ligand-bearing, long-circulating PEGylated nanocarriers, initially liposomes [89, 94]. The first method involved the modification of preformed liposomes, while a second method involved mixing the ligand–PEG–lipid conjugate with other liposomal matrix-forming components, and then creating unilamellar vesicles [94, 95]. A third approach involved the use of a spontaneous insertion strategy, whereby ligand–PEG–PE conjugates were incubated with preformed loaded liposomes [94].

Multiple targeting ligands were attached to PEGylated pharmaceutical nanocarriers to achieve an improved delivery [96]. Thus, HER2-overexpressing tumors were targeted using anti-HER2 PEGylated doxorubicin-loaded liposomes [92, 97]. When attached to PEGylated liposomes, the antibody CC52 against rat colon adenocarcinoma CC531 provided a specific accumulation of liposomes in a rat model of metastatic CC531 [98]. A nucleosome-specific monoclonal antibody (mAb 2C5) which was capable of recognizing various tumor cells via the tumor cell surface-bound nucleosomes led to a significant improvement in Doxil® targeting to tumor cells, and also increased its cytotoxicity [99] both *in vitro* and *in vivo* in different test systems [99, 100] that included an intracranial human brain U-87 tumor xenograft in nude mice [101]. The same antibody was also used to effectively target long-circulating PEG-liposomes loaded with an agent for tumor photodynamic therapy (PDT), both to multiple cancer cells *in vitro* and to experimental tumors *in vivo*. A significantly enhanced tumor cell killing was also provided under the conditions of PDT [102].

PEGylated gold nanoparticles were also conjugated to monoclonal F19 antibodies as targeted labeling agents for human pancreatic carcinoma tissues [103]. Similarly, PEGylated CNTs were modified with antibodies that specifically targeted the cytoplasmic compartment [104].

Since transferrin receptors (TfRs) are overexpressed on the surface of many tumor cells, antibodies against TfR as well as transferrin itself are among popular ligands for targeting various nanoparticulate drug carriers, including liposomes to tumors and inside tumor cells [105]. In order to enhance transferrin interaction with TfR on TfR-overexpressing tumor cells and to increase the efficiency of transfection, transferrin could be also attached to the surface of nanoparticles via the PEG spacer [106]. Many studies have involved the coupling of transferrin to PEG on PEGylated liposomes, in order to combine longevity and targetability for drug delivery into solid tumors [107]. Nanoparticles composed of PLA were surface-modified with PEG and with an anti-TfR monoclonal antibody to produce PEGylated immunoparticles which were about 120 nm in size [108]. A similar approach was applied to deliver antitumor agents for PDT [109, 110] and for the intracellular delivery of cisplatin into gastric cancer [111]. Transferrin-coupled, doxorubicin-loaded PEGylated liposomes demonstrated an increased binding and toxicity against tumors in rats [112]. Transferrin–PEG–liposomes have also been used for the delivery of borocaptate to malignant glioma for neutron capture therapy [113]. Interestingly, an increase in the expression of TfRs was also identified in the post-ischemic cerebral endothelium, and used to deliver transferrin-modified PEG–liposomes to the post-ischemic brain in rats [114].

As the folate receptor (FR) is frequently overexpressed in many tumor cells, the targeting of tumors with folate-modified nanocarriers that include liposomes represents a popular approach. Early studies demonstrated the possibility of delivering first macromolecules [115], and then liposomes [116], into living cells utilizing FR endocytosis, which enabled any multidrug resistance to be bypassed. Subsequently, numerous folate-targeted liposomal systems have been described for the treatment of cancer (for reviews, see Refs [117, 118]). Folate was also attached to the surface of cyanoacrylate-based nanoparticles via activated PEG blocks [119]. Similarly, PEG–PCL-based particles which had been surface-modified with folate and then loaded with paclitaxel demonstrated an increased cytotoxicity [120]. Superparamagnetic magnetite nanoparticles which had been modified with folate (with or without a PEG spacer) demonstrated a better uptake by cancer cells, that could be used for both diagnostic [e.g., as magnetic resonance imaging (MRI) agents] and therapeutic purposes [121, 122]. Folate-modified, PEG-grafted hyperbranched PEI was used for tumor-targeted gene therapy [123], while docetaxel-loaded PEGylated PLGA nanoparticles were modified with folate to increase their cellular uptake and anti-cancer cytotoxicity [124]. Folate-modified PEG-dendrimers loaded with 5-fluorouracil also demonstrated an enhanced activity in tumor-bearing mice [125]. Folate was also used to target PEGylated gold particles to tumor cells [126]; folate-modified PEGylated QDs were also used for the visualization of receptor-mediated endocytosis [127].

5.3
Stimuli-Sensitivity and Intracellular Targeting

Multifunctional long-circulating and targeted pharmaceutical nanocarriers can be functionalized still further by imparting them with additional sensitivity to various stimuli, whether characteristic for the target pathological zone or applied externally (see Table 5.1). Those stimuli characteristic for pathological tissues include pH and redox conditions, whereas temperature can serve as a local stimulus both within the tissue (inflammation and local hyperthermia) and from the outside. For example, the intratumoral pH value in tumors may fall to 6.5 (i.e., approximately one pH unit lower than in normal blood) because of hypoxia and massive cell death inside the tumor [128, 129], and may fall still further inside the cells (especially in endosomes) to 5.5 and even below [130]. At the same time, the intracellular concentration of glutathione (i.e., redox potential) in cancer cells may be significantly higher (several hundred-fold) than the normal extracellular levels [131]. Stimuli such as ultrasound and (electro)magnetic fields may be applied only "artificially," and mainly from the outside.

5.3.1
pH-Sensitive Systems

pH sensitivity was used to modify, in a desired way, the drug/DDS behavior in pathological areas where there was a decreased pH value, including tumors, inf-

Table 5.1 Stimuli utilized to control a drug-delivery system (DDS).

Stimuli	Stimuli origin
pH	Internal–decreased pH in pathological areas, such as tumors, infarcts, and inflammations, because of hypoxia and massive cell death; relatively low pH in endosomes and cytoplasm
Redox conditions	Internal–increased concentration of glutathione inside many pathological cells
Temperature	Internal–inflammation-provoked hyperthermia External–temperature can be increased in target tissues by locally applied ultrasound, or by accumulating magnetic nanoparticles in the target with subsequent action of a high-frequency electromagnetic field
Ultrasound	External–sonication can be applied to the body to facilitate DDS penetration into cells and drug/gene release
Magnetic field	External–magnetic field can concentrate magneto-sensitive DDS in required areas

arcts, and inflammations. Thus, pH-sensitive liposomes have been created to destabilize and release the incorporated load at a lowered pH-value. Such liposomes contained phospholipids capable of protonation or the formation of nonbilayered structures at decreased pH, and destabilizing liposomal or liposomal and endosomal membranes with the subsequent drug/DNA release from liposomes and endosomes into the cytoplasm [132, 133]. In addition to membrane-destabilizing lipid components, there exists a large family of membrane-destabilizing anionic polymers that also can enhance the endosomal escape of various drugs and biomacromolecules [134]. This family includes carboxylated polymers, copolymers of acrylic and methacrylic acids, copolymers of maleic acid, and polymers and copolymers of N-isopropylacrylamide, which demonstrate a lower critical solution (solubility/insolubility switch) at physiological temperatures; moreover, when they precipitate they tend to destabilize the biomembranes with which they are interacting [135].

Despite a decreased pH-sensitivity, multifunctional long-circulating PEGylated pH-sensitive liposomes can still deliver their contents very effectively into the cytoplasm (for a recent review, see Ref. [36]). Serum-stable, long-circulating PEGylated pH-sensitive liposomes were also prepared by using a combination of PEG and a pH-sensitive, terminally alkylated copolymer of N-isopropylacrylamide and methacrylic acid [37]. The additional modification of pH-sensitive liposomes with an antibody results in pH-sensitive immunoliposomes. The main advantages of antibody-bearing, pH-sensitive liposomes include cytoplasmic delivery, targetability, and a facilitated uptake (i.e., an improved intracellular availability) via receptor-mediated endocytosis. The successful application of pH-sensitive immunoliposomes has been demonstrated for the delivery of a variety of molecules, including fluorescent dyes, antitumor drugs, proteins, and DNA [136]. A combination of liposome pH-sensitivity and specific ligand targeting for cytosolic drug delivery, utilizing decreased endosomal pH values, was described for both folate and transferrin-targeted liposomes [137].

In the case of *polyplexes*, which cannot directly destabilize the endosomal membrane, the mechanism of DNA escape from endosomes is associated with the ability of polymers (such as PEI) to strongly protonate under the acidic pH inside endosome, and to create a charge gradient that eventually provokes a water influx, followed by endosomal swelling and disintegration [138]. The attachment of nuclear localization sequences to plasmid DNA may enhance its nuclear translocation and transfection efficiency [139].

Polymeric micelles can also demonstrate pH-sensitivity and an ability to escape from endosomes. Thus, micelles prepared from PEG–poly(aspartate hydrazone adriamycin) not only easily release an active drug at the lowered pH values typical for endosomes, but also facilitate its cytoplasmic delivery and toxicity against cancer cells [140]. Alternatively, micelles for the intracellular delivery of antisense oligonucleotides (ODNs) were prepared from ODN–PEG conjugates complexed with a cationic fusogenic peptide, KALA, and provided a much higher intracellular delivery of the ODN than could be achieved with free ODN [141]. Compensation of the negative charge of PEG–lipid micelles [142] by the addition of positively charged lipids to PEG–PE micelles caused a marked improvement in the uptake by cancer cells of a drug-loaded mixed PEG–PE/positively charged lipid micelles, and their escape from the endosomes. This approach was used to increase the intracellular delivery and, consequently, the anticancer activity of the micellar paclitaxel by preparing paclitaxel-containing micelles from the mixture of PEG–PE and positively charged lipids [143]. Multifunctional polymeric micelles capable of pH-dependent dissociation and drug release, when loaded with doxorubicin and supplemented with biotin as a cancer cell-interacting ligand, have also been described [144]. Paclitaxel was also loaded into mixed micelles that could undergo dissociation into unimers leading to drug liberation, even above the critical micelle concentration (CMC) value due to an ionization of their components at certain pH values [145].

Polymeric components with pH-sensitive (pH-cleavable) bonds are used to produce stimuli-responsive DDSs that are stable in the circulation or in normal tissues, but which acquire an ability to degrade and to release the entrapped drugs in body areas or cell compartments with a lowered pH, such as tumors, infarcts, inflammation zones, cell cytoplasm, or endosomes [36, 37, 146]. Serum-stable, long-circulating PEGylated pH-sensitive liposomes were also prepared using a combination of PEG and pH-sensitive terminally alkylated copolymer of *N*-isopropylacrylamide and methacrylic [37] on the same liposome. In this case, attachment of the pH-sensitive polymer to the liposomes surface might facilitate liposome destabilization and drug release in compartments with decreased pH values.

The combination of liposome pH-sensitivity and specific ligand targeting for cytosolic drug delivery utilizing decreased endosomal pH values was described for folate- and transferrin-targeted liposomes [147, 148]. For this, PEG–dendrimer combinations were used to build pH-sensitive micelles capable of releasing micelle-incorporated doxorubicin at acidic pH values [149].

The stimuli-sensitivity of PEG coats can also allow for the preparation of multifunctional DDSs with temporarily "hidden" functions that, under normal circumstances, would be "shielded" by the protective PEG coat, but become exposed after PEG detachment. Such systems would require that multiple functions be attached to the surface of the nanocarrier and function in a certain coordinated manner.

5.3.2
Temperature-Sensitive Systems

The idea of using temperature-sensitive nanocarriers was derived initially from the fact that many pathological areas demonstrate distinct hyperthermia. Additionally, there exist various means to heat the required area in the body. Initially, it was found that a significantly greater fraction of intravenously administered liposomes and other nanocarriers would accumulate in a tumor mass upon heating to 42 °C in human ovarian carcinoma xenograft model; likewise, a higher concentration and efficacy was observed for doxorubicin delivered into tumors in temperature-sensitive liposomes [150, 151]. Temperature-sensitive liposomes frequently include dipalmitoylphosphatidylcholine (DPPC) as the key component, since liposomes usually become "leaky" at a gel-to-liquid crystalline phase transition which, in the case of DPPC, takes place at 41 °C [152]. Liposomes can also be made temperature-sensitive by incorporating (by grafting) certain polymers that display a lower critical solution temperature (LCST), slightly above the physiological value [153, 154]. Because these polymers are soluble below the LCST, and precipitate when the temperature increases above the LCST, they can cause the liposomal membrane to be damaged during precipitation, and allow for drug release [155]. The most usual representative of this class of polymers is poly(N-isopropylacrylamide) (NIPAM) [156]. Similarly, polymeric micelles can be rendered temperature-sensitive by assembling them from amphiphilic copolymers, in which one of the blocks demonstrate properties similar to those of NIPAM [157].

5.3.3
Magnetically Sensitive Systems

Drug carriers, such as microcapsules, can be loaded not only with the drug alone but also with magnetic nanoparticles; this allows for the manipulation of such capsules in a magnetic field. Alternatively, the inclusion of metallic nanoparticles is possible; these can respond to an external electromagnetic field, allowing the rate of drug release to be controlled either by oscillating or heating the carrier [158]. In nanomedicine, iron oxide nanoparticles with particle size of approximately 4–10 nm (superparamagnetic iron oxide nanoparticles, SPIONs) have recently attracted much interest (e.g., see Refs [159–165]). The concept of magnetic drug targeting (MDT), whereby magnetically susceptible particles are guided towards the intended pathological site under the influence of external magnets, was first introduced by Widder *et al.* [166] and has recently attracted much attention following advances in nanotechnology. For example, Gang *et al.* [167] have demonstrated the targeting of magnetic PCL nanoparticles loaded with gemcitabine in a pancreatic cancer xenograft mouse model, by using external magnets. Cinteza *et al.* have also reported the coloading of polymeric micelles of diacylphospholipid–PEG with the photosensitizing drug 2-[1-hexyloxyethyl]-2-devinyl pyropheophorbide-a, and magnetic SPION for magnetic drug targeting *in vitro*. Alexiou *et al.* have employed mitoxantrone-loaded SPION, and targeted them to VX2 squamous cell carcinoma in rabbits by using external magnets [168, 169]. MDT has also been used to improve localized drug delivery to interstitial tumor targets; in particular, it is currently being developed to improve drug delivery to tumor vessels. Thus, magnetite (MAG-C) was loaded into cationic liposomes together with etoposide and dacarbazine [170]. It is very important, that the co-incorporation of SPION and drugs into the same nanocarriers to only a minimal extent, influences the efficacy of drug loading [171]. By using magnetic fluid depots, the concept of magnetic hyperthermia has been demonstrated by Johannsen *et al.* [172]. Likewise, Wust *et al.* have conducted feasibility and tolerance studies to test the applicability of whole-body magnetic field applicators and iron oxide nanoparticles in human patients. Notably, the results of the study illustrated the great potential for SPION-based hyperthermia [173].

5.3.4
Ultrasound-Sensitive Systems

Whilst the application of external ultrasound to control drug delivery and release from nanocarriers is a relatively novel approach, the first reports of this topic date back to 1998, when acoustically active liposphers were described that contained paclitaxel [174]. This whole concept is based on the creation of DDSs that, upon accumulation in required areas, can be made "leaky" by a locally applied external ultrasound, so as to liberate any incorporated drugs or genes. Much promising data on drug and gene delivery via ultrasound-sensitive drug carriers are already in existence, and some of these are reviewed in Refs [175, 176]. Acoustically active liposomes containing a small quantity of a certain gas (air) or a perfluorated hydrocarbon, and initially developed as an ultrasound contrast agent, can be loaded with various drugs that will be released when the liposomes are damaged by applied ultrasound [177, 178]. Polymeric micelles have also been prepared, that are capable of incorporating various drugs (e.g., doxorubicin) and releasing them after ultrasonication; these micelles can also assist in the delivery of DDSs inside the cells [179, 180]. A similar approach was used for the local release of thrombolytic enzymes (e.g., tissue plasminogen activator) from echogenic liposomes for the treatment of blood clot formation [181]. Studies on ultrasound-sensitive formulations, and the mechanisms that control drug release from such formulations, represent an important area of research into stimuli-sensitive nanocarriers [182, 183].

5.3.5
Redox-Sensitive Systems

A high redox potential difference, which exists between the reducing intracellular space and oxidizing extracellular space, can also be utilized for the construction of stimuli-sensitive DDSs [131]. With this in mind, drug or DNA can be loaded into the nanocarrier, the structure of which is normally maintained by the presence of disulfide bonds. When such bonds are reduced to thiol groups, due to the presence of high glutathione levels inside the cell, the integrity of the carrier is compromised such that drug or DNA can be released. Previously, polymers with positively charged and thiol groups incorporated into the polymer structure have been used to complex DNA (via a positive charge) and to form a polymeric network (via disulfide bridges formed from the groups) [184]. When reduced, the disulfide bridges are converted back to thiols, while the polymeric carrier will disintegrate and facilitate DNA release. The intracellular delivery of plasmid DNA was also performed using thiolated gelatin nanoparticles [185, 186]. Transfection efficacy was also enhanced by using DNA condensed with thiolated PEI [187]. Redox-responsive liposomes have also been prepared from standard phospholipids with the addition of a small quantity of a lipid, in which the head and tail are linked by the disulfide bond [188]. Long-circulating, redox-responsive liposomes with a detachable PEG coat have also been described [189]. The general scheme of stimuli-sensitive DDSs is shown in Figure 5.4.

5.3.6
Intracellular Delivery of Pharmaceutical Nanocarriers

The intracellular transport of different biologically active molecules represents a key problem in drug delivery in general, as many pharmaceutical agents – including various large molecules (e.g., proteins, enzymes, antibodies) and even drug-loaded pharmaceutical nanocarriers – must be delivered intracellularly in order to exert their therapeutic action inside the cytoplasm, or onto the nucleus or other specific organelles, such as the lysosomes, mitochondria, or endoplasmic reticulum. In addition, an intracytoplasmic drug delivery in cancer treatment may overcome important obstacles in anti-cancer chemotherapy, such as multidrug resistance. Unfortunately, the cell membrane is problematic, in that it prevents large molecules (e.g., peptides, proteins, DNA) from spontaneously entering the cells. Moreover, even when the drugs have been safely delivered into the cell cytoplasm, they must still find their way to specific organelles (e.g., nuclei, lysosomes, mitochondria). Some approaches to the intracellular delivery of pharmaceutical nanocarriers have been discussed earlier in connection with pH-sensitive functions.

The addition of a positive charge to the nanocarrier can significantly enhance its uptake by cells and, indeed, the use of cationic lipids and cationic polymers as transfection vectors for the intracellular delivery of DNA was first proposed many years ago [190, 191]; for a recent review, see Ref. [192]. Complexes between cationic

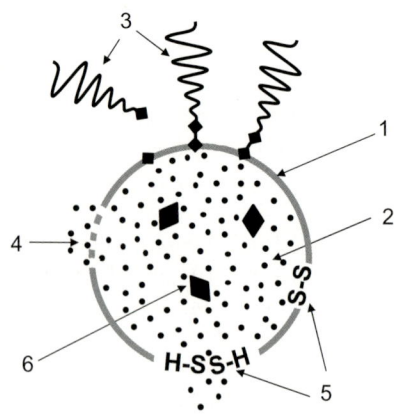

Figure 5.4 Schematic of the different stimuli acting on the stimuli-sensitive nanocarrier and expected responses. (**1**) Nanocarrier; (**2**) Drug; (**3**) The protective polymeric coating attached to the surface of the nanocarrier via pH-sensitive bonds could be detached and removed by the action of a lowered pH in certain pathological areas (tumors) or inside cellular compartments (cytoplasm, endosome); (**4**) Temperature-sensitive coating or components of the carrier, which can be influenced by the heat (hyperthermia in certain pathological areas, or the heat brought upon by an external source) to destabilize the carrier and allow for drug release; (**5**) Redox-sensitive coating or components of the carrier, which can be influenced by changing redox conditions (increased glutathione), for example by transforming –S–S–bonds into thiol groups, and allowing for drug release; (**6**) Particles of magnetosensitive material (SPION), which can allow the whole nanocarrier to be transported to required site under the action of an external magnetic field.

lipids (such as Lipofectin®) and DNA (lipoplexes), and complexes between cationic polymers, such as PEI [193] and DNA (polyplexes) are formed because of the strong electrostatic interactions between the positively charged carrier and negatively charged DNA. A slight net positive charged of already-formed lipoplexes and polyplexes is believed to facilitate their interaction with negatively charged cells, and improve their transfection efficiency [194]. *Endocytosis* represents the main mechanism of lipoplex/polyplex internalization employed by cells [195].

A relatively recent approach for intracellular drug delivery is based on the modification of drugs and drug carriers with certain proteins and peptides that demonstrate a unique ability to penetrate into cells (the "transduction" phenomenon). The cell-penetrating peptide (CPP)-mediated intracellular delivery of large molecules and nanoparticles was proved to proceed via the energy-dependent macropinocytosis, with subsequent enhanced escape from endosome into the cell cytoplasm [196]. In contrast, the individual CPPs or CPP-conjugated small molecules were shown to penetrate cells via electrostatic interactions and hydrogen bonding, and seemed not to depend on the energy requirements [197].

It has been shown that CPPs are capable of internalizing nanosized particles into the cells [198, 199]. For example, SPION – when conjugated with TATp and fluorescein isothiocyanate – were taken up quickly by the T cells, B cells and macrophages, followed by a migration of the conjugates, primarily to the cytoplasm, where they could be tracked easily by using MRI [200]. Dextran-coated SPIONs derivatized with TATp were internalized into lymphocytes by over 100-fold more efficiently than non-modified particles [201]. Even relatively large particles, such as liposomes, could be delivered into the various cells by multiple TATp or other CPP molecules being attached to their surfaces [202, 203]. The translocation of TATp-liposomes (both plain and PEGylated) into cells required the direct interaction of the liposomal TATp with the cell surface [204]. Complexes of TATp-liposomes with a plasmid encoding for the green fluorescent protein (GFP), were used for the successful *in vitro* transfection of various tumor and normal cells, as well as for the *in vivo* transfection of tumor cells in mice bearing Lewis lung carcinoma [205] (the combination of positive charge for DNA complexation and cell-penetrating functions). Antp and TATp, when coupled to small unilamellar liposomes, were accumulated within tumor cells and dendritic cells more effectively than unmodified control liposomes [206]. The coupling of TATp to the outer surface of the liposomes was also described, which resulted in an enhanced binding and endocytosis of the liposomes in ovarian carcinoma cells [207]. Anti-liposomes have been also considered as a carrier system for an enhanced cell-specific delivery of liposome-entrapped molecules [206]. The octamer of arginine (R8), attached to the surface of siRNA-loaded liposomes, provided their effective intracellular delivery and silencing of the targeted gene [208].

Cell-penetrating function could be beneficially combined with the stimuli-sensitivity discussed earlier. Thus, a nanoparticulate DDS should be able: (i) to accumulate in the required organ or tissue, and; (ii) then penetrate inside the target cells, where they deliver their load (e.g., drug or DNA). Organ or tissue (tumor, infarct) accumulation could be achieved by the passive targeting via the EPR effect, or by antibody-mediated active targeting, while an intracellular delivery could be mediated by certain internalizable ligands (folate, transferrin) or by CPPs, such as TAT or polyArg [209, 210]. When in the blood, the cell-penetrating function should be temporarily inactivated (shielded) to prevent a non-specific drug delivery into non-target cells. However, by already being inside the target the delivery system should lose its protective coat, thus exposing the cell-penetrating function and providing intracellular drug delivery [211]. This is especially important for CPP-bearing nanocarriers, as all CPPs are highly nonselective and can lead their cargo to any cells, including many nontarget cells.

Recently, targeted long-circulating PEGylated liposomes and PEG–PE-based micelles possessing several functionalities [212, 213] have been described. Such systems are capable of targeting a specific cell or organ by attaching the monoclonal antibody (infarct-specific antimyosin antibody 2G4 or cancer-specific antinucleosome antibody 2C5) to their surface. On the other hand, these nanocarriers could be additionally modified with TATp moieties attached to the surface of the nanocarrier via the short spacer. PEG–PE used for liposome surface modification or for micelle preparation, was rendered degradable by inserting the pH-sensitive hydrazone bond between PEG and PE (PEG–Hz–PE). Under normal pH values, TATp functions on the surface of nanocarriers were "shielded" by long protecting

PEG chains (pH-degradable PEG_{2000}–PE or PEG_{5000}–PE), or by long pNP–PEG–PE moieties used to attach antibodies to the nanocarrier (non-pH-degradable PEG_{3400}–PE or PEG_{5000}–PE). At pH 7.5–8.0, both liposomes and micelles demonstrated a high specific binding with antibody substrates, but a very limited internalization by cells. However, upon brief incubation at lower pH values (pH 5.0–6.0), the nanocarriers lost their protective PEG shell because of acidic hydrolysis of PEG–Hz–PE, and were effectively internalized by cells via TATp moieties. In vivo, TATp-modified pGFP-loaded liposomal preparations have been administered via an intratumoral route in tumor-bearing mice, and the efficacy of tumor cell transfection followed after 72 h. The administration of pGFP–TATp–liposomes with a non-pH-sensitive PEG coating resulted in only a minimal transfection of tumor cells, mainly because of steric hindrances for the liposome-to-cell interaction created by the PEG coat. In contrast, the administration of pGFP–TATp–liposomes with the low pH-detachable PEG resulted in a highly efficient transfection, since the removal of PEG under the action of the decreased intratumoral pH led to an exposure of the liposome-attached TATp residues, an enhanced penetration of the liposomes inside the tumor cells, and an effective intracellular delivery of the pGFP [51].

Interesting multifunctional envelope-type devices have been recently described for the cytoplasmic delivery of proteins, DNA, and oligonucleotides [214]. Nano particles have been formed by condensation of the substances to be delivered inside sells with lipid derivatives of CPPs (such as polyarginine), and were efficiently internalized by the cells before releasing their cargo into the cytosol.

5.4
A New Challenge: Theranostics

One recent challenge in the area of multifunctional pharmaceutical nanocarriers has been that of *theranostics*–that is, the use pharmaceutical nanocarriers for diagnostic/imaging purposes simultaneously with their therapeutic use. This should allow the real-time biodistribution and target accumulation of these materials to be followed simultaneously. With this in mind, the contrast reporter moieties can be added to multifunctionalized nanocarriers. Currently used medical imaging modalities include: gamma-scintigraphy, MRI, CT, and ultrasonography, together with some modifications of those methods. Because each modality has its own advantages and disadvantages, it is highly desirable to have more than one type of reporter group associated with the same nanocarrier.

Liposomes, polymeric micelles, dendrimers, and iron oxide nanoparticles are frequently used as templates for engineering theranostic carriers. In the case of liposomes for gamma-scintigraphy and MRI, when heavy-metal atoms are used as contrast moieties the reporter metal can first be chelated into a soluble chelator (e.g., diethylene triamine pentaacetic acid, DTPA), and then incorporated into the interior of a liposome [215]. Alternatively, a chelating compound can be chemically modified with a hydrophobic group, which causes it to be anchored onto the liposome surface [216]; this is especially beneficial for MRI contrast systems when, in order to achieve a better signal the reporter metal must be exposed to water [217–219]. Clearly, this is true for liposomes (or other nanocarriers) that already bear various functions (such as longevity, targetability, stimuli-sensitivity, intracellular penetration). Different chelators and different hydrophobic anchors have been examined for the preparation of ^{111}In, ^{99m}Tc, Mn-, and Gd-loaded liposomes [220–224]. The amphiphilic chelating probes (paramagnetic Gd–DTPA–PE and radioactive ^{111}In-DTPA–SA) can also be incorporated into PEG(5 kDa)–PE micelles and used for *in vivo* MRI and scintigraphy [225]. Polychelating amphiphilic polymers (PAPs) were synthesized which consisted of the main chain, but with multiple side chelating groups capable of firmly binding many reporter metal atoms, and also hydrophobic terminal groups which allowed for polymer adsorption onto hydrophobic nanoparticles or incorporation into hydrophobic domains of liposomes or micelles, thus providing a sharp increase in the number of bound reporter metal atoms per particle and, thus, of the image signal intensity [226]. In case of MRI, metal atoms chelated into polymer side groups are directly exposed to the water environment that enhances the relaxivity of the paramagnetic ions [222, 227, 228]. Such PAP-nanoparticles were used for *in vivo* MRI of lymphatic system components with Gd-loaded nanocarriers. The performance of Gd–PAP–liposomes or –micelles could be further improved in case of the co-incorporation of amphiphilic PEG onto the liposome membrane or micelle surface. This effect can be explained by an increased relaxivity of PEG–Gd–liposomes, due to the presence of an increased amount of PEG-associated water protons in the close vicinity of chelated Gd ions [229, 230]. Gd–PAP–PEG-liposomes that additionally were modified with the cancer-specific monoclonal antibody, demonstrated a rapid and specific tumor accumulation, and could serve as effective contrast agents for tumor MRI [231, 232].

The combination of drug loading, longevity, targetability, and contrast properties will, most likely, result in a new generation of multifunctional nanopharmaceuticals. Thus, long-circulating PEGylated liposomes loaded with doxorubicin and additionally decorated with a tumor-specific antibody and contrast moieties [233, 234], demonstrated an increased therapeutic activity *in vivo*, while their target accumulation could be easily followed using either gamma-scintigraphy (Figure 5.5) or MRI. Multifunctional nanocarriers for image-guided drug delivery, which combine therapeutic and imaging agents merged into one preparation, have also been described [179, 235]. In the former of these studies, ultrasonic tumor imaging was combined with a targeted therapy, using doxorubicin.

This combination of drug (chemotherapeutic) and contrast agent load on the same nanocarrier allows for the development of a personalized nanotherapeutic approach. This is because the efficiency of nanocarrier accumulation in the tumor (as determined by the "leakiness" of the blood vessels and traced by the contrast agent, for example, by mammography) can assist in the correct dose of therapeutic agent to treat individual tumors [235, 236]. Such systems may, in future, be developed on the basis of various pharmaceutical nanocarriers [237], including lipid-based [238] and polymeric particles [239].

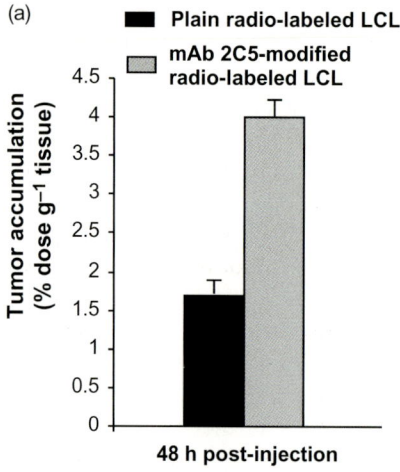

Figure 5.5 Combination of the longevity, targetability, and contrast function. (a) Radiolabeled (^{111}In) long-circulating PEGylated liposomes (LCL) modified with the cancer-specific monoclonal antibody 2C5 demonstrate an enhanced tumor accumulation; (b) This method can also be used for the rapid and specific tumor visualization by gamma-scintigraphy (in mice).

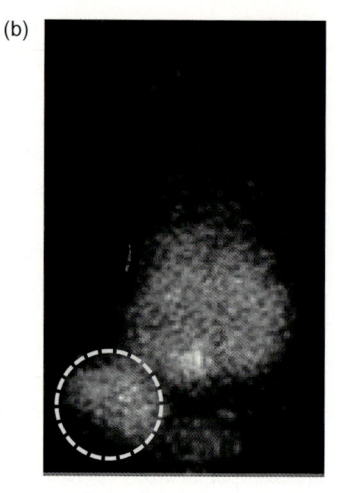

Iron oxide-based nanopreparations have attracted much attention over the past few year, and have provided a combined opportunity of magnetically mediated drugs targeting and MR imaging (for reviews, see Refs [159, 240–242]). Thus, diacyllipid micelle-based nanopreparations have been suggested for the delivery of drugs in PDT [171], while PEG-modified magnetic nanoparticles with immobilized methotrexate have been suggested for MRI and drug delivery [243]. Similarly, chlorotoxin-conjugated iron oxide nanoparticles were prepared for the specific delivery of therapeutic and MRI contrast agents to tumors, *in vitro* and *in vivo* [240]. Both, drug- and dye-loaded iron oxide nanoparticles can serve as therapeutic agents and dual optical/MR contrast agents [244]. In being modified with specific ligands, such as recombinant peptide containing the amino-terminal fragment of urokinase-type plasminogen activator (uPA), magnetic particles can be used to target breast cancer tumors overexpressing uPA receptors, for enhanced imaging and delivery [245]. In a similar fashion, folic acid-modified Pluronic F127-coated magnetic nanoparticles can be used for the targeting, diagnosis, and therapy of tumors which overexpress the folate receptor [246].

As mentioned above, specific interest is currently being shown in the development of dual modality imaging systems, such as positron emission tomography (PET)/CT [237], with the imaging system serving a simultaneous therapeutic role in the case of radionuclides [237]. Multimodal imaging systems frequently utilize liposomes which, as noted above, already have a well-established clinical reputation and have been successfully loaded with a wide variety of contrast agents (e.g., iron oxide nanoparticles, QDs, radionuclides, MRI contrast agents) [247]. A liposomal vascular contrast agent which was suited simultaneously for CT and MRI-based applications demonstrated a good efficacy in mice and rabbits [248]. Similarly, a variety of polymeric nanoparticles has been described which was designed to serve as imaging agents for two different modalities. Thus, multifunctional chitosan nanoparticles have been prepared encapsulating QDs and Gd–DTPA, and shown to be suitable for both optical and magnetic resonance imaging [249]. Nanoparticles which can serve for magnetic and fluorescent or magnetic, fluorescent and PET imaging, have also been reviewed [250–252]. Such nanoprobes for dual-modality magnetofluorescent imaging have been additionally modified with PEG for longevity, and with folate for tumor targeting [253].

To conclude, over a relatively short period of time the area of multifunctional pharmaceutical nanocarriers has undergone a dramatic development. At the experimental level, there is now available a broad selection of multifunctional systems providing various combinations of longevity, targetability, stimuli-sensitivity, cell-penetrating ability, and contrast properties. The evident success of such systems in multiple animal experiments promises the development of clinically acceptable systems for theranostic applications in the foreseeable future.

References

1 Torchilin, V. (ed.) (2006) *Nanoparticulates as Drug Carriers*, Imperial College Press, London, UK.

2 Thassu, D., Deleers, M. and Pathak, Y. (eds) (2007) *Nanoparticulate Drug Delivery Systems*, Informa Healthcare USA, New York.

3 Torchilin, V. (2008) Multifunctional and stimuli-sensitive pharmaceutical nanocarriers. *European Journal of Pharmaceutics and Biopharmaceutics*, **71**, 432–44.

4 van Vlerken, L.E. and Amiji, M.M. (2006) Multi-functional polymeric nanoparticles for tumour-targeted drug delivery. *Expert Opinion on Drug Delivery*, **3**, 205–16.

5 Torchilin, V.P. (2006) Multifunctional nanocarriers. *Advanced Drug Delivery Reviews*, **58**, 1532–55.

6 Lasic, D.D. and Martin, F.J. (eds) (1995) *Stealth Liposomes*, CRC Press, Boca Raton.

7 Torchilin, V.P. and Trubetskoy, V.S. (1995) Which polymers can make nanoparticulate drug carriers long-circulating? *Advanced Drug Delivery Reviews*, **16**, 141–55.

8 Moghimi, S.M. and Szebeni, J. (2003) Stealth liposomes and long circulating nanoparticles: critical issues in pharmacokinetics, opsonization and protein-binding properties. *Progress in Lipid Research*, **42**, 463–78.

9 Maeda, H., Wu, J., Sawa, T., Matsumura, Y. and Hori, K. (2000) Tumor vascular permeability and the EPR effect in macromolecular therapeutics: a review. *Journal of Controlled Release*, **65**, 271–84.

10 Gabizon, A.A. (1995) Liposome circulation time and tumor targeting: implications for cancer chemotherapy. *Advanced Drug Delivery Reviews*, **16**, 285–94.

11 Kommareddy, S., Tiwari, S.B. and Amiji, M.M. (2005) Long-circulating polymeric nanovectors for tumor-selective gene delivery. *Technology in Cancer Research and Treatment*, **4**, 615–25.

12 Klibanov, A.L., Maruyama, K., Torchilin, V.P. and Huang, L. (1990) Amphipathic polyethyleneglycols effectively prolong the circulation time of liposomes. *FEBS Letters*, **268**, 235–7.

13 Papahadjopoulos, D., Allen, T.M., Gabizon, A., Mayhew, E., Matthay, K., Huang, S.K., Lee, K.D., Woodle, M.C., Lasic, D.D., Redemann, C. *et al.* (1991) Sterically stabilized liposomes: improvements in pharmacokinetics and

antitumor therapeutic efficacy. *Proceedings of the National Academy of Sciences of the United States of America*, **88**, 11460–4.

14 Allen, T.M. (1994) The use of glycolipids and hydrophilic polymers in avoiding rapid uptake of liposomes by the mononuclear phagocyte system. *Advanced Drug Delivery Reviews*, **13**, 285–309.

15 Needham, D., McIntosh, T.J. and Lasic, D.D. (1992) Repulsive interactions and mechanical stability of polymer-grafted lipid membranes. *Biochimica et Biophysica Acta*, **1108**, 40–8.

16 Torchilin, V.P. and Papisov, M.I. (1994) Why do polyethylene glycol-coated liposomes circulate so long? *Journal of Liposome Research*, **4**, 725–39.

17 Torchilin, V.P., Omelyanenko, V.G., Papisov, M.I., Bogdanov, A.A., Jr, Trubetskoy, V.S., Herron, J.N. and Gentry, C.A. (1994) Poly(ethylene glycol) on the liposome surface: on the mechanism of polymer-coated liposome longevity. *Biochimica et Biophysica Acta*, **1195**, 11–20.

18 Senior, J.H. (1987) Fate and behavior of liposomes in vivo: a review of controlling factors. *Critical Reviews in Therapeutic Drug Carrier Systems*, **3**, 123–93.

19 Gabizon, A. and Papahadjopoulos, D. (1992) The role of surface charge and hydrophilic groups on liposome clearance in vivo. *Biochimica et Biophysica Acta*, **1103**, 94–100.

20 Zalipsky, S. (1995) Chemistry of polyethylene glycol conjugates with biologically active molecules. *Advanced Drug Delivery Reviews*, **16**, 157–82.

21 Yamaoka, T., Tabata, Y. and Ikada, Y. (1994) Distribution and tissue uptake of poly(ethylene glycol) with different molecular weights after intravenous administration to mice. *Journal of Pharmaceutical Sciences*, **83**, 601–6.

22 Torchilin, V.P. (2002) Strategies and means for drug targeting: an overview, in *Biomedical Aspects of Drug Targeting* (eds V. Muzykantov and V.P. Torchilin), Kluwer Academic Publishers, Boston, pp. 3–26.

23 Torchilin, V.P. (1996) How do polymers prolong circulation times of liposomes. *Journal of Liposome Research*, **9**, 99–116.

24 Torchilin, V.P., Trubetskoy, V.S., Whiteman, K.R., Caliceti, P., Ferruti, P. and Veronese, F.M. (1995) New synthetic amphiphilic polymers for steric protection of liposomes in vivo. *Journal of Pharmaceutical Sciences*, **84**, 1049–53.

25 Torchilin, V.P., Levchenko, T.S., Whiteman, K.R., Yaroslavov, A.A., Tsatsakis, A.M., Rizos, A.K., Michailova, E.V. and Shtilman, M.I. (2001) Amphiphilic poly-N-vinylpyrrolidones: synthesis, properties and liposome surface modification. *Biomaterials*, **22**, 3035–44.

26 Woodle, M.C., Engbers, C.M. and Zalipsky, S. (1994) New amphipathic polymer-lipid conjugates forming long-circulating reticuloendothelial system-evading liposomes. *Bioconjugate Chemistry*, **5**, 493–6.

27 Klibanov, A.L., Maruyama, K., Beckerleg, A.M., Torchilin, V.P. and Huang, L. (1991) Activity of amphipathic poly(ethylene glycol) 5000 to prolong the circulation time of liposomes depends on the liposome size and is unfavorable for immunoliposome binding to target. *Biochimica et Biophysica Acta*, **1062**, 142–8.

28 Maruyama, K., Takizawa, T., Yuda, T., Kennel, S.J., Huang, L. and Iwatsuru, M. (1995) Targetability of novel immunoliposomes modified with amphipathic poly(ethylene glycol)s conjugated at their distal terminals to monoclonal antibodies. *Biochimica et Biophysica Acta*, **1234**, 74–80.

29 Sou, K., Endo, T., Takeoka, S. and Tsuchida, E. (2000) Poly(ethylene glycol)-modification of the phospholipid vesicles by using the spontaneous incorporation of poly(ethylene glycol)-lipid into the vesicles. *Bioconjugate Chemistry*, **11**, 372–9.

30 Maeda, H. (2001) The enhanced permeability and retention (EPR) effect in tumor vasculature: the key role of tumor-selective macromolecular drug targeting. *Advances in Enzyme Regulation*, **41**, 189–207.

31 Gabizon, A. and Papahadjopoulos, D. (1988) Liposome formulations with

prolonged circulation time in blood and enhanced uptake by tumors. *Proceedings of the National Academy of Sciences of the United States of America*, **85**, 6949–53.

32. Boman, N.L., Masin, D., Mayer, L.D., Cullis, P.R. and Bally, M.B. (1994) Liposomal vincristine which exhibits increased drug retention and increased circulation longevity cures mice bearing P388 tumors. *Cancer Research*, **54**, 2830–3.

33. Allen, T.M., Mehra, T., Hansen, C. and Chin, Y.C. (1992) Stealth liposomes: an improved sustained release system for 1-beta-D-arabinofuranosylcytosine. *Cancer Research*, **52**, 2431–9.

34. Rose, P.G. (2005) Pegylated liposomal doxorubicin: optimizing the dosing schedule in ovarian cancer. *Oncologist*, **10**, 205–14.

35. Allen, T.M., Hansen, C.B. and de Menezes, D.E.L. (1995) Pharmacokinetics of long-circulating liposomes. *Advanced Drug Delivery Reviews*, **16**, 267–84.

36. Simoes, S., Moreira, J.N., Fonseca, C., Duzgunes, N. and de Lima, M.C. (2004) On the formulation of pH-sensitive liposomes with long circulation times. *Advanced Drug Delivery Reviews*, **56**, 947–65.

37. Roux, E., Passirani, C., Scheffold, S., Benoit, J.P. and Leroux, J.C. (2004) Serum-stable and long-circulating, PEGylated, pH-sensitive liposomes. *Journal of Controlled Release*, **94**, 447–51.

38. Guo, X. and Szoka, F.C., Jr (2001) Steric stabilization of fusogenic liposomes by a low-pH sensitive PEG-diortho ester-lipid conjugate. *Bioconjugate Chemistry*, **12**, 291–300.

39. Zhang, J.X., Zalipsky, S., Mullah, N., Pechar, M. and Allen, T.M. (2004) Pharmaco attributes of dioleoylphosphatidyletholamine/ cholesterylhemisuccinate liposomes containing different types of cleavable lipopolymers. *Pharmacological Research*, **49**, 185–98.

40. Leroux, J., Roux, E., Le Garrec, D., Hong, K. and Drummond, D.C. (2001) N-isopropylacrylamide copolymers for the preparation of pH-sensitive liposomes and polymeric micelles. *Journal of Controlled Release*, **72**, 71–84.

41. Roux, E., Stomp, R., Giasson, S., Pezolet, M., Moreau, P. and Leroux, J.C. (2002) Steric stabilization of liposomes by pH-responsive N-isopropylacrylamide copolymer. *Journal of Pharmaceutical Sciences*, **91**, 1795–802.

42. Lee, E.S., Na, K. and Bae, Y.H. (2003) Polymeric micelle for tumor pH and folate-mediated targeting. *Journal of Controlled Release*, **91**, 103–13.

43. Lee, E.S., Shin, H.J., Na, K. and Bae, Y.H. (2003) Poly(L-histidine)-PEG block copolymer micelles and pH-induced destabilization. *Journal of Controlled Release*, **90**, 363–74.

44. Jones, M.C., Ranger, M. and Leroux, J.C. (2003) pH-sensitive unimolecular polymeric micelles: synthesis of a novel drug carrier. *Bioconjugate Chemistry*, **14**, 774–81.

45. Suzawa, T., Nagamura, S., Saito, H., Ohta, S., Hanai, N., Kanazawa, J., Okabe, M. and Yamasaki, M. (2002) Enhanced tumor cell selectivity of adriamycin-monoclonal antibody conjugate via a poly(ethylene glycol)-based cleavable linker. *Journal of Controlled Release*, **79**, 229–42.

46. Potineni, A., Lynn, D.M., Langer, R. and Amiji, M.M. (2003) Poly(ethylene oxide)-modified poly(beta-amino ester) nanoparticles as a pH-sensitive biodegradable system for paclitaxel delivery. *Journal of Controlled Release*, **86**, 223–34.

47. Yoo, H.S., Lee, E.A. and Park, T.G. (2002) Doxorubicin-conjugated biodegradable polymeric micelles having acid-cleavable linkages. *Journal of Controlled Release*, **82**, 17–27.

48. Cheung, C.Y., Murthy, N., Stayton, P.S. and Hoffman, A.S. (2001) A pH-sensitive polymer that enhances cationic lipid-mediated gene transfer. *Bioconjugate Chemistry*, **12**, 906–10.

49. Venugopalan, P., Jain, S., Sankar, S., Singh, P., Rawat, A. and Vyas, S.P. (2002) pH-sensitive liposomes: mechanism of triggered release to drug and gene delivery prospects. *Pharmazie*, **57**, 659–71.

50 Kale, A.A. and Torchilin, V.P. (2007) "Smart" drug carriers: PEGylated TATp-modified pH-sensitive liposomes. *Journal of Liposome Research*, **17**, 197–203.

51 Kale, A.A. and Torchilin, V.P. (2007) Enhanced transfection of tumor cells in vivo using "Smart" pH-sensitive TAT-modified pegylated liposomes. *Journal of Drug Targeting*, **15**, 538–45.

52 Xu, H., Deng, Y., Chen, D., Hong, W., Lu, Y. and Dong, X. (2008) Esterase-catalyzed dePEGylation of pH-sensitive vesicles modified with cleavable PEG-lipid derivatives. *Journal of Controlled Release*, **130**, 238–45.

53 Gref, R., Minamitake, Y., Peracchia, M.T., Trubetskoy, V., Torchilin, V. and Langer, R. (1994) Biodegradable long-circulating polymeric nanospheres. *Science*, **263**, 1600–3.

54 Harper, G.R., Davies, M.C., Davis, S.S., Tadros, T.F., Taylor, D.C., Irving, M.P. and Waters, J.A. (1991) Steric stabilization of microspheres with grafted polyethylene oxide reduces phagocytosis by rat Kupffer cells in vitro. *Biomaterials*, **12**, 695–700.

55 Shan, X., Yuan, Y., Liu, C., Tao, X., Sheng, Y. and Xu, F. (2009) Influence of PEG chain on the complement activation suppression and longevity in vivo prolongation of the PCL biomedical nanoparticles. *Biomedical Microdevices*, **11**, 1187–94.

56 Gindy, M.E., Ji, S., Hoye, T.R., Panagiotopoulos, A.Z. and Prud'homme, R.K. (2008) Preparation of poly(ethylene glycol) protected nanoparticles with variable bioconjugate ligand density. *Biomacromolecules*, **9**, 2705–11.

57 Shan, X., Liu, C., Yuan, Y., Xu, F., Tao, X., Sheng, Y. and Zhou, H. (2009) In vitro macrophage uptake and in vivo biodistribution of long-circulation nanoparticles with poly(ethylene-glycol)-modified PLA (BAB type) triblock copolymer. *Colloids and Surfaces B, Biointerfaces*, **72**, 303–11.

58 Betancourt, T., Byrne, J.D., Sunaryo, N., Crowder, S.W., Kadapakkam, M., Patel, S., Casciato, S. and Brannon-Peppas, L. (2009) PEGylation strategies for active targeting of PLA/PLGA nanoparticles. *Journal of Biomedical Materials Research. Part A: Early View*, **91**, 263–76.

59 Cu, Y. and Saltzman, W.M. (2009) Controlled surface modification with poly(ethylene)glycol enhances diffusion of PLGA nanoparticles in human cervical mucus. *Molecular Pharmaceutics*, **6**, 173–81.

60 Chan, J.M., Zhang, L., Yuet, K.P., Liao, G., Rhee, J.W., Langer, R. and Farokhzad, O.C. (2009) PLGA-lecithin-PEG core-shell nanoparticles for controlled drug delivery. *Biomaterials*, **30**, 1627–34.

61 Illum, L., Church, A.E., Butterworth, M.D., Arien, A., Whetstone, J. and Davis, S.S. (2001) Development of systems for targeting the regional lymph nodes for diagnostic imaging: in vivo behaviour of colloidal PEG-coated magnetite nanospheres in the rat following interstitial administration. *Pharmaceutical Research*, **18**, 640–5.

62 Gupta, A.K. and Curtis, A.S. (2004) Surface modified superparamagnetic nanoparticles for drug delivery: interaction studies with human fibroblasts in culture. *Journal of Materials Science: Materials in Medicine*, **15**, 493–6.

63 Gou, M.L., Qian, Z.Y., Wang, H., Tang, Y.B., Huang, M.J., Kan, B., Wen, Y.J., Dai, M., Li, X.Y., Gong, C.Y. and Tu, M.J. (2008) Preparation and characterization of magnetic poly(epsilon-caprolactone)-poly(ethylene glycol)-poly(epsilon-caprolactone) microspheres. *Journal of Materials Science: Materials in Medicine*, **19**, 1033–41.

64 Zhang, F., Kang, E.T., Neoh, K.G. and Huang, W. (2001) Modification of gold surface by grafting of poly(ethylene glycol) for reduction in protein adsorption and platelet adhesion. *Journal of Biomaterials Science. Polymer Edition*, **12**, 515–31.

65 Kah, J.C., Wong, K.Y., Neoh, K.G., Song, J.H., Fu, J.W., Mhaisalkar, S., Olivo, M. and Sheppard, C.J. (2009) Critical parameters in the pegylation of gold nanoshells for biomedical applications: an in vitro macrophage study. *Journal of Drug Targeting*, **17**, 181–93.

66 Niidome, T., Yamagata, M., Okamoto, Y., Akiyama, Y., Takahashi, H., Kawano, T., Katayama, Y. and Niidome, Y. (2006) PEG-modified gold nanorods with a stealth character for in vivo applications. *Journal of Controlled Release*, **114**, 343–7.

67 Boisselier, E., Salmon, L., Ruiz, J. and Astruc, D. (2008) How to very efficiently functionalize gold nanoparticles by "click" chemistry. *Chemical Communications*, **44**, 5788–90.

68 Maus, L., Spatz, J.P. and Fiammengo, R. (2009) Quantification and reactivity of functional groups in the ligand shell of PEGylated gold nanoparticles via a fluorescence-based assay. *Langmuir*, **25**, 7910–17.

69 von Maltzahn, G., Park, J.H., Agrawal, A., Bandaru, N.K., Das, S.K., Sailor, M.J. and Bhatia, S.N. (2009) Computationally guided photothermal tumor therapy using long-circulating gold nanorod antennas. *Cancer Research*, **69**, 3892–900.

70 Gu, Y.J., Cheng, J., Lin, C.C., Lam, Y.W., Cheng, S.H. and Wong, W.T. (2009) Nuclear penetration of surface functionalized gold nanoparticles. *Toxicology and Applied Pharmacology*, **237**, 196–204.

71 Kim, D., Park, S., Lee, J.H., Jeong, Y.Y. and Jon, S. (2007) Antibiofouling polymer-coated gold nanoparticles as a contrast agent for in vivo X-ray computed tomography imaging. *Journal of the American Chemical Society*, **129**, 7661–5.

72 Lee, H. and Larson, R.G. (2009) Molecular dynamics study of the structure and interparticle interactions of polyethylene glycol-conjugated PAMAM dendrimers. *Journal of Physical Chemistry B*, **113**, 13202–7.

73 Kaminskas, L.M., Kelly, B.D., McLeod, V.M., Boyd, B.J., Krippner, G.Y., Williams, E.D. and Porter, C.J. (2009) Pharmacokinetics and tumor disposition of PEGylated, methotrexate conjugated poly-l-lysine dendrimers. *Molecular Pharmaceutics*, **6**, 1190–204.

74 Kojima, C., Regino, C., Umeda, Y., Kobayashi, H. and Kono, K. (2010) Influence of dendrimer generation and polyethylene glycol length on the biodistribution of PEGylated dendrimers. *International Journal of Pharmaceutics*, **383**, 293–6.

75 Kaminskas, L.M., Wu, Z., Barlow, N., Krippner, G.Y., Boyd, B.J. and Porter, C.J. (2009) Partly-PEGylated poly-L-lysine dendrimers have reduced plasma stability and circulation times compared with fully PEGylated dendrimers. *Journal of Pharmaceutical Sciences*, **98**, 3871–5.

76 Wang, W., Xiong, W., Wan, J., Sun, X., Xu, H. and Yang, X. (2009) The decrease of PAMAM dendrimer-induced cytotoxicity by PEGylation via attenuation of oxidative stress. *Nanotechnology*, **20**, 105103.

77 Lopez, A.I., Reins, R.Y., McDermott, A.M., Trautner, B.W. and Cai, C. (2009) Antibacterial activity and cytotoxicity of PEGylated poly(amidoamine) dendrimers. *Molecular BioSystems*, **5**, 1148–56.

78 Liu, Z., Tabakman, S.M., Chen, Z. and Dai, H. (2009) Preparation of carbon nanotube bioconjugates for biomedical applications. *Nature Protocols*, **4**, 1372–82.

79 Murray, A.R., Kisin, E., Leonard, S.S., Young, S.H., Kommineni, C., Kagan, V.E., Castranova, V. and Shvedova, A.A. (2009) Oxidative stress and inflammatory response in dermal toxicity of single-walled carbon nanotubes. *Toxicology*, **257**, 161–71.

80 Cheng, J., Fernando, K.A., Veca, L.M., Sun, Y.P., Lamond, A.I., Lam, Y.W. and Cheng, S.H. (2008) Reversible accumulation of PEGylated single-walled carbon nanotubes in the mammalian nucleus. *ACS Nano*, **2**, 2085–94.

81 Murakami, T., Sawada, H., Tamura, G., Yudasaka, M., Iijima, S. and Tsuchida, K. (2008) Water-dispersed single-wall carbon nanohorns as drug carriers for local cancer chemotherapy. *Nanomedicine*, **3**, 453–63.

82 Delogu, L.G., Magrini, A., Bergamaschi, A., Rosato, N., Dawson, M.I., Bottini, N. and Bottini, M. (2009) Conjugation of antisense oligonucleotides to PEGylated carbon nanotubes enables efficient knockdown of PTPN22 in T lymphocytes. *Bioconjugate Chemistry*, **20**, 427–31.

83 Schipper, M.L., Iyer, G., Koh, A.L., Cheng, Z., Ebenstein, Y., Aharoni, A.,

Keren, S., Bentolila, L.A., Li, J., Rao, J., Chen, X., Banin, U., Wu, A.M., Sinclair, R., Weiss, S. and Gambhir, S.S. (2009) Particle size, surface coating, and PEGylation influence the biodistribution of quantum dots in living mice. *Small*, **5**, 126–34.

84 Susumu, K., Mei, B.C. and Mattoussi, H. (2009) Multifunctional ligands based on dihydrolipoic acid and polyethylene glycol to promote biocompatibility of quantum dots. *Nature Protocols*, **4**, 424–36.

85 Warnement, M.R., Tomlinson, I.D., Chang, J.C., Schreuder, M.A., Luckabaugh, C.M. and Rosenthal, S.J. (2008) Controlling the reactivity of amphiphilic quantum dots in biological assays through hydrophobic assembly of custom PEG derivatives. *Bioconjugate Chemistry*, **19**, 1404–13.

86 Daou, T.J., Li, L., Reiss, P., Josserand, V. and Texier, I. (2009) Effect of poly(ethylene glycol) length on the in vivo behavior of coated quantum dots. *Langmuir*, **25**, 3040–4.

87 Blume, G., Cevc, G., Crommelin, M.D., Bakker-Woudenberg, I.A., Kluft, C. and Storm, G. (1993) Specific targeting with poly(ethylene glycol)-modified liposomes: coupling of homing devices to the ends of the polymeric chains combines effective target binding with long circulation times. *Biochimica et Biophysica Acta*, **1149**, 180–4.

88 Torchilin, V.P., Levchenko, T.S., Lukyanov, A.N., Khaw, B.A., Klibanov, A.L., Rammohan, R., Samokhin, G.P. and Whiteman, K.R. (2001) p-Nitrophenylcarbonyl-PEG-PE-liposomes: fast and simple attachment of specific ligands, including monoclonal antibodies, to distal ends of PEG chains via p-nitrophenylcarbonyl groups. *Biochimica et Biophysica Acta*, **1511**, 397–411.

89 Zalipsky, S., Gittelman, J., Mullah, N., Qazen, M.M. and Harding, J.A. (1998) Biologically active ligand-bearing polymer-grafted liposomes, in *Targeting of Drugs 6: Strategies for Stealth Therapeutic Systems* (ed. G. Gregoriadis), Plenum Press, New York, pp. 131–9.

90 Torchilin, V.P., Lukyanov, A.N., Gao, Z. and Papahadjopoulos-Sternberg, B. (2003) Immunomicelles: targeted pharmaceutical carriers for poorly soluble drugs. *Proceedings of the National Academy of Sciences of the United States of America*, **100**, 6039–44.

91 Klibanov, A.L. (1998) Antibody-mediated targeting of PEG-coated liposomes, in *Long Circulating Liposomes : Old Drugs, New Therapeutics* (eds M.C. Woodle and G. Storm), Springer, Berlin, p. 269.

92 Park, J.W., Kirpotin, D.B., Hong, K., Shalaby, R., Shao, Y., Nielsen, U.B., Marks, J.D., Papahadjopoulos, D. and Benz, C.C. (2001) Tumor targeting using anti-her2 immunoliposomes. *Journal of Controlled Release*, **74**, 95–113.

93 Benhar, I., Padlan, E.A., Jung, S.H., Lee, B. and Pastan, I. (1994) Rapid humanization of the Fv of monoclonal antibody B3 by using framework exchange of the recombinant immunotoxin B3(Fv)-PE38. *Proceedings of the National Academy of Sciences of the United States of America*, **91**, 12051–5.

94 Zalipsky, S., Mullah, N., Harding, J.A., Gittelman, J., Guo, L. and DeFrees, S.A. (1997) Poly(ethylene glycol)-grafted liposomes with oligopeptide or oligosaccharide ligands appended to the termini of the polymer chains. *Bioconjugate Chemistry*, **8**, 111–18.

95 Wong, J.Y., Kuhl, T.L., Israelachvili, J.N., Mullah, N. and Zalipsky, S. (1997) Direct measurement of a tethered ligand-receptor interaction potential. *Science*, **275**, 820–2.

96 Brannon-Peppas, L. and Blanchette, J.O. (2004) Nanoparticle and targeted systems for cancer therapy. *Advanced Drug Delivery Reviews*, **56**, 1649–59.

97 Gao, J., Zhong, W., He, J., Li, H., Zhang, H., Zhou, G., Li, B., Lu, Y., Zou, H., Kou, G., Zhang, D., Wang, H., Guo, Y. and Zhong, Y. (2009) Tumor-targeted PE38KDEL delivery via PEGylated anti-HER2 immunoliposomes. *International Journal of Pharmaceutics*, **374**, 145–52.

98 Kamps, J.A., Koning, G.A., Velinova, M.J., Morselt, H.W., Wilkens, M., Gorter, A., Donga, J. and Scherphof, G.L. (2000) Uptake of long-circulating

immunoliposomes, directed against colon adenocarcinoma cells, by liver metastases of colon cancer. *Journal of Drug Targeting*, **8**, 235–45.

99 Lukyanov, A.N., Elbayoumi, T.A., Chakilam, A.R. and Torchilin, V.P. (2004) Tumor-targeted liposomes: doxorubicin-loaded long-circulating liposomes modified with anti-cancer antibody. *Journal of Controlled Release*, **100**, 135–44.

100 ElBayoumi, T.A. and Torchilin, V.P. (2009) Tumor-targeted nanomedicines: enhanced antitumor efficacy in vivo of doxorubicin-loaded, long-circulating liposomes modified with cancer-specific monoclonal antibody. *Clinical Cancer Research*, **15**, 1973–80.

101 Gupta, B. and Torchilin, V.P. (2007) Monoclonal antibody 2C5-modified doxorubicin-loaded liposomes with significantly enhanced therapeutic activity against intracranial human brain U-87 MG tumor xenografts in nude mice. *Cancer Immunology, Immunotherapy: CII*, **56**, 1215–23.

102 Roby, A., Erdogan, S. and Torchilin, V.P. (2007) Enhanced in vivo antitumor efficacy of poorly soluble PDT agent, meso-tetraphenylporphine, in PEG-PE-based tumor-targeted immunomicelles. *Cancer Biology and Therapy*, **6**, 1136–42.

103 Eck, W., Craig, G., Sigdel, A., Ritter, G., Old, L.J., Tang, L., Brennan, M.F., Allen, P.J. and Mason, M.D. (2008) PEGylated gold nanoparticles conjugated to monoclonal F19 antibodies as targeted labeling agents for human pancreatic carcinoma tissue. *ACS Nano*, **2**, 2263–72.

104 Cato, M.H., D'Annibale, F., Mills, D.M., Cerignoli, F., Dawson, M.I., Bergamaschi, E., Bottini, N., Magrini, A., Bergamaschi, A., Rosato, N., Rickert, R.C., Mustelin, T. and Bottini, M. (2008) Cell-type specific and cytoplasmic targeting of PEGylated carbon nanotube-based nanoassemblies. *Journal of Nanoscience and Nanotechnology*, **8**, 2259–69.

105 Hatakeyama, H., Akita, H., Maruyama, K., Suhara, T. and Harashima, H. (2004) Factors governing the in vivo tissue uptake of transferrin-coupled polyethylene glycol liposomes in vivo. *International Journal of Pharmaceutics*, **281**, 25–33.

106 Li, Y., Ogris, M., Wagner, E., Pelisek, J. and Ruffer, M. (2003) Nanoparticles bearing polyethyleneglycol-coupled transferrin as gene carriers: preparation and in vitro evaluation. *International Journal of Pharmaceutics*, **259**, 93–101.

107 Ishida, O., Maruyama, K., Tanahashi, H., Iwatsuru, M., Sasaki, K., Eriguchi, M. and Yanagie, H. (2001) Liposomes bearing polyethyleneglycol-coupled transferrin with intracellular targeting property to the solid tumors in vivo. *Pharmaceutical Research*, **18**, 1042–8.

108 Olivier, J.C., Huertas, R., Lee, H.J., Calon, F. and Pardridge, W.M. (2002) Synthesis of pegylated immunonanoparticles. *Pharmaceutical Research*, **19**, 1137–43.

109 Derycke, A.S. and De Witte, P.A. (2002) Transferrin-mediated targeting of hypericin embedded in sterically stabilized PEG-liposomes. *International Journal of Oncology*, **20**, 181–7.

110 Gijsens, A., Derycke, A., Missiaen, L., De Vos, D., Huwyler, J., Eberle, A. and de Witte, P. (2002) Targeting of the photocytotoxic compound AlPcS4 to HeLa cells by transferrin conjugated PEG-liposomes. *International Journal of Cancer*, **101**, 78–85.

111 Iinuma, H., Maruyama, K., Okinaga, K., Sasaki, K., Sekine, T., Ishida, O., Ogiwara, N., Johkura, K. and Yonemura, Y. (2002) Intracellular targeting therapy of cisplatin-encapsulated transferrin-polyethylene glycol liposome on peritoneal dissemination of gastric cancer. *International Journal of Cancer*, **99**, 130–7.

112 Li, X., Ding, L., Xu, Y., Wang, Y. and Ping, Q. (2009) Targeted delivery of doxorubicin using stealth liposomes modified with transferrin. *International Journal of Pharmaceutics*, **373**, 116–23.

113 Doi, A., Kawabata, S., Iida, K., Yokoyama, K., Kajimoto, Y., Kuroiwa, T., Shirakawa, T., Kirihata, M., Kasaoka, S., Maruyama, K., Kumada, H., Sakurai, Y., Masunaga, S., Ono, K. and Miyatake, S. (2008) Tumor-specific targeting of sodium borocaptate (BSH) to malignant glioma by transferrin-PEG liposomes: a

modality for boron neutron capture therapy. *Journal of Neuro-Oncology*, **87**, 287–94.

114 Omori, N., Maruyama, K., Jin, G., Li, F., Wang, S.J., Hamakawa, Y., Sato, K., Nagano, I., Shoji, M. and Abe, K. (2003) Targeting of post-ischemic cerebral endothelium in rat by liposomes bearing polyethylene glycol-coupled transferrin. *Neurological Research*, **25**, 275–9.

115 Leamon, C.P. and Low, P.S. (1991) Delivery of macromolecules into living cells: a method that exploits folate receptor endocytosis. *Proceedings of the National Academy of Sciences of the United States of America*, **88**, 5572–6.

116 Lee, R.J. and Low, P.S. (1994) Delivery of liposomes into cultured KB cells via folate receptor-mediated endocytosis. *Journal of Biological Chemistry*, **269**, 3198–204.

117 Lu, Y. and Low, P.S. (2002) Folate-mediated delivery of macromolecular anticancer therapeutic agents. *Advanced Drug Delivery Reviews*, **54**, 675–93.

118 Gabizon, A., Shmeeda, H., Horowitz, A.T. and Zalipsky, S. (2004) Tumor cell targeting of liposome-entrapped drugs with phospholipid-anchored folic acid-PEG conjugates. *Advanced Drug Delivery Reviews*, **56**, 1177–92.

119 Stella, B., Arpicco, S., Peracchia, M.T., Desmaele, D., Hoebeke, J., Renoir, M., D'Angelo, J., Cattel, L. and Couvreur, P. (2000) Design of folic acid-conjugated nanoparticles for drug targeting. *Journal of Pharmaceutical Sciences*, **89**, 1452–64.

120 Park, E.K., Lee, S.B. and Lee, Y.M. (2005) Preparation and characterization of methoxy poly(ethylene glycol)/poly(epsilon-caprolactone) amphiphilic block copolymeric nanospheres for tumor-specific folate-mediated targeting of anticancer drugs. *Biomaterials*, **26**, 1053–61.

121 Zhang, Y., Kohler, N. and Zhang, M. (2002) Surface modification of superparamagnetic magnetite nanoparticles and their intracellular uptake. *Biomaterials*, **23**, 1553–61.

122 Choi, H., Choi, S.R., Zhou, R., Kung, H.F. and Chen, I.W. (2004) Iron oxide nanoparticles as magnetic resonance contrast agent for tumor imaging via folate receptor-targeted delivery. *Academic Radiology*, **11**, 996–1004.

123 Liang, B., He, M.L., Xiao, Z.P., Li, Y., Chan, C.Y., Kung, H.F., Shuai, X.T. and Peng, Y. (2008) Synthesis and characterization of folate-PEG-grafted-hyperbranched-PEI for tumor-targeted gene delivery. *Biochemical and Biophysical Research Communications*, **367**, 874–80.

124 Esmaeili, F., Ghahremani, M.H., Ostad, S.N., Atyabi, F., Seyedabadi, M., Malekshahi, M.R., Amini, M. and Dinarvand, R. (2008) Folate-receptor-targeted delivery of docetaxel nanoparticles prepared by PLGA-PEG-folate conjugate. *Journal of Drug Targeting*, **16**, 415–23.

125 Singh, P., Gupta, U., Asthana, A. and Jain, N.K. (2008) Folate and folate-PEG-PAMAM dendrimers: synthesis, characterization, and targeted anticancer drug delivery potential in tumor bearing mice. *Bioconjugate Chemistry*, **19**, 2239–52.

126 Dixit, V., Van den Bossche, J., Sherman, D.M., Thompson, D.H. and Andres, R.P. (2006) Synthesis and grafting of thioctic acid-PEG-folate conjugates onto Au nanoparticles for selective targeting of folate receptor-positive tumor cells. *Bioconjugate Chemistry*, **17**, 603–9.

127 Song, E.Q., Zhang, Z.L., Luo, Q.Y., Lu, W., Shi, Y.B. and Pang, D.W. (2009) Tumor cell targeting using folate-conjugated fluorescent quantum dots and receptor-mediated endocytosis. *Clinical Chemistry*, **55**, 955–63.

128 Vaupel, P., Kallinowski, F. and Okunieff, P. (1989) Blood flow, oxygen and nutrient supply, and metabolic microenvironment of human tumors: a review. *Cancer Research*, **49**, 6449–65.

129 Wike-Hooley, J.L., Haveman, J. and Reinhold, H.S. (1984) The relevance of tumour pH to the treatment of malignant disease. *Radiotherapy and Oncology: Journal of the European Society for Therapeutic Radiology and Oncology*, **2**, 343–66.

130 Gerweck, L.E. and Seetharaman, K. (1996) Cellular pH gradient in tumor versus normal tissue: potential exploitation for the treatment of cancer. *Cancer Research*, **56**, 1194–8.

131 Saito, G., Swanson, J.A. and Lee, K.D. (2003) Drug delivery strategy utilizing conjugation via reversible disulfide linkages: role and site of cellular reducing activities. *Advanced Drug Delivery Reviews*, **55**, 199–215.

132 Litzinger, D.C. and Huang, L. (1992) Phosphatidylethanolamine liposomes: drug delivery, gene transfer and immunodiagnostic applications. *Biochimica et Biophysica Acta*, **1113**, 201–27.

133 Torchilin, V.P., Zhou, F. and Huang, L. (1993) pH-sensitive liposomes. *Journal of Liposome Research*, **3**, 201–55.

134 Yessine, M.A. and Leroux, J.C. (2004) Membrane-destabilizing polyanions: interaction with lipid bilayers and endosomal escape of biomacromolecules. *Advanced Drug Delivery Reviews*, **56**, 999–1021.

135 Yessine, M.A., Lafleur, M., Meier, C., Petereit, H.U. and Leroux, J.C. (2003) Characterization of the membrane-destabilizing properties of different pH-sensitive methacrylic acid copolymers. *Biochimica et Biophysica Acta*, **1613**, 28–38.

136 Geisert, E.E., Jr, Del Mar, N.A., Owens, J.L. and Holmberg, E.G. (1995) Transfecting neurons and glia in the rat using pH-sensitive immunoliposomes. *Neuroscience Letters*, **184**, 40–3.

137 Xu, L., Huang, C.C., Huang, W., Tang, W.H., Rait, A., Yin, Y.Z., Cruz, I., Xiang, L.M., Pirollo, K.F. and Chang, E.H. (2002) Systemic tumor-targeted gene delivery by anti-transferrin receptor scFv-immunoliposomes. *Molecular Cancer and Therapeutics*, **1**, 337–46.

138 Boussif, O., Lezoualc'h, F., Zanta, M.A., Mergny, M.D., Scherman, D., Demeneix, B. and Behr, J.P. (1995) A versatile vector for gene and oligonucleotide transfer into cells in culture and in vivo: polyethylenimine. *Proceedings of the National Academy of Sciences of the United States of America*, **92**, 7297–301.

139 Branden, L.J., Mohamed, A.J. and Smith, C.I. (1999) A peptide nucleic acid-nuclear localization signal fusion that mediates nuclear transport of DNA. *Nature Biotechnology*, **17**, 784–7.

140 Bae, Y., Nishiyama, N., Fukushima, S., Koyama, H., Yasuhiro, M. and Kataoka, K. (2005) Preparation and biological characterization of polymeric micelle drug carriers with intracellular pH-triggered drug release property: tumor permeability, controlled subcellular drug distribution, and enhanced in vivo antitumor efficacy. *Bioconjugate Chemistry*, **16**, 122–30.

141 Jeong, J.H., Kim, S.W. and Park, T.G. (2003) Novel intracellular delivery system of antisense oligonucleotide by self-assembled hybrid micelles composed of DNA/PEG conjugate and cationic fusogenic peptide. *Bioconjugate Chemistry*, **14**, 473–9.

142 Lukyanov, A.N., Hartner, W.C. and Torchilin, V.P. (2004) Increased accumulation of PEG-PE micelles in the area of experimental myocardial infarction in rabbits. *Journal of Controlled Release*, **94**, 187–93.

143 Wang, J., Mongayt, D. and Torchilin, V.P. (2005) Polymeric micelles for delivery of poorly soluble drugs: preparation and anticancer activity in vitro of paclitaxel incorporated into mixed micelles based on poly(ethylene glycol)-lipid conjugate and positively charged lipids. *Journal of Drug Targeting*, **13**, 73–80.

144 Lee, E.S., Na, K. and Bae, Y.H. (2005) Super pH-sensitive multifunctional polymeric micelle. *Nano Letters*, **5**, 325–9.

145 Shim, W.S., Kim, S.W., Choi, E.K., Park, H.J., Kim, J.S. and Lee, D.S. (2006) Novel pH sensitive block copolymer micelles for solvent free drug loading. *Macromolecular Bioscience*, **6**, 179–86.

146 Roux, E., Francis, M., Winnik, F.M. and Leroux, J.C. (2002) Polymer based pH-sensitive carriers as a means to improve the cytoplasmic delivery of drugs. *International Journal of Pharmaceutics*, **242**, 25–36.

147 Turk, M.J., Reddy, J.A., Chmielewski, J.A. and Low, P.S. (2002) Characterization of a novel pH-sensitive peptide that enhances drug release from folate-targeted liposomes at endosomal pHs. *Biochimica et Biophysica Acta*, **1559**, 56–68.

148 Shi, G., Guo, W., Stephenson, S.M. and Lee, R.J. (2002) Efficient intracellular drug and gene delivery using folate receptor-targeted pH-sensitive liposomes composed of cationic/anionic lipid combinations. *Journal of Controlled Release*, **80**, 309–19.

149 Gillies, E.R., Jonsson, T.B. and Fréchet, J.M.J. (2004) Stimuli-responsive supramolecular assemblies of linear-dendritic copolymers. *Journal of the American Chemical Society*, **126**, 11936–43.

150 Meyer, D.E., Shin, B.C., Kong, G.A., Dewhirst, M.W. and Chilkoti, A. (2001) Drug targeting using thermally responsive polymers and local hyperthermia. *Journal of Controlled Release*, **74**, 213–24.

151 Ponce, A.M., Vujaskovic, Z., Yuan, F., Needham, D. and Dewhirst, M.W. (2006) Hyperthermia-mediated liposomal drug delivery. *International Journal of Hyperthermia*, **22**, 205–13.

152 Yatvin, M.B., Weinstein, J.N., Dennis, W.H. and Blumenthal, R. (1978) Design of liposomes for enhanced local release of drugs by hyperthermia. *Science*, **202**, 1290–3.

153 Kono, K. (2001) Thermosensitive polymer-modified liposomes. *Advanced Drug Delivery Reviews*, **53**, 307–19.

154 Schild, H.G. (1992) Poly(N-isopropylacrylamide): experiment, theory and application. *Progress in Polymer Science*, **17**, 163–249.

155 Kono, K., Nakai, R., Morimoto, K. and Takagishi, T. (1999) Thermosensitive polymer-modified liposomes that release contents around physiological temperature. *Biochimica et Biophysica Acta*, **1416**, 239–50.

156 Kono, K., Yoshino, K. and Takagishi, T. (2002) Effect of poly(ethylene glycol) grafts on temperature-sensitivity of thermosensitive polymer-modified liposomes. *Journal of Controlled Release*, **80**, 321–32.

157 Yoshida, R., Uchida, K., Kaneko, Y., Sakai, K., Kikuchi, A., Sakurai, Y. and Okano, T. (1995) Comb-type grafted hydrogels with rapid deswelling response to temperature changes. *Nature*, **374**, 240–2.

158 Sukhorukov, G.B., Rogach, A.L., Garstka, M., Springer, S., Parak, W.J., Munoz-Javier, A., Kreft, O., Skirtach, A.G., Susha, A.S., Ramaye, Y., Palankar, R. and Winterhalter, M. (2007) Multifunctionalized polymer microcapsules: novel tools for biological and pharmacological applications. *Small*, **3**, 944–55.

159 Gupta, A.K., Naregalkar, R.R., Vaidya, V.D. and Gupta, M. (2007) Recent advances on surface engineering of magnetic iron oxide nanoparticles and their biomedical applications. *Nanomedicine*, **2**, 23–39.

160 Gupta, A.K. and Gupta, M. (2005) Synthesis and surface engineering of iron oxide nanoparticles for biomedical applications. *Biomaterials*, **26**, 3995–4021.

161 Ito, A., Shinkai, M., Honda, H. and Kobayashi, T. (2005) Medical application of functionalized magnetic nanoparticles. *Journal of Bioscience and Bioengineering*, **100**, 1–11.

162 Shinkai, M. and Ito, A. (2004) Functional magnetic particles for medical application. *Advances in Biochemical Engineering/Biotechnology*, **91**, 191–220.

163 Wagner, S., Schnorr, J., Pilgrimm, H., Hamm, B. and Taupitz, M. (2002) Monomer-coated very small superparamagnetic iron oxide particles as contrast medium for magnetic resonance imaging: preclinical in vivo characterization. *Investigative Radiology*, **37**, 167–77.

164 Anzai, Y. and Prince, M.R. (1997) Iron oxide-enhanced MR lymphography: the evaluation of cervical lymph node metastases in head and neck cancer. *Journal of Magnetic Resonance Imaging*, **7**, 75–81.

165 Ngaboni Okassa, L., Marchais, H., Douziech-Eyrolles, L., Cohen-Jonathan, S., Souce, M., Dubois, P. and Chourpa, I. (2005) Development and characterization of sub-micron poly(D,L-lactide-*co*-glycolide) particles loaded with magnetite/maghemite nanoparticles. *International Journal of Pharmaceutics*, **302**, 187–96.

166 Widder, K.J., Senyel, A.E. and Scarpelli, G.D. (1978) Magnetic microspheres: a model system of site specific drug

delivery in vivo. *Proceedings of the Society for Experimental Biology and Medicine*, **158**, 141–6.
167 Gang, J., Park, S.B., Hyung, W., Choi, E.H., Wen, J., Kim, H.S., Shul, Y.G., Haam, S. and Song, S.Y. (2007) Magnetic poly epsilon-caprolactone nanoparticles containing Fe_3O_4 and gemcitabine enhance anti-tumor effect in pancreatic cancer xenograft mouse model. *Journal of Drug Targeting*, **15**, 445–53.
168 Alexiou, C., Arnold, W., Klein, R.J., Parak, F.G., Hulin, P., Bergemann, C., Erhardt, W., Wagenpfeil, S. and Lubbe, A.S. (2000) Locoregional cancer treatment with magnetic drug targeting. *Cancer Research*, **60**, 6641–8.
169 Alexiou, C., Jurgons, R., Schmid, R.J., Bergemann, C., Henke, J., Erhardt, W., Huenges, E. and Parak, F. (2003) Magnetic drug targeting–biodistribution of the magnetic carrier and the chemotherapeutic agent mitoxantrone after locoregional cancer treatment. *Journal of Drug Targeting*, **11**, 139–49.
170 Dandamudi, S. and Campbell, R.B. (2007) The drug loading, cytotoxicity and tumor vascular targeting characteristics of magnetite in magnetic drug targeting. *Biomaterials*, **28**, 4673–83.
171 Cinteza, L.O., Ohulchanskyy, T.Y., Sahoo, Y., Bergey, E.J., Pandey, R.K. and Prasad, P.N. (2006) Diacyllipid micelle-based nanocarrier for magnetically guided delivery of drugs in photodynamic therapy. *Molecular Pharmaceutics*, **3**, 415–23.
172 Johannsen, M., Gneveckow, U., Eckelt, L., Feussner, A., Waldofner, N., Scholz, R., Deger, S., Wust, P., Loening, S.A. and Jordan, A. (2005) Clinical hyperthermia of prostate cancer using magnetic nanoparticles: presentation of a new interstitial technique. *International Journal of Hyperthermia*, **21**, 637–47.
173 Wust, P., Gneveckow, U., Johannsen, M., Bohmer, D., Henkel, T., Kahmann, F., Sehouli, J., Felix, R., Ricke, J. and Jordan, A. (2006) Magnetic nanoparticles for interstitial thermotherapy–feasibility, tolerance and achieved temperatures. *International Journal of Hyperthermia*, **22**, 673–85.
174 Unger, E.C., McCreery, T.P., Sweitzer, R.H., Caldwell, V.E. and Wu, Y. (1998) Acoustically active liposheres containing paclitaxel: a new therapeutic ultrasound contrast agent. *Investigative Radiology*, **33**, 886–92.
175 Unger, E.C., Porter, T., Culp, W., Labell, R., Matsunaga, T. and Zutshi, R. (2004) Therapeutic applications of lipid-coated microbubbles. *Advanced Drug Delivery Reviews*, **56**, 1291–314.
176 Liu, Y., Miyoshi, H. and Nakamura, M. (2006) Encapsulated ultrasound microbubbles: therapeutic application in drug/gene delivery. *Journal of Controlled Release*, **114**, 89–99.
177 Huang, S.L. and MacDonald, R.C. (2004) Acoustically active liposomes for drug encapsulation and ultrasound-triggered release. *Biochimica et Biophysica Acta*, **1665**, 134–41.
178 Tartis, M.S., McCallan, J., Lum, A.F., LaBell, R., Stieger, S.M., Matsunaga, T.O. and Ferrara, K.W. (2006) Therapeutic effects of paclitaxel-containing ultrasound contrast agents. *Ultrasound in Medicine and Biology*, **32**, 1771–80.
179 Rapoport, N., Gao, Z. and Kennedy, A. (2007) Multifunctional nanoparticles for combining ultrasonic tumor imaging and targeted chemotherapy. *Journal of the National Cancer Institute*, **99**, 1095–106.
180 Husseini, G.A., Diaz de la Rosa, M.A., Gabuji, T., Zeng, Y., Christensen, D.A. and Pitt, W.G. (2007) Release of doxorubicin from unstabilized and stabilized micelles under the action of ultrasound. *Journal of Nanoscience and Nanotechnology*, **7**, 1028–33.
181 Tiukinhoy-Laing, S.D., Buchanan, K., Parikh, D., Huang, S., MacDonald, R.C., McPherson, D.D. and Klegerman, M.E. (2007) Fibrin targeting of tissue plasminogen activator-loaded echogenic liposomes. *Journal of Drug Targeting*, **15**, 109–14.
182 Fang, J.Y., Hung, C.F., Liao, M.H. and Chien, C.C. (2007) A study of the formulation design of acoustically active liposheres as carriers for drug delivery. *European Journal of Pharmaceutics and Biopharmaceutics*, **67**, 67–75.

183 Schroeder, A., Avnir, Y., Weisman, S., Najajreh, Y., Gabizon, A., Talmon, Y., Kost, J. and Barenholz, Y. (2007) Controlling liposomal drug release with low frequency ultrasound: mechanism and feasibility. *Langmuir*, **23**, 4019–25.

184 Cavallaro, G., Campisi, M., Licciardi, M., Ogris, M. and Giammona, G. (2006) Reversibly stable thiopolyplexes for intracellular delivery of genes. *Journal of Controlled Release*, **115**, 322–34.

185 Kommareddy, S. and Amiji, M. (2005) Preparation and evaluation of thiol-modified gelatin nanoparticles for intracellular DNA delivery in response to glutathione. *Bioconjugate Chemistry*, **16**, 1423–32.

186 Kommareddy, S. and Amiji, M. (2007) Poly(ethylene glycol)-modified thiolated gelatin nanoparticles for glutathione-responsive intracellular DNA delivery. *Nanomedicine*, **3**, 32–42.

187 Carlisle, R.C., Etrych, T., Briggs, S.S., Preece, J.A., Ulbrich, K., Seymour, L.W. (2004) Polymer-coated polyethylenimine/DNA complexes designed for triggered activation by intracellular reduction. *Journal of Gene Medicine*, **6**, 337–44.

188 Kevin, R.W. and Otto, S. (2005) Reversible covalent chemistry in drug delivery. *Current Drug Discovery Technologies*, **2**, 123–60.

189 Kirpotin, D., Hong, K.L., Mullah, N., Papahadjopoulos, D. and Zalipsky, S. (1996) Liposomes with detachable polymer coating: destabilization and fusion of dioleoylphosphatidylethanolamine vesicles triggered by cleavage of surface-grafted poly(ethylene glycol). *FEBS Letters*, **388**, 115–18.

190 Xu, Y. and Szoka, F.C., Jr (1996) Mechanism of DNA release from cationic liposome/DNA complexes used in cell transfection. *Biochemistry*, **35**, 5616–23.

191 Wu, G.Y. and Wu, C.H. (1987) Receptor-mediated in vitro gene transformation by a soluble DNA carrier system. *Journal of Biological Chemistry*, **262**, 4429–32.

192 Elouahabi, A. and Ruysschaert, J.M. (2005) Formation and intracellular trafficking of lipoplexes and polyplexes. *Molecular Therapy*, **11**, 336–47.

193 Kunath, K., von Harpe, A., Fischer, D., Petersen, H., Bickel, U., Voigt, K. and Kissel, T. (2003) Low-molecular-weight polyethylenimine as a non-viral vector for DNA delivery: comparison of physicochemical properties, transfection efficiency and in vivo distribution with high-molecular-weight polyethylenimine. *Journal of Controlled Release*, **89**, 113–25.

194 Sakurai, F., Inoue, R., Nishino, Y., Okuda, A., Matsumoto, O., Taga, T., Yamashita, F., Takakura, Y. and Hashida, M. (2000) Effect of DNA/liposome mixing ratio on the physicochemical characteristics, cellular uptake and intracellular trafficking of plasmid DNA/cationic liposome complexes and subsequent gene expression. *Journal of Controlled Release*, **66**, 255–69.

195 Ogris, M., Steinlein, P., Carotta, S., Brunner, S. and Wagner, E. (2001) DNA/polyethylenimine transfection particles: influence of ligands, polymer size, and PEGylation on internalization and gene expression. *AAPS PharmSci*, **3**, E21.

196 Wadia, J.S., Stan, R.V. and Dowdy, S.F. (2004) Transducible TAT-HA fusogenic peptide enhances escape of TAT-fusion proteins after lipid raft macropinocytosis. *Nature Medicine*, **10**, 310–15.

197 Rothbard, J.B., Jessop, T.C., Lewis, R.S., Murray, B.A. and Wender, P.A. (2004) Role of membrane potential and hydrogen bonding in the mechanism of translocation of guanidinium-rich peptides into cells. *Journal of the American Chemical Society*, **126**, 9506–7.

198 Josephson, L., Tung, C.H., Moore, A. and Weissleder, R. (1999) High-efficiency intracellular magnetic labeling with novel superparamagnetic-Tat peptide conjugates. *Bioconjugate Chemistry*, **10**, 186–91.

199 Torchilin, V.P. (2008) Tat peptide-mediated intracellular delivery of pharmaceutical nanocarriers. *Advanced Drug Delivery Reviews*, **60**, 548–58.

200 Kaufman, C.L., Williams, M., Ryle, L.M., Smith, T.L., Tanner, M. and Ho, C. (2003) Superparamagnetic iron oxide particles transactivator protein-fluorescein isothiocyanate particle

labeling for in vivo magnetic resonance imaging detection of cell migration: uptake and durability. *Transplantation*, **76**, 1043–6.
201 Zhao, M., Kircher, M.F., Josephson, L. and Weissleder, R. (2002) Differential conjugation of tat peptide to superparamagnetic nanoparticles and its effect on cellular uptake. *Bioconjugate Chemistry*, **13**, 840–4.
202 Tseng, Y.L., Liu, J.J. and Hong, R.L. (2002) Translocation of liposomes into cancer cells by cell-penetrating peptides penetratin and tat: a kinetic and efficacy study. *Molecular Pharmacology*, **62**, 864–72.
203 Gorodetsky, R., Levdansky, L., Vexler, A., Shimeliovich, I., Kassis, I., Ben-Moshe, M., Magdassi, S. and Marx, G. (2004) Liposome transduction into cells enhanced by haptotactic peptides (Haptides) homologous to fibrinogen C-termini. *Journal of Controlled Release*, **95**, 477–88.
204 Levchenko, T.S., Rammohan, R., Volodina, N. and Torchilin, V.P. (2003) Tat peptide-mediated intracellular delivery of liposomes. *Methods in Enzymology*, **372**, 339–49.
205 Torchilin, V.P., Levchenko, T.S., Rammohan, R., Volodina, N., Papahadjopoulos-Sternberg, B. and D'Souza, G.G. (2003) Cell transfection in vitro and in vivo with nontoxic TAT peptide-liposome-DNA complexes. *Proceedings of the National Academy of Sciences of the United States of America*, **100**, 1972–7.
206 Marty, C., Meylan, C., Schott, H., Ballmer-Hofer, K. and Schwendener, R.A. (2004) Enhanced heparan sulfate proteoglycan-mediated uptake of cell-penetrating peptide-modified liposomes. *Cellular and Molecular Life Sciences*, **61**, 1785–94.
207 Fretz, M.M., Koning, G.A., Mastrobattista, E., Jiskoot, W. and Storm, G. (2004) OVCAR-3 cells internalize TAT-peptide modified liposomes by endocytosis. *Biochimica et Biophysica Acta*, **1665**, 48–56.
208 Zhang, C., Tang, N., Liu, X., Liang, W., Xu, W. and Torchilin, V.P. (2006) siRNA-containing liposomes modified with polyarginine effectively silence the targeted gene. *Journal of Controlled Release*, **112**, 229–39.
209 Gupta, B., Levchenko, T.S. and Torchilin, V.P. (2005) Intracellular delivery of large molecules and small particles by cell-penetrating proteins and peptides. *Advanced Drug Delivery Reviews*, **57**, 637–51.
210 Lochmann, D., Jauk, E. and Zimmer, A. (2004) Drug delivery of oligonucleotides by peptides. *European Journal of Pharmaceutics and Biopharmaceutics*, **58**, 237–51.
211 Sawant, R.M., Hurley, J.P., Huang, Z., Szoka, F.C. and Torchilin, V.P. (2005) 32nd International Symposium on Controlled Release, Controlled Release Society, Inc., Miami, poster #406.
212 Sawant, R.M., Hurley, J.P., Salmaso, S., Kale, A., Tolcheva, E., Levchenko, T.S. and Torchilin, V.P. (2006) "SMART" drug delivery systems: double-targeted pH-responsive pharmaceutical nanocarriers. *Bioconjugate Chemistry*, **17**, 943–9.
213 Kale, A.A. and Torchilin, V.P. (2007) Design, synthesis, and characterization of pH-sensitive PEG-PE conjugates for stimuli-sensitive pharmaceutical nanocarriers: the effect of substitutes at the hydrazone linkage on the pH stability of PEG-PE conjugates. *Bioconjugate Chemistry*, **18**, 363–70.
214 Suzuki, R., Yamada, Y. and Harashima, H. (2007) Efficient cytoplasmic protein delivery by means of a multifunctional envelope-type nano device. *Biological and Pharmaceutical Bulletin*, **30**, 758–62.
215 Tilcock, C., Unger, E., Cullis, P. and MacDougall, P. (1989) Liposomal Gd-DTPA: preparation and characterization of relaxivity. *Radiology*, **171**, 77–80.
216 Kabalka, G.W., Davis, M.A., Holmberg, E., Maruyama, K. and Huang, L. (1991) Gadolinium-labeled liposomes containing amphiphilic Gd-DTPA derivatives of varying chain length: targeted MRI contrast enhancement agents for the liver. *Magnetic Resonance Imaging*, **9**, 373–7.
217 Unger, E., Cardenas, D., Zerella, A., Fajardo, L.L. and Tilcock, C. (1990) Biodistribution and clearance of

liposomal gadolinium-DTPA. *Investigative Radiology*, **25**, 638–44.

218 Putz, B., Barsky, D. and Schulten, K. (1994) Mechanisms of liposomal contrast agents in magnetic resonance imaging. *Journal of Liposome Research*, **4**, 771–808.

219 Schwendener, R.A. (1994) Liposomes as carriers for paramagnetic gadolinium chelates as organ specific contrast agents for magnetic resonance imaging (MRI). *Journal of Liposome Research*, **4**, 837–55.

220 Phillips, W.T. and Goins, B. (1995) Targeted delivery of imaging agents by liposomes, in *Handbook of Targeted Delivery of Imaging Agents* (ed. V. Torchilin), CRC Press, Boca Raton, pp. 149–73.

221 Schwendener, R.A., Wuthrich, R., Duewell, S., Wehrli, E. and von Schulthess, G.K. (1990) A pharmacokinetic and MRI study of unilamellar gadolinium-, manganese-, and iron-DTPA-stearate liposomes as organ-specific contrast agents. *Investigative Radiology*, **25**, 922–32.

222 Torchilin, V.P. (1997) Surface-modified liposomes in gamma- and MR-imaging. *Advanced Drug Delivery Reviews*, **24**, 301–13.

223 Kabalka, G.W., Davis, M.A., Moss, T.H., Buonocore, E., Hubner, K., Holmberg, E., Maruyama, K. and Huang, L. (1991) Gadolinium-labeled liposomes containing various amphiphilic Gd-DTPA derivatives: targeted MRI contrast enhancement agents for the liver. *Magnetic Resonance in Medicine*, **19**, 406–15.

224 Grant, C.W., Karlik, S. and Florio, E. (1989) A liposomal MRI contrast agent: phosphatidylethanolamine-DTPA. *Magnetic Resonance in Medicine*, **11**, 236–43.

225 Torchilin, V.P. (2007) Micellar nanocarriers: pharmaceutical perspectives. *Pharmaceutical Research*, **24**, 1–16.

226 Torchilin, V.P. (2000) Polymeric contrast agents for medical imaging. *Current Pharmaceutical Biotechnology*, **1**, 183–215.

227 Trubetskoy, V.S. and Torchilin, V.P. (1994) New approaches in the chemical design of Gd-containing liposomes for use in magnetic resonance imaging of lymph nodes. *Journal of Liposome Research*, **4**, 961–80.

228 Torchilin, V.P. (1999) Novel polymers in microparticulate diagnostic agents. *Chemical Technology*, **29**, 27–34.

229 Torchilin, V.P., Trubetskoy, V.S., Narula, J. and Khaw, B.A. (1995) PEG-modified liposomes for gamma- and magnetic resonance imaging, in *Stealth Liposomes* (eds D.D. Lasic and F.J. Martin), CRC Press, Boca Raton, pp. 225–31.

230 Trubetskoy, V.S., Cannillo, J.A., Milshtein, A., Wolf, G.L. and Torchilin, V.P. (1995) Controlled delivery of Gd-containing liposomes to lymph nodes: surface modification may enhance MRI contrast properties. *Magnetic Resonance Imaging*, **13**, 31–7.

231 Erdogan, S., Roby, A. and Torchilin, V.P. (2006) Enhanced tumor visualization by gamma-scintigraphy with ^{111}In-labeled polychelating-polymer-containing immunoliposomes. *Molecular Pharmaceutics*, **3**, 525–30.

232 Erdogan, S., Medarova, Z.O., Roby, A., Moore, A. and Torchilin, V.P. (2008) Enhanced tumor MR imaging with gadolinium-loaded polychelating polymer-containing tumor-targeted liposomes. *Journal of Magnetic Resonance Imaging*, **27**, 574–80.

233 Elbayoumi, T.A. and Torchilin, V.P. (2006) Enhanced accumulation of long-circulating liposomes modified with the nucleosome-specific monoclonal antibody 2C5 in various tumours in mice: gamma-imaging studies. *European Journal of Nuclear Medicine and Molecular Imaging*, **33**, 1196–205.

234 Elbayoumi, T.A., Pabba, S., Roby, A. and Torchilin, V.P. (2007) Antinucleosome antibody-modified liposomes and lipid-core micelles for tumor-targeted delivery of therapeutic and diagnostic agents. *Journal of Liposome Research*, **17**, 1–14.

235 Koning, G.A. and Krijger, G.C. (2007) Targeted multifunctional lipid-based nanocarriers for image-guided drug delivery. *Anti-cancer Agents in Medicinal Chemistry*, **7**, 425–40.

236 Sajja, H.K., East, M.P., Mao, H., Wang, Y.A., Nie, S. and Yang, L. (2009)

Development of multifunctional nanoparticles for targeted drug delivery and noninvasive imaging of therapeutic effect. *Current Drug Discovery Technologies*, **6**, 43–51.

237 Mitra, A., Nan, A., Line, B.R. and Ghandehari, H. (2006) Nanocarriers for nuclear imaging and radiotherapy of cancer. *Current Pharmaceutical Design*, **12**, 4729–49.

238 Karathanasis, E., Chan, L., Balusu, S.R., D'Orsi, C.J., Annapragada, A.V., Sechopoulos, I. and Bellamkonda, R.V. (2008) Multifunctional nanocarriers for mammographic quantification of tumor dosing and prognosis of breast cancer therapy. *Biomaterials*, **29**, 4815–22.

239 Pridgen, E.M., Langer, R. and Farokhzad, O.C. (2007) Biodegradable, polymeric nanoparticle delivery systems for cancer therapy. *Nanomedicine*, **2**, 669–80.

240 Sun, C., Fang, C., Stephen, Z., Veiseh, O., Hansen, S., Lee, D., Ellenbogen, R.G., Olson, J. and Zhang, M. (2008) Tumor-targeted drug delivery and MRI contrast enhancement by chlorotoxin-conjugated iron oxide nanoparticles. *Nanomedicine*, **3**, 495–505.

241 McCarthy, J.R. and Weissleder, R. (2008) Multifunctional magnetic nanoparticles for targeted imaging and therapy. *Advanced Drug Delivery Reviews*, **60**, 1241–51.

242 Gao, J., Gu, H. and Xu, B. (2009) Multifunctional magnetic nanoparticles: design, synthesis, and biomedical applications. *Accounts of Chemical Research*, **42**, 1097–107.

243 Kohler, N., Sun, C., Fichtenholtz, A., Gunn, J., Fang, C. and Zhang, M. (2006) Methotrexate-immobilized poly(ethylene glycol) magnetic nanoparticles for MR imaging and drug delivery. *Small*, **2**, 785–92.

244 Santra, S., Kaittanis, C., Grimm, J. and Perez, J.M. (2009) Drug/dye-loaded, multifunctional iron oxide nanoparticles for combined targeted cancer therapy and dual optical/magnetic resonance imaging. *Small*, **5**, 1862–8.

245 Yang, L., Peng, X.H., Wang, Y.A., Wang, X., Cao, Z., Ni, C., Karna, P., Zhang, X., Wood, W.C., Gao, X., Nie, S. and Mao, H. (2009) Receptor-targeted nanoparticles for in vivo imaging of breast cancer. *Clinical Cancer Research*, **15**, 4722–32.

246 Lin, J.J., Chen, J.S., Huang, S.J., Ko, J.H., Wang, Y.M., Chen, T.L. and Wang, L.F. (2009) Folic acid-Pluronic F127 magnetic nanoparticle clusters for combined targeting, diagnosis, and therapy applications. *Biomaterials*, **30**, 5114–24.

247 Al-Jamal, W.T. and Kostarelos, K. (2007) Liposome-nanoparticle hybrids for multimodal diagnostic and therapeutic applications. *Nanomedicine*, **2**, 85–98.

248 Zheng, J., Liu, J., Dunne, M., Jaffray, D.A. and Allen, C. (2007) In vivo performance of a liposomal vascular contrast agent for CT and MR-based image guidance applications. *Pharmaceutical Research*, **24**, 1193–201.

249 Tan, W.B. and Zhang, Y. (2007) Multi-functional chitosan nanoparticles encapsulating quantum dots and Gd-DTPA as imaging probes for bio-applications. *Journal of Nanoscience and Nanotechnology*, **7**, 2389–93.

250 Mulder, W.J., Griffioen, A.W., Strijkers, G.J., Cormode, D.P., Nicolay, K. and Fayad, Z.A. (2007) Magnetic and fluorescent nanoparticles for multimodality imaging. *Nanomedicine*, **2**, 307–24.

251 Cheon, J. and Lee, J.H. (2008) Synergistically integrated nanoparticles as multimodal probes for nanobiotechnology. *Accounts of Chemical Research*, **41**, 1630–40.

252 Wehrl, H.F., Judenhofer, M.S., Wiehr, S. and Pichler, B.J. (2009) Pre-clinical PET/MR: technological advances and new perspectives in biomedical research. *European Journal of Nuclear Medicine and Molecular Imaging*, **36** (Suppl. 1), S56–68.

253 Ke, J.H., Lin, J.J., Carey, J.R., Chen, J.S., Chen, C.Y. and Wang, L.F. (2010) A specific tumor-targeting magnetofluorescent nanoprobe for dual-modality molecular imaging. *Biomaterials*, **31**, 1707–15.

6
Polymersomes and Their Biomedical Applications

Giuseppe Battaglia

6.1
Introduction

The chemical and physical transformation of molecules and macromolecules such as proteins, nucleic acids, carbohydrates, and lipids, represents the "central dogma" of life. All such transformations must take place in highly organized compartments, with biological compartmentalization being achieved by exploiting the formation of membranes that have thicknesses of a few nanometers that enclose aqueous environments in various structures, the sizes of which range from a few tens of nanometers (e.g., trafficking vesicles) up to even meters (e.g., motoneuron plasma membranes) [1]. These membranes result from the self-assembly of amphiphilic molecules known as *phospholipids*. Amphiphiles have a dual nature that comprises both hydrophilic (water-soluble) regions and hydrophobic (water-insoluble) regions. In the presence of water, the hydrophobic regions tend to minimize the contact with water, by attracting each other, while the hydrophilic parts prefer to make contact with water and so repel each other. Thus, the membrane assembly is clearly the perfect building block for generating confined aqueous compartments. Indeed, this two-dimensional (2-D) structure increases the efficiency considerably, and opens the opportunity for irreversible charge separation and the transitory storage of energy in the form of chemical potential gradients [1]. This modular design has clearly been a crucial evolutionary step in the production of some order within the cell, and to control such a complex machine [2]. Numerous different molecules are distributed among the organized cellular subspaces. The differences in concentration and composition between the subcellular compartments generate a composition and concentration gradient that is vital for the directed flow of freshly synthesized material, from one intracellular organelle to other organelles, or to the plasma membrane.

The extraordinary ability of Mother Nature to engineer functional and dynamic supramolecular structures with a tunable size has been mimicked by synthetic amphiphiles. In particular, recent new advances in controlled polymerization techniques [3–5] have allowed the design of a new class of amphiphilic membrane, based on block copolymers [6]. The latter are macromolecules that comprise two

Nanomaterials for the Life Sciences Vol.10: Polymeric Nanomaterials. Edited by Challa S. S. R. Kumar
Copyright © 2011 WILEY-VCH Verlag GmbH & Co. KGaA, Weinheim
ISBN: 978-3-527-32170-4

or more different polymers, whereby the combination of different polymer backbones provides a range of properties that is defined by the designed molecules. Polymeric amphiphiles can be considered as "super" amphiphiles, as they have molecular weights up to several orders of magnitude higher than that of phospholipids. As a consequence of such a macromolecular nature [7], the amphiphiles are able to self-assemble into highly entangled membranes, providing the final structure with improved mechanical properties and stability compared to those of a membrane which is based on phospholipids [8–10]. The simplest structure that membranes can form is a core–shell sphere, known as vesicle or "polymersome" (where "some" derives from the Greek, meaning "body of"). Polymersomes can be formed with sizes ranging from tens of nanometers to tens of micrometers, and with a relatively high control on the size distribution. This flexibility has attracted the attention of many scientists and engineers, who have investigated polymersomes for their different biomedical applications, which span from drug delivery [11, 12] to diagnostics [13, 14].

6.2
The Chemistry of Polymersomes

Block copolymers are macromolecules that comprise two or more different polymers, with the combination of different polymer backbones providing a range of properties that is defined by the designed molecules. The final form of copolymers can be controlled by their synthesis; for example, it is possible to prepare diblock, triblock, multiblock, random block, star, and graft copolymers, with the possibilities for molecular design being limited only by the chemist's imagination and skills. In particular, amphiphilic block copolymers are formed by a combination of hydrophilic and hydrophobic portions which tend to self-assemble in contact with water due to a *hydrophobic effect* [15]. Consequently, depending on the hydrophilic fraction, different ordered structures can be formed, such as spherical micelles, cylindrical micelles, and vesicles. To be precise, reducing the hydrophilic fraction will lead to the formation of cylindrical rather than spherical micelles, while further decrease will cause the formation of vesicles [9]. The chemical structure of the polymersomes forming block copolymers can be either simple or very complex, and this gives the corresponding polymersomes their peculiar properties. A traditional example is the commercially available and Food and Drugs Administration (FDA)-approved poly(ethylene oxide)–poly(propylene oxide) (PEO–PPO) [16], while other simple structures that have been reported include:

- Poly(acrylic acid)–polystyrene (PAA–PS) [17–19], poly(ethylene oxide)–polystyrene (PEO–PS) [20], and poly(acrylic acid)–polybutadiene (PAA–PBD) [21], all of which are able to form large compound vesicular aggregates.

- PBD–PEO [8, 22] and PEO–poly(ethyl ethylene) (PEO–PEE) [8, 22], the corresponding vesicles of which present mechanical toughness and biological inertness.

- Polyethers such as poly(butylene oxide)–poly(ethylene oxide) (PBO–PEO) [23], which form vesicles with selective permeability [24].

- PEO–poly(lactic acid) (PEO–PLA), PEO–poly(ε-caprolactone) (PEO–PCL), and poly(ethylene oxide)–poly(trimethylene carbonate) (PEO–PTC) that self assemble into biodegradable vesicles [25].

More complicated block copolymer structures have been synthesized, giving rise to vesicles that show:

- pH-sensitivity: these include PBD–(poly-L-glutamate) [26], poly(2-methacryloxyethyl phosphorylcholine)–poly(2-(diisopropylamino)ethyl methacrylate) [27], poly(L-arginine)-poly(L-leucine) [28], poly(2-vinylpyridine)–PEO [29] and PAA–polystyrene-poly(4-vinyl pyridine) [30].

- pH-tunable membrane permeability: such as PEO–poly(2-(diethylamino)ethyl methacrylate-*stat*-3-(trimethoxysilyl) propyl methacrylate) [31] and PAA–poly(distearin acrylate) [32].

- UV light sensitivity: such as PEO–poly(methylphenylsilane) and azobenzene-containing poly(methacrylate)–PAA (PMA–PAA) [33].

- Sensitivity to oxidation: such as PEO–poly(propylene sulfide) (PPS) [34].

- Sensitivity to reduction: such as PEO–SS–PPS [34, 35].

- An ability to incorporate membrane proteins, such as: poly(2-methyloxazoline)–poly(dimethylsiloxane)–poly(2-methyloxazoline) [36] and poly(2-ethyl-2-oxazoline)–poly(dimethylsiloxane)–poly(2-ethyl-2-oxazoline) [37].

- Selective permeability: such as polystyrene-poly(L-isocyanoalanine(2-thiophen-3-yl-ethyl)amide) [38].

- Degradability and controllable vesicle diameter: such as poly(L-lysine)–poly(L-leucine and poly(L-glutamic acid)–poly(L-leucine) [39].

- An ability to form vesicles with a single emulsion technique such as: PBD–poly(N-methyl-4-vinyl-pyridinium iodide [40]. Vesicle-forming polystyrene–poly(propylene imine) dendrimer [41], helical block copolypeptides such as polystyrene–poly(isocyano-L-alanine-L-alanine), polystyrene–poly(isocyano-L-alanine-L-histidine) [42] and PBD–poly(L-lysine) [42–44] have also been reported, as well as a combination of the two features such as dendritic helical (L-lysine)–poly(γ-benzyl-L-glutamate) block copolypeptides [45].

Depending on the molecular architecture of the copolymer, the membrane will have different conformations. For example, tri-block hydrophobic–hydrophilic–hydrophobic copolymers are similar to diblock copolymers, in that there is only one molecular conformation that can lead to membrane formation. Namely, the hydrophobic chain ends must assemble into a membrane, and the hydrophilic block must form a loop (U-shape). On the other hand, hydrophilic–hydrophobic–hydrophilic copolymers can have two possible conformations: (i) the hydrophobic block can either form a loop so that the hydrophilic chains are on the same side of the membrane (U-shape); or (ii) they can stretch to form a monolayer, with the two hydrophilic blocks at the opposite sides of the membrane (I-shape) [23, 46–49]. A similar conformation can be formed by pentablock copolymers [50]. This also applies when a third (or even a fourth) chemically different block is added.

Nonetheless, the overall geometry is the same, and multiblock copolymers have an extra level of control within the polymersomes, introduced by the extra interaction between the blocks. ABC copolymers, where A and C are hydrophilic and B is hydrophobic, assemble into asymmetric "Janus" (as in the double-faced Roman god) particles [51]. These ABC copolymers assemble into asymmetric membranes [48, 52, 53] that form polymersomes, the internal and external surface chemistries of which are different one from another. Brannan et al. [54] have also shown that ABCA tetrablock copolymers, where A is hydrophilic and B and C are both hydrophobic, assemble into vesicles, the membrane of which has an internal morphology that changes from lamellae to cylinders, thus altering the volume fraction between B and C.

6.3
Polymersomes: Physico-Chemical Properties

6.3.1
Membrane Conformations

The synthetic nature of block copolymers allows the application of different compositions and functionalities over a limitless range of molecular weights and, consequently, of membrane thicknesses [7]. Polymersome membranes comprise a highly entangled hydrophobic layer with thickness t, and two hydrophilic polymer brushes, with thickness d, that stabilize the membrane (Figure 6.1). The electron

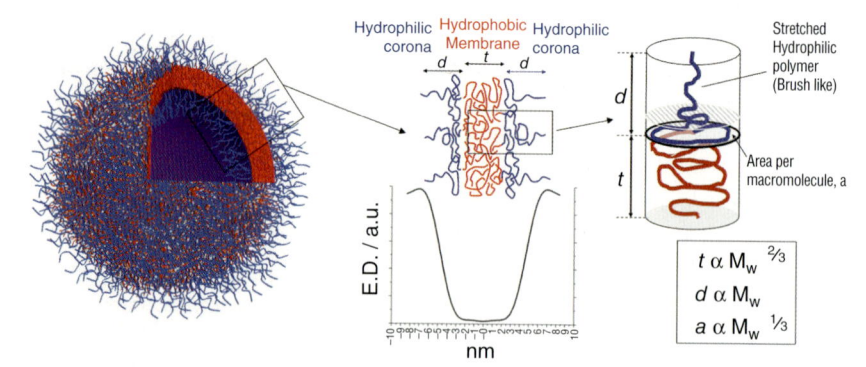

Figure 6.1 Schematic of a polymersome, including the electron density distribution across the polymersome membrane and the single macromolecule conformation.

$t \propto M_w^{2/3}$
$d \propto M_w$
$a \propto M_w^{1/3}$

density distribution of PEO-based polymersomes has recently been calculated by combining small-angle X-ray scattering and electron microscopy, to show that both the hydrophilic brush and the membrane have similar sizes [61]. However, as shown in Figure 6.1, whereas the hydrophilic brush thickness scales almost linearly with the copolymer molecular weight [61], which is typical of dense polymer brushes in solution [62, 63], the hydrophobic membrane follows a power law with an exponent smaller than the unity [7, 64], which is typical of bulk strongly segregated systems [65].

Different studies have contributed to highlight the fact that high molecular weights are related to thicker membranes; in particular, the polymersomes membrane thickness, d, ranges approximately between 2 and 30 nm [8, 66–69], while lipid membranes present $d \approx 3–5$ nm [67, 68]. The power law, $d \sim M_w^b$, has been proven, both experimentally and theoretically, to be applicable for relating the amphiphile's molecular weight to the membrane thickness [7, 64, 70–75]. The two boundary conditions for the exponent are $b = 1$ for fully stretched [7] and $b = 1/2$ for random Gaussian coil conformation [66, 75]. The former case is the theoretical limit approached by phospholipids, while the latter is known as "weak segregation" in block copolymer melts. In the intermediate condition, as in the polymersomes case, the exponent was found to be $b = 2/3$ [7, 76]. This results from an "interdigitation" of hydrophobic copolymer chains that has been shown to occur at the bilayer midplane, also computationally [76]. The elastic bending, stretching, and shear of polymersomes have each been measured using micropipette aspiration, and compared to the liposomes' corresponding values; both liposomes and polymersomes present similar elastic properties, but the latter are able to undergo a higher tension and area strain before rupture [8, 70]. In fact, the low thickness values make liposomes more susceptible to defects and fluctuations with respect to polymersomes. Nevertheless, micromanipulation techniques performed on high-molecular-weight polymersomes contributed to highlight a quadratic increase of bending rigidity with bilayer thickness [9], due to the higher viscosity caused by an increased interdigitation of macromolecules in the midplane of the membrane. Lateral diffusivity measurements [70, 77–79] showed that membrane fluidity decreases with the increasing M_w of block copolymers; in particular, polymeric membranes present diffusion coefficients ($D \sim 0.1\,\mu m^2 s^{-1}$ and less) that are at least one order of magnitude lower than those of lipid membranes ($D \sim 1\,\mu m^2 s^{-1}$). Another feature of polymersomes is their reduced permeability compared to liposomes [8, 24, 80]; it has been shown that the membrane crossing rates depend on the thickness, as predicted by first Fick's law [81]. This represents an important goal, as one of the fundamental properties of amphiphilic membranes is their partitioning ability of aqueous volumes with different compositions and concentrations. Furthermore, depending on the chemistry of the amphiphilic block copolymers used, it is possible to achieve a selectivity of polymeric membranes towards determined molecules [24], which makes polymersomes particularly suitable for controlled release applications.

The fine-tuning of the polymersomes' properties, such as membrane mechanical properties [24], membrane permeability [66], and the polymersomes' interaction with plasma proteins, improves their circulation half-time in living systems [24]. Compared to other vesicular structures, polymersomes have been shown to present a higher stability and content retention [82]. Unlike block copolymers, traditional surfactants such as phospholipids usually have molecular weights below 1 kDa, imparting biomembranes with "soft" properties that are crucial for cells functions such as endocytosis, cell division, and clustering [7].

Polymersomes can be thought of as colloidal dispersions [68] and, as such, the system can be perturbed by altering the balance between the forces stabilizing the dispersion. In a colloidal dispersion, the dispersed particles have a huge surface area that gives rise to a large interfacial energy; however, due to a complex balance between the attractive interfacial and van der Waals forces with a repulsive electrostatic double layer forces, the particles do not aggregate as might be expected [83, 84]. Rather, when particles do collide in solution they stick irreversibly, such that the aggregate size in a colloidal dispersion will increase over time. A similar fusion process occurs in both lipid and polymeric vesicle dispersions. In Nature, fusion and fission are important processes within the cell, as they enable the transport of chemicals from one organelle to another in a controlled fashion; for example, membrane fission is notably common in cells during the process of endocytosis. Thus, the study of membrane fusion and fission is important to understand the complex processes involved within the cell. When both Menger *et al.* and Nomura *et al.* reported on fusion and fission in giant lipid vesicles [85–88], it was found that the injection of additives to vesicle dispersions caused a variety of effects, including aggregation, fusion, and budding. In particular, vesicle budding with the addition of octyl glucoside was observed, whereas sodium cholate addition caused a mass fusion between vesicles to the point of decay [89]. The reversible adhesion between oppositely charged lipid vesicles has also been observed [86, 88]. In this case, the separation of adhered vesicles was induced by internal membrane dislocations, although it was not clear whether the separation of biological cells transiently adhered together operated via the same mechanism. These processes are rapid in the presence of lipid vesicles, but have been observed on a longer time scale when using polymersomes [89–91]. Zhou *et al.* observed the membrane fusion [9] and fission [92] of giant polymeric vesicles in real-time; notably, subjecting the dispersions to ultrasound induced vesicle–vesicle fusion so as to produce larger polymersomes. The fusion of these polymeric vesicles takes place on the minute time scale, which is much slower than for liposomes [89, 93] (microseconds/seconds) and biological membranes [94, 95] (milliseconds). On this slower time scale, four clear intermediate stages have been observed as the membranes adheres and fuses, and the central pore slowly increases in size to give a single larger vesicle. By contrast, the same authors induced fission in giant polymersomes by adding glucose to the dispersion. Here, the proposed mechanism involved the generation an osmotic shock to the semi-permeable vesicle membrane, which then budded off to form smaller vesicles as the water forced its way through the membrane. Eisenberg and colleagues reported the fusion and fission of polystyrene-*b*-poly(acrylic acid) (PS–PAA) vesicles by altering the water : organic solvent ratio [96]. In particular, when increasing the water content of the PS–PAA vesicle dispersion the system minimized the total interfacial energy by increasing the polymersomes size (fusion). In contrast, when decreasing the water content, polymersomes fission was observed. Furthermore, for smaller polymersomes the interfacial area was more affected by the size, leading to a narrow size distribution, whereas in the presence of larger polymersomes a wider size distribution was observed. The water content also affected the kinetics of the fusion process of PS–PAA vesicles [91]. Recently, it has been observed [90] that the addition of a PEO homopolymer (and no other water-soluble polymers) to PEO polymersome dispersions induced polymersomes fusion and gelation, depending on the homopolymer's molecular weight. The control of fusion can, in turn, allow the effective control of the topology of the polymersomes. Similarly, the active control of vesicular gels allows the formation of gels with three levels of supramolecular control; that is, copolymer chemistry, membrane structure and, finally, polymersomes aggregation.

6.3.2
Responsive Polymersomes

One advantage of polymers compared to their small-molecular-weight counterparts is that the properties of the polymer solution are strictly controlled by the side group chemistry. A polymeric chain comprises both soluble and insoluble parts, the balance of which dictates the overall solubility; consequently, polymer solubility can be affected by the application of external stimuli such as temperature, pH, ionic strength, and light. This approach has been largely exploited in polymer engineering, and several devices based on polymer solubility have been devised as a result (T. Smart *et al.*, unpublished results). In the case of polymersomes, these properties can be employed to generate structures capable of dissolving under certain environmental conditions. The simplest approach has been to couple hydrophilic polymers with hydrolyzable polymers such as PLA [25, 55, 57, 97, 98]. In the proximity of water, these polymers will degrade, leading to a non-reversible disassembly of the polymersome and hence the release of its content. The ability to degrade after a definite time, or upon stimulation, seems critical for the design of new carriers, as does a controllable vesicle diameter. Similarly, non-reversible transitions can be achieved by using block copolymers based on PPS as the hydrophobic block [49, 58, 99]. In this case, the change in hydrophilic/hydrophobic balance is due to the oxidation of PPS into more hydrophilic sulfoxides and sulfones upon exposure to oxidative environments, such as H_2O_2, leading first to the transition of vesicles to cylindrical micelles, spherical micelles and, eventually, into nonassociating unimers. Such an approach, in combination with specific enzymes, can open new perspectives in the field of biosensors for the detection of molecules of interest, such as glucose and inflammatory signals. An alternative approach to creating oxidation-sensitive polymersomes was reported by Cerritelli *et al.* [34], who synthesized block copolymers comprising the hydrophilic PEO and the hydrophobic PPS, linked via a disulfide bond. A more reversible disassembly can be obtained by combining hydrophilic polymers with polymers for which the solubility is heavily dependent on pH, temperature, and UV light. For example, pH-responsive polymersomes based on poly(L-glutamic acid)–poly(L-lysine) have been reported [35] that can be formed reversibly in weak acidic or basic aqueous solutions. Armes and coworkers [100] demonstrated the spontaneous formation of biocompatible vesicles when the pH of the solution was increased from 2 to 7, by using a diblock copolymer made from poly(2-(methacryloyloxy) ethyl phosphorylcholine) (PMPC), a hydrophilic and highly biocompatible block, and poly(2-(diisopropylamino)ethyl methacrylate) (PDPA), a hydrophobic block capable of being protonated. The protonation of the PDPA part at a pH-vale below its pK_a value turned the polymer hydrophilic, and led to a dissolution of the vesicles [101]. Similarly poly(2-vinylpyridine-b-ethylene oxide) (P2VP–PEO) polymersomes have been reported [102] to undergo a similar transition at pH about 5. A similar responsiveness can be achieved by having a hydrophilic block, such as weak polyacids and polybases, that change their macromolecular conformation as a function of the pH. As the acid or the basic groups become protonated and deprotonated, the polyelectrolytes will undergo the transition from a hypercoiled to a stretched configuration [29]. When attached to a hydrophobic block, this transition alters the assembly properties and results in the formation of vesicles and micelles, depending on the pH of the solution [17, 103, 104]. Similar transitions have been observed as a function of temperature using poly[N-(3-aminopropyl)-methacrylamide hydrochloride]–poly(N-isopropylacrylamide) (PAPMA-PNIPAM) diblock copolymers. McCormick and coworkers [105] reported that the polymerosomes formed reversibly at temperatures in excess of the lower critical solution temperature of the pure PNIPAM block. A similar approach was used using PEO–PNIPAM temperature-responsive polymersomes [106]. Rank and coworkers have also shown that temperature can be exploited to trigger the transition from micelles to vesicles [60]. Polymersomes from PEO–poly(2-vinyl pyridine) (PEO–P2VP) block copolymers were converted reversibly into cylindrical micelles when the temperature was lowered to 4 °C. This temperature-sensitive behavior was attributed to the effect of temperature on PEO solubility in polar solvents [107]. Finally, a similar transition can be induced by using UV-sensitive groups such as PEO–poly(methylphenylsilane) (PEO–PMPS) and azobenzene-containing poly(methacrylate)–poly(acrylic acid) (PAA–PAzoMA) [107]. More recently, Mabrouk et al. [108] demonstrated asymmetric polymersomes, the membrane of which had only one leaflet composed of UV-sensitive liquid-crystalline copolymer poly(ethylene glycol)-co-poly(4-butyloxy-2-(4-(methacryloyloxy)butyloxy)-4-(4-butyloxybenzoyloxy)azobenzene) (PEG-b-PMAazo444). Once exposed to UV light, these polymersome underwent bursting due to changes in the spontaneous curvature of their membranes [109].

6.3.3
Surface Chemistry of the Polymersomes

The polymersomes membrane results from the self-assembly in water of amphiphilic block copolymers and, as discussed above, almost any chemistry can be used provided that the necessary hydrophilic/hydrophobic conditions for the assembly are maintained. The hydrophilic blocks assemble into a dense brush configuration that, depending on the polymer nature (i.e., cationic, anionic, or neutral), will control the surface characteristics of the polymersome and hence its interaction with the environment. For example, PEO-containing block copolymers will assemble into polymersomes with a highly hydrated and yet neutral polymer brush [109] which has very limited interaction with proteins [110, 111]. Such nonfouling nature allowed the PEO–polymersomes to withstand biological fluids almost without interacting with the immune system [112]. This ability to "hide" from the immune system – known as "stealth" – has been largely exploited to enhance the circulation times of liposomes and other nanoparticles [113, 119]. The brush thickness and conformation is crucial for the degree of stealth of the polymersomes; this was demonstrated by Photos et al., who reported different circulation times for PEO polymersomes of various molecular weights [115]. PEG and other nonfouling polymers are highly desirable for applications involving extended times in contact with biological fluids. The translation of these design principles can provide significant help in the synthesis of many ad hoc chemistries for new specific applications. In addition, it has recently been shown that the polymersomes' surface can be further modified to introduce domains of different chemistry. This can be achieved by mixing different polymersome-forming copolymers so as to induce a polymersome-confined polymer–polymer phase separation. As shown in Figure 6.2a, the different surface topologies can be identified as a function of the PMPC–PDPA/PEG–PDPA ratio [82]. Furthermore, studies on polymersomes phase separation (Figure 6.2b) have been recently conducted by Discher and coworkers on PAA–PBD/PEO–PBD copolymer couples [117]. Here, the ability of phase separation was combined with the responsive nature of the PAA, showing ion- and pH-induced transitions from patchy polymersomes to patchy micelles.

On a second level of complexity, the polymersomes' surfaces can be decorated with active moieties such as proteins, antibodies, vitamins, and carbohydrates (Figure 6.3). One of the simplest methods to decorate polymersomes is by exploiting the biotin–avidin complex [116]. The linkage of biotin moieties to the hydrophilic block of the copolymer is used to attach avidin groups to preformed polymersomes, which in turn can bind to biotinylated targeting ligands, as each avidin group has four binding sites for biotin [118–121]. Broz et al. have taken advantage of biotin–avidin chemistry to attach polyG, specific to the scavenger receptor A1 (SRA1) of macrophage cells to polymersomes formed from an "ABA" PMOXA–PDMS–PMOXA triblock copolymer [122]. By using a similar approach, polymersomes conjugated with the anti-ICAM-1 antibody were tested to treat endothelial cells undergoing inflammation (in such cells, the ICAM-1 molecule is overexpressed during intracellular adhesion). These studies highlight the importance of tag-orientation on the polymersomes surface [122]. Other conjugation strategies have been followed by coupling thiolated antibodies with maleimide-functionalized PEO–poly(caprolactam) copolymers [120]. These antibody-functionalized polymersomes were shown to target the OX21 receptor expressed on the blood–brain barrier, thus facilitating the delivery of therapeutic peptides within the brain [123]. A similar chemical approach was used by Christian et al.

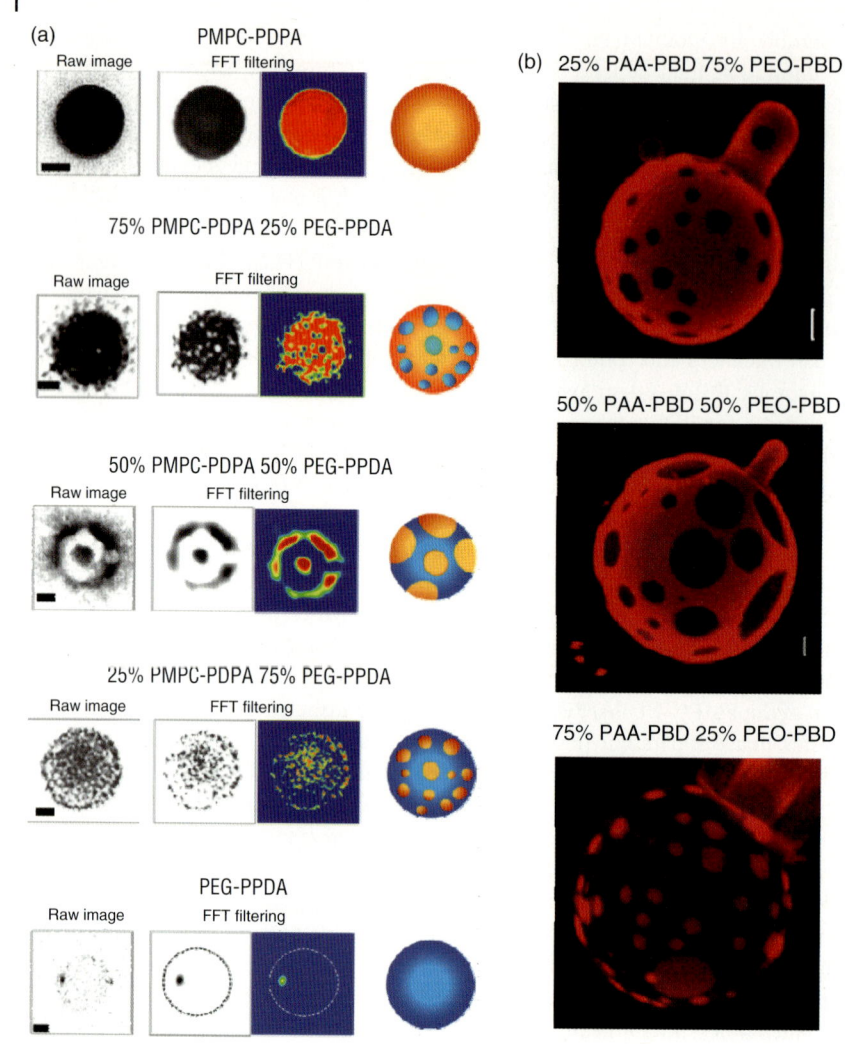

Figure 6.2 (a) Nanometer-sized polymersomes imaged by transmission electron microscopy (TEM) and analyzed using fast Fourier transform (FFT) filtering. Phase segregation generated by binary different ratio mixtures of PMPC–PDPA and PEG–PDPA copolymers forming patchy polymersomes that display surface domains highlighted by selective staining of the PMPC chain by phosphotungstenic acid (selective for ester groups). Scale bar=50 nm [116]; (b) Micrometer-sized polymersomes phase separation at the generated by mixing different ratio of PEO–PBD and PAA–PBD copolymers and imaged by confocal microscopy during micropipette aspiration. The red fluorescence is from the tetraethyl rhodamine-5 carbonyl azide chemically conjugated to the PEO–PBD copolymer. Scale bar=2 μm [117].

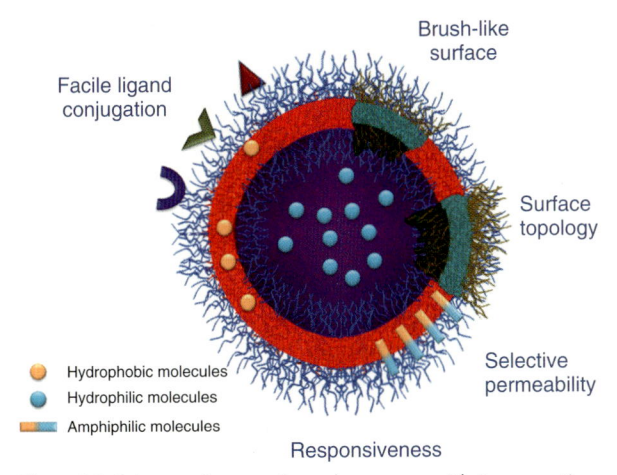

Figure 6.3 Scheme of responsive polymersome with its properties.

to functionalize polymersomes with the cell-permeable peptide, TAT, to enhance cellular uptake [123]. Upadhyay *et al.* have also shown that the targeting moiety can be engineered so as to act as hydrophilic block of the copolymer. In this case, a hyaluronan-*b*-poly(γ-benzyl-L-glutamate) copolymer was synthesized by exploiting the hyaluronal ability to target the type 1 transmembrane glycoprotein CD44 which is upregulated in certain tumors [124].

6.4
Polymersomes Formation and Preparation

Polymersomes can be prepared by dissolving the block copolymer in an organic solvent suitable for all blocks, followed by the addition of water that is a poor solvent for the hydrophobic block(s) (the "phase-inversion" technique) [17, 125]. Alternatively, they can be prepared by the direct dissolution of block copolymers in aqueous solution [7, 126, 127], or by film rehydration or electroformation ("solvent-free" techniques) [8, 54, 59, 128]. In the former case, water is added to a polymeric film in the presence (or not) of an energy source, such as shear, ultrasound, or alternating current [22]. When the phase-inversion technique is used, the transition from spherical micelles to cylindrical micelles, and finally to vesicles, has often been observed during the addition of water to the solution [129], as a consequence of the increased interfacial tension [130]. Vesicles formed via the phase-inversion technique may contain considerable amounts of organic solvent residues, however, which may limit their application in the biomedical field. Although, the preparation methods are basically the same as used for liposomes [9, 131–133], the formation of polymersomes can be more difficult and slower, depending on the flexibility of the block copolymer chains. When a solvent-free rehydration of vesicle-forming amphiphiles films is carried out, different phase sequences have been reported when the water concentration is increased. At high copolymer concentrations, lamellar phases are first visible (connected membranes) whereas, upon the addition of water, bicontinuous phases are reported to be formed, followed by vesicular gels and, eventually, vesicles [133–136]. Previously, the phase diagram of polymersomes forming copolymers as a function of the molecular weight and concentration in water has been reported [137]. In this case, the size of the amphiphile has a significant effect on the formation of both the lyotropic phases and dispersed vesicles. The evolution from bulk solid to lyotropic liquid shows that the low-molecular-weight copolymer assembles first into an inverted hexagonal structure and then lamellae, whereas the high-molecular-weight dissolves directly to the lamellar phase. The transition from lamellae to the sponge phase is essentially unaffected by the molecular weight of the amphiphiles. In contrast, however, evolution of the sponge phase into vesicles is qualitatively different, depending on the size of the amphiphilic block copolymer. Smaller copolymers form dispersed vesicles at quite high concentrations, whereas high-molecular-weight copolymers initially form peculiar vesicular gel clusters, which eventually break up into dispersed vesicles. The amphiphile molecular weight affects also the nature of the vesicles formed, with the larger amphiphiles forming exclusively unilamellar vesicles on dilution, and the smaller amphiphiles giving rise to multilamellar vesicles, still in a lyotropic solution. Furthermore, membrane undulations and unbinding depend heavily on the stiffness of the membrane which, in turn, is governed by the amphiphilic block copolymer molecular weight. In particular, more flexible membranes lead to earlier unbinding and multilamellar vesicles, while unbinding occurs at lower concentrations as the membrane flexibility is reduced forming unilamellar vesicles. Polymersomes formation under solvent-free conditions is dictated by the mutual diffusion of water into the bulk block copolymer, and *vice versa*. It has been shown that the amphiphilic block copolymers swell in water, allowing two qualitatively different growth regimes [138]. Initially, water and copolymer diffuse into each other following a subdiffusional growth as a result of the molecular-level arrangement of the amphiphilic membranes that comprise the swollen copolymer. After a critical time, which is exponential in polymer molecular weight, the amphiphilic membranes reach their equilibrium morphology and, consequently, the growth starts to follow a Fickian diffusion. In vesicle formation under solvent-free conditions, the driving force for vesicles formation is the concentration gradient between the copolymer front, diffusing in water, and the water front, diffusing in the copolymer. When an extra energy source is supplied (mechanical mixing, sonication, AC electrical field, etc.) in order to keep this gradient constant, the mutual diffusion is enhanced. Thus, the lyotropic lamellae swell and very soon are placed under conditions of complete and drastic unbinding; the amphiphilic membranes have plenty of time to evolve into the most stable structure. Under simple hydration, on the other hand, the concentration gradient decreases almost linearly with time. As the concentration gradient changes, the lamellar structures may not have sufficient time to unbind completely [139]; consequently, the average dimension of polymersomes is considerably affected by the method of preparation. In particular, it has been observed that both phase inversion [139] and electroformation [129, 140] usually lead to micrometer-sized vesicles, while rehydration with vigorous mixing, sonication, or extrusion, leads to nanometer-sized vesicles [141, 142]. Thus, the final size of the formed vesicles in a given solution is dictated more by kinetics aspects than by thermodynamics [143]. By using crystal growth arguments, it has been shown that the dimensions of polymersomes are inversely proportional to the concentration gradient [9]; thus, under conditions of mixing, where the concentration gradient is kept high and almost constant, vesicles are shown to have diameter on the orders of nanometers, whereas under mild diffusion conditions the diameter may be of decades of micrometers. During hydration of the vesicle-forming amphiphiles, tubular structures similar to natural myelins can grow. These are kinetic instabilities occurring at the solid/liquid interface in the absence of an extra energy source [139]. Such tubular structures have recently been observed and studied also for amphiphilic polymers [144]; in particular, it was found that an oriented growth of such myelins could be achieved under a particular concentration gradient. In analogy to natural membranes, where the presence of cholesterol enhances myelination processes (e.g., in nerve cells), for polymeric membranes it was found that myelin growth was favored by the presence of chloroform in the swelling solution, acting as a fluidizing agent [145].

6.5
Biomedical Applications

Some important applications of polymersomes span from medical imaging [145], to the encapsulation of proteins, DNA and anticancer drugs for delivery applications.

6.5.1
Medical Imaging

With the advent of new imaging techniques, improvements in current intracellular labeling procedures became fundamental. As well as providing an excellent delivery system, polymersomes are also extremely useful for imaging applications. As an example, Hammer and coworkers recently demonstrated the application of a simple diblock copolymer system such as PEO-b-PCL and PEO-b-poly(γ-methyl-ε-caprolactone) (PMCL), to create polymersomes with special emissive properties [146–150]. This was achieved by incorporating porphyrin-based fluorophores within the polymeric membrane, such that the final structure showed optical properties similar to those of quantum dots (QDs), from the visible to the infrared. Such near-infrared (NIR)-polymersomes, when conjugated to a cell-permeable peptide (TAT), have been used by Christian et al. [151] to provide an efficient intracellular delivery for future dendritic cell tracking with these optical probes.

With similar purpose, the PMPC–PDPA polymersomes-mediated delivery of fluorophores has been recently explored by the present author's group. In detail, fluorescent amphiphilic molecules have been incorporated into the polymersomes' membrane and delivered within cells in order to generate a noncytotoxic, nonimmunogenic cellular tracking system [124].

In addition to the incorporation of fluorophores within polymersomes, the potential of those carriers to the field of imaging has been further exemplified by the encapsulation of magnetic resonance imaging (MRI) contrast agents within porous PEO–PBD polymersomes, in a study conducted by Cheng and Tsourkas [152]. Here, the polymersomes were prepared by the hydration of a thin film of the PEO–PBD diblock copolymer mixed with 1-palmitoyl-2-oleoyl-sn-glycero-3-phosphocholine (POPC) phospholipid at a molar ratio of 85:15. A chelated gadolinium (Gd) covalently attached to a dendrimer was loaded within the polymersome lumen, and the diblock copolymer was then crosslinked by free-radical polymerization via the addition of a chemical initiator, to increase the stability of the polymersome membrane (see Figure 6.4a). Upon the addition of a surfactant (e.g., Triton X-100), the phospholipid component of the polymersomal membrane was removed, generating holes in the membrane for a facile influx of water (see Figure 6.4a). The strength of the signal provided by the Gd ions for imaging purposes is dependent on rapid exchange kinetics between the Gd-bound water molecules and the surrounding bulk water [146], which is facilitated by the presence of pores in the polymersomal membrane. Conjugation of the Gd chelates to dendrimers inhibited their escape via the polymersome pores. Their encapsulation within polymersomes has the advantage of greatly increasing the signal intensity generated by the Gd ions compared to free Gd, since virtually 44 000 Gd ions were loaded into each polymersome on average. Furthermore, these polymersomes offer the potential to encapsulate both hydrophilic and hydrophobic active agents and to add targeting ligands to their exterior (as discussed above), rendering the generation of targeted nanocarriers for "combined drug delivery and MR imaging," a viable area of research. Similarly, Yang and coworkers [146] utilized a binary mixture of amphiphilic folate-poly(ethylene glycol)-poly(D,L-lactide) (folate–PEG–PDLLA) and NH_2–PEG–PDLLA to generate polymersomes in order to encapsulate hydrophilic superparamagnetic iron oxide nanoparticles (SPIONs) via double emulsion. Those particles have a broad range of applications spanning from MRI, drug delivery, therapy and cell labeling, but unfortunately they are hydrophobic and therefore would have a limited application *in vivo* without a proper carrier [154]. Results obtained have shown that the encapsulation enables a high loading level of SPION particles, followed by their clustering within the aqueous core. The resulting superparamagnetic properties enable an effective contrast for cancer diagnosis and treatment [155].

Another imaging application of PEO–PBD copolymers has been reported by the Maskos research group, whereby QDs were encapsulated within polymersome membranes [154]. Although QD nanocrystals have been investigated mainly on the basis of their broad imaging applications, unfortunately their inorganic nature induces a cellular toxicity that limits their biological application compared to the more biocompatible, organic-based fluorophores. This was confirmed by studies

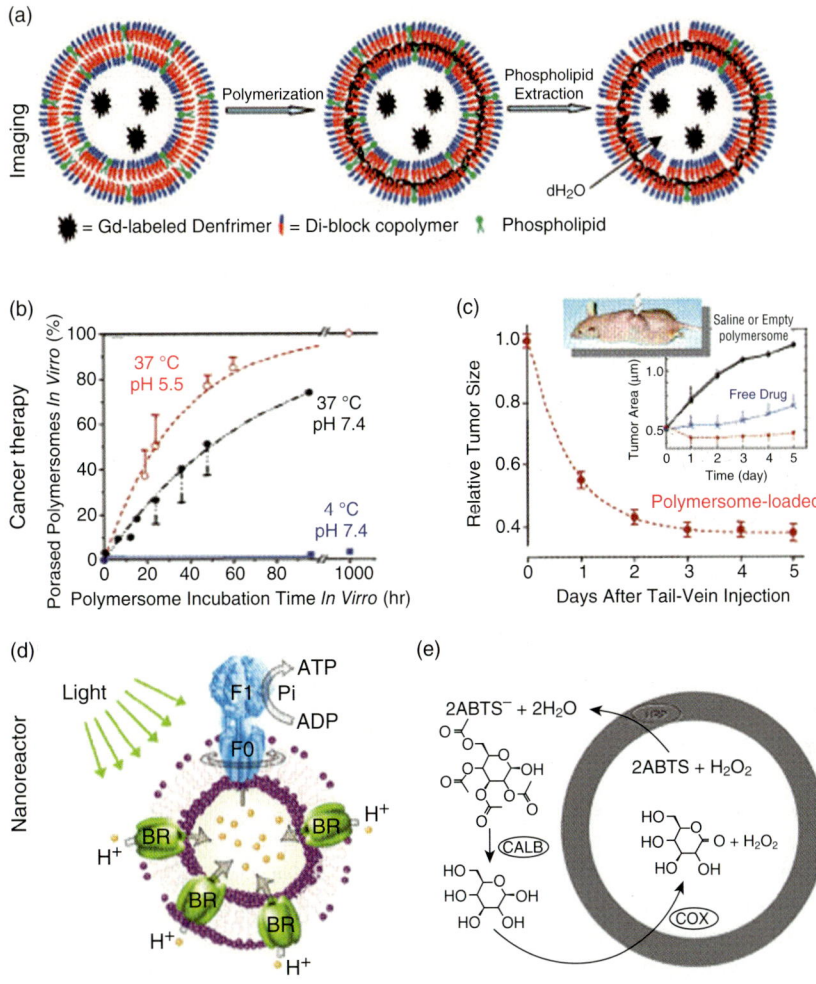

Figure 6.4 (a) Schematic representation of the generation of porous polymersomes loaded with a Gd-chelated dendrimer for magnetic resonance imaging applications [153]; (b) Rate of poration of polymersomes formed from a 3 : 1 mol : mol mixture of PEO–PBD and PEO–PLA, following their incubation at both 37 °C and at 4 °C with buffered saline solutions. At 37 °C, the effect of pH was also investigated, by applying conditions of both physiological pH and mildly acidic pH conditions, analogous to the endocytotic environment; (c) Reduction in the size of a human breast tumor implanted into nude mice following successful delivery of the polymersomes loaded with both paclitaxel and doxorubicin anticancer drugs into the tumor tissue. In control experiments, when mice were injected with saline, unloaded polymersomes, or the free drug, the tumor growth was not impeded [146]; (d) Schematic representation of the "proteopolymersome" nanoreactor, showing that the bacteriorhodopsin (BR) channels present within the polymersome membrane can transport H^+ ions into the polymersome interior in response to light, triggering the conversion of ADP to ATP in the presence of inorganic phosphate (Pi) by the action of the F0F1–ATP synthase rotary motor protein, which is also incorporated within the polymersome membrane [98]; (e) Schematic representation of a three-step cascade reaction for the conversion of ABTS to ABTS•+, catalyzed by enzymes positioned within different polymersome compartments [37].

conducted with murine melanoma cells treated with polymersomes encapsulating porphyrin-based fluorophores [156].

6.5.2
Cancer Therapy

Cancer therapy is another emerging application of polymersomes, based on their ability to load both hydrophilic and hydrophobic anticancer drugs within their lumen and membrane cores, respectively. Both, hydrophilic (doxorubicin) and hydrophobic (paclitaxel) anticancer drugs have been loaded simultaneously within PEO–PLA and PEO–PCL polymersomes (generated by mixing these hydrolytically degradable block copolymers 1:3 mol:mol with inert PEO–PBD) by Ahmed and coworkers within the Discher research group [38, 55, 98]. The polymersomes were able to deliver these drugs into a human breast tumor implanted into mice, and growth retardation and a reduction in tumor area were also observed (Figure 6.4c). At five days after an intravenous injection of the polymersomes into mice, the tumor shrank to less than 50% of its original size. Due to the biodegradability of the ester blocks comprising the polymersomes, and the blending of a degradable block copolymer with a chemically inactive block copolymer, the drug release rate could be controlled, with the mildly acidic environment within cellular endosomes triggering a faster ester hydrolysis and polymersome poration compared to the extracellular environment (Figure 6.4b). Temperature was also found to have an effect on the rate of polymersome poration (see Figure 6.4c).

6.5.3
Polymersomes as Delivery Vectors

The potentials of PEO–PBD and PMPC–PDPA as delivery vectors have undergone extensive investigation, with both having been applied to the encapsulation of different proteins. Arifin and Palmer demonstrated the ability of PEO–PBD to encapsulate bovine hemoglobin (Hb), without perturbing its affinity for oxygen, thus mimicking the red blood cells' capability. Such an effect can, in turn, be used to generate alternative *in vivo* oxygen therapeutics [56].

PMPC–PDPA have been studied by the present author's group for the delivery of primary human antibodies, in order to achieve not only "in live" imunolabeling (by using fluorescently labeled antibodies), but also to exploit the therapeutic potential of the antibodies [128].

6.5.4
Nanoreactors

Extensive studies have also been undertaken by Meier and coworkers in the field of protein incorporation within polymersome walls, to yield nanoreactors [36, 157, 158]. An example of this is the incorporation of ionophores (i.e., lipophilic molecules used to carry ions across membranes, for example, by forming channels to avoid contact of the ions with the hydrophobic membrane core, or by binding to the ions, thereby reducing their polarity) within poly(2-ethyl-2-oxazoline)–poly(dimethylsiloxane)–poly(2-ethyl-2-oxazoline) (PMOXA–PDMS–PMOXA) triblock copolymer polymersome membranes [159], to allow a controlled calcium phosphate precipitation to occur within the polymersome aqueous cores. For this, anionic phosphates were encapsulated within the polymersomes, and the ion-carrier channels successfully used to transport Ca^{2+} across the polymer membrane, which allowed calcium phosphate precipitation to occur in the polymersome lumen.

6.5.5
Artificial Cells and Organelles

Various research groups have attempted to generate artificial cells/cell organelles by the incorporation of enzymes [36] or protein channels [38, 160, 161] within polymersome membranes, thereby creating nanoreactors. Montemagno and coworkers have developed a multiprotein polymersome system for ATP synthesis [162]. These polymersomes, which were formed from an ABA triblock copolymer, poly(2-ethyl-2-oxazoline)–poly(dimethylsiloxane)–poly(2-ethyl-2-oxazoline) (PEtOz–PDMS–PEtOz), incorporated two different protein channels within the polymersome membrane: (i) ATP synthetase, a rotary motor protein; and (ii) bacteriorhodopsin, which can act as a proton pump in response to light (Figure 6.4d), resulting in an influx of hydrogen ions (H^+) into the polymersome lumen. Through coupled reactions involving both protein channels, ATP was successfully synthesized in the presence of ADP and inorganic phosphate (Pi), providing this "hybrid proteopolymersome" with potential use in numerous applications, including the "... *in vitro* investigation of cellular metabolism" [37].

Van Hest and coworkers have also investigated the incorporation of enzymes within PS–polyisocyanoalanine(2-thiophene-3-yl-ethyl)amide (PIAT) polymersomes for the generation of polymersome "nanoreactors" [37, 38, 163]. These block copolymers comprised a mobile hydrophobic block (PS) and a rigid hydrophilic block (PIAT), and yielded polymersomes that contained pores in the membrane, allowing the diffusion of small solutes between the polymersome lumen and the external aqueous milieu, while larger biomacromolecules remained entrapped [170]. Similarly, Van Hest and colleagues succeeded in encapsulating enzymes within these polymersomes, and were also able to impart a high degree of control over the location of the enzymes within the polymersomes [38, 163, 164]. As a consequence, three different enzymes involved in a multistep reaction – *Candida antarctica* lipase B (CALB), horseradish peroxidase (HRP), and glucose oxidase (GOx) – were located outside the polymersomes, within the polymersome lumen, and within the polymersome membrane, respectively. The coupled catalytic activity of these enzymes was found to be 100-fold higher when located inside the polymersome compartments than when free in solution. A schematic representation of the nanoreactor and the catalytic conversion of glucose into the lactone, releasing hydrogen peroxide for catalytic reaction with 2,2′-azinobis

(3-ethylbenzothiazoline-6-sulfonic acid (ABTS), to yield ABTS•+, is shown in Figure 6.4e.

6.5.6
Gene Therapy

Finally, gene therapy is an area of research that is currently attracting significant attention, based on its potential to replace or "switch off" the expression of particular genes. The intracellular delivery of naked genes lacks efficiency due to their negative charge and size, as they experience a repulsive interaction from the negatively charged cell plasma membrane. Various methods to either avoid interaction with the cell membrane (e.g., by the direct injection of genes into the cell using electroporation techniques), or to use a vector to transport genes into the cell, have been investigated, and some of these have involved the use of polymersomes.

Korobko and coworkers [164, 165] encapsulated DNA in polymersomes prepared from a cationic amphiphilic diblock copolymer poly-(butadiene-b-N-methyl-4-vinyl pyridinium) (PBD–P4VPQI). Subsequently, DNA accumulation appeared both in the polymersome core and brush, due to the preparation protocol followed; however, the final formulation was very stable and thus suitable for air purification and redispersion in aqueous PEG solution. An intracellular delivery was achieved by endosomal rupture due to the polymer charge [166].

Poly(amino acid) (poly(AA))-based polymersomes have also been considered for transfection by Brown and coworkers, who used a triblock copolymer generated by attaching methoxy-PEG and palmitic acid to a poly-L-lysine (PLL) or poly-L-ornithine (PO) backbone [166]. These slightly anionic polymersomes have shown improved results in transfection efficiency both *in vitro* and *in vivo* compared to a basic poly(AA)/DNA complex or naked DNA but, unfortunately, they proved to be cytotoxic at therapeutic concentrations [167]. In contrast, the neutral system proved to be more effective as a gene delivery vector, without generating any significant morbidity. An example was degradable PEG–PLA polymersomes, which displayed an encapsulation efficiency and delivery of small interfering RNA (siRNA) *in vitro* which was comparable to levels achieved with the commonly used lipoplex of siRNA and Lipofectamine 2000 (LF2K) [168]. More detailed polymers were dissolved in dimethyl sulfoxide (DMSO) and added to an aqueous solution of fluorescently labeled siRNA to visually assess the encapsulation and delivery. Furthermore, a siRNA, specific for the silencing of lamin A/C, was added. By using an emulsion technique, polymersomes (93 ± 7 nm) were formed and subsequently purified by dialysis. Results obtained with these polymersomes in A549 cell lines confirmed a general 30% knockdown of lamin expression. Although the knockdown results were similar to those obtained with LF2K- or lipoplex-mediated delivery, the PEG–PLA polymersomes proved to be less cytotoxic and therefore more suitable for *in vivo* applications [169].

In emphasizing the importance of stealth, PMPC–PDPA polymersomes were able to physically encapsulate DNA with high transfection efficiencies in both cell lines and primary cells [59, 169], modestly affecting the viability of the cells. Furthermore, the presence of biocompatible PMPC chains on the external corona, coupled with the 200–400 nm vesicle dimensions, are likely to promote relatively long *in vivo* circulation times.

6.6
Conclusions

Polymersomes are the biomimetic analogues of natural phospholipid vesicles based on amphiphilic block copolymers. Their macromolecular nature imparts polymersomes with intrinsic and unique chemical and mechanical stabilities, as well as an amazing versatility that allows their properties to be easily tuned simply by adjusting either the block copolymers' chemistry and/or their molecular weight. In particular, the selection of hydrophilic blocks dictates the polymersomes' surface chemistry and, thus, their interaction with the environment, while the molecular weight will tune their mechanical properties and permeability. The possibility to finely design polymersomes will allow their use in many different applications, ranging from classical drug delivery to the generation of biomimetic nanoreactors. Although polymersomes are the simplest structures that amphiphilic membranes can form, in analogy with biological systems more complex synthetic structures can be created. Indeed, several examples of block copolymer membrane-enclosed structures have been reported, such as complex multilamellar structures [148–150], tubular myelin-like structures [110, 146, 147], and tissue-like structures [151]. Parameters such as spontaneous curvature, for example, can induce remarkable modifications in the morphology of complex structures, such as perforated vesicles, as shown by [94] Polymersomes may, then, be seen simply as a starting point on the scientific pathway towards the perspective of new biomimetic, intricate structures capable of conducting complex functions, such as the Golgi apparatus or neural axons. One day – perhaps not too far away – it might even become possible to prepare a review entitled, "From polymersomes to *polymercells*."

References

1 Alberts, B. *et al.* (2002) *Molecular Biology of the Cell*, 4th edn, Garland Science.
2 Mansy, S.S. *et al.* (2008) Template-directed synthesis of a genetic polymer in a model protocell. *Nature*, 454, 122–5.
3 Matyjaszewski, K. and Spanswick, J. (2005) Controlled/living radical polymerization. *Materials Today*, 8, 26–33.
4 Matyjaszewski, K. and Xia, J. (2001) Atom transfer radical polymerization. *Chemical Reviews*, 101, 2921–90.
5 Dove, A.P. (2008) Controlled ring-opening polymerisation of cyclic esters: polymer blocks in self-assembled nanostructures. *Chemical Communications*, 6446–70.
6 LoPresti, C., Lomas, H., Massignani, M., Smart, T. and Battaglia, G. (2009) Polymersomes: nature inspired nanometer sized compartments. *Journal of Materials Chemistry*, 19, 3576–90.
7 Battaglia, G. and Ryan, A.J. (2005) Bilayers and interdigitation in block copolymer vesicles. *Journal of the*

American Chemical Society, **127**, 8757–64.
8 Discher, B.M. et al. (1999) Polymersomes: tough vesicles made from diblock copolymers. *Science*, **284**, 1143–6.
9 Discher, D.E. and Eisenberg, A. (2002) Polymer vesicles. *Science*, **297**, 967–73.
10 Discher, D.E. et al. (2007) Emerging applications of polymersomes in delivery: from molecular dynamics to shrinkage of tumors. *Progress in Polymer Science*, **32**, 838–57.
11 Discher, D.E. and Ahmed, F. (2006) Polymersomes. *Annual Review of Biomedical Engineering*, **8**, 323–41.
12 Christian, D.A. et al. (2009) Polymersome carriers: from self-assembly to siRNA and protein therapeutics. *European Journal of Pharmaceutics and Biopharmaceutics*, **71**, 463–74.
13 Levine, D.H. et al. (2008) Polymersomes: a new multi-functional tool for cancer diagnosis and therapy. *Methods*, **46**, 25–32.
14 Onaca, O., Enea, R., Hughes, D.W. and Meier, W. (2009) Stimuli-responsive polymersomes as nanocarriers for drug and gene delivery. *Macromolecular Bioscience*, **9**, 129–39.
15 Tanford, C. (1979) Interfacial free energy and the hydrophobic effect. *Proceedings of the National Academy of Sciences of the United States of America*, **76**, 4175–6.
16 Schillén, K., Bryskhe, K. and Mel'nikova, Y.S. (1999) Vesicles formed from poly(ethylene oxide) poly(propylene oxide)-poly(ethylene oxide) triblock copolymer in dilute aqueous solution. *Macromolecules*, **32**, 6885–8.
17 Zhang, L. and Eisenberg, A. (1995) Multiple morphologies of "crew-cut" aggregates of polystyrene-b-poly(acrylic acid) block copolymers. *Science*, **268**, 1728–31.
18 Kabanov, A.V., Bronich, T.K., Kabanov, V.A., Yu, K. and Eisenberg, A. (1998) Spontaneous formation of vesicles from complexes of block ionomers and surfactants. *Journal of the American Chemical Society*, **120**, 9941–2.
19 Shen, H. and Eisenberg, A. (2000) Control of architecture in block-copolymer vesicles. *Angewandte Chemie International Edition*, **39**, 3310–12.
20 Yu, K. and Eisenberg, A. (1998) Bilayer morphologies of self-assembled crew-cut aggregates of amphiphilic PS-b-PEO diblock copolymers in solution. *Macromolecules*, **31**, 3509–18.
21 Geng, Y., Ahmed, F., Bhasin, N. and Discher, D.E. (2005) Visualizing worm micelle dynamics and phase transitions of a charged diblock copolymer in water. *Journal of Physical Chemistry B*, **109**, 3772–9.
22 Lee, J.C.-M. et al. (2001) Preparation, stability, and in vitro performance of vesicles made with diblock copolymers. *Biotechnology and Bioengineering*, **73**, 135–45.
23 Zhulina, E.B. and Halperin, A. (1992) Lamellar mesogels amd mesophases: a self-consistent-field theory. *Macromolecules*, **25**, 5730–41.
24 Battaglia, G., Ryan, A.J. and Tomas, S. (2006) Polymeric vesicle permeability: a facile chemical assay. *Langmuir*, **22**, 4910–13.
25 Meng, F., Hiemstra, C., Engbers, G.H.M. and Feijen, J. (2003) Biodegradable polymersomes. *Macromolecules*, **36**, 3004–6.
26 Kukula, H., Schlaad, H., Antonietti, M. and Förster, S. (2002) The formation of polymer vesicles or "peptosomes" by polybutadiene-block-poly(L-glutamate)s in dilute aqueous solution. *Journal of the American Chemical Society*, **124**, 1658–63.
27 Du, J., Tang, Y., Lewis, A.L. and Armes, S.P. (2005) pH-sensitive vesicles based on a biocompatible zwitterionic diblock copolymer. *Journal of the American Chemical Society*, **127**, 17982–3.
28 Holowka, E.P., Sun, V.Z., Kamei, D.T. and Deming, T.J. (2007) Polyarginine segments in block copolypeptides drive both vesicular assembly and intracellular delivery. *Nature Materials*, **6**, 52–7.
29 Borchert, U. et al. (2006) pH-induced release from P2VP-PEO block copolymer vesicles. *Langmuir*, **22**, 5843–7.
30 Liu, F. and Eisenberg, A. (2003) Preparation and pH-triggered inversion of vesicles from poly(acrylic acid)-block-polystyrene-block-poly(4-vinyl pyridine).

Journal of the American Chemical Society, **125**, 15059–64.

31 Du, J. and Armes, S.P. (2005) pH-responsive vesicles based on a hydrolytically self-cross-linkable copolymer. *Journal of the American Chemical Society*, **127**, 12800–1.

32 Chiu, H.-C., Lin, Y.-W., Huang, Y.-F., Chuang, C.-K. and Chern, C.-S. (2008) Polymer vesicles containing small vesicles within interior aqueous compartments and pH-responsive transmembrane channels. *Angewandte Chemie International Edition*, **47**, 1875–8.

33 Wang, G., Tong, X. and Zhao, Y. (2004) Preparation of azobenzene-containing amphiphilic diblock copolymers for light-responsive micellar aggregates. *Macromolecules*, **37**, 8911–17.

34 Napoli, A., Valentini, M., Tirelli, N., Martin, M. and Hubbell, J.A. (2004) Oxidation-responsive polymeric vesicles. *Nature Materials*, **3**, 183–9.

35 Cerritelli, S., Velluto, D. and Hubbell, J.A. (2007) PEG-SS-PPS: reduction-sensitive disulfide block copolymer vesicles for intracellular drug delivery. *Biomacromolecules*, **8**, 1966–72.

36 Sauer, M., Haefele, T., Graff, A., Nardin, C. and Meier, W. (2001) Ion-carrier controlled precipitation of calcium phosphate in giant ABA triblock copolymer vesicles. *Chemical Communications*, 2452–3.

37 Choi, H.J. and Montemagno, C.D. (2005) Artificial organelle: ATP synthesis from cellular mimetic polymersomes. *Nano Letters*, **5**, 2538–42.

38 Vriezema, D.M. *et al.* (2007) Positional assembly of enzymes in polymersome nanoreactors for cascade reactions. *Angewandte Chemie International Edition*, **46**, 7378–82.

39 Holowka, E.P., Pochan, D.J. and Deming, T.J. (2005) Charged polypeptide vesicles with controllable diameter. *Journal of the American Chemical Society*, **127**, 12423–8.

40 Korobko, A.V., Backendorf, C. and Van der Maarel, J.R.C. (2006) Plasmid DNA encapsulation within cationic diblock copolymer vesicles for gene delivery. *Journal of Physical Chemistry B*, **110**, 14550–6.

41 Van Hest, J.C.M., Delnoye, D.A.P., Baars, M.W.P.L., von Genderen, M.H.P. and Mejer, E.W. (1995) Polystyrene-dendrimer amphiphilic block copolymers with a generation-dependent aggregation. *Science*, **268**, 1592–5.

42 Cornelissen, J.J.L.M., Fischer, M., Sommerdijk, N.A.J.M. and Nolte, R.J.M. (1998) Helical superstructures from charged poly(styrene)-poly(isocyanodipeptide) block copolymers. *Science*, **280**, 1427–30.

43 Gebhardt, K.E., Ahn, S., Venkatachalam, G. and Savin, D.A. (2008) Role of secondary structure changes on the morphology of polypeptide-based block copolymer vesicles. *Journal of Colloid and Interface Science*, **317**, 70–6.

44 Sigel, R., Losik, M. and Schlaad, H. (2007) pH responsiveness of block copolymer vesicles with a polypeptide corona. *Langmuir*, **23**, 7196–9.

45 Kim, K.T., Winnik, M.A. and Manners, I. (2006) Synthesis and self-assembly of dendritic-helical block copolypeptides. *Soft Matter*, **2**, 957–65.

46 Dai, L.M. and Toprakcioglu, C. (1992) End-adsorbed triblock copolymer chains at the liquid-solid interface: bridging effects in a good solvent. *Macromolecules*, **25**, 6000–6.

47 Levicky, R., Koneripalli, N., Tirrell, M. and Satija, S.K. (1998) Concentration profiles in densely tethered polymer brushes. *Macromolecules*, **31**, 3731–4.

48 Stoenescu, R., Graff, A. and Meier, W. (2004) Asymmetric ABC-triblock copolymer membranes induce a directed insertion of membrane proteins. *Macromolecular Bioscience*, **4**, 930–5.

49 Napoli, A., Tirelli, N., Wehrli, E. and Hubbell, J.A. (2002) Lyotropic behavior in water of amphiphilic ABA triblock copolymers based on poly(propylene sulfide) and poly(ethylene glycol). *Langmuir*, **18**, 8324–9.

50 Sommerdijk, N.A.J.M., Holder, S.J., Hiorns, R.C., Jones, R.G. and Nolte, R.J.M. (2000) Self-assembled structures from an amphiphilic multiblock copolymer containing rigid semiconductor segments. *Macromolecules*, **33**, 8289–94.

51 Walther, A. and Muller, A.H.E. (2008) Janus particles. *Soft Matter*, **4**, 663–8.
52 Wittemann, A., Azzam, T. and Eisenberg, A. (2007) Biocompatible polymer vesicles from biamphiphilic triblock copolymers and their interaction with bovine serum albumin. *Langmuir*, **23**, 2224–30.
53 Blanazs, A., Massignani, M., Battaglia, G., Armes, S.P. and Ryan, A.J. (2009) Tailoring macromolecular expression at block copolymer vesicle surface. *Advanced Functional Materials*, **19**, 2906–14.
54 Brannan, A.K. and Bates, F.S. (2004) ABCA tetrablock copolymer vesicles. *Macromolecules*, **37**, 8816–19.
55 Ahmed, F. and Discher, D.E. (2004) Self-porating polymersomes of PEG-PLA and PEG-PCL: hydrolysis-triggered controlled release vesicles. *Journal of Controlled Release*, **96**, 37–53.
56 Ahmed, F. et al. (2006) Shrinkage of a rapidly growing tumor by drug-loaded polymersomes: pH-triggered release through copolymer degradation. *Molecular Pharmaceutics*, **3**, 340–50.
57 Najafi, F. and Sarbolouki, M.N. (2003) Biodegradable micelles/polymersomes from fumaric/sebacic acids and poly(ethylene glycol). *Biomaterials*, **24**, 1175–82.
58 Napoli, A., Boerakker, M.J., Tirelli, N., Nolte, R.J.M., Sommerdijk, N.A.J.M. and Hubbell, J.A. (2004) Glucose-oxidase based self-destructing polymeric vesicles. *Langmuir*, **20**, 3487–91.
59 Lomas, H. et al. (2008) Non-cytotoxic polymer vesicles for rapid and efficient intracellular delivery. *Faraday Discussions*, **139**, 143–59.
60 Qin, S., Geng, Y., Discher, D.E. and Yang, S. (2006) Temperature-controlled assembly and release from polymer vesicles of poly(ethylene oxide)-block-poly(N-isopropylacrylamide). *Advanced Materials*, **18**, 2905–9.
61 Smart, T., Mykhaylyk, O.O., Ryan, A.J. and Battaglia, G. (2009) Polymersomes hydrophilic brush scaling relations. *Soft Matter*, **5**, 3607–10.
62 de-Gennes, P.G. (1980) Conformations of polymers attached to an interface. *Macromolecules*, **13**, 1069–75.
63 de-Gennes, P.G. (1980) *Scaling Concepts in Polymer Physics*, Cornell University Press.
64 Jain, S. and Bates, F.S. (2004) Consequences of nonergodicity in aqueous binary PEO-PB micellar dispersions. *Macromolecules*, **37**, 1511–23.
65 Förster, S. et al. (1999) Fusion of charged block copolymer micelles into toroid networks. *Journal of Physical Chemistry B*, **103**, 6657–68.
66 Bermudez, H., Brannan, A.K., Hammer, D.A., Bates, F.S. and Discher, D.E. (2002) Molecular weight dependence of polymersome membrane structure, elasticity, and stability. *Macromolecules*, **35**, 8203–8.
67 Cevc, G. and Marsh, D. (1987) *Phospholipid Bilayers: Physical Principles and Models*, vol. 5, John Wiley & Sons. ISBN-10: 047109255X.
68 Disher, D.E. and Ahmed, F. (2006) Polymersomes. *Annual Review of Biomedical Engineering*, **8**, 323–41.
69 Aranda-Espinoza, H., Bermudez, H., Bates, F.S. and Disher, D.E. (2001) Electromechanical limits of polymersomes. *Physical Review Letters*, **87**, 208301.
70 Srinivas, G., Discher, D.E. and Klein, M. (2004) Self-assembly and properties of diblock copolymers by coarse-grain molecular dynamics. *Nature Materials*, **3**, 6338–644.
71 Hadziioannou, G. and Skoulios, A. (1982) Molecular weight dependence of lamellar structure in styrene/isoprene two- and three-block copolymers. *Macromolecules*, **15**, 258–62.
72 Hashimoto, T., Shibayama, M. and Kawai, H. (1980) Domain-boundary structure of styrene-isoprene block copolymer films cast from solution. 4. Molecular-weight dependence of lamellar microdomains. *Macromolecules*, **13**, 1237–47.
73 Richards, R.W. and Thomason, J.L. (1983) Small-angle neutron scattering study of block copolymer morphology. *Macromolecules*, **16**, 982–92.
74 Ryan, A.J., Mai, S.-M., Fairclough, J.P.A., Hamley, I.W. and Booth, C. (2001) Ordered melts of block copolymers of

ethylene oxide and 1,2-butylene oxide. *Physical Chemistry Chemical Physics*, **3**, 2961–71.

75 Ortiz, V. et al. (2005) Dissipative particle dynamics simulations of polymersomes. *Journal of Physical Chemistry B*, **109**, 17708–14.

76 Matsen, M.W. and Bates, F.S. (1995) Testing the strong-stretching assumption in a block copolymer microstructure. *Macromolecules*, **28**, 8884–6.

77 Bermùdez, H., Hammer, D.A. and Disher, D.E. (2004) Effect of bilayer thickness on membrane bending rigidity. *Langmuir*, **20**, 540–3.

78 Dalhaimer, P., Bates, F.S., Arada-Espinoza, H. and Disher, D.E. (2003) Synthetic cell elements from block copolymers – hydrodynamic aspects. *Comptes Rendus Physique*, **4**, 251–8.

79 Lodge, T.P. and Dalvi, M.C. (1995) Mechanisms of chain diffusion in lamellar block copolymers. *Physical Review Letters*, **75**, 657–60.

80 Lee, J.C.M., Santore, M., Bates, F.S. and Disher, D.E. (2002) From membranes to melts, rouse to reptation: diffusion in polymersome versus lipid bilayers. *Macromolecules*, **35**, 323–6.

81 Mecke, A., Dittrich, C. and Meier, W. (2006) Biomimetic membranes designed from amphiphilic block copolymers. *Soft Matter*, **2**, 751–9.

82 Photos, P.J., Bacakova, L., Discher, B., Bates, F.S. and Discher, D.E. (2003) Polymer vesicles in vivo: correlations with PEG molecular weight. *Journal of Controlled Release*, **90**, 323–34.

83 Jones, R.A.L. (2004) *Soft Condensed Matter*, Oxford University Press.

84 Belloni, L. (2000) Colloidal interactions. *Journal of Physics: Condensed Matter*, **12**, R549–87.

85 Israelachvili, J.N. (2002) *Intermolecular and Surface Forces*, 9th edn, Elsevier Science Imprint.

86 Menger, F.M. and Gabrielson, K. (1994) Chemically-induced birthing and foraging in vesicle systems. *Journal of the American Chemical Society*, **116**, 1567–8.

87 Menger, F.M. and Balachander, N. (1992) Chemically-induced aggregation, budding, and fusion in giant vesicles: direct observation by light microscopy. *Journal of the American Chemical Society*, **114**, 5862–3.

88 Menger, F.M. and Seredyuk, V.A. (2003) Internally catalyzed separation of adhered lipid membranes. *Journal of the American Chemical Society*, **125**, 11800–1.

89 Nomura, F. et al. (2004) Microscopic observations reveal that fusogenic peptides induce liposome shrinkage prior to membrane fusion. *Proceedings of the National Academy of Sciences of the United States of America*, **101**, 3420–5.

90 Choucair, A.A., Kycia, A.H. and Eisenberg, A. (2003) Kinetics of fusion of polystyrene-b-poly(acrylic acid) vesicles in solution. *Langmuir*, **19**, 1001–8.

91 Luo, L. and Eisenberg, A. (2001) Thermodynamic size control of block copolymer vesicles in solution. *Langmuir*, **17**, 6804–11.

92 Zhou, Y. and Yan, D. (2005) Real-time membrane fusion of giant polymer vesicles. *Journal of the American Chemical Society*, **127**, 10468–9.

93 Zhou, Y. and Yan, D. (2005) Real-time membrane fission of giant polymer vesicles. *Angewandte Chemie International Edition*, **44**, 3223–6.

94 Haluska, C.K. et al. (2006) Time scales of membrane fusion revealed by direct imaging of vesicle fusion with high temporal resolution. *Proceedings of the National Academy of Sciences of the United States of America*, **103**, 15841–6.

95 Shillcock, J.C. and Lipowsky, R. (2005) Tension-induced fusion of bilayer membranes and vesicles. *Nature Materials*, **4**, 225–8.

96 Lei, G. and MacDonald, R.C. (2003) Lipid bilayer vesicle fusion: intermediates captured by high-speed microfluorescence spectroscopy. *Biophysical Journal*, **85**, 1585–99.

97 de Las Heras Alarcon, C., Pennadam, S. and Alexander, C. (2005) Stimuli responsive polymers for biomedical applications. *Chemical Society Reviews*, **34**, 276–85.

98 Ahmed, F. et al. (2006) Biodegradable polymersomes loaded with both paclitaxel and doxorubicin permeate and shrink tumors, inducing apoptosis in proportion to accumulated drug. *Journal of Controlled Release*, **116**, 150–8.

99 Ghoroghchian, P.P. et al. (2006) Bioresorbable vesicles formed through spontaneous self-assembly of amphiphilic poly(ethylene oxide)-block-polycaprolactone. *Macromolecules*, **39**, 1673–5.

100 Rodriguez-Hernandez, J., Babin, J., Zappone, B. and Lecommandoux, S. (2005) Preparation of shell cross-linked nano-objects from hybrid-peptide block copolymers. *Biomacromolecules*, **6**, 2213–20.

101 Du, J., Tang, Y., Lewis, A.L. and Armes, S.P. (2005) pH-sensitive vesicles based on a biocompatible zwitterionic diblock copolymer. *Journal of the American Chemical Society*, **127**, 17982–3.

102 Lomas, H. et al. (2008) Non-cytotoxic polymer vesicles for rapid and efficient intracellular delivery. *Faraday Discussions*, **139**, 143–59.

103 Ruiz-Pérez, L. et al. (2008) Conformation of poly(methacrylic acid) chains in dilute aqueous solution. *Macromolecules*, **41**, 2203–11.

104 Choucair, A. and Eisenberg, A. (2003) Control of amphiphilic block copolymer morphologies using solution conditions. *The European Physical Journal. E, Soft Matter*, **10**, 37–44.

105 Fernyhough, C., Ryan, A.J. and Battaglia, G. (2009) pH controlled assembly of a polybutadiene–poly(methacrylic acid) copolymer in water: packing considerations and kinetic limitations. *Soft Matter*, **5**, 1674–82.

106 Li, Y., Lokitz, B.S. and McCormick, C.L. (2006) Thermally responsive vesicles and their structural locking through polyelectrolyte complex formation. *Angewandte Chemie International Edition*, **45**, 5792–5.

107 Rank, A., Hauschild, S., Forster, S. and Schubert, R. (2009) Preparation of monodisperse block copolymer vesicles via a thermotropic cylinder-vesicle transition. *Langmuir*, **25**, 1337–44.

108 Wang, G., Tong, X. and Zhao, Y. (2004) Preparation of azobenzene-containing amphiphilic diblock copolymers for light-responsive micellar aggregates. *Macromolecules*, **37**, 8911–17.

109 Mabrouk, E., Cuvelier, D., Brochard-Wyart, F., Nassoy, P. and Li, M.H. (2009) Bursting of sensitive polymersomes induced by curling. *Proceedings of the National Academy of Sciences of the United States of America*, **106**, 7294–8.

110 Smart, T.P., Mykhaylyk, O.O., Ryan, A.J. and Battaglia, G. (2009) Polymersomes hydrophilic brush scaling relations. *Soft Matter*, **5**, 3607–10.

111 Alcantar, N.A., Aydil, E.S. and Israelachvili, J.N. (2000) Polyethylene glycol-coated biocompatible surfaces. *Journal of Biomedical Materials Research*, **51**, 343–51.

112 Israelachvili, J.N. and Wennerstrom, H. (1996) Role of hydration and water structure in biological and colloidal interactions. *Nature*, **379**, 219–25.

113 Vonarbourg, A., Passirani, C., Saulnier, P. and Benoit, J.-P. (2006) Parameters influencing the stealthiness of colloidal drug delivery systems. *Biomaterials*, **27**, 4356–73.

114 Lasic, D.D. (1994) Sterically stabilized vesicles. *Angewandte Chemie International Edition*, **33**, 1685–98.

115 Papahadjopoulos, D. et al (1991) Sterically stabilized liposomes: improvements in pharmacokinetics and antitumor therapeutic efficacy. *Proceedings of the National Academy of Sciences of the United States of America*, **88**, 11460–4.

116 Christian, D.A. et al. (2009) Spotted vesicles, striped micelles and Janus assemblies induced by ligand binding. *Nature Materials*, **8**, 843–9.

117 Massignani, M. et al. (2009) Controlling cellular uptake by surface chemistry, size, and surface topology at the nanoscale. *Small*, **5**, 2424–32.

118 Kuntz, I.D., Chen, K., Sharp, K.A. and Kollman, P.A. (1999) The maximal affinity of ligands. *Proceedings of the National Academy of Sciences of the United States of America*, **96**, 9997–10002.

119 Hammer, D.A. et al. (2008) Leuko-polymersomes. *Faraday Discussions*, **139**, 129–41.

120 Lin, J.J., Ghoroghchian, P., Zhang, Y. and Hammer, D.A. (2006) Adhesion of antibody-functionalized polymersomes. *Langmuir*, **22**, 3975–9.

121 Lin, J.J. et al. (2004) The effect of polymer chain length and surface

density on the adhesiveness of functionalized polymersomes. *Langmuir*, **20**, 5493–500.

122 Broz, P. et al. (2005) Cell targeting by a generic receptor-targeted polymer nanocontainer platform. *Journal of Controlled Release*, **102**, 475–88.

123 Pang, Z.Q. et al. (2008) Preparation and brain delivery property of biodegradable polymersomes conjugated with OX26. *Journal of Controlled Release*, **128**, 120–7.

124 Christian, N.A. et al. (2007) Tat-functionalized near-infrared emissive polymersomes for dendritic cell labeling. *Bioconjugate Chemistry*, **18**, 31–40.

125 Upadhyay, K.K. et al. (2009) Biomimetic doxorubicin loaded polymersomes from hyaluronan-block-poly(gamma-benzyl glutamate) copolymers. *Biomacromolecules*, **10**, 2802–8.

126 Luo, L. and Eisenberg, A. (2001) Thermodynamic stabilization of mechanism of block copolymer vesicles. *Journal of the American Chemical Society*, **123**, 1012–13.

127 Bryskhe, K., Jansson, J., Topgaard, D., Shillén, K. and Olsson, U. (2004) Spontaneous vesicle formation in a block copolymer system. *Journal of Physical Chemistry B*, **108**, 9710–19.

128 Arifin, D.R. and Palmer, A.F. (2005) Polymersome encapsulated hemoglobin: a novel type of oxygen carrier. *Biomacromolecules*, **6**, 2172–81.

129 Angelova, M.I. and Dimitrov, D.S. (1986) Liposome electroformation. *Faraday Discussions*, **81**, 303.

130 Shen, H. and Eisenberg, A. (1999) Morphological phase diagram for a ternary system of block copolymer PS_{310}-b-PAA_{52}/Dioxane/H_2O. *Journal of Physical Chemistry B*, **103**, 9473–87.

131 Pozo Navas, B., Lohner, K., Deutsch, G., Sevcsik, E., Riske, K.A., Dimova, R., Garidel, P. and Pabst, G. (2005) Composition dependence of vesicle morphology and mixing properties in a bacterial model membrane system. *Biochimica et Biophysica Acta Biomembranes*, **1716**, 40–8.

132 Dimova, R. et al. (2006) A practical guide to giant vesicles. Probing the membrane nanoregime via optical microscopy. *Journal of Physics: Condensed Matter*, **18**, S1151.

133 Lipowsky, R. and Leibler, S. (1986) Unbinding transitions of interacting membranes. *Physical Review Letters*, **56**, 2541–4.

134 Lipowsky, R. and Lipowsky, R. (1995) From bunches of membranes to bundles of strings. *Zeitschrift für Physik B: Condensed Matter*, **97**, 193–203.

135 Lipowsky, R. (1991) The conformation of membranes. *Nature*, **349**, 475–81.

136 Roux, D., Codon, C. and Cates, M.E. (1992) Sponge phases in surfactant solutions. *Journal of Physical Chemistry*, **96**, 4174–87.

137 Helfrich, W. (1994) Lyotropic lamellar phases. *Journal of Physics: Condensed Matter*, **6**, A79–92.

138 Battaglia, G. and Ryan, A.J. (2006) Effect of amphiphile size on the transformation from a lyotropic gel to a vesicular dispersion. *Macromolecules*, **39**, 798–805.

139 Battaglia, G. and Ryan, A.J. (2006) Pathways of polymeric vesicle formation. *Journal of Physical Chemistry B*, **110**, 10272–9.

140 Moscho, A., Orwar, O., Chiu, D.T., Modi, B.P. and Zare, R.N. (1996) Rapid preparation of giant unilamellar vesicles. *Proceedings of the National Academy of Sciences of the United States of America*, **93**, 11443–7.

141 Menger, F.M. and Angelova, M.I. (1998) Giant vesicles: imitating the cytological processes of cell membranes. *Accounts of Chemical Research*, **31**, 789–97.

142 Gregoriadis, G. (1984) *Preparation of Liposomes*, CRC press.

143 Olson, F., Hunt, C.A., Szoka, F.C., Vail, W.J. and Papahadjopoulos, D. (1979) Preparation of liposomes of defined size distribution by extrusion through polycarbonate membrane. *Biochimica et Biophysica Acta*, **9**, 557.

144 Buchanan, M., Egelhaaf, S.U. and Cates, M.E. (2000) Dynamics of interface instabilities in nonionic lamellar phases. *Langmuir*, **16**, 3718–26.

145 Battaglia, G. and Ryan, A.J. (2006) Neuron-like tubular membranes made of diblock copolymer amphiphiles.

Angewandte Chemie International Edition, 45, 2052–6.
146 Cheng, Z.L. and Tsourkas, A. (2008) Paramagnetic porous polymersomes. *Langmuir*, 24, 8169–73.
147 Ghoroghchian, P.P. *et al.* (2007) Controlling bulk optical properties of emissive polymersomes through intramembranous polymer-fluorophore interactions. *Chemistry of Materials*, 19, 1309–18.
148 Ghoroghchian, P.P., Therien, M.J. and Hammer, D.A. (2009) In vivo fluorescence imaging: a personal perspective. *Wiley Interdisciplinary Reviews: Nanomedicine and Nanobiotechnology*, 1, 156–67.
149 Duncan, T.V., Ghoroghchian, P.P., Rubtsov, I.V., Hammer, D.A. and Therien, M.J. (2008) Ultrafast excited-state dynamics of nanoscale near-infrared emissive polymersomes. *Journal of the American Chemical Society*, 130, 9773–84.
150 Ghoroghchian, P.P. *et al.* (2005) Broad spectral domain fluorescence wavelength modulation of visible and near-infrared emissive polymersomes. *Journal of the American Chemical Society*, 127, 15388–90.
151 Ghoroghchian, P.P. *et al.* (2005) Near-infrared-emissive polymersomes: self-assembled soft matter for in vivo optical imaging. *Proceedings of the National Academy of Sciences of the United States of America*, 102, 2922–7.
152 Massignani, M., Canton, I., Sun, T., Hearnden, V., MacNeil, S., Blanazs, A., Armes, S.P. and Battaglia, A.L.G. (2010) Enhanced fluorescence imaging of live cells by effective cytosolic delivery of probes. *PLoS One*, 5 (5), e10459.
153 Wu, S.P. *et al.* (2005) Near-infrared optical imaging of B16 melanoma cells via low-density lipoprotein-mediated uptake and delivery of high emission dipole strength tris[(porphinato)zinc(II)] fluorophores. *Bioconjugate Chemistry*, 16, 542–50.
154 Yang, X.Q. *et al.* (2009) Tumor-targeting, superparamagnetic polymeric vesicles as highly efficient MRI contrast probes. *Journal of Materials Chemistry*, 19, 5812–17.
155 Baghi, M. *et al.* (2005) The efficacy of MRI with ultrasmall superparamagnetic iron oxide particles (USPIO) in head and neck cancers. *Anticancer Research*, 25, 3665–70.
156 Mueller, W. *et al.* (2009) Hydrophobic shell loading of PB-b-PEO vesicles. *Macromolecules*, 42, 357–61.
157 Massignani, M., Canton, I., Patikarnmonthon, N., Warren, N., Armes, S., Lewis, A. and Battaglia, G. (2010) Cellular delivery of antibodies: effective targeted subcellular imaging and new therapeutic tool. Available from *Nature Proceedings*. Available at: http://hdl.handle.net/10101/npre.2010.4427.1.
158 Nardin, C., Thoeni, S., Widmer, J., Winterhalter, M. and Meier, W. (2000) Nanoreactors based on (polymerized) ABA-triblock copolymer vesicles. *Chemical Communications*, 1433–4.
159 Stoenescu, R. and Meier, W. (2002) Vesicles with asymmetric membranes from amphiphilic ABC triblock copolymers. *Chemical Communications*, (24), 3016–17.
160 Nallani, M. *et al.* (2006) A nanophosphor-based method for selective DNA recovery in synthosomes. *Biotechnology Journal*, 1, 828–34.
161 Onaca, O., Nallani, M., Ihle, S., Schenk, A. and Schwaneberg, U. (2006) Functionalized nanocompartments (Synthosomes): limitations and prospective applications in industrial biotechnology. *Biotechnology Journal*, 1, 795–805.
162 Nallani, M. *et al.* (2006) A nanocompartment system (Synthosome) designed for biotechnological applications. *Journal of Biotechnology*, 123, 50–9.
163 Nallani, M. *et al.* (2007) Polymersome nanoreactors for enzymatic ring-opening polymerization. *Biomacromolecules*, 8, 3723–8.
164 van Dongen, S.F.M., Nallani, M., Cornelissen, J.L.L.M., Nolte, R.J.M. and van Hest, J.C.M. (2009) A three-enzyme cascade reaction through positional assembly of enzymes in a polymersome nanoreactor. *Chemistry: A European Journal*, 15, 1107–14.

165 Korobko, A.V., Backendorf, C. and van der Maarel, J.R. (2006) Plasmid DNA encapsulation within cationic diblock copolymer vesicles for gene delivery. *Journal of Physical Chemistry B*, **110**, 14550–6.

166 Korobko, A.V., Jesse, W. and van der Maarel, J.R. (2005) Encapsulation of DNA by cationic diblock copolymer vesicles. *Langmuir*, **21**, 34–42.

167 Brown, M.D. *et al.* (2000) Preliminary characterization of novel amino acid based polymeric vesicles as gene and drug delivery agents. *Bioconjugate Chemistry*, **11**, 880–91.

168 Brown, M.D. *et al.* (2003) In vitro and in vivo gene transfer with poly(amino acid) vesicles. *Journal of Controlled Release*, **93**, 193–211.

169 Kim, Y.H. *et al.* (2009) Polymersome delivery of siRNA and antisense oligonucleotides. *Journal of Controlled Release*, **134**, 132–40.

170 Vriezema, D.M., Hoogboom, J., Velonia, K., Takazawa, K., Christianen, P.C.M., Maan, J.C., Rowan, A.E. and Nolte, R.J.M. (2003) Vesicles and polymerized vesicles from thiophene-containing rod–coil block copolymers. *Angewandte Chemie International Edition*, **42**, 772–6.

Part Two
Nanoparticles

Nanomaterials for the Life Sciences Vol.10: Polymeric Nanomaterials. Edited by Challa S. S. R. Kumar
Copyright © 2011 WILEY-VCH Verlag GmbH & Co. KGaA, Weinheim
ISBN: 978-3-527-32170-4

7
Synthetic Approaches to Organic Nanoparticles
Stefan Köstler and Volker Ribitsch

7.1
Introduction

Over the past two decades, research in the general fields of nanoscience and nanotechnology has experienced a tremendous growth such that, today, both are well-established scientific disciplines with ever-increasing importance. One major area of research in this rapidly growing discipline is concerned with the development of synthetic methods for nanoparticulate materials, their study, and the development of their applications. This is in particular, related to the high expectations that were – and still are – set for these materials with applications in the fields of microelectronics, photonics and catalysis, and also in the life sciences. In 2009, the number of known chemical substances amounted to more than 50 millions,[1] the vast majority of which are classified as organic compounds. Although, the number of organic compounds is much higher than that of inorganic ones, until now nanotechnology research has been mainly focused on inorganic materials. A huge number of synthetic methods for different inorganic nanoparticles (such as metals, oxides, chalcogenides) have been developed, and many of these have become well established for applications such as biological labeling, medical imaging, catalysis, optoelectronics, or coatings technology [1–3]. In comparison, much less effort has been put into the development of synthetic methods for organic nanoparticles, which contribute a relatively small share to the total field of nanoparticle research. This is reflected by the number of reports related to nanoparticles; for example, a search for the simple query "nanoparticle?" in the scientific search engine Scopus (www.scopus.com) resulted in more than 86 000 hits. Yet, only slightly more than 700 hits were found in a similar search using a more specific query[2] related to organic nanoparticles.

---

1) From CAS REGISTRY (http://www.cas.org/newsevents/releases/50millionth090809.html) (accessed on 14 December 2009).

2) Query: ("organic nanoparticle?" OR "polymer nanoparticle?" OR "polymer nanosuspension?" OR "organic nanosuspension?") search on www.scopus.com (accessed 14 December 2009).

Nanomaterials for the Life Sciences Vol.10: Polymeric Nanomaterials. Edited by Challa S. S. R. Kumar
Copyright © 2011 WILEY-VCH Verlag GmbH & Co. KGaA, Weinheim
ISBN: 978-3-527-32170-4

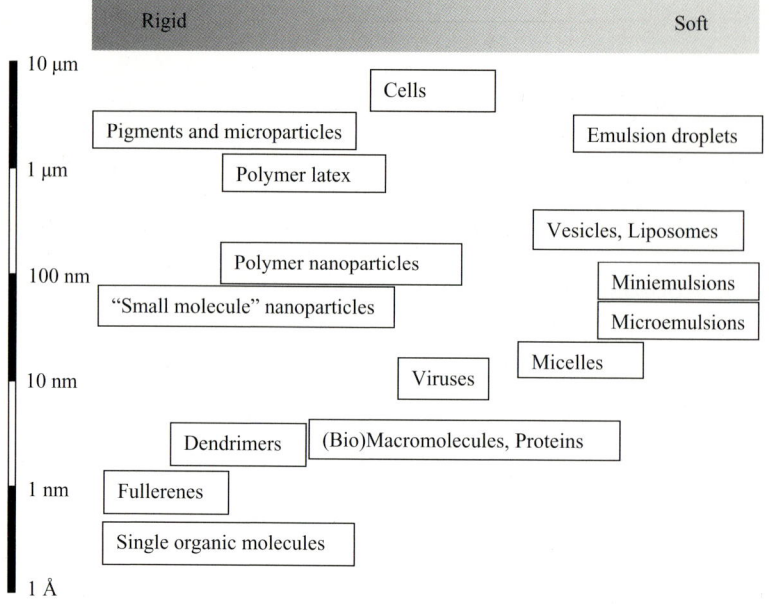

Figure 7.1 Schematic overview on the dimensions of organic molecules, supramolecular assemblies, and particles.

7.1.1
Types of Organic Particle and Scope of This Chapter

When considering the different types of organic particle and supramolecular structure that can be found in the size range spanning from nanometer to micrometer dimensions, a broad range of different materials is found (see Figure 7.1).

At the upper end of the scale are found many industrially used organic particles (e.g., dye particles, pigments, drug microparticles), classical emulsion droplets, and biological cells which are in the micrometer range. In contrast, the lower end of the scale represents most typical "small" organic molecules, which are of subnanometer dimensions in homogeneous solution. Between these two groups can be found almost a "whole world" of different macromolecules, supramolecular structures, and assemblies of molecules or macromolecules with dimensions in the nanometer scale. Many macromolecules, such as high-molecular-weight polymers and proteins in solution, have dimensions in the nanometer range, yet most of them are not commonly referred to as "nanoparticles." Exceptions to this may be dendrimers or fullerenes, which are considered to be single-molecule nanoparticles. Some of the organic microstructures and nanostructures listed in Figure 7.1 are formed by essentially liquid particles (such as emulsion droplets), or can be regarded as rather "soft" structures (e.g., micelles). These "soft" particles or structures are characterized by being permanently in dynamic equilibrium with the surrounding liquid. This means that they may change their composition to a large degree, or even completely disintegrate upon the (sometimes slight) adjustment of external parameters, such as concentration, temperature, and ionic strength. Most micelles formed from amphiphiles or block-copolymers belong to this category, although block copolymers in particular often form so-called "frozen micelles," which are kinetically stabilized structures [4]. In this chapter, attention is focused on "rigid" or solid organic particles; these are assemblies of organic molecules or macromolecules which are in a kinetically stabilized state after their formation. Consequently, such particles could, in principle, be isolated as solid powders or transferred from one medium into another. Organic nanoparticles in the notation used here, are organic particles which are in the nanometer scale in all three dimensions. Nano-objects with significantly larger extensions in one or two dimensions (e.g., nanofibers, nanotubes, nanosheets) will not be considered here, as they often require the application of quite different synthesis strategies.

Several reviews and monographs which provided a rather comprehensive overview on organic nanoparticle science, including their properties and synthesis efforts, were produced several years ago [5–7]. More recent reviews on this topic are mostly focused on special application areas (e.g., polymeric drug delivery [8], drug particles [9], food additives [10]), or on special synthesis methods (e.g., supercritical fluid techniques [11]). The aim of this chapter is to provide an overview of the most important current methods for the synthesis of organic nanoparticles. This includes classical methods, such as emulsion-based syntheses, as well as newly developed techniques such as those in which lithographic or microfluidic technologies are applied.

7.1.2
Characteristics of Organic Nanoparticles

In inorganic nanoparticles, the basic building blocks are atoms which are arranged in the form of clusters or crystal lattices. In contrast, the basic building blocks of organic nanoparticles are either "small organic molecules" (having molecular masses up to a few hundred Da) or macromolecules (with molecular masses up to several hundred thousand Da). As a consequence of the larger building blocks, the number of molecules in organic nanoparticles is typically lower than the number of atoms in inorganic nanoparticles of comparable size. A few special organic nanoparticles, such as fullerenes and dendrimers, are composed of only a single molecule. Several of the well-known optical and electronic size effects of inorganic nanoparticles are due to so-called "quantum confinement" effects. When particles become similar to or smaller than the Bohr radius of Wannier excitons, many electronic properties are determined by particle size, and hence strong size effects are observed. For organic nanoparticles such quantum confinement effects are not expected due to the different type of electronic processes and the smaller radii of Frenkel excitons typical for organic materials [12]. Rather than depending on quantum confinement effects, size-dependent effects in organic nanoparticles are more likely caused by the different energetic state of molecules located at the surface of particles [13]. Organic molecules located at surfaces and interfaces are known to have different electronic structures and properties than molecules in the bulk of a molecular crystal or particle. In nanoparticulate materials, a considerable part of the molecules are located at the interface, and this can be expected to have strong effects on different materials' properties, ranging from optical properties [13] to bioavailability [14, 15].

Organic nanoparticles can, in principle, be prepared from a huge, almost infinite, number of different organic compounds, including small molecules as well as macromolecules. Depending on the particle size and molecular interactions within the particle, the properties of organic nanoparticles may differ significantly from those of the bulk phase of the respective compound. Often, organic nanoparticles can be assumed to somehow represent an intermediate stage between bulk solid organic materials and homogeneous molecular solutions. Therefore, organic nanoparticles offer immense potential for the creation and design of materials with unique properties for different fields of application. These may range from pharmaceuticals and food science, imaging and sensors, to organic electronics and photonics.

7.2
Methods of Organic Nanoparticle Preparation

In this section, an overview is provided of the broad spectrum of current methods that have been developed and described for the preparation of nanoparticles from small organic molecules or organic polymers. In nanoscience and nanotechnology, it is common to categorize structuring and synthetic methods to belong either to the class of "top-down" or "bottom-up" approaches. In this notation, "top-down" usually means the miniaturization of systems or the physical division of larger, macroscopic systems into nanosized structures or materials. This involves typically engineering-based approaches and physical structuring techniques such as lithographic methods. In contrast, "bottom-up" represents the preparation of nanostructures and nanomaterials by assembling single atoms, molecules, or macromolecules. These approaches are usually based on either chemical synthesis, directed assembly, or the self-assembly of materials. As with many other nanotechnologies, the different synthetic methods used to prepare organic nanoparticles can also conveniently be subdivided into these two categories.

7.2.1
Top-Down Approaches to Organic Nanoparticles

7.2.1.1 Reduction of Particle Size by Mechanical Forces
Probably, the most obvious "top-down" route to organic nanoparticles is the application of mechanical milling methods, for grinding macroscopic or microparticles down to the nanoscale. Mechanical milling is a well-known unit operation in

process engineering, and different milling methods are successfully used in numerous industrial processes. Unfortunately, mechanical milling has several inherent limitations for the preparation of nanoparticles. In any case, high-energy milling techniques are necessary to overcome the huge increase in surface energy related to nanoparticle formation. Furthermore, with decreasing particle size, it becomes increasingly difficult to use the mechanical impact and shear forces efficiently for particle breakup, without simultaneously inducing agglomeration [16]. Contamination of the nanosized product by abrasion and wear stemming from milling media and mill parts can also never be completely avoided. Further limitations arise for the grinding of organic materials, as these are often quite soft materials and have a limited thermal stability. Despite these limitations, milling methods have found widespread practical application for organic particle formation in two important branches of industry, namely pigment formulation and pharmaceutical compound formulation. An overview on the mechanical devices mostly used for milling pigments (e.g., bead mills, attritors, ball mills) has been provided in Ref [17]. In many cases, such as paints and most printing inks, pigment particle sizes in the lower micron range are sufficient for the desired applications. For other applications, such as inkjet inks, real nanoparticle dispersions with particle sizes below 100 nm are required, and these can be obtained by using prolonged wet bead milling techniques [18]. As the bioavailability of hydrophobic, poorly soluble, active compounds in pharmaceutical formulations can be significantly enhanced by a reduction of particle size to the submicron and nanometer range [15], nanoparticle preparation and formulation have also been identified as important key technologies by the pharmaceutical industry. Among the first methods developed for drug nanoparticle preparation were attrition methods such as wet ball milling, as introduced by Liversidge *et al.* [19, 20]. Wet milling techniques can now be considered a mature technology, and much effort has been put into upscaling in order to reach industrial drug production scales. For example, wet milling technology has now been commercialized by Elan Drug Technologies, Ireland (www.elandrugtechnologies.com), under the name NanoCrystal® technology [21].

In contrast to these upscaling efforts, an important current research topic in the field of wet-milling deals with miniaturization. Nanoparticle formulations are often a prerequisite for the application of hydrophobic drug compounds. Therefore, it would be desirable to prepare them at a very early stage of drug discovery and screening [22]. Since, in these early development stages, usually only minute amounts of new compounds are available, correspondingly miniaturized nanoparticle preparation and formulation techniques need to be developed. An important step in this direction was made by Van Eerdenbrugh *et al.*, who developed milling techniques in 96-micro-well plates or small glass vials, requiring only 10 mg of material [23]. With both methods, nanoparticle suspensions suitable for further characterization could be obtained. Nevertheless, the approach using glass vials seems more appropriate, as the milling technique inside the 96-well plates suffered from considerable well deformation and abrasive wear [23]. Other recent research topics have been focused on more systematic studies concerning the effectiveness of different milling additives used as stabilizers for the formed nanoparticles. Several attempts to correlate the observed stabilization capacities with physico-chemical properties, identified rather complicated dependencies involving parameters such as molecular weight, hydrophobicity, surface activity, and solution viscosity [24–27]. Besides the wet attrition processes described to date, dry grinding processes applicable to organic materials such as pharmaceuticals were developed. Several reports have been made describing nanoparticle formation by the co-grinding of pharmaceutical compounds with additives such as polymer/surfactant mixtures [28, 29] or cyclodextrins [30]. Another interesting dry-grinding approach aimed at minimal product contamination was developed by Juhnke and Weichert [31, 32]. This dry-grinding process was carried out at cryogenic temperatures in order to facilitate brittle fracture of the particles, and used solid carbon dioxide and/or water ice particles as the milling media. The application of these special milling media avoids contamination of the product by abrasive wear from the milling media. Furthermore, they can easily be removed from the product simply by warming up in the case of carbon dioxide, or by lyophilization for water-ice.

Another group of technologies based on the mechanical reduction of particle size is based on the so-called "high-pressure homogenization" principle. While in dry- and wet-grinding techniques the primary mechanical forces are impact and shear, high-pressure homogenization relies on cavitation. The classical method of high-pressure homogenization for organic nanoparticle formation is based on a piston-gap instrument [33, 34]. A premilled (by ball- or jet-milling) suspension consisting of organic microparticles in an aqueous stabilizer solution is forced through a small gap of approximately 25 µm by applying pressure differences of up to 150 MPa. Inside the narrow gap, the fluid is strongly accelerated, leading to a highly increased dynamic fluid pressure. Consequently, the static pressure on the fluid has to decrease and falls below the vapor pressure of water, which leads to the formation of vapor bubbles. After leaving the gap, the ambient pressure is reached instantaneously and the bubbles implode. Cavitation forces, created during vapor bubble implosion, are mainly responsible for disintegration of the microparticles and the formation of nanoparticles. Usually, several consecutive homogenization cycles are required in order to obtain the desired particle size distributions of typically a few hundred nanometers; nevertheless, the process times are usually shorter than for mechanical grinding. Other advantages are the ease of upscaling and reduced contamination by the erosion of instrument parts [33, 35]. For certain examples, the size and stability of drug nanoparticle suspensions prepared by such a cavitation-based high-pressure process were reported to be comparable with the bottom-up synthesis [36]. This technology has also already been commercialized, under the trade name DissoCubes® technology, by SkyePharma (www.skyepharma.com/), while a variant which uses nonaqueous solvents, called Nanopur®, has been developed by PharmaSol (www.pharmasol-berlin.de). More recently, high-pressure systems based on a modular nozzle design instead of the classical piston-gap geometry were developed [37]. This instrument geometry allowed an enhanced control over the individual mechanical forces of cavitation, impact, and shear that act on the particles.

7.2.1.2 Lithographic Methods

Lithographic methods such as photo-, electron-beam or X-ray lithography have long been used in microelectronics and semiconductor technology, mainly for the fabrication of two-dimensional (2-D) nanostructures and microstructures. Despite their widespread use, these methods are in general not very suitable for the preparation of colloidal particles. Notably, they either require very expensive equipment (e.g., electron-beam or X-ray lithography) or they are rather limited in their resolution (visible light lithography) and are generally rather slow to operate. Moreover, they often lack compatibility with organic or biological materials due to radiation damage and the requirement for plasmas, corrosive gases, or solvents in the etching and lift-off steps. Nevertheless, polymer colloids and microparticles have been fabricated by using photolithography [38], holographic lithography [39], and a combination of photolithography and microfluidics [40]. Apart from these classical methods, another lithographic method termed nanoimprint lithography (NIL) is gaining increasing importance. NIL is based on the replication of structures by imprinting a master into a polymer or liquid polymer precursor solution. As a consequence, features with nanometer resolution can be fabricated using this technique, and it can also be scaled-up for large-volume productions by using a roll-to-roll technology [41]. Based on the NIL process, deSimone and coworkers developed a very versatile technology for the preparation of polymer nanoparticles, called "Particle Replication in Nonwetting Templates" (PRINT®) [42–44]. This is based on the use of a nonwetting fluoropolymer material for the mold, thus avoiding the formation of a wetting film of polymer precursor solution on the substrate between the mold cavities. Thus, in contrast to conventional NIL, no residuum layer is formed after curing, which ultimately allows for an easy harvesting of the particles (see Figure 7.2).

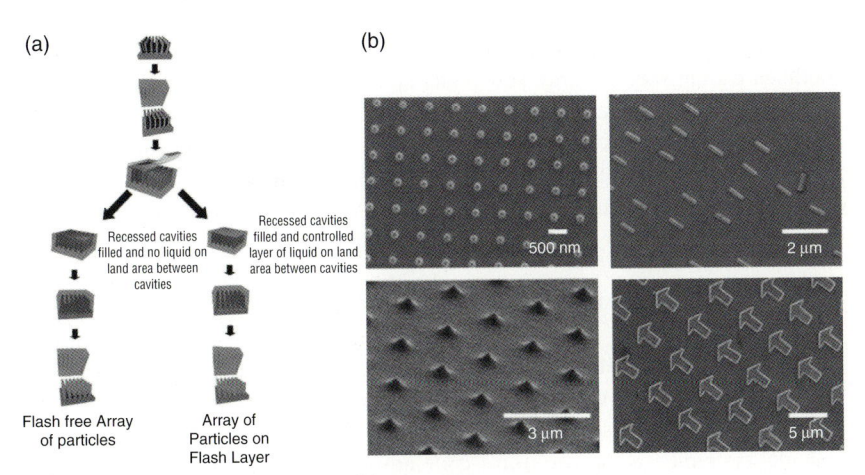

Figure 7.2 (a) Schematic illustration of the PRINT® process of particle formation without and with the flash layer; (b) Examples of polymer particles of different shapes prepared by the PRINT® process. Reproduced with permission from Refs [43, 44]; © 2005 and 2008, American Chemical Society.

The PRINT® process enables the preparation of a broad variety of particles down to the nanometer size, while the materials that have been used in the process range from photocurable and photocrosslinkable polymers to proteins [45]. Furthermore, many different and biologically active substances such as fluorophores, drugs, proteins, and DNA can be incorporated into the particles. Particles prepared via the PRINT® process are monodisperse, and can be engineered at will into almost any arbitrary shape (Figure 7.2); this has been demonstrated by the replication of self-assembled nano-objects, such as polymer micelles and virus particles [46]. This extremely versatile technology of particle formation shows great potential for nanomedical applications, and is currently commercialized by the start-up company Liquidia Technologies (www.liquidia.com). In similar approaches, other variations of NIL techniques have recently been used for polymer nanoparticle engineering. For example, the use of a so-called step-and-flash imprint lithography (S-FIL) led to the formation of crosslinked peptide nanoparticles that could easily be harvested by rinsing with aqueous buffer solutions, due to the presence of a water-soluble sacrificial transfer layer on the substrate [47]. In another example, high-pressure molding with etched silicon masters was demonstrated for the preparation and release of polymeric nanorods, using a bilayer NIL process [48].

7.2.2
Bottom-Up Approaches to Organic Nanoparticles

Whilst many of the top-down approaches used for nanoparticle synthesis (see above) are employed in large-scale industrial particle production, most organic nanoparticles used for fundamental research and development are prepared using bottom-up approaches. The reason for this is that, bottom-up approaches are better suited for small-scale, batch particle synthesis, and allow a rapid and easy systematic variation of the preparation conditions. Many bottom-up particle synthesis methods do not require large and costly special equipment (such as milling or lithography tools), and are therefore easier to implement in research laboratories. Another important advantage is that, in most cases, bottom-up synthesis approaches allow for a better control of properties such as particle size, size distribution, and composition. For this reason, bottom-up methods are increasingly used also for large-scale, industrial nanoparticle syntheses.

7.2.2.1 Solution-Based Bottom-Up Methods

7.2.2.1.1 **Emulsion Methods** An important group of methods, within almost innumerable examples of organic nanoparticle preparation, is based on the application of different types of emulsion for nanoparticle formation. The *emulsion*, a colloidal system based on two immiscible liquids, is used either as a microscopic reaction vessel or as an intermediate stage for particle formation.

Emulsion polymerization is a well-known, classical, and widely applied method for the synthesis of polymer microspheres and nanospheres. Although emulsion polymerization is of enormous scientific and technological importance, it will not be discussed here in detail (due to limitations of space), and the reader is referred to a specialized review [49]. The same applies to more recent variants of the classical emulsion polymerization technique, such as polymerization in oil-in-oil emulsions and miniemulsion polymerization. The use of oil-in-oil instead of oil-in-water emulsions extends the range of suitable materials to water-sensitive monomers or catalysts, and also allows the use of polyaddition and polycondensation reactions (for a recent review, see Ref. [50]).

Miniemulsion polymerization is an extraordinarily versatile method for the synthesis of polymer and complex hybrid nanoparticles. It is based on reactions occurring in small emulsion droplets stabilized by a surfactant, with a costabilizer acting as an osmotic pressure agent [51] (more detailed information is provided in a recent review [52]). Apart from applying emulsion droplets as the reaction containers for polymerization, emulsions can be used as intermediate-stage formulations in the precipitation of small organic molecules or polymers in poor solvents (such as water). In such a process, the organic material is first dissolved in an organic solvent, after which an emulsion is formed by dispersing the solution in water containing either surfactants or polymeric stabilizers. The size of the emulsion droplets can, to a certain degree, be controlled by the type and concentration of the stabilizers. In order to convert the formed emulsion into an aqueous dispersion of organic nanoparticles, the organic solvent must first be removed. In most cases, this is accomplished either by evaporation or diffusion out of the emulsion droplets (see Figure 7.3).

Numerous different examples of this basic process have been described, mainly for pharmaceutical, food, and photographic applications [5, 7]. Horn and Rieger assigned different variants of the process into several subcategories, depending on polarity of the solvent used and the type of organic compound to be dispersed [5]. If lipophilic solvents are used, emulsions are usually prepared by mechanical homogenization, after which the solvent can be removed by evaporation or extensive washing of the emulsion. Both, particle size and size distribution depend on the droplet sizes and solution concentration. The organic compounds forming the particles can be either small organic molecules, polymers, or a mixture of these. In variants using partially water-miscible, amphiphilic solvents, emulsions are first formed but these are then destabilized by the addition of further water. This causes the solvent to diffuse out of – and water to diffuse into – the droplets, such that the solution becomes supersaturated and the organic compounds are precipitated inside the droplets [53]. This general method has also been applied to water-dilutable microemulsions [54].

Microemulsions are thermodynamically stable systems that form spontaneously upon mixing and stirring of the components, and do not require high-shear instruments for emulsification. Current research examples of emulsification–evaporation and emulsification–diffusion techniques are often focused on the systematic investigation of preparation conditions and the application of chemometric techniques [55, 56]. Other research has been aimed at enhancing the biocompatibility of nanoparticles and the synthetic process; this may be achieved by avoiding organic

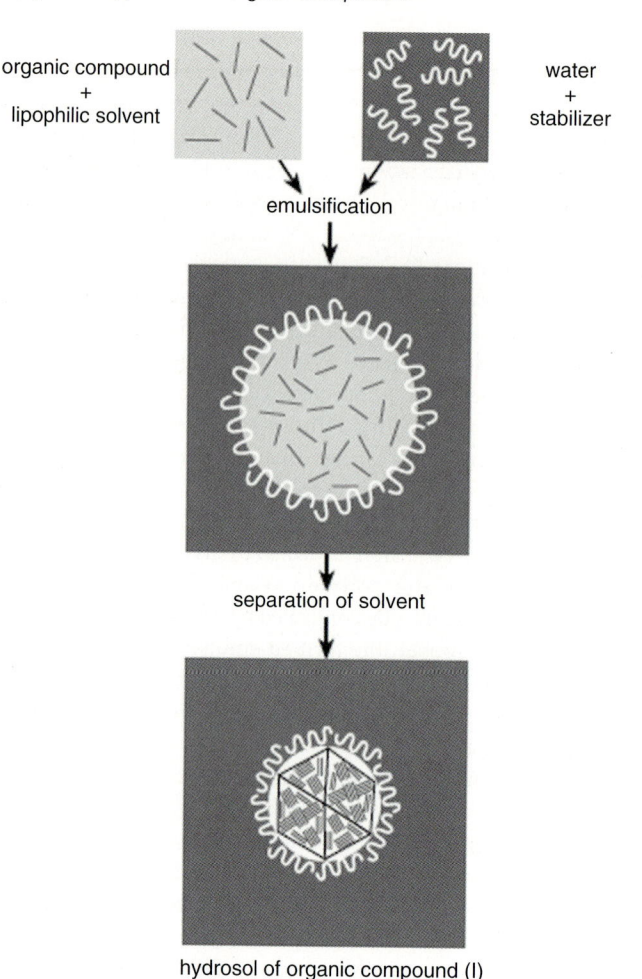

Figure 7.3 Principle of the preparation of organic nanoparticles by emulsion methods. Reproduced with permission from Ref. [5]; © 2001, John Wiley & Sons.

solvents, using the melt emulsion method [57]. In this case, emulsification is carried out at temperatures above the melting point of the active compound, such that the active compound itself can form the oil phase of the emulsion. Another possibility would be to use biocompatible substances, such as common food proteins, as emulsifiers and particle stabilizers [58].

Apart from these classical emulsion methods, several more recent developments in emulsion-based organic nanoparticle synthesis have been reported, some of which will be discussed here. A completely stabilizer-free emulsion-based organic nanoparticle synthesis method has been described [59] in which, as a first step, a solution of the organic material in a hot organic solvent is injected into an equally hot aqueous medium to form an emulsion. Next, particle nucleation and crystal growth are induced by cooling the emulsion to room temperature. Finally, the emulsion is broken and the organic nanoparticles are transferred and dispersed in the aqueous phase by adding surface-active agents and/or sonication [59]. Nanoparticle dispersions obtained using this method were reported to be highly stable and to have a narrow size distribution; moreover, the particle size can be controlled by the solution concentration [60].

In a slight variation from the above-described approaches, a nanoparticle synthetic method based on water-in-oil microemulsions has also been applied to organic nanoparticles. Such water-in-oil microemulsion methods are used exhaustively in inorganic nanoparticle synthesis [61, 62], and can be adapted for the synthesis of organic compound nanoparticles [63, 64]. In this case, an empty microemulsion is first prepared which consists of small, surfactant-stabilized water droplets dispersed in an oil phase. A solution of the organic compound in a suitable solvent is then added dropwise to the microemulsion, followed by ultrasonication or mechanical stirring. The process of nanoparticle formation involves several stages. First, the solution of the organic compound in the solvent penetrates inside the aqueous cores by crossing the interfacial film, after which the organic compound is precipitated in the aqueous cores, due to its insolubility in water. The nuclei thus formed can grow because of an exchange of the organic compound between the aqueous cores, leading to a dispersion of organic nanoparticles stabilized by a surfactant shell [62]. In analogy to many examples of inorganic nanoparticle synthesis, water-in-oil microemulsions can further be used as microcontainments for reactants. As one example, nanoparticles of the hydrophilic polymer chitosan have recently been synthesized by mixing two different microemulsions, respectively containing chitosan and a crosslinker molecule in the aqueous cores [65].

A novel, generic and versatile process for the emulsion-mediated synthesis of dry organic nanoparticle powders has been described by Zhang *et al.* [66]. The process is based on a combination of emulsification and freeze-drying, whereby an oil-in-water emulsion is formed by using a volatile organic solvent containing the dissolved organic compound as the oil phase, and an aqueous surfactant or polymer stabilizer solution as the water phase. After having frozen the emulsion by using cryogenic liquids (mostly liquid nitrogen), both the water and the organic solvent are evaporated by freeze-drying. This step results in a porous, solid mixture consisting of nanosized particles of the previously dissolved organic compound and water-soluble surfactants or polymers. These porous solids can be easily redispersed in water to yield stable nanoparticle dispersions (Figure 7.4). This method is currently being commercialized by the start-up company Iota NanoSolutions Ltd (www.iotanano.com/), and has been demonstrated with several hydrophobic organic compounds such as dyes, biocides, and pharmaceuticals [66, 67]. A very similar technique, based on the formulation of a spontaneously forming microemulsion and subsequent evaporation of all solvents by freeze-drying, has been described for the synthesis of sub-100 nm-sized drug nanoparticles [68]. This method was recently further elaborated in order to avoid the lyophilization step;

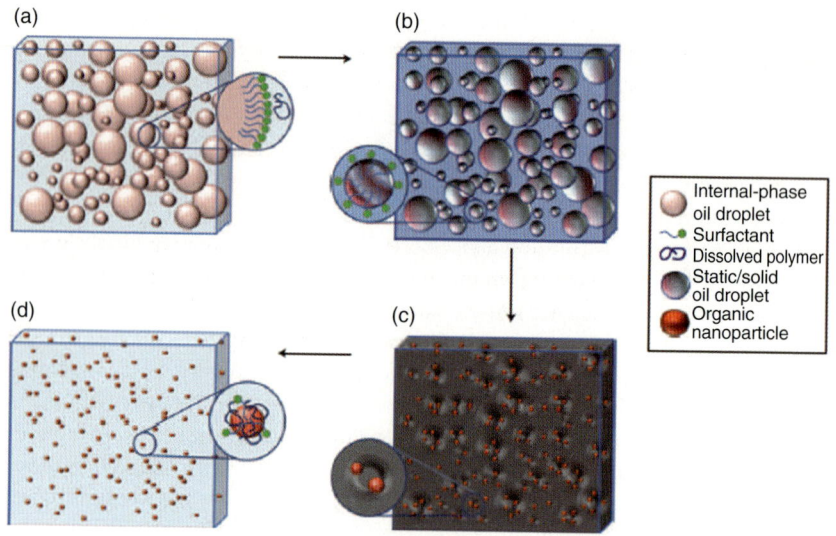

Figure 7.4 Schematic representation of nanoparticle formation during freeze-drying. (a) Oil-in-water emulsion (organic compound dissolved in oil droplets); (b) Frozen oil-in-water emulsion; (c) Emulsion-templated porous solid after freeze-drying; (d) Organic nanodispersion after addition of water. Reproduced with permission from Ref. [66]; © 2008, Macmillan Publishers Ltd.

and solid, easily redispersible drug nanoparticles were produced by the spray-drying of a drug-loaded oil-in-water microemulsion, using easily evaporable solvents [69].

7.2.2.1.2 Reprecipitation Methods

The different emulsion-based methods for nanoparticle synthesis described in Section 7.2.2.1.1 are two-step procedures that comprise an initial emulsion formation and, subsequently, an actual nanoparticle formation via a process such as evaporation or diffusion. In contrast, reprecipitation methods for organic nanoparticle synthesis are single-step procedures based on the mixing of a solution of the organic compound with a poor or nonsolvent. The solvent used to dissolve the organic compound must be completely and well miscible with the poor solvent (usually water), in order to achieve high levels of supersaturation almost instantaneously after mixing. A high supersaturation is important in order to promote rapid nucleation and to limit the growth of nanoparticles. Usually, the solution of an organic compound in a miscible solvent (e.g., alcohol, acetone, tetrahydrofuran) is added to an excess of a poor solvent (e.g., water), although the reverse process of adding the poor solvent to the solution is also possible. The reprecipitation method was first applied to polymer nanoparticles by Fessi et al. [70], and to small-molecule organic nanoparticles by Horn [5, 71] and Kasai et al. [72, 73]. Due to its simplicity and versatility, the reprecipitation method has been applied to the synthesis of nanoparticles of a large number of different materials. Among polymer nanoparticles prepared using this technique, the majority of studies have been targeted at drug-delivery applications, whereby a broad variety of pharmaceutically active compounds have been incorporated into nanoparticles of different biocompatible polymers, by reprecipitation. Examples of such polymeric nanoparticles include lactic acid polymers and copolymers [74, 75], polysaccharides [76], conventional synthetic polymers [77], and proteins [78]. Other examples of polymer nanoparticles include porous nanoparticles for use as dielectrics [79], or fluorescent polymer nanoparticles for sensor and biomarker applications. Such polymeric sensor nanoparticles may be loaded with luminescent indicator dyes [80] or semiconductor quantum dots (QDs) [81], or be formed from conjugated, intrinsically fluorescent polymers [82, 83]. With regards to the reprecipitation of small molecules, drug delivery is an important application, and many examples of the nanoparticle formation of poorly soluble hydrophobic drugs have been reported [84, 85]. The formulation of drug nanoparticles based on a reprecipitation technique, known as NanoMorph® technology, is also offered commercially by SOLIQS at Abott GesmbH & Co. KG [86]. Likewise, many other small-molecule organic compounds have been used to prepare nanoparticles by reprecipitation. Among these are included dyes [13, 87], polycyclic aromatic hydrocarbons [88–90], porphyrins [91], or carotenes [92]. Often, it may be desirable to remove the organic solvent from the nanodispersion after synthesis since, depending on the solvent type and concentration, the residual organic solvent can give rise to Ostwald ripening of particles, or it can interfere in the desired application. Such solvent removal can be accomplished by dialysis, vacuum evaporation or precipitation–redispersion, or perhaps by using a new technique such as flash-evaporation [93]. Whilst some of the above-described nanoparticles were prepared by reprecipitation in pure nonsolvent, in most cases additives such as surfactants and polymers were added to the nonsolvent in order to stabilize the nanoparticles formed. Besides acting as stabilizers, the ionic and polymeric additives were found in many cases also to affect particle growth and, in turn, the particles' size, shape, and structure [94, 95]. This is believed to be mainly due to adsorption during particle growth, or preferential adsorption to certain crystal facets. The results of a recent study indicated that additives may also affect organic nanoparticle size by acting as "seeds" or nucleation sites for nanocrystal formation [96]. Another elegant method has been to directly control the intermolecular forces responsible for nanoparticle formation by modifying the side groups attached to the molecule's backbone [83]. A different–but very effective–way to tune the optical properties of organic nanoparticles is by the coprecipitation of two dye molecules in order to prepare doped nanoparticles. Here, the emission spectrum can be tuned gradually due to a resonance energy transfer between the two different dye components, rather than by particle size [97].

Based on the classical reprecipitation method, several variants and modifications have been developed, the main aim being to improve the efficiency of the method, and to gain a better control or extend its applicability. One quite obvious modification has been the application of ultrasound during reprecipitation, which generally accelerates the mixing and increases nucleation rates in the crystallization of organic materials, leading to a decrease in particle size [98, 99]. When several different organic nanoparticles were prepared by reprecipitation under sonication, the particle sizes were usually reduced and the colloidal stability enhanced compared to simple stirring [100, 101].

One shortcoming of the classical reprecipitation method is that simply dropping or injecting the solution into a poor solvent does not allow any control over the mixing process. The reprecipitation proceeds via dispersion of the organic solution in the poor solvent in the form of small droplets, followed by liquid counterdiffusion, while particle formation occurs either by nucleation or spinodal decomposition [87]. Therefore, the initial drop size and the mixing times represent key parameters for defining nucleation and particle growth kinetics. In the classical reprecipitation method, these mixing parameters are poorly defined and cannot be varied systematically. One approach to better define the particle formation process would be to carry out the reprecipitation in so-called "membrane contactors" or "membrane mixers." These are common crossflow filtration systems in which one liquid is added to the other through the pores of an ultrafiltration membrane. The technique not only allows adjustments to be made to the liquid addition rate, but also leads to the generation of very small solution droplets of a uniform size, with a subsequent rapid interdiffusion of solvents and improved particle size distributions [102]. Nanoparticles of polymers, as well as of small molecules, have been prepared by reprecipitation in such membrane contactors [102–104].

Another approach to the problem of mixing and particle formation kinetics in organic nanoparticle reprecipitation is based on a mixer device that allows tuning of the mixing times. Such a mixer was realized in a so-called "Confined Impinging Jet" geometry [105, 106], where two high-velocity liquid jets – of organic solution and poor solvent, respectively – are directed against each other and impinge in a mixing chamber to form a highly turbulent mixing zone. The mixing times obtained were in the millisecond range, and could be adjusted by varying the jet velocities. In this way, the mixing times could be tuned to be faster than the induction time of nanoparticle nucleation and growth, thus eliminating the effect of mixing on particle formation and allowing for the preparation of a narrow particle size distribution. Besides the synthesis of drug and carotene nanoparticles [106], the technique was shown also to be applicable to the creation of multifunctional nanoparticles containing drugs and fluorescent indicators [107]. The Confined Impinging Jet technique is interesting for larger-scale organic nanoparticle synthesis, as it is a continuous process and is also easy to upscale. Hence, considerable effort has been expended into the optimization of reactor geometry and the exact description of the hydrodynamic situation in the mixing zone, using computational fluid dynamics (CFD) simulation methods [108–110] (Figure 7.5). The flexibility of the method has been further enhanced by the advancement to the so-called Vortex Jet Mixer geometry, which allows the simultaneous mixing of more than two liquid jets [107, 111].

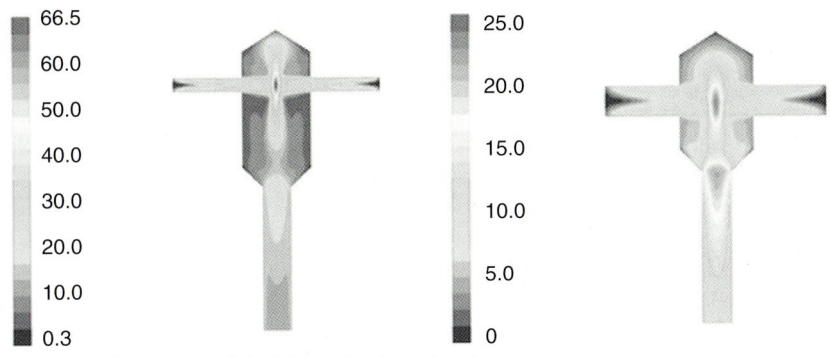

Figure 7.5 Computational fluid dynamics-derived contour plots for the turbulent intensity (%) in two Confined Impinging Jet Reactors of different geometry. Reproduced with permission from Ref. [109]; © 2009, Elsevier.

Several other process variants for the reprecipitation of organic materials have been developed, which target the improvement of mixing and mass transfer by applying additional centrifugal forces. In the "spinning disk process," the liquids are spread as thin films on a rapidly rotating disk [112], whereas in "high-gravity-controlled precipitation" a "rotating packed bed" is used on which the feed liquid streams are spread into thin films and droplets [113, 114]. In contrast to the above-described instrumental approaches, a "chemical approach" of spatially controlled reprecipitation has been developed to achieve a better control of the reprecipitation process. This is based on the spatially confined reprecititation of organic material inside the pores of sol–gel materials [115–117], and allows the preparation of organic–inorganic hybrid materials that consist of almost monodispersed organic nanoparticles within silica xerogels.

All of the examples of organic nanoparticle synthesis by reprecipitation discussed to date have relied on a sudden change in organic compound solubility upon changing the environment from a good to poor solvent. However, this classical – often-called "solvent shifting" – method is not the only mechanism by which the solubility of an organic compound can be changed. Solvent quality can be strongly affected by a change in pH, as many organic molecules and polymers contain dissociable charged groups. Shifting the pH of a solution can create a state of supersaturation of an organic compound, and can thus be used for the synthesis of organic nanoparticles [7]. Nanoparticles of both, small molecules and polymers, have been synthesized by this pH-shifting method, which can be seen as a variant of the reprecipitation process [7, 118]. Recently, the basic concept of the pH-shifting method has been generalized to the more generic "ion-association" approach [119, 120]. Whereas, in the pH-shifting method the solubility of an organic compound is changed by the association or dissociation of H^+-ions, in the "ion-association" method a hydrophobic ion-pair is formed by the addition of an aqueous solution containing a hydrophobic counter-ion (e.g., tetraphenylborate) to an aqueous solution of a charged organic compound. The hydrophobic ion-pair then can precipitate in the form of organic nanoparticles. The "ion-association method" has been applied to the synthesis of different porphyrin and cyanine dye nanoparticles [119–122]. Whilst these last examples rely on ion-exchange reactions, the technique has been further extended by using a real chemical reaction for particle formation. Kang *et al.* described, for the first time, the formation of monodispersed perylene nanoparticles by the reduction of perylene cations in acetonitrile [123]. This well-defined homogeneous solution process can easily be upscaled, and allows almost monodisperse particles to be obtained, based on the simple separation of nucleation and particle growth steps. The latter method is similar to the approach used for the synthesis of many inorganic nanomaterials and could, in principle, be applied to a vast number of other organic chemical reactions.

7.2.2.1.3 Supercritical Fluid Techniques

Different supercritical fluid techniques have become an important set of methods in chemical synthesis, and also in the synthesis of a broad range of nanomaterials [124]. Their common characteristic is that they employ a fluid at temperature and pressure conditions above the critical point, a so-called "supercritical fluid." These supercritical fluids have unique properties, such as tunable density and special solvency behavior, that can be exploited in the synthesis of nanomaterials. The most widely used supercritical fluid is carbon dioxide, because of its low critical point and low toxicity. Many supercritical fluid techniques used in the synthesis of organic materials can be seen as a special form of the reprecipitation method. The supercritical fluid can act either as the good or the poor solvent for the organic compound. Consequently, most supercritical fluid techniques can be categorized as belonging either to the group of "solvent" or "anti-solvent" techniques [11]. The important so-called "rapid expansion of supercritical solutions" (RESS) method belongs to the first group of "solvent" methods. This process involves the rapid expansion of a near-saturated solution of the organic compound in a supercritical fluid into a low-pressure chamber through a nozzle. During expansion, a rapid nucleation and subsequent aggregation leads to the precipitation of small particles. A slightly modified version of the classical process was developed by expanding the supercritical solvent into another liquid solvent [125]. This revised technique was reported to yield exclusively nanoparticles of polymers and other materials, while in the classical RESS process often microparticle products were obtained due to aggregation.

In the second group of supercritical fluid techniques, the organic compound is dissolved in a liquid solvent that is miscible with the supercritical fluid, which is used as a poor solvent. In this method, which is commonly called the "supercritical anti-solvent" (SAS) process, mixing of the two fluids is usually accomplished by spraying through a nozzle or orifice. In principle, either the addition of a liquid to the supercritical fluid, or the reverse process, is possible. Many different types of organic nanoparticle, ranging from drug or dye molecules to polymers, have been prepared using this technique [126]. Several reviews on the application of different types of supercritical fluid technique for the preparation of nanoparticles – especially in the pharmaceutical field – have been published recently [11, 124, 126, 127]. Besides the numerous examples of drug and polymer nanoparticles, the synthesis of organic semiconductor nanoparticles for the fabrication of light-emitting diodes was also reported using a modified RESS technique [128].

7.2.2.1.4 Precipitation by Polyelectrolyte Complex Formation

The synthesis of nanoparticles by the formation of complexes between oppositely charged polyelectrolytes is somewhat similar to the "ion-association" method mentioned above (see Section 7.2.2.1.2). The mixing of oppositely charged polyelectrolyte solutions leads to an association of the polymers by electrostatic interactions, to a partial charge neutralization, and to complex formation. Such complexes can be formed using either synthetic polyelectrolytes or charged biomacromolecules (e.g., peptides, proteins, polysaccharides). Under certain conditions, the polyelectrolyte complexes can precipitate as nanosized particles. Depending on which of the two polyelectrolytes is used in excess, the formed nanoparticles can be tuned to be either positively or negatively charged (for a review, see Ref. [129]).

This method of nanoparticle synthesis has been especially investigated, using charged polysaccharides as polyelectrolytes. The polysaccharide can either be complexed with an oppositely charged polymer, this being another polysaccharide, a synthetic polyelectrolyte, or another charged biomacromolecule. Among cationic polysaccharides, *chitosan* being the most frequently used material, whereas many different anionic polysaccharides such as carboxymethylcellulose, dextran sulfate, hyaluronic acid or heparin, have been used for nanoparticle preparation [130, 131]. The sizes and structures of nanoparticles from polyelectrolyte complexes differ widely, depending on parameters such as the mode of addition, the molecular weight, chain flexibility, and the molar ratio of the polyelectrolytes. Consequently, different mechanisms were proposed for the different cases. One mechanism – the "scrambled-egg" model – describes the formation of highly aggregated complexes with a 1:1 stoichiometry of the polyions. This model commonly applies to complexes formed from polyions, with strong ionic groups and similar molecular weight. For the association of such similar molecular weight polyelectrolytes, it was found that lower molecular weights gave better, more stable complexes, whereas complexes of high-molecular-weight polyions were more susceptible to agglomeration [132]. For the association of polyelectrolytes with different molecular weights, a relationship between the nanoparticle size and the chain length ratio of the used polyelectrolytes was established. This relation could be explained by a host/guest concept, in which the polyelectrolyte with the higher molar weight acts as host for the other one [133]. Different mechanisms were found, depending on which of the two polyelectrolytes was used in excess (see Figure 7.6 for one example). This can be attributed to the different reactivities of the respective ions, and the different flexibility and conformation of the chains [134].

In addition to the two polymeric species responsible for complexation, these particles may contain additional components such as other polymer or drug molecules. Many polysaccharide-derived particles have been explored as biocompatible and biodegradable nanoparticles for controlled drug delivery (see , Refs [129, 131,

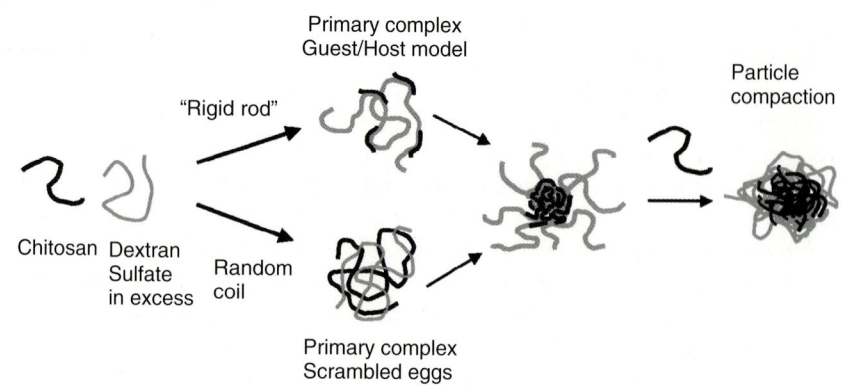

Figure 7.6 Mechanisms suggested for the formation of polyelectrolyte complexes from chitosan and dextran sulfate in excess. Reproduced with permission from Ref. [134]; © 2007, American Chemical Society.

135, 136]). An especially important and wide field of research into controlled drug delivery using polyelectrolyte complexes relates to the complexation of nucleic acids such as DNA and RNA, whereby several polycations can lead to nanosized precipitates that are useful for the direct delivery of drugs into living cells. These nanoparticles can be seen as a type of nonviral vector for DNA transfection in applications such as gene therapy [137, 138]. Due to their inherently good biocompatibility, polysaccharides such as chitosan are again among the preferred polycations used for DNA complexation in these applications [138]. In other examples, specialized peptides (such as protamine) were used as biocompatible polycations well-suited for DNA or oligonucleotide delivery [139]. These so-called "proticles" were shown to be capable of effectively crossing the blood–brain barrier after having first been coated with apolipoprotein [140].

Recently, it has been demonstrated that, in principle, nonelectrostatic interactions can also be applied to the formation of nanoparticles by the complexation of macromolecules. Nanoparticles with a diameter of about 130 nm were formed by host–guest interactions between a dextran containing grafted alkyl chains on the one hand, and a poly-beta-cyclodextrin on the other hand [141, 142].

7.2.2.1.5 Organic Nanoparticle Synthesis in Microfluidic Devices Today, both microfluidics – which relates to the manipulation of fluid streams in microchannels – and "Lab-on-a-Chip" technologies are increasingly applied in a variety of fields in chemistry and the life sciences. Clearly, such systems are very advantageous, and are applied extensively in situations and for applications where there is a need to work with small volumes of liquid; typical examples include bioanalytical systems [143] and single-cell studies [144]. In addition to the ability to handle very small volumes, microfluidic systems also provide rigorous control and well-defined flow conditions that are important for many chemical reactions and processes. It is for just such reasons that, during the past decade, many microfluidic systems have been investigated and developed for roles in organic syntheses [145]. Subsequently, the advantages of well-defined reaction and flow conditions in microfluidic devices have also been exploited for the synthesis of inorganic nanoparticles and polymer microparticles [146–148]. During the past few years, microfluidics has shown great promise for organic nanoparticle syntheses, with many of the underlying principles – such as the nucleation of organic material and further growth – having been studied in detail for protein crystallization in microfluidic devices [149]. Whilst much of this knowledge can be transferred to organic nanoparticle synthesis applications, it should, in principle, be possible to realize most of the solution-based nanoparticle synthesis methods discussed above inside microfluidic channels. Yet, very few examples of the microfluidic synthesis of organic nanoparticles have been reported to date.

Microfluidic devices are highly versatile instruments for the precisely controlled formation of emulsion-like fluid streams. These consist of well-defined liquid droplets in a stream of an immiscible continuous medium, with either simple T-mixers or so-called "hydrodynamic flow focusing" being used for the formation of well-defined droplets in microfluidic channels. The formation of organic nanoparticles in such droplets has been reported on occasion. For example, Su *et al.* described the precipitation of an organic compound inside such droplets on mixing two different droplet streams of different solvent quality in a simple T-mixer set-up [150], while Rondeau *et al.* reported on the formation of biopolymer nanoparticles by interdiffusion, using droplet streams of partially miscible solvents [151]. Beside mixing, other process steps and unit operations in nanoparticle synthesis can also be carried out in microfluidic devices. For example, the cooling of emulsions in a microfluidic system in order to induce crystallization of the droplets using a micro heat exchanger has been demonstrated [152].

Apart from an ability to form droplets, the properties of almost exclusively laminar flow and rapid diffusion, which are unique to microfluidic devices, have also been exploited for organic nanoparticle formation. As a result of the low Reynolds numbers in microchannels, the flow patterns are laminar; this means that two streams of miscible liquids can flow in parallel, and that mixing will occur by diffusion. This, on the one hand, allows for rigorous control of mixing, while on the other hand it means that due to the small dimensions, the diffusion lengths are extremely short and consequently diffusion can be very rapid. As discussed in Section 7.2.2.1.2, the mixing and mixing times are crucial for nanoparticle synthesis by reprecipitation, and therefore microfluidic synthesis approaches could be highly advantageous for this method. Indeed, it has been shown that, by using a simple Y-shaped channel geometry, the micromixing times can be below 1 ms, and thus below the typical induction times for nucleation [153]. This allows for a rapid and homogeneous nucleation that would result in a smaller particle size and a narrower size distribution than in macroscopic batch reactors [153]. The synthesis of different hydrophobic drug nanoparticles by reprecipitation in microfluidic devices has been demonstrated to yield nanoparticles of a narrow size distribution, and with a tunable size [153, 154]. Indeed, it was found that the particle size could

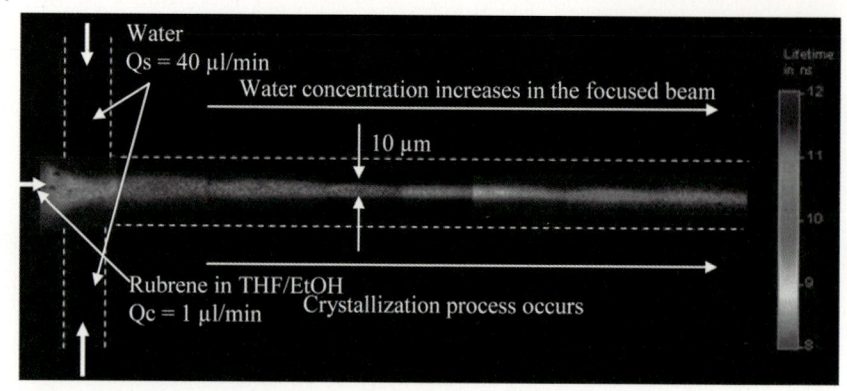

Figure 7.7 Fluorescence lifetime microscopy image reconstitution of the lifetime evolution of rubrene along the channel. The change in fluorescence lifetime can be used to follow the crystallization of rubrene in the first 3 mm of the channel. Reproduced with permission from Ref. [156]; © 2007, Elsevier.

effectively be changed from 80 to 450 nm by varying the channel geometry and flow rates [153, 154]. However, the most effective parameter for tuning the particle size was the flow rate of the poor solvent, and this was confirmed by computational modeling of the process [155].

In order to improve the precipitation set-up in microfluidic devices, the diffusion times can be further decreased by so-called "hydrodynamic flow focusing" (see Figure 7.7). For this, a central liquid stream is sheathed by two adjacent streams of a higher flow rate, such that the central stream is squeezed (or focused) into a very narrow stream. This narrow stream allows for an accelerated mixing with the side streams, by diffusion. The synthesis of organic nanoparticles in such microfluidic devices by reprecipitation has been demonstrated for small aromatic hydrocarbons [156] and block-copolymers [157]. In addition to enhanced control and preparation conditions, microfluidic devices allow for an *in situ* observation of the particle formation process using, for example, fluorescence microscopy [156] (Figure 7.7). Although all of the microfluidic devices discussed so far operate in the laminar flow regime, it has been shown recently that turbulent flow conditions in microfluidic reactors can be applied to nanoparticle synthesis simply by a downscaling of the Confined Impinging Jet-type reactors (see Section 7.2.2.1.2.) to the microscale [158].

7.2.3
Vapor Condensation-Based Methods

In analogy to the situation that occurs during the precipitation of an organic compound from solution, both nucleation and subsequent particle growth may also occur by condensation from the gas phase. In principle, a broad range of organic materials can easily be transferred into the gas phase, without decomposition, at reduced pressure. Consequently, several different methods for the synthesis of organic nanoparticles based on vapor-phase condensation have been developed. In a method first introduced by Toyotama, the material is evaporated in an inert gas atmosphere under reduced pressure, and deposited in the form of aggregates or small particles onto a cooled substrate. Indeed, nanoparticles of a wide variety of organic materials, ranging from aromatic hydrocarbons, over many drug molecules, to hormones, and even polymers, have been prepared using this technique [159]. Further developments of this basic evaporation process include the partial or total evaporation and subsequent condensation of organic materials in an aerosol-flow reactor [160], or the direct condensation of an organic vapor inside a liquid dispersion medium [161].

Another method described for the preparation of organic nanoparticles is based on the pulsed-laser ablation of large, several-micrometer-sized, organic crystals suspended in a liquid [162, 163]. These organic crystals absorb the laser light, which leads to a local increase in temperature and the evaporation of a small amount of material from the crystal surface. This vaporized material is rapidly cooled by the surrounding liquid to form nanoparticles. To date, such laser ablation methods have been mainly applied to the preparation of aromatic hydrocarbon and dye (especially phthalocyanine) nanoparticles [162].

7.3 Application of Organic Nanoparticles

From an application-oriented point of view, the procedure of organic nanoparticle syntheses does not end following actual particle formation and stabilization against further agglomeration. Irrespective of the intended application areas, which may range from organic electronics to pharmaceuticals, a number of further processing steps are usually required. Typically, the initially synthesized nanoparticle dispersions must be transformed into formulations that are practically useful for the subsequent fabrication of, for example, organic nanoparticle-based coatings, bioassays, electronic- and sensor-devices, or dosage forms. Such further processing steps might include the concentration, purification, surface modification, labeling, or sterilization of organic nanoparticles.

One of the most frequent limitations, with respect to further processing, is related to the particle concentration. Very few synthetic methods can provide nanoparticle dispersions with a narrow size distribution and well-defined composition in sufficiently high concentrations for subsequent processing steps. A concentration of nanoparticle dispersions can, in some cases, be achieved by simple evaporation of the dispersion medium, or even by drying or freeze-drying the particles. Often, concentration can be combined with common methods of nanoparticle purification such as centrifugation, dialysis, and/or micro- and ultrafiltration techniques. For an overview on the different nanoparticle post-processing techniques, and efforts to scale-up organic nanoparticle synthesis, the reader is referred to a recent review [8]. A second review provides a good overview of post-processing techniques for the solidification of nanoparticle dispersions in order to obtain solid dosage forms for pharmaceutical applications (e.g., spray-drying, lyophilization) [164].

Provided that adequate techniques for nanoparticle synthesis and the necessary post-treatment are found, organic nanoparticles could be expected to be introduced successfully in various high-value applications and products. Currently, the formulation of active compounds in the form of nanosized particle dosage forms represents the most deeply explored field of application for organic nanoparticles. However, such nanoparticle formulations continue to open new possibilities in pharmaceutical technology, such as the enhanced drug delivery of poorly soluble compounds [14, 15, 22], the controlled release of drugs from polymer nanoparticles, or targeted delivery to special organs, perhaps by applying the ability of nanoparticles to cross the blood–brain barrier [78, 140]. Similarly, the principles of increased bioavailability and controlled release of nanoparticle systems can be used in related fields such as food technology (e.g., nanostructured food additives [10, 165]).

In contrast, applications that exploit special optical or electrical properties of organic nanoparticles remain in a nascent state, despite many organic nanoparticles showing a broad spectrum of interesting optical effects that range from nonlinear-optical properties [166] to photoswitchable fluorescence [167]. Fluorescent organic nanoparticles have, for example, been investigated as biological labels and probes for DNA [168, 169], proteins [170], cells [82] and for immunolabeling [171]. Polymer nanoparticles loaded with luminescent indicator dyes can also serve as versatile nanosized chemo- and biosensorparticles [80]. Furthermore, the use of nanoparticles of certain aromatic hydrocarbons as chemosensors and biosensors was proposed by exploiting the resonance energy transfer between nanoparticles and organic dyes [172]. Based on this principle, thin sol–gel films with incorporated organic nanoparticles were suggested for applications such as chemical sensor arrays [173] and biochips [174].

7.4
Summary and Future Perspectives

A wide variety of methods has been developed for the synthesis of organic nanoparticles, forming a type of "toolbox" from which the most appropriate method can be selected depending on material and application. This toolbox spans from classical top-down approaches, such as milling, to novel bottom-up synthetic methods in microfluidic devices. An overview of the most important techniques for organic nanoparticle synthesis, as reviewed in this chapter, is provided in Table 7.1.

A key task for the future development of nanotechnology applications using organic materials will be the development of approaches for the controllable synthesis of organic nanomaterials. This means, on the one hand, that a rigorous

Table 7.1 Overview on the different nanoparticle synthesis methods.

	Nanoparticle preparation method		s/p[a]	Typical material(s)	Reference(s)
Top-down	Mechanical size reduction		s	Drugs, pigments	[20, 33]
	Nanoimprint lithography		p	(Bio)polymers	[43]
Bottom-up	Solution-based methods	Emulsion methods	p/s	(Bio)polymers, dyes, drugs	[5, 8]
		Reprecipitation	p/s	(Bio)polymers, dyes, drugs, proteins	[5, 7, 8]
		Supercritical fluid techniques	p/s	(Bio)polymers, dyes, drugs	[11, 124, 126]
		Polyelectrolyte complex formation	p	Polyelectrolytes, polysaccharides, DNA	[129, 131]
		Microfluidic synthesis	p/s	Polymers, drugs, dyes	–
	Vapor phase-based methods		s	Dyes, drugs, pigments	[159, 162]

a s, small organic molecules; p, polymers.

control must be exercised over the size and polydispersity of the particles, whilst on the other hand it must be possible to control the morphology and structural properties of the particles. On occasion, variation of the molecular structure of organic molecules, by employing different side groups or substitution patterns, may represent a successful tool to control which type of nanostructure is obtained. For example, Tian et al. [175] described the controlled formation of either spherical nanoparticles or nanorods from an organic dye, depending on the type of side groups present. Other methods used to control the morphology of organic nanomaterials include the adjustment of preparation conditions (e.g., solvent, concentration, temperature) or template-assisted syntheses in micelles or vesicles (see, e.g., Refs [176, 177]). The addition of different macromolecules or stabilizers in the reprecipitation of small molecules has also frequently been described to allow for a certain control over particle morphology [95]. Besides the particle size and morphology, future approaches in organic nanoparticle synthesis should be aimed at controlling the crystallinity and phase of the particles. For many organic materials, the order and orientation of molecules in the solid state are, to a large extent, responsible for their properties. The formation of different types of molecular aggregate (J- and H- aggregates) in organic nanoparticles may have a strong effect on changing their optical [5, 92] or electrical properties [178]. The chemical and pharmacokinetic properties of organic nanoparticles also depend on crystallinity and phase with, in general, crystalline particles being more stable while amorphous particles represent a higher energy state and hence an increased solubility. Thus, the controlled preparation of amorphous nanoparticles of hydrophobic, poorly soluble drugs may be desired for some pharmaceutical applications [179].

One emerging field of increasing future importance will be the controlled synthesis of internally and chemically structured nanoparticles. Such structured organic nanoparticles would be expected to show special optical and electronic properties, and can also form new, unusual superstructures. The ability to synthesize such structured nanoparticles is also a prerequisite for the preparation of so-called "multifunctional nanoparticles," which combine multiple functionalities such as drug-delivery, specific site targeting, controlled release, imaging, sensing, and controlled biodegradability, within a single-particle structure. Multifunctional nanoparticles come close to the idea of smart–almost autonomous–nanodevices, and somehow represent the ultimate goal of nanoparticle synthetic approaches. To date, most approaches for the preparation of multifunctional organic nanoparticles have been aimed at the integration of drug-delivery and imaging functions (e.g., Refs [81, 107].). Nevertheless, investigations of multifunctional organic nanoparticles remain at a nascent stage, and will surely be one of the most important future fields in nanobiotechnology.

References

1 Cushing, B.L., Kolesnichenko, V.L. and O'Connor, C.J. (2004) Recent advances in the liquid-phase syntheses of inorganic nanoparticles. *Chemical Reviews*, **104** (9), 3893–946.

2 Kwon, S.G. and Hyeon, T. (2008) Colloidal chemical synthesis and formation kinetics of uniformly sized nanocrystals of metals, oxides, and chalcogenides. *Accounts of Chemical Research*, **41** (12), 1696–709.

3 Schmid, G. (2004) *Nanoparticles*, Wiley-VCH Verlag GmbH, Weinheim.

4 Nagarajan, R. (2008) "Frozen" micelles: polymer nanoparticles of controlled size by self-assembly, in *Nanoparticles: Synthesis, Stabilization, Passivation, and Functionalization*, vol. 996 (eds R. Nagarajan and T.A. Hatton), ACS Symposium Series, American Chemical Society, pp. 341–56.

5 Horn, D. and Rieger, J. (2001) Organic nanoparticles in the aqueous phase–theory, experiment, and use. *Angewandte Chemie International Edition*, **40** (23), 4330–61.

6 Masuhara, H., Nakanishi, H. and Sasaki, K. (2003) *Single Organic Nanoparticles*, Springer, Berlin, p. 402.

7 Texter, J. (2001) Precipitation and condensation of organic particles. *Journal of Dispersion Science and Technology*, **22** (6), 499–527.

8 Vauthier, C. and Bouchemal, K. (2009) Methods for the preparation and manufacture of polymeric nanoparticles. *Pharmaceutical Research*, **26** (5), 1025–58.

9 Date, A.A. and Patravale, V.B. (2004) Current strategies for engineering drug nanoparticles. *Current Opinion in Colloid and Interface Science*, **9** (3-4), 222–35.

10 Acosta, E. (2009) Bioavailability of nanoparticles in nutrient and nutraceutical delivery. *Current Opinion in Colloid and Interface Science*, **14**, 3–15.

11 Yasuji, T., Takeuchi, H. and Kawashima, Y. (2008) Particle design of poorly water-soluble drug substances using supercritical fluid technologies. *Advanced Drug Delivery Reviews*, **60** (3), 388–98.

12 Schwoerer, M. and Wolf, H.C. (2007) *Organic Molecular Solids*, Wiley-VCH Verlag GmbH, Weinheim.

13 Patra, A., Hebalkar, N., Sreedhar, B., Sarkar, M., Samanta, A. and Radhakrishnan, T.P. (2006) Tuning the size and optical properties in molecular nano/microcrystals. *Small*, **2** (5), 650–9.

14 Kipp, J.E. (2004) The role of solid nanoparticle technology in the parenteral delivery of poorly water-soluble drugs.

International Journal of Pharmaceutics, **284** (1-2), 109–22.

15 Merisko-Liversidge, E.M. and Liversidge, G.G. (2008) Drug nanoparticles: formulating poorly water-soluble compounds. *Toxicologic Pathology*, **36** (1), 43–8.

16 Schönert, K. and Steier, K. (1971) Die Grenze der Zerkleinerung bei kleinen Korngrößen. *Chemie Ingenieur Technik*, **43** (13), 773–7.

17 Doroszkowski, A. (1994) Paints, in *Technological Applications of Dispersions*, vol. 52 (ed. R.B. McKay), Marcel Decker, Inc., New York, pp. 1–67.

18 Bishop, J.F. and Czekai, D.A. (1997) Ink jet inks containing nanoparticles of organic pigments. US Patent 5, 679, 138 1997.

19 Liversidge, G.G. and Cundy, K.C. (1995) Particle size reduction for improvement of oral bioavailability of hydrophobic drugs: I. Absolute oral bioavailability of nanocrystalline danazol in beagle dogs. *International Journal of Pharmaceutics*, **125** (1), 91–7.

20 Merisko-Liversidge, E.M., Liversidge, G.G. and Cooper, E.R. (2003) Nanosizing: a formulation approach for poorly-water-soluble compounds. *European Journal of Pharmaceutical Sciences*, **18** (2), 113–20.

21 NanoCrystal® Technology (2009) *Commercialised Products*. http://www.elandrugtechnologies.com/nanocrystal_technology/commercialised (accessed 31 October 2009).

22 Kesisoglou, F., Panmai, S. and Wu, Y. (2007) Nanosizing–oral formulation development and biopharmaceutical evaluation. *Advanced Drug Delivery Reviews*, **59** (7), 631–44.

23 Van Eerdenbrugh, B., Stuyven, B., Froyen, L., Van Humbeeck, J., Martens, J.A., Augustijns, P. and Van den Mooter, G. (2009) Downscaling drug nanosuspension production: processing aspects and physicochemical characterization. *AAPS PharmSciTech*, **10** (1), 44–53.

24 Choi, J.-Y., Park, C.H. and Lee, J. (2008) Effect of polymer molecular weight on nanocomminution of poorly soluble drug. *Drug Delivery*, **15** (5), 347–53.

25 Choi, J.-Y., Yoo, J.Y., Kwak, H.-S., Nam, B.U. and Lee, J. (2005) Role of polymeric stabilizers for drug nanocrystal dispersions. *Current Applied Physics*, **5** (5), 472–4.

26 Lee, J., Choi, J.-Y. and Park, C.H. (2008) Characteristics of polymers enabling nano-comminution of water-insoluble drugs. *International Journal of Pharmaceutics*, **355** (1-2), 328–36.

27 Van Eerdenbrugh, B., Vermant, J., Martens, J.A., Froyen, L., Van Humbeeck, J., Augustijns, P. and Van den Mooter, G. (2009) A screening study of surface stabilization during the production of drug nanocrystals. *Journal of Pharmaceutical Sciences*, **98** (6), 2091–103.

28 Pongpeerapat, A., Wanawongthai, C., Tozuka, Y., Moribe, K. and Yamamoto, K. (2008) Formation mechanism of colloidal nanoparticles obtained from probucol/PVP/SDS ternary ground mixture. *International Journal of Pharmaceutics*, **352** (1-2), 309–16.

29 Wanawongthai, C., Pongpeerapat, A., Higashi, K., Tozuka, Y., Moribe, K. and Yamamoto, K. (2009) Nanoparticle formation from probucol/PVP/sodium alkyl sulfate co-ground mixture. *International Journal of Pharmaceutics*, **376** (1-2), 169–75.

30 Wongmekiat, A., Tozuka, Y., Moribe, K., Oguchi, T. and Yamamoto, K. (2007) Preparation of drug nanoparticles by co-grinding with cyclodextrin: formation mechanism and factors affecting nanoparticle formation. *Chemical and Pharmaceutical Bulletin*, **55** (3), 359–63.

31 Juhnke, M. and Weichert, R. (2005) Nanoparticles of soft materials by high-energy milling at low temperatures. 7th World Congress of Chemical Engineering, Glasgow.

32 Juhnke, M. and Weichert, R. (2005) Zerkleinerung weicher Materialien ohne Verunreinigung der Produkte durch die Mahlkörper. *Chemie Ingenieur Technik*, **77** (1-2), 90–4.

33 Müller, R.H., Jacobs, C. and Kayser, O. (2001) Nanosuspensions as particulate drug formulations in therapy: rationale for development and what we can expect for the future. *Advanced Drug Delivery Reviews*, **47** (1), 3–19.

34 Müller, R.H. and Peters, K. (1998) Nanosuspensions for the formulation of poorly soluble drugs: I. Preparation by a size-reduction technique. *International Journal of Pharmaceutics*, **160** (2), 229–37.

35 Krause, K.P., Kayser, O., Mäder, K., Gustl, R. and Müller, R.H. (2000) Heavy metal contamination of nanosuspensions produced by high-pressure homogenisation. *International Journal of Pharmaceutics*, **196** (2), 169–72.

36 Verma, S., Gokhale, R. and Burgess, D.J. (2009) A comparative study of top-down and bottom-up approaches for the preparation of micro/nanosuspensions. *International Journal of Pharmaceutics*, **380** (1-2), 216–22.

37 Shah, U., Vemavarapu, C., Askins, V., Lodaya, M., Elzinga, P. and Mollan, M.J. (2006) Nanoparticle formation: a modular high pressure system to enhance biopharmaceutical properties of poorly soluble drugs. *Pharmaceutical Technology*, **4**, article no. 311239.

38 Hernandez, C.J. and Mason, T.G. (2007) Colloidal alphabet soup: monodisperse dispersions of shape-designed LithoParticles. *Journal of Physical Chemistry C*, **111** (12), 4477–80.

39 Moon, J.H., Kim, A.J., Crocker, J.C. and Yang, S. (2007) High-throughput synthesis of anisotropic colloids via holographic lithography. *Advanced Materials*, **19** (18), 2508–12.

40 Dendukuri, D., Pregibon, D.C., Collins, J., Hatton, T.A. and Doyle, P.S. (2006) Continuous-flow lithography for high-throughput microparticle synthesis. *Nature Materials*, **5** (5), 365–9.

41 Ahn, S.H. and Guo, L.J. (2009) Large-area roll-to-roll and roll-to-plate nanoimprint lithography: a step toward high-throughput application of continuous nanoimprinting. *ACS Nano*, **3** (8), 2304–10.

42 Euliss, L.E., DuPont, J.A., Gratton, S. and DeSimone, J. (2006) Imparting size, shape, and composition control of materials for nanomedicine. *Chemical Society Reviews*, **35**, 1095–104.

43 Gratton, S.E.A., Williams, S.S., Napier, M.E., Polhaus, P.D., Zhou, Z., Wiles, K.B., Maynor, B.W., Shen, C., Olafsen, T., Samulski, E.T. and DeSimone, J.M. (2008) The pursuit of a scalable nanofabrication platform for use in material and life science applications. *Accounts of Chemical Research*, **41** (12), 1685–95.

44 Rolland, J.P., Maynor, B.W., Euliss, L.E., Exner, A.E., Denison, G.M. and de Simone, J.M. (2005) Direct fabrication and harvesting of monodisperse, shape-specific nanobiomaterials. *Journal of the American Chemical Society*, **127** (28), 10096–100.

45 Kelly, J.Y. and de Simone, J.M. (2008) Shape-specific, monodisperse nano-molding of protein particles. *Journal of the American Chemical Society*, **130** (16), 5439.

46 Maynor, B.W., LaRue, I., Hu, Z., Rolland, J.P., Pandya, A., Fu, Q., Liu, J., Spontak, R.J., Sheiko, S.S., Samulski, E.T. and De Simone, J.M. (2007) Supramolecular nanomimetics: replication of micelles, viruses, and other naturally occurring nanoscale objects. *Small*, **3** (5), 845–9.

47 Glangchai, L.C., Mary, C.-M., Shi, L. and Roy, K. (2008) Nanoimprint lithography based fabrication of shape-specific, enzymatically-triggered smart nanoparticles. *Journal of Controlled Release*, **125** (3), 263–72.

48 Buyukserin, F., Aryal, M., Gao, J. and Hu, W. (2009) Fabrication of polymeric nanorods using bilayer nanoimprint lithography. *Small*, **5** (14), 1632–6.

49 Thickett, S.C. and Gilbert, R.G. (2007) Emulsion polymerization: state of the art in kinetics and mechanisms. *Polymer*, **48** (24), 6965–91.

50 Klapper, M., Nenov, S., Haschick, R., Müller, K. and Müllen, K. (2008) Oil-in-oil emulsions: a unique tool for the formation of polymer nanoparticles. *Accounts of Chemical Research*, **41** (9), 1190–201.

51 Landfester, K. (2001) The generation of nanoparticles in miniemulsions. *Advanced Materials*, **13** (10), 765–8.

52 Landfester, K. (2009) Miniemulsion polymerization and the structure of polymer and hybrid nanoparticles. *Angewandte Chemie International Edition*, **48** (26), 4488–507.

53 Trotta, M., Gallarate, M., Pattarino, F. and Morel, S. (2001) Emulsions containing partially water-miscible solvents for the preparation of drug nanosuspensions. *Journal of Controlled Release*, **76** (1-2), 119–28.

54 Trotta, M., Gallarate, M., Carlotti, M.E. and Morel, S. (2003) Preparation of griseofulvin nanoparticles from water-dilutable microemulsions. *International Journal of Pharmaceutics*, **254** (2), 235–42.

55 Ambrus, R., Kocbek, P., Kristl, J., Sibanc, R., Rajko, R. and Szabo-Revesz, P. (2009) Investigation of preparation parameters to improve the dissolution of poorly water-soluble meloxicam. *International Journal of Pharmaceutics*, **381** (2), 153–9.

56 Zeng, H., Li, X., Zhang, G. and Dong, J. (2009) System investigation of the formation of beta-cypermethrin nanosuspension: influence of the formulation variables. *Journal of Dispersion Science and Technology*, **30** (1), 76–82.

57 Kocbek, P., Baumgartner, S. and Kristl, J. (2006) Preparation and evaluation of nanosuspensions for enhancing the dissolution of poorly soluble drugs. *International Journal of Pharmaceutics*, **312** (1-2), 179–86.

58 Chu, B.-S., Ichikawa, S., Kanafusa, S. and Nakajima, M. (2007) Preparation of protein-stabilized β-carotene nanodispersions by emulsification–evaporation method. *Journal of the American Oil Chemists Society*, **84** (11), 1053–62.

59 Kwon, E., Oikawa, H., Kasai, H. and Nakanishi, H. (2007) A fabrication method of organic nanocrystals using stabilizer-free emulsion. *Crystal Growth and Design*, **7** (4), 600–2.

60 Ujiiye-Ishii, K., Kwon, E., Kasai, H., Nakanishi, H. and Oikawa, H. (2008) Methodological features of the emulsion and reprecipitation methods for organic nanocrystal fabrication. *Crystal Growth and Design*, **8** (2), 369–71.

61 Capek, I. (2004) Preparation of metal nanoparticles in water-in-oil (w/o) microemulsions. *Advances in Colloid and Interface Science*, **110** (1-2), 49–74.

62 Destree, C. and Nagy, J.B. (2006) Mechanism of formation of inorganic and organic nanoparticles from microemulsions. *Advances in Colloid and Interface Science*, **123–126**, 353–67.

63 Debuigne, F., Jeunieau, L., Wiame, M. and Nagy, J.B. (2000) Synthesis of organic nanoparticles in different w/o microemulsions. *Langmuir*, **16** (20), 7605–11.

64 Destree, C., George, S., Champagne, B., Guillaume, M., Ghijsen, J. and Nagy, J.B. (2009) J-complexes of retinol formed within the nanoparticles prepared from microemulsions. *Colloid and Polymer Science*, **286** (1), 15–30.

65 Tallury, P., Kar, S., Bamrungsap, S., Huang, Y.-F., Tan, W. and Santra, S. (2009) Ultra-small water-dispersible fluorescent chitosan nanoparticles: synthesis, characterization and specific targeting. *Chemical Communications*, **17**, 2347–9.

66 Zhang, H., Wang, D., Butler, R., Campbell, N.L., Long, J., Tan, B., Duncalf, D.J., Foster, A.J., Hopkinson, A., Taylor, D., Angus, D., Cooper, A.I. and Rannard, S.P. (2008) Formation and enhanced biocidal activity of water-dispersible organic nanoparticles. *Nature Nanotechnology*, **3**, 506–11.

67 Rannard, S. (2008) Opportunities for organic nanoparticles. *Specialty Chemicals Magazine*, **6**, 40–1.

68 Margulis-Goshen, K. and Magdassi, S. (2009) Formation of simvastatin nanoparticles from microemulsion. *Nanomedicine*, **5** (3), 271–81.

69 Margulis-Goshen, K., Netivi, H.D., Major, D.T., Gradzielski, M., Raviv, U. and Magdassi, S. (2010) Formation of organic nanoparticles from volatile microemulsions. *Journal of Colloid and Interface Science*, **342** (2), 283–92.

70 Fessi, H., Puisieux, F., Devissaguet, J.P., Ammoury, N. and Benita, S. (1989) Nanocapsule formation by interfacial polymer deposition following solvent displacement. *International Journal of Pharmaceutics*, **55** (1), R1–4.

71 Horn, D. (1989) Preparation and characterization of microdisperse bioavailable carotenoid hydrosols.

Angewandte Makromolekulare Chemie, **166** (1), 139–53.

72 Kasai, H., Nalwa, H.S., Oikawa, H., Okada, S., Matsuda, H., Minami, N., Kakuta, A., Ono, K., Mukoh, A. and Nakanishi, H. (1992) Novel preparation method of organic microcrystals. *Japanese Journal of Applied Physics Part 2: Letters*, **31** (8A), L1132–4.

73 Nakanishi, H. and Oikawa, H. (2003) Reprecipitation method for organic nanocrystals, in *Single Organic Nanoparticles* (eds H. Masuhara, H. Nakanishi and K. Sasaki), Springer, Berlin, pp. 17–31.

74 Legrand, P., Lesieur, S., Bochot, A., Gref, R., Raatjes, W., Barratt, G. and Vauthier, C. (2007) Influence of polymer behaviour in organic solution on the production of polylactide nanoparticles by nanoprecipitation. *International Journal of Pharmaceutics*, **344** (1-2), 33–43.

75 Bilati, U., Allémann, E. and Doelker, E. (2005) Development of a nanoprecipitation method intended for the entrapment of hydrophilic drugs into nanoparticles. *European Journal of Pharmaceutical Sciences*, **24** (1), 67–75.

76 Hornig, S. and Heinze, T. (2008) Efficient approach to design stable water-dispersible nanoparticles of hydrophobic cellulose esters. *Biomacromolecules*, **9** (5), 1487–92.

77 Hornig, S., Heinze, T., Becer, C.R. and Schubert, U.S. (2009) Synthetic polymer nanoparticles by nanoprecipitation. *Journal of Materials Chemistry*, **19**, 3838–40.

78 Zensi, A., Begley, D., Pontikis, C., Legros, C., Mihoreanu, L., Wagner, S., Büchel, C., von Briesen, H. and Kreuter, J. (2009) Albumin nanoparticles targeted with Apo E enter the CNS by transcytosis and are delivered to neurones. *Journal of Controlled Release*, **137** (1), 78–86.

79 Zhao, G., Ishizaka, T., Kasai, H., Oikawa, H. and Nakanishi, H. (2007) Fabrication of unique porous polyimide nanoparticles using a reprecipitation method. *Chemistry of Materials*, **19** (8), 1901–5.

80 Borisov, S.M., Mayr, T., Mistlberger, G., Waich, K., Koren, K., Chojnacki, P. and Klimant, I. (2009) Precipitation as a simple and versatile method for preparation of optical nanochemosensors. *Talanta*, **79** (5), 1322–30.

81 Nehilla, B.J., Allen, P.G. and Desasis, T.A. (2008) Surfactant-free, drug-quantum-dot coloaded poly(lactide-*co*-glycolide) nanoparticles: towards multifunctional nanoparticles. *ACS Nano*, **2** (3), 538–44.

82 Green, M., Howes, P., Berry, C., Argyros, O. and Thanou, M. (2009) Simple conjugated polymer nanoparticles as biological labels. *Proceedings of the Royal Society A*, **465** (2109), 2751–9.

83 Wang, F., Han, M.-Y., Mya, K.Y., Wang, Y. and Lai, Y.-H. (2005) Aggregation-driven growth of size-tunable organic nanoparticles using electronically altered conjugated polymers. *Journal of the American Chemical Society*, **127** (29), 10350–5.

84 Dong, Y., Ng, W.K., Hu, J., Shen, S. and Tan, R.B.H. (2010) A continuous and highly effective static mixing process for antisolvent precipitation of nanoparticles of poorly water-soluble drugs. *International Journal of Pharmaceutics*, **386** (1–2), 256–61.

85 Matteucci, M.E., Hotze, M.A., Johnston, K.P. and Williams, III, R.O. (2006) Drug nanoparticles by antisolvent precipitation: mixing energy versus surfactant stabilization. *Langmuir*, **22** (21), 8951–9.

86 *NanoMorph: Technology for Amorphous Nanoparticles* http://www.soliqs.com/NanoMorph-R.20.0.html (accessed 23 January 2010).

87 Brick, C.M., Palmer, H.J. and Whitesides, T.H. (2003) Formation of colloidal dispersions of organic materials in aqueous media by solvent shifting. *Langmuir*, **19** (16), 6367–80.

88 Kim, H.Y., Bjorklund, T.G., Lim, S.H. and Bardeen, C.J. (2003) Spectroscopic and photocatalytic properties of organic tetracene nanoparticles in aqueous solution. *Langmuir*, **19** (9), 3941–6.

89 Latterini, L., Roscini, C., Carlotti, B., Aloisi, G.G. and Elisei, F. (2006) Synthesis and characterization of

perylene nanoparticles. *Physica Status Solidi A*, **203** (6), 1470–5.

90 Lim, S.-H., Bjorklund, T.G., Spano, F.C. and Bardeen, C.J. (2004) Exciton delocalization and superradiance in tetracene thin films and nanoaggregates. *Physical Review Letters*, **92** (10), 107402.

91 Gong, X., Milic, T., Xu, C., Battes, J.D. and Drain, C.M. (2002) Preparation and characterization of porphyrin nanoparticles. *Journal of the American Chemical Society*, **124** (48), 14290–1.

92 Auweter, H., Haberkorn, H., Heckmann, W., Horn, D., Lüddecke, E., Rieger, J. and Weiss, H. (1999) Supramolecular structure of precipitated nanosize-carotene particles. *Angewandte Chemie International Edition*, **38** (15), 2188–91.

93 Kumar, V. and Prud'homme, R.K. (2009) Nanoparticle stability: processing pathways for solvent removal. *Chemical Engineering Science*, **64** (6), 1358–61.

94 Bertorelle, F., Lavabre, D. and Fery-Forgues, S. (2003) Dendrimer-tuned formation of luminescent organic microcrystals. *Journal of the American Chemical Society*, **125**, 6244–53.

95 Fery-Forgues, S., Abyan, M. and Lamere, J.-F. (2008) Nano- and microparticles of organic fluorescent dyes. *Annals of the New York Academy of Sciences*, **1130**, 272–9.

96 Oliveira, D., Baba, K., Mori, J., Miyashita, Y., Kasai, H., Oikawa, H. and Nakanishi, H. (2010) Using an organic additive to manipulate sizes of perylene nanoparticles. *Journal of Crystal Growth*, **312** (3), 431–6.

97 Peng, A.-D., Xiao, D.-B., Ma, Y., Yang, W.-S. and Yao, J.-N. (2005) Tuneable emission from doped 1,3,5-triphenyl-2-pyrazoline organic nanoparticles. *Advanced Materials*, **17**, 1073–2070.

98 Dalvi, S.V. and Dave, R.N. (2009) Controlling particle size of a poorly water-soluble drug using ultrasound and stabilizers in antisolvent precipitation. *Industrial and Engineering Chemistry Research*, **48** (16), 7581–93.

99 Dennehy, R.D. (2003) Particle engineering using power ultrasound. *Organic Process Research and Development*, **7** (6), 1002–6.

100 Al-Kaysi, R., Müller, A.M., Ahn, T.-S., Lee, S. and Bardeen, C.J. (2005) Effects of sonication on the size and crystallinity of stable zwitterionic organic nanoparticles formed by reprecipitation in water. *Langmuir*, **21** (17), 7990–4.

101 Kang, P., Chen, C., Hao, L., Zhu, C., Hu, Y. and Chen, Z. (2004) A novel sonication route to prepare anthracene nanoparticles. *Materials Research Bulletin*, **39**, 545–51.

102 Jia, Z., Xiao, D., Yang, W., Ma, Y., Yao, J.-N. and Liu, Z. (2004) Preparation of perylene nanoparticles with a membrane mixer. *Journal of Membrane Science*, **241** (2), 387–92.

103 Charcosset, C. and Fessi, H. (2005) Preparation of nanoparticles with a membrane contactor. *Journal of Membrane Science*, **266** (1-2), 115–20.

104 Jia, Z., Yang, W., Zhou, Z., Yao, J.-N. and Liu, Z. (2006) Preparation of pyrazoline nanoparticles with a membrane mixer. *Colloids and Surfaces A: Physicochemical and Engineering Aspects*, **276** (1-3), 22–7.

105 Johnson, B.K. and Prud'homme, R.K. (2003) Generic method of preparing multifunctional fluorescent nanoparticles using flash NanoPrecipitation. *Australian Journal of Chemistry*, **56** (10), 1021–4.

106 Johnson, B.K., Saad, W. and Prud'homme, R.K. (2006) Nanoprecipitation of pharmaceuticals using mixing and block copolymer stabilization, in *Polymeric Drug Delivery II*, vol. 924 (ed. S. Svenson), ACS Symposium Series, American Chemical Society, pp. 278–91.

107 Akbulut, M., Ginart, P., Gindy, M.E., Theriault, C., Chin, K.H., Soboyejo, W. and Prud'homme, R.K. (2009) Generic method of preparing multifunctional fluorescent nanoparticles using flash NanoPrecipitation. *Advanced Functional Materials*, **19** (5), 718–25.

108 Gavi, E., Marchisio, D.L. and Barresi, A.A. (2007) CFD modelling and scale-up of Confined Impinging Jet Reactors. *Chemical Engineering Science*, **62** (8), 2228–41.

109 Lince, F., Marchisio, D.L. and Barresi, A.A. (2009) Smart mixers and reactors for the production of pharmaceutical

nanoparticles: proof of concept. *Chemical Engineering Research and Design*, **87** (4), 543–9.
110 Marchisio, D.L., Rivautella, L. and Barresi, A.A. (2006) Design and scale-up of chemical reactors for nanoparticle precipitation. *AIChE Journal*, **52** (5), 1877–87.
111 Liu, Y., Cheng, C., Liu, Y., Prud'homme, R.K. and Fox, R.O. (2008) Mixing in a multi-inlet vortex mixer (MIVM) for flash nano-precipitation. *Chemical Engineering Science*, **63** (11), 2829–42.
112 Anantachoke, N., Makha, M., Raston, C.L., Reutrakul, V., Smith, N.C. and Saunders, M. (2006) Fine tuning the production of nanosized β-carotene particles using spinning disk processing. *Journal of the American Chemical Society*, **128** (42), 13847–53.
113 Chen, J.-F., Zhou, M.-Y., Shao, L., Wang, Y.-Y., Yun, J., Chew, N.Y.K. and Chan, H.-K. (2004) Feasibility of preparing nanodrugs by high-gravity reactive precipitation. *International Journal of Pharmaceutics*, **269** (1), 267–74.
114 Hu, T.-T., Wang, J.-X., Shen, Z.-G. and Chen, J.-F. (2008) Engineering of drug nanoparticles by HGCP for pharmaceutical applications. *Particuology*, **6** (4), 239–51.
115 Ibanez, A., Maximov, S., Guiu, A., Chaillout, C. and Baldeck, P.L. (1998) Controlled nanocrystallization of organic molecules in sol-gel glasses. *Advanced Materials*, **10** (18), 1540–3.
116 Monnier, V., Sanz, N., Botzung-Appert, E., Bacia, M. and Ibanez, A. (2006) Confined nucleation and growth of organic nanocrystals in sol-gel matrices. *Journal of Materials Chemistry*, **16**, 1401–9.
117 Sanz, N., Gaillot, A.-C., Usson, Y., Baldeck, P.L. and Ibanez, A. (2000) Organic nanocrystals grown in sol-gel coatings. *Journal of Materials Chemistry*, **10**, 2723–6.
118 Sheibat-Othman, N., Burne, T., Charcosset, C. and Fessi, H. (2008) Preparation of pH-sensitive particles by membrane contactor. *Colloids and Surfaces A: Physicochemical and Engineering Aspects*, **315** (1-3), 13–22.
119 Ou, Z.-M., Yao, H. and Kimura, K. (2006) Organic nanoparticles of porphyrin without self-aggregation. *Chemistry Letters*, **35** (7), 782.
120 Yao, H., Ou, Z. and Kimura, K. (2005) Ion-based organic nanoparticles: synthesis, characterization, and optical properties of pseudoisocyanine dye nanoparticles. *Chemistry Letters*, **34** (8), 1108.
121 Ou, Z.-M., Yao, H. and Kimura, K. (2007) Organic nanoparticles of cyanine dyes in aqueous solution. *Bulletin of the Chemical Society of Japan*, **80** (2), 295–302.
122 Ou, Z., Yao, H. and Kimura, K. (2007) Preparation and optical properties of organic nanoparticles of porphyrin without self aggregation. *Journal of Photochemistry and Photobiology A: Chemistry*, **189** (1), 7–14.
123 Kang, L., Wang, Z., Cao, Z., Ma, Y., Fu, H. and Yao, J. (2007) Colloid chemical reaction route to the preparation of nearly monodispersed perylene nanoparticles: size-tunable synthesis and three-dimensional self organization. *Journal of the American Chemical Society*, **129** (23), 7305–12.
124 Byrappa, K., Ohara, S. and Adschiri, T. (2008) Nanoparticles synthesis using supercritical fluid technology – towards biomedical applications. *Advanced Drug Delivery Reviews*, **60** (3), 299–327.
125 Sun, Y.-P., Meziani, M.J., Pathak, P. and Qu, L. (2005) Polymeric nanoparticles from rapid expansion of supercritical fluid solution. *Chemistry: A European Journal*, **11** (5), 1366–73.
126 Reverchon, E., De Marco, I. and Torino, E. (2007) Nanoparticles production by supercritical antisolvent precipitation: a general interpretation. *Journal of Supercritical Fluids*, **43** (1), 126–38.
127 Moribe, K., Tozuka, Y. and Yamamoto, K. (2008) Supercritical carbon dioxide processing of active pharmaceutical ingredients for polymorphic control and for complex formation. *Advanced Drug Delivery Reviews*, **60** (3), 328–38.
128 Jagannathan, R., Irvin, G., Blanton, T. and Jagannathan, S. (2009) Organic nanoparticles: preparation, self-assembly,

and properties. *Advanced Functional Materials*, **16** (6), 747–53.

129 Hartig, S.M., Greene, R.R., Dikov, M.M., Prokop, A. and Davidson, J.M. (2007) Multifunctional nanoparticulate polyelectrolyte complexes. *Pharmaceutical Research*, **24** (12), 2353–69.

130 Boddohi, S., Moore, N., Johnson, P.A. and Kipper, M.J. (2009) Polysaccharide-based polyelectrolyte complex nanoparticles from chitosan, heparin, and hyaluronan. *Biomacromolecules*, **10** (6), 1402–9.

131 Liu, Z., Jiao, Y., Wang, Y., Zhou, C. and Zhang, Z. (2008) Polysaccharides-based nanoparticles as drug delivery systems. *Advanced Drug Delivery Reviews*, **60** (15), 1650–62.

132 Hartig, S.M., Carlesso, G., Davidson, J.M. and Prokop, A. (2007) Development of improved nanoparticulate polyelectrolyte complex physicochemistry by nonstoichiometric mixing of polyions with similar molecular weights. *Biomacromolecules*, **8** (1), 265–72.

133 Schatz, C., Domard, A., Viton, C., Pichot, C. and Delair, T. (2004) Versatile and efficient formation of colloids of biopolymer-based polyelectrolyte complexes. *Biomacromolecules*, **5** (5), 1882–92.

134 Drogoz, A., David, L., Rochas, C., Domard, A. and Delair, T. (2007) Polyelectrolyte complexes from polysaccharides: formation and stoichiometry monitoring. *Langmuir*, **23** (22), 10950–8.

135 Oyarzun-Ampuero, F.A., Brea, J., Loza, M.I., Torres, D. and Alonso, M.J. (2009) Chitosan–hyaluronic acid nanoparticles loaded with heparin for the treatment of asthma. *International Journal of Pharmaceutics*, **381** (2), 122–9.

136 Zheng, Y., Yang, W., Wang, C., Hu, J., Fu, S., Dong, L., Wu, L. and Shen, X. (2007) Nanoparticles based on the complex of chitosan and polyaspartic acid sodium salt: preparation, characterization and the use for 5-fluorouracil delivery. *European Journal of Pharmaceutics and Biopharmaceutics*, **67** (3), 621–31.

137 Kabanov, A.V. and Kabanov, V.A. (1995) DNA complexes with polycations for the delivery of genetic material into cells. *Bioconjugate Chemistry*, **6** (1), 7–20.

138 Fernandes, J.C., Tiera, M.J. and Winnik, F.M. (2006) Chitosan nanoparticles for non-viral gene therapy, in *Polysaccharides for Drug Delivery and Pharmaceutical Applications*, vol. 934 (eds R.H. Marchessault, F. Ravenelle and X.X. Zhu), ACS Symposium Series, American Chemical Society, pp. 177–200.

139 Junghans, M., Kreuter, J. and Zimmer, A. (2000) Antisense delivery using protamine–oligonucleotide particles. *Nucleic Acids Research*, **28** (10), e45.

140 Kratzer, I., Wernig, K., Panzenboeck, U., Bernhart, E., Reicher, H., Wronski, R., Windisch, M., Hammer, A., Malle, E., Zimmer, A. and Sattler, W. (2007) Apolipoprotein A-I coating of protamine–oligonucleotide nanoparticles increases particle uptake and transcytosis in an in vitro model of the blood–brain barrier. *Journal of Controlled Release*, **117** (3), 301–11.

141 Daoud-Mahammed, S., Ringard-Lefebvre, C., Razzuoq, N., Rosilio, V., Gillet, B., Couvreur, P., Amiel, C. and Gref, R. (2007) Spontaneous association of hydrophobized dextran and poly-β-cyclodextrin into nanoassemblies: formation and interaction with a hydrophobic drug. *Journal of Colloid and Interface Science*, **307** (1), 83–93.

142 Gref, R., Amiel, C., Molinard, K., Daoud-Mahammed, S., Sebille, B., Gillet, B., Beloeil, J.-C., Ringard, C., Rosilio, V., Poupaert, J. and Couvreur, P. (2006) New self-assembled nanogels based on host–guest interactions: characterization and drug loading. *Journal of Controlled Release*, **111** (3), 316–24.

143 West, J., Becker, M., Tombrink, S. and Manz, A. (2008) Micro total analysis systems: latest achievements. *Analytical Chemistry*, **80** (12), 4403–19.

144 El-Ali, J., Sorger, P. and Jensen, K.F. (2006) Cells on chips. *Nature*, **442** (7101), 403–11.

145 deMello, A.J. (2006) Control and detection of chemical reactions in microfluidic systems. *Nature*, **442** (7101), 394–402.

146 Dendukuri, D. and Doyle, P.S. (2009) The synthesis and assembly of polymeric

microparticles using microfluidics. *Advanced Materials*, **21** (41), 4071–86.

147 Hung, L.-H. and Lee, A.P. (2007) Microfluidic devices for the synthesis of nanoparticles and biomaterials. *Journal of Medical and Biological Engineering*, **27** (1), 1–6.

148 Jahn, A., Reiner, J.E., Vreeland, W.N., DeVoe, D.L., Locasio, L.E. and Gaitan, M. (2008) Preparation of nanoparticles by continuous-flow microfluidics. *Journal of Nanoparticle Research*, **10** (6), 925–34.

149 Leng, J. and Salmon, J.-B. (2009) Microfluidic crystallization. *Lab on a Chip*, **9** (1), 24–34.

150 Su, Y.-F., Kim, H., Kovenklioglu, S. and Lee, W.Y. (2007) Continuous nanoparticle production by microfluidic-based emulsion, mixing and crystallization. *Journal of Solid-State Chemistry*, **180** (9), 2625–9.

151 Rondeau, E. and Cooper-White, J.J. (2008) Biopolymer microparticle and nanoparticle formation within a microfluidic device. *Langmuir*, **24** (13), 6937–45.

152 Jasch, K., Barth, N., Fehr, S., Bunjes, H., Augustin, W. and Scholl, S. (2009) A microfluidic approach for a continuous crystallization of drug carrier nanoparticles. *Chemical Engineering and Technology*, **32** (11), 1806–14.

153 Zhao, H., Wang, J.-X., Wang, Q.-A., Chen, J.-F. and Yun, J. (2007) Controlled liquid antisolvent precipitation of hydrophobic pharmaceutical nanoparticles in a microchannel reactor. *Industrial and Engineering Chemistry Research*, **46** (24), 8229–35.

154 Ali, H.S.M., York, P. and Blagden, N. (2009) Preparation of hydrocortisone nanosuspensions through a bottom-up nanoprecipitation technique using microfluidic reactors. *International Journal of Pharmaceutics*, **375** (1-2), 107–13.

155 Ali, H.S.M., Blagden, N., York, P., Amani, A. and Brook, T. (2009) Artificial neural networks modelling the prednisolone nanoprecipitation in microfluidic reactors. *European Journal of Pharmaceutical Sciences*, **37** (3-4), 514–22.

156 Desportes, S., Yatabe, Z., Baumlin, S., Genot, V., Lefevre, L.-P., Ushiki, H., Delaire, J.A. and Pansu, R.B. (2007) Fluorescence lifetime imaging microscopy for in situ observation of the nanocrystallization of rubrene in a microfluidic set-up. *Chemical Physics Letters*, **446**, 212–16.

157 Karnik, R., Gu, F., Basto, P., Cannizzaro, C., Dean, L., Kyei-Manu, W., Langer, R. and Farokhzad, O.C. (2008) Microfluidic platform for controlled synthesis of polymeric nanoparticles. *Nano Letters*, **8** (9), 2906–12.

158 Liu, Y., Olsen, M.G. and Fox, R.O. (2009) Turbulence in a microscale planar confined impinging-jets reactor. *Lab on a Chip*, **9**, 1110–18.

159 Toyotama, H. (1997) Organic compound ultra-fine particles, in *Ultra-Fine Particles: Exploratory Science and Technology* (eds C. Hayashi, R. Uyeda and A. Tasaki), Noyes Publications, Westwood, New Jersey, pp. 286–92.

160 Raula, J., Kuivanen, A., Lähde, A., Jiang, H., Antopolsky, M., Kansikas, J. and Kauppinen, E.I. (2007) Synthesis of L-leucine nanoparticles via physical vapor deposition at varying saturation conditions. *Journal of Aerosol Science*, **38**, 1172–84.

161 Köstler, S., Rudorfer, A., Haase, A., Satzinger, V., Jakopic, G. and Ribitsch, V. (2009) Direct condensation method for the preparation of organic nanoparticle dispersions. *Advanced Materials*, **21** (24), 2505–10.

162 Asahi, T., Sugiyama, T. and Masuhara, H. (2008) Laser fabrication and spectroscopy of organic nanoparticles. *Accounts of Chemical Research*, **41** (12), 1790–8.

163 Masuhara, H. and Asahi, T.. (2003) Laser ablation method for organic nanoparticles, in *Single Organic Nanoparticles* (eds H. Masuhara, H. Nakanishi and K. Sasaki), Springer, Berlin, pp. 32–43.

164 Van Eerdenbrugh, B., Van den Mooter, G. and Augustijns, P. (2008) Top-down production of drug nanocrystals: nanosuspension stabilization, miniaturization and transformation into solid products. *International Journal of Pharmaceutics*, **364** (1), 64–75.

165 Velikov, K. and Pelan, E. (2008) Colloidal delivery systems for micronutrients and nutraceuticals. *Soft Matter*, **4** (10), 1964–80.

166 Yi, T., Clément, R., Haut, C., Catala, L., Gacoin, T., Tancrez, N., Ledoux, I. and Zyss, J. (2005) J-aggregated dye-MnPS3 hybrid nanoparticles with giant quadratic optical nonlinearity. *Advanced Materials*, **17** (3), 335–8.

167 Tian, Z., Wu, W. and Li, A.D.Q. (2009) Photoswitchable fluorescent nanoparticles: preparation, properties and applications. *ChemPhysChem*, **10** (15), 2577–91.

168 Jinshui, L., Lun, W., Feng, G., Yongxing, L. and Yun, W. (2003) Novel fluorescent colloids as a DNA fluorescent probe. *Analytical and Bioanalytical Chemistry*, **377** (2), 346–9.

169 Wang, L., Xia, T., Wang, L., Chen, H., Dong, L. and Bian, G. (2005) Preparation and application of a novel core-shell organic nanoparticle as a fluorescent probe in the determination of nucleic acids. *Microchimica Acta*, **149** (3-4), 267–72.

170 Wang, L., Wang, L., Dong, L., Bian, G., Xia, T. and Chen, H. (2005) Direct fluorimetric determination of γ-globulin in human serum with organic nanoparticle biosensor. *Spectrochimica Acta, Part A*, **61**, 129–33.

171 Kim, H.-J., Lee, J., Kim, T.-H., Lee, T.S. and Kim, J. (2008) Highly emissive self-assembled organic nanoparticles having dual color capacity for targeted immunofluorescence labeling. *Advanced Materials*, **20** (6), 1117–21.

172 Botzung-Appert, E., Monnier, V., Ha Duong, T., Pansu, R. and Ibanez, A. (2004) Polyaromatic luminescent nanocrystals for chemical and biological sensors. *Chemistry of Materials*, **16** (9), 1609–11.

173 Botzung-Appert, E., Zaccaro, J., Gourgon, C., Usson, Y., Baldeck, P.L. and Ibanez, A. (2005) Spatial control of organic nanocrystal nucleation in sol-gel thin films for 3-D optical data storage devices or chemical multi-sensors. *Journal of Crystal Growth*, **283** (3-4), 444–9.

174 Dubuisson, E., Monnier, V., Sanz-Menez, N., Boury, B., Usson, Y., Pansu, R.B. and Ibanez, A. (2009) Brilliant molecular nanocrystals emerging from sol–gel thin films: towards a new generation of fluorescent biochips. *Nanotechnology*, **20**, 316301.

175 Tian, Z., Chen, Y., Yang, W., Yao, J., Zhu, L. and Shuai, Z. (2004) Low-dimensional aggregates from stilbazolium-like dyes. *Angewandte Chemie International Edition*, **43** (31), 4060–3.

176 Pramod, P., Thomas, K.G. and George, M.V. (2009) Organic nanomaterials: morphological control for charge stabilization and charge transport. *Chemistry: An Asian Journal*, **4** (6), 806–23.

177 Zhao, Y.S., Fu, H., Peng, A.-D., Ma, Y., Xiao, D. and Yao, J.-N. (2008) Low-dimensional nanomaterials based on small organic molecules: preparation and optoelectronic properties. *Advanced Materials*, **20** (15), 2859–976.

178 Maltsev, E.I., Lypenko, D.A., Bobinkin, V.V., Tameev, A.R., Shapiro, B.I., Schoo, H.F.M. and Vannikov, A.V. (2002) Near-infrared electroluminescence in polymer composites based on organic nanocrystals. *Applied Physics Letters*, **81** (16), 3088–90.

179 Cooper, E.R. (2010) Nanoparticles: a personal experience for formulating poorly water soluble drugs. *Journal of Controlled Release*, **141** (3), 300–2.

8
Organic Nanoparticles Using Microfluidic Technology for Drug-Delivery Applications

Wei Cheng, Lorenzo Capretto, Martyn Hill and Xunli Zhang

8.1
Introduction

Over the past decade, nanomedicine has emerged as a new field of medicine where nanoscale materials are used to deliver a wide range of pharmaceutically active organic compounds such as drugs, genes, and imaging agents [1]. Challenges in synthesizing and formulating organic compound-based nanostructured materials remain, although the synthesis of inorganic nanomaterials has been extensively studied over decades, with good control of particle shape and size [2–4]. Consequently, special formulation techniques are required to disperse solid organic materials into water, to maintain the dispersion for a certain time period, and to functionalize organic nanoparticles.

Microfluidics is a new and emerging science and technology field dealing with microscale systems that process, or manipulate, small amounts of materials (typically 10^{-18} to 10^{-9} liters [5]. The main feature of such microsystems is the microscale channel network with a channel width of about 100 μm (roughly the diameter of a human hair), where fluids are brought together using a variety of pumping techniques for mixing, reaction, separation, or analysis. Relevant to the formulation of organic nanoparticles, the advantages of microfluidic and lab-on-a-chip technologies will allow not only the creation of particles of a well-controlled size distribution, but also the integration of a number of measurement systems into the microreactor to monitor the process when the particles are generated and, more importantly, to institute real-time feedback.

Having established that microfluidic and lab-on-a-chip technologies not only support complex chemical processes but also offer distinct operational advantages, they provide a promising new and direct route to obtaining colloidal systems and nanostructures in a continuous-flow format. As a result, some attempts to form inorganic nanoparticles (e.g., TiO_2 and CdSe) using microreactors have been reported by research groups such as that of deMello at Imperial College London [6] and of Ismagilov at the University of Chicago [7]. The results obtained have demonstrated the advantages of microfluidic methodology in terms of particle size control. In addition, other groups, including those of Barrow at Cardiff University [8] and of Weitz at Harvard University [9], have used microreactors to produce microsized particles or droplets with a potential application in drug delivery. However, studies on the formation of organic nanoparticles are in the early stages in terms of synthesis versus functionality (even in conventional batch systems). Moreover, there is to date very little significant experimental evidence available on the early stage of organic nanoparticle formation, especially during or immediately after the mixing of reactants.

The characteristics of a microfluidic environment make it one of the most attractive fields for the production of organic nanoparticles for drug-delivery systems. These characteristics allow the production of organic particles with controlled size and size distribution and the spatial and temporal investigation of the formation of organic nanoparticles, together with the process scale-out based on the concept of parallel processing. However, at present few reports have comprehensively discussed microfluidic techniques for the formation of organic nanoparticles. Hence, the aim of this chapter is to discuss the advantages of microfluidic and lab-on-a-chip technologies for the development and understanding of organic nanoparticle production, and its current or potential application in drug discovery. Despite being a newly emerging field, it is anticipated that microfluidic theory and practice for organic nanoparticles will benefit the future research and development of the processes envisioned for the continuous-flow production of a wide range of organic nanostructures with fine-tunable sizes and shapes, via control of the nanostructures' nucleation and growth.

In this chapter, the most recent research and development of organic nanoparticles using microfluidic technology for drug-delivery applications will be reviewed, the aim being to elucidate the importance of organic nanoparticles in formulation science and the development of new strategies for new and better products. The conventional methodology for the synthesis of organic nanoparticles is also discussed, and macroscale and microscale syntheses compared by outlining the necessity to, and the advantages of, scaling down in operational space dimensions. The unique physical characteristics and theory of the microfluidic environment will then be discussed in the context of organic particle formation, which includes the omnipresence of laminar flow, well-controlled and faster heat and mass transfer, with the opportunity of integrating processes and measurement systems into a single technology platform. A perspective will also be provided on future studies for the use of microfluidics to create organic nanoparticles, including the controlled synthesis of organic nanoparticles and/or the spatial and temporal investigation of nanoparticle formation, by integrating advanced analytical techniques.

8.1.1
Batch Synthesis of Organic Nanoparticles

Nanodispersed particles can be obtained by either mechanical milling of the raw material using wet- or dry-milling processes ("top-down" methods), or by the precipitation or condensation of the products or educts dissolved in solvents, with subsequent separation of the antisolvent (comminution method). Although milling processes have been widely used in the formulation of poorly soluble active compounds, they are generally considered unsuitable. This applies particularly to the production of nanoparticles of narrow distribution, because mechanical energy in the form of shearing and cavitation forces for particle milling are applied, which this makes it difficult to reduce the size of the particles without simultaneously inducing particle agglomeration [10]. In addition, abrasion during the milling process can cause contamination of the end product which is difficult to separate, especially in active compound formulations, while it may be difficult to produce nanosized organic particles consistently when using solid-particle milling [11]. Hence, milling processes will not be discussed in this chapter.

The production of organic colloidal nanoparticles has involved the investigation of several conventional preparation methods, including reprecipitation [12], microemulsion, and evaporation [13, 14]. In these processes, a nonaqueous solution of the molecularly dissolved organic compounds is first prepared, followed by precipitation under supersaturation conditions. The supersaturation conditions can be created by various methods, such as mixing the organic solution with a nonsolvent (e.g., water) [15], diluting a micellar solution containing the organic compounds as a solubilizate below the critical micelle concentration (CMC) [14], or by rapidly changing (generally decreasing) the temperature [15]. To control the nanoparticle size and stabilize the dispersion, the addition of suitable stabilizers is generally required at an appropriate point during the precipitation process [16]. Over the past decade, some excellent reviews have focused on the conventional synthesis of organic nanoparticles [17, 18].

8.1.2
Specifications of Reactors: Macroscale versus Microscale Syntheses

As nanoparticle synthesis moves to controllable production, there are opportunities for both scientific and technological innovation. Typically, a precise concentration and temperature control over short lengths and timescales is required to produce well-defined and controllable nanoscale structures through synthetic chemistry. Those challenges that remain largely unresolved include the elucidation of the general physical mechanisms of nanostructure formation and growth, and the sensitivity of the particle morphology, composition, and size to the processing conditions. These challenges are difficult to achieve when using conventional macroscale synthetic chemistry techniques.

Batch-scale nanoparticle syntheses are typically carried out in stirred flasks, where stirring is conventionally used to mix the reactants rapidly and to maintain the growing particles in suspension (Figure 8.1a). In terms of the synthesis of core–shell particles, a controlled addition of the secondary reactants is required. In such cases, the addition rate and mixing speed often determine the presence or absence of secondary particle nucleation, the homogeneity of shell growth, and the state of aggregation of the final mixture. The homogeneity of concentration and temperature are also crucial for monodisperse particle size distributions,

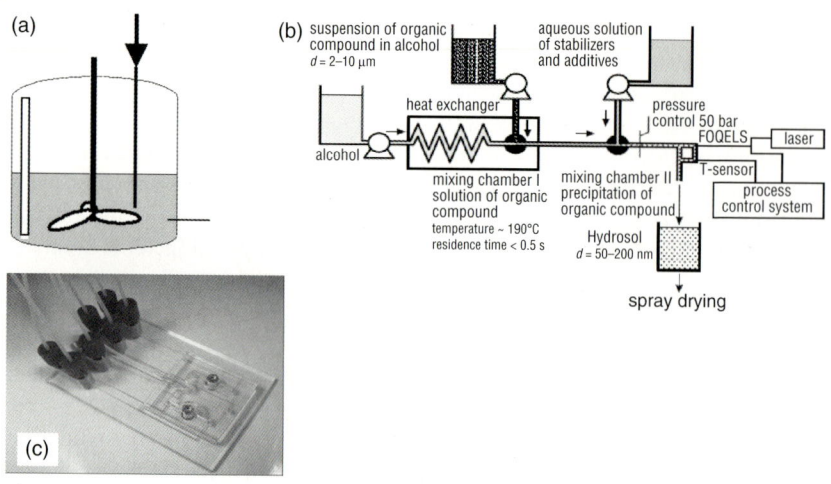

Figure 8.1 (a) Conventional batch reactor; (b) Macroscale continuous-flow reactor [17]; (c) Microreactor.

mainly because of the sensitivity of colloidal nucleation and morphology to local temperature and composition. However, such batch-scale syntheses often involve the heterogeneous spatial and temporal distributions of concentration and temperature, as well as uncontrollable additions and mixing rates. Furthermore, the batch-to-batch variations that always occur in batch-scale syntheses will further complicate the scale-up of batch-scale stirred-flask syntheses when producing greater quantities of product.

Although most studies on the production of nanoparticles have been conducted using stirring batch processes, various continuous-flow reactors have recently been developed. An example is the turbulent jet mixing chamber developed by Prud'homme and coworkers, whereby an organic solution is rapidly mixed with an antisolvent (water), followed by precipitation at a lower temperature in the presence of stabilizing surfactants [19]. In this case, by varying the jet velocity, which in turn determines the mixing time, different nanoparticle sizes and distributions have been obtained (Figure 8.1b). Haberkorn et al. have used a rapid mixing nozzle connected to a reaction tube to investigate a range of precipitation reactions, including organic pigment nanoparticle formation [15]. Similarly, by changing the flow velocity and length of the reaction tube, both the mixing time and the precipitation reaction time were varied, and this resulted in different nanoparticle sizes and distributions. By coupling the reaction tube with small-angle X-ray scattering (SAXS) detection, online information was obtained regarding the structural inhomogeneities within the reacting systems, at early stages of particle formation. Alison and coworkers designed a continuous-flow crystallizer for the simultaneous collection of both SAXS and wide-angle X-ray diffraction (WAXD) data during the nucleation and crystal growth of a small organic molecule from solution [20].

Although continuous-flow chemical reactors overcome some of the drawbacks of stirred-flask synthesis, such as ensuring the spatial homogeneity of the concentration and temperature, mixing in such macroscale flow reactors remains an issue. In the aforementioned studies, reagent mixing was carried out over a short time period using turbulent flow, but unfortunately turbulent mixing coupled with rapid aggregation leads to the formation of complex, time-dependent patterns of spatially localized regions of supersaturation, which in turn results in particle nuclei that contain different amounts of the components. Hence, depending on the fluidic conditions, the different regions of the reactor will contain particles at different stages of the nucleation, growth and stabilization processes. As a result, poor quality products may be created in terms of nanoparticle size, size distribution, and stability. In addition, the difficulty (if not impossibility) of characterizing the precise conditions for a specific flow unit within the bulk flow, under turbulent flow conditions, represents a significant challenge for modeling and predicting such processes in terms of fluidics and the mass transfer from both spatial and temporal aspects.

The recent global development of microfluidic and lab-on-a-chip technologies, however, has demonstrated conclusively that such miniaturized systems offer many advantages over conventional macroscale reactors in achieving controllable, information-rich, high-throughput and environment-friendly processes [21–23]. This can be largely attributed to the key feature of microscale channel networks within such microdevices (Figure 8.1c). When scaling down in operation space dimension (compared to the conventional macroscale system), they not only reduce the sample volume but also bring unique characteristics to the microscale fluidic environment, where the spatial and temporal control of reagents is achieved under a nonturbulent, diffusive mixing regime.

Compared to conventional macroscale batch systems, the microscale synthesis of nanoparticles using microfluidic technology has unique operating characteristics and advantages:

- It offers a laminar flow environment where diffusion dominates mass transfer. Laminar flow, as the simplest flow phase, enables the flow to be modeled and controlled precisely. In addition, backmixing can be eliminated in the diffusion-dominated mass transfer regime; this is in contrast to the hydrodynamic, poorly controllable batch processes where a secondary particle agglomeration can occur.

- A short and highly controlled mixing time can be easily achieved in the microscale synthesis system. In a microchannel with a width of a few tens of microns, where the mass transfer is dominated by diffusion, sub-second mixing times can be achieved based on a molecule's diffusion across the channel. Such sub-second mixing times are believed to be important in the formation of nanoparticles, especially in the early stages. However, by carefully selecting the flow rate and geometry, this mixing can be precisely controlled.

- The microscale system provides a high surface-to-volume ratio which, in association with a small volume, offers rapid heat transfer to achieve thermal homogeneity throughout the entire reaction volume, a situation which is crucial to the monodisperse particle size distributions.

- The ability to manipulate reagent concentrations and reaction interfaces within the channel network of a microreactor, in both space and time, provides a high level of reaction control that is unattainable in conventional bulk reactors.

8.1.3
Properties and Application of Organic Nanoparticles for Drug Delivery

Drug delivery is a multidisciplinary science in which the methods or processes of administering pharmaceutical compounds to achieve a therapeutic effect, whether in humans or animals, is studied. Whilst the conventional formulations used to administer drugs rely (preferably) on noninvasive pills, sprays, and topical formulations, an alternative route involves the intravenous administration of solutions which provide a much faster appearance of any drug effect, but are often much less patient-compliant. The use of different types of pharmaceutical formulation not only implies different routes of administration, but also provides the potential to modify the drug-release profile in terms of the absorption, distribution, and elimination of the drug(s). In the past, this possibility has led to a rapid expansion in the field of drug delivery and the development of advanced drug-delivery formulations. In fact, by controlling the concentration level (modified drug release) and/or location of the drug in the body (targeted drug delivery), adverse side effects may be reduced and lower doses often needed, which in turn has the clear benefits of improving not only product efficacy and safety but also patient compliance. Moreover, by using an appropriate formulation it is possible to formulate even drugs with very poor water-solubility, for therapeutic purposes.

A major part of the current drug-delivery system development is devoted to so-called "nanomedicine." Nanostructured materials can be used to modulate the biodistribution and metabolism of drugs, so as to "hit" the target but without damaging healthy tissues. Nanoscale materials used for drug delivery include liposomes, polymer–drug conjugates, polymer–DNA complexes, and polymeric micelles. Because of their small dimensions, these drug-delivery systems can be injected directly into the bloodstream; this causes the drug to be released in a modified manner, and also provides a means of targeting the drug or genes towards a specific tissue. By using a specific moiety for active targeting, stimuli-responsive release, or by levering the peculiar biodistribution of nanoparticles in the body, these delivery systems can preferentially release the drug into a specific tissue, without damaging any healthy tissues. During the early twentieth century, Paul Ehrlich proposed the use of drugs as so-called "magic bullets"; today – a century later – this concept is becoming reality, based on the development of targeted nanomedicines [24]. Although the synthesis of inorganic nanomaterials has been extensively studied, with a good control of particle shape and size, the synthesis and formulation of organic compound-based nanostructured materials remains a challenge for a variety of reasons [25, 26]:

- Although most pharmaceutically active organic compounds are hydrophobic, poorly soluble in water, or even water-insoluble, the medical applications require their administration into an aqueous environment, and at a sufficient concentration.
- The nanomedicine product should be formulated to be stable over a certain period (the "shelf-life") at different temperatures.
- Ideally, site avoidance and reduced toxicity should be provided by a minimal uptake in sensitive tissues, and by facilitating a sustained release.
- The nanostructures should have a uniform size, or a narrow distribution.

Thus, special formulation techniques are required to disperse the solid organic materials into water, to maintain the dispersion for a certain time period, and to functionalize the organic nanoparticles.

8.2
Microfluidic Synthesis of Organic Nanoparticles

The size, shape, and crystal structure characteristics of nanoparticles strongly affect their physical and chemical properties. Consequently, it is not surprising that a superior control of the process parameters is desirable when creating nanoparticles in order to produce the required material. In this respect, microfluidic reactors offer a series of potential advantages in the production of nanoparticles, because they enable the fine control and manipulation of fluid and fluid interfaces. In microfluidic devices, the reactions are carried out in small reaction channels with diameters ranging between a few and several hundreds of microns. These small dimensions, and the resultant large area-to-volume ratio, enable a rapid and more uniform heat and mass transfer that can lead to dramatic improvements in the yield and size distribution of nanoparticles, while reducing the formation of undesirable byproducts. In addition, the possible use of solvent-recycling and integrated separation techniques may well provide cost-effective and environment-friendly technologies for future production.

8.2.1
Overview: Unique Features of Microfluidic Reactors for the Controlled Synthesis of Organic Nanoparticles

A series of features make microfluidic reactors particularly appealing for the production and investigation of nanoparticles. Most of these features derive from the unique characteristics of the microchannel flow and the general microfluidic environment, such as the omnipresence of a highly predictable laminar flow and a large surface-to-volume ratio. Such characteristics include:

- An efficient, fast and controllable mixing which is dominated by diffusion under continuous-flow conditions, and results in a homogeneous reaction environment.
- A more efficient temperature control and heat transfer.
- *In situ* monitoring of the progress of nanoparticle formation, through spatial resolution.
- The spatial and temporal control of reactions by adding reagent at precise time intervals during the reaction progress.
- The control of the nanoparticles' characteristics by controlling the kinetics of the process.
- A high-throughput screening of various formulations by varying the process parameters online.
- The opportunity to integrate post-synthesis processes and measurement systems into a single technology platform.
- The possible scale-up of a process, simply by increasing the number of microreactors involved.

These unique characteristics reveal the significant potential of this technique to transform current classical batch technology for the production of nanoparticles into a continuous, microfluidic process. Unfortunately, this area of research is still clearly in its infancy, and further studies are required to highlight the superiority of microfluidic processes over the conventional batch processes.

8.2.2
Microfluidic Reactors for Organic Nanoparticles

During the past decade, tremendous interest has been shown in the development of microfluidic methods for the chemical synthesis of inorganic nanoparticles, and in the production of nanocrystalline semiconductors with uniform and tunable size distributions [2]. Although much fewer studies have been conducted on microfluidic methods for organic nanoparticle synthesis, attention is focused at this point on recent studies of the microfluidic fabrication of organic nanoparticles, including emulsions, nanoprecipitations, and liposomes. The production of organic nanoparticles using microfluidic methods is compared to the traditional bulk fabrication approach, by using specific examples, to highlight the benefits that microfluidic technology can bring to organic nanoparticle formation.

8.2.2.1 Emulsions
Water droplets have been shown to serve as reactors for biological and chemical reactions [27]. Although emulsions have been used to formulate organic nanoparticles, the conventional emulsion-based methods of manufacture lead to the production of particles with a wide range of diameters (and kinetics of release) in each

8.2 Microfluidic Synthesis of Organic Nanoparticles

batch. In order to create droplet emulsions with narrow size distributions, production methods have been developed which use microfluidic systems. Indeed, emulsions generated in microfluidic systems have been used for applications that include bioanalysis, fluid optics, and organic syntheses [28–30].

T-junctions and flow-focusing nozzles are two broad classes of device used to generate emulsions in microfluidic platforms, and these make it possible to generate monodisperse particles and offer flexibility with regards to the size of emulsion produced. In a recent study, a systematic effort was made to characterize the kinetics of drug release from particles generated by microfluidic devices [31]. Such a flow-focusing approach has also been employed to fabricate polymer microparticles, by combining the generation of monodisperse droplets of solvent containing poly(lactic-co-glycolic acid) (PLGA) and the drug, with the subsequent removal of solvent to form PLGA–drug particles (Figure 8.2). The particles generated were almost monodisperse (polydispersity index 3.9%). When a model amphiphilic drug (bupivacaine) was incorporated within the biodegradable matrix of the particles, a kinetic analysis showed its release from these monodisperse particles was slower than when using conventional methods of the same average size but with a broader distribution of sizes. Most importantly, a significantly lower initial burst of drug release was exhibited than was observed with conventional particles. This difference in the initial kinetics of drug release was attributed to the uniform distribution of the drug inside the particles generated by microfluidic methods. Notably, the results confirmed the value of microfluidics for generating homogeneous systems of particles for drug delivery.

A versatile microfluidic technique for fabricating monodisperse biocompatible polymersomes with biocompatible and biodegradable diblock copolymers for the efficient encapsulation of actives has recently been described [32] (Figure 8.3). In this case, a double emulsion was used as a template for the assembly of amphiphilic

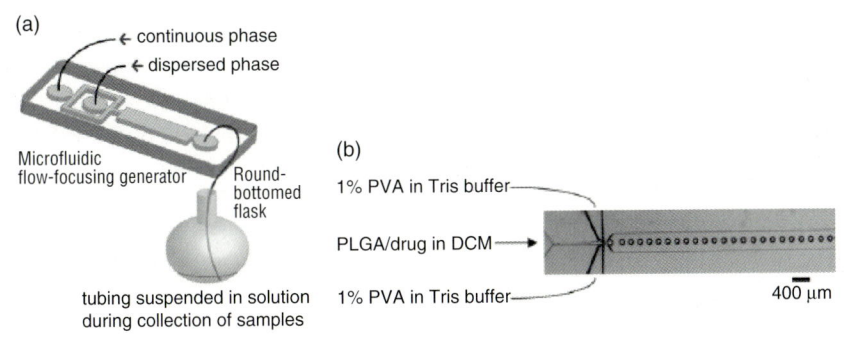

Figure 8.2 The use of microfluidic flow-focusing devices to fabricate monodisperse drug-loaded particles from biodegradable polymers and the drug-delivery properties of those particles. (a) Schematic illustration of the procedure; (b) Optical microscopy image showing the orifice of the flow-focusing region generating droplets of dichloromethane in water. Data taken and readapted from Ref. [31].

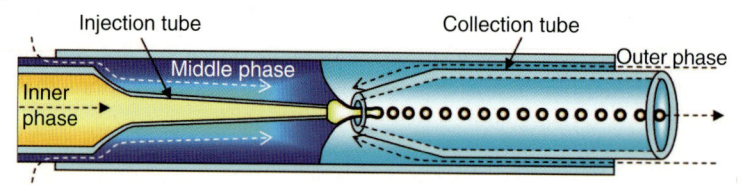

Figure 8.3 Schematic of the microcapillary geometry for generating double emulsions. The geometry requires the outer phase to be immiscible with the middle phase, which is in turn immiscible with the inner phase. Data taken and readapted from Ref. [32].

diblock copolymers poly(ethylene-glycol)-*b*-polylactic acid (PEG-*b*-PLA) into vesicular structures during solvent evaporation. The polymersomes can be used to encapsulate small hydrophilic solutes and, when triggered by an osmotic shock, will break and release the solutes, thus providing a simple and effective release mechanism. The same technique can also be applied to diblock copolymers with different hydrophilic-to-hydrophobic block ratios, or to mixtures of diblock copolymers and hydrophobic homopolymers. The ability to produce polymer vesicles with copolymers of different block ratios, and to incorporate different homopolymers into polymersomes, should allow the properties of the polymersomes, such as membrane thickness, mechanical response, permeability and thermal stability, to be fine-tuned.

The ability to produce monodisperse particles for drug delivery also has several practical advantages. A list of recent studies in which microfluidic technology was used to generate emulsion particles is shown in Table 8.1. The reduced shear stresses used to prepare particles in a microfluidics device, compared to a conventional emulsion, can help to maintain the bioactivity of shear-sensitive biomolecular drugs (i.e., protein therapeutics) released from biodegradable particles. Given that the conventional emulsion approach typically produces aggregates that must be removed by filtration, particles prepared using the microfluidic method can be produced with higher yields than are typical via the conventional approach; this is a significant advantage, particularly when very expensive drugs are involved. Furthermore, the use of monodisperse particles enables the injection of larger (i.e., slower-releasing) particles, because the larger particles on the upper tail of the particle size distribution (which increase the probability of needle clogging) are absent. This flow-focusing approach has wide potential uses in the controlled production of microspheres for pharmaceutical application.

8.2.2.2 Nanoprecipitation

Unlike an emulsion formation, which relies on hydrodynamic instability to break up an immiscible polymeric solution into droplets and subsequently form small particles through crosslinking or solvent evaporation, nanoprecipitation involves the formation of nanoparticles through the self-assembly of block copolymers by rapidly mixing miscible polymer solutions with water. Nanoprecipitation offers a

8.2 Microfluidic Synthesis of Organic Nanoparticles

Table 8.1 Recent studies involving the use of microfluidic technology for organic nanoparticle formation.

Slot no.	Materials of reactor	Type of microfluidic reactor T-junction	Synthesis methods	Produced organic particles 2,20-dipyridylamine (DPA)	Year of publication	Reference
1	Teflon	Micromixer	Emulsion	Nanoparticles	2007	[30]
2	PDMS/glass	Flow-focusing	Nanoprecipitation	PLGA–PEG	2008	[34]
3	Teflon	Flow-focusing	Nanoprecipitation	Hydrocortisone	2009	[33]
4	PDMS	Flow-focusing	Emulsion	PLGA–drug particles	2009	[31]
5	Glass	Glass microcapillary	Emulsion (double)	PEG–PLA polymersomes	2008	[32]
6	Silicon	Flow-focusing	Supersaturation	Liposomes	2007	[37]
7	Steel plate	Flow-focusing	Supersaturation	Solid lipid nanoparticles	2008	[39]
8	Steel plate	Flow-focusing	Supersaturation	Solid lipid nanoparticles	2009	[40]
9	Glass	Glass capillaries	Supersaturation	Solid lipid nanoparticles	2008	[41]
10	PDMS	Axisymmetric flow-focusing	Solidification	Alginate particles	2008	[51]

PDMS, poly(dimethylsiloxane).

simple and gentle formulation under ambient conditions, without the use of chemical additives or harsh formulation processes. However, the typical synthesis of nanoparticles by nanoprecipitation involves the dropwise addition of a polymer–organic solvent solution into a larger quantity of water, which results in a slow and uncontrolled mixing. Although small PLGA-based nanoparticles with diameters <100 nm can be obtained through batch nanoprecipitation, they have a poor drug encapsulation and a rapid drug release, compounded with a high polydispersity.

However, nanoprecipitation through rapid and controlled mixing using microfluidics can enable the formation of more homogeneous nanoparticles, and also provide a better control of the nanoparticle properties such as size, surface characteristics, and drug loading. The creation of a relatively stable aqueous hydrocortisone nanosuspension using microfluidic reactors has recently been examined [33] (Figure 8.4). This involved a study of the parameters of the microfluidic precipitation process that affect the size of the generated drug particles. Such parameters included the flow rates of drug solutions and antisolvents, microfluidic

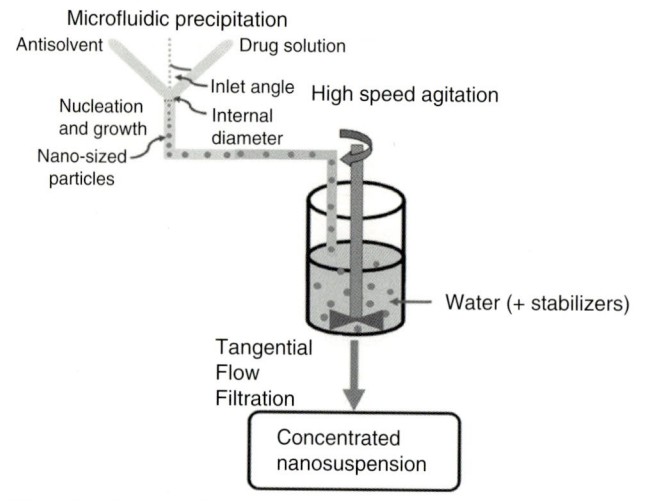

Figure 8.4 Diagram of microreactor set-up for nanosuspension preparation. Data taken and readapted from Ref. [33].

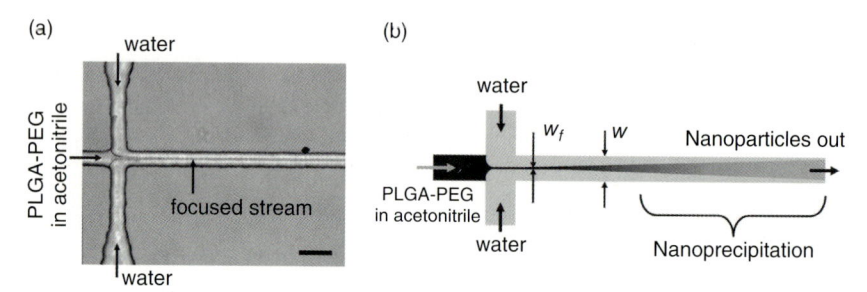

Figure 8.5 Illustration of nanoprecipitation by hydrodynamic flow focusing.
(a) A microfluidic device for the hydrodynamic flow focusing of polymeric nanoparticles in water. Scale bar = 50 μm; (b) The process of mixing can be carried out in a microfluidic device using hydrodynamic flow focusing, where the polymer stream is focused into a thin stream between two water streams with higher flow rates. Data taken and readapted from Ref. [34].

channel diameters, microreactors inlet angles, and drug concentrations. The results revealed that hydrocortisone nanosized dispersions in the range of 80 to 450 nm could be obtained, and that the mean particle size could be changed by modifying the experimental parameters and design of the microreactors. The nanosized particles generated from a microreactor were rapidly introduced into an aqueous solution of stabilizers stirred at high speeds with a propeller mixer.

Most recently, a rapid and tunable microfluidic device has been demonstrated to synthesize drug-encapsulated biodegradable polymeric PLGA–PEG nanoparticles with a defined size, a lower polydispersity, and a higher drug loading with slower release [34] (Figure 8.5a). The PLGA–PEG nanoparticles are synthesized in a microfluidic channel by rapidly mixing polymer–acetonitrile solutions and water, using a hydrodynamic flow focusing in a controlled nanoprecipitation process.

In hydrodynamic flow focusing, the fluid stream to be mixed flows along the central channel, and meets two adjacent streams flowing at higher flow rates (Figure 8.5b). At low Reynolds numbers, the central stream is squeezed into a narrow stream between the two adjacent streams; the narrow width of the focused stream then enables rapid mixing through diffusion.

In this study, the nanoprecipitation synthesis of smaller and more homogeneous nanoparticles has been demonstrated using microfluidic technology. Microfluidics enables a greater control over the rate of mixing and, in conjunction with the controlling precursor composition, can be used to tune the nanoparticle size, homogeneity, and drug loading and release. The study results also suggested that the microfluidic synthesis of nanoparticles via the self-assembly of polymeric precursors can enable a better control over the physico-chemical properties of the nanoparticles, which may prove beneficial in the emerging field of nanomedicine.

8.2.2.3 Liposomes

Liposomes are interesting as transport vehicles for *in vivo* applications such as drug delivery, where they are thought to achieve a selective and sufficiently high localization of active drugs at the disease site [35]. The bulk hydration of lipids in aqueous buffer is used to yield large polydisperse and multilamellar liposomes. This method, as well as other traditional liposome production bulk methods such as freeze–thaw cycling, film hydration and reversed-phase evaporation, often lead to heterogeneous and uncontrolled chemical and mechanical conditions during liposome formation, and can produce liposomes that are polydisperse in both size and lamellarity.

Consequently, it is challenging to produce liposome formulations with a defined size for the specific application and with little size variation in their population. Jahn *et al.* first reported a microfluidic channel network for the controlled formation of manometer-sized liposomes through hydrodynamic focusing [36]. In this case, the liposomes were formed by a diffusively driven process when a stream of lipids dissolved in an organic solvent such as isopropyl alcohol (IPA) was hydrodynamically sheathed between two oblique buffer streams in a microfluidic channel. The laminar flow conditions facilitated diffusive mixing at the two miscible liquid interfaces, predictably diluting the alcohol concentration below the solubility limit of lipids and initiating lipid self-assembly into small unilamellar vesicles. In another recent report, the same group examined the mechanism that controls liposome size and homogeneity by modifying the microfluidic design [37] (Figure 8.6a). Here, the deep channels of a higher aspect ratio with a rectangular cross-sectional area led to a more homogeneous velocity profile across the channel height, and reduced the impact of surface effects at the bottom and top of the channel. The use of microfluidic techniques to produce liposome formulations of monodisperse distributions, the size of which could be controlled by adjusting the

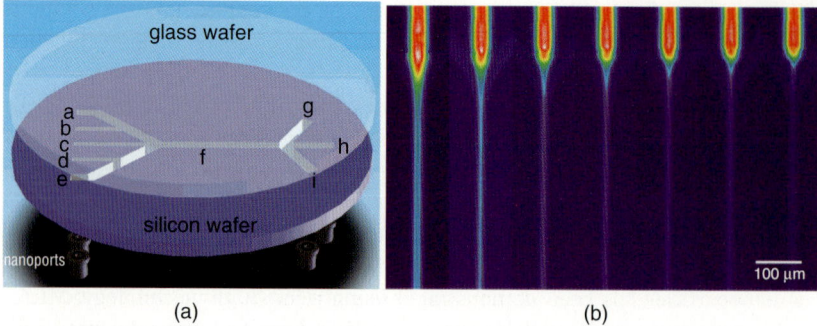

Figure 8.6 (a). Schematic of the microfluidic device. An exploded view showing the fluid ports attached to the reverse side of the silicon wafer, the channel network etched into silicon with five inlet channels (designated a–e) on the left and three outlet channels (designated g–i) on the right and the sealing with a glass wafer via anodic bonding; (b). False color confocal microscope images showing the hydrodynamic focusing of an isopropyl alcohol (IPA) stream by two adjacent aqueous buffer streams (not visible). The focused IPA stream containing sulforhodamine B for visualizing purposes enters from the top. Seven different flow rate ratios (FRRs) are shown, increasing from 5 to 35 in increments of 5 from left to right at a total constant volumetric flow rate (VFR) of 100 µl min^{-1}. Data taken and readapted from Ref. [37].

fluid flow rates in the microfluidic network, was further demonstrated (Figure 8.6b).

The results of these studies showed clearly that nanometer-sized liposomes with a very narrow size distribution can be produced using microfluidics, and could potentially open applications for on-demand liposome-mediated delivery of point-of-care personalized therapeutics. Solid lipid nanoparticles (SLNs) are considered an alternative drug-carrier system to conventional emulsions, liposomes, and polymeric nanoparticles. Among the main advantages of the SLN delivery system can be included a controlled release, long-term stability, good tolerability, and the ability to prevent the loaded drugs from being degraded; moreover, the SLN system can be administered easily via the oral, parenteral, dermal, or pulmonary route [38]. Recently, SLNs were produced using microreactors based on a simple flow-focusing configuration [39], and combined with a shaped branch channel to displace gas so as to prevent the deposit and blockage of SLNs and ensure a continuous process [40]. In both cases, the process demonstrated control of the dimensions of the nanoparticles produced, simply by varying the volumetric flow rates (VFRs) of the three inlet streams, using microfluidics.

8.2.3
Controlled Operating Parameters of Microfluidic Reactors

Among the different types of microfluidic reactor, the use of a focusing enhanced mixer has attracted most attention (Table 8.1). This type of microfluidic reactor not only provides an easy means of controlling and varying the mixing time, but also includes very easy design and fabrication procedures. Polymeric micelles (PMs) represent a class of polymeric nanoparticles with a core–shell structure that is usually formed spontaneously by the self-assembly of block copolymer unimers in a liquid [41]. Recently, PMs have been shown to serve as a drug-delivery system for carrying bioactive molecules to the target site in the human body [42]. Because the rate of drug release, accumulation site and kinetics of elimination from the body of these systems depend on the size and size distribution of the nanoparticles, it is easy to understand the possible impact that microfluidic production methods might have in this field. In this section, using a model of the synthesis of nanosized PMs using hydrodynamic flow focusing, the parameters of the microfluidic precipitation process that affect the size of the generated organic nanoparticles and characterize the kinetics of their synthesis, are investigated.

8.2.3.1 Flow Velocity, Microfluidic Dimension, and Mixing Time

The production of PMs made from a Pluronic® triblock copolymer was investigated using a microfluidic approach. The Pluronic® PMs can be prepared in a microfluidic reactor by rapidly mixing polymer–dimethylsulfoxide (DMSO) solutions and water, using hydrodynamic flow focusing in a controlled nanoprecipitation process (Figure 8.7). The microreactors were fabricated in glass, using the microfabrication techniques, and had a semi-circular shape.

In hydrodynamic flow focusing the central stream is squeezed into a narrow stream between the two sheath streams [43–45]. Because the mixing time is inversely proportional to the square of the diffusion path length (in this case

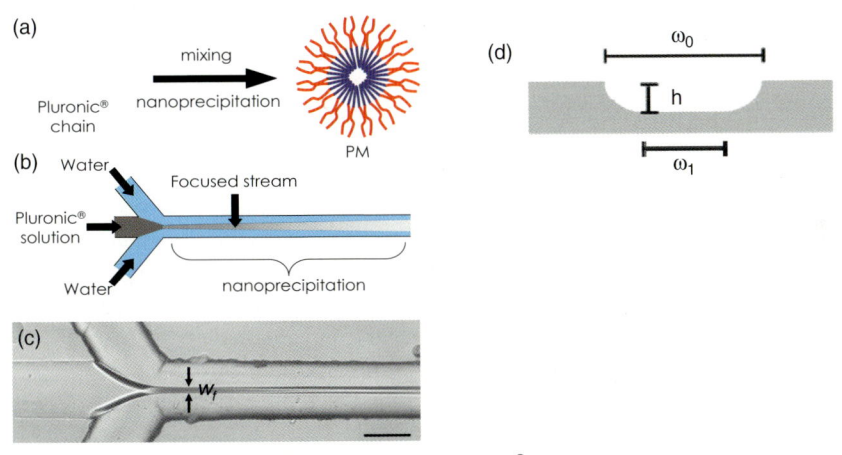

Figure 8.7 Schematic of the nanoprecipitation of Pluronic® block copolymer. (a–c) Pluronic® chains self-assemble into polymeric micelles (PM) when a water-miscible solution of the polymers is mixed with water, in which the polymer is partially soluble; (d) The channel cross-section shape of the microfluidic channel.

represented by the focused stream width), decreasing the stream width results in a faster mixing. A simple mass flow balance within the microchannel can provide a theoretical model to estimate the width of the focused central stream. Taking into account the shape of the cross-section of the microchannel, the width of the focused stream w_f can be derived as follows:

$$w_f = \frac{Q_I}{Q_E + Q_I} \cdot \left(w_b + \frac{\pi}{2}h\right) \cdot \alpha \tag{8.1}$$

where Q_I and Q_E are the VFR for the central stream (polymer solution) and sheath stream (non-solvent solution), respectively, and w_b and h are the width and height of the outlet channel as reported in Figure 8.7. The experimental factor α is introduced to improve agreement between the theoretical data of the focused stream width and the experimental data. The mixing time (τ_{mix}) can now be estimated from the diffusion time scale as:

$$\tau_{mix} \approx \frac{w_f^2}{4D} \approx \frac{\left(w_b + \frac{\pi}{2}h\right)^2 \cdot \alpha^2}{4D\left(1 + \frac{1}{R}\right)^2} \tag{8.2}$$

where D is the diffusion coefficient of the solvent and R is the ratio of the flow rate of the polymer-containing stream and the total flow rate of the non-solvent.

8.2.3.2 Mixing Time, Aggression Time, and the Damkohler Number

In a microfluidic reactor (Figure 8.7), the PMs can be synthesized by rapidly mixing a polymeric solution with water, using hydrodynamic flow focusing. Water, in which the polymer is partially soluble, triggers nanoprecipitation of the block copolymer unimers and their aggregation to form PMs. The self-assembly of the block copolymer nanoparticles during nanoprecipitation is believed to occur in three stages [34]: (i) nucleation of the block copolymer unimers; (ii) fusion of the existing particles; and (iii) the formation of an overlapping brush corona that results in kinetically frozen nanoparticles (Figure 8.8). The solvent quality affects the size of the particles by changing the crucial aggregation size; therefore, the nanoparticle size is expected to depend on the time scale associated with mixing in the solvent τ_{mix}. The relationship of the mixing time to the timescale associated with block copolymer aggregation τ_{agg} is also significant. The ratio between the two timescales is expressed as the Damkohler number (Da):

$$Da = \tau_{mix} / \tau_{agg} \tag{8.3}$$

When $Da < 1$, mixing occurs faster than the time scale associated with the nanoparticle nucleation, and therefore the nanoparticle size would be expected to be independent of the mixing time and the polymer concentration. The result is that nanoparticles with a characteristic dimension are represented by the critical size, which corresponds to the creation of an overlapping brush corona, and thereby

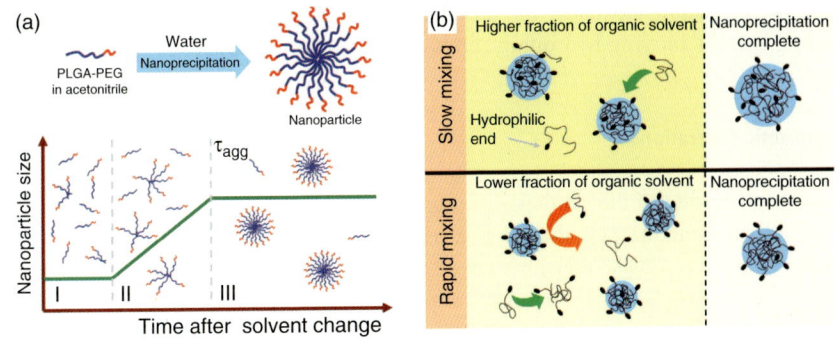

Figure 8.8 Nanoprecipitation of polymeric micelles (PMs) (taking PLGA–PEG diblock copolymers as an example). (a) PLGA–PEG diblock copolymers self-assemble into nanoparticles when a water-miscible solution of the polymers is mixed with water, in which the PLGA block is poorly soluble; (b) The mechanism of self-assembly of nanoparticles during nanoprecipitation [34].

nanoparticles are expected to be more homogeneous than those produced with a slower mixing. Conversely, for $Da > 1$, the size of the produced nanoparticle increases with either an increase in mixing time or the initial polymer concentration.

Slow mixing ($\tau_{mix} > \tau_{agg}$) results in the aggregation of polymers to form nanoparticles when mixing is incomplete; this occurs in the presence of a higher fraction of organic solvent in the solution. Under these conditions, the polymers are easily adsorbed onto the nanoparticle aggregates, burying the hydrophilic end groups and leading to the formation of larger nanoparticles. Rapid mixing ($\tau_{mix} < \tau_{agg}$) results in the aggregation of polymers when mixing is nearly complete; this occurs in the presence of a lower fraction of organic solvent in the solution. Polymers cannot easily be adsorbed onto or inserted into nanoparticles, and hence a greater proportion of nanoparticles that are smaller in size is nucleated, such that fewer hydrophilic end groups are buried inside the nanoparticle.

Under rapid microfluidic mixing, the solvent exchange is complete even before the polymers begin to aggregate. Therefore, nanoparticle assembly occurs in solvent conditions that more closely match the final solvent; that is, water with a small fraction (5%) of acetonitrile. However, under bulk mixing the time scale of the nanoparticle assembly is smaller than the timescale of solvent exchange. Consequently, nanoparticle assembly occurs in solvent conditions that are very different from the final solvent condition; that is, assembly occurs when the fraction of acetonitrile in the solution is larger than that in the final solvent. Various studies have suggested that a mechanism in which the barrier to the insertion of polymers from the solution into nanoparticles is lower when solvent exchange is incomplete, leading to an increased nanoparticle size. In addition to increasing the barrier for the insertion of polymers into the nanoparticle, incomplete mixing during nanoprecipitation facilitates the adsorption of the polymers onto existing polymers, thereby burying hydrophilic ends inside the nanoparticle.

8.2.4
Synthetic Operations

Although studies of the synthesis of organic nanoparticles in microfluidic devices are still in their infancy, there has already been a clear demonstration of their superiority over "flask" reactions (as discussed above). In fact, most studies have focused on inorganic nanoparticles, which indicates that there is a potential for obtaining a better control in the size, size distribution, and shape of the nanomaterials. Exactly how the crucial features of synthetic operations are utilized to control the synthesis of organic nanoparticles is discussed in the following subsections.

8.2.4.1 Micromixing

The framework of classical nucleation and crystallization theory provides a useful model to describe the formation of the colloidal system via the wet method, and to understand the role of efficient mixing in the dimensional characteristic of the nanoparticles produced [46]. The model involves an initial nucleation phase in which seed particles, called *nuclei*, precipitate spontaneously, and a subsequent growth phase in which the initial seeds capture the remaining dissolved solute. The nucleation phase occurs when the concentration of the solute reaches a supersaturated concentration, whereby a shower of nuclei is formed. Thus, if the concentration of the solute is lower than the critical nucleation threshold, no new nuclei can form. However, the concentration is still sufficient to allow growth of the nuclei already formed. The growth phase proceeds until the concentration of the still-dissolved material has fallen to the equilibrium concentration. In the nucleation phase, nucleation and growth occur concurrently; therefore, the earlier the nuclei are formed the larger the resulted nanoparticles will grow. In order to obtain nanoparticles batches with a narrow size distribution, it is important to tune the process to ensure that the nucleation occurs more quickly than the growth phase. In addition, to obtain a monodisperse batch, a homogeneous environment is required in terms of temperature and concentration, during both the nucleation and growth phases [47].

Microfluidic mixers can provide both homogeneous and fast mixing, and have the potential to represent a method to obtain nanoparticles with excellent dimensional characteristics. Figure 8.9 shows two examples of flow strategies for mixing organic molecular solutions. One strategy is based on co-flowing laminar streams with interfaces in both the vertical and horizontal planes, and the other on sequenced injections of plugs of the different components. Variations in channel designs and dimensions, flow rates and flow patterns enable a high level of spatial and temporal control of contacts between the different components.

Alongside the laminar flow microfluidic reactors, different groups have presented alternative approaches to the chemical synthesis, based on segmented flow

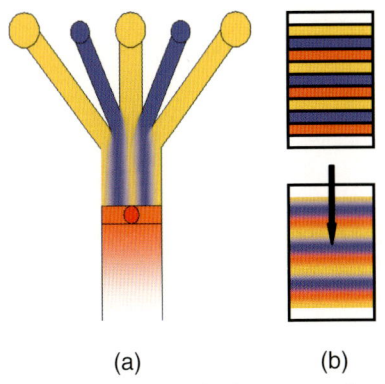

Figure 8.9 Two microfluidic strategies for mixing organic molecular solution (blue), nonsolvent (yellow) and stabilizing agent (red). (a) The blue and yellow streams co-flow in parallel, with the red component added continuously downstream in a horizontal sheet flow; (b) Sequenced plugs of each component are injected.

or multiphase flow microfluidic reactors [48, 49]. In the chemical synthesis of nanoparticles, one important parameter that strongly affects the monodispersity of the produced samples is the residence time distribution (RTD), which reflects the mean time that a particle spends inside the reactor. In laminar flow microfluidic reactors, the parabolic flow profile (fluid moving slower near the channel wall than in the center) and the associated axial dispersion cause a variation in residence times, which in turn leads to a wider size distribution of the synthesized nanoparticles [50]. Problems with laminar flow reactors can be avoided with segmented and droplet flow microreactors, and this leads to a better control of nanoparticle size distribution because the droplets or slugs function as a microsized reactor that flows along the channel, with a time determined by the flow rate.

8.2.4.2 Online Process of Various Reactants

The wet chemical synthesis of nanomaterials using microfluidic reactors can also take advantage of another unique ability of the microfluidic reactor to operate within continuous-flow regimes that allow additional reagents to be added downstream as required. Such a feature allows for the pre- and post-treatment and multistep synthesis in a single continuous-flow regime (Figure 8.10), which is required for nanoparticle formation and characterization. The reagent input streams (organic molecular solution and precipitating solvent) are initially mixed to provide precipitation and particle formation. Further inputs along the main channel enable the sequenced addition of one or more stabilizers or capping reagents and, if appropriate, reagents to link and functionalize the particles. Reagent flow can be achieved by either pressure-driven forces using syringe pumps, or electrokinetic forces by applying high voltages along the channel.

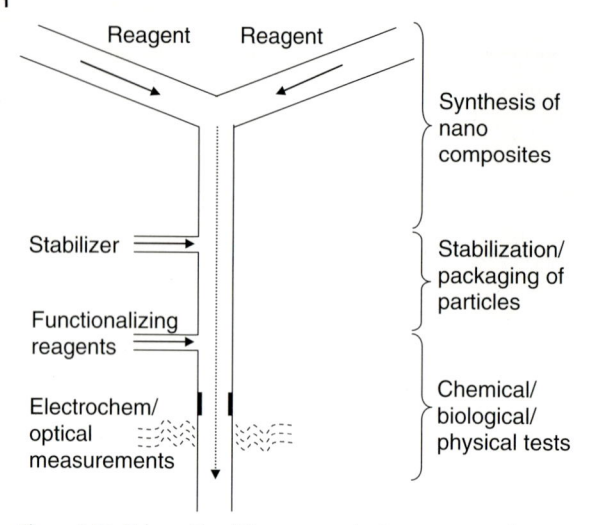

Figure 8.10 Schematic of the proposed microreactor, with integrated processes.

Nanomaterial synthesis in microfluidic devices also takes advantage of the unique ability of the microfluidic channel to work in a continuous-flow regime, allowing the spatial and temporal control of reactions by adding reagents at precise time intervals during the reaction progress. This allows microfluidic reactors to carry out pre- and post-treatment in the same reactor. One example of the multi-step formation of nanoparticles using microfluidics includes a novel microfluidic method.

For the production of crosslinked alginate microparticles and nanoparticles, monodisperse droplets are generated by extruding an aqueous alginate solution using an axisymmetric flow-focusing design [51] (Figure 8.11). As it flows downstream in the channel, because of water and the continuous phase being partially miscible, the water diffuses very slowly out of the polymeric droplets into the transport fluid, which shrinks the drops and condenses the polymer phase. The resulting size of the solid particles depends on the polymer concentration and the ensuing balance between the kinetics of the crosslinking reaction and the volume loss due to solvent diffusion. The study details a single-step microfluidic technique for the formation of alginate microparticles of sizes ranging from 1 to 50 μm via near-equilibrium solvent diffusion within a microfluidic device, and a two-step method which was shown to generate biopolymer nanoparticles of sizes ranging from 10 to 300 nm. The methodologies presented in this study demonstrate its flexibility to be extended to the preparation of nanoparticles from a wide range of mixed synthetic and biologically derived polymers.

8.2.4.3 Thermal Control and Heat Transfer

High surface-to-volume ratios are key parameters in defining fluid flow characteristics at the microscale. Of equal importance is the effect of these ratios on

8.2 Microfluidic Synthesis of Organic Nanoparticles

(a)

A- Alginate solution
B- Crosslinking agents solution
C- Organic solvent

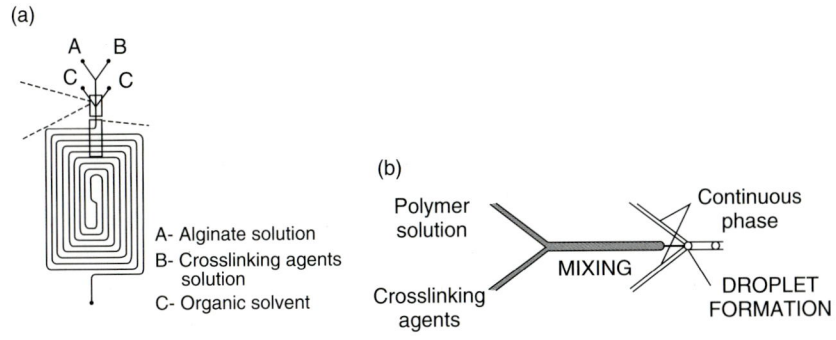

(b) Polymer solution / Crosslinking agents → MIXING → Continuous phase → DROPLET FORMATION

Figure 8.11 Schematic of the channel layout, showing a Y-shaped channel and two co-flowing channels. Polymer solution and crosslinking agents are injected via the two Y-shaped inlet channels, which join together to form the main flow channel. Dimethyl carbonate (DMC) is injected further downstream via the secondary Y-shaped channels to generate alginate droplets at their junction with the main flow channel. As the droplets flow downstream, the water diffuses from the drops into the DMC [51].

diffusion-mediated mass and heat transfer in reactive processes. For example, typical microfluidic devices exhibit high thermal transfer efficiencies because of their reduced thermal masses and high surface-to-volume ratios, and this allows exothermic and/or high-temperature reactions to be performed in an efficient and controllable (isothermal) manner [21]. Microfluidic environments have been shown to provide an efficient temperature and thereby reaction control in continuous-flow reactors for multicomponent reactions [52] and nitrations [53].

The large surface-to-volume ratio of microreactors also offers the possibility of an accelerated heat exchange and accurate temperature control for nanoparticle synthesis. To control the microreactor temperature, a temperature-regulated reactor holder can be constructed using an in-house-modified microscopic hot-stage Peltier device. Capillary tubing in hot oil baths has been used for the efficient and fine temperature control of microreactors [54, 55]. By varying the temperature and the flow rate ratio (FRR), nanoparticles with different sizes can be produced. This increased accuracy in temperature and reaction time controls a high reproducibility of particle size distribution. A method to control the volume and velocity of drops generated in a flow-focusing device, both dynamically and independently, has recently been reported [56]. This involves the simultaneous tuning of the temperature of the nozzle of the device and the flow rate of the continuous phase, and requires a continuous-phase liquid which has a viscosity that varies steeply with temperature (Figure 8.12). Increasing the temperature of the flow-focusing nozzle from 0 to 80 °C increases the volume of the drops by almost two orders of magnitude, while tuning the temperature and flow rate independently controls the drop volume and drop velocity. This method can be implemented in on-chip applications, where thermal management is already incorporated into the system, such as DNA amplification using the polymerase chain reaction and nanoparticle synthesis.

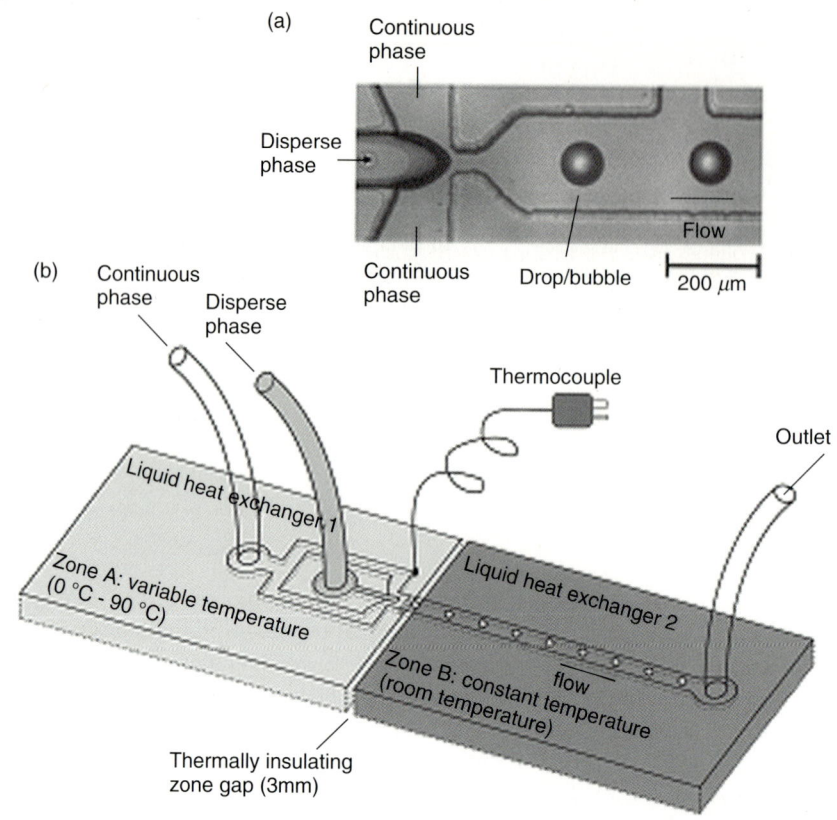

Figure 8.12 Illustration of the experimental set-up. (a) Top view of the microfluidic flow-focusing generator in operation. Drops of water were generated in a continuous phase of mineral oil. The device also had a side channel that was kept plugged for the experiments described here; (b) The device was placed in contact with two copper liquid heat exchangers that made up the two constant temperature zones, A and B. A thermocouple was used to measure the temperature of the nozzle [56].

8.2.4.4 Spatial and Temporal Kinetic Control

In addition to providing an efficient control of the dimensional characteristics of the nanomaterials produced, microreactors can also be used to investigate and control the fundamental reaction processes in nanoparticle formation. Microfluidic devices provide a platform for the *in situ* monitoring of nanoparticles formation through the ability to spatially resolve the nucleation and growth phases in the reactor during synthesis. Using a poly(methyl methacrylate) (PMMA) microreactor, cobalt nanoparticle formation was probed at three different positions using synchrotron radiation-based X-ray absorption spectroscopy (XRS) [57], together with reference spectra of the precursor. The final product collected at the end of microfluidic system showed that the time resolution of the reaction (in the order of milliseconds) was obtained by spatial resolution within the microreactor.

Sounart et al. [58] recently undertook a spatially resolved investigation of nanoparticle growth during synthesis in a microfluidic reaction channel, and demonstrated the spatially resolved, on-chip monitoring of semiconductor nanoparticle synthesis in a continuous-flow microfluidic reactor.

Although most studies have paid attention to the kinetics of the synthesis of quantum dot (QD) nanocrystals, their results have provided a direct insight into a fundamental component of microreaction technology, with reactions that are initiated on contact in micromixers beginning by diffusional mixing between laminar reagent streams. Likewise, the nanoprecipitation of organic nanoparticles can also be spatially observed using these spatially resolved spectroscopic imaging techniques and the spectroscopy of the microreactor, where kinetic and mechanistic data on nanoparticle nucleation and growth can be spatially acquired within the reaction/diffusion zone between two laminar flowing reagent streams at a steady state. Kinetics information that would be difficult to observe in a batch reaction can be elucidated in the configuration, because the diffusional mixing and added information from analysis of the product spatial distribution (such as diffusion-limited homogeneous reaction and particle nucleation) are controlled using microfluidic technology.

8.2.4.5 Self-Assembly Mechanism and Competitive Reaction

Recently, significant interest has been expressed in developing a process to produce nanoparticles of organic compounds at a high solid concentration and a low colloidal stabilizer content, with the goal of tailoring the surface properties of the nanoparticles through the formation of unique composite organic and block copolymer nanoparticles [59]. Synthetic carotenoids represent a group of extensively studied organic actives, in which bioavailability is improved by hydrosol formation [60]. Organic nanoparticle formation has been developed based on the hydrosol formation of the water-insoluble organic compounds β-carotene with the amphiphilic diblock copolymer polystyrene-b-poly(ethylene oxide) (PS-b-PEO).

As discussed above, in order to achieve a Damkohler number less than 1, the mixing time must be less than the induction time for the formation of a block copolymer nanoparticle, τ_{agg}, and the induction time for the nucleation and growth, τ_{ng}, of organic particles less than 1 μm. There are two precipitation times for the self-assembly of block copolymers and competitive formation of organic particles: the ratio of the induction times can be balanced effectively by changing the active concentration or the molecular architecture of the diblock copolymer. When the two precipitation times are properly balanced to match one another, the insoluble portion of the protective colloid (the block copolymer) is deposited on the surface of the growing organic particle, to freeze the size distribution at that desired for a particular formulation. The precipitation process can be further subdivided into the characteristic induction times for copolymer aggregation, τ_{agg}, and active organic nucleation and growth, τ_{ng}, (Figure 8.13a). The "reactions" compete such that, when the times are matched, the block copolymer can interact with the growing active particle to alter nucleation and growth and offer colloidal stabilization.

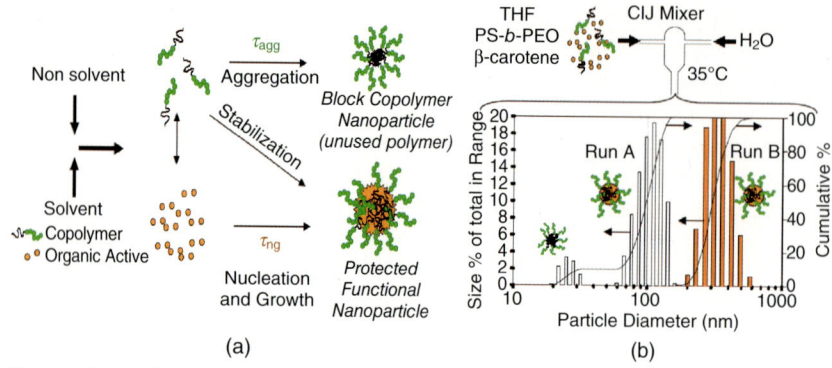

Figure 8.13 (a) Illustration showing that an organic active and an amphiphilic diblock copolymer are molecularly dissolved in an organic phase and mixed rapidly with a miscible antisolvent for the active and one block of the copolymer; (b) Nanoprecipitation of β-carotene and polystyrene (10 monomers)-block-poly(ethylene oxide) (68 monomers) (PS10-b-PEO68) [59].

If either the block copolymer or the organic active precipitates well before the other, the two processes do not interact sufficiently to yield nanoparticles using a confined impinging jets mixer (Figure 8.13b). However, by balancing the aggregation time for the copolymers and the nucleation and growth time for the organic active, the size distribution can be controlled. In the case of very fast mixing, as when using microfluidic reactors, mixing occurs within a period of milliseconds, which is faster than the induction time for each precipitation ($Da < 1$). This has enabled the process to be run under a "homogeneous" starting condition, where the effect of mixing is not convoluted with the role of the precipitation times. It is to be expected that such microscale reactors for the formation of water-insoluble organic compounds will attract further interest, and in time be used to understand the competitive precipitation of both block copolymers and organic active compounds. When using a microfluidic reactor, with the mixing rate sufficiently fast and the metastable zone width exceeded for both compounds, an homogeneous competitive kinetics dictates the resultant product.

8.3
Microfluidic-Related Organic Nanoparticles for Drug Delivery

Nanoscale materials have been used to deliver a wide range of pharmaceutically active organic compounds, such as drugs, genes, and imaging agents [1]. Most pharmaceutically active organic compounds are hydrophobic, poorly soluble in water, or even water-insoluble. The nanosized particles of organic materials increase the solubility of the organic compounds, which enables them to be administered into the aqueous environment, at sufficient concentration. Today, there is an increasing interest in the development of new aqueous formulations based on nanosized particles of organic materials for two main reasons [61]:

- The efficacy (bioavailability) of many of these types of product depends on the (low) molecular solubility of the organic species in water. The solubility—and thereby the intrinsic efficacy—depends on the particle size, and can be significantly enhanced for nanosized particles relative to the micron sizes produced, for example, by milling processes.

- A small particle size can dramatically increase the accessibility, and hence the performance, of a pharmaceutically active species. For example, the blood concentration of β-carotene is five- to 10-fold higher when a single dose of $6\,\text{mg}\,\text{kg}^{-1}$ is administered.

As discussed above, microfluidic reactors provide unique characteristics for the synthesis and formulation of organic compound-based nanostructured materials, with good control of the particle shape and size. Thus, properties such as the solubility, stability and bioavailability of these organic nanoparticles are tunable to a required condition for pharmaceutical applications. In this section, microfluidic-related organic nanoparticles for drug delivery will be discussed, by providing specific samples to address the benefits that microfluidic technology may bring to the field of drug discovery.

8.3.1
Drug Encapsulation and Release

In drug formulation, nanodispersed polymeric systems are attracting significant interest with regard to the control of the spatial and temporal kinetics of drug encapsulation and release at the site of action. Such effects are key to achieving an optimal pharmokinetic effect, such as modulating the biodistribution and metabolism of drugs to hit their target(s), without damaging healthy tissues [62]. However, the polydispersity and batch-to-batch variation in both the size and morphology of conventional nanoparticles produce undesirable variations in the rate of particle degradation, the stability of the drug, and the kinetics of its release. In this regard, microfluidic technology can offer the potential to develop polymer-based transport systems for the controlled nanodispersed formulation of the active material during storage and administration [63].

A microfluidic device has been developed to assemble drug-encapsulated PLGA–PEG nanoparticles by tuning the nanoparticle composition to increase drug loading, without adversely affecting the size and size distribution [34]. When docetaxel (Dtxl) was used as a model therapeutic agent, it was found that the loading of Dtxl could be maximized and its release rate prolonged by the addition of hydrophobic PLGA to the PLGA–PEG nanoparticle. As a result, the encapsulation efficiency and drug loading of the nanoparticles obtained by hydrodynamic flow focusing was almost doubled following PLGA addition, whereas nanoprecipitation on microfluidic reactors resulted in a higher encapsulation and loading than with bulk nanoprecipitation. These results confirmed that rapid and tunable microfluidic mixing could be used to synthesize drug-encapsulated biodegradable polymeric PLGA–PEG nanoparticles with a defined size, a lower polydispersity, and a higher drug loading, yet with a slower rate of release.

The emulsion method was developed for incorporating bupivacaine (a local anesthetic) by dissolving the free-base form of the drug into a solution of dichloromethane and PLGA in a microfluidic reactor [31]. For comparison, particles with a similar average size were also prepared using the conventional single-emulsion technique, and assayed to release the drug. This process demonstrated that monodisperse particles prepared using microfluidics would release the drug more slowly than conventional polydispersed particles of a similar average size. However, the initial burst release of drug was significantly less than that observed with the corresponding conventional polydispersed particles. A further kinetics analysis led to the conclusion that faster mixing in microfluidic devices is more homogeneous, and leads to a more uniform drug distribution within the particles. Particles prepared using the conventional emulsion method may have drug-rich domains at or near the particle surface, and this can account for the more rapid rate of drug release from particles fabricated using this method. Besides polymeric nanoparticles as a carrier, SLNs are currently undergoing intensive investigation as a new generation of drug carrier systems for pharmaceutical applications [39–41]. SLNs are prepared by the melt-homogenization of a matrix lipid in surfactant-containing aqueous media; this requires the use of high, well-defined cooling rates, and offers interesting new possibilities for the manufacture of such drug-carrier systems. In a recent study, a microfluidic device for the continuous-melt crystallization of SLN suspensions was established [64], allowing for high and well-defined cooling rates. Because of their small volumes and superior heat and mass transfer performance, the microfluidic devices ensure a precise setting and control of the optimum process conditions.

8.3.2
Stimuli-Responsive Release

Polymeric nanoparticles that release the drug as a response to specific chemical or physical stimuli have been proposed to enhance therapeutic efficiency and reduce adverse side effects. Such drug-delivery systems are designed to release the drug in response to external stimuli. Consequently, in recent years stimuli-responsive polymeric nanoparticles have attracted much attention, and excellent progress has been made in their development [65–68]. The most frequently applied stimuli are pH, temperature, redox potential, magnetic field, light, and ultrasound. The results of a recent study showed that, although the selectivity and specificity of attaching a target molecule to a magnetic carrier are related to the surface functionalization, the efficiency of the magnetic manipulation in microfluidics involves an interplay of various other parameters, such as the inlet velocity of the fluid containing the magnetic nanoparticles, the size of the nanoparticles, the

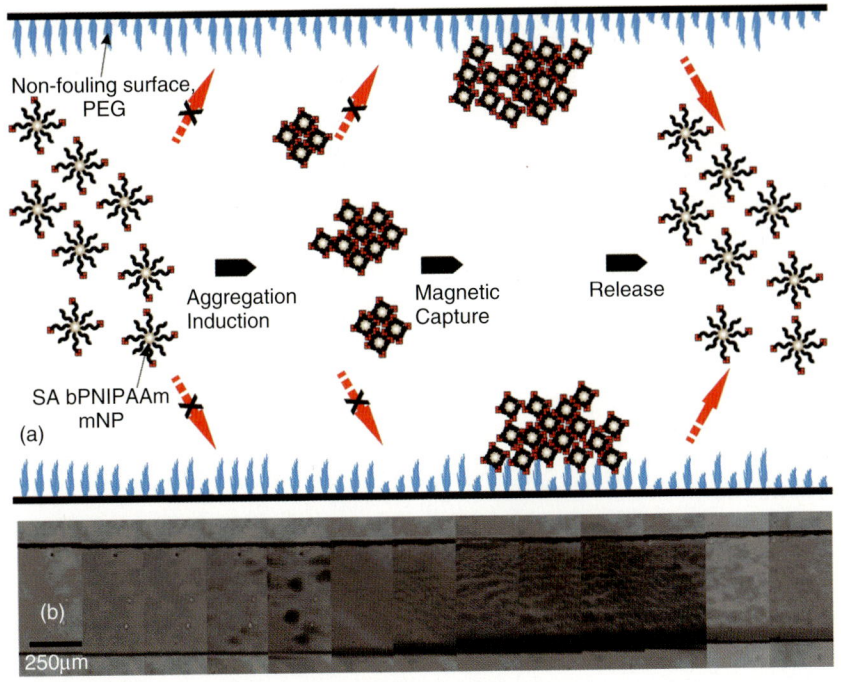

Figure 8.14 Illustration of the particle capture and release in PEGylated poly(dimethylsiloxane) (PDMS) microfluidic channels [69].

magnetic field strength and its orientation, and the geometry of the device [68]. Nonetheless, this technique has shown much promise for the controllable manipulation of these smart polymeric nanoparticles, by combining the microfluidic approach to manipulate chemical or biological target drug molecules by attaching them to the carrier.

A stimuli-responsive magnetic nanoparticle system for the diagnostic target capture and concentration has been developed using a microfluidic device [69] (Figure 8.14). In this case, telechelic poly(N-isopropylacrylamide) (PNIPAAm) polymer chains were synthesized with dodecyl tails at one end and a reactive carboxylate at the opposite end, by using a reversible addition fragmentation transfer (RAFT) technique. These PNIPAAm chains self-associated into nanoscale micelles that were used as dimensional confinements to synthesize magnetic nanoparticles. The resultant superparamagnetic nanoparticles exhibited a gamma-Fe_2O_3 core (approximately 5 nm) with a layer of carboxylate-terminated PNIPAAm chains as a corona on the surface. Magnetic nanoparticles are able to associate with biotinylated targets as individual particles, after which the application of a combined temperature increase and magnetic field can be used to magnetically separate the aggregated particles onto the PEG-modified polydimethylsiloxane channel walls of a microfluidic device. When the magnetic field is turned off and the temperature reversed, the captured aggregates redisperse into the channel flow stream for further downstream processing. However, by integrating the dual magnetic- and temperature-responsive process into the microfluidic device, this approach can be used to capture diagnostic targets at a controlled time point and microfluidic channel position. Using this approach, nanoparticles can then be isolated and released after capturing target molecules, thus overcoming the problem of the low magnetophoretic mobility of the individual particle, while retaining the advantages of a high surface-to-volume ratio and faster diffusive properties during target capture.

8.3.3
Nanomedicine Delivery to Target Cells

Nanoparticles have been used for targeted delivery to interact with antigens that are differentially expressed by a subset of cells or tissues [70]. Commonly, *in vivo* animal experiments are used to evaluate the consequence of each change on the biodistribution of particles; however, these require a large number of animals and are costly and time-consuming.

The development of *in vitro* systems to optimize various parameters that influence cell–particle interactions are attracting growing interest. Microfluidic devices provide such benefit by controllably manipulating the cells as well as the nanoparticles by integrating an external force, such as a magnetic force, to achieve a targeted delivery [71–73]. One of the pioneering efforts included the development of a simple microscale device to use as a model of microcirculation to examine the targeting efficacy of polymeric nanoparticle and microparticle delivery vehicles *in vitro* [71]. This model could be used to evaluate the interaction of polymeric nanoparticles conjugated to aptamers that recognized the transmembrane prostate-specific membrane antigen (PSMA), with cells seeded in microchannels. The binding of particles to cells that expressed (or not) the PSMA was evaluated with respect to changes in fluid shear stress, PSMA expression on target cells, and particle size. It was shown that microfluidic devices offered the advantages of scalability, low cost, reproducibility and high-throughput capability, and could also be used for a wide array of cell–particle systems, before *in vivo* experiments were conducted.

To enable *in vivo* therapeutic delivery to cells without their rapid disappearance into the circulatory system following injection, magnetic nanoparticles can be used for the localization of therapeutic materials and targeting of cells by using an externally applied magnetic field. Whilst this can be an effective and noninvasive method for directing materials and cells to a specific site, it is unclear whether magnetically labeled cells can withstand the significant mechanical forces exerted on them. It also remains unclear as to whether the magnetic force is sufficiently strong to overcome the blood flow in arteries, such that the particles are directed to the target site. Microfluidic devices can be used for this type of study by modeling the physiological flow conditions of the bloodstream *in vitro* to produce well-defined flow rates, which can then be related to shear stress levels with a

8.3 Microfluidic-Related Organic Nanoparticles for Drug Delivery

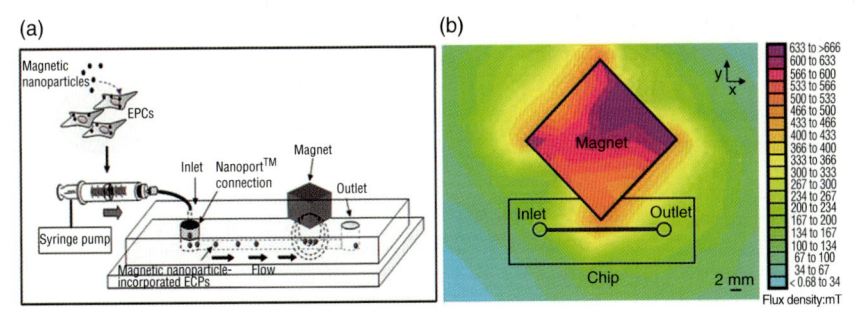

Figure 8.15 (a) A schematic illustration of the microfluidic system; (b) the two-dimensional (2-D) magnetic field distribution around the microchannel [72].

measurable biological response. In a recent study, a microfluidic system was used to create and control the flow rate and to serve as a model of the *in vivo* human blood system [72], for assessing the magnetically mediated incorporation of endothelial progenitor cells (EPCs) for biomedical applications (Figure 8.15). In spite of the weak magnetic field used (400 mT), those cells containing magnetic nanoparticles were able to withstand the high flow rate within the microchannel, and were directed to the desired site in the microchannel. These results confirmed that the microfluidic system would indeed serve as a useful tool for understanding the behavior of magnetic nanoparticle-incorporated cells within the human circulatory system.

Nanoparticle-based drug delivery relies heavily on the in-depth study of cellular responses to treatments and their microenvironments [73]. Drug-delivery efficiency, subcellular targeting and therapeutic efficacy were all analyzed by observing, microscopically, the morphological alterations and viability of treated HeLa cells. In this regard, the dynamic time-lapsed, long-term observation of cell-based assays, especially the multiplexed, high-throughput observation of the interaction between nanoparticle delivery and targeting cells, is desired. During the past decade, significant interest has been expressed in developing a microfluidic platform for high-throughput cell-based studies. Microfluidics offers the possibility for the high-throughput, high-magnification observation of real-time, cell-based study and functional assays [74, 75]. Recently, a microfluidic cell culture platform for the real-time *in vitro* microscopic observation and evaluation of cellular functions and responses to nanoparticles has been reported [76]. Within this system, microheaters, a microtemperature sensor and micropumps are integrated to achieve a self-contained, perfusion-based, cell culture microenvironment, the key feature of which includes an ultra-thin culture chamber that allows real-time, high-resolution cellular imaging by combining bright-field and fluorescent optics to visualize nanoparticle–cell/organelle interactions. The system offers a universal platform for the high-throughput analysis of organic nanoparticles and targeting cells [77].

8.4
Conclusions and Prospective Study

8.4.1
Materials, Design, and Fabrication

Glass, silicon, poly(dimethylsiloxane) (PDMS) and SU8 (epoxy-based photoresists) have each been used to construct microfluidic reactors, and each has been shown to have their own advantages and disadvantages. In organic nanoparticle application, although widely employed as a substrate for the microfluidic synthesis of organic nanoparticles [31, 34, 51], PDMS has been found to swell in the presence of many common organic solvents, particularly alkane and aromatic solvents, and is limited for solvent-based synthesis. Glass can be the favored substrate material for microfluidic solvent-based syntheses, because it presents good optical properties, heat-dissipating efficiency and, most importantly, has a high resistance to mechanical and chemical stresses, although its fabrication is both time-consuming and expensive. Thus, there is constant interest in developing an alternative inexpensive rapid prototyping of solvent-resistant microfluidic devices [78–80]. Such examples include microfluidic devices produced from a photocurable perfluoropolyether, which has a high resistance to swelling by using methylene chloride [78], and a rapid prototyping technique based on a commercially available thiolene-based optical adhesive to fabricate more solvent-resistant microfluidic devices [79]. A variety of microfluidic reactors have been designed for the different syntheses of organic nanoparticles, including emulsions [32], nanoprecipitation [33], and the production of liposomes [37]. Flow-focusing microfluidic reactors, because of their feasible design and ease of fabrication, have been utilized for all applications. In laminar flow microfluidic reactors, the parabolic flow profile (fluid moving slower near the channel wall than in the center) and the associated axial dispersion lead to a wider size distribution of synthesized nanoparticles. Although very few reports have been made on organic nanoparticle formation based on segmented flow or multiphase flow microfluidic reactors, more complicated microfluidic reactors are likely to be used to provide an understanding of the more complicated synthesis information, by offering a better control over nanoparticle size distribution.

Currently, photolithography, followed by etching, is the standard process to fabricate microfluidic reactors. Recently, the concept of nanofluidic devices presented exciting opportunities to build a system to operate or manipulate single molecules or particles, with the early advances of applying the principles of nanotechnology to medicine [80]. In addition, multiple fabrication techniques, such as electron-beam lithography, focused ion-beam milling and nanoimprint lithography, have been used to construct nanostructures for manipulating and measuring the different types of nanoparticle in solution [81]. Further fabrication concepts could be applied when more delicate components such as micropumps, microvalves and microheaters are required to be integrated within microfluidic reactors

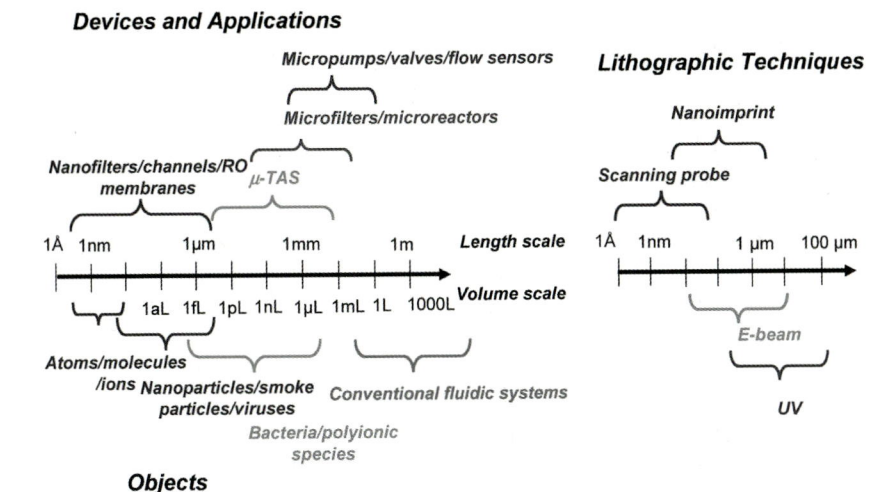

Figure 8.16 Schematic description of the length scales and lithographic techniques used in micro- and nanosystems [80].

for the controlled synthesis and manipulation of single organic nanoparticles (Figure 8.16).

8.4.2
High-Throughput Microfluidic Processes

When microfluidics shows the promise of the controlled synthesis of organic nanoparticles, the scale-up of nanoparticle synthesis using the microfluidic process, such as stacking thousands of microfluidic reactors together, is necessary to meet industry requirements [82]. However, one of the major challenges is that to achieve an efficient scale-up of nanoparticle production in a microfluidic channel, the same flow rate must be assured in all the arrayed microchannels. As explained above, the flow rate across the channel can affect different parameters that are important for controlling the reaction conditions, such as heat and mass transfer and residence time. The use of pumps for each channel can ensure a uniform distribution of the flow, but this is expensive. With regards to the high-throughput controlled synthesis of inorganic nanoparticles, a manifold PDMS microreactor with the controlled multipoint addition and mixing of a reactant to a primary feed was successfully utilized for silica-coated titania nanoparticles [83]. Moreover, an integrated stacked microreactor system was utilized to synthesize metallic nanoparticles, although the high-throughput synthesis of organic nanoparticles has not yet been reported [2].

High-throughput production processes also rely on an efficient way to characterize the produced nanoparticles. In this respect, there remains a lack of any general method to enable the high throughput of individual nanoparticles, although recent developments on high-throughput Raman flow spectrometry based on surface-enhanced resonant Raman scattering (SERRS) has shown some promise, not only for single nanoparticle sensitivity but also for high-throughput high spectral resolution [84]. To date, very few studies have been conducted on microfluidics for the high-throughput characterization of nanoparticles. Recently, however, a microfluidic channel was used to simulate a nanoparticle focusing lens to estimate the path of randomly moving nanoparticles through a focusing lens, using an immersed finite-element method [85]. It is anticipated that such approaches, as well as significant improvements, will be made in the future either through modeling or properly designed experiments.

8.4.3
Controlled Synthesis of Organic Nanoparticles

It remains a challenge to control all the critical features of nanoparticles (e.g., size, size distribution, shape, structure) simultaneously in a single experiment. Fewer studies have conducted with organic nanoparticles synthesized using microfluidic approaches, although inorganic metal nanoparticles – particularly QDs – have been studied in detail. However, it is anticipated that microfluidics will be utilized as a powerful platform in the coming years for the continuous-flow production of a wide range of organic nanostructures with fine-tunable sizes and shapes, via the controllable processes of nanostructure nucleation and growth. The consideration of laminar flow conditions will lead to broader velocity and residence-time distributions. It is also expected that segmented flow microreactors will be largely involved in the future development of new designs of microreactors for a superior control of the size and size distribution of organic particles.

In organic nanoparticle precipitation processes, the control of the local level of supersaturation is crucial for tailoring product characteristics, while mixing is also of primary importance when determining local supersaturation levels. Microfluidic devices need to be optimized for the physical and chemical characteristics of the nanomaterial to be synthesized. In particular, both reactor design and operation parameters such as reactor geometries, feed input mode, feed rate, feed point locations and reagent concentrations, will have a significant effect on the resulting particle properties. Thus, it is crucial to know the three-dimensional (3-D) and time-dependent fluidics and reagent concentration distributions for each component during and after mixing. The microfluidic modeling of the synthesis process aids the design and operation of the microfluidic delivery system with varying computer-controlled flow pattern switching procedures for organic nanoparticle formation. Thus, further investigations and research are foreseen in this direction, which will not only improve the comprehension of the nanoparticle precipitation process but also explain how reactor geometry, flow and mixing and the reagent concentrations can affect the precipitation process and the particle characteristics in the microreactor. In addition, by taking advantage of chip-based technologies for the real-time sizing of nanoparticles during nanoparticle synthesis [2], it is anticipated that further research and investigation will bring control to the output.

8.4.4
Spatial and Temporal Kinetics Investigation of Nanoparticles

Studies on the formation of organic nanoparticles are still at an early stage. For instance, there is hardly any significant experimental evidence available concerning the early stage of organic nanoparticle formation, especially during or immediately after mixing the reactants. Although microfluidic devices provide a platform for *in situ* monitoring of the progress of nanoparticle formation through the ability to spatially resolve the nucleation and growth phases in the reactor during synthesis, few reports exist relating to the on-chip monitoring of the mechanistic principles in the formation of such organic nanoparticles, partly because of the lack of directly measurable optically characteristics from organic nanoparticles. By deliberately incorporating fluorescent/luminescent molecules into the core–shell-type organic nanoparticles, on-chip spectroscopic detection can be performed during the synthesis of organic nanoparticles to monitor their formation kinetics. For example, depending on the particle size and aggregate structure, the β-carotene hydrosol can exhibit different colors, visibly ranging from yellow to orange to red, in such way that its kinetics of formation could be detected with a high degree of precision using a simple microscopic spectrophotometer. It is anticipated that, by using similar types of organic nanoparticle models, further research and studies will conducted on the kinetics investigation of organic nanoparticles using the microfluidic platform.

In addition, the highly temporal monitoring of the kinetics of organic particle formation, to some extent, relies heavily on the development of methodologies to characterize single nanoparticles. A combination of lasers, confocal and scanning probe microscopes, and highly sensitive optical detection systems are indispensable to achieve this. Recent reports have described the use of confocal laser spectroscopy to demonstrate that nanoparticle engineering and optimization is guided by a rapid optical characterization of large numbers of individual metal nanoparticles freely diffusing in colloidal solutions [86]. Furthermore, the kinetics of atomic deposition onto a single gold nanocrystal has been directly observed, using surface plasma spectroscopy [87]. It is expected that such advanced methodologies will be applied to investigations of the kinetics of organic nanoparticle formation, with high temporal resolution.

References

1 Peer, D., Karp, J.M., Hong, S., Farokhzad, O.C., Margalit, R. and Langer, R. (2007) Nanocarriers as an emerging platform for cancer therapy. *Nature Nanotechnology*, **2**, 751–60.

2 Song, Y., Hormes, J. and Kumar, C.S.S.R. (2008) Microfluidic synthesis of nanomaterials. *Small*, **4**, 698–711.

3 Chang, C.H., Paul, B.K., Remcho, V.T., Atre, S. and Hutchison, J.E. (2008) Synthesis and post-processing of nanomaterials using microreaction technology. *Journal of Nanoparticle Research*, **10**, 965–80.

4 Song, Y., Modrow, H., Henry, L.L., Saw, C.K., Doomes, E., Hormes, J. and Kumar,

C.S.S. (2006) Microfluidic synthesis of cobalt nanoparticles. *Chemistry of Materials*, **18**, 2817–27.
5 Whitesides, G.M. (2006) The origins and the future of microfluidics. *Nature*, **442**, 368–73.
6 Cottam, B.F., Krishnadasan, S., deMello, A.J., deMello, J.C. and Shaffer, K.S. (2007) Accelerated synthesis of titanium oxide nanostructures using microfluidic chip. *Lab on a Chip*, **7**, 167–9.
7 Song, H., Chen, D.L. and Ismagilov, R.F. (2006) Reactions in droplets in microfluidic channels. *Angewandte Chemie International Edition*, **45**, 7336–56.
8 Barrow, D.A., Harries, N., Jones, T.G. and Bouris, K. (2004) Novel microfluidic geometries and processing methodologies enabling the precision formation of micro- and nano-particles. European Patent WO2004/043598.
9 Kim, J.W., Utada, A.S., Fernández-Nieves, A., Hu, Z. and Weitz, D.A. (2007) Fabrication of monodisperse gel shells and functional microgels in microfluidic devices. *Angewandte Chemie International Edition*, **46**, 1819–22.
10 Peters, D. (1996) Ultrasound in material chemistry. *Journal of Materials Chemistry*, **6**, 1605–18.
11 Khammana, O., Chaisana, W., Yimniruna, R. and Ananta, S. (2007) Effect of vibro-milling time on phase formation and particle size of lead zirconate nanopowders. *Materials Letters*, **61** (13), 2822–6.
12 Jagannathan, R., Irvin, G., Blanton, T. and Jagannathan, S. (2006) Organic nanoparticles: preparation, self-assembly, and properties. *Advanced Functional Materials*, **16** (6), 747–53.
13 Debuigne, F., Jeunieau, L., Wiame, M. and Nagy, J.B. (2000) Synthesis of organic nanoparticles in different w/o microemulsions. *Langmuir*, **16** (20), 7605–11.
14 Shaw, D. and David, B. (eds) (2007) *Handbook of Micro and Nanoparticle Science and Technology*, Kluwer Academic Publishers, Dordrecht, The Netherlands.
15 Haberkorn, H., Franke, D., Frechen, T., Goesele, W. and Rieger, J. (2003) Early stages of particle formation in precipitation reactions-quinacridone and boehmite as generic examples. *Journal of Colloid and Interface Science*, **259** (1), 112–26.
16 Fu, H., Xiao, D., Yao, J. and Yang, G. (2003) Nanofibers of 1,3-diphenyl-2-pyrazoline induced by cetyltrimethylammonium bromide micelles. *Angewandte Chemie International Edition*, **42**, 2883–6.
17 Horn, D. and Rieger, J. (2001) Organic nanoparticles in the aqueous phase-theory, experiment, and use. *Angewandte Chemie International Edition*, **40** (23), 4330–61.
18 Masuhara, H., Nakanishi, H. and Sasaki, K. (eds) (2003) *Single Organic Nanoparticles*, Springer.
19 Zhu, Z., Anacker, J.L., Ji, S., Hoye, T.R., Macosko, C.W. and Prud'homme, R.K. (2007) Formation of block copolymer-protected nanoparticles via reactive impingement mixing. *Langmuir*, **23**, 10499–504.
20 Alison, H.G., Davey, R.J., Garside, J., Tiddy, G.J.T., Clarke, D.T. and Jones, G.R. (2003) Using a novel plug flow reactor for the in situ, simultaneous, monitoring of SAXS and WAXD during crystallisation from solution. *Physical Chemistry Chemical Physics*, **5**, 4998–5000.
21 deMello, A.J. (2006) Control and detection of chemical reactions in microfluidic systems. *Nature*, **442**, 394–402.
22 Quiram, D.J., Jensen, K.F., Schmidt, M.A., Mills, P.L., Ryley, J.F., Wetzel, M.D. and Kraus, D.J. (2007) Integrated microreactor system for gas-phase catalytic reactions. 2. Microreactor packaging and testing. *Industrial and Engineering Chemistry Research*, **46** (25), 8306–18.
23 Zhang, X. and Haswell, S.J. (2006) Materials matter in microfluidic devices. *American Materials Research Society Bulletin*, **31**, 95–9.
24 Plank, C. (2009) Silence the target. *Nature Nanotechnology*, **4**, 544–5.
25 Hosokawa, M. (ed.) (2007) *Nanoparticle Technology Handbook*, Elsevier, Amsterdam, The Netherlands.
26 Rapoport, N. (2007) Physical stimuli-responsive polymeric micelles for anti-cancer drug delivery. *Progress in Polymer Science*, **32**, 962–90.

27. Kelly, B.T. et al. (2007) Miniaturizing chemistry and biology in microdroplets. *Chemical Communications*, **18**, 1773–88.
28. Hatakeyama, T., Chen, D.L.L. and Ismagilov, R.F. (2006) Microgram-scale testing of reaction conditions in solution using nanoliter plugs in microfluidics with detection by MALDI-MS. *Journal of the American Chemical Society*, **128**, 2518–19.
29. Hashimoto, M., Mayers, B., Garstecki, P. and Whitesides, G.M. (2006) Flowing lattices of bubbles as tunable, self-assembled diffraction gratings. *Small*, **2**, 1292–8.
30. Su, Y.-F., Kim, H., Kovenklioglu, S. and Lee, W.Y. (2007) Continuous nanoparticle production by microfluidic-based emulsion, mixing and crystallization. *Journal of Solid State Chemistry*, **180**, 2625–9.
31. Qiaobing, X., Hashimoto, M., Dang, T.T., Hoare, T., Kohane, D.S., Whitesides, G.H., Langer, R. and Anderson, D.G. (2009) Preparation of monodisperse biodegradable polymer microparticles using a microfluidic flow-focusing device for controlled drug delivery. *Small*, **5** (13), 1575–81.
32. Shum, H.C., Kim, J.W. and Weitz, D.A. (2008) Microfluidic fabrication of monodisperse biocompatible and biodegradable polymersomes with controlled permeability. *Journal of the American Chemical Society*, **130**, 9543–9.
33. Hany, S.M.A., Yorka, P. and Blagdena, N. (2009) Preparation of hydrocortisone nanosuspension through a bottom-up nanoprecipitation technique using microfluidic reactors. *International Journal of Pharmaceutics*, **375**, 107–13.
34. Karnik, R., Gu, F., Basto, P., Cannizzaro, C., Dean, L., Kyei-Manu, W., Langer, R. and Farokhzad, O.C. (2008) Microfluidic platform for controlled synthesis of polymeric nanoparticles. *Nano Letters*, **9**, 2906–12.
35. Jahn, A., Reiner, J.E., Vreeland, W.N., DeVoe, D.L., Locascio, L.E. and Gaitan, M. (2008) Preparation of nanoparticles by continuous-flow microfluids. *Journal of Nanoparticle Research*, **10**, 925–34.
36. Jahn, A., Vreeland, W.N., Gaitan, M. and Locascio, L.E. (2004) Controlled vesicle self-assembly in microfluidic channels with hydrodynamic focusing. *Journal of the American Chemical Society*, **126** (9), 2674–5.
37. Jahn, A., Vreeland, W.N., DeVoe, D.L., Locascio, L.E. and Gaitan, M. (2007) Microfluidic directed formation of liposomes of controlled size. *Langmuir*, **23** (11), 6289–93.
38. Almeida, A.J. and Souto, E. (2007) Solid lipid nanoparticles as a drug delivery system for peptides and proteins. *Advanced Drug Delivery Reviews*, **59**, 478–90.
39. Zhang, S.H., Yun, J.X., Shen, S.C., Chen, Z., Yao, K.J., Chen, J.Z. and Chen, B.B. (2008) Formation of solid lipid nanoparticles in a microchannel system with a cross-shaped junction. *Chemical Engineering Science*, **63**, 5600–5.
40. Yun, J., Zhang, S., Shen, S., Chen, Z., Yao, K. and Chen, J. (2009) Continuous production of solid lipid nanoparticles by liquid flow-focusing and gas displacing method in microchannels. *Chemical Engineering Science*, **64**, 4115–22.
41. Zhang, S.H., Shen, S.C., Chen, Z., Yun, J.X., Yao, K., Chen, B. and Chen, J.Z. (2008) Preparation of solid lipid nanoparticles in co-flowing microchannels. *Chemical Engineering Science*, **144**, 324–8.
42. Smart, T., Lomas, H., Massignani, M., Flores-Merino, M.V., Perez, L.R. and Battaglia, G. (2008) Block copolymer nanostructures. *Nano Today*, **3**, 38–46.
43. Ismagilov, R.F., Stroock, A.D., Kenis, P.J.A. and Whitesides, G. (2000) Experimental and theoretical scaling laws for transverse diffusive broadening in two-phase laminar flows in microchannels. *Applied Physics Letters*, **76**, 2376–8.
44. Nguyen, N.T. and Huang, X. (2005) Mixing in microchannels based on hydrodynamic focusing and time-interleaved segmentation: modelling and experiment. *Lab on a Chip*, **5**, 1320–6.
45. Yang, A.S. and Hsieh, W.H. (2007) Hydrodynamic focusing investigation in a micro-flow cytometer. *Biomedical Microdevices*, **9**, 113–22.
46. Jan, S., Miroslav, S., Vaccaroc, A. and Morbidelli, M. (2006) Effects of mixing on aggregation and gelation of nanoparticles.

Chemical Engineering and Processing, **45**, 936–43.

47 La Mer, V.K. and Dinegar, R.H. (1950) Theory, production and mechanism of formation of monodispersed hydrosols. *Journal of the American Chemical Society*, **72**, 4847–54.

48 Chan, E.M., Alivisatos, A.P. and Mathies, R.A. (2005) High-temperature microfluidic synthesis of CdSe nanocrystals in nanoliter droplets. *Journal of the American Chemical Society*, **127**, 13854–61.

49 Shestopalov, I.A., Tice, J.D. and Ismagilov, R.F. (2004) Multi-step synthesis of nanoparticles performed on millisecond time scale in a microfluidic droplet-based system. *Lab on a Chip*, **4**, 316–21.

50 Khan, S.A., Gunther, A., Schmidt, M.A. and Jensen, K.F. (2004) Microfluidic synthesis of colloidal silica. *Langmuir*, **20**, 8604–11.

51 Rondeau, E. and Cooper-White, J.J. (2008) Biopolymer microparticle and nanoparticle formation within a microfluidic device. *Langmuir*, **24** (13), 6937–45.

52 Mitchell, M.C., Spikmans, V. and Mello, A.J. (2001) Microchip-based synthesis and analysis: control of multicomponent reaction products and intermediates. *Analyst*, **126**, 24–7.

53 Kawaguchi, T., Miyata, H., Ataka, K., Mae, K. and Yoshida, J. (2005) Room-temperature Swern oxidations by using a microscale flow system. *Angewandte Chemie International Edition*, **44**, 2413–16.

54 Nakamura, H., Yamaguchi, Y., Miyazaki, M., Maeda, H., Uehara, M. and Mulvaney, P. (2002) Preparation of CdSe nanocrystals in micro-flow-reactor. *Chemical Communications*, **23**, 2844.

55 Nakamura, H., Tashiro, A., Yamaguchi, Y., Miyazaki, M., Watari, T., Shimizua, H. and Maeda, H. (2004) Application of a microfluidic reaction system for CdSe nanocrystal preparation: their growth kinetics and photoluminescence analysis. *Lab on a Chip*, **4**, 237–40.

56 Stan, C.A., Tang, S.K. and Whitesides, G.M. (2009) Independent control of drop size and velocity in microfluidic flow-focusing generators using variable temperature and flow rate. *Analytical Chemistry*, **81** (6), 2399–402.

57 Svetlana, Z., Rohini, D., Russell, D.L., Proyag, D., Kumar, C.S.S., Goettert, J. and Hormes, J. (2007) The wet chemical synthesis of Co nanoparticles in a microreactor system: a time-resolved investigation by X-ray absorption spectroscopy. *Nuclear Instruments and Methods in Physics Research Section A – Accelerators Spectrometers Detectors and Associated Equipment*, **582**, 239.

58 Sounart, T.L., Safier, P.A., Voigt, J.A., Hoyt, J., Tallant, D.R., Matzke, C.M. and Michalske, T.A. (2007) Spatially-resolved analysis of nanoparticle nucleation and growth in a microfluidic reactor. *Lab on a Chip*, **7**, 908.

59 Brian, K., Johnson, A.C. and Prud'homme, R.K. (2003) Flash nanoprecipitation of organic actives and block copolymers using a confined impinging jets mixer. *Australian Journal of Chemistry*, **56**, 1021–4.

60 Stahl, W., Ale-Agha, N. and Polidori, M.C. (2002) Non-antioxidant properties of carotenoids. *Biological Chemistry*, **383**, 553–8.

61 Brick, M.C., Palmer, H.J. and Whitesides, T.H. (2003) Formation of colloidal dispersions of organic materials in aqueous media by solvent shifting. *Langmuir*, **19**, 6367–80.

62 Farokhzad, O.C. and Langer, R. (2009) Impact of nanotechnology on drug delivery. *ACS Nano*, **3** (1), 16–20.

63 Chin, C.D., Linder, V. and Sia, S.K. (2007) Lab-on-a-chip devices for global health: past studies and future opportunities. *Lab on a Chip*, **7**, 41–57.

64 Jasch, K., Barth, N., Fehr, S., Bunjes, H., Augustin, W. and Scholl, S. (2009) A microfluidic approach for a continuous crystallization of drug carrier nanoparticles. *Chemical Engineering and Technology*, **32**, 1806–14.

65 Meng, F., Zhong, Z. and Feijen, J. (2009) Stimuli-responsive polymersomes for programmed drug delivery. *Biomacromolecules*, **10** (2), 197–209.

66 Rijcken, C.J.F., Soga, O., Hennink, W.E. and van Nostrum, C.F. (2007) Triggered destabilisation of polymeric micelles and vesicles by changing polymers polarity: an

attractive tool for drug delivery. *Journal of Controlled Release*, **120**, 131–48.
67 Schmaljohann, D. (2006) Thermo- and pH-responsive polymers in drug delivery. *Advanced Drug Delivery Reviews*, **58**, 1655–70.
68 Wu, L., Zhang, Y., Palaniapan, M. and Roy, P. (2009) Magnetic nanoparticle migration in microfluidic two-phase flow. *Journal of Applied Physics*, **105**, 123909.
69 Lai, J.J., Hoffman, J.M., Ebara, M., Hoffman, A.S., Estourne, C., Wattiaux, A. and Stayton, P.S. (2007) Dual magnetic-/temperature-responsive nanoparticles for microfluidic separations and assays. *Langmuir*, **23**, 7385–91.
70 Mykhaylyk, O., Antequera, Y.S., Vlaskou, D. and Plank, C. (2007) Generation of magnetic nonviral gene transfer agents and magnetofection in vitro. *Nature Protocols*, **2**, 2391–411.
71 Farokhzad, O.C., Khademhosseini, A., Jon, S., Hermmann, A., Cheng, J., Chin, C., Kiselyuk, A., Teply, B., Eng, G. and Langer, R. (2005) Microfluidic system for studying the interaction of nanoparticles and microparticles with cells. *Analytical Chemistry*, **77** (17), 5453–9.
72 Kim, J.A., Lee, H.J., Kang, H.J. and Park, T.H. (2009) The targeting of endothelial progenitor cells to a specific location within a microfluidic channel using magnetic nanoparticles. *Biomedical Microdevices*, **11** (1), 287–96.
73 Tseng, C.L., Wang, T.W., Dong, C.C., Wu, S.Y.H., Young, T.H., Shieh, M.J., Lou, P.J. and Lin, F.H. (2007) Development of gelatin nanoparticles with biotinylated EGF conjugation for lung cancer targeting. *Biomaterials*, **28**, 3996–4005.
74 West, J., Becker, M., Tombrink, S. and Manz, A. (2008) Micro total analysis systems: latest achievements. *Analytical Chemistry*, **80**, 4403–19.
75 El-Ali, J., Sorger, P.K. and Jensen, K.F. (2006) Cells on chips. *Nature*, **442**, 403–11.
76 Hsieh, C.C., Huang, S.B., Wu, P.C., Shieh, D.B. and Lee, G.B. (2009) A microfluidic cell culture platform for real-time cellular imaging. *Biomedical Microdevices*, **11** (4), 903–13.
77 Weibell, D.B. and Whitesides, G.M. (2006) Applications of microfluidics in chemical biology. *Current Opinion in Chemical Biology*, **10** (6), 584–91.
78 Rolland, J.P., Van Dam, R.M., Schorzman, D.A., Quake, S.R. and DeSimone, J.M. (2004) Solvent-resistant photocurable "liquid teflon" for microfluidic device fabrication. *Journal of the American Chemical Society*, **126** (8), 2322–3.
79 Harrison, C., Cabral, J.T., Stafford, C.M., Karim, A. and Amis, E.J. (2004) A rapid prototyping technique for the fabrication of solvent-resistant structures. *Journal of Micromechanics and Microengineering*, **14**, 153–8.
80 Prakash, S., Piruska, A., Gatimu, E.N., Bohn, P.W., Sweedler, J.V. and Shannon, M.A. (2008) Nanofluidics: systems and applications. *IEEE Sensors Journal*, **8** (5), 441–50.
81 Stavis, S.M. et al. (2009) Nanofluidic structures with complex three-dimensional surfaces. *Nanotechnology*, **20**, 165302–9.
82 Dendukuri, D., Pregibon, D.C., Collins, J., Hatton, T.A. and Doyle, P.S. (2006) Continuous-flow lithography for high-throughput microparticle synthesis. *Nature Materials*, **5**, 365–9.
83 Khan, S.A. and Jensen, K.F. (2007) Microfluidic synthesis of titania shells on colloidal silica. *Advanced Materials*, **19**, 2556–60.
84 Sebba, D.S., Watson, D.A. and Nolan, J.P. (2009) High throughput single nanoparticle spectroscopy. *ACS Nano*, **3** (6), 1477–84.
85 Lee, T.R., Chang, Y.S., Choi, J.B., Liu, W.K. and Kim, Y.J. (2009) Numerical simulation of a nanoparticle focusing lens in a microfluidic channel by using immersed finite element method. *Journal of Nanoscience and Nanotechnology*, **5**, 7407–11.
86 Becker, J., Schubert, O. and Sonnichsen, C. (2007) Gold nanoparticle growth monitored *in situ* using a novel fast optical single-particle spectroscopy method. *Nano Letters*, **7**, 1664–9.
87 Novo, C., Funston, A.M. and Mulvaney, P. (2008) Direct observation of chemical reactions on single gold nanocrystals using surface plasmon spectroscopy. *Nature Nanotechnology*, **3**, 598–60.

9
Lipid–Polymer Nanomaterials

Corbin Clawson, Sadik Esener and Liangfang Zhang

9.1
Introduction

During recent years, nanomaterials have attracted increasing interest from both the academic and industrial arenas, because of the unique physico-chemical properties of materials at the nanoscale, including ultra-small dimensions, vast surface area-to-volume ratios, and high reactivities [1]. These properties have been intensively studied and used to improve the treatment and diagnosis of diseases [2]. Whilst the history of applying nanomaterials to medicine and the life sciences can be traced back to many decades ago, tremendous progress and success have only been made during the past two decades [3–5]. Although a number of nanomaterial-based therapeutic and diagnostic products are now approved for clinical use, the search for new nanoscale materials for medical and biological applications remains as a cutting-edge research topic, and promises great advances. As advances come, more complex nanoscale systems will be developed that can provide safer and more effective strategies to cure diseases. Currently, many research groups are combining a variety of natural and synthetic molecules to create hybrid supramolecular structures not previously possible. Lipids and polymers represent two extremely useful nanomaterial building blocks that are invaluable to the design and engineering of such nanoscale systems.

Lipids are a large group of amphiphilic or hydrophobic small molecules that have many biological functions, including the structural components of cell membranes, cellular signaling molecules, and high-density energy storage. In general, there are eight basic classes of lipids, including fatty acyls, glycerolipids, glycerophospholipids, sphingolipids, sterol lipids, prenol lipids, saccharolipids, and polyketides [6]. Amphiphilic lipids have long been known to self-assemble into structures such as membranes, micelles, and vesicles [7]. Because of their unique properties, they are increasingly being recruited in the rational design and synthesis of nanomaterials for use in a broad range of applications.

Polymers are large molecules consisting of repeating structural subunits or monomers. They can be simply divided into two groups, natural polymers and synthetic polymers. Natural polymers, which include proteins, polynucleotides

Nanomaterials for the Life Sciences Vol.10: Polymeric Nanomaterials. Edited by Challa S. S. R. Kumar
Copyright © 2011 WILEY-VCH Verlag GmbH & Co. KGaA, Weinheim
ISBN: 978-3-527-32170-4

(such as DNA and RNA) and polysaccharides, are essential to life and form the bulk of organic materials. Synthetic polymers are common in everyday life, and are important engineering and design materials. Some common synthetic polymers include polyacrylates, polyamides, polyesters, polycarbonates, polyimides, and polystyrenes. Natural and synthetic polymers are used extensively in nanomaterials for their unique properties. However, finding polymers suitable for use in the medical and life sciences limits the possibilities to biocompatible, nontoxic polymers in most cases [8]. Whether designing a drug-delivery nanoparticle to administer chemotherapeutics, or a surface coating of an implantable device that will reduce host rejection, polymers play a large role in the fields of bio- and nanomaterials.

Biology utilizes both lipids and polymers to orchestrate the complex symphony of life. In order to overcome the limitations of traditional approaches to medicine and engineering, researchers are applying biomimetic approaches to solve problems, gain new insights, and develop new therapeutics. Both lipids and polymers are complex molecules with a wide range of characteristics and properties that a designer may choose from. Using both lipids and polymers in a rational design increases the inherent complexity of the system, but it also opens up a much larger toolkit that the designer can draw from. As the fundamental understanding of biology, biochemistry, and molecular dynamics increases, an increasing ability to successfully utilize these two important classes of molecules – lipids and polymers – together in new systems will open new doors in science, medicine, and technology. In this chapter, a comprehensive review is provided of the current status of lipid–polymer hybrid nanomaterials with a length scale ranging from molecular level to nanostructures, and to microstructures. Specifically, the synthesis, characterization and application of lipid–polymer conjugates (i.e., lipopolymers), lipid–polymer hybrid nanoparticles, and lipid–polymer films and coatings will be reviewed. The chapter is completed with a summary and a prospective outlook on the future of lipid–polymer nanomaterials.

9.2
Lipopolymers

Lipopolymers are a class of molecules consisting of a polymer moiety and a lipid moiety that are covalently linked together. Natural lipopolymers produced by cells, such as glycolipids, lipopolysaccharides and lipoproteins, have innumerable functions that include cell metabolism, enzymatic activity, blood transport, transmembrane transport, and a host of others. It is beyond the scope of this chapter to discuss the structures, functions or importance of the variety of natural lipopolymers. Instead, attention will be focused on synthetic lipopolymers which are defined as either synthetic or natural lipids covalently bound to either synthetic or natural polymers made through a deliberate, synthetic reaction.

Synthetic lipopolymers are useful in a host of applications, from creating long-circulating nanoparticles to artificial cartilage surfaces for joint replacements. The synthesis and fabrication of lipopolymers, their properties and characterization, and their application in the life sciences and medicine are discussed in the following subsections. The types of lipopolymers discussed include phospholipids conjugated to either hydrophilic polymers or to peptides and proteins, and lipids or lipid derivatives conjugated as side chains to polymer backbones.

9.2.1
Synthesis and Fabrication of Lipopolymers

The synthesis of lipid–polymer conjugates can be implemented by using a variety of chemical reactions. Lipids are commercially available with modified headgroups, such that their functionalization and conjugation to other molecules is straightforward. Functionalized headgroups on commercially available phospholipids include amine, glycol, cyanuric chloride, carboxylic acid, thiol and protected thiols, maleimide, and biotin. With this toolkit, many different molecules can easily be conjugated to lipid headgroups, either directly or through the use of another linker. The typical conjugation reactions include:

- **Carboxyl–amine reaction:** 1-ethyl-3-(3-dimethylaminopropyl) carbodiimide hydrochloride (EDC) is a zero-length crosslinker that is often used with N-hydroxysulfosuccinimide (NHS) to couple carboxyl groups to primary amines. Polymers with carboxyl end groups are easily conjugated to lipids with amine-modified headgroups, and *vice versa*. Cyanuric chloride groups are also amine-reactive, and can be conjugated to polymers under mild basic conditions, without having to prederivatize the polymers [9].

- **Thiol–thiol reaction:** Thiols can be conjugated to other thiols to form disulfide bonds. In this case, a reducing agent would be used to break any existing disulfide bridges in order to allow new disulfide bonds to be formed between the lipid and the added polymer molecule. Protected thiol headgroups are used to prevent the lipids from forming disulfide bridges and crosslinking with other lipids. A thiol group will also readily react with the carbon double bond of a maleimide lipid head group to form a stable carbon–sulfur bond. Peptides and proteins that contain the amino acid cysteine are often easily conjugated to lipids in this manner, taking advantage of the thiol in the cysteine residue.

- **Biotin–avidin reaction:** Biotin creates a very strong ligand–receptor bond with the protein avidin. The dissociation constant of avidin from biotin is measured to be $K_D \approx 10^{-15}$ M, and is recognized as one of the strongest noncovalent bonds known [10]. Avidin can simultaneously bind four biotin molecules with high affinity and specificity. However, with a molecular weight of 66–69 kDa, avidin is large compared to most lipids. This limits its usefulness in subsequent applications, where steric hindrance may cause problems.

Many different polymers can be conjugated to lipids using the aforementioned conjugation techniques. Most often, hydrophilic polymers are conjugated to phospholipids in order to produce an amphiphilic molecule that can easily be incorporated into phospholipid membranes. A few of the many possible hydrophilic polymers capable of being conjugated to lipids include poly(ethylene glycol) (PEG), poly(acryloylmorpholine), and poly(vinylpyrrolidone) (PVP) [11]. One of the first classes of hydrophilic polymer molecules to be synthetically conjugated to phospholipid headgroups for use in nanotechnology was polysaccharide, an important class of carbohydrate biomolecules [12].

Glycolipids are naturally occurring biomolecules that consist of a lipid and an attached carbohydrate. These natural glycolipids serve as the inspiration for synthetically conjugating polysaccharides to phospholipid headgroups. *Dextran*, a polysaccharide made from repeating glucose molecules, was first conjugated to lipids and incorporated into liposomes by Pain *et al.* in 1984 [12]. Shortly thereafter, the hydrophilic polymer PEG was conjugated to various phospholipids to create amphiphilic lipid–polymer conjugates [13]. PEG has since become the most popular hydrophilic polymer to conjugate to phospholipids, and lipid–PEG complexes are commercially available with a variety of molecular weights of PEG chains, lipid types, and functional terminal groups on the PEG polymer. A space-filling model and a stick structure of a lipid–PEG conjugate are shown in Figure 9.1.

Besides synthetic polymers, other natural biopolymers such as proteins and peptides can also be conjugated to lipids. Many of the conjugation techniques used to attach biopolymers to lipids are similar to the methods previously discussed. *Antibodies* are commonly conjugated to lipids via disulfide linkages or through a maleimide–thiol reaction. Peptides with cysteine residues are equally easy to conjugate to phospholipids via the thiol functional group on the cysteine amino acid. Amide bonds are also easily formed between a carboxylated phospholipid headgroup and the amine of the N-terminus of a protein or peptide, or *vice versa*. Ester bonds, which are less stable than amide bonds, can also be formed between a hydroxyl functionalized lipid and the carboxyl end group of a peptide.

Another class of lipopolymers consists of lipids or lipid derivatives conjugated to a polymer backbone. Phosphorylcholine is the zwitterionic headgroup of many phospholipids. Poly(2-(methacryloyloxy)-ethyl phosphorylcholine) (PMPC) is a polymer chain based on a methacrylic phosphorylcholine monomer; this polymer

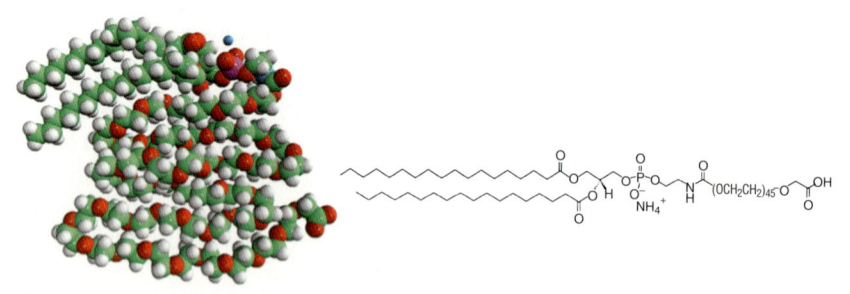

Figure 9.1 Space-filling model and stick structure of a common lipid–PEG molecule, 1,2-distearoyl-sn-glycero-3-phosphoethanolamine-N-(carboxy(polyethylene glycol)-2000) (ammonium salt), also known as DSPE-PEG(2000) carboxylic acid.

is extremely hydrophilic and biocompatible due to an abundance of the zwitterionic phosphorylcholine groups.

Although chemical conjugation is the most popular method of synthesizing lipopolymers, polymerization represents an alternative method. The basic strategy is to attach a monomer or polymer-initiating moiety to the lipid and to proceed with polymerization *in situ*. For this, common methods of polymerization can be used, including ring-opening and radical polymerization [14].

Purification to remove unreacted lipids and polymers is an inevitable step after the completion of the synthesis reaction. As a first-pass purification of the lipid–polymer conjugates, a solvent precipitation or solvent extraction is commonly used, with further purification often accomplished using chromatography. In synthesis methods involving *in situ* polymerization on the lipid, dialysis is a common method of purification.

9.2.2
Properties and Characterization of Lipopolymers

The characterization of a lipid–polymer conjugate can be achieved with a variety of analytical tools. *Chromatography* is commonly used to verify a change in molecular weight or hydrophobicity resulting from the conjugation:

- High-performance liquid chromatography (HPLC) is frequently used both as an analytical and as a purification tool for sample sizes ranging from the microgram to milligram scale.

- Thin-layer chromatography (TLC) can be used to visualize the reaction products and verify that the reaction has gone to completion. Recovering a sample from TLC is less straightforward than with HPLC, but does not require an expensive equipment set-up.

- Gel-permeation chromatography (GPC) can also be used to verify a change in molecular weight of a reaction product in order to confirm successful conjugation of the polymer to the lipid.

- Mass spectrometry is another high-resolution method of determining the molecular weight or elemental composition of a product.

- Nuclear magnetic resonance spectroscopy (NMR), both ^{1}H and ^{13}C, as well as infrared spectroscopy (IR) are often used to confirm the presence or absence of functional groups and the number and type of chemical entities in order to confirm the completion of a reaction.

9.2.3
Applications of Lipopolymers in the Life Sciences

Lipopolymers are useful in a wide variety of applications relevant to the life sciences, including self-assembly, biocompatibility, decreasing toxicity, and cellular targeting.

Lipopolymers can be designed to self-assemble, forming useful structures. Lipids are able to self-assemble into bilayers, vesicles, and micelles due to their amphiphilic nature. Conjugating hydrophilic polymers to these lipids capitalizes on their ability to self-assemble. For example, PEG is a hydrophilic, neutrally charged, biocompatible polymer which can be conjugated to a lipid, such that the resultant lipid–PEG lipopolymer is amphiphilic and retains the ability to self-assemble. The PEG chain also adds steric stability to lipid layers, preventing the adsorption of other macromolecules and proteins. When incorporated onto a nanoparticle surface, the added steric stability reduces aggregation and the uptake of particles by the reticuloendothelial system (RES) [15]. This effectively increases the *in vivo* circulation half-life of the particles, which is an important characteristic for drug-delivery and imaging systems. This strategy, as it relates to liposomes and nanoparticles, is discussed further in Section 9.3.

Lipopolymers can also be designed to increase biocompatibility. PMPC and other phosphorylcholine-based polymers are extremely hydrophilic, with an estimated 22 water molecules being associated with each polar phosphorylcholine monomer [16]. Because phosphorylcholine is an essential component in many biological membranes, this class of lipopolymers is also very biocompatible [13]. Phosphorylcholine copolymers are widely used in applications requiring high blood compatibility, low biofouling and protein adsorption, or low cell adhesion [17, 18]. Biomedical applications include drug-eluting stents, catheters, and artificial joint surfaces [19–21].

Lipopolymers present a less toxic alternative to traditional polymers. Polyethyleneimine (PEI) is a cationic polymer and a common transfection reagent, but is cytotoxic. Subsequently, lipopolymers consisting of low-molecular-weight PEI conjugated to cholesterol were shown to efficiently transfect myocardial cells while also reducing the cytotoxicity [22]. When the cholesterol–PEI molecules were complexed with DNA encoding for the vascular endothelial growth factor protein (VEGF), the lipopolymer complexes were shown to enter the cell via the cholesterol uptake pathway, confirming the benefits of the lipopolymer strategy. In another study, by replacing the ε-NH$_2$ groups of PEI with fatty acid tails of various lengths, the transfection of bone marrow stromal cells was increased by 20–25%, and was positively correlated with the extent of fatty acid substitution [23].

Lipopolymers are also used to incorporate targeting ligands into lipid vesicles and membranes. Antibodies, peptides, proteins, and aptamers conjugated to lipids can function as targeting moieties presented on the surfaces of vesicles, liposomes, or other particles [24, 25]. Cell-penetrating peptides such as oligoarginine can aid in the lipopolymer gaining access to the intracellular space [26].

9.3
Lipid–Polymer Hybrid Nanoparticles

Nanoparticles are used to deliver therapeutic agents to tissues, as imaging agents, and as biosensors and diagnostic tools [2, 27, 28]. Many advantages of using nanoparticles to deliver drugs have been recognized during the past decades. For example, nanoparticles can:

- improve the solubility of poorly water-soluble drugs;
- prolong the half-life of drugs in the systemic circulation by reducing immunogenicity;
- release drugs at a sustained rate or in an environmentally responsive manner and thus lower the frequency of administration;
- deliver drugs in a targeted manner to minimize systemic side effects; and
- deliver two or more drugs simultaneously for combination therapy to generate a synergistic effect and suppress drug resistance [1, 3, 29, 30].

As a result, a few pioneering nanoparticle-based therapeutic products have been introduced into the pharmaceutical market, while numerous ensuing products are currently undergoing clinical trials or are entering the pipeline. Polymeric nanoparticles and liposomes are two widely used drug-delivery platforms [1, 5].

Liposomes are lipid vesicles that form through the self-assembly of amphiphilic lipids into bilayer membranes. Liposomes are a versatile drug-delivery platform owing to their easy surface modification chemistry, low toxicity and, with the incorporation of hydrophilic polymers, a long circulation half-life. Doxil® (liposomal doxorubicin) was the first liposomal drug formulation to be approved by the US Food and Drug Administration (FDA) in 1995, for the treatment of AIDS-related Kaposi's sarcoma. Other liposomal drug formulations to receive FDA approval since then include DaunoXome® (daunorubicin liposomes), DepotDur® (morphine liposomes), Visudyne® (verteporfin liposomes), and Ambisome® (amphotericin B liposomes) [4]. Conjugating PEG molecules to proteins had previously been shown to increase the circulation half-life of the proteins, and the strategy was consequently applied to liposomes and other lipid vesicles. The resulting liposomes showed an increased circulation half-life in mice, from <30 min for bare liposomes to 5 h [15].

Polymeric nanoparticles are used to encapsulate drugs and deliver them in a controlled manner, providing an optimal dose over an extended period of time. This is especially useful when administering highly toxic, poorly water-soluble drugs, or drugs and biomolecules that are prone to degradation. Combinatorial drug therapy is also possible by encapsulating multiple therapeutic modalities into one nanoparticle system [2, 31]. Polymeric nanoparticles can form by self-assembly when copolymers consisting of blocks of polymers with different hydrophobicities are used. The polymers self-assemble, with the more hydrophobic blocks in the core and the hydrophilic blocks forming a shell, and interfacing with the aqueous solution. The use of amphiphilic polymers allows hydrophobic molecules to be encapsulated in the particle core, while the hydrophilic shell reduces protein adsorption and biofouling of the particle, which results in an extended systemic stability [8, 32].

The advantages of nanocarrier systems can be best utilized by combining lipids and polymers to broaden the toolkit available to nanoengineers. Lipid–polymer hybrid nanoparticles consisting of both lipids and polymers aim to capitalize on the advantages of both liposomal and polymeric nanoparticle systems, while reducing the intrinsic limitations of each. In general, there are two types of lipid–polymer hybrid nanoparticle: polymer-protected lipidic nanoparticles, and lipid-coated polymeric nanoparticles.

9.3.1
Synthesis and Fabrication of Lipid–Polymer Hybrid Nanoparticles

9.3.1.1 Synthesis of Polymer-Protected Lipidic Nanoparticles

The liposome was first described in 1965, when the British hematologist Dr Alec Bangham used an electron microscope to examine dry phospholipids and observed lipid vesicles with a bilayer conformation [7]. Liposomes generally exist as one of three varieties:

- Large multilamellar vesicles (LMV)
- Large unilamellar vesicles (LUV)
- Small unilamellar vesicles (SUV).

Multilamellar vesicles have many layers of lipids, while unilamellar vesicles have only a single bilayer membrane. Large vesicles (which may range from several hundred nanometers to several microns in diameter) are useful for creating artificial cell models and for studying membrane dynamics, while small vesicles (20–200 nm diameter) are more useful as nanocarrier delivery vehicles.

Among the various methods used to manufacture liposomes, most employ high shear forces to create unilamellar vesicles. Such techniques include thin-film hydration, sonication, extrusion, cyclical freeze–thaw, reverse-phase evaporation, and ether injection [33].

For thin-film hydration method, liposomes are made by drying a lipid/lipopolymer film and then rehydrating the film with an aqueous solution. Generally, a mixture of lipids and lipopolymers is dissolved in an organic solvent such as chloroform, which is then removed using either rotary evaporation or applying a stream of inert gas, such that only the lipid film remains. An aqueous solution is then added to the film and the lipids rehydrated to form vesicles. Large unilamellar vesicles can be produced by gently hydrating a thin lipid mixture film on a polytetrafluoroethylene (PTFE) substrate and collecting the "clouds" that form just above the substrate. In this case, a correct lipid mixture and concentration are critical to produce consistently large vesicles.

Sonication or extrusion through a polycarbonate membrane with pores characteristic of the desired size of the liposomes reduces multilamellar liposomes to unilamellar vesicles. The extrusion method also has the benefit of producing a much lower size polydispersity [33].

In the cyclic freeze–thaw method, hydrated lipids and lipopolymers are repeatedly frozen and heated above their phase-transition temperature, in order to break up the multilamellar vesicles and produce unilamellar vesicles.

Reverse-phase evaporation techniques for preparing liposomes involve dissolving a mixture of lipids and lipopolymers in an organic solvent that is immiscible with water. The solvent is typically less dense than water, such as diethyl ether or isopropyl ether. The organic phase is first added to an aqueous solution and then allowed to evaporate, usually under sonication or reduced pressure. As the solvent evaporates, the lipids form reverse micelles and then bilayer structures encapsulating the aqueous phase. Encapsulation efficiencies as high as 57% of the total aqueous phase have been reported [34].

Although liposomes first became useful tools for studying bilayer membranes and as artificial cell models, it was not until polymers were conjugated to lipids and incorporated into liposomes that the circulation half-life of liposomes became sufficiently long to allow for meaningful therapeutic and other *in vivo* uses [15]. In 1984, Pain *et al.* conjugated dextran to the surface of liposomes in order to increase their circulation half-life [12]. Gangliosides were also conjugated to liposomes and showed an increase in circulation time [35]. The term "stealth" was applied to liposomes and other nanocarriers that were able to avoid uptake by the RES and spleen. PEG showed an improvement over gangliosides and dextran in increasing the circulation half-life of liposomes [15], and soon became the standard lipopolymer complex used to impart "stealth" characteristics to lipid-based nanocarriers [36]. The mechanism by which PEG increases the circulation half-life of nanocarriers likely includes a decrease in the biofouling and protein adsorption to the lipid surface, as well as an increase in steric stability, preventing membrane fusion with other liposomes and cell membranes.

Proteins such as immunoglobulins (antibodies) and their fragments are commonly conjugated to liposomes, as tumor-targeting ligands and cell-penetrating ligands. These surface-attached ligands are able to bind to receptors on the surface of cells, which allows the liposomes to accumulate in the tissue of interest. Without the steric protection of a polymer coating such as PEG, most targeted liposome systems do not circulate for a sufficiently long period to allow accumulation in the targeted tissue [4].

The steric stability achieved by PEGylated liposomes is beneficial while en route to the tissue site of interest; however, it can also prevent uptake of the liposome by the cell, thus reducing the effectiveness of agents that act on intracellular components. One method developed to overcome this limitation has involved the use of pH-sensitive liposomes, which become fusogenic in mildly acidic environments. Poly(glycidol)-modified liposomes show an ability to fuse with endosomal or lysosomal membranes upon uptake through the endocytic pathway [37]. The contents of the poly(glycidol)-modified liposomes were observed to enter the cytosol of HeLa cells, indicating a fusion of the liposomal and cellular membranes. A cartoon of the poly(glycidol)-modified liposomes is shown in Figure 9.2; in this case, when the poly(glycidol) polymer was modified with 3-methyl-glutaryl, the efficiency of the fusion was increased. Other strategies to develop pH-sensitive liposomes have involved incorporating a cleavable PEG chain onto the surface of the liposome; such chains are cleaved free in response to environmental stimuli such as enzymes, pH changes, or hydrolysis [38].

Lipidic nanoparticles can also be manufactured without an aqueous interior. Solid lipid nanoparticles are typically prepared from lipids that are solid at room

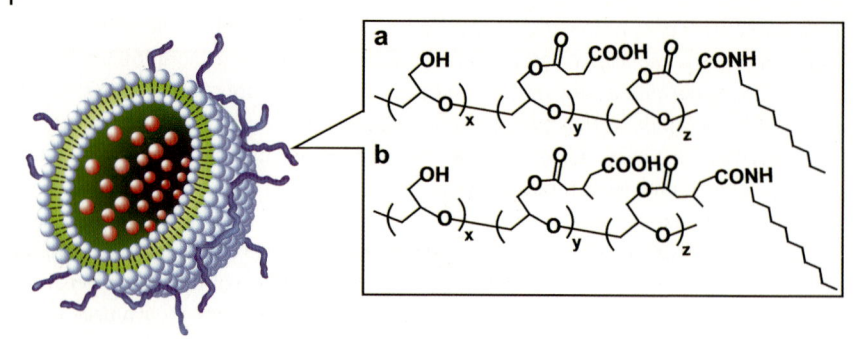

Figure 9.2 Schematic of a pH-sensitive liposome modified with poly(glycidol) polymers and the chemical structures of two different poly(glycidol) polymers. (a) Succinylated poly(glycidol); (b) 3-Methyl-glutarylated poly(glycidol) [37].

temperature, along with a surfactant or polar lipid shell to act as a stabilizing coating. The manufacturing techniques typically used are either hot or cold high-pressure homogenization:

- In *hot homogenization*, the lipids are melted and emulsified in an aqueous surfactant solution, and then homogenized by forcing the emulsification through a high-pressure nozzle. The droplets undergo very high shear stress, breaking them into smaller droplets; the homogenized emulsification is then allowed to cool and recrystallize.

- *Cold homogenization* involves the mechanical milling of crystallized lipids below (usually well below) the melting temperature of the lipids. The resulting microparticles are then dispersed in a surfactant solution and homogenized through a high-pressure nozzle at or below room temperature.

Common lipids used to form the solid core include triglycerides, hard fats, and waxes. Some common surfactants include lecithin, phospholipids, and poloxamers (triblock copolymers of poly(propylene oxide) and PEG) [39]. Functionalization of the surface of solid lipid nanoparticles is accomplished in much the same way as for liposomes. Lipid–PEG complexes can be used to stabilize solid lipid nanoparticles, as with many other nanoparticle systems [40].

Other specialized lipid-based nanoparticle systems include niosomes (nonionic surfactant vesicles) [41], ethesomes (soft, malleable vesicles made from phospholipids and ethanol) [42], and archaeosomes (vesicles made from polar lipids extracted from archaea membranes) [43]. These nanoparticle systems can likewise benefit from polymer stabilization with PEG or other hydrophilic polymers. They can also be targeted through the attachment of proteins, peptides, or other polymer-based targeting ligands. These systems will not be discussed at length at this point.

9.3.1.2 Synthesis of Lipid-Coated Polymeric Nanoparticles

Lipid-coated polymeric nanoparticles are here defined as nanoparticles consisting of a polymer core and a lipid shell, where the polymers and the lipids associate through hydrophobic interactions, van der Waals attractions, electrostatic interactions, or some other forces, but are not covalently bound to each other. An additional polymer layer, such as a hydrophilic PEG layer, may be used to stabilize the particles, and often is covalently bound to the lipid shell.

The earliest lipid-coated polymeric nanoparticles were made by fusing liposomes with polymeric nanoparticles. First, polymer core particles are prepared using one of several methods, including an emulsion or double-emulsion method, high-pressure homogenization, or nanoprecipitation. In the double-emulsion method, the polymer is dissolved in a water-immiscible solvent such as chloroform. If a water-soluble molecule is to be encapsulated, a primary emulsion is made of the aqueous solution containing the drug in the polymer solution; this is termed the water-in-oil emulsion (or simply the w/o emulsion). The w/o emulsion is then added to a second aqueous solution containing stabilizing surfactants such as poly(vinyl alcohol) (PVA). This mixture is also emulsified through sonication or other means, thus creating a secondary emulsion. This double emulsion is referred to as a water-in-oil-in-water emulsion (or a w/o/w emulsion).

The single-emulsion method can be used if the agent to be encapsulated is hydrophobic. In that case, the agent is simply added to the oil phase with the polymer and a single emulsion is created through sonication or otherwise. The organic solvent is then allowed to evaporate, leaving behind solidified polymeric nanoparticles.

High-pressure homogenization is used to produce polymeric nanoparticles in similar fashion as it is used to make solid lipid nanoparticles. The polymer is broken into nanoparticles through very high shear stress as it passes through a very narrow nozzle or gap at very high pressures.

The nanoprecipitation method relies on two miscible solvents; one is a good solvent for the polymer, and one is a poor polymer solvent. Typical solvent choices are acetonitrile and water, tetrahydrofuran and water, or dimethylformamide and water. The polymer is dissolved in the good solvent and subsequently added to the poor solvent. The good solvent then diffuses into the poor solvent, leaving the polymer to precipitate into nanoparticles. Depending on the protocol, the mixing of the solvents can occur during stirring, dropwise, or during sonication. All of these parameters, as well as polymer concentration and solvent volumes, may affect particle size and distribution.

After preparing the polymeric core nanoparticles, liposomes are also prepared using techniques previously described, such as thin-film hydration and sonication. Mixing the polymeric nanoparticles and the liposomes together with vigorous vortexing results in a fusion of the liposomes with the polymeric core [44]. As the main driving force of the fusion of the liposomes and the polymeric core particles is thought to be an electrostatic interaction between the polymer core and the liposomes [44], it is important to choose polymer and lipid systems that are

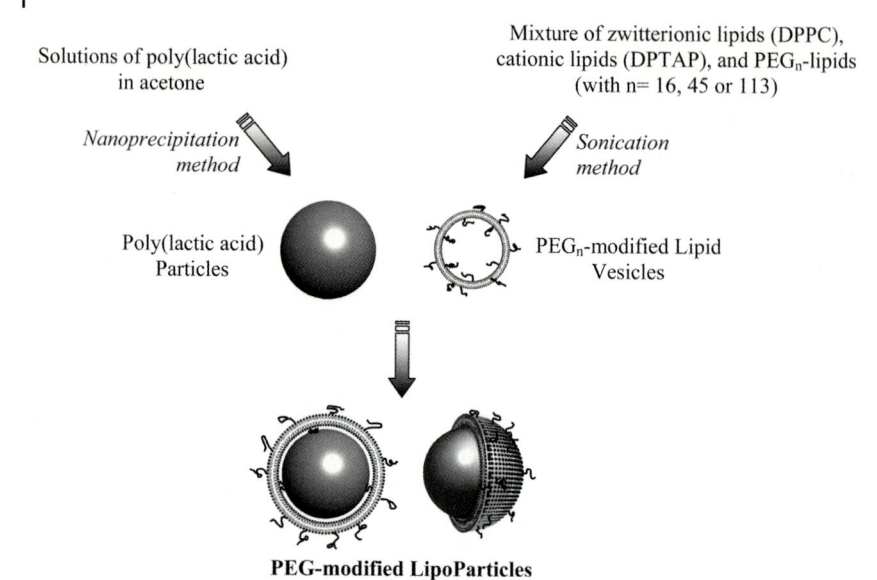

Figure 9.3 Schematic of the process of preparing lipid-coated polymeric nanoparticles, or lipoparticles [44].

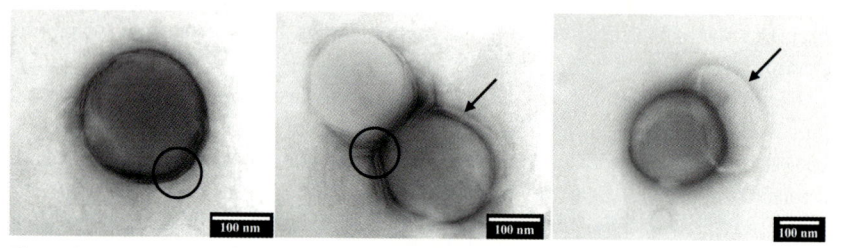

Figure 9.4 TEM images of lipoparticles (polymeric nanoparticles fused with liposomes). The circles highlight multilamellar lipids; the arrows point to lipid vesicles that did not fully fuse with the polymer core [44].

compatible with this method. One such compatible combination is carboxylated poly(lactic acid) and a mixture of dipalmitoyl phosphatidylcholine (DPPC) and dipalmitoyl trimethylammonium-propane (DPTAP). This combination results in an anionic polymer core and cationic liposomes; a schematic of the process of creating these hybrid lipoparticles is shown in Figure 9.3, while transmission electron microscopy (TEM) images of the lipid-coated polymeric nanoparticles are shown in Figure 9.4. When using this method, some vesicles do not completely fuse with the polymer core (as highlighted by the arrows in Figure 9.4). Multiple layers of lipids can also fuse with the polymer core, resulting in inconsistent lipid shells. During fusion of the liposome with the polymer particle, a small part of the aqueous interior of the liposome remains unexpelled, located between the polymer core and the lipid bilayer. Advantageously, this volume may contain dissolved hydrophilic molecules that cannot be encapsulated inside the hydrophobic polymer core or within the hydrophobic lipid bilayer membrane [44].

Another method used to form lipid–polymer hybrid nanoparticles employs self-assembly in order to create a more uniform lipid shell around the polymer core. In this case, phospholipids and phospholipid–PEG complexes are dissolved in an aqueous solution (a small amount of ethanol can be added to help solubilize the lipids in aqueous solution if they are not readily soluble in water). The polymer and any drugs to be encapsulated are dissolved in a water-miscible solvent, as in preparation for nanoprecipitation. The polymer should be hydrophobic in order to provide the driving force for lipid self-assembly; alternatively, oppositely charged polymer and lipids can be used to provide the driving force. The lipid solution is heated above the phase-transition temperature of the lipids, after which the polymer solution is added to the aqueous lipid solution. The polymer solvent then diffuses into the aqueous solution, allowing the polymer to precipitate into the nanoparticles. The lipids and lipopolymers self-assemble on the particle core with their hydrophobic tails sticking to the hydrophobic surface of the polymer and their hydrophilic headgroups extending into the aqueous solution. Thus, a monolayer of lipids is formed around the polymer core in a one-step self-assembly process [45]. The hydrophilic lipid–PEG lipopolymer forms a sterically stabilizing layer, preventing aggregation or biofouling and resulting in an increased stability, both *in vitro* and *in vivo*. In this procedure, poly(lactic acid) (PLA), poly(glycolic acid) (PGA), poly(lactic-*co*-glycolic acid) (PLGA), and poly(ε-caprolactone) (PCL) are the polymers most often used, because they are FDA approved for clinical use and are completely biodegradable. They are also hydrophobic, providing a good platform for the self-assembly process. A schematic of the constituent parts required for the nanoprecipitation self-assembly process is shown in Figure 9.5. The lipid monolayer or the outer PEG layer can be functionalized with targeting ligands, using the techniques previously described.

The benefits of using the self-assembly nanoprecipitation method include the ability to scale-up production easily, as it is a simple one-step process. Fewer processing steps can also lead to major cost savings, especially when all of the steps must be performed in a controlled environment in order to maintain sterility and avoid contamination, for example, for *in vivo* use.

9.3.2
Characterization of Lipid–Polymer Hybrid Nanoparticles

The characterization of lipid–polymer hybrid nanoparticles is important to understand their pharmacokinetics, pharmacodynamics, and general fate and efficacy *in vivo*. Size is an important characteristic of any nanoparticle system designed for clinical use. As shown in Figure 9.6, physiological barriers such as hepatic filtration, renal excretion and tissue extravasation must be considered when optimizing the size of a nanoparticle system [46]. Particles of <10 nm are prone to be cleared

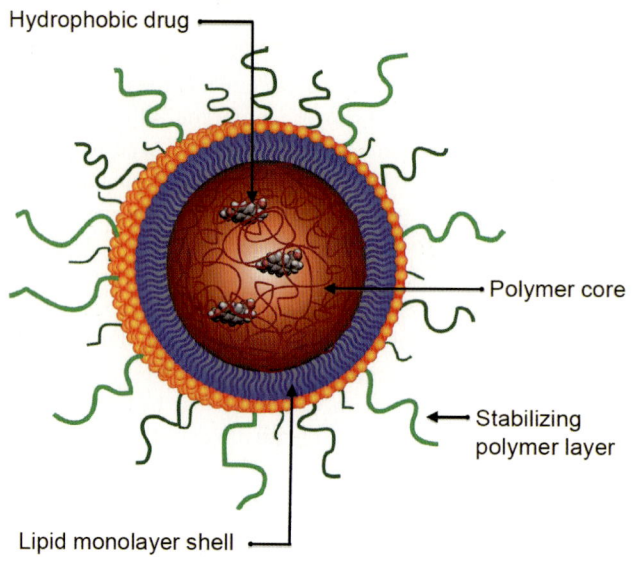

Figure 9.5 Through a one-step nanoprecipitation process, a lipid-coated polymeric nanoparticle forms with a hydrophobic polymer core encapsulating hydrophobic drugs, a lipid monolayer shell, and an outer stabilizing lipid–PEG layer.

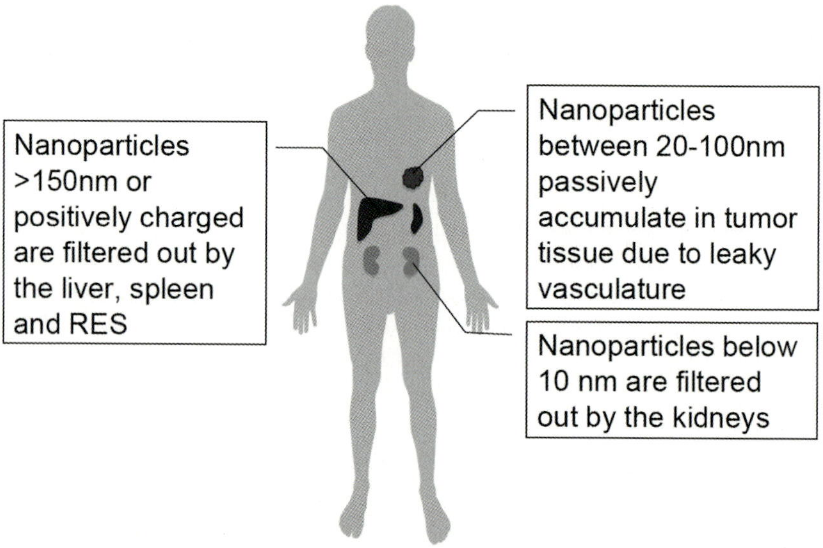

Figure 9.6 Nanoparticle size and charge are important characteristics. Nanoparticles smaller than 10 nm are quickly filtered out by the kidneys, while those larger than 150 nm are filtered out by the liver or spleen. The "sweet spot" for increased circulation half-life, as well as passive tumor targeting, lies between 20 and 100 nm. Positively charged particles are also quickly filtered out by the liver and reticuloendothelial system (RES).

by the kidneys [47], reducing the amount of time available for accumulation in the tissue of interest, while those >150 nm are mostly filtered out by the liver and spleen on the first pass [46]. Protein adsorption on the particle surface was also shown to have a significant positive correlation with particle size. Nanoparticles that were 80 nm in diameter resulted in a 6% serum protein adsorption (by weight), while nanoparticles of 171 nm and 243 nm, of the same formulation, resulted in 23% and 34% protein adsorption, respectively [46].

Nanoparticle size is often measured using dynamic light scattering (DLS), whereby a laser is shone through a solution in which the nanoparticles are suspended. The auto-correlation of the diffraction pattern of the laser is used to estimate the diffusion characteristics of the particles, and the size is calculated using the Stokes equation. DLS instruments can also be used to measure the polydispersity of a population of particles, providing more information than just the average size of the population. The zeta potential of the particles can also be measured using DLS, by applying an oscillating electric field and monitoring the movement of the particles as they are attracted and repulsed by the electric field. The zeta potential is a measure of the electrokinetic potential between the bulk fluid and the layer of ions strictly associated with the particle surface. Measuring the zeta potential of a particle system is important because it affects the blood residence time and uptake by the RES. Particles with primary amines on the surface, and therefore a net positive charge, were shown to be taken up more readily through phagocytosis [46]. In order to avoid hepatic and reticuloendothelial filtration, nanoparticles should have a neutral or slightly negative surface charge.

Because DLS monitors the rate of diffusion or movement through a fluid, the actual size being measured is the hydrodynamic size; this may differ from the true physical size, especially if the particle is not spherical. Hence, it is often beneficial to use electron microscopy to verify the particle size, shape, and morphology. Stains can be used to increase electron contrast in order to highlight and quantify specific particle components. For example, uranyl acetate can be used to increase the electron contrast of lipids and roughly quantify the amount of lipids associated with the polymer core when using TEM. A TEM image of lipid-coated polymeric nanoparticles synthesized using the self-assembly nanoprecipitation process is shown in Figure 9.7 [45]. The correlation between the lipid/polymer ratio and the size and charge of the nanoparticles is also shown in Figure 9.7; the size of the particles is inversely correlated with the amount of lipid used, presumably because of the reduced effect of the stabilizing PEG layer. The charge of the particles becomes more negative with increasing lipid concentration, due to the negative carboxyl groups on the PEG chains. The size and charge of the particles also show a correlation with the molecular weight or viscosity of the polymer used in the nanoprecipitation process.

Another method used to quantify the lipid concentration associated with hybrid lipid–polymer nanoparticles involves the use of phospholipase enzymes that degrade the phospholipids, yielding byproducts (e.g., hydrogen peroxide) that can be monitored directly through titration reactions [44].

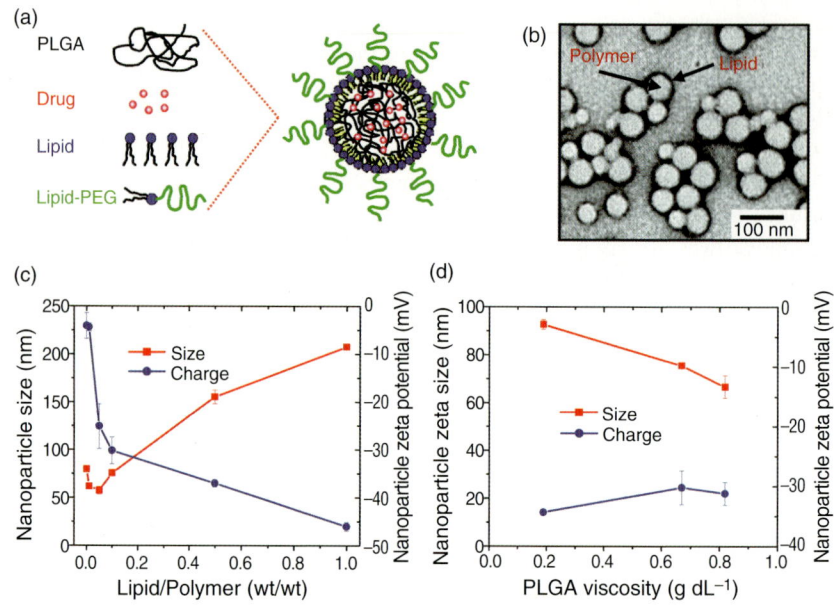

Figure 9.7 (a) Schematic of the self-assembled lipid-coated polymeric nanoparticles used in the studies represented by panels (b–d); (b) A TEM image of lipid-coated polymeric nanoparticles. The lipids were stained with uranyl acetate to increase electron contrast, and subsequently were visualized as a dark ring around a light polymer core; (c) The effects of varying the lipid/polymer weight ratio on particle size and zeta potential; (d) The effects of varying polymer molecular weight (indicated by polymer viscosity) on particle size and zeta potential [45].

For many diseases, including cancer, the efficacy of a drug treatment can be enhanced by maintaining the local drug concentration within the therapeutic window for an extended period of time. For this reason, the drug-release kinetics of lipid–polymer nanoparticles have undergone extensive investigation. One benefit that lipid–polymer nanoparticles have over bare polymeric nanoparticles is a slower drug release rate, due to the hydrophobic barrier that is presented to the encapsulated drug by the lipid shell. The lipid shell also reduces water penetration into the polymeric core, thus reducing polymer degradation and subsequent drug release. The drug encapsulation efficiencies and drug-release kinetics for lipid-coated polymeric nanoparticles with PLGA as the core, and with a shell of phospholipids and lipid–PEG lipopolymers, are shown in Figure 9.8. In this study, the drug encapsulation efficiency was increased from 19% for PLGA–PEG nanoparticles to 59% for PLGA–lipid–PEG particles, while the drug release half-life was increased from 10 h to 20 h, respectively [45]. These changes represent a significant increase in drug encapsulation and a significant slowing of drug release, the result being a nanoparticle system capable of an efficient and sustained delivery of a drug dose.

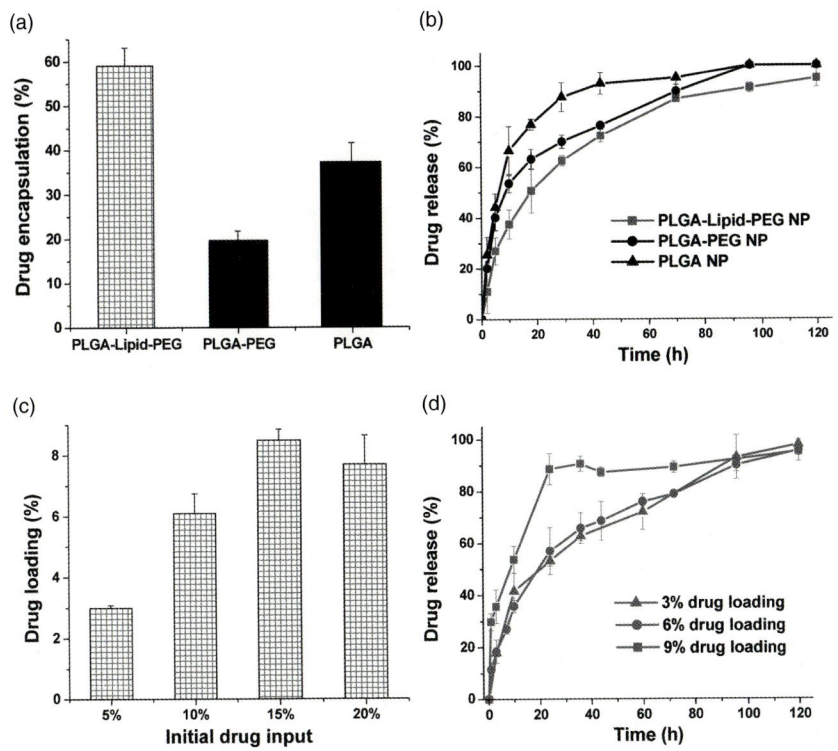

Figure 9.8 Loading and release profiles of nanoparticles (NP) loaded with docetaxel, a chemotherapeutic agent. (a) Drug encapsulation efficiency of lipid-coated PLGA nanoparticles compared to PLGA–PEG and bare PLGA nanoparticles; (b) Drug-release profiles for the same nanoparticles over 120 h; (c) Drug loading yield of lipid-coated PLGA nanoparticles with various initial drug inputs; (d) Drug-release profiles for lipid-coated PLGA nanoparticles loaded with varying amounts of drug.

9.3.3
Applications of Lipid–Polymer Hybrid Nanoparticles

Lipid–polymer hybrid nanoparticles show the most promise in improving the delivery of therapeutic, diagnostic, and imaging agents, selectively, to a tissue of interest. In order to accomplish this goal, nanoparticles must have a sufficiently long circulation half-life in order to allow accumulation in the tissue(s), and the targeting of the nanoparticles, whether passive or active, must be effective enough to prevent accumulation in unwanted areas. The drug encapsulation and release characteristics must also lie within a suitable range in order to maintain drug levels during the optimal therapeutic window.

The steric stabilization offered by hydrophilic polymer layers on the surface of lipidic and polymeric nanoparticles provides the nanoparticles with the opportunity to reach a targeted destination before delivering their payload. The use of hydrophilic polymers – and PEG in particular – has significantly increased the circulation half-life of lipid–polymer nanoparticles from minutes to hours, or even days [48, 49], allowing the targeting properties of lipid–polymer nanoparticles sufficient time to function. Targeted drug delivery has special significance in the battle against cancer, where highly toxic, systemically administered chemotherapeutic agents cause undesirable, and often severe, side effects. Thus, the targeted delivery of anti-cancer agents, if effective, should lead to a significant reduction in the severity of these side effects.

The tumor vasculature has also been shown to be leaky and irregularly formed, with large numbers of fenestrations not present in normal blood vessel walls; this results in a naturally occurring mechanism for passive targeting. Typically, a leaky vasculature would allows particles of between 50 and 150 nm to exit the blood system and accumulate in the extracellular space. This passive targeting effect has been exploited for the delivery of chemotherapeutics and imaging agents [36, 50, 51].

Lipid–polymer hybrid nanoparticles can be formulated to efficiently encapsulate and deliver a wide variety of compounds. For example, polymer-protected lipidic nanoparticles can encapsulate water-soluble drugs and biomolecules, while cationic liposomes have been used for gene therapy or small interfering RNA (siRNA) delivery in diseases including cancer [52], glucose metabolism disorders [53], cystic fibrosis [54], central nervous system injury [55], and even systemic gene delivery [56]. Cationic liposomes show an electrostatic attraction towards the negatively charged polynucleotides, so that the loading efficiencies of the liposomes is very high [57]. Cationic liposomes have also been investigated as drug-delivery vehicles for chemotherapeutics, based on the natural affinity of cationic vesicles for the tumor microvasculature [58]. Unfortunately, cationic liposomes have been shown to be cytotoxic, and their clinical use has not become widespread [59].

Lipid-coated polymeric nanoparticles have several advantages over traditional liposomes. For example, the polymeric core of lipid-coated polymeric nanoparticles allows hydrophobic drugs to be encapsulated efficiently, increasing their bioavailability following systemic administration. The polymeric core also adds mechanical stability to the particles, especially after procedures such as lyophilization, where the mechanical integrity of lipidic particles may be compromised and the effective encapsulation efficiency reduced.

Lipid–polymer hybrid nanoparticles offer significant advantages in the sustained release of encapsulated agents over administration of the free compound. Polymer-protected lipidic particles such as liposomes allow for the slow diffusion of encapsulated compounds across their bilayer membrane, increasing the therapeutic efficacy of many drugs. However, such particles may often suffer from "burst release," where the interior contents are released in a single burst upon rupture or fusion of the lipid membrane. Although the core of a lipid-coated polymeric nanoparticle allows the encapsulated drug to diffuse only slowly from the polymer matrix, eventually the core will swell as it is penetrated by water, allowing more of the drug to diffuse out. Finally, the polymer will undergo bulk erosion, releasing the remainder of the encapsulated material. Alternatively, the release rate may be modified by providing an environmental stimulus, such as a change in pH or temperature. The drug-release kinetics can also be tuned by selecting polymers with varying degrees of density, hydrophobicity, or degradation rate.

Lipid–polymer nanoparticles represent the ideal platform for drug delivery, based on their long circulation half-life, targeting abilities, encapsulation characteristics, and sustained-release kinetics. Lipid–polymer nanoparticles can also be used in a variety of other areas of research and manufacture, including cell mimetic systems [60], cosmetics [61], and the food sciences [62].

9.4
Lipid–Polymer Films and Coatings

Lipid–polymer nanomaterials can be used to fabricate supramolecular structures that approach the macro scale, but with thicknesses on the order of nanometers, while still retaining their unique design properties as nanomaterials. Whilst lipopolymers are widely used to fabricate films and coatings, their unique properties of self-assembly, easy modification through known chemistry, and their amphiphilic nature may offer benefits in research investigations, biomedical devices, sensors, and many other applications.

Lipid–polymer films have been used in investigations of lipid membrane dynamics, and also as a model system to study complex cellular membrane interactions [63]. The essential functions and properties of cellular membranes, including ion channel function, immune reactions, cell adhesion and migration, in addition to the functions of many membrane-associated proteins, have been investigated using lipopolymer films [64]. Lipid–polymer films are useful not only as a model membrane system, but also as superlubricious coatings on replacement joints, where they mimic the role of natural cartilage [20], and also as highly sensitive chemical odorant detectors [65] and nanofilters [66].

9.4.1
Synthesis of Lipid–Polymer Films and Coatings

During several decades, solid-supported lipid membranes have become a standard component in the study of cell membranes, mainly because they are easily fabricated by depositing phospholipids on a solid substrate and hydrating the space between the lipids and the substrate with distilled H_2O. However, as the distance separating the lipid from the substrate is usually between 0.5 and 2 nm, this may alter the conformation or function of membrane-associated proteins, and also limit the diffusion of molecules across the membrane. In order to overcome these limitations, polymer-cushioned and polymer-tethered membranes have been developed. A schematic of these different membrane support options is shown in Figure 9.9 [67].

The *polymer cushion* maintains a lubricated layer between the membrane and the solid substrate, and also separates the membrane from the substrate

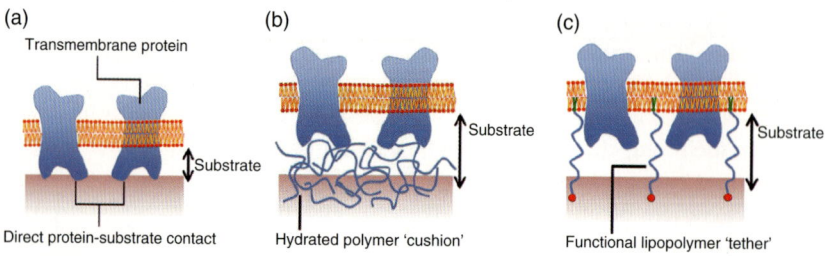

Figure 9.9 Schematic illustrations. (a) Solid-supported lipid membrane; (b) Polymer-cushioned lipid membrane; (c) Polymer-tethered lipid membrane [67].

sufficiently to avoid van der Waals interactions between the two. The typical thickness of the polymer cushion layer is between 5 and 100 nm. In particular, regenerated cellulose may serve as an excellent polymer to cushion a lipid membrane film, especially as it can be manufactured as films as thin as 5 nm and with tunable wetting properties [68].

Another lipid–polymer tool that has been used to separate lipid membranes from their solid substrate is the *lipid–polymer tether*. Lipid–PEG molecules can be incorporated into the bilayer membrane where the PEG chain acts as a tether, keeping the membrane at a consistent distance from the solid support [67]. Other tethers used in this way have included oligopeptides with thiol groups [69] and poly(2-oxazoline) [70]. By adjusting the tether length and lateral density, the lateral diffusivity of the lipid membrane can be altered, revealing how this may affect membrane protein function [67].

9.4.2
Characterization of Lipid–Polymer Films and Coatings

The characteristic properties of lipid–polymer films and coatings include their thickness, roughness, and dynamic properties, all of which can be characterized using specific techniques.

The thickness of the lipid polymer film can be quantified by using ellipsometry, in which a change in the polarization of light reflected from the thin film is correlated with the film's thickness. In this way, films thinner than the wavelength of the reflected light, even down to single atom layers, can be accurately measured using nondestructive ellipsometry [71, 72].

The roughness of the film can be measured using atomic force microscopy (AFM), especially *in situ* AFM. Indeed, the morphology of the lipid–polymer film can be accurately characterized, with true atomic resolution, in this way. The fact that *in situ* AFM allows the structures to be probed under ambient conditions, or even in a liquid, represents a major advantage for studying biologically relevant systems [73, 74].

The dynamic properties such as diffusion activities can be measured using a variety of fluorescence techniques, including florescence recovery after photobleaching (FRAP) [75, 76] and fluorescence correlation spectroscopy (FCS) [77, 78]. In FRAP, a small section of a fluorescently labeled film is photobleached; the lateral diffusion is then measured by tracking the spread of the photobleached area over time.

9.4.3
Applications of Lipid–Polymer Films and Coatings

Lipid–polymer films and coatings have been used in a variety of life sciences applications, including model membrane platforms, chemical sensors, and bioimplant surface coatings.

Both tethered and cushioned lipid membranes have been used extensively to study cell membrane functions and interactions with intracellular and extracellular networks. Native cell membranes can be spread onto polymer cushions in order to probe the membrane-associated proteins in their native conformation, and at native concentrations [67]. Lipid polymers have been used to study lipid membrane microdomains, such as *lipid rafts* [63]; these are sphingolipid- and cholesterol-rich regions of cellular membranes that are important in many cellular processes, including signal transduction, apoptosis, cell migration, and protein sorting [79]. Lipid rafts have also been suggested as the locations of entry for bacteria, viruses, and toxins. A tethered lipid–polymer film substrate was developed to study lipid raft dynamics by patterning the substrate with nanoarrays of lipid–polymers inside nanowells, and then depositing the bilayer membrane containing lipid raft components, so as to allow an easy probing of the lipid rafts. Tethered lipopolymer membranes were also shown to be useful as nanoscale filters for the purification of proteins [66]. The control of lateral diffusivity, by altering the length and density of the polymer tethers supporting the bilayer membrane, allowed proteins to aggregate selectively in membrane microdomains.

Lipid–polymer films were also developed to function as a chemical sensor. For this, disulfide lipids and lipid–PEG molecules were chemisorbed onto electrodes of a quartz crystal microbalance, creating a support on which further layers were physisorbed. Chemical odorants in the air that come into contact with the lipid–polymer films can be analyzed and recognized by the functional groups present on the odorants. The lipid–polymer films enhance the sensitivity and selectivity of the quartz crystal microbalance odorant sensor [65].

Deterioration of the articulating surface of artificial joints can lead to a requirement for joint replacement surgery, especially if the recipient is young and active. Lipid–polymer coatings were shown to create a superlubricious surface on an artificial joint metal substrate, decreasing the frictional forces responsible for surface wear. The deposition of a 2-methacryloyloxyethyl phosphorylcholine (MPC) polymer onto the surface of a cobalt–chromium–molybdenum (Co–Cr–Mo) alloy produced a highly hydrophilic surface that mimicked the superlubricious properties of articular cartilage.

Lipid–polymer films and coatings have a wide variety of applications, especially in areas where self-assembly is needed, complex biological membrane interactions are being explored, or where the unique properties of lipopolymers can offer benefits.

9.5
Summary and Future Perspective

In summary, lipid–polymer nanomaterials represent a new class of functional material that consists of both lipids and polymers in a rational design. In recent years, these materials have become an increasingly popular area of research, and have also demonstrated promising applications in the life sciences, medicine, and many other areas of scientific advancement. Lipid–polymer conjugates (lipopolymers) are essential building blocks in the world of nanomaterials and nanotechnologies, as they may be applied to numerous systems ranging from unique drug-delivery vehicles to the creation of artificial biomembranes for the study of cell membrane functions. Lipid–polymer hybrid nanoparticles have also attracted much attention as a robust drug-delivery platform, showing great promise for curing diseases and improving the health of individuals. The major advantages of lipid–polymer hybrid nanoparticles include their tunable size and surface charge, high drug loading yields, sustained and controllable drug-release kinetics, long circulation lifetimes, negligible cytotoxicity, and their ability to target the delivery of drugs. Lipid–polymer films and coatings represent another application of lipid–polymer nanomaterials, in that they provide an excellent platform for mimicking biomembranes and creating effective anti-fouling surfaces.

In looking to the future, as both knowledge of the complexity of biomolecular interactions and the ability to engineer nanoscale systems increase, so too will the ability to harness the molecular tools of lipids and polymers in ways that will open new doors to science, medicine, and technology. At this point, special attention should be drawn to some key aspects in the development of lipid–polymer nanomaterials for the life sciences and medical applications:

- A systematic testing of the stability and sterility of lipid–polymer hybrid nanomaterials will determine the feasibility and safety of using new nanostructures that consist of both lipids and polymers, in the clinical situation.

- An evaluation of the physiological behavior and therapeutic efficacy of lipid–polymer hybrid nanomaterials in true animal models. Many new nanostructures consisting of both lipids and polymers have been synthesized and shown promising profiles *in vitro*. However, the testing of these nanostructures in animal models is essential, in order to evaluate their possible use in clinical practice.

- An exploration of the systematic approaches to scaling-up lipid–polymer hybrid nanomaterials for high-level production. Recently, the large-scale production of therapeutic nanostructures has become rather a "bottleneck" when translating these nanoscale systems for practical use. Indeed, some nanostructures with potent functionalities have proven to be non-scalable due to their structural complexity or systemic toxicity, which has rendered them less useful. Nonetheless, others can – at least potentially – be produced in large quantities, although their physico-chemical properties may be changed during large-scale fabrication processes.

References

1 Zhang, L., Gu, F.X., Chan, J.M., Wang, A.Z., Langer, R.S. and Farokhzad, O.C. (2007) Nanoparticles in medicine: therapeutic applications and developments. *Clinical Pharmacology and Therapeutics*, **83**, 761–9.

2 Farokhzad, O.C. and Langer, R. (2006) Nanomedicine: developing smarter therapeutic and diagnostic modalities. *Advanced Drug Delivery Reviews*, **58**, 1456–9.

3 Davis, M.E., Chen, Z.G. and Shin, D.M. (2008) Nanoparticle therapeutics: an emerging treatment modality for cancer. *Nature Reviews Drug Discovery*, **7**, 771–82.

4 Torchilin, V.P. (2005) Recent advances with liposomes as pharmaceutical carriers. *Nature Reviews Drug Discovery*, **4**, 145–60.

5 Wagner, V., Dullaart, A., Bock, A.-K. and Zweck, A. (2006) The emerging nanomedicine landscape. *Nature Biotechnology*, **24**, 1211–17.

6 Fahy, E., Subramaniam, S., Brown, H.A., Glass, C.K., Merrill, A.H., Jr, Murphy, R.C. *et al.* (2005) A comprehensive classification system for lipids. *Journal of Lipid Research*, **46**, 839–62.

7 Bangham, A.D., Standish, M.M. and Watkins, J.C. (1965) Diffusion of univalent ions across lamellae of swollen phospholipids. *Journal of Molecular Biology*, **13**, 238.

8 Tong, R. and Cheng, J.J. (2007) Anticancer polymeric nanomedicines. *Polymer Reviews*, **47**, 345–81.

9 Bendas, G., Krause, A., Bakowsky, U., Vogel, J. and Rothe, U. (1999) Targetability of novel immunoliposomes prepared by a new antibody conjugation technique. *International Journal of Pharmaceutics*, **181**, 79–93.

10 Green, N.M. (1963) Avidin. 1. The use of [14C] biotin for kinetic studies and for assay. *Biochemical Journal*, **89**, 585–91.

11 Vladimir, P.T., Vladimir, S.T., Kathleen, R.W., Paolo, C., Paolo, F. and Francesco, M.V. (1995) New synthetic amphiphilic polymers for steric protection of liposomes in vivo. *Journal of Pharmaceutical Sciences*, **84**, 1049–53.

12 Pain, D., Das, P., Ghosh, P. and Bachhawat, B. (1984) Increased circulatory half-life of liposomes after conjunction with dextran. *Journal of Biosciences*, **6**, 811–16.

13 Hayward, J.A. and Chapman, D. (1984) Biomembrane surfaces as models for polymer design – the potential for hemocompatibility. *Biomaterials*, **5**, 135–42.

14 Patrick, T., Rudolf, Z. and Simona, S. (2002) Synthesis of end-functionalized lipopolymers and their characterization with regard to polymer-supported lipid membranes. *Macromolecular Bioscience*, **2**, 387–94.

15 Klibanov, A.L., Maruyama, K., Torchilin, V.P. and Huang, L. (1990) Amphipathic polyethyleneglycols effectively prolong the circulation time of liposomes. *FEBS Letters*, **268**, 235–7.

16 Feng, W., Nieh, M.P., Zhu, S., Harroun, T.A., Katsaras, J. and Brash, J.L. (2007) Characterization of protein resistant, grafted methacrylate polymer layers bearing oligo(ethylene glycol) and phosphorylcholine side chains by neutron reflectometry. *Biointerphases*, **2**, 34–43.

17 Ahmad, H., Dupin, D., Armes, S.P. and Lewis, A.L. (2009) Synthesis of biocompatible sterically-stabilized poly(2-(methacryloyloxy) ethyl phosphorylcholine) latexes via dispersion

polymerization in alcohol/water mixtures. *Langmuir*, **25**, 11442–9.

18 Ishihara, K., Aragaki, R., Ueda, T., Watenabe, A. and Nakabayashi, N. (1990) Reduced thrombogenicity of polymers having phospholipid polar groups. *Journal of Biomedical Materials Research*, **24**, 1069–77.

19 Ho, S.P., Nakabayashi, N., Iwasaki, Y., Boland, T. and LaBerge, M. (2003) Frictional properties of poly(MPC-co-BMA) phospholipid polymer for catheter applications. *Biomaterials*, **24**, 5121–9.

20 Kyomoto, M., Moro, T., Iwasaki, Y., Miyaji, F., Kawaguchi, H., Takatori, Y. *et al.* (2009) Superlubricious surface mimicking articular cartilage by grafting poly(2-methacryloyloxyethyl phosphorylcholine) on orthopaedic metal bearings. *Journal of Biomedical Materials Research. Part A: Early View*, **91A**, 730–41.

21 Lewis, A.L., Willis, S.L., Small, S.A., Hunt, S.R., O'Byrne, V. and Stratford, P.W. (2004) Drug loading and elution from a phosphorylcholine polymer-coated coronary stent does not affect long-term stability of the coating in vivo. *Bio-Medical Materials and Engineering*, **14**, 355–70.

22 Lee, M., Rentz, J., Han, S.O., Bull, D.A. and Kim, S.W. (2003) Water-soluble lipopolymer as an efficient carrier for gene delivery to myocardium. *Gene Therapy*, **10**, 585–93.

23 Incani, V., Lin, X.Y., Lavasanifar, A. and Uludag, H. (2009) Relationship between the extent of lipid substitution on poly(L-lysine) and the DNA delivery efficiency. *ACS Applied Materials and Interfaces*, **1**, 841–8.

24 Backer, M.V. and Backer, J.M. (2010) A "dock and lock" approach to preparation of targeted liposomes. *Methods in Molecular Biology*, **605**, 257–66.

25 Manuela, V. and Maya, S. (2002) Designing of "intelligent" liposomes for efficient delivery of drugs. *Journal of Cellular and Molecular Medicine*, **6**, 465–74.

26 Maitani, Y. and Hattori, Y. (2009) Oligoarginine-PEG-lipid particles for gene delivery. *Expert Opinion on Drug Delivery*, **6**, 1065–77.

27 Ferrari, M. (2007) Cancer nanotechnology: opportunities and challenges. *Nature Nanotechnology*, **2**, 37–47.

28 Langer, R. (1998) Drug delivery and targeting. *Nature*, **392**, 5–10.

29 Gu, F.X., Karnik, R., Wang, A.Z., Alexis, F., Levy-Nissenbaum, E., Hong, S. *et al.* (2007) Targeted nanoparticles for cancer therapy. *Nano Today*, **2**, 14–21.

30 Peer, D., Karp, J.M., Hong, S., Farokhzad, O.C., Margalit, R. and Langer, R. (2007) Nanocarriers as an emerging platform for cancer therapy. *Nature Nanotechnology*, **2**, 751–60.

31 Farokhzad, O.C., Cheng, J., Teply, B.A., Sherifi, I., Jon, S., Kantoff, P.W. *et al.* (2006) Targeted nanoparticle-aptamer bioconjugates for cancer chemotherapy in vivo. *Proceedings of the National Academy of Sciences of the United States of America*, **103**, 6315–20.

32 Cheng, J., Teply, B.A., Sherifi, I., Sung, J., Luther, G., Gu, F.X. *et al.* (2007) Formulation of functionalized PLGA-PEG nanoparticles for in vivo targeted drug delivery. *Biomaterials*, **28**, 869–76.

33 Mozafari, M.R. and Khosravi-Darani, K. (2007) *An Overview of Liposome-Derived Nanocarrier Technologies*, Springer, pp. 113–23.

34 Szoka, F. and Papahadjopoulos, D. (1978) Procedure for preparation of liposomes with large internal aqueous space and high capture by reverse phase evaporation. *Proceedings of the National Academy of Sciences of the United States of America*, **75**, 4194–8.

35 Allen, T.M., Ryan, J.L. and Papahadjopoulos, D. (1985) Gangliosides reduce leakage of aqueous-space markers from liposomes in the presence of human-plasma. *Biochimica et Biophysica Acta*, **818**, 205–10.

36 Mozafari, M.R., Pardakhty, A., Azarmi, S., Jazayeri, J.A., Nokhodchi, A. and Omri, A. (2009) Role of nanocarrier systems in cancer nanotherapy. *Journal of Liposome Research*, **19**, 310–21.

37 Yuba, E., Kojima, C., Harada, A., Tana, Watarai, S. and Kono, K. (2009) pH-Sensitive fusogenic polymer-modified liposomes as a carrier of antigenic proteins for activation of cellular immunity. *Biomaterials*, **31**, 943–51.

38 Xu, H., Deng, Y.-H. and Chen, D.-W. (2008) Recent advances in the study of cleavable PEG-lipid derivatives modifying

liposomes. *Yaoxue Xuebao*, **43**, 18–22.
39. Schwarz, C. (1999) Solid lipid nanoparticles (SLN) for controlled drug delivery II. drug incorporation and physicochemical characterization. *Journal of Microencapsulation*, **16**, 205–13.
40. Rai, S., Paliwal, R., Gupta, P.N., Khatri, K., Goyal, A.K., Vaidya, B. et al. (2008) Solid lipid nanoparticles (SLNs) as a rising tool in drug delivery science: one step up in nanotechnology. *Current Nanoscience*, **4**, 30–44.
41. Uchegbu, I.F. and Vyas, S.P. (1998) Non-ionic surfactant based vesicles (niosomes) in drug delivery. *International Journal of Pharmaceutics*, **172**, 33–70.
42. Paolino, D., Lucania, G., Mardente, D., Alhaique, F. and Fresta, M. (2005) Ethosomes for skin delivery of ammonium glycyrrhizinate: in vitro percutaneous permeation through human skin and in vivo anti-inflammatory activity on human volunteers. *Journal of Controlled Release*, **106**, 99–110.
43. Benvegnu, T., Lemiegre, L. and Cammas-Marion, S. (2009) New generation of liposomes called archaeosomes based on natural or synthetic archaeal lipids as innovative formulations for drug delivery. *Recent Patents on Drug Delivery and Formulation*, **3**, 206–20.
44. Thevenot, J., Troutier, A.-L., David, L., Delair, T. and Ladavière, C. (2007) Steric stabilization of lipid/polymer particle assemblies by poly(ethylene glycol)-lipids. *Biomacromolecules*, **8**, 3651–60.
45. Zhang, L., Chan, J.M., Gu, F.X., Rhee, J.-W., Wang, A.Z., Radovic-Moreno, A.F. et al. (2008) Self-assembled lipid-polymer hybrid nanoparticles: a robust drug delivery platform. *ACS Nano*, **2**, 1696–702.
46. Alexis, F., Pridgen, E., Molnar, L.K. and Farokhzad, O.C. (2008) Factors affecting the clearance and biodistribution of polymeric nanoparticles. *Molecular Pharmacology*, **5**, 505–15.
47. Choi, H.S., Liu, W., Liu, F., Nasr, K., Misra, P., Bawendi, M.G. et al. (2010) Design considerations for tumour-targeted nanoparticles. *Nature Nanotechnology*, **5**, 42–7.
48. Moghimi, S.M. and Szebeni, J. (2003) Stealth liposomes and long circulating nanoparticles: critical issues in pharmacokinetics, opsonization and protein-binding properties. *Progress in Lipid Research*, **42**, 463–78.
49. Whiteman, K.R., Subr, V., Ulbrich, K. and Torchilin, V.P. (2001) Poly(HPMA)-coated liposomes demonstrate prolonged circulation in mice. *Journal of Liposome Research*, **11**, 153–64.
50. Parveen, S. and Sahoo, S.K. (2008) Polymeric nanoparticles for cancer therapy. *Journal of Drug Targeting*, **16**, 108–23.
51. Stevanovic, M. and Uskokovic, D. (2009) Poly(lactide-co-glycolide)-based micro and nanoparticles for the controlled drug delivery of vitamins. *Current Nanoscience*, **5**, 1–14.
52. Schatzlein, A.G. (2001) Non-viral vectors in cancer gene therapy: principles and progress. *Anti-Cancer Drugs*, **12**, 275–304.
53. Giannoukakis, N. and Trucco, M. (2003) Gene therapy technology applied to disorders of glucose metabolism: promise, achievements, and prospects. *Biotechniques*, **35**, 122–45.
54. Lee, T.W.R., Matthews, D.A. and Blair, G.E. (2005) Novel molecular approaches to cystic fibrosis gene therapy. *Biochemical Journal*, **387**, 1–15.
55. Yang, K., Clifton, G.L. and Hayes, R.L. (1997) Gene therapy for central nervous system injury: the use of cationic liposomes: an invited review. *Journal of Neurotrauma*, **14**, 281–97.
56. Liu, F. and Huang, L. (2002) Development of non-viral vectors for systemic gene delivery. *Journal of Controlled Release*, **78**, 259–66.
57. De Rosa, G. and La Rotonda, M.I. (2009) Nano and microtechnologies for the delivery of oligonucleotides with gene silencing properties. *Molecules*, **14**, 2801–23.
58. Campbell, R.B., Ying, B., Kuesters, G.M. and Hemphill, R. (2009) Fighting cancer: from the bench to bedside using second generation cationic liposomal therapeutics. *Journal of Pharmaceutical Sciences*, **98**, 411–29.
59. Lv, H.T., Zhang, S.B., Wang, B., Cui, S.H. and Yan, J. (2006) Toxicity of

cationic lipids and cationic polymers in gene delivery. *Journal of Controlled Release*, **114**, 100–9.

60 Antunes, F.E., Marques, E.F., Miguel, M.G. and Lindman, B. (2009) Polymer-vesicle association. *Advances in Colloid and Interface Science*, **147–148**, 18–35.

61 Choi, M.J. and Maibach, H.I. (2005) Elastic vesicles as topical/transdermal drug delivery systems. *International Journal of Cosmetic Science*, **27**, 211–21.

62 Taylor, T.M., Davidson, P.M., Bruce, B.D. and Weiss, J. (2005) Liposomal nanocapsules in food science and agriculture. *Critical Reviews in Food Science and Nutrition*, **45**, 587–605.

63 Lee, B.K., Lee, H.Y., Kim, P., Suh, K.Y. and Kawai, T. (2009) Nanoarrays of tethered lipid bilayer rafts on poly(vinyl alcohol) hydrogels. *Lab on a Chip*, **9**, 132–9.

64 Sackmann, E. and Tanaka, M. (2000) Supported membranes on soft polymer cushions: fabrication, characterization and applications. *Trends in Biotechnology*, **18**, 58–64.

65 Wyszynski, B., Somboon, P. and Nakamoto, T. (2008) Mixed self-assembled lipopolymers with spacer lipids enhancing sensitivity of lipid-derivative QCMs for odor sensors. *Sensors and Actuators B*, **134**, 72–8.

66 Albertorio, F., Daniel, S. and Cremer, P.S. (2006) Supported lipopolymer membranes as nanoscale filters: simultaneous protein recognition and size-selection assays. *Journal of the American Chemical Society*, **128**, 7168–9.

67 Tanaka, M. and Sackmann, E. (2005) Polymer-supported membranes as models of the cell surface. *Nature*, **437**, 656–63.

68 Schaub, M., Wenz, G., Wegner, G., Stein, A. and Klemm, D. (1993) Ultrathin films of cellulose on silicon-wafers. *Advanced Materials*, **5**, 919–22.

69 Bunjes, N., Schmidt, E.K., Jonczyk, A., Rippmann, F., Beyer, D., Ringsdorf, H. *et al.* (1997) Thiopeptide-supported lipid layers on solid substrates. *Langmuir*, **13**, 6188–94.

70 Purrucker, O., Fortig, A., Ludtke, K., Jordan, R. and Tanaka, M. (2005) Confinement of transmembrane cell receptors in tunable stripe micropatterns. *Journal of the American Chemical Society*, **127**, 1258–64.

71 Porter, M.D., Bright, T.B., Allara, D.L. and Chidsey, C.E.D. (1987) Spontaneously organized molecular assemblies .4. Structural characterization of normal-alkyl thiol monolayers on gold by optical ellipsometry, infrared-spectroscopy, and electrochemistry. *Journal of the American Chemical Society*, **109**, 3559–68.

72 Sukhishvili, S.A. and Granick, S. (2000) Layered, erasable, ultrathin polymer films. *Journal of the American Chemical Society*, **122**, 9550–1.

73 Xie, A.F., Yamada, R., Gewirth, A.A. and Granick, S. (2002) Materials science of the gel to fluid phase transition in a supported phospholipid bilayer. *Physical Review Letters*, **89**, 246103.

74 Zhang, L., Spurlin, T.A., Gewirth, A.A. and Granick, S. (2006) Electrostatic stitching in gel-phase supported phospholipid bilayers. *Journal of Physical Chemistry B*, **110**, 33–5.

75 Pucadyil, T.J., Mukherjee, S. and Chattopadhyay, A. (2007) Organization and dynamics of NBD-labeled lipids in membranes analyzed by fluorescence recovery after photobleaching. *Journal of Physical Chemistry B*, **111**, 1975–83.

76 Ratto, T.V. and Longo, M.L. (2002) Obstructed diffusion in phase-separated supported lipid bilayers: a combined atomic force microscopy and fluorescence recovery after photobleaching approach. *Biophysical Journal*, **83**, 3380–92.

77 Zhang, L. and Granick, S. (2005) Lipid diffusion compared in outer and inner leaflets of planar supported bilayers. *Journal of Chemical Physics*, **123**, 211104.

78 Zhang, L. and Granick, S. (2005) Slaved diffusion in phospholipid bilayers. *Proceedings of the National Academy of Sciences of the United States of America*, **102**, 9118–21.

79 Simons, K. and Toomre, D. (2000) Lipid rafts and signal transduction. *Nature Reviews Molecular Cell Biology*, **1**, 31–9.

10
Core–Shell Polymeric Nanomaterials and Their Biomedical Applications

Ziyad S. Haidar and Maryam Tabrizian

10.1
Introduction

The development of revolutionary materials of biological origin for use in the diagnosis and treatment of devastating human diseases has become a vital area of research during the past few decades [1, 2], with attention focused on novel biomaterials produced at the nanometer scale [3]. These materials offer impressive solutions when applied to medical challenges such as infectious diseases, diabetes, cancer, and cardiovascular or skeletal problems. By considering the complex design of the human body's natural systems, and taking advantage of their special physical characteristics, certain nanomaterials can stimulate, respond to, and interact with target cells and tissues in *controlled* ways, to induce desired physiological responses but with a minimum of undesirable effects [4, 5]. Among the plethora of size-dependent properties of nanomaterials, the optical [6] and magnetic [7] effects continue to be the most often applied for biological purposes, with significant advances having been made in the fields of drug delivery, gene therapy, novel drug synthesis, bioimaging and the detection of cancerous cells [1, 2]. Recent progress, especially in the case of polymer-, polyelectrolyte- or polysaccharide-based nanostructures (nanoparticles, nanocapsules and nanofibers), has promised much for the controlled delivery of therapeutic agents and diagnostics. The aim of this chapter is to describe recent advances in the use of polymeric-based core–shell nanoparticulate systems for key biomedical applications, such as bioimaging, tissue engineering, controlled drug delivery, and oncology. However, as the number of reports and reviews related to the theory and practice of nanoparticles increases on a daily basis, the reader is advised to examine the original data for any in-depth discussions of specific polymer-based core–shell nanostructures not included here.

10.2
Core–Shell Nanomaterials of Biomedical Interest

In general, nanostructures used for nanomedicine may be either: (i) one-dimensional (1-D) thin-surface films and coatings; (ii) two-dimensional (2-D) mono- or multi-layered carbon nanotubes (CNTs) and colloidal assemblies; or (iii) three-dimensional (3-D) semiconductor spherical nanocrystals (1–10 nm, quantum dots; QDs), dendrimers and micelles (1–100 nm), organic spherical cages (C_{60} fullerenes), nanoshells, and/or core–shell nanoparticles, depending on the method of synthesis, material composition, and intended purpose/mode of use [4, 8–11] (see Figure 10.1). Given the wide array of currently available and possible material combinations, Sounderya and Zhang [12] have, for reasons of simplicity, recently classified core–shell nanoparticles based broadly on the materials from which the core and shell are constructed. The important point to consider is that the need to shift from traditional bulk biomaterials, microparticles and nanopolymers to core–shell structured nanoparticles is, in essence, to improve the overall optical, magnetic, and/or agent delivery properties, to enhance particle morphology and monodispersity, to prevent agglomeration, to reduce toxicity and, as a result, to attempt to spy, orchestrate, and mimic the complex biological scenarios involved [9]. This enhanced surface reactivity can be further modified by coating the nanoparticles with various materials to achieve a better stability and biocompatibility [11, 13]. In addition, by dictating the type and number of layers that surround the core, *multifunctional* core–shell nanostructures can be formulated where the magnetic and luminescent layers can be combined to detect and manipulate the particles *in situ*. Thus, core–shell-designed nanoparticles, specifically, are considered to be advanced nanostructures that offer distinct and superior advantages compared to traditional nanomaterials, providing significance and need for their use [11–13].

One common formula for the construction of core–shell nanostructures has been to use silica for the core, and a second "sticky" compound to adhere gold particles, usually to the outside surface, to create the shell. An example of the use

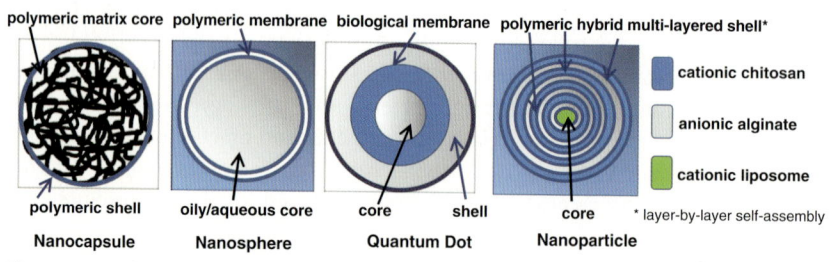

Figure 10.1 Schematic representation of the most commonly investigated polymeric core–shell nanostructures, ranging from nanocapsules to layer-by-layer self-assembled multilayered nanoparticles.

of nanoshells of this type was in the killing of cancer cells in mice [12, 14]; the injection of such nanoshells into a tumor, followed by exposure to radiation, led to the nanoshells being heated to a sufficient level to kill the tumor cells [14]. In contrast, a traditional nanosized drug carrier may be used to solubilize poorly water-soluble anticancer drugs, not only prolonging their circulation time in the blood but also reducing or diminishing any adverse side effects. When further functionalized or coated, core–shell nanoparticles are capable of performing a targeted and controlled drug delivery [11], and also to serve as superior biomarkers for tumors and cancer cells, based on their enhanced properties of permeability and retention. The design, therefore, will not only enhance the nanoparticles' thermal and chemical stability but also improve their solubility and cytocompatibility, allowing the conjugation of other bioactive molecules [15]. Likewise, semiconductor cores composed of QDs can be surrounded with shells made from semiconductor materials, metal oxides or polymeric materials, such as silica, for applications in biology and medicine as fluorescent bioimaging agents [16]. Today, QDs are most commonly conjugated to proteins, nucleic acids or other biologically relevant molecules to serve as molecular *targeting* devices in diagnosis and molecular detection, profiling, and signal transduction studies [17]. Such QDs have also been used to map the lymphatic system and sentinel lymph nodes, which are important structures in cancer therapy [18, 19]. The results of other studies have revealed that PEGylated QDs [poly(ethylene glycol) (PEG) is an FDA-approved biocompatible and biodegradable copolymer] are capable of escaping the reticuloendothelial system (RES), emphasizing the favorable effect of the nanoshell and providing a unique opportunity for their extensive use in animal models [20]. Hence, the most promising core–shell nanostructures are those which have a polymeric core and/or a polymeric shell, and can be dispersed in a matrix of any class of material, the property of which is to be modified or enhanced; these are termed *polymeric-based core–shell nanostructures*. For example, polymeric-based micellar systems that contain a hydrophobic core surrounded by hydrophilic polymers and typically are assembled from amphiphilic diblock or triblock copolymers, have been widely investigated as carriers for hydrophobic drugs [21]. Such materials offer a relatively simple preparation, an efficient drug loading, and controlled release kinetics, and as a result several formulations of chemotherapeutic agents, including doxorubicin [22] and paclitaxel [23], are currently undergoing clinical trials. It is these promising systems that form the focal point of this chapter.

10.3
Core–Shell Polymeric Nanoparticles

Polysaccharide-based nanoparticles offer interesting alternatives to the nanostructures described above, and have undergone increasing scrutiny as the delivery vehicles of choice. These particles can take various forms, including solid spheres, capsules, and matrices [20]:

- Solid particles are generally present where the core and surface are continuous. Unfortunately, this limits the entrapment or encapsulation of the therapeutic agents, so that their use in drug-delivery systems is rare.
- Capsules comprise a polymeric outer membrane surrounding a hollow core, similar in basic structure to liposomes.
- Matrix systems are formed by the interaction between several polymer chains, which results in the formation of a "net-like" complex with inter-chain pores (a sponge-like structure also referred to as "nanosponges") suitable for the gradual and sustained release of the load [24].

Nanoparticle synthesis and preparation (as summarized in Table 10.1) will, therefore, be dependent on many factors, the most important of which is the polymer(s) used, there being an abundance to choose from according to the intended application and purpose of use [25]. "Smart" or *stimuli-responsive polymer-based systems*, in the form of crosslinked or physical gels covalently linked onto surfaces that reversibly alter their physico-chemical characteristics in response to their environment [26], have been at the center of recent intensive explorations. A recent comprehensive review, provided by Alves and Mano [27], describes the details of such promising injectable and bioinspired self-assembling systems.

10.3.1
Biodegradable Core–Shell Polymeric Nanoparticles

While traditional nanoparticles with hydrophobic surfaces are rapidly opsonized and cleared by the body's RES and mononuclear phagocytic system, targeted therapeutic delivery requires persistence of the nanoparticulate systems within the systemic circulation [28]. For that reason, among others, the modification and coating of surfaces with different cationic materials and hydrophilic polymers, in order to promote nanoparticulate–cell interaction and cellular internalization, has been deemed desirable [23]. Coatings (nanoshells) create a cloud of chains at the nanoparticle surface which repels the plasma proteins [29]. Typical biodegradable that have been polymers used and intensively investigated in the formulation of polymeric-based core–shell nanoparticles, mainly as drug-, protein- and gene-delivery vehicles, include alginate, chitosan, gelatin, hyaluronic acid, poly(lactic-co-glycolic acid) (PLGA), polylactic acid (PLA), polycaprolactone (PCL), and poly-alkyl-cyanoacrylates (PAC) [20, 30–33]. The bioavailability, enhanced encapsulation, controlled release and cytocompatibility and biocompatibility of these materials improves the therapeutic value of nanoencapsulated agents. The impact of various significant nanomedicine biomolecules and nanotherapy drugs upon encapsulation in such biodegradable nanosystems was recently highlighted by Kumari and coworkers [33]. A recent review of the current challenges in biodegradable nanoparticulate systems, mainly aimed at the delivery of growth factors such as bone morphogenetic proteins, has also been produced by the present authors [34, 35]. The performance of nanoparticles *in vivo* is influenced by their size and morphological characteristics, their surface charge, and the chemistries and

Table 10.1 Synthesis methods for polymeric core–shell nanomaterials.

Method	Description	Examples
Sol–gel (ionic gelation)*	Used for synthesis of metal or polymer (core) metal oxide (shell) nanoparticles in inorganic matrix (silica) forming a gel. Also for some semiconductor nanoparticles (heat-treated gelation) forming an oxide shell on top of metal or polymer core	Iron/iron oxide and copper/copper oxide in silica matrix polypyrrole/iron oxide and CdSe/CdS nanoparticles *Ionic gelation (also known as complex coacervation)
Reverse micelle (emulsification)	For controlling nanoparticle size and morphology by performing synthesis in emulsions or in solutions that form micelles (by mixing aqueous reactant with suitable surfactant) followed by oxidation polymerization and shell coating (sol–gel)	Iron (core) gold (shell) and cobalt (core) platinum (shell) Polymeric nanoparticles can be synthesized if emulsion of monomer is thermodynamically stable
Mechano-chemical	*Sonochemical synthesis* (chemical reaction for nanoparticle synthesis and sonication to improve speed of reaction and particle dispersion) and *Electrodeposition* (formation of nanoparticle shell with charged polymers to control nanoparticle size) are only two of the techniques used mainly for synthesizing core–shell nanoparticles with magnetic properties. Others include: pulverization, colloidal chemical synthesis, mechanical attrition, layer deposition, and reduction techniques	**Metallic shell nanocomposites:** Iron oxide (core) gold (shell) and iron (core) cobalt (shell) **Non-metallic shell nanoparticles:** Silica (core) PAPBA (shell) arid CuS (core) PVA (shell)
Polymerization	Free radical polymerization (FRP) or atom transfer radical polymerization (ATRP) have been used with ATRP showing superiority in terms of molecular weight and particle size control. ATRP also facilitates the coating of polymers on silica cores. A third polymerization method is known as chemical oxidation polymerization (COP), where a monomer is polymerized by adding the appropriate oxidizing agents. COP is mostly conducted for metallic nanoparticles and core–shell nanocomposites formed by a chemical reaction and reduction	**ATRP method:** Gold (core) silica (shell) nanoparticles–biomimetic approach **COP method:** Silver (core) polyaniline (shell), silver (core) polypyrrole (shell) and PbS (core) polypyrrole (shell) nanocomposites
Electrospinning (nanorods/nanowires)	Used to fabricate core–shell nanowires (co-electrospinning) where the nanofiber core material (by traditional electrospinning) has been employed in conjunction with coating processes such as chemical or physical vapor deposition, spin coating, or spraying to coat the core with a layered polymer shell. A new template wetting technique referred to as a template-assisted sequential solution wetting was demonstrated to fabricate core-shell polymer nanowires with adjustable shell thicknesses	**Metallic nanocables (nanoelectric devices):** Silver (conducting core) silica (insulating shell) **Nanofibers (core–shell polymeric devices):** Polyaniline (core) schizophyllan (coat) in biological sensor applications

molecular weights of the polymers involved [36, 37]. Surface-modified liposomes with alternate coats of alginate and chitosan, by virtue of the extended configuration on the particle surface acting as a barrier or a spacer, reduced the extent of agglomeration and possibly also clearance by hepatic macrophages upon administration *in vivo* [11, 37]. The core–shell structure also allowed a modulation of the mechanism of release from the nanoparticles, by controlling the number of layers that formed the shell [11, 36, 38]. An additional control might also be gained by modulating the molecular weight of the polymer(s) used; indeed, it has been shown that the higher the molecular weight of a polymer, the slower will be the *in vitro* release of drugs [38]. Although surface properties will be discussed briefly later in the chapter, some examples of polymeric-colloids with advanced core–shell structures are presented first.

10.3.2
Core–Shell Colloids

Polymeric colloids are small particles that range from 100 to 400 nm in diameter, that have been used for a variety of applications dating back to the 1970s. Amphiphilic copolymers are good examples of polymers that self-assemble into a core–shell structure, where the hydrophilic core can serve as a reservoir for genes, enzymes, and also a variety of drugs with various characteristics [39]. For example, Ranjan *et al.* [40] recently demonstrated the antibacterial efficacy of amphiphilic core–shell nanostructures encapsulating gentamicin against an *in vivo* intracellular *Salmonella* model. Novel colloidal systems have been introduced as another means (depending on the crosslinker used) by which core–shell nanoparticles can be synthesized and their final properties, such as elasticity and biodegradability, either modified or improved. A PCL core which is grafted and surrounded by a hydrophilic dextran shell is a stable colloidal core–shell (triphasic release) system aimed at biomedical applications (Table 10.2), while PLA, PLGA, and PCL nanoparticles coated with PEG have all been extensively studied for protein adsorption [41]. In this case, the PEG chains in the shell provide a clear added advantage of a requisite longer blood circulation time than do particles without this layer. Likewise, as such particles are not recognized by cells of the mononuclear phagocyte system, they can serve as a persistent drug-delivery system, with the potential for *in vivo* administration [33]. This "camouflage" method creates a hydrophilic protective layer or shell around the nanoparticles that is able to repel the absorption of opsonin proteins via steric repulsion forces, thus blocking and delaying the first step in the opsonization process [42]. This is the main reason why smaller nanoparticles (<100 nm in diameter) are able to avoid recognition by cellular defense systems following surface modification and, as a consequence, are able to circulate for longer periods in the body. Yang *et al.* [20] have reviewed the most recent advances in the development of polymeric-based nanoparticles, while Finne-Wistrand and Albertsson [42] have conducted extensive studies of the polymeric structure and use of polymers in the design of resorbable core–shell structures that have been evaluated for biomedical applications. Additional relevant data are

Table 10.2 Polysaccharide-based nanoparticulate delivery systems.

Class of nanoparticles (NPs) and synopsis	Examples (s)
Covalently crosslinked polysaccharide NPs *Use of biocompatible intermolecular crosslinkers with the aid of water-soluble condensation agents*	Glutaraldehyde-crosslinked chitosan-based NPs (but cytotoxic). Today, succinic acid, malic acid, tartaric acid and citric acid are used for intermolecular crosslinking of chitosan NPs
Ionically crosslinked polysaccharide NPs *Milder preparation conditions than covalent crosslinking. Polyanions and polycations act as ionic crosslinkers for polycationic and polyanionic polysaccharides, respectively*	Most widely used polyanion crosslinker is tripolyphosphate (TPP). TPP-crosslinked chitosan NPs are common, forming a gel by ionic interaction between positively charged amino groups of chitosan and negatively charged counterions of TPP. Alginate-coated chitosan NPs have also been formulated for gene delivery
Polysaccharide NPs by polyelectrolyte complexation *Feasible with any oppositely charged polymers by intermolecular electrostatic interactions, but restricted to water-soluble and biocompatible polymers for cytocompatibility and biocompatibility purposes. Most common with chitosan*	*Anionic polysaccharides and peptides*: carboxymethyl cellulose–chitosan NPs (containing plasmid DNA) for potential genetic immunization and poly-v-glutamic acid–chitosan NPs for enhancing intestinal paracellular transport in Caco-2 cell monolayers and enhanced penetration depth in skin (for gene expression)
Self-assembly of hydrophobically modified polysaccharides *Amphiphilic copolymers are spontaneously synthesized when hydrophilic polymeric chains are grafted with hydrophobic segments in an aqueous environment*	Comb-like NPs (<200 nm) of co-polymers; dextran and poly(ε-caprolactone) were prepared to incorporate bovine serum albumin and lectin with hemagglutinating properties, suggesting the possible application of this type of surface-modified NP for targeted oral administration

Chemical structure of a polymeric core-shell nanoparticle

incorporating a hydrophobic core (PCL) and a hydrophilic shell (dextran) also available on the preparation of colloid core–shell particle systems (mainly via crosslinking mechanisms), and their use in solving a wide array of biomedical and pharmaceutical problems [20, 33, 42].

10.3.3
Surface Properties and Modification Techniques

In core–shell-designed nanostructures and nanoparticles, the therapeutic agents, biomolecules and/or drugs are normally either conjugated to the surface, entrapped within the shell, or encapsulated and protected inside the core [42]. The unique size of the nanoparticles (the smaller the better) is amenable to surface functionalization or modification to achieve the desired characteristics for a specific application [42]. Various methods have been employed to enhance the load or cargo retention time, to reduce the nonspecific distribution and to target tissues or specific cell-surface antigens with targeting ligands, peptides and antibodies. In targeted drug delivery, nanoparticle surface modification remains a very important approach to release drugs in the bloodstream with "stealth" invisibility against the natural defense system of the body (opsonization), as noted above [28]. Logically, a longer circulation time would also increase the probability of the nanoparticles and their respective loads reaching their intended targets. However, it is noteworthy that the surface modification, synthesis and choice of polymeric material for formulating nanometric core–shell carriers are entirely dependent on the nature of the drugs, as well as the delivery period, stability, permeability and drug-release profile [43]. Thus, if a sustained systemic circulation is required, then the surface of the hydrophobic nanoparticles must be modified in order that they are not entrapped rapidly in the liver or spleen [37]. Such surface modification may be achieved by coating the core with a static hydrophilic protective layer, such as PEG, poly-vinyl pyrrolidone (PVP) or chitosan derivatives that prevent macromolecules from interacting with the nanoparticle, even at a low surface coverage, so that they release their load at a specific site [28, 42]. The accumulation of drug-loaded nanoparticles at the target site is more important than their circulation and retention in the circulatory system of the body, and the details of both passive and active targeting techniques have been described elsewhere [28, 42, 44]. In *passive targeting*, the nanoparticles accumulate at the tumor sites as a results of the tumor's leaky vasculature, whereas in *active targeting* the nanoparticles must carry the targeting molecules on their surfaces before they can interact with the surrounding tissues. An example of an active surface modification is the use of an immunoglobulin G (IgG) shell to enhance the immunoresponse of the core [45]. In contrast, hydrophilic polymers can be tailored to the nanoparticle surfaces in two ways, namely via the adsorption of surfactants, or through the use of branched copolymers. As shown in Table 10.2 (insert), dextran was used to coat the surface of PCL nanoparticles in order to inhibit protein adsorption [42]. Furthermore, a poly(N-isopropylacrylamide) copolymer, when introduced onto the surface of PCL nanoparticles, formed a hydrogel layer and provided an additional diffusion barrier. The grafting of a ligand onto the surface/shell of a nanoparticle in order to specifically target receptors has been investigated, with most such ligands belonging to the lectin family and being well recognized for their involvement in cell-recognition and -adhesion processes [42–46].

10.4
Biomedical Applications of Core–Shell Polymeric Nanostructures

The three main developments currently foreseen in this arena of nanobiotechnology are: *sensorization* (biosensors); *diagnosis* (biomarkers and bioimaging); and *drug delivery* (controlled-release systems). Advances in applications such as bioimaging and cell-labeling, agent delivery, targeting, cancer, tissue engineering and immunoassays, along with details of the major limitations and possible future advances, are outlined in the following subsections.

10.4.1
Bioimaging, Biological, and Cellular Labeling

Luminescent core–shell nanostructures such as QDs, magnetic nanoparticles and other nanoscaled biomarkers have been tested in most biotechnological applications that involve fluorescence and luminescence, including DNA array technology, immunofluorescence assays, magnetic resonance imaging (MRI) and cell and animal biology (vascular and lymphatic systems, tumor cell labeling). As they tend to target and penetrate cells very easily, and also enjoy a better spin-lattice relaxation time, biocompatible core–shell magnetic nanoparticles with an iron oxide or cobalt core have been used to enhance MRI contrast and subsequent imaging quality [47]. Recently, thermoresponsive polymer-coated magnetic cores were loaded with anticancer drugs (doxorubicin) for magnetic drug targeting, followed by simultaneous hyperthermia and drug release (multimodal treatment of cancer), with good results [48]. Nanocomposite biomarkers also have the ability to operate at over wider pH and temperature ranges [49]. The high-detection benefits of using QD-based labels in (sandwich immunocomplex) immunoassays [50] were realized by exploiting their spectral multiplexing capability to excite and detect several labeled species simultaneously, using a single light source [51]. Many studies have shown the promising potential of using QDs as new probes; these include *in vitro* and multicellular animal models that allow the real-time, continuous and prolonged (over periods of hours to days) monitoring and tracking of single and/or multiple cellular components and biological processes. This was not possible to achieve with luminescent particles, which function in the visible or near-infrared (NIR) range [48]. The modern trend is therefore, to benefit from novel core–shell nanostructures that emit in the ultraviolet (UV)-visible range with low signal-to-noise ratios (SNRs) suitable for a wider and more specific cellular detection [44, 52]. The most recent applications of QDs for multiple organelle and subcellular imaging are shown in Figure 10.2, where modification of the surfaces of currently available QDs has been performed in parallel. Brighter, multifunctional core–shell

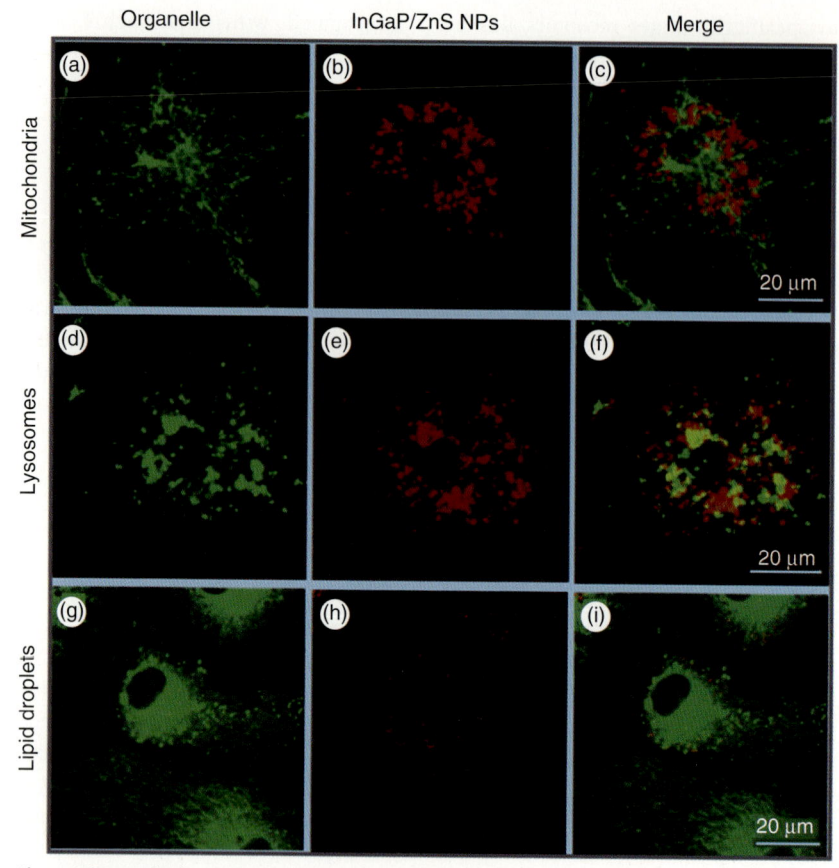

Figure 10.2 Multiple organelle labeling and subcellular location of InGap/ZnS nanoparticles in live primary cortical cells. Confocal images showing stained organelles with vital dyes (a and c); lysosomes (d and f); lipid droplets (g and i) and 10 nm InGap/ZnS nanoparticles (b, e, and h).
Reproduced with permission from Ref. [52]; © 2009, J Nanomedicine published by Future Medicine.

nanoparticles (with magnetic and luminescent capabilities for cellular labeling and separation) that provide anti-Stokes emission, such as semiconductor nanoparticles, lanthanide-based nanoparticles and gold-coated silica nanoparticles (to name but a few), have recently been attracting increasing interest [12, 17], although much caution as to the choice of core and shell material has been deemed necessary to obtain clearer 3-D images. For example, QDs conjugated to secondary antibodies have been shown to improve the specificity of Western blot assays when compared to expensive and traditional radioactive labels used for the ultrasensitive detection of tracer proteins [53, 54]. This, along with the potential to greatly reduce the background autofluorescence of biological samples, would appear to represent a promising advance towards the next generation of inexpensive and improved flow cytometry systems, probes, assays, and diagnostic set-ups. The separation and purification of biological molecules and cells also represent areas in which such particles with well-defined and controlled properties can be used [44, 54]. In the case of medical diagnostics, the aim is to prepare novel core–shell nanomaterials that can detect and bind antibodies on a covalent basis [49, 50, 53, 55].

10.4.1.1 Cytotoxicity of QDs

One major concern with regards to the use of QDs has been their cytotoxic properties. Following studies of QD cytotoxicity [49, 56], the size, charge, concentration, and outer coating bioactivity (e.g., capping material and functional groups), as well as the oxidative, photolytic, and mechanical stability have each been implicated as determining factors in QD toxicity [49, 56]. Whilst it has been well established that QDs contain cytotoxic elements (notably cadmium ions), it was announced by Evident Technologies, in 2006, that QDs containing no heavy metals were commercially available. Nonetheless, QD-related cytotoxicity has been mainly reported to be exacerbated by their large exposed surfaces (photolysis or oxidation). Consequently, the most important aspect of QD toxicity is their stability, not only *in vivo* but also during their synthesis and storage [49, 56]. In an effort to avoid such shortcomings when employing QDs in biological fluids, the modulation of the chemical nature of the QD surfaces, as well as altering their solubility and adding further chemical functionalities, have been explored [49, 56, 57]. To this end, different coating technologies of the biological surfaces, polymeric nanoshells and traditional layers consisting of ZnS compounds have been incorporated into QDs [57]. The aim here is to obtain not only water-soluble QDs but also functionalized nanocrystals capable of fulfilling several tasks, such as receptor targeting or the sensing of low-molecular-weight substances. Polymers have also been shown to enhance QD stability and function [57]; for example, Susumu *et al.* [58, 59] applied a polyethylene surface coat to QDs (in this case, a biotin-coated surface was reacted with streptavidin-coated QDs) and demonstrated an enhanced fluorescence intensity [58], stability, and biological functionality [59]. Elsewhere, when the effects of PEG-coated cationic CdSe/ZnS QDs were demonstrated in live cells, the PEG shell was seen to enhance the long-term stability of the nanoparticles over a broad pH-range, while the amine-functionalized core revealed an efficient intracellular labeling [60]. When, in nude mice, hyaluronic acid (HA)-conjugated QDs were administered subcutaneously and tracked in real-time, an NIR wavelength of 800 nm was detectable for up to two months, according to the real-time bioimaging. Polymer functionalization is, therefore, necessary to ensure biocompatibility [61, 62]. Developments in QD synthesis/bioconjugation, the effects of polymer coating of QDs via different strategies (ligand-exchange and -capping), as well as their applications in medicine and future directions, have been reviewed elsewhere [60–63].

10.4.2
Glucose Monitoring and Biosensing in Diabetes Mellitus

Nanobiosensors are currently undergoing investigation for a possible role in the life-time management of diabetes mellitus, by measuring glucose levels accurately in the blood, in noninvasive fashion [64, 65]. For example, glucose oxidase (GOx), when coupled to core–shell nanotubes, has been recently shown to serve as a catalytic biomolecular sensor [66, 67]. A single-molecule detection to study molecular diversity in diabetes pathology has also been explored [64]. Further, the feasibility of a novel ultrasensitive protein nanoprobe system based on self-assembled supramolecular protein nanoparticles (10–15 nm) that would bind GAD65 (an antibody which serves as an early marker of Type I diabetes) during the early phase of pancreatic β-cell destruction was investigated. This system has the potential to identify high-risk individuals several years before the clinical onset of Type I diabetes [68]. Duong and Rhee [69] also recently synthesized CdSe/ZnS core–shell QDs and conjugated them with specific enzymes, namely GOx and horseradish peroxidase, for use as probes to biosense or detect levels of blood glucose.

10.4.3
Drug Delivery

During the past decade, one of the main advances of nanobiotechnology has been in drug delivery, with bifunctional core–shell nanostructures composed of a luminescent core and a shell conjugated with a biomolecule/drug having been used for targeting/drug delivery and imaging. Likewise, drug-eluting contact lenses have benefited from similar advances, with polymeric-based hydrogels introduced as ophthalmic drug-delivery systems to treat glaucoma [70, 71]. Although, to date, many nanoparticles have been investigated for drug delivery, very few have obtained FDA approval [72]. This has been due predominantly to their potential toxicity, poor biocompatibility, instability *in vivo*, low drug-loading capacity, and/or batch-to-batch inconsistencies in their bio-physico-chemical properties [61, 72]. Today, polymeric-based core–shell nanoparticles find increasing applications in the biomedical and pharmaceutical areas, mainly to overcome the limitations of poorly-soluble drugs or to produce long-acting injectable formulations and specific drug-targeting options. In particular, they can provide controlled-release properties due to the biodegradability, pH, ion and/or temperature sensibility of the polymers (stimuli-responsive polymers) [73–75].

Polysaccharide-based core–shell nanoparticles can also improve the utility of encapsulated drugs, and reduce any toxic side effects [11, 76]. Given the design, agents are entrapped into their interior structures and/or absorbed onto their exterior surfaces. At present, this type of nanoparticle has been widely used to deliver drugs, polypeptides, proteins, vaccines, nucleic acids, and genes [11, 35, 75, 76]. Modern trends in nanoparticulate drug-delivery systems research is focused on:

- a combination of polymer materials to obtain suitable and proper drug release profiles and kinetics;
- optimization of drug encapsulation efficiencies and loading capacities;
- shell modification to enhance specific targeting capabilities; and
- optimization of preparatory techniques for clinical use and industrial production.

Numerous polymeric materials have been investigated, including PGA, PLA, chitosan, alginates, the poly(acrylic acid) family, proteins, or polypeptides (e.g., gelatin). Among these, polysaccharides continue to be the most popular polymeric material for the preparation of core–shell nanoparticles for drug-delivery applications [11, 35, 73–76], with the material choice determining the multifunctional nature of the resultant particles. Recently, Liu *et al.* [77] reviewed an exhaustive range of polymers as potential polysaccharide-based nanoparticulate drug-delivery systems, while Mano [78] described the general aspects of the different types of stimulus that can be used to modulate biological response, by using stimuli-responsive polymeric (natural and synthetic) systems and investigating their biomedical applications. Furthermore, Van Tomme and colleagues [75] examined several examples of the commonly investigated *in situ* gelling and photopolymerizable systems, and their potential in biomedical applications [75].

10.4.3.1 Natural Polymer-Based Drug-Delivery Systems: Layer-by-Layer Self-Assembly

The idea of polyelectrolyte coatings obtained by the alternate deposition of polyanions and polycations emerged as a novel way to functionalize surfaces [79, 80]. This was quickly applied to the drug-delivery field, where the layer-by-layer (LbL) technique was extended from the build-up of multilayered polyelectrolytic films on macroscopic flat substrates [81] to the construction of core–shell particles on various spherical templates and colloidal particles [82]. Unfortunately, particle flocculation or aggregation was difficult to overcome initially, and thus far the adsorption of only a single layer of biopolymer, such as polyvinyl alcohol (PVA) [83] or chitosan [84] on a charged liposomal surface has been reported. In addition to being nontoxic, biocompatible, biodegradable and hydrophilic, it has been shown that biomolecules can be assembled and entrapped within polyelectrolyte layers, hence maintaining their bioactivity [79, 85]. A series of alginate–chitosan nanosized polyionic complexes was also designed for gene therapy [86]. Besides the known advantages, which include the size property, a longer shelf-life and an ability to entrap more drugs [87], nanosized systems reside longer in the circulation and therefore greatly extend the macromolecular biological activity when compared to microparticles [88]. Recent findings have led to a pioneering use of the LbL self-assembly technique as a simple and effective way to enhance the bioactivity of encapsulated drugs and growth factors, as well as QDs/MRI contrasts. The incorporation of growth factors such as bone morphogenetic proteins (BMPs) within polymer-based core–shell nanoparticulate systems, for example, has resulted in advanced positive effects. Indeed, such potent proteins have been shown to induce tissue regeneration [34–38, 89] and, when used in injectable delivery systems, to provide a localized, predictable and metered release (as deemed necessary to maintain the bioactivity of such morphogens) of the growth factor [34–38]. The advantages of nanoscaled liposomes and natural polysaccharide-based polymers (namely sodium alginate and chitosan) were combined with those of core–shell nanoparticles and hydrogels to formulate novel biocompatible and biodegradable drug- and gene-delivery systems, with applications in the field of bone tissue engineering, cardiovascular disease, and beyond [11, 36–38].

Ionic gelation is a method particularly suited to water-soluble polymers (for other techniques, see Table 10.1), and which can result in an injectable gel [36, 37] or be applied with the LbL self-assembly of polyelectrolytes to develop multifunctional 2-D and 3-D drug-delivery systems. Examples of this include paclitaxel nanocoatings on stents (drug-eluting stents) to protect against abrasion and to prevent thrombogenesis, restenosis and to control the arterial healing processes [90], and alginate–chitosan nanoparticles for their potential application as DNA carriers [86, 91]. The effect of the core–shell design (Figure 10.3) in protecting encapsulated loads from biodegradation, and the benefit of polymers such as chitosan in localizing, modulating and sustaining agent release and any likely side effects (such as tissue distress), have been demonstrated in diverse *in vitro* and *in vivo* studies utilizing small- and large-animal models over prolonged periods of 70 days [36, 37]. The results obtained have demonstrated an attractive potential for clinical application in skeletal indications, such as craniofacial distraction osteogenesis and fracture healing.

10.4.3.2 Synthetic and Composite Drug-Delivery Systems: Functionalized Nanoshells

It has been also shown that nanoparticulate systems help to reduce the adverse effects of chemotherapy in cancer studies [92–94]. These systems are prepared by forming drug–polymer complexes in which the drug is uniformly dispersed, or by creating nanoscaled vesicular cores (such as liposomes and micelles) to entrap the drug molecules [95, 96]. The ability of liposomal-based formulations to encapsulate hydrophilic therapeutic agents (e.g., the chemotherapeutic docetaxel) at high loading efficiencies, and to shield the encapsulated drugs from external conditions, can be further enhanced by the application of a shell composed of inert and biocompatible polymers, such as alginate, chitosan and PEG [95]. This significantly reduces their systemic clearance rates and prolongs the circulation half-life *in vivo*, as described earlier. PEG-end groups in the shell may also be functionalized with specific ligands for targeting to specific sites of the cells, tissues, and organs of interest [20]. Recently, Chan *et al.* [97] have developed self-assembled lipid core–polymer shell hybrid nanoparticles for potential use in controlled docetaxel delivery. The delivery system was created via a modified nanoprecipitation method combined with the self-assembly of three biomaterials: biodegradable PLGA as a hydrophobic core; soybean phosphatidylcholine (lecithin) as a monolayer surrounding the core; and 1,2-distearoyl-sn-glycero-3-phosphoethanolamine-N-carboxy-carboxy(PEG) to form a PEG shell providing electrostatic and steric stabilizations, a longer circulation half-life *in vivo*, and also functional-end groups for the attachment of targeting ligands such as antibodies, peptides, and aptamers [97]. Further, phosphate-functional core–shell nanoparticles composed of a polystyrene core and a shell of crosslinked polymers (alkenyl phosphate binding) were recently formulated for the bioactive release of vascular endothelial growth factor (VEGF) in potential angiogenesis applications [2]. The controlled release properties

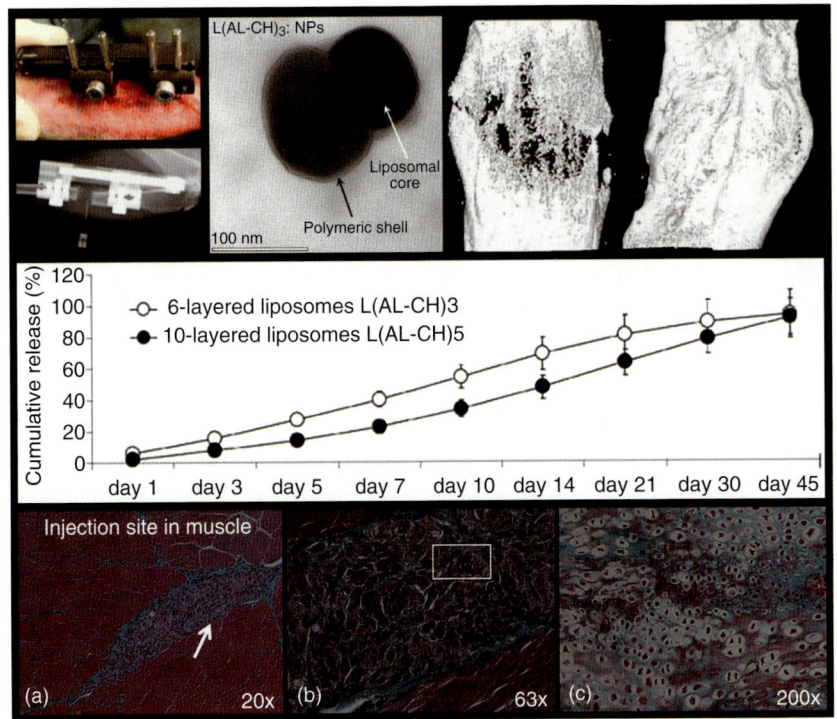

Figure 10.3 An overview of studies involving polymeric-based core–shell nanoparticles in a drug-delivery application. **Top row:** In a rabbit model of long bone distraction osteogenesis (tibial lengthening), a single injection of a low-dose (0.5 µg) OP-1, loaded into hybrid core–shell nanoparticles (three-bilayered alginate/chitosan shell surrounding a liposomal core) was sufficient to accelerate and enhance new bone regeneration when compared to a free protein injection. **Middle row:** Linear and controlled (metered) release profile exhibited from the delivery system by varying the number of polymeric layers in the shell (a slower release profile is associated with a thicker and denser shell). **Bottom row:** Safety and biocompatibility of the core–shell nanoparticles demonstrated in a rat toxicity model where timely blood (complete blood count) and organ function analysis was conducted over a period of 70 days. The histological images shown represent the injection site in the thigh muscle of rats, and demonstrate the safety and localization effect of the nanoparticles (composition and release effects), where no tissue distress was noted in any of the harvested samples. The clinical photographs, radiographs, TEM and Faxitron images, release data plots and muscle tissue histomorphometry were adapted and modified from Refs [36–38].

of synthetic polymers has been combined with the biocompatibility of natural polymers in recent years, by forming polymeric composites. Examples of this include PLGA–gelatin composites, collagen–PLG–alginate composites, and hyaluronan-impregnated PLA sponges [98, 99]. PLGA nanospheres immobilized onto prefabricated nanofibrous poly(L-lactic acid) (PLLA) scaffolds were used to load one BMP (BMP-7, also known as osteogenic protein-1; OP-1), and to promote *in vivo* bone regeneration [100]. However, a significant failure of bone induction was observed due to loss of the bioactivity of the loaded protein and rapid release from the scaffolds following subcutaneous implantation into rats [100]. In a recent review [35], a classification was created for the carrier biomaterials (particularly natural and synthetic polymers) and their combinations in different formats for the delivery of growth factors to preclinical and clinical sites of bone regeneration and repair. Applications of core–shell nanoparticulate systems allowing injectability, for example, will extend into a much broader range of orthopedic as well as craniofacial and orodental indications, including bone, cartilage, and tendon/ligament (and periodontal) tissue regeneration and repair.

Surface charge is another important parameter in determining how nanoparticles interact with cells, the membranes of which are usually negatively charged. Further, the morphology, hydrophobicity and size allows drugs to accumulate in solid tumors, which are characterized by extensive angiogenesis, a defective vascular architecture, an impaired lymphatic drainage, and an increased production of permeability factors [101, 102]. The fundamental physico-chemical properties for efficacious and safe nanoparticulate drug delivery specifically within a tumor microenvironment, depending on the cancer type, the stage of disease, the site of implantation, and also on the host species, were recently investigated by Adiseshaiah *et al.* [103].

10.4.4
Cancer

Although polymer colloids have been used previously in cancer therapy (as described above), one major disadvantage of anticancer drugs is their lack of selectivity towards tumor tissues, which may lead to severe side effects and low cure rates [104]. Photodynamic therapy (PDT) represents an effective and selective means of destroying the diseased tissue, without damaging adjacent healthy regions. In this case, the nanoparticles are actively taken up by the tumor cells, while subsequent irradiation with visible light results in an irreversible destruction of the impregnated cells [104]. In attempting to address some of these issues, several nanosized liposomal and polymer conjugate-based drugs have been developed and are currently undergoing clinical trials [105] (Table 10.3). Thus, in minimizing the toxic effects of the anticancer drugs, the nanoparticles will also reduce the devastating and adverse effects of chemotherapy [92–94].

Polymeric-based core–shell nanoparticles may also serve as multifunctional therapeutic agents, rather than be used simply as passive carriers of the drug cargo. Such "smart" or "intelligent" systems can, potentially, carry drugs to a target tissue, image that tissue, and release the drug either in response to a signal or on reaching the appropriate cellular compartment [92–94]. Targeted nanotherapy may also be achieved by coupling a specific antibody or a small-molecular-weight ligand (e.g., folic acid) to the surface of a nanoparticle that recognizes a protein which is selectively expressed on the tumor cells [93, 94]. To this end, magnetic iron oxide particles or polymeric-based nanotubes may be used as core nanostructures, whereby imaging can be accomplished via MRI with iron oxide nanoparticles or fluorescence methods with QDs, while targeting is achieved by the functionalized

Table 10.3 Examples of different nanoparticles used in cancer therapy.

Nanostructure	Synopsis
Nanocapsules Drug within core is surrounded by a polymeric membrane	Stability of the **cisplatin**-containing core is optimized by varying the lipid composition of the bilayer shell
Nanospheres Drug is physically and uniformly dispersed within a matrix-based system	Bovine serum albumin nanospheres containing **5-fluorouracil** demonstrated a greater tumor inhibition than the free drug
Micelles In an aqueous solution of amphiphilic block co-polymers containing the drug	Micelle delivery of **doxorubicin** increases cytotoxicity to prostate carcinoma cells
Liposomes Drug within natural phospholipids and cholesterol artificial spherical vesicles	Radiation-guided drug delivery of liposomal **cisplatin** to tumor blood vessels results in improved tumor growth delay
Dendrimers Drug within a series of polymer branches surrounding an inner core	Targeted delivery within dendrimers improved the cytotoxic response of the cells to **methotrexate** 100-fold over free drug
Polymeric core-shell nanoparticles (Fe_3O_4 magnetic core and a dextran thermosensitive shell)	Targeted **doxorubicin** delivery system with longer circulation time, reduced side effects and controlled drug release in response to the change in external temperature

Updated from Ref. [105].

shell. An example of this was the tumor-targeting antibody, Herceptin, which was prepared against the HER2 receptor amplified in breast tumors. The coupling of Herceptin to gold nanoshells enabled targeting of the HER2-overexpressing tumor cells which, when irradiated, were specifically killed [106, 107]. Likewise, when nanoshells of ultrathin gold layers surrounding a core of silica were investigated, an adjustment of the core and shell thickness allowed the nanoshells to absorb and scatter light at a desired wavelength [108]. Nanoshells for cancer therapeutic purposes have been designed to have a peak optical absorption in the NIR, as this wavelength provides the optimal tissue penetration, while the metal shell converts the absorbed light into heat, with great efficiency and stability. Due to their small size, the nanoshells become concentrated preferentially in cancer cells as a result of an enhanced permeation and retention (EPR) effect. If required, a supplementary specificity can be engineered by attaching antigens onto the nanoshells, as these will be recognized specifically by the cancer cells [107]. The ability of PEG-based C_{60} fullerenes to generate highly reactive singlet oxygen has also been tested as a potential approach to PDT in tumor cells [12].

Thus, the concept of using multifunctional nanoparticles for cancer therapy has been validated in several experimental systems, and has shown great potential for major advances in the future [77, 78, 107].

Finally, although the bulk of this chapter has been dedicated to injectable and implantable modes of administration, a number of biodegradable nanoparticles have recently been developed as insulin carriers for the oral and transdermal delivery routes, as an alternative to injection. An example of this involved the development of biocompatible and biodegradable pH-sensitive alginate nanospheres which released insulin over extended periods in the less-acidic intestinal environment (but not in the highly acidic gastric environment), and are currently undergoing development [42, 109].

10.4.5
Miscellaneous Applications

Polymeric-based core–shell nanocomposites, whether polymer–polymer or polymer–metal combinations, are currently being used in dental braces and resins, as well as in joint-, hip-, and knee-replacements [12]. Ultra-high-molecular-weight polyethylene–silver and polymer-coated titanium are such examples [110, 111]. An artificial hybrid material was also prepared from 15–18 nm ceramic nanoparticles coated with a poly(methyl methacrylate) copolymer [112] for applications in restorative and conservative dentistry, given the scratch-resistant property of the material. Using a tribology approach, the viscoelastic behavior (healing) of human dentition was demonstrated [113].

Other nanostructures, such as colloidal silver and titania, are now being used in medical dressings for quicker wound healing, and also in the design of filter materials for the better and efficient separation of components, respectively [12, 113].

It should be noted, however, that these applications are not totally restricted to biological areas, as the enhancement of catalytic activity is today a highly dynamic field [114, 115]. For example, when enzymes were immobilized onto core–shell nanoparticles or catalysts and encapsulated suitably into nanoparticles, they demonstrated much higher activities than had previously been seen, with such benefits considered to have potential applications for biosensing [114, 115]. Recently, a shell composed of a poly(N-isopropylacrylamide) network was grafted onto a solid polystyrene core, creating a thermosensitive microgel onto which β-D-glucosidase (an almond-derived enzyme) was then adsorbed. An observed enhancement of enzymatic activity was considered to have resulted from the strength of the interactions between the enzyme and the microgel by hydrogen bonding, as was revealed via Fourier-transform infrared spectroscopy [114, 115].

10.5
Future Prospects

Today, the clear and outstanding potential of core–shell nanoparticles is considered to stem from an ability to produce structures with combinations of properties that none of the individual materials possesses. Whilst the majority of commercial natural polymer-incorporated nanoparticulate applications in medicine are geared towards bioactive drug delivery, a much greater potential exists for many other high-demand applications, including antitumor therapy, protein therapy, gene therapy, AIDS treatments, and radiotherapy. Currently, in the biosciences, nanoparticles are continuing to replace organic dyes in applications that require not only a high photostability but also high multiplexing capabilities. It is to be expected that, in future, nanomaterials will have a radical effect on the way in which disease is diagnosed, treated, and prevented. For example, developments have been made in the directing, targeting and remote control of the functions of nanoprobes, such as conducting magnetic nanoparticles to a tumor, where they can either release their drug load or simply be heated indirectly so as to destroy the surrounding tissues. A major trend in the further development of nanomaterials is to render them multifunctional and controllable by the application of external signals or by the local environment, effectively converting them into "smart" or "intelligent" nanodevices. Notably, nanoparticulate drug-delivery systems have clear advantages over conventional strategies, in that they can increase the bioavailability, solubility and permeability of many potent drugs, and also of short-lived biological morphogens. The use of both localized and release-controlled polymeric core–shell nanoparticles will also lead to a reduction in drug dosage frequency and, in turn, improve patient compliance. Moreover, the unique problems associated with some of these drugs can be minimized by safeguarding their stability and preserving their structure. Hybrid nanoparticles represent an ingenious approach to treatment by enabling targeted delivery and controlled release, with the further potential of combining diagnosis and therapy emerging as a major tool in nanomedicine. The main goals are to improve the stability of drugs within the biological environment, to mediate the biodistribution of active compounds, and to improve drug loading, targeting, transport, release, and interaction with biological barriers. Of note, whilst major problems relating to the cytotoxicity and degradation byproducts of the drugs persist, improvements in biocompatibility remain a logical concern and the focus of future research. Protective layers, whether constructed as single or multiple assemblies, as offered by polymeric core–shell nanostructures, appear to show great promise in tackling these restrictive issues.

Acknowledgments

These studies were supported by the National Research and Engineering Council of Canada (NSERC) and the Canadian Institutes for Health Research (CIHR)–Regenerative Medicine/Nanomedicine in the framework of funding grants to the CBB.

References

1 Sahoo, S.K. and Labhasetwar, V. (2003) Nanotech approaches to drug delivery and imaging. *Drug Discovery Today*, 8, 1112–20.

2 Gilmore, J.L., Yi, X., Quan, L. and Kabanov, A.V. (2008) Novel nanomaterials for clinical neuroscience. *Journal of Neuroimmune Pharmacology*, **3**, 83–94.

3 Ramsden, J.J. (2005) What is nanotechnology? *Nanotechnology Perceptions*, **1**, 3–17.

4 Murray, C.B., Kagan, C.R. and Bawendi, M.G. (2000) Synthesis and characterisation of monodisperse nanocrystals and close-packed nanocrystals assemblies. *Annual Review of Materials Research*, **30**, 545–610.

5 Whitesides, G.M. (2003) The "right" size in nanobiotechnology. *Nature Biotechnology*, **2**, 1161–5.

6 Parak, W.J., Gerion, D., Pellegrino, T., Zanchet, D., Micheel, C., Williams, C.S., Boudreau, R., Le Gros, M.A., Larabell, C.A. and Alivisatos, A.P. (2003) Biological applications of colloidal nanocrystals. *Nanotechnology*, **14**, R15–27.

7 Pankhurst, Q.A., Connolly, J., Jones, S.K. and Dobson, J. (2003) Applications of magnetic nanoparticles in biomedicine. *Journal of Physics D: Applied Physics*, **36**, R167–81.

8 Boscovic, B.O. (2007) Carbon nanotubes and nanofibres. *Nanotechnology Perceptions*, **3**, 141–58.

9 Zandonella, C. (2003) The tiny toolkit. *Nature*, **423**, 10–12.

10 Mazzola, L. (2003) Commercializing nanotechnology. *Nature Biotechnology*, **21**, 1137–43.

11 Haidar, Z.S., Hamdy, R.C. and Tabrizian, M. (2008) Protein release kinetics for core-shell hybrid nanoparticles based on the layer-by-layer assembly of alginate and chitosan on liposomes. *Biomaterials*, **29**, 1207–15.

12 Sounderya, N. and Zhang, Y. (2008) Use of core/shell structured nanoparticles for biomedical applications. *Recent Patents on Biomedical Engineering*, **1**, 34–42.

13 Burda, C., Chen, X., Narayana, R. and El Sayed, M.A. (2005) Chemistry and properties of nanocrystals of different shapes. *Chemical Reviews*, **105**, 1025–102.

14 Brigger, I., Dubernet, C. and Couvreur, P. (2002) Nanoparticles in cancer therapy and diagnosis. *Advanced Drug Delivery Reviews*, **54**, 631–51.

15 Papadimitriou, S. and Bikiaris, D. (2009) Novel self-assembled core–shell nanoparticles based on crystalline amorphous moieties of aliphatic copolyesters for efficient controlled drug release. *Journal of Controlled Release*, **138**, 177–84.

16 Schreder, B., Schmidt, T., Ptatschek, V., Spanhel, L., Materny, A. and Kiefer, W. (2000) Raman characterization of CdTe/CdS-core-shell-clusters in colloids and films. *Journal of Crystal Growth*, **214**, 782–6.

17 Chu, M.Q., Cheng, D.L. and Zhu, J. (2006) Preparation of quantum dot-coated magnetic polystyrene nanospheres for cancer cell labelling and separation. *Nanotechnology*, **17**, 3268–73.

18 Zhou, M. and Gho, I. (2006) Quantum dots and peptides: a bright future together. *Peptide Science*, **88**, 325–39.

19 Sandros, M.G., Shete, V. and Benson, D.E. (2006) Selective, reversible, reagentless maltose biosensing with core–shell semiconducting nanoparticles. *Analyst*, **131**, 229–35.

20 Yang, Y.Y., Wang, Y., Powell, R. and Chan, P. (2006) Polymeric core-shell nanoparticles for therapeutics. *Clinical and Experimental Pharmacology and Physiology*, **33**, 557–62.

21 Kang, D.Y., Kim, M.J., Kim, S.T., Oh, K.S., Yuk, S.H. and Lee, S. (2008) Size characterization of drug-loaded polymeric core/shell nanoparticles using asymmetrical flow field-flow fractionation. *Analytical and Bioanalytical Chemistry*, **390**, 2183–8.

22 Chilcott, J., Lloyd Jones, M. and Wilkinson, A. (2009) Docetaxel for the adjuvant treatment of early node-positive breast cancer: a single technology appraisal. *Health Technology Assessment*, **13**, 1–7.

23 Wei, Z., Hao, J., Yuan, S., Li, Y., Juan, W., Sha, X. and Fang, X. (2009) Paclitaxel-loaded Pluronic P123/F127 mixed polymeric micelles: formulation, optimization and in vitro characterization. *International Journal of Pharmaceutics*, **376**, 176–85.

24. Borini, S., D'Auria, S., Rossi, M. and Rossi, A.M. (2005) Writing 3D protein nanopatterns onto a silicon nanosponge. *Lab on a Chip*, **5**, 1048–52.
25. Gupta, P., Vermani, K. and Garg, S. (2002) Hydrogels: from controlled release to pH-responsive drug delivery. *Drug Discovery Today*, **7**, 569–79.
26. Kojima, C. (2010) Design of stimuli-responsive dendrimers. *Expert Opinion on Drug Delivery*, **7** (3), 307–19.
27. Alves, N.M. and Mano, J.F. (2008) Chitosan derivatives obtained by chemical modifications for biomedical and environmental applications. *International Journal of Biological Macromolecules*, **43**, 401–14.
28. Owens, D.E. III and Peppas, N.A. (2006) Opsonization, biodistribution, and pharmacokinetics of polymeric nanoparticles. *International Journal of Pharmaceutics*, **307**, 93–102.
29. Preston, T.C. and Signorell, R. (2009) Growth and optical properties of gold nanoshells prior to the formation of a continuous metallic layer. *ACS Nano*, **3**, 3696–706.
30. Park, J.H., Ye, M. and Park, K. (2005) Biodegradable polymers for microencapsulation of drugs. *Molecules*, **10**, 146–61.
31. Azzam, T. and Eisenberg, A. (2007) Monolayer-protected gold nanoparticles by the self-assembly of micellar poly(ethylene oxide)-b-poly(epsilon-caprolactone) block copolymer. *Langmuir*, **23**, 2126–32.
32. Ramzi, A., Rijcken, C.J., Veldhuis, T.F., Schwahn, D., Hennink, W.E. and van Nostrum, C.F. (2008) Core-shell structure of degradable, thermosensitive polymeric micelles studied by small-angle neutron scattering. *Journal of Physical Chemistry B*, **112**, 784–92.
33. Kumari, A., Yadav, S.K. and Yadav, S.C. (2010) Biodegradable polymeric nanoparticles based drug delivery systems. *Colloids and Surfaces B, Biointerfaces*, **75**, 1–18.
34. Haidar, Z.S., Hamdy, R.C. and Tabrizian, M. (2009) Delivery of recombinant bone morphogenetic proteins for bone regeneration and repair. Part A: current challenges in BMP delivery. *Biotechnology Letters*, **31**, 1817–24.
35. Haidar, Z.S., Hamdy, R.C. and Tabrizian, M. (2009) Delivery of recombinant bone morphogenetic proteins for bone regeneration and repair. Part B: delivery systems for BMPs in orthopaedic and craniofacial tissue engineering. *Biotechnology Letters*, **31**, 1825–35.
36. Haidar, Z.S., Tabrizian, M. and Hamdy, R.C. (2010) A hybrid rhOP-1 delivery system enhances new bone regeneration and consolidation in a rabbit model of distraction osteogenesis. *Growth Factors*, **28**, 44–55.
37. Haidar, Z.S., Hamdy, R.C. and Tabrizian, M. (2010) Biocompatibility and safety of a hybrid core-shell nanoparticulate OP-1 delivery system intramuscularly administered in rats. *Biomaterials*, **31**, 2746–54.
38. Haidar, Z.S., Azari, F., Hamdy, R.C. and Tabrizian, M. (2009) Modulated release of OP-1 and enhanced preosteoblast differentiation using a core-shell nanoparticulate system. *Journal of Biomedical Materials Research. Part A: Early View*, **91**, 919–28.
39. Lee, W.F. and Cheng, T.S. (2009) Synthesis and drug-release behavior of porous biodegradable amphiphilic co-polymeric hydrogels. *Journal of Biomaterials Science. Polymer Edition*, **20**, 2023–37.
40. Ranjan, A., Pothayee, N., Seleem, M.N., Tyler, R.D. Jr, Brenseke, B., Sriranganathan, N., Riffle, J.S. and Kasimanickam, R. (2009) Antibacterial efficacy of core-shell nanostructures encapsulating gentamicin against an in vivo intracellular *Salmonella* model. *International Journal of Nanomedicine*, **4**, 289–97.
41. Ydens, I., Degee, P., Nouvel, C., Dellacherie, E., Six, J.L. and Dubois, P. (2005) Surfactant-free stable nanoparticles from biodegradable and amphiphilic poly(ε-caprolactone)-grafted dextran copolymers. *e-Polymers*, **46**, 1–11.
42. Finne-Wistrand, A. and Albertsson, A.C. (2006) The use of polymer design in resorbable colloids. *Annual Review of Materials Research*, **36**, 369–95.

43 Zhang, J. and Misra, R.D.K. (2007) Magnetic drug-targeting carrier encapsulated with thermosensitive smart polymer: core–shell nanoparticle carrier and drug release response. *Acta Biomaterialia*, **3**, 838–50.

44 Surendiran, A., Sandhiya, S., Pradhan, S.C. and Adithan, C. (2009) Novel applications of nanotechnology in medicine. *Indian Journal of Medical Research*, **130**, 689–701.

45 Santander-Ortega, M.J., Bastos-González, D. and Ortega-Vinuesa, J.L. (2007) Electrophoretic mobility and colloidal stability of PLGA particles coated with IgG. *Colloids and Surfaces B, Biointerfaces*, **60**, 80–8.

46 Sant, S., Poulin, S. and Hildgen, P. (2008) Effect of polymer architecture on surface properties, plasma protein adsorption, and cellular interactions of pegylated nanoparticles. *Journal of Biomedical Materials Research. Part A: Early View*, **87**, 885–95.

47 Lee, P.W., Hsu, S.H., Wang, J.J., Tsai, J.S., Lin, K.J., Wey, S.P., Chen, F.R., Lai, C.H., Yen, T.C. and Sung, H.W. (2010) The characteristics, biodistribution, magnetic resonance imaging and biodegradability of superparamagnetic core-shell nanoparticles. *Biomaterials*, **31**, 1316–24.

48 Purushotham, S., Chang, P.E., Rumpel, H., Kee, I.H., Ng, R.T., Chow, P.K., Tan, C.K. and Ramanujan, R.V. (2009) Thermoresponsive core-shell magnetic nanoparticles for combined modalities of cancer therapy. *Nanotechnology*, **20**, 305101.

49 Choi, H.S., Liu, W., Liu, F., Nasr, K., Misra, P., Bawendi, M.G. and Frangioni, J.V. (2010) Design considerations for tumour-targeted nanoparticles. *Nature Nanotechnology*, **5**, 42–7.

50 Zhu, X., Duan, D. and Publicover, N.G. (2010) Magnetic bead based assay for C-reactive protein using quantum-dot fluorescence labeling and immunoaffinity separation. *Analyst*, **135**, 381–9.

51 Xu, H., Sha, M.Y., Wong, E.Y., Uphoff, J., Xu, Y., Treadway, J.A., Truong, A., O'Brien, E., Asquith, S., Stubbins, M., Spurr, N.K., Lai, E.H. and Mahoney, W. (2003) Multiplexed SNP genotyping using the Qbead system: a quantum dot-encoded microsphere-based assay. *Nucleic Acids Research*, **31**, 43.

52 Behrendt, M., Sandros, M.G., McKinney, R.A., McDonald, K., Przybytkowski, E., Tabrizian, M. and Maysinger, D. (2009) Imaging and organelle distribution of fluorescent InGaP/ZnS nanoparticles in glial cells. *Nanomedicine*, **4**, 747–61.

53 Kim, M.J., Park, H.Y., Kim, J., Ryu, J., Hong, S., Han, S.J. and Song, R. (2008) Western blot analysis using metal-nitrilotriacetate conjugated CdSe/ZnS quantum dots. *Analytical Biochemistry*, **379**, 124–6.

54 Shiohara, A., Hanada, S., Prabakar, S., Fujioka, K., Lim, T.H., Yamamoto, K., Northcote, P.T. and Tilley, R.D. (2010) Chemical reactions on surface molecules attached to silicon quantum dots. *Journal of the American Chemical Society*, **132**, 248–53.

55 Zhang, H., Sachdev, D., Wang, C., Hubel, A., Gaillard-Kelly, M. and Yee, D. (2009) Detection and downregulation of type I IGF receptor expression by antibody-conjugated quantum dots in breast cancer cells. *Breast Cancer Research and Treatment*, **114**, 277–85.

56 Sadik, O.A., Zhou, A.L., Kikandi, S., Du, N., Wang, Q. and Varner, K. (2009) Sensors as tools for quantitation, nanotoxicity and nanomonitoring assessment of engineered nanomaterials. *Journal of Environmental Monitoring*, **11**, 1782–800.

57 Hauck, T.S., Anderson, R.E., Fischer, H.C., Newbigging, S. and Chan, W.C. (2010) In vivo quantum-dot toxicity assessment. *Small*, **6**, 138–4.

58 Susumu, K., Uyeda, H.T., Medintz, I.L. and Mattoussi, H. (2007) Design of biotin-functionalized luminescent quantum dots. *Journal of Biomedicine and Biotechnology*, **7**, 90651.

59 Susumu, K., Uyeda, H.T., Medintz, I.L., Pons, T., Delehanty, J.B. and Mattoussi, H. (2007) Enhancing the stability and biological functionalities of quantum dots via compact multifunctional ligands. *Journal of the American Chemical Society*, **129**, 13987–96.

60 Lee, J., Kim, J., Park, E., Jo, S. and Song, R. (2008) PEG-ylated cationic CdSe/ZnS QDs as an efficient intracellular labeling agent. *Physical Chemistry Chemical Physics*, **10**, 1739–42.

61 Jia, H. and Titmuss, S. (2009) Polymer-functionalized nanoparticles: from stealth viruses to biocompatible quantum dots. *Nanomedicine*, **4**, 951–66.

62 Aillon, K.L., Xie, Y., El-Gendy, N., Berkland, C.J. and Forrest, M.L. (2009) Effects of nanomaterial physicochemical properties on in vivo toxicity. *Advanced Drug Delivery Reviews*, **61**, 457–66.

63 Hotz, C.Z. (2005) Applications of quantum dots in biology: an overview. *Methods in Molecular Biology*, **303**, 1–17.

64 Pickup, J.C., Zhi, Z.L., Khan, F., Saxl, T. and Birch, D.J. (2008) Nanomedicine and its potential in diabetes research and practice. *Diabetes/Metabolism Research and Reviews*, **24**, 604–10.

65 Chang, A., Orth, A., Le, B., Menchavez, P. and Miller, L. (2009) Performance analysis of the OneTouch® UltraVue blood glucose monitoring system. *Journal of Diabetes Science and Technology*, **3**, 1158–65.

66 Wang, J. and Musameh, M. (2003) Enzyme-dispersed carbon-nanotube electrodes: a needle microsensor for monitoring glucose. *Analyst*, **128**, 1382–5.

67 Cai, C. and Chen, J. (2004) Direct electron transfer of glucose oxidase promoted by carbon nanotubes. *Analytical Biochemistry*, **332**, 75–83.

68 Lee, S.H., Lee, H., Park, J.S., Choi, H., Han, K.Y., Seo, H.S., Ahn, K.Y., Han, S.S., Cho, Y., Lee, K.H. and Lee, J. (2007) A novel approach to ultrasensitive diagnosis using supramolecular protein nanoparticles. *FASEB Journal*, **21**, 1324–34.

69 Duong, H.D. and Rhee, J.I. (2007) Use of CdSe/ZnS core-shell quantum dots as energy transfer donors in sensing glucose. *Talanta*, **73**, 899–905.

70 Kapoor, Y., Thomas, J.C., Tan, G., John, V.T. and Chauhan, A. (2009) Surfactant-laden soft contact lenses for extended delivery of ophthalmic drugs. *Biomaterials*, **30**, 867–78.

71 Xinming, L., Yingde, C., Lloyd, A.W., Mikhalovsky, S.V., Sandeman, S.R., Howel, C.A. and Liewen, L. (2008) Polymeric hydrogels for novel contact lens-based ophthalmic drug delivery systems: a review. *Contact Lens and Anterior Eye*, **31**, 57–64.

72 Lü, J.M., Wang, X., Marin-Muller, C., Wang, H., Lin, P.H., Yao, Q. and Chen, C. (2009) Current advances in research and clinical applications of PLGA-based nanotechnology. *Expert Review of Molecular Diagnostics*, **9**, 325–41.

73 Mahmud, A., Xiong, X.B., Aliabadi, H.M. and Lavasanifar, A. (2007) Polymeric micelles for drug targeting. *Journal of Drug Targeting*, **15**, 553–84.

74 Panda, J.J., Mishra, A., Basu, A. and Chauhan, V.S. (2008) Stimuli responsive self-assembled hydrogel of a low molecular weight free dipeptide with potential for tunable drug delivery. *Biomacromolecules*, **9**, 2244–50.

75 Van Tomme, S.R., Storm, G. and Hennink, W.E. (2008) In situ gelling hydrogels for pharmaceutical and biomedical applications. *International Journal of Pharmaceutics*, **355**, 1–18.

76 Mincheva, R., Bougard, F., Paneva, D., Vachaudez, M., Manolova, N., Rashkov, I. and Dubois, P. (2009) Natural polyampholyte-based core-shell nanoparticles with N-carboxyethylchitosan-containing core and poly(ethylene oxide) shell. *Biomacromolecules*, **10**, 838–44.

77 Liu, Z., Jiao, Y., Wang, Y., Zhou, C. and Zhang, Z. (2008) Polysaccharides-based nanoparticles as drug delivery systems. *Advanced Drug Delivery Reviews*, **60**, 1650–62.

78 Mano, J.F. (2008) Stimuli-responsive polymeric systems for biomedical applications. *Advanced Engineering Materials*, **10**, 515–27.

79 Thierry, B., Winnik, F.M., Merhi, Y., Silver, J. and Tabrizian, M. (2003) Bioactive coatings of endovascular stents based on polyelectrolyte multilayers. *Biomacromolecules*, **4**, 1564–71.

80 Quinn, J.F. and Caruso, F. (2004) Facile tailoring of film morphology and release properties using layer-by-layer assembly

of thermoresponsive materials. *Langmuir*, **20**, 20–2.
81 Decher, G. (1997) Fuzzy nanoassemblies: toward layered polymeric multicomposites. *Science*, **277**, 1232–7.
82 Yap, H.P., Quinn, J.F., Ng, S.M., Cho, J. and Caruso, F. (2005) Colloid surface engineering via deposition of multilayered thin films from polyelectrolyte blend solutions. *Langmuir*, **21**, 4328–33.
83 Takeuchi, H., Kojima, H., Yamamoto, H., Toshitada, T., Hidekazu, T. and Tomaoki, H. (1998) Physical stability of size controlled small unilamellar liposomes coated with a modified polyvinyl alcohol. *International Journal of Pharmaceutics*, **164**, 103–11.
84 Galovic, R.R., Barisic, K., Pavelic, Z., Zanic, G.T., Cepelak, I. and Filipovic-Grcic, J. (2002) High efficiency entrapment of superoxide dismutase into mucoadhesive chitosan-coated liposomes. *European Journal of Pharmaceutical Sciences*, **15**, 441–8.
85 Hillberg, A.L. and Tabrizian, M. (2006) Biorecognition through layer-by-layer polyelectrolyte assembly: in-situ hybridization on living cells. *Biomacromolecules*, **7**, 2742–50.
86 Douglas, K.L. and Tabrizian, M. (2005) Effect of experimental parameters on the formation of alginate-chitosan nanoparticles and evaluation of their potential application as DNA carrier. *Journal of Biomaterials Science. Polymer Edition*, **16**, 43–56.
87 Gref, R., Minamitake, Y., Perracchia, M.T., Trubeskoy, V., Torchilin, V. and Langer, R. (1994) Biodegradable long-circulating polymeric nanospheres. *Science*, **263**, 1600–3.
88 Desai, M.P., Labhasetwar, V., Amidon, G.L. and Levy, R.J. (1996) Gastrointestinal uptake of biodegradable microparticles: effect of particle size. *Pharmaceutical Research*, **13**, 1838–45.
89 Kim, H.D. and Valentini, R.F. (2002) Retention and activity of BMP-2 in hyaluronic acid-based scaffolds in vitro. *Journal of Biomedical Materials Research*, **59**, 573–84.
90 Sydow-Plum, G., Haidar, Z.S., Merhi, Y. and Tabrizian, M. (2008) Modulating the release kinetics of paclitaxel from membrane-covered stents using different loading strategies. *Materials*, **1**, 25–43.
91 Douglas, K.L., Piccirillo, C.A. and Tabrizian, M. (2006) Effects of alginate inclusion on the vector properties of chitosan-based nanoparticles. *Journal of Controlled Release*, **27**, 354–61.
92 Saravanakumar, G., Kim, K., Park, J.H., Rhee, K. and Kwon, I.C. (2009) Current status of nanoparticle-based imaging agents for early diagnosis of cancer and atherosclerosis. *Journal of Biomedical Nanotechnology*, **5**, 20–35.
93 Ozpolat, B., Sood, A.K. and Lopez-Berestein, G. (2010) Nanomedicine based approaches for the delivery of siRNA in cancer. *Journal of Internal Medicine*, **267**, 44–53.
94 Wu, W., Aiello, M., Zhou, T., Berliner, A., Banerjee, P. and Zhou, S. (2010) In-situ immobilization of quantum dots in polysaccharide-based nanogels for integration of optical pH-sensing, tumor cell imaging, and drug delivery. *Biomaterials*, **31**, 3023–31.
95 Wu, S.Y. and McMillan, N.A. (2009) Lipidic systems for in vivo siRNA delivery. *AAPS Journal*, **11**, 639–52.
96 Franzen, S. and Lommel, S.A. (2009) Targeting cancer with "smart bombs": equipping plant virus nanoparticles for a "seek and destroy" mission. *Nanomedicine*, **4**, 575–88.
97 Chan, J.M., Zhang, L., Yuet, K.P., Liao, G., Rhee, J.W., Langer, R. and Farokhzad, O.C. (2009) PLGA lecithin-PEG core-shell nanoparticles for controlled drug delivery. *Biomaterials*, **30**, 1627–34.
98 Kenley, R., Marden, L., Turek, T., Jin, L., Ron, E. and Hollinger, J.O. (1994) Osseous regeneration in the rat calvarium using novel delivery systems for recombinant human bone morphogenetic protein-2 (rhBMP-2). *Journal of Biomedical Materials Research*, **28**, 1139–47.
99 Higuchi, T., Kinoshita, A., Takahashi, K., Oda, S. and Ishikawa, I. (1999) Bone regeneration by recombinant human bone morphogenetic protein-2 in rat

mandibular defects. An experimental model of defect filling. *Journal of Periodontology*, **70**, 1026–31.

100 Wei, G., Jin, Q., Giannobile, W.V. and Ma, P.X. (2007) The enhancement of osteogenesis by nano-fibrous scaffolds incorporating rhBMP-7 nanospheres. *Biomaterials*, **28**, 2087–96.

101 Corsi, F., De Palma, C., Colombo, M., Allevi, R., Nebuloni, M., Ronchi, S., Rizzi, G., Tosoni, A., Trabucchi, E., Clementi, E. and Prosperi, D. (2009) Towards ideal magnetofluorescent nanoparticles for bimodal detection of breast-cancer cells. *Small*, **5**, 2555–64.

102 Chung, Y.I., Kim, J.C., Kim, Y.H., Tae, G., Lee, S.Y., Kim, K. and Kwon, I.C. (2010) The effect of surface functionalization of PLGA nanoparticles by heparin- or chitosan-conjugated Pluronic on tumor targeting. *Journal of Controlled Release*, **143** (3), 374–82.

103 Adiseshaiah, P.P., Hall, J.B. and McNeil, S.E. (2010) Nanomaterial standards for efficacy and toxicity assessment. *Wiley Interdisciplinary Reviews: Nanomedicine and Nanobiotechnology*, **2**, 99–112.

104 Haigron, P., Dillenseger, J.L., Luo, L. and Coatrieux, J.L. (2010) Image-guided therapy: evolution and breakthrough [a look at]. *IEEE Engineering in Medicine and Biology Magazine*, **29**, 100–4.

105 Orive, G., Hernández, R.M., Gascón, A.R. and Pedraz, J.L. (2005) Micro and nano drug delivery systems in cancer therapy. *Cancer Therapy*, **3**, 131–8.

106 Lissett, B.R., Chang, J., Fu, K., Sun, J., Hu, Y., Gobin, A., Yu, T. and Drezek, R.A. (2008) Evaluation of immunotargeted gold nanoshells as rapid diagnostic imaging agents for HER2-overexpressing breast cancer cells: a time-based analysis. *Nanobiotechnology*, **4**, 1–8.

107 Bickford, L.R., Agollah, G., Drezek, R. and Yu, T.K. (2010) Silica-gold nanoshells as potential intraoperative molecular probes for HER2-overexpression in ex vivo breast tissue using near-infrared reflectance confocal microscopy. *Breast Cancer Research and Treatment*, **120** (3), 547–55.

108 Wang, C., Chen, J. and Talavage, T. (2009) Gold nanorod/Fe_3O_4 nanoparticle "nano-pearl-necklaces" for simultaneous targeting, dual-mode imaging, and photothermal ablation of cancer cells. *Angewandte Chemie International Edition*, **48**, 2759–63.

109 Ramesan, R.M. and Sharma, C.P. (2009) Challenges and advances in nanoparticle-based oral insulin delivery. *Expert Review of Medical Devices*, **6**, 665–76.

110 Yu, B., Ahn, J.S., Lim, J.I. and Lee, Y.K. (2009) Influence of TiO_2 nanoparticles on the optical properties of resin composites. *Dental Materials*, **25**, 1142–7.

111 Kim, K.H. and Ramaswamy, N. (2009) Electrochemical surface modification of titanium in dentistry. *Dental Materials Journal*, **28**, 20–36.

112 de la Isla, A., Brostow, W., Bujard, B., Estevez, M., Rodriguez, J.R., Vargas, S. and Castano, V.M. (2003) Nanohybrid scratch resistant coating for teeth and bone viscoelasticity manifested in tribology. *Materials Research Innovations*, **7**, 110–14.

113 Berube, D.M. (2006) The magic of nano. *Nanotechnology Perceptions*, **2**, 249–55.

114 Ballauff, M. and Lu, Y. (2007) "Smart" nanoparticles: preparation, characterization and applications. *Polymer*, **48**, 1815–23.

115 Welsch, N., Wittemann, A. and Ballauff, M. (2009) Enhanced activity of enzymes immobilized in thermoresponsive core-shell microgels. *Journal of Physical Chemistry B*, **113**, 16039–45.

11
Polymer Nanoparticles and Their Cellular Interactions

Volker Mailänder and Katharina Landfester

11.1
Introduction

Polymeric particles in the submicrometer size range are used extensively in biomedical applications [1–8]. During the past few years, many research groups have conducted investigations how cells of human origin might interact with nanomaterials. This is of particular interest as cellular therapy can offer great opportunities for regenerative medicine, especially for the repair of tissue function following organ damage; indeed, the addition of such nanomaterials might even enhance the ability of these cellular products to repair the damaged tissues. In this situation, nanomaterials carrying specific molecules could influence the fate of differentiation of the cells, or enable the detection of their migration and homing.

Nanomaterials are characterized by their size, which is well below the micrometer range and the size range of mammalian cells. Nonetheless, nanomaterials are larger than single molecules (small molecules) or even proteins, ranging in size from several nanometers to some hundreds of nanometers.

For many years, nanotechnology – and the nanomaterials thus produced – has promised to employ the specific properties of such supramolecular assemblies and materials, so that hitherto inaccessible effects could be exploited for new applications. Many of these properties are not demonstrated by single molecules or assemblies in the micrometer range; rather, some effects that are found only within this nanosize range are in the field of physics, including the optical properties of nanoparticles (gold nanoparticles of different colors, depending on their size, fluorescent "quantum dots"), the superparamagnetism of small magnetic nanoparticles [9], or the supercooling of fluids in confined geometries [10]. Other examples are the high surface areas required for catalysis [11] and adsorption [12]. In biology and medicine, superparamagnetic iron oxide nanoparticles have been employed for cell selection [13] and as magnetic resonance imaging (MRI) contrast agents. During the past few years it has also become clear that the uptake of nanoparticles is possible in a wide variety of cells – an effect which appears, in cell biology, to be specific for materials in the range of 50 to 200 nm [14, 15].

Nanomaterials for the Life Sciences Vol. 10: Polymeric Nanomaterials. Edited by Challa S. S. R. Kumar
Copyright © 2011 WILEY-VCH Verlag GmbH & Co. KGaA, Weinheim
ISBN: 978-3-527-32170-4

While the labeling and selection of cells represent possible applications, targeted nanoparticulate drug delivery is today considered to be one of the most promising techniques for increasing the efficiency of drug administration [16, 17]. Following their administration, the degradation or metabolism of drugs can be effectively reduced by the drug being incorporated into, or adsorbed onto, a carrier system. In this way, it is possible that the oral application of insulin or heparin [18, 19] or the transdermal delivery of drugs [20, 21] might serve as novel routes for drug administration.

Although, in the past, the development of many drugs has been abandoned due to their inaccessibility to intracellular targets, or their sensitivity to extracellular degradation (e.g., the enzymatic digestion of nucleic acids), nanoparticles might prove to be of great value for the delivery of such drugs. For example, some small molecules may not be taken up by certain cell types, or may be actively removed from the cytoplasm of the cell, perhaps by multi-drug resistance-related proteins [22]. Hence, in such a situation the ability to deliver a drug directly into the cell would clearly be highly advantageous.

The information provided in this chapter will relate to stem cells and differentiated mammalian cells. Whilst hematopoetic stem cell populations have been collected, processed and manipulated for clinical reapplication in malignant and non-malignant diseases for more than three decades, further applications have been investigated in recent years. Today, regenerative therapy using cell preparations of different origin to repair damaged tissue is a rapidly evolving field in medicine. However, in order to understand and harness the possible use of nanoparticles for (stem) cell therapy, their interactions with cells must be investigated. Thus, the chapter will provide details of investigations into the effects of polymeric materials on the interaction with cells, especially with regards to the influence of the polymer itself and of the other components (e.g., surfactants) used in the preparation of nanoparticles.

Such interactive behavior may be modified deliberately, by using monomers with substituted side groups; alternatively, the surfaces of the nanoparticles, which serve as the interaction partners with the cell surface and other cellular compartments and proteins, can be altered intentionally by modifying parameters such as molecular charge or the nature of the amino acids present. In the case of surface functionalization, which can be achieved via copolymerization, the influence of the density of side groups per square nanometer (at least for some nanoparticles) must be investigated to establish a density–uptake relationship, in addition to the uptake kinetics. Likewise, subcellular localization should be determined since, in order to use nanoparticles for DNA delivery, they must first be transported into the cell nucleus.

Pharmacological inhibitors of defined endocytotic pathways may also be used to investigate and dissect the uptake mechanisms by which the nanoparticles are endocytosed by cells.

The value of various reporters is also demonstrated, where nanoparticles can serve either as marker systems for future *in vivo* studies (e.g., investigations into the homing and trafficking of (stem) cells), or in a more complex form with encapsulated substances as possible drug-delivery systems.

11.1.1
Nanoparticle Synthesis by Miniemulsion

Today, polymeric particles in the sub-micrometer size range are used extensively in biomedical applications. Compared to other nanoparticulate systems (e.g., liposomes, micelles), nanoparticles show an increased colloidal stability, a better chemical resistance, and their creation is usually much more easily achieved. Notably, when nanoparticles are envisaged as specific carriers for therapeutic, contrasting or imaging agents, their development is pursued with attention focused clearly on their ability to permeate through different tissues.

While other synthetic processes result in distinct types of nanoparticles, the miniemulsion process has proved to be the most versatile, as it permits the use of a wide variety of materials in the synthesis of nanoparticles. In particular, the miniemulsion process enables defined modifications to be made of the nanoparticles' parameters, even in the presence of a complex structure. The principle of miniemulsion is described in detail in Ref. [10], and shown schematically in Figure 11.1. Following the mixing of an oily phase with an aqueous phase, oil droplets are formed by stirring; these droplets are comparatively large (in the micrometer range) and their sizes differ widely. In order to obtain smaller particles (in the nanometer range) and to provide monodisperse nanodroplets, a high shear stress is applied by using either ultrasound or high-pressure homogenization. When the nanodroplets are formed they can be stabilized by adding a surfactant.

If the oily phase consists of monomers, they can be polymerized such that the geometry and size of the droplets is preserved beyond the stability of the oily droplets. The intriguing point here is that any substances present within the nanoparticles are enclosed by the same process. In order to remove any nonreacted monomers and most of the surfactant, the nanoparticles are dialyzed for use in cell experiments. The use of a different surfactant, as well as a change in the

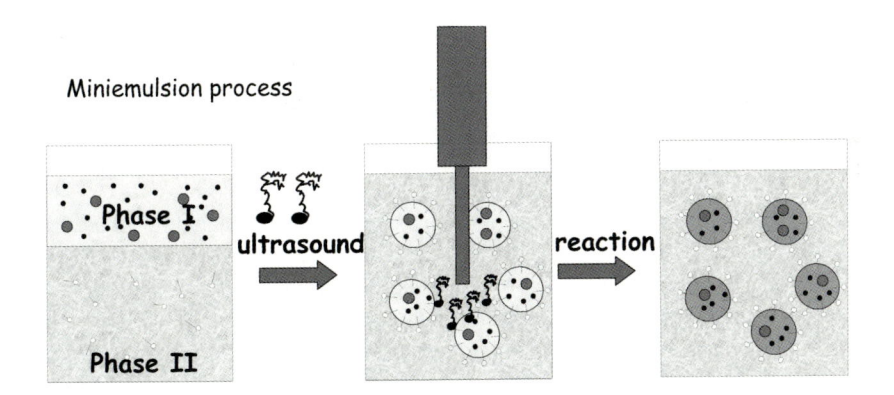

Figure 11.1 Schematic of the miniemulsion process.

water:oil ratio leads to the creation of aqueous nanodroplets in an oily continuous phase (the so-called "inverse miniemulsion").

It should be noted that the principle of miniemulsion allows a wide variety of polymers, enclosed substances (reporters, drugs, etc.), surfactants, geometries (nanoparticles, nanocapsules, etc.), each with different surface functionalities, to be synthesized.

11.2
Nanoparticles as Labeling Agents for Cellular Therapeutics

11.2.1
Experimental Polystyrene Nanoparticles with Iron Oxide (Magnetite) and a Fluorescent Dye as Reporter

Commercially available MRI contrast agents (e.g., Feridex®, Resovist®) are based on superparamagnetic iron oxide nanoparticles which typically are approximately 10 nm in size. The superparamagnetic effect of these nanoparticles leads to their use not only as interesting contrast agents for magnetic resonance tomography [23], but also as nonviral vehicles for gene therapy [24, 25], drug delivery [17, 26, 27], immunization [19, 28–30], and detoxification [12].

In order to improve the properties of magnetite nanoparticles with regards to their aggregation, coagulation, and iron leakage, the proposal was made that they should be efficiently encapsulated in a hydrophobic polymer shell, rather than in a biologically unstable dextran coating. This was recently achieved by Landfester et al., using the miniemulsion process [31, 32]. With a high magnetite content, it is possible to achieve a uniform distribution in the polymer [32]; moreover, the use of nonbiodegradable polystyrene (PS) as the polymer would allow the nanoparticles to be used for long-term studies in animals, but not in humans.

In order that the behavior of nanoparticles could be monitored more easily, the decision was taken to create "dual-reporter nanoparticles"; such nanoparticles can be detected in different ways, typically using fluorescence or MRI [33]. Thus, a series of magnetic PS particles encapsulating magnetite nanoparticles (10–12 nm) in a hydrophobic poly(styrene-co-acrylic acid) shell in which the fluorescent dye PMI (N-(2,6-diisopropylphenyl)perylene-3,4-dicarbonacidimide) was encapsulated, was synthesized via a three-step miniemulsion process [33]. This allowed the incorporation of a large amount of iron oxide (magnetite) – typically 30–40% (w/w) – while polymerization of the styrene monomer led to the production of nanoparticles of 45–70 nm in size; these findings were subsequently confirmed by others [34, 35]. Furthermore, the copolymerization of styrene with the hydrophilic acrylic acid led to varying amounts of carboxyl groups being present on the surface, which in turn enabled the creation of quite complex nanoparticles.

Surface modification with carboxylic groups (by copolymerization with acrylic acid) can be used for coupling lysine, glutamine, or asparagine to the carboxylic

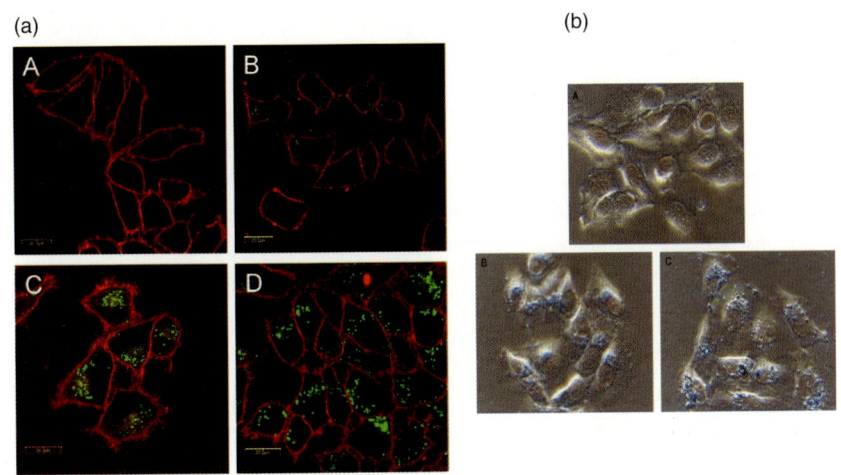

Figure 11.2 (a) Laser scanning microscopy of HeLa cells. The cell membranes are stained with RH414 in red (contours) and the particles in green (dots in the cells). A, Untreated control; B, VHMPM2 (2% of acrylic acid); C, VHMPM5 (5% of acrylic acid); D, VHMPM10 (10% of acrylic acid). Reproduced with permission from Ref. [33]; © 2006; IOP Publishing; (b) Prussian blue staining of VH3ka and VH3kaLys in HeLa cells. While there are no blue spots in A (control cells), there are bluish spots in VH3ka when PLL (B) was added and VH3kaLys (C) even without the addition of a transfection agent. Holzapfel, V., et al. (2006) Synthesis and biomedical applications of functionalized fluorescent and magnetic dual reporter nanoparticles as obtained in the miniemulsion process; © (2006), IOP Publishing.

groups via 1-ethyl-3-(3-dimethylaminopropyl)carbodiimide hydrochloride (EDC) coupling [36].

For biomedical evaluation, the nanoparticles were incubated with different cell types. The introduction of carboxyl groups onto the particle surfaces led to an enhanced cellular uptake of nanoparticles, as indicated by detection of the fluorescent signal using fluorescence-activated cell sorting; FACS [33] and laser scanning microscopy (LSM) (Figure 11.2a).

Whilst the nanoparticles where readily detectable by using FACS and LSM, at no point did the uptake of iron exceed 1 pg per cell (data not shown). The quantity of iron oxide required to be present in cells for most biomedical applications (e.g., detection by MRI) must be significantly higher (ca. 10–20 pg Fe per cell) [37]. A further increase of uptake can be accomplished by using transfection agents such as poly-L-lysine or other positively charged polymers [33, 38, 39] [33], and this will also enable detection by Prussian blue staining (Figure 11.2b). The addition of poly-L-lysine to the nanoparticles as a separate reagent has been described previously [37, 38, 40, 41]; this functionality can also be engrafted onto the surface of the nanoparticles by the covalent coupling of lysine to carboxyl groups.

The amount of iron that could be transfected was even higher than when a transfection agent was added to the nanoparticles, and was adsorbed only physically [33]. Furthermore, the subcellular localization of these nanoparticles was demonstrated, using transmission electron microscopy (TEM), to be clustered in the endosomal compartments [33].

While coagulation and clumping proved to be a major concern with the commercially available dextran nanoparticles Resovist® or Feridex®, no coagulation or clumping was observed with the carboxyl-functionalized nanoparticles in the cell culture media. Indeed, even at the highest concentrations tested, all particles that were not removed by washing were found to have become located inside the tested cell types, and not simply adsorbed onto the cell surfaces [33].

11.2.2
Gadolinium as a Reporter in Nanoparticles

Among several reports describing the combination of gadolinium molecules, Lipinski et al. used Gd-containing immunomicelles within the range of 85 to 130 nm, with between four and nine Gd atoms per micelle [42]. Although micelles demonstrate a limited stability after injection into the blood, mixing them with diacetylene, followed by a photolytic polymerization, led to an increased stability [43]. Wooley et al. synthesized gadolinium-labeled crosslinked nanoparticles with hydrodynamic diameters of approximately 40 nm by using diblock copolymers [poly(acrylic acid) (PAA) and poly(methacrylic acid) (PMA)] micelles, which were then covalently crosslinked by amidation with ethyl amines. Gadolinium-coordinated penta-acetic acids were then located at the surface of the micelles, allowing a rapid water exchange and thus high relaxivities [44]. Yu et al. conducted an intensive investigation into the synthesis of fibrin-targeted contrast agents to detect fibrin in human thrombus *in vitro* [45], while lipid-encapsulated perfluorocarbon nanoparticles with numerous Gd–diethylenetriaminepenta-acetic acid (DTPA) complexes incorporated onto the outer surface were synthesized by other groups for use as a contrast agent [45, 46]. In this case, more than 90 000 Gd[3+] could be attached on particles of 250 nm diameter, which resulted in a 20–40% decrease in T_1 relaxation.

Likewise, the encapsulation of gadolinium complexes inside apoferritin spheres resulted in approximately 10 units of the hydrophilic gadolinium chelate complex being encapsulated per apoferritin sphere [47].

An alternative approach was developed by Reynolds et al., who synthesized core–shell nanoparticles loaded with a gadolinium salt in an intermediate layer. The high porosity of the shell permitted a rapid exchange of water, such that a reduction in relaxation time could be determined [48].

These data point to the fact that, in order to maintain the relaxation properties after encapsulation of these lanthanides, the most important concern after encapsulation is to ensure a sufficient water exchange between the encapsulated contrast agent and the surrounding ^1H protons, and as a consequence of this the shell of the nanocapsule must be adjusted. Moreover, the gadolinium complex must not be able to diffuse out, while the ^1H protons should be able to enter and leave the interior of the nanocapsule very easily.

Nanocapsules of sizes between 80 and 400 nm with a hydrophobic core and a polymer shell were synthesized by Tiarks et al., using a one-step procedure, via the direct aqueous miniemulsion technique [49]. An inverse miniemulsion process allows the synthesis of nanocapsules with an aqueous core, with polyurea, polythiourea, or polyurethane shells being obtained via interfacial polyaddition at the water/oil interface [50].

Nanocapsules containing the hydrophilic gadolinium complexes Magnevist® and Gadovist® (for structural formulas, see Ref. [33]) were obtained via an inverse miniemulsion process [51], with the shell thickness being adjusted by modifying the monomer concentration; in this way, the shell can be rendered impermeable to the Gd complex. The hallmark of this procedure is a high degree of flexibility; indeed, the diffusion of water molecules through the capsule walls can be ensured due to the porosity of the polymeric shell, and nanocapsule wall thicknesses of 20–40 nm could be obtained [51].

These nanoparticles were first tested for their relaxivity in cyclohexane as the continuous phase, where they showed a low T_1 relaxivity. This could be easily explained by bearing in mind that, due to a limited water content inside the capsules and an absence of water molecules outside the nanocapsules, the number of protons would be by far insufficient for local strong T_1 shortening. However, the T_1 relaxivity was not inferior compared to that for nonencapsulated Magnevist® solution, after having transferred the nanocapsules to water as a continuous phase [51].

The unchanged T_1 relaxivities provide proof of a sufficient water/proton exchange through the polymeric shell. As might be expected, the T_1 relaxivity was also found to depend on the thickness of the polymer shell, which in turn affects the efficiency of the water exchange.

Both, blood cells and proteins, may cause alterations in the exchange rate of water from the outside to the inside, for example, by coating the nanoparticles and thereby increasing the wall thickness or porosity. Hence, experiments were conducted which simulated physiological conditions (isotonic NaCl solution, human blood) [33], the results of which showed clearly that, in all cases, the proton exchange was sufficiently high. The results of these experiments also demonstrated the clear potential for using nanocapsules as new contrast agent materials for MRI.

More recently, functionalization of the nanocapsules has become possible, and this should allow the targeting of defined molecular structures for the detection of specific diseases.

11.3
Uptake of Polymeric Nanoparticles into Cells

The uptake and interaction of nanoparticles with cells may be influenced by several parameters, which can be categorized as follows:

- The structure and morphology of the polymer in the nanoparticles themselves. Both, the hydrophilic and hydrophobic polymeric surface properties of nanoparticles are known to influence cell adhesion during the uptake process.

It has been reported that, with increasing hydrophobicity of the polymer, the attachment on cells and subsequent internalization is enhanced [52].

- Surface groups that are covalently bound to the nanoparticles by copolymerization should influence cellular uptake, as the cell membrane has not only a negative net charge but also consists of areas with positively charged proteins.
- Amphiphilic polymers can be used in order to enhance cellular uptake. These polymers are physically adsorbed onto the nanoparticles' surface (as a transfection agent).

11.3.1
Influence of Polymer on Uptake

As the polymer comprises up most of the nanoparticles, it might be expected that the uptake of nanoparticles could be altered by the type of polymer used. The other properties (e.g., size) must be kept stable, while the surface density of functionalization, and of the surfactant, should not cover the entire surface in dense fashion. In miniemulsions, the surface density of the functionalization can be adjusted by the amount of comonomers used; the density of the surfactant is minimized in miniemulsions compared to microemulsions, and is further reduced by dialysis after synthesis.

11.3.1.1 Polystyrene

Styrene can be easily emulsified and polymerized in miniemulsions, and many modifications are possible. For example, copolymerization with a monomer that produces surface functionalization is easily achieved (see Refs [14, 33, 53]). Within the context of a cellular environment, PS does not degrade significantly; however, whilst for long-term experiments in cell cultures and animals this is a required property, it would be undesirable for use in humans. The nonbiodegradability of a nanoparticles would result in its accumulation, the negative side effects of which would be of major concern in humans.

The uptake of unmodified PS nanoparticles was only minimal by the cell lines tested [54]. If PS is modified by charged comonomers, then the uptake would mostly be enhanced. The results of the cell experiments are described in detail in Section 11.3.2.1.

11.3.1.2 Polyisoprene

It has been speculated that the uptake behavior of nanoparticles into cells might be facilitated if the polymers used showed a high degree of similarity to natural structures. The terpene structure, which is found widely in nature, occurs in essential oils and pheromones, and in polymer science is known as the monomer *isoprene*. Thus, it was suggested that polyisoprene (PI) nanoparticles would be well suited to the interaction with the lipophobic portion of the cell membrane, and for cell uptake. The remaining double bonds in the PI latex may also be used to functionalize the polymeric particles [55, 56].

11.3 Uptake of Polymeric Nanoparticles into Cells

Fluorescent PI nanoparticles have been synthesized, via the miniemulsion technique, as marker particles for cells in what emerged as the first report on PI nanoparticles. In this case, an adherent cell line (HeLa) and a suspension cell line (Jurkat) were used to study the uptake of the nonfunctionalized PI nanoparticles, in the absence of any transfection agents.

11.3.1.2.1 Concentration-Dependence of Particles on Uptake
When increasing the concentration of nanoparticles in the cell medium, the occurrence of a plateau should be expected as the cell most likely has only a limited capacity for nanoparticle uptake. In this analysis, the HeLa and Jurkat cell lines were incubated with increasing concentrations of PI nanoparticles (ranging from 37.5 to 4800 µg) in the culture medium for 24 h, in the absence of any transfection agents. Both cell lines (Figure 11.3) showed a steady and rapid increase in uptake over the lower particle concentration range, but higher concentrations resulted in a flattening of the curve. Such a state of saturation might be caused by a depletion of the

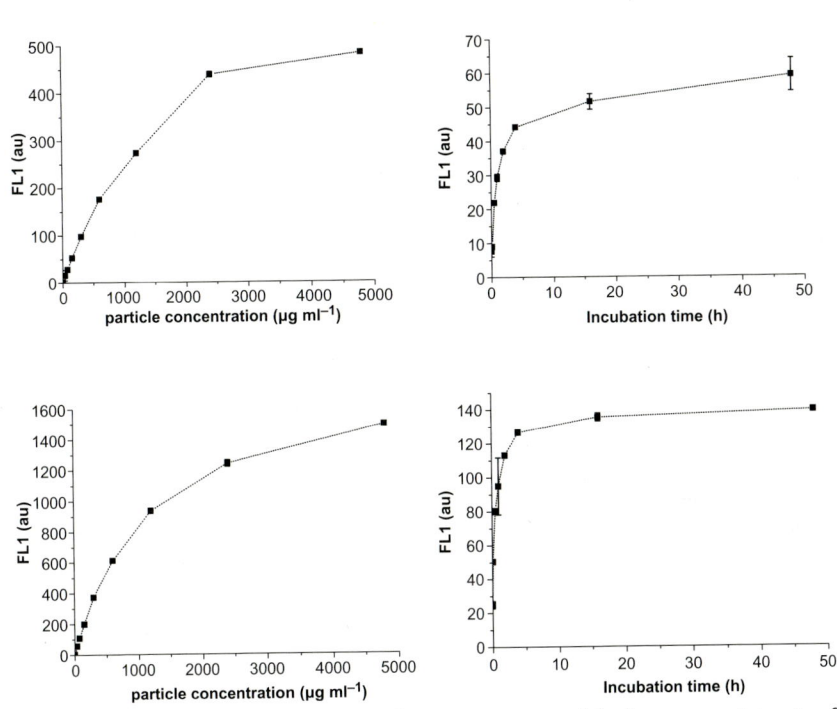

Figure 11.3 Left: FACS analysis of Jurkat cells (lower row) and HeLa cells (upper row) incubated with different concentrations of dialyzed PI particles overnight for 24 h. Saturation of the curves is near-complete after incubation of 2400 mg of solid content of the particles per ml of medium; Right: FACS measurements of the fluorescence intensity of Jurkat cells (lower row) and HeLa cells (upper row) after incubation of fluorescent PI nanoparticles for different time periods (0, 10, 30 min, 1, 2, 4, 16, and 48 h). In both cell types, the half-maximum uptake was achieved within 1 h [57].

endocytotic mechanisms, or saturation of the cells' storage capacity. As a consequence, a particle concentration of 75 µg ml^{-1} was used for further experiments, as this was well below the saturation level.

11.3.1.2.2 Kinetics of the Uptake

The Jurkat and HeLa cells were incubated in the culture medium in the presence of 75 µg ml^{-1} PI nanoparticles for up to 48 h to evaluate the uptake kinetics (see Figure 11.3). Both cell lines showed a rapid uptake during the first hour; in the case of the PI nanoparticles, particle internalization was seen to commence during the first minutes of incubation, with the uptake kinetic reaching a plateau and resulting in saturation. With the 48 h value set as a maximum, the half-maximum was reached at 1 h for HeLa cells, and at less than 1 h for Jurkat cells.

The PI nanoparticles were taken up much more rapidly during the first 4 h of incubation when compared to PS particles, in the presence of the transfection agent poly-L-lysine (PLL) or amino-functionalized PS nanoparticles (see Ref. [57], PS nanoparticles with PLL; and Ref. [57], 0.21 NH$_2$ groups per nm^2). The functionalized PS nanoparticles showed an absence of saturation, even after a 24 h period of incubation, while the half-maximum value was well above 6 h.

The difference in uptake kinetics between the PI and PS nanoparticles may be accounted for by two distinct mechanisms. It was speculated that, in the case of the rather high initial uptake of the PI particles, a receptor-mediated process might be suspected, although no specific receptor has been identified. A receptor-mediated internalization process normally occurs much faster than does constitutional uptake [15, 58–60].

In conclusion, the PI nanoparticles were internalized by different cell lines relevant to biomedical applications. In particular, the Jurkat cells – as a model for T cells – demonstrated a fair degree of uptake when compared to PS nanoparticles or to Feridex® or Resovist®. Thus, PI nanoparticles could be used to label these difficult-to-transfect cells efficiently, if a marker were to be incorporated into the particles. Moreover, as PI is not biodegradable, the particles should be suited to long-term applications in animal studies.

11.3.1.3 Nanoparticles Composed of Different Proportions of Polystyrene and Polyisoprene

As (unfunctionalized) PS particles show a lower uptake rate than PI nanoparticles, the uptake rates could be tuned by the proportion of PS in the PI/PS copolymer particles. It has been shown previously [57] that there is a trend towards a higher uptake when the proportion of PI is increased, thereby confirming the results obtained with pure PS or PI particles.

11.3.1.4 Poly (n-butylcyanoacrylate) (PBCA) Nanoparticles

While the polymers discussed above are nonbiodegradable, the poly (alkylcyanoacrylate) (PACA) nanoparticles are both biocompatible and biodegradable, and capable of absorbing or entrapping bioactive compounds. As a consequence, they can serve as ideal tools in applications where the nanoparticles are administered repeatedly, for example, as a drug-carrier system. Many different compounds have been used as the "payload," including inorganic crystallites (e.g., magnetite [61]), various drugs (e.g., methotrexate [62], doxorubicin [63–65]), and even oligopeptides (e.g., dalargin [66, 67]) or proteins (e.g., insulin [18, 68, 69]).

When Couvreur added the alkylcyanoacrylate monomer dropwise into a HCl solution containing a polymeric, nonionic surfactant, the described particles showed a broad size distribution that ranged from less than 100 nm to over 1 μm [70]. In this case, the particle size, stability of the dispersion and molar masses of the polymer were shown to depend heavily on the pH of the continuous phase [68, 71–73], and also on the type and concentration of the surfactant [74, 75].

Despite the extensive application of emulsion polymerization with nonionic or polymeric surfactants for the preparation of PACA nanoparticles, there remain several limitations. One problem relates to the low polymer content of the dispersions (ca. 1 wt%), while another involves the large amount of surfactant compared to monomer (the surfactant:monomer ratio may be 1:1, or even higher; see Refs [66, 76].). Modification of the poly(n-butyl cyanoacrylate) (PBCA) particle surface is achieved by the choice of surfactant, which is physically adsorbed (as with polysorbates [66, 67]) or chemically bonded via, for example, a hydroxyl group of dextran [74, 77] or poly(ethylene glycol) (PEG) [78] to the particles' surface. Some reports have been made that these modifications with polysorbates allow the particles to permeate through the blood–brain barrier (see Chapter 13). Moreover, PEGylated particles show a greater persistence in the circulatory system.

Fluorescent dye-labeled unfunctionalized and functionalized PBCA nanoparticles with a narrow size distribution were prepared by applying the miniemulsion technique, whereby the acid used as the continuous phase and the applied initiator solution determined the molar mass distribution of the polymer [79]. In this case, the initiator is incorporated onto the particles' surface and thus serves to functionalize the particles, whereas unfunctionalized particles are obtained after initiation with NaOH solution. Amines, amino acids, or PEGs used as initiators lead to surface-functionalized particles with these molecules.

As PBCA nanoparticles are relatively toxic in cell culture assays, concentrations of only 0.75 to 75 μg ml^{-1} were investigated. The finding that the uptake was almost linear over this range confirmed that the mechanisms involved had not been saturated [79].

Whilst the molar mass of the polymer determines the onset and extent of apoptosis, the total uptake is not dependent on the molar mass. Indeed, different uptake kinetics have been demonstrated with HeLa and Jurkat cells following incubation with the same particle batch [79]. The use of confocal laser scanning microscopy failed to demonstrate any significant differences in intracellular particle distribution for both cell lines, nor for the particle batches [79].

11.3.1.5 Polyester Nanoparticles: Poly(ε-Caprolactone), Poly(D,L-Lactide), Poly(D,L-lactide-co-Glycolide)

A further class of biodegradable polymers, in addition to the PACAs (such as PBCA), would be polyesters such as poly(ε-caprolactone) (PCL) or poly(D,L-lactide)

(PLLA). These materials are the first choice for sustained-release applications [19, 69, 80–84], and are also biocompatible as they demonstrate a low level of toxicity *in vivo*. Consequently, they represent ideal candidates for the controlled release of pharmacologically active substances [85–88].

For other nanoparticles (such as PS, noted above), the monomer in the nanodroplets was first emulsified and then polymerized. In contrast, the polymerization of L-lactide is difficult to achieve in miniemulsions, as the chemical reactions (polycondensation or ring-opening) cannot be carried out easily in water, or in close contact with water. Consequently, the process was modified so that the emulsification of an organic phase containing the dissolved preformed polymer that had been dissolved in an organic solvent and was within an aqueous solution containing the surfactant, could be performed. In this way, monodisperse nanodroplets could be obtained and tailored to a size range of 50 to 500 nm [49]. Precipitation of the polymer is initiated in the second stage as the organic solvent is removed from the system. In addition, the process allows the entrapment of hydrophobic components (emulsion/solvent evaporation process). In this way, nanoparticles were prepared from a series of biodegradable polymers that included PLLA, PCL, and poly(D,L-lactide-*co*-glycolide) (PLGA). Following the incorporated also of a fluorescent dye, these particles could be used as platform particles for the encapsulation of various hydrophobic materials, such as drugs.

11.3.1.5.1 Polymer Degradation

As the mechanism of aliphatic polyester degradation involves hydrolysis of the backbone ester groups, it was hypothesized that most hydrophobic and crystalline PCL and PLLA nanoparticles would degrade slower than PLGA, based on the physico-chemical characteristics of the polymers [89–94]. However, as the use of ultrasound during miniemulsion formation led to the induction of polymer degradation [95], the molecular weight average of the final particles prepared without ultrasonication was also investigated [96]. Subsequently, the degradation of polymers with a high initial molecular weight was found to be more affected by ultrasonication than was that of shorter chain-length polymers.

After formulation, the polymer molecular weight fell rapidly as a result of random hydrolytic cleavage of the ester bonds, with the degradation rate of the amorphous PLGA particles being slightly faster than for other polymers due to the higher permeability of the material for water molecules. Over a period of five months, the molecular weights of the PLLA, PCL, and PLGA nanoparticles prepared with sodium dodecylsulfate (SDS) were decreased by 40%, 50%, and 47%, respectively. In contrast, when the same particles were stabilized with Lutensol AT50, the loss in molecular weight was 20%, 30%, and 27%, respectively [96]. This confirmed that the presence of ionic groups on the particle surface (by the action of SDS) would accelerate degradation of the polymer due to an improved water access [97] although, interestingly the particle size did not change significantly when the polymer was hydrolyzed. These results were in agreement with the findings of Vert *et al.* [93]

11.3.1.5.2 Cellular Uptake of Biodegradable Nanoparticles: Flow Cytometry

For both Jurkat and HeLa cells, the uptake patterns of the various particle variants were similar [96]. Typically, the molecular weights of the polymers (H = high-molecular-weight; L = low-molecular-weight) had no influence on the uptake of either the PLLA or PCL particles. Neither could any consistent correlation between the ζ-potential or particle size and cellular uptake be detected. Within this series of nanoparticles, only the PLGA nanoparticles show a greater degree of uptake into the cells.

Except for PCL particle uptake in Jurkat cells, the surfactant located on the surface of the nanoparticles had a clear influence on cellular uptake. In most cases, anionically stabilized SDS particle variants were taken up to a greater extent than the Lutensol AT50 (nonionically stabilized) analogues, with this effect being most pronounced for the uptake of PLLA particles [96] but much weaker for PLGA and PCL [14, 54].

11.3.1.5.3 Cellular Uptake of Biodegradable Nanoparticles: Confocal Laser Scanning Microscopy

The intracellular location of the nanoparticles was confirmed using confocal laser scanning microscopy (cLSM) [96], with the respective types of polymer particle stabilized with Lutensol AT50 being identified after a 24 h period of incubation with both Jurkat and HeLa cells. Notably, the images recorded following uptake of the SDS variants showed similar results (although, due to the different PMI contents of the particles [96] the results cannot be compared directly with those obtained with fluorescence microscopy). The staining patterns were similar in the Jurkat and HeLa cells. Moreover, it was noticeable (but not explained) that the uptake of both PCL (e.g., PCLH-16) and PLGA (e.g., PLGA-18) particles resulted in a diffuse background staining with some bright spots, whereas the background staining was very weak following PLLA (e.g., PLLAL-15) particle uptake [96].

The TEM images of HeLa cells acquired after 24 h incubation showed the PLLA particles as bright spots in the vicinity of cells, and also intracellularly (Figure 11.4), while filopod-like membrane extensions appeared to enclose the particles. In most cases, multiple particles were localized in endosomes of varying sizes, a situation which was comparable to SPIOs or PS but was in contrast to PI. Although a number of the individual particles appeared to have escaped from the compartments (arrows in Figure 11.4), further investigations will be required to determine if the particles are truly outside the endosomes.

Win and Feng [98], using Caco-2 cells, reported the nuclear uptake of vitamin E (D-alpha-tocopheryl polyethylene)-coated PLGA particles of about 200 nm. These results contrasted with those obtained not only for PLGA but also all other nanoparticles tested. Indeed, no particles were ever identified in the nucleus, during either cLSM or TEM imaging.

The uptake kinetics was studied for PLLAL-15 Lut and PCLH-16 Lut, both of which were formulated with the nonionic (PEG-containing) surfactant Lutensol AT50. As shown previously [96], the initial rates of uptake for both particles were

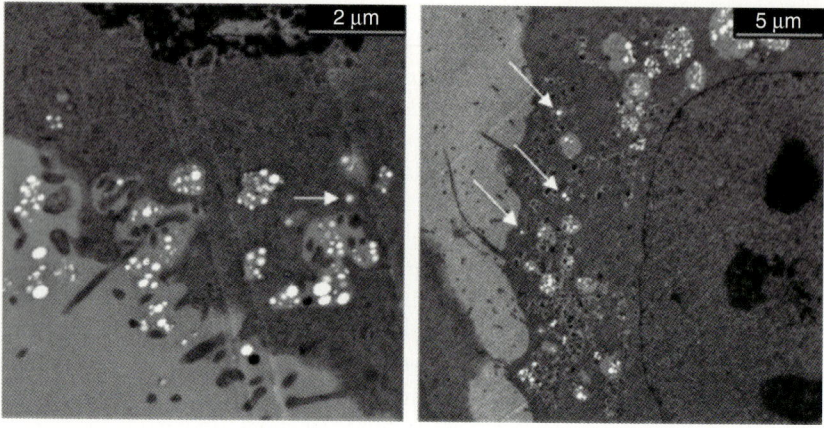

Figure 11.4 TEM images of HeLa cells incubated for 24 h with Lutensol stabilized PLLAH Lut particles. The arrows indicate particles which seem to have escaped from endosomal compartments. Reproduced with permission from Ref. [96]; © 2008, Wiley VCH.

rather high during the first few hours of incubation and subsequently reached a plateau, although the uptake of PCLH-16 Lut occurred more rapidly than that of PLLAL-15 Lut. Recently, the uptake of similar amino-functionalized PS particles that had been stabilized with a nonionic Lutensol AT50 surfactant, was investigated in HeLa and Jurkat cell lines under similar conditions [14, 54]. In this case, whereas the half-maximum amount for PCLH-16 Lut and PLLAH-1 Lut was achieved in less than 2 h and less than 4 h, respectively, the amino-functionalized PS particles showed a more steady increase of cellular uptake, with no plateau, within 24 h. Clearly, different endocytotic mechanisms may again be involved in the uptake of these particulate materials.

A small, but detectable, quantity of particles was found to have localized inside HeLa cells (as detected by cLSM) after only a 15 min period of incubation with PCL particles, thus confirming the results obtained with FACS studies [96]. The more diffuse staining of cytoplasm that occurred after 24 h differed from a more clustered staining after 2 h. It was suggested that such an effect might be caused by distribution events such as endosomal escape, although particle degradation might also have been involved.

11.3.2
Influence of Transfection Agents: Surface Modifications of Nanoparticles by Covalently Linked Groups

11.3.2.1 Functionalization of Nanoparticle Surfaces by Carboxylic and Amino Side Groups

An understanding of the interactions of nanoparticles with cells is crucial for improving their interaction *in vivo* and *in vitro*. Either internalization into cells is the aspired goal (e.g., for transfection or labeling), or it should be avoided (e.g., for blood pool contrast agents or nanoparticles for the slow release of extracellular components). In particular, the uptake of nanoparticles into cells and their degradation in intracellular compartments is of major importance for many applications.

At this point, attention is focused on ways to increase the rate of intracellular uptake of polymeric particles. It might be hypothesized that positively charged nanoparticles should be taken up more efficiently than negatively charged nanoparticles, as the cell membranes are negatively charged. Although, consequently, positively charged (cationic) transfection agents [99] were used by many research groups [38], these transfection agents were found generally to be toxic and hence were not approved for clinical applications. Thus, therapeutic interventions cannot be performed in humans by using these materials.

Previous attempts to examine the influence of extent of surface charge and charge density of nanoparticles on cellular uptake were hampered, mainly because series of nanoparticles with different quantities of surface charge could not be produced under the same conditions.

11.3.2.1.1 Copolymerization of Styrene and Monomers with Charged Side Groups and Characterization of Uptake into Cells

In order to investigate the influence of surface charge on the uptake of nanoparticles into cells, a series of fluorescent polystyrene latex particles with carboxyl and amino functionalities on their surfaces was synthesized, using the miniemulsion technique.

The miniemulsion allows for a specific modification of the surface by inserting functional groups, which play an important role in further applications. This also enabled, in a second step, the biomolecules to be coupled covalently to the nanoparticles (e.g., amino acids peptides, proteins, DNA), an approach which proved to be much more efficient and stable than physical adsorption [100, 101].

In order to obtain negatively charged surfaces containing carboxylic groups, acrylic acid (AA) and styrene (St) were co-emulsified, whereas for positively charged surfaces, styrene and aminoethyl methacrylate hydrochloride (AEMH) were co-polymerized. In this case, the fluorescent dye N-(2,6-diisopropylphenyl)perylene-3,4-dicarboximide (PMI) was incorporated as reporter into the copolymer nanoparticles. The subsequent determination of the amount of PMI per gram polymeric dispersion permitted a quantitative readout of the particle uptake. The particle size ranged from 100 to 175 nm, and was dependent on the amount and nature of the functional co-monomer. All latexes were characterized using TEM, dynamic light scattering (DLS), UV-visible spectroscopy, and ζ-potential measurements, while an electrolyte titration was used to determine the numbers of surface functional groups. Any correlation between nanoparticle uptake and the surface charge was assessed using semi-quantitative FACS measurements.

Slight differences were observed between poly(St-co-AA) and poly(St-co-AEMH) particles with regards to the particles themselves. First, the size of the nanoparticles was influenced by the co-monomer over a relatively narrow size range. Second, increasing the AA fraction produced larger particles [54], whereas an increase in the amount of AEMH led to the production of smaller particles. Finally, composite particles with a size range of 100–150 nm for AA as a co-monomer and of 120–175 nm for AEMH were synthesized. However, these size variations should not have any effect on cellular uptake as only particles much smaller (<50 nm) or much larger (>300 nm) have been reported to be taken up in a different manner. Thus, such variation should be considered irrelevant during the following cell experiments. Following the surface modification of PS nanoparticles with carboxyl or amino groups, an altered uptake behavior was observed in nonphagocytotic HeLa, Jurkat, and mesenchymal stem cells (MSCs). In all of the cell types studied, with the exception of MSCs, a clear correlation of surface charge and fluorescence intensity was demonstrated (Figure 11.5).

By using surface-functionalized polymeric nanoparticles with a range of carboxyl group densities, a density–uptake relationship was established. For an efficient uptake of nanoparticles, a negatively charged surface was sufficient and no transfection agents were needed (see Figure 1 in Ref. [14]). The optimum density of carboxyl groups for particles was about 0.5 groups per nm^2, while a decrease in

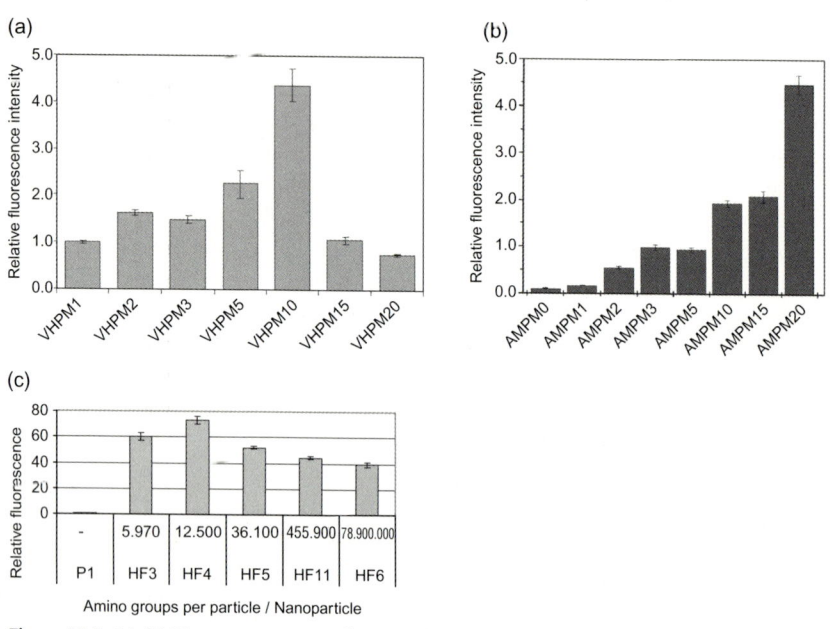

Figure 11.5 (a) FACS measurements of HeLa cells incubated for 24 h with fluorescent poly(St-co-AA) nanoparticles. Each bar shows the average normalized fluorescence intensity (nFL). Each condition was tested in triplicate; (b) FACS measurements of HeLa cells incubated for 24 h with fluorescent amino-functionalized nanoparticles. Each condition was tested in triplicate. Reproduced with permission from Ref. [54]; © 2008, Wiley VCH; (c) FACS measurements of MSCs incubated for 24 h with amino functionalized fluorescent nanoparticles. The diagram shows no clear correlation between surface charge and relative fluorescence intensity for MSCs. Reproduced with permission from Ref. [14]; © 2006, Elsevier.

uptake occurred if the density of the carboxyl groups was increased [54]. It was also shown that the intracellular uptake depended not only on the charge of the polymeric model nanoparticle, but also on the type of target cell. For MSCs, the presence of carboxyl groups increased cellular uptake compared to a nonfunctionalized particle, whereas the density of carboxyl groups was less relevant for MSCs than for HeLa cells, with only a small quantity being sufficient to induce cell uptake [102]. This may explain why differences were detected in the uptake of Resovist® (composed of carboxydextran) and of Feridex® (composed of dextran and citric acid). It is likely that the citric acid may account for extra carboxyl groups on the surface of the Feridex® particles, and this may be sufficient to induce intracellular uptake in MSCs. However, the opposite effect may occur in HeLa cells, which require a higher density of carboxyl groups to achieve an effective intracellular uptake.

11.3.2.1.2 Comparison of Amino- and Carboxy-Functionalized Nanoparticles
In order to compare different experiments, an unfunctionalized nanoparticle (termed VHPM1 or P1) was synthesized. Likewise, in order to be able to compare the different experiments, the uptake of the nanoparticle was normalized and the unfunctionalized nanoparticles were set as "1."

In the case of HeLa cells, it was found that while carboxyl-functionalized PS nanoparticles of 100 nm showed only a slight increase in total particle uptake after 24 h compared to unfunctionalized PS particles of the same size (about sixfold better than P1), the amino-functionalized particles showed a greater than 40-fold increase in uptake [102].

The different uptake behaviors for surface-modified nanoparticles (either covalently attached or physically absorbed moieties) can be explained for positively charged nanoparticles, at least in part, by the fact that the cell membrane has a negative net charge. Therefore, positively charged nanoparticles should be attracted to cell membranes. The reason why negatively charged nanoparticles also show an enhanced uptake is less easily explained, but interactions with surface molecules (e.g., proteins) may be responsible for the adhesion and uptake. Phosphatidylserine present in the cell membrane has also been linked to fusion events of cationic agents with cell membranes [103].

11.3.2.1.3 Subcellular Distribution of Charged Nanoparticles: TEM and cLSM
Previously, both cLSM and TEM have been used to reveal differences in the subcellular localization of nanoparticles. In both MSCs and HeLa cells, the nanoparticles were mostly located inside cellular compartments which resembled endosomes (Figure 11.6a and b), whereas a cytoplasmic localization was not observed in TEM studies. Neither were nanoparticles detected in the mitochondria, the cell nucleus, or the Golgi apparatus.

Interestingly, in Jurkat and KG1a cells, the nanoparticles were located predominantly in clusters on or near the cell surface, as shown using cLSM (Figure 11.6c and d). Studies using scanning electron microscopy (SEM) showed that the microvilli were also involved in the process of adherence of nanoparticles to KG1a and

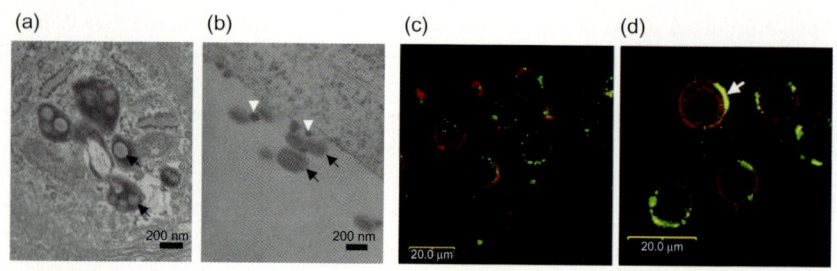

Figure 11.6 (a,b) TEM analyses of MSCs. (a) Incubation for 24 h with particle. The particles (arrows) can be found in groups in intracellular compartments which resemble endosomes; (b) MSCs incubated wit for 24 h. Particles are attached to the cell membrane (black arrows). Electron-dense material can be seen between the attached particles (white arrowheads); (c,d) cLSM of KG1a cells. (c) HF4 and (d) HF5 (green channel) where the green signal is located in clusters at the cell membrane (arrow). Reproduced with permission from Ref. [14]; © 2006, Elsevier.

Jurkat cells [14], while an increased uptake was detected with increasing surface charge when using cLSM. Whilst no aggregation was observed in carboxy-functionalized nanoparticles, a tendency towards aggregation in the medium and on the cell surface was observed when very large amounts of amino groups were present [14].

Thus, it can be concluded that the attachment of particles to the cell membrane as an initial step seems mostly to be affected by the surface charge of the particles. Furthermore, differences in the intracellular localization of particles between various cell lines (HeLa, MSCs, KG1a, and Jurkat) can be explained by the different endocytotic/pinocytotic properties of these cells.

11.3.3
Size Dependency of Cellular Uptake

Conflicting data have been presented on the size dependency of particles. For example, when Müller et al. [104] studied the uptake of negatively charged PS particles of different sizes (100, 200, and 500 nm) and different surface hydrophobicity in HL60 cells, the hydrophobic particles were seen to be taken up into the cells much more efficiently than their hydrophilic counterparts. Moreover, an increase in cellular uptake was also observed with increasing particle size. Similar uptake studies of PS particles (50, 100, 200, 500, and 1000 nm) in Caco-2 cells were also reported by Win and Feng, but had a different outcome [98]. Typically, the results showed that the particle uptake decreased with an increase in particle size, the exception being the 50 nm particles, which showed the lowest uptake efficiency. This indicated that there may be a limit below which particles are no longer efficiently taken up, and are treated by the cell just like any other macromolecule, in that a specific receptor is required for their internalization; they cannot induce endocytosis by themselves. Notably, all of the particles were negatively charged, in the range of -18 to -37 mV.

A possible explanation for the contrasting results obtained for the uptake of same-sized particles, as reported by Müller et al. [104] and Win and Feng [98], might be the differences in cell type used in these studies. A size- and cell-type-dependent particle uptake was reported by Zauner et al. [105], who used different-sized negatively charged polystyrene particles (20, 93, 220, 560, and 1010 nm) for *in vitro* uptake studies in six different cell lines (HUVEC, ECV 304, HNX 14C, KLN 205, HepG2 and Hepa 1–6). The different cell types appeared to have different cut-off values for the size of particles that could be taken up, with particles in the size range of 100 to 220 nm being taken up by all cell lines in an efficient manner. In a subsequent study, the uptake and distribution of colored PS microspheres of much larger sizes (0.75, 2, and 4.5 μm) was examined in intestinal lymphoid tissues [106]. Although particles of all sizes were identified in Peyer's patches, the density of the 0.75 μm microspheres was much higher than that of the other particles. The interaction of PS particles with the inflamed tissue of an experimental model of colitis, as a function of particle size, was investigated by Lamprecht et al. [107]. In this case, the particles possessed a negatively charged surface, and had average diameters of 100 nm, 1000 nm, and 10 μm. Whilst a higher deposition rate was observed for the smaller particles, it was also proposed that particles possessing a negative charge might adhere more readily to inflamed tissues. This would apply especially to the stomach [108], which expresses high concentrations of positively charged proteins, such that the affinity towards negatively charged substances may be increased.

11.4
Influence of Nanoparticles on (Stem) Cell Differentiation

If nanoparticles are to be used in mammalian (stem) cells, it must be ensured that the nanoparticles, together with all of their components, will not alter the specific functions of the target cells (as in diagnostic use), or they will alter the biological functions in only a defined manner. Unless being used as chemotherapeutic agents in tumor therapy, nanomaterials should be nontoxic when in contact with cells (i.e., not only stem cells but also differentiated cells). Neither should the nanoparticles affect the growth kinetics of the cells. Indeed, differentiated cells should still perform their regular functions (e.g., synthesis of extracellular matrix, production and release of cytokines, metabolic functions such as the degradation of products, signaling to neighboring cells, etc.).

To date, investigations with stem cells have been conducted only infrequently [109]. For example, the ability of the stem cells to migrate and interact with other cells – whether via surface-expressed receptors or by paracrine hormone secretion – should not be altered in any undesired way [110]. In stem cell biology, further functions should not be distorted; this applies especially to the ability for self-renewal – that is, the capacity to divide, such that one cell of the two daughter cells is a true undifferentiated stem cell, while the other one loses this ability and gives rise to differentiated cells. This is also termed asymmetric cell division [111].

It is also important that the variation in differentiation potential must be assessed, since there remains the possibility that the differentiation potential of stem cells will be altered unintentionally by the influence of the nanomaterial, even when – after exposure to nanomaterials – undifferentiated stem cells are still present after several passages *in vivo* or *in vitro*. For example, conflicting data exist as to whether MSCs lose their chondrogenic differentiation potential after incubation with superparamagnetic iron oxide (SPIO) particles formulated with dextran [112, 113]. Interestingly, this has also been demonstrated for quantum dots (QDs) composed of cadmium and selenium [114]. This might point towards the possibility that this effect on the chondrogenesis of MSCs may not depend on the type of the reporter molecule, nor on its assembly within the nanoparticles. Rather, other hitherto unidentified components or properties may be effective, and – at least potentially – this could be exploited to steer the differentiation of MSCs away from the chondrogenic pathway if this was not the desired differentiated cell population.

When the multi-lineage potential of MSCs was investigated in the presence of phosphonate-functionalized PS nanoparticles, incubation in the presence of such particles failed to disturb the osteogenic, adipogenic, or chondrogenic differentiation [115]. This was demonstrated qualitatively by the presence of a mineralized matrix in the case of osteogenic differentiation, by fat droplets in the case of adipogenic differentiation, and proteoglycans in chondrogenic pellets [115]. Moreover, these findings were confirmed by the expression of a broad spectrum of relevant marker genes, which was unaffected by the presence of the particles (Figure 11.7). Although, several authors have examined the effect of different particles on MSC differentiation, the latter parameter was investigated only qualitatively or was restricted to certain lineages. Whereas, fluorescence-labeled biocompatible mesoporous silica nanoparticles showed no effect on osteogenic and

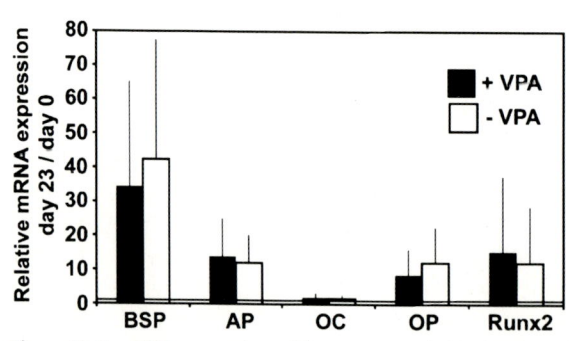

Figure 11.7 mRNA expression of the osteogenic differentiation marker genes bone sialoprotein (BSP), alkaline phosphatase (AP), osteocalcin (OC), osteopontin (OP) and the transcription factor Runx2 in cells cultured under osteogenic conditions with VPA particles (black bars) and without particles (white bars) at day 23 of the experiment. Levels of mRNA were normalized to GAPDH mRNA levels and related to the expression at day 0 (mRNA expression at day 0 was set at 1), $n = 3$. Reproduced with permission from Ref. [115]; © 2010, Elsevier.

adipogenic MSC differentiation [116, 117], others identified a negative influence on osteogenic differentiation following treatment with biphasic calcium phosphate particles, which was related to particle release from the bioceramics [118]. In particular, titanium or cobalt–chromium–molybdenum particles, produced through the wearing of metallic implants, were found to suppress osteogenic differentiation [119, 120].

11.5
Endocytosis

As shown previously, nanoparticles are taken up by a variety of cells, with one of the most interesting questions being the route that they take during the uptake process. This is especially important in order to manipulate the uptake of such nanomaterials into cells; indeed, such manipulation may even permit the targeting of a cargo to a specific subcellular compartment.

For small molecules and macromolecules such as proteins, the routes of uptake into cells have been well described [121, 122]. However, particles ranging from a few nanometers to several hundred nanometers are much larger, and the conditions and mechanisms of uptake – which involve several possible mechanisms such as pinocytosis, nonspecific endocytosis, receptor-mediated endocytosis and, for larger particles, phagocytosis – have not been studied thoroughly [123]. Whilst for receptor-mediated endocytosis, biological activators such as opsonin and lectin, as well as scavenger receptors, have been identified [124], no receptors have been identified for the endocytosis of nanoparticles. However, it is possible to learn from the mechanisms of endocytosis for biologically active molecules, despite them being much smaller than many nanoparticles. "Particles" of similar size to the present nanoparticles have been synthesized and their uptake behavior has been studied extensively [125–128].

The first of the routes to be characterized was the clathrin-coated pit pathway, a specialized protein assembly of the triskelion clathrin that forms a net-like basket [129]. A second class of vesicles, coated with the protein caveolin (and thus termed caveolae) are derived from lipid rafts [130], which are in turn characterized as rigid membrane microdomains enriched with phospholipids, sphingolipids, and cholesterol [131]. Uncoated vesicles may also be formed (also by macropinocytosis) from the lipid rafts; these vesicles are larger than their coated counterparts, and may allow the endocytosis of larger particles (>150 nm). Different receptors are known to be involved in endocytosis when specific cargos, such as transferrin, are to be taken up. Some of these receptors are concentrated into clathrin-coated pits, or in the lipid raft domain (for a review, see Ref. [132]).

In order to study the specific mechanisms of uptake, a series of model substances and inhibitors have been used [15, 133], with many of the components (e.g., dynamin) or cytoskeletal components having been shown to be crucial for the different pathways [132]. Endocytotic pathways should perhaps be regarded as a network of different properties which can be combined with each other. For example, the small GTPase dynamin, which pinches off the endocytotic sack from the cell membrane, is essential during the internalization of clathrin or caveolin-coated vesicles. Consequently, the inhibition of dynamin would affect both pathways.

The redundancy of the endocytotic mechanisms and molecular machinery involved can be demonstrated, for example, by using dynamin knockout mice that survive because an alternative molecular machinery can provide a means of endocytosis [134]. The different cell types may also have different uptake mechanisms, the less prominent of which may be upregulated when the standard pathway is blocked.

As noted above, surface charge is an important factor for endocytotic uptake into different cell lines (see Section 11.3.2.1), and consequently it should be expected that uptake mechanisms may also differ, depending on the surface properties. In recent studies, an evaluation was made as to which endocytotic mechanisms were involved in the uptake of well-defined positively and negatively charged 100 nm fluorescent PS nanoparticles into HeLa cells. It was shown that, independent of the particle charge (cationic or anionic), endocytosis is an energy-dependent process and highly reliant on dynamin and F-actin, as shown by an inhibition of uptake in the presence of dynasore and cytochalasin D, respectively (Figure 11.8). As genistein inhibits particle uptake, tyrosine-specific protein kinases (located in lipid rafts) must become involved in the uptake [53]. Whilst these factors suggest the need for a dynamin- and lipid raft-dependent uptake mechanism for the uptake of nanoparticles into HeLa cells, cholesterol sequestration or depletion by filipin or methyl-β-cyclodextrin, respectively, did not hinder the particle uptake [53].

The particles' surface charge also impacts any further uptake mechanisms in HeLa cells. For example, macropinocytosis appears to be an important mechanism in the uptake of positively charged nanoparticles into HeLa cells, as demonstrated by a strong inhibition of positively charged nanoparticles by 5-(N-ethyl-N-isopropyl) amiloride (EIPA) [53]. The fact that uptake was also hindered in the presence of nocodazole and indomethacin also points to an involvement of the microtubule network, and to the presence of cyclooxygenases in the uptake process of positively charged particles in HeLa cells (Figure 11.8). Likewise, in HeLa cells clathrin-coated pits play only a minor role in the uptake of positively charged nanoparticles, having no effect on the endocytosis of negatively charged nanoparticles (see below for data relating to MSCs). However, approximately 20% of the endocytosis of positively charged particles is inhibited by chlorpromazine [53]. Subsequent TEM studies also indicated only a minor involvement of clathrin-coated pits, as several nanoparticles appeared to be engulfed by a structure (Figure 11.8b) that appeared to be macropinocytic. It should be noted, however, that the rigid clathrin structure cannot be dilated to any significant degree [15].

Unfortunately, the results of these studies failed to answer the question regarding the role of lipid rafts. Cholesterol depletion and/or sequestration should affect lipid rafts very effectively, but the uptake was largely independent of the cholesterol depletion/sequestration. Consequently, these results would indicate that lipid rafts

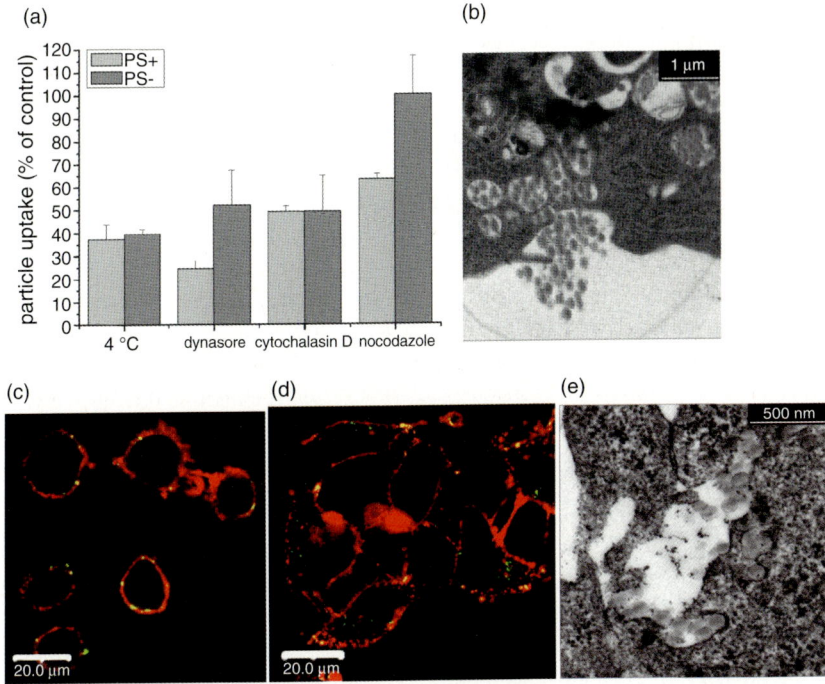

Figure 11.8 (a) Uptake of PS+ or PS- particles into HeLa cells under different inhibitory conditions. HeLa cells were preincubated for 30 min with the respective drugs, and particle uptake within 1 h was determined by flow cytometry. Mean values and standard deviations of two independent triplicate experiments are given. After incubation with filipin or MBCD, the fluorescence signal was never significantly lower than the respective positive control; (b) HeLa cells were incubated with positively charged polystyrene nanoparticles. The image shows a HeLa cell during the endocytosis of nanoparticles; (c–e) HeLa cells after 1 h incubation with PS+ particles (green) in the presence of nocodazole, or (d) as a positive control without any drugs. The cell membranes were stained with RH414 (red). The colocalization of particles and the cell membrane is shown in yellow. Particles were detected as not yet endocytosed by the cell and remained near the outer membrane (e, TEM image). Reproduced with permission from Ref. [53]; © 2008, Wiley-VCH.

are not involved in the uptake process. On the other hand, the inhibition of lipid raft-associated proteins showed a significant effect.

Thus, it can be concluded that, depending on the surface charge of the nanoparticles, differences in their uptake and/or in the intracellular trafficking of the endosomes may occur in HeLa cells. Interestingly, negatively charged nanoparticles were less inhibited by a dynamin inhibitor, which suggested the possibility that a hitherto unidentified dynamin-independent process may contribute to the uptake of negatively charged nanoparticles.

As the details of an uptake mechanism may depend on the cell type studied, MSCs were selected next as an alternative cell type. In the case of aminofunctionalized nanoparticles and their uptake into MSCs, a predominance of the clathrin-coated pit pathway could be demonstrated, whereby the amino groups on the nanoparticles enabled (or, at least, substantially enhanced) clathrin-mediated uptake by the MSCs [135]. The influence of other components of the nanoparticles (such as the surfactant used during synthesis) could be excluded by demonstrating differences between nanoparticles synthesized in exactly the same manner as amino-functionalized nanoparticles, but only with an amino-functionalized co-monomer. Subsequent quantitative analyses of confocal images of the clathrin-mediated and clathrin-independent internalization of nanoparticles by MSCs revealed the clathrin-independent processes to be slower, and significantly delayed. This was confirmed first by using flow cytometry, which enabled the analysis of several thousand cells within minutes, and subsequently by using LSM. Quantitative confocal imaging also enabled the acquisition of additional, detailed insights into the uptake kinetics and the subcellular localization of the nanoparticles [135].

11.6
Summary

Beginning with experiments using commercially available SPIO nanoparticles intended for the labeling of (stem) cell preparations, the primary question was how these nanomaterials would interact with cells (for a graphical representation of the possible applications, see Figure 11.9). In an effort to provide an answer, special attention was paid to determine the crucial parameters on both the nanoparticles and the cells, so that the former would be endocytosed in the latter. It

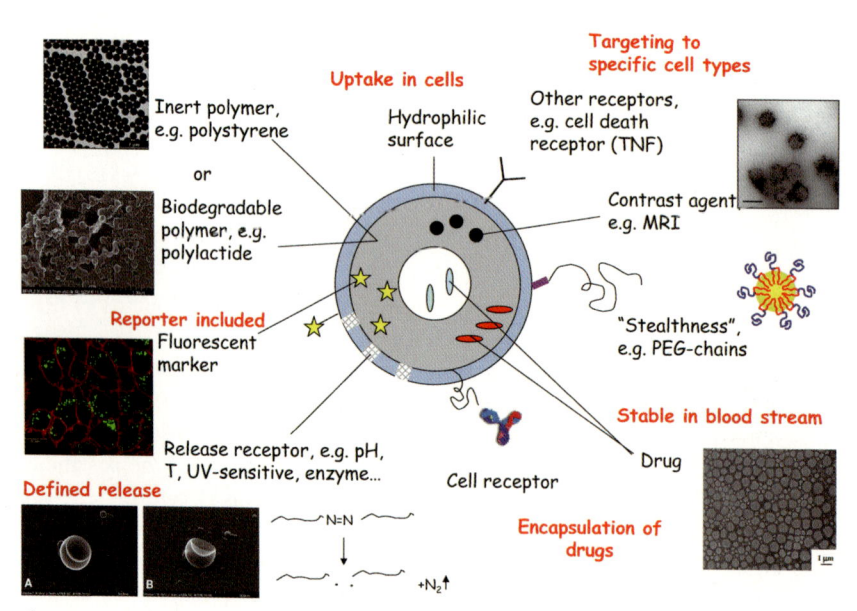

Figure 11.9 Scheme of possible applications of nanoparticles and nanocapsules.

has been shown how the polymeric material can influence cellular uptake, and that the uptake kinetics is much faster for PI nanoparticles as compared to PS. Nanoparticles can be synthesized from a range of polymers used for technical applications, as well as their biocompatible and biodegradable counterparts, with all processes conducted via the miniemulsion technique. Moreover, cell uptake can be "tuned" simply by altering the relative amounts of the monomers when two monomers (e.g., PI and PS) are copolymerized. Surface modifications, such as ensuring a positive or negative charge on the surface of the nanoparticles, will enhance cellular uptake. Furthermore, such an effect can be "titrated" by using a series of nanoparticles with a range of densities of effective side groups. Cellular uptake can also be enhanced by the use of transfection agents; for example, if uptake is not desired, then PEG functionalization will cause the particles to be retained within the extracellular space.

In order to detect the presence of nanoparticles in the cells, different reporters were encapsulated (e.g., fluorescent dyes), to conduct *in vitro*, and possibly also *in vivo*, studies. Among these can be included SPIO and gadolinium, for MRI applications. Interestingly, when more than one of these reporters is included in the nanoparticle, the resulting material is termed a "dual-reporter" nanoparticle.

With regards to their subcellular distribution, nanoparticles were found to locate mostly in the endosomes, but were not able to enter the cell nucleus. Consequently, an application such as gene delivery would need to ensure that the DNA would be transported by, and released from, the nanoparticle in order to gain access to the cell nucleus.

References

1 Moghimi, S.M., Hunter, A.C. and Murray, J.C. (2005) Nanomedicine: current status and future prospects. *FASEB Journal*, **19**, 311–30.

2 Allemann, E., Leroux, J.C. and Gurny, R. (1998) Polymeric nano- and microparticles for the oral delivery of peptides and peptidomimetics. *Advanced Drug Delivery Reviews*, **34**, 171–89.

3 Panyam, J. and Labhasetwar, V. (2003) Biodegradable nanoparticles for drug and gene delivery to cells and tissue. *Advanced Drug Delivery Reviews*, **55**, 329–47.

4 Matuszewski, L. *et al.* (2005) Cell tagging with clinically approved iron oxides: feasibility and effect of lipofection, particle size, and surface coating on labeling efficiency. *Radiology*, **235**, 155–61.

5 Gupta, A.K. and Curtis, A.S.G. (2004) Surface modified superparamagnetic nanoparticles for drug delivery: interaction studies with human fibroblasts in culture. *Journal of Materials Science: Materials in Medicine*, **15**, 493–6.

6 Shenoy, D.B. and Amiji, M.M. (2005) Poly(ethylene oxide)-modified poly(var epsilon-caprolactone) nanoparticles for targeted delivery of tamoxifen in breast cancer. *International Journal of Pharmaceutics*, **293**, 261–70.

7 Cegnar, M. *et al.* (2004) Poly(lactide-co-glycolide) nanoparticles as a carrier system for delivering cysteine protease inhibitor cystatin into tumor cells. *Experimental Cell Research*, **301**, 223–31.

8 Delie, F. and Blanco-Príeto, M.J. (2005) Polymeric particulates to improve oral bioavailability of peptide drugs. *Molecules*, **10**, 65–80.

9 Jun, Y.W., Jang, J.T. and Cheon, J. (2007) Magnetic nanoparticle assisted molecular MR imaging. *Advances in Experimental Medicine and Biology*, **620**, 85–106.
10 Landfester, K. (2006) Synthesis of colloidal particles in miniemulsions. *Annual Review of Materials Research*, **36**, 231–79.
11 Narayanan, R. and El-Sayed, M.A. (2005) FTIR study of the mode of binding of the reactants on the Pd nanoparticle surface during the catalysis of the Suzuki reaction. *Journal of Physical Chemistry B*, **109**, 4357–60.
12 Fallon, M.S. et al. (2004) A physiologically-based pharmacokinetic model of drug detoxification by nanoparticles. *Journal of Pharmacokinetics and Pharmacodynamics*, **31**, 381–400.
13 Miltenyi, S. (1997) CD34+ selection: the basic component for graft engineering. *Oncologist*, **2**, 410–13.
14 Lorenz, M.R. et al. (2006) Uptake of functionalized, fluorescent-labeled polymeric particles in different cell lines and stem cells. *Biomaterials*, **27**, 2820–8.
15 Rejman, J. et al. (2004) Size-dependent internalization of particles via the pathways of clathrin- and caveolae-mediated endocytosis. *Biochemical Journal*, **377**, 159–69.
16 Bender, A.R. et al. (1996) Efficiency of nanoparticles as a carrier system for antiviral agents in human immunodeficiency virus-infected human monocytes/macrophages in vitro. *Antimicrobial Agents and Chemotherapy*, **40**, 1467–71.
17 Wang, G. and Uludag, H. (2008) Recent developments in nanoparticle-based drug delivery and targeting systems with emphasis on protein-based nanoparticles. *Expert Opinion on Drug Delivery*, **5**, 499–515.
18 Sullivan, C.O. and Birkinshaw, C. (2004) In vitro degradation of insulin-loaded poly(n-butylcyanoacrylate) nanoparticles. *Biomaterials*, **25**, 4375–82.
19 Jiao, Y. et al. (2002) In vitro and in vivo evaluation of oral heparin-loaded polymeric nanoparticles in rabbits. *Circulation*, **105**, 230–5.
20 Alvarez-Roman, R. et al. (2004) Enhancement of topical delivery from biodegradable nanoparticles. *Pharmaceutical Research*, **21**, 1818–25.
21 Alvarez-Roman, R. et al. (2004) Skin penetration and distribution of polymeric nanoparticles. *Journal of Controlled Release*, **99**, 53–62.
22 Kourti, M. et al. (2007) Expression of multidrug resistance 1 (MDR1), multidrug resistance-related protein 1 (MRP1), lung resistance protein (LRP), and breast cancer resistance protein (BCRP) genes and clinical outcome in childhood acute lymphoblastic leukemia. *International Journal of Hematology*, **86**, 166–73.
23 Lewin, M. et al. (2000) Tat peptide-derivatized magnetic nanoparticles allow in vivo tracking and recovery of progenitor cells. *Nature Biotechnology*, **18**, 410–14.
24 Cohen, H. et al. (2000) Sustained delivery and expression of DNA encapsulated in polymeric nanoparticles. *Gene Therapy*, **7**, 1896–905.
25 Panyam, J. and Labhasetwar, V. (2003) Biodegradable nanoparticles for drug and gene delivery to cells and tissue. *Advanced Drug Delivery Reviews*, **55**, 329–47.
26 Gupta, A.K. and Curtis, A.S. (2004) Surface modified superparamagnetic nanoparticles for drug delivery: interaction studies with human fibroblasts in culture. *Journal of Materials Science: Materials in Medicine*, **15**, 493–6.
27 Chawla, J.S. and Amiji, M.M. (2002) Biodegradable poly(epsilon-caprolactone) nanoparticles for tumor-targeted delivery of tamoxifen. *International Journal of Pharmaceutics*, **249**, 127–38.
28 Elamanchili, P. et al. (2004) Characterization of poly(D,L-lactic-co-glycolic acid) based nanoparticulate system for enhanced delivery of antigens to dendritic cells. *Vaccine*, **22**, 2406–12.
29 Russell-Jones, G.J. (2000) Oral vaccine delivery. *Journal of Controlled Release*, **65**, 49–54.
30 Damge, C. et al. (1988) New approach for oral administration of insulin with

polyalkylcyanoacrylate nanocapsules as drug carrier. *Diabetes*, **37**, 246–51.
31 Landfester, K. and Ramirez, L. (2003) Encapsulated magnetite particles for biomedical applications. *Journal of Physics: Condensed Matter*, **204**, 22–31.
32 Ramirez, L. and Landfester, K. (2003) Magnetic polystyrene nanoparticles with a high magnetite content obtained by miniemulsion processes. *Macromolecular Chemistry and Physics*, **206**, 2440–9.
33 Holzapel, V. et al. (2006) Synthesis and biomedical applications of functionalized fluorescent and magnetic dual reporter nanoparticles as obtained in the miniemulsion process. *Journal of Physics: Condensed Matter*, **18**, 2581–94.
34 Liu, X. et al. (2004) Surface modification and characterization of magnetic polymer nanospheres prepared by miniemulsion polymerization. *Langmuir*, **20**, 10278–82.
35 Qian, Z., Zhang, Z. and Chen, Y. (2008) A novel preparation of surface-modified paramagnetic magnetite/polystyrene nanocomposite microspheres by radiation-induced miniemulsion polymerization. *Journal of Colloid and Interface Science*, **327**, 354–61.
36 Hua, X.F. et al. (2006) Characterization of the coupling of quantum dots and immunoglobulin antibodies. *Analytical and Bioanalytical Chemistry*, **386**, 1665–71.
37 Frank, J.A. et al. (2004) Methods for magnetically labeling stem and other cells for detection by in vivo magnetic resonance imaging MR techniques for in vivo molecular and cellular imaging. *Cytotherapy*, **6**, 621–5.
38 Arbab, A.S. et al. (2003) Intracytoplasmic tagging of cells with ferumoxides and transfection agent for cellular magnetic resonance imaging after cell transplantation: methods and techniques. *Transplantation*, **76**, 1123–30.
39 Kraitchman, D.L. et al. (2003) In vivo magnetic resonance imaging of mesenchymal stem cells in myocardial infarction. *Circulation*, **107**, 2290–3.
40 Arbab, A.S. et al. (2004) Comparison of transfection agents in forming complexes with ferumoxides, cell labeling efficiency, and cellular viability. *Molecular Imaging*, **3**, 24–32.
41 Frank, J.A. et al. (2002) Magnetic intracellular labeling of mammalian cells by combining (FDA-approved) superparamagnetic iron oxide MR contrast agents and commonly used transfection agents. *Academic Radiology*, **9**, S484–7.
42 Lipinski, M.J. et al. (2006) MRI to detect atherosclerosis with gadolinium-containing immunomicelles targeting the macrophage scavenger receptor. *Magnetic Resonance in Medicine*, **56**, 601–10.
43 Vuu, K. et al. (2005) Gadolinium-rhodamine nanoparticles for cell labeling and tracking via magnetic resonance and optical imaging. *Bioconjugate Chemistry*, **16**, 995–9.
44 Turner, J.L. et al. (2005) Synthesis of gadolinium-labeled shell-crosslinked nanoparticles for magnetic resonance imaging applications. *Advanced Functional Materials*, **15**, 1248–54.
45 Yu, X. et al. (2000) High-resolution MRI characterization of human thrombus using a novel fibrin-targeted paramagnetic nanoparticle contrast agent. *Magnetic Resonance in Medicine*, **44**, 867–72.
46 Morawski, A.M. et al. (2004) Targeted nanoparticles for quantitative imaging of sparse molecular epitopes with MRI. *Magnetic Resonance in Medicine*, **51**, 480–6.
47 Aime, S., Frullano, L. and Crich, S.G. (2002) Compartmentalization of a gadolinium complex in the apoferritin cavity: a route to obtain high relaxivity contrast agents for magnetic resonance imaging. *Angewandte Chemie International Edition*, **41**, 1017–19.
48 Reynolds, C.H. et al. (2000) Gadolinium-loaded nanoparticles: new contrast agents for magnetic resonance imaging. *Journal of the American Chemical Society*, **122**, 8940–5.
49 Landfester, K. (2001) Polyreactions in miniemulsions. *Macromolecular Rapid Communications*, **22**, 896–936.
50 Crespy, D. et al. (2007) Polymeric nanoreactors for hydrophilic reagents synthesized by interfacial

polycondensation on miniemulsion droplets. *Macromolecules*, **40**, 3122–35.

51 Jagielski, N. et al. (2007) Nanocapsules synthesized by miniemulsion technique for application as new contrast agent materials. *Macromolecular Chemistry and Physics*, **208**, 2229–41.

52 Hu, Y. et al. (2006) Effect of PEG conformation and particle size on the cellular uptake efficiency of nanoparticles with the HepG2 cells. *Journal of Controlled Release*, **6**, 6.

53 Dausend, J. et al. (2008) Uptake mechanism of oppositely charged fluorescent nanoparticles in HeLa cells. *Macromolecular Bioscience*, **8**, 1135–43.

54 Holzapfel, V. et al. (2005) Preparation of fluorescent carboxyl and amino functionalized polystyrene particles by miniemulsion polymerization as markers for cells. *Macromolecular Chemistry and Physics*, **205**, 2440–9.

55 Derouet, D., Mulder-Houdayer, S. and Brosse, J.-C. (2005) Chemical modification of polydienes in latex medium: study of epoxidation and ring opening of oxiranes. *Journal of Applied Polymer Science*, **95**, 39–52.

56 Phinyocheep, P. et al. (2005) Chemical degradation of epoxidized natural rubber using periodic acid: preparation of epoxidized liquid natural rubber. *Journal of Applied Polymer Science*, **95**, 6–15.

57 Lorenz, M.R. et al. (2008) Synthesis of fluorescent polyisoprene nanoparticles and their uptake into various cells. *Macromolecular Bioscience*, **8**, 711–27.

58 Falcone, S. et al. (2006) Macropinocytosis: regulated coordination of endocytic and exocytic membrane traffic events. *Journal of Cell Science*, **119**, 4758–69.

59 Lai, S.K. et al. (2007) Privileged delivery of polymer nanoparticles to the perinuclear region of live cells via a non-clathrin, non-degradative pathway. *Biomaterials*, **28**, 2876–84.

60 Nabi, I.R. and Le, P.U. (2003) Caveolae/raft-dependent endocytosis. *Journal of Cell Biology*, **161**, 673–7.

61 Arias, J.L. et al. (2001) Synthesis and characterization of poly(ethyl-2cyanoacrylate) nanoparticles with

magnetic core. *Journal of Controlled Release*, **77**, 309–21.

62 Reddy, L.H. and Murthy, R.R. (2004) Influence of polymerization technique and experimental variables on the particle properties and release kinetics of methotrexate from poly(butylcyanoacrylate) nanoparticles. *Acta Pharmaceutica*, **54**, 103–18.

63 Kattan, J. et al. (1992) Phase I clinical trial and pharmacokinetic evaluation of doxorubicin carried by polyisohexylcyanoacrylate nanoparticles. *Investigational New Drugs*, **10**, 191–9.

64 Steiniger, S.C.J. et al. (2004) Chemotherapy of glioblastoma in rats using doxorubicin-loaded nanoparticles. *International Journal of Cancer*, **109**, 759–67.

65 Gulyaev, A.E. et al. (1999) Significant transport of doxorubicin into the brain with polysorbate 80-coated nanoparticles. *Pharmaceutical Research*, **16**, 1564–9.

66 Alyautdın, R. et al. (1995) Analgesic activity of the hexapeptide dalargin adsorbed on the surface of polysorbate 80-coated poly(butyl cyanoacrylate) nanoparticles. *European Journal of Pharmaceutics and Biopharmaceutics*, **41**, 44–8.

67 Olivier, J.-C. et al. (1999) Indirect evidence that drug brain targeting using polysorbate 80-coated polybutylcyanoacrylate nanoparticles is related to toxicity. *Pharmaceutical Research*, **16**, 1836–42.

68 Behan, N. and Birkinshaw, C. (2001) Preparation of poly(butyl cyanoacrylate) nanoparticles by aqueous dispersion polymerisation in the presence of insulin. *Macromolecular Rapid Communications*, **22**, 41–3.

69 Couvreur, P. (1988) Polyalkylcyanoacrylates as colloidal drug carrier. *CRC Critical Reviews in Therapeutic Drug Carrier Systems*, **5**, 1–17.

70 Couvreur, P. et al. (1979) Adsorption of antineoplastic drugs to polyalkylcyanoacrylate nanoparticles and their release in calf serum. *Journal of Pharmaceutical Sciences*, **68**, 1521–4.

71 Lescure, F. et al. (1992) Optimization of polyalkylcyanoacrylate nanoparticle

preparation: influence of sulfur dioxide and pH on nanoparticle characteristics. *Journal of Colloid and Interface Science*, **154**, 77–86.
72. El-Egakey, M.A., Bentele, V. and Kreuter, J. (1983) Molecular weights of polycyanacrylate nanoparticles. *International Journal of Pharmaceutics*, **13**, 349–52.
73. Douglas, S.J. et al. (1984) Particle size and size distribution of poly(butyl-2-cyanoacrylate) nanoparticles. I. Influence of physicochemical factors. *Journal of Colloid and Interface Science*, **101**, 149–58.
74. Douglas, S.J., Illum, L. and Davis, S.S. (1985) Particle size and size distribution of poly(butyl 2-cyanoacrylate) nanoparticles. II. Influence of stabilizers. *Journal of Colloid and Interface Science*, **103**, 154–63.
75. Vasnick, L. et al. (1985) Molecular weights of free and drug-loaded nanoparticles. *Pharmaceutical Research*, **2**, 36–41.
76. Seijo, B. et al. (1990) Design of nanoparticles of less than 50 nm diameter: preparation characterization and drug loading. *International Journal of Pharmaceutics*, **62**, 1–7.
77. Chauvierre, C. et al. (2004) Evaluation of the surface properties of dextran-coated poly(isobutylcyanoacrylate) nanoparticles by spin-labelling coupled with electron resonance spectroscopy. *Colloid and Polymer Science*, **282**, 1016–25.
78. Peracchia, M.T. et al. (1997) Complement consumption by poly(ethylene glycol) in different conformations chemically coupled to poly(isobutyl 2-cyanoacrylate) nanoparticles. *Life Sciences*, **61**, 749–61.
79. Weiss, C.K. et al. (2007) Cellular uptake behavior of unfunctionalized and functionalized PBCA particles prepared in a miniemulsion. *Macromolecular Bioscience*, **7**, 883–96.
80. Göpferich, A. (1996) Polymer degradation and erosion: mechanisms and applications. *European Journal of Pharmaceutics and Biopharmaceutics*, **42**, 1–11.
81. Vauthier, C. et al. (2003) Poly(alkylcyanoacrylates) as biodegradable materials for biomedical applications. *Advanced Drug Delivery Reviews*, **55**, 519–48.
82. Kreuter, J. (2001) Nanoparticulate systems for brain delivery of drugs. *Advanced Drug Delivery Reviews*, **47**, 65–81.
83. Gref, R. et al. (1994) Biodegradable long-circulating polymeric nanospheres. *Science*, **263**, 1600–3.
84. Gurny, R. et al. (1981) Development of biodegradable and injectable lattices for controlled release of potent drugs. *Drug Development and Industrial Pharmacy*, **7**, 1–25.
85. Duvvuri, S., Gaurav Janoria, K. and Mitra, A.K. (2006) Effect of polymer blending on the release of ganciclovir from PLGA microspheres. *Pharmaceutical Research*, **23**, 215–23.
86. Lai, M.K. and Tsiang, R.C.C. (2004) Encapsulating acetaminophen into poly(L-lactide) microcapsules by solvent-evaporation technique in an O/W emulsion. *Journal of Microencapsulation*, **21**, 307–16.
87. Mehta, R.C. et al. (1994) Biodegradable microspheres as depot system for parenteral delivery of peptide drugs. *Journal of Controlled Release*, **29**, 375–84.
88. Benny, O. et al. (2005) Continuous delivery of endogenous inhibitors from poly(lactic-*co*-glycolic acid) polymeric microspheres inhibits glioma tumor growth. *Clinical Cancer Research*, **11**, 768–76.
89. Tracy, M.A. et al. (1999) Factors affecting the degradation rate of poly(lactide-*co*-glycolide) microspheres in vivo and in vitro. *Biomaterials*, **20**, 1057–62.
90. Lemoine, D. et al. (1996) Stability study of nanoparticles of poly(ε-caprolactone), poly(D,L-lactide) and poly(D,L-lactide-*co*-glycolide). *Biomaterials*, **17**, 2191–7.
91. Park, T.G. (1994) Degradation of poly(-lactic acid) microspheres: effect of molecular weight. *Journal of Controlled Release*, **30**, 161–73.
92. Park, T.G. (1995) Degradation of poly(lactic-*co*-glycolic acid) microspheres: effect of copolymer composition. *Biomaterials*, **16**, 1123–30.
93. Li, S. and Vert, M. (1994) Morphological changes resulting from the hydrolytic degradation of stereocopolymers derived

from L- and DL-lactides. *Macromolecules*, **27**, 3107–10.
94 Makino, K., Arakawa, M. and Kondo, T. (1985) Preparation and in vitro degradation properties of polylactide microcapsules. *Chemical and Pharmaceutical Bulletin*, **33**, 1195–201.
95 Reich, G. (1998) Ultrasound-induced degradation of PLA and PLGA during microsphere processing: influence of formulation variables. *European Journal of Pharmaceutics and Biopharmaceutics*, **45**, 165–71.
96 Musyanovych, A. *et al.* (2008) Preparation of biodegradable polymer nanoparticles by miniemulsion technique and their cell interactions. *Macromolecular Bioscience*, **8**, 127–39.
97 Coffin, M.D. and McGinity, J.W. (1992) Biodegradable pseudolatexes: the chemical stability of poly(D,L-lactide) and poly(ε-caprolactone) nanoparticles in aqueous media. *Pharmaceutical Research*, **9**, 200–5.
98 Win, K.Y. and Feng, S.-S. (2005) Effects of particle size and surface coating on cellular uptake of polymeric nanoparticles for oral delivery of anticancer drugs. *Biomaterials*, **26**, 2713–22.
99 Gershon, H. *et al.* (1993) Mode of formation and structural features of DNA-cationic liposome complexes used for transfection. *Biochemistry*, **32**, 7143–51.
100 Gibanel, S. *et al.* (2001) Monodispersed polystyrene latex particles functionalized by the macromonomer technique. II. Application in immunodiagnosis. *Polymers for Advanced Technologies*, **12**, 494–9.
101 Ortegavinuesa, J.L. and Hidalgoalvarez, R. (1995) Sequential adsorption of F(Ab') (2) and BSA on negatively and positively charged polystyrene latexes. *Biotechnology and Bioengineering*, **47**, 633–9.
102 Mailander, V. *et al.* (2008) Carboxylated superparamagnetic iron oxide particles label cells intracellularly without transfection agents. *Molecular Imaging and Biology*, **10**, 138–46.
103 Stebelska, K., Dubielecka, P.M. and Sikorski, A.F. (2005) The effect of PS content on the ability of natural membranes to fuse with positively charged liposomes and lipoplexes. *Journal of Membrane Biology*, **206**, 203–14.
104 Müller, R.H. *et al.* (1997) Influence of fluorescent labelling of polystyrene particles on phagocytic uptake, surface hydrophobicity, and plasma protein adsorption. *Pharmaceutical Research*, **14**, 18–24.
105 Zauner, W., Farrow, N.A. and Haines, A.M.R. (2001) In vitro uptake of polystyrene microspheres: effect of particle size, cell line and cell density. *Journal of Controlled Release*, **71**, 39–51.
106 Hoshi, S. *et al.* (1999) Uptake of orally administered polystyrene latex and poly(D,L-lactic/glycolic acid) microspheres into intestinal lymphoid tissues in chickens. *Veterinary Immunology and Immunopathology*, **70**, 33–42.
107 Lamprecht, A., Schäfer, U. and Lehr, C.M. (2001) Size-dependent bioadhesion of micro- and nanoparticulate carriers to the inflamed colonic mucosa. *Pharmaceutical Research*, **18**, 788–93.
108 Nagashima, R. (1981) Mechanisms of action of sucralfate. *Journal of Clinical Gastroenterology*, **3**, 117–27.
109 Amsalem, Y. *et al.* (2007) Iron-oxide labeling and outcome of transplanted mesenchymal stem cells in the infarcted myocardium. *Circulation*, **116**, I38–45.
110 Schafer, R. *et al.* (2007) Transferrin receptor upregulation: in vitro labeling of rat mesenchymal stem cells with superparamagnetic iron oxide. *Radiology*, **244**, 514–23.
111 Braydich-Stolle, L. *et al.* (2005) In vitro cytotoxicity of nanoparticles in mammalian germline stem cells. *Toxicological Sciences*, **88**, 412–19.
112 Kostura, L. *et al.* (2004) Feridex labeling of mesenchymal stem cells inhibits chondrogenesis but not adipogenesis or osteogenesis. *NMR in Biomedicine*, **17**, 513–17.
113 Arbab, A.S. *et al.* (2005) Labeling of cells with ferumoxides-protamine sulfate complexes does not inhibit function or differentiation capacity of hematopoietic or mesenchymal stem cells. *NMR in Biomedicine*, **18**, 553–9.

114 Hsieh, S.C. et al. (2006) The internalized CdSe/ZnS quantum dots impair the chondrogenesis of bone marrow mesenchymal stem cells. *Journal of Biomedical Materials Research. Part B: Applied Biomaterials*, **79**, 95–101.

115 Tautzenberger, A. et al. (2010) Effect of functionalised fluorescence-labelled nanoparticles on mesenchymal stem cell differentiation. *Biomaterials*, **31**, 2064–71.

116 Liu, H.-M. et al. (2008) Mesoporous silica nanoparticles improve magnetic labeling efficiency in human stem cells. *Small*, **4**, 619–26.

117 Chung, T.-H. et al. (2007) The effect of surface charge on the uptake and biological function of mesoporous silica nanoparticles in 3T3-L1 cells and human mesenchymal stem cells. *Biomaterials*, **28**, 2959–66.

118 Saldaña, L. et al. (2009) Calcium phosphate-based particles influence osteogenic maturation of human mesenchymal stem cells. *Acta Biomaterialia*, **5**, 1294–305.

119 Okafor, C.C. et al. (2006) Particulate endocytosis mediates biological responses of human mesenchymal stem cells to titanium wear debris. *Journal of Orthopaedic Research*, **24**, 461–73.

120 Schofer, M.D. et al. (2008) The role of mesenchymal stem cells in the pathogenesis of Co-Cr-Mo particle induced aseptic loosening: an in vitro study. *Bio-Medical Materials and Engineering*, **18**, 395–403.

121 Fischer, R. et al. (2002) A quantitative validation of fluorophore-labelled cell-permeable peptide conjugates: fluorophore and cargo dependence of import. *Biochimica et Biophysica Acta*, **1564**, 365–74.

122 Jensen, K.D. et al. (2003) Cytoplasmic delivery and nuclear targeting of synthetic macromolecules. *Journal of Controlled Release*, **87**, 89–105.

123 Pratten, M.K. and Lloyd, J.B. (1986) Pinocytosis and phagocytosis: the effect of size of a particulate substrate on its mode of capture by rat peritoneal macrophages cultured in vitro. *Biochimica et Biophysica Acta*, **881**, 307–13.

124 Schulze, E. et al. (1995) Cellular uptake and trafficking of a prototypical magnetic iron oxide label in vitro. *Investigative Radiology*, **30**, 604–10.

125 Sieczkarski, S.B. and Whittaker, G.R. (2002) Influenza virus can enter and infect cells in the absence of clathrin-mediated endocytosis. *Journal of Virology*, **76**, 10455–64.

126 Szymkiewicz, I., Shupliakov, O. and Dikic, I. (2004) Cargo- and compartment-selective endocytic scaffold proteins. *Biochemical Journal*, **383**, 1–11.

127 Stuart, A.D. and Brown, T.D. (2006) Entry of feline calicivirus is dependent on clathrin-mediated endocytosis and acidification in endosomes. *Journal of Virology*, **80**, 7500–9.

128 Llorente, A. et al. (2000) Apical endocytosis of ricin in MDCK cells is regulated by the cyclooxygenase pathway. *Journal of Cell Science*, **113** (Pt 7), 1213–21.

129 Alberts, B. et al. (1989) *Molecular Biology of the Cell*, Garland Publishing, Inc., New York.

130 Anderson, R.G. (1998) The caveolae membrane system. *Annual Review of Biochemistry*, **67**, 199–225.

131 Nichols, B. (2003) Caveosomes and endocytosis of lipid rafts. *Journal of Cell Science*, **116**, 4707–14.

132 Kirkham, M. and Parton, R.G. (2005) Clathrin-independent endocytosis: new insights into caveolae and non-caveolar lipid raft carriers. *Biochimica et Biophysica Acta*, **1746**, 349–63.

133 Qaddoumi, M.G. et al. (2003) Clathrin and caveolin-1 expression in primary pigmented rabbit conjunctival epithelial cells: role in PLGA nanoparticle endocytosis. *Molecular Vision*, **9**, 559–68.

134 Klingauf, J. (2007) Synaptic vesicle dynamics sans dynamin. *Neuron*, **54**, 857–8.

135 Jiang, X. et al. (2010) Specific effects of surface amines on polystyrene nanoparticles in their interactions with mesenchymal stem cells. *Biomacromolecules*, **11**, 748–53.

12
Radiopaque Polymeric Nanoparticles for X-Ray Medical Imaging

Shlomo Margel, Anna Galperin, Hagit Aviv, Soenke Bartling and Fabian Kiessling

12.1
Introduction

X-ray imaging is a well-known and extremely valuable tool for the detection and diagnosis of various disease states in the human body, as first indicated by G.G. Liversidge *et al.* [1]. Briefly, in X-ray imaging, transmitted radiation is used to produce a radiograph that is based on the overall tissue attenuation characteristics. X-rays pass through various tissues and are attenuated by scattering – that is, by reflection, refraction, or energy absorption. Unfortunately, some body organs, vessels and anatomic sites exhibit such minimal absorption of X-ray radiation that radiographs of these body portions are difficult to obtain. In order to overcome this problem, radiologists routinely introduce an X-ray-absorbing medium containing a contrast agent into such body organs, vessels and anatomic sites.

Recent reports concerning novel biomaterials have indicated an increasing interest in the development of radiopaque polymeric particles as contrast agents for X-ray imaging. Contrast agents composed of iodinated polymers may be used for various applications, including imaging of the blood pool or of certain body organs, the detection of various disease states, and monitoring embolization processes. Each application requires optimal particles with different physical and chemical properties. For example, in order to image a certain body organ by intravenous injection, it is essential that the particles are of nanometer size, with a narrow size distribution. In contrast, for embolization purposes the particles should be of micro- or macro-meter-sized diameter. X-ray contrast agents may be used not only for medical imaging, but also for other applications; examples include high-refractive-index materials and the detection of various materials such as plastics, inks, cloth stents, and dental compositions [2–4].

Contrast agent research began as early as 1895, in the same year that W.C. Röntgen discovered X-rays. Subsequently, elements with high atomic numbers, such as bromine, iodine, bismuth and lead, were soon found to be useful as contrast enhancers. For example, strontium bromide and lithium and sodium iodides were each studied as the first water-soluble contrast agents, although iodine later became the leading contrast-giving atom for X-ray techniques.

Nanomaterials for the Life Sciences Vol.10: Polymeric Nanomaterials. Edited by Challa S. S. R. Kumar
Copyright © 2011 WILEY-VCH Verlag GmbH & Co. KGaA, Weinheim
ISBN: 978-3-527-32170-4

Today, all intravascular administrations require iodinated contrast agents. In order to visualize the gastrointestinal tract barium sulfate is used, while in rare cases xenon gas may be inhaled for delineation of the pulmonary system. The main reasons that iodine took over the leading role in contrast agent synthesis included its high absorption coefficient for radiation, its chemical versatility, and relative inertness. Contrast agents are requested to be pharmacologically "invisible" – that is, they should not exhibit any pharmacological activity in the organism; their only role is to absorb X-rays. Accordingly, whilst sodium iodide should be ideal in terms of its absorption (85% of the molecule is iodine), the same is not true from a toxicity perspective, as free iodide ions affect not only the thyroid glands in particular but also other organs and tissues. Organically bound (i.e., stable) iodine – particularly if shielded by hydrophilic groups – is much better tolerated. Other factors which contribute to contrast agent toxicity include the hydrophilicity of the molecule, and the osmolality and viscosity of the preparation. The historic development of X-ray contrast agents was a logical consequence of these considerations, moving from inorganic iodide to organic mono-iodine compounds (Uroselectan A), bis-iodine (Uroselectan B) and tris-iodine substances (diatrizoate); from lipophilic to hydrophilic agents; from ionic (diatrizoate) to nonionic drugs (iopromide); from monomers (iopromide) to dimers (iotrolan), and then to water-soluble and insoluble polymers [5–27].

Despite being used for X-ray imaging, water-soluble iodinated organic compounds and polymers have significant drawbacks; typically, their inherent hyperosmolality can cause extremely severe side effects when used as a gastrointestinal contrast agent in dehydrated patients. For this reason, these materials have become rather controversial, and some radiologists no longer use them in the gastrointestinal tract [28].

Water-insoluble, iodinated organic polymers in different shapes – that is, micelles, dendrimers, and liposomes – have also been developed as contrast agents for X-ray imaging [9, 29]. Particulate contrast agents were developed because the particles make almost no contribution to the osmotic pressure of a medium. Accordingly, the use of particulate contrast agents affords the possibility of administering high-iodine-content materials that lack the fluid and ion shifts associated with water-soluble agents. An additional advantage of particulate contrast agents is based on their ability to bind and/or encapsulate desired drugs within the particulate matrix, for controlled release or targeting purposes. Unfortunately, the development of suitable particulate contrast agents for *in vivo* X-ray imaging has not yet been achieved, mainly because of the difficulties involved in their development. The size and shape of the particles must be such that they do not cause embolism in the capillaries (unless used for this purpose), while *in vivo* aggregation must be avoided to prevent the occurrence of embolism. The particles should also be biocompatible, show a reduced toxicity, and be of the correct size in order that they are to be phagocytosed into the cellular space of an organ. One significant obstacle to the intravenous use of particles bonded with a homing agent for targeting a desired tissue is the rapid clearance of these particles from the bloodstream by the macrophages of the reticuloendothelial system (RES). For example, polystyrene particles as small as 60 nm in diameter are cleared from the blood within 2–3 min; however, it is well known that by coating the particles with appropriate surfactants (e.g., poly(ethylene glycol); PEG), their half-lives in the blood will be significantly increased [30]. Thus, for targeting a specific tissue (e.g., breast tissue) for X-ray imaging via intravenous administration, it is necessary to have X-ray contrast agent nanoparticles less than about 200 nm in size; they must also be bonded with an appropriate homing agent, coated with an appropriate material, and capable of fulfilling certain other essential requirements (e.g., nontoxic, biocompatible). It should be noted that present-day X-ray contrast agents are not suitable for targeting purposes, and very few studies have been conducted dealing with this issue. However, Popovtzer *et al.* recently demonstrated *in vitro* the possibility of using targeted gold nanoparticles for the diagnosis of head and neck cancer, using computed tomography (CT) [31].

In this chapter, attention is focused on the synthesis, characterization and X-ray-imaging capabilities of iodinated nanoparticles of narrow size distributions and various sizes, prepared by different heterogeneous polymerizations of the monomer 2-methacryloyloxyethyl (2,3,5-triiodobenzoate), MAOETIB. The polymer thus obtained, P(MAOETIB), was selected for these studies as it is expected to be both nontoxic and biodegradable. In fact, in the body it would be expected to decompose only by the action of various enzymes (e.g., chymotrypsin and esterases) to form triiodobenzoic acid (a common useful water-soluble X-ray contrast agent) and poly(2-hydroxyethyl methacrylate) (polyHEMA; a nontoxic water-soluble polymer). The synthesis and imaging capabilities of the following uniform iodinated radiopaque nanoparticles are described in this chapter [32–37]:

1. P(MAOETIB), poly(2-methacryloyloxyethyl(2,3,5-triiodobenzoate), nanoparticles of narrow size distribution prepared by the emulsion polymerization of MAOETIB [32].

2. P(MAOETIB-GMA), poly(2-methacryloyloxyethyl(2,3,5-triiodobenzoate) – glycidyl methacrylate), copolymeric nanoparticles of narrow size distribution prepared by the emulsion copolymerization of MAOETIB and GMA [33].

3. γ-Fe$_2$O$_3$/P(MAOETIB) core–shell nanoparticles of narrow size distribution prepared by the seeded emulsion polymerization of MAOETIB in the presence of γ-Fe$_2$O$_3$ nanoparticles dispersed in an aqueous continuous phase [34]. (*Note:* These details were excluded from the chapter, for reasons of limited space.)

12.2
Synthesis of the Monomer MAOETIB

The monomer MAOETIB was synthesized according to Scheme 12.1 [23]:
Briefly, 2,3,5-triiodobenzoic acid (50 g, 0.10 mol), HEMA (15 g, 0.11 mol), N,N'-dicyclohexylcarbodiimide (DCC; 23 g, 0.11 mol) and 4-pyrrolidinopyridine (1.5 g, 0.01 mol) were dispersed in ether (500 ml) and stirred at room temperature for

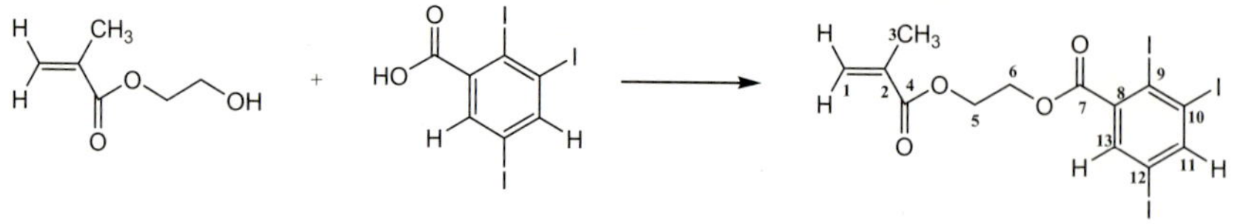

Scheme 12.1

18 h. The solid was filtered off, and the residue washed with fresh ether. The ether solution was then washed with HCl (2 M), saturated NaHCO$_3$, and brine, after which the organic phase was dried over MgSO$_4$, filtered, and evaporated to produce an orange solid. Pure white crystals of MAOETIB (m.p. 95 °C) were obtained by recrystallization of the orange solid twice from ethyl acetate.

12.3
Radiopaque Iodinated P(MAOETIB) Nanoparticles

12.3.1
Synthesis of the P(MAOETIB) Nanoparticles

P(MAOETIB) nanoparticles of 30.6 ± 5.0 nm were prepared by the emulsion polymerization of MAOETIB [38]. Briefly, 5 ml of a toluene solution containing 400 mg MAOETIB (2%, w/v$_{H2O}$) were introduced into a vial containing 20 ml of 1% (w/ v$_{H2O}$) sodium dodecyl sulfate (SDS) aqueous solution and 0.05% (w/v$_{H2O}$) of the initiator potassium persulfate (PPS). The mixture was than shaken at 73 °C for 12 h. The organic phase containing the toluene and excess monomer was then extracted from the aqueous phase. Excess SDS was removed from the aqueous dispersion by extensive dialysis against water. Dried radiopaque P(MAOETIB) nanoparticles were subsequently obtained by lyophilization. The effect of various polymerization parameters, such as monomer, initiator and surfactant concentrations, on the size and size distribution of the particles has also been elucidated.

12.3.2
Characterization of the P(MAOETIB) Nanoparticles

A typical transmission electron microscopy (TEM) image of the dried P(MAOETIB) nanoparticles, prepared according to the description in Section 12.3.1, is shown in Figure 12.1a. Although, after evaporation of the aqueous medium (as required for TEM analysis), the iodinated radiopaque polymeric nanoparticles formed agglomerates, those agglomerates composed of nanoparticles with an average diameter of 28.9 ± 6.3 nm were easily visualized. In contrast, P(MAOETIB) nanoparticles dispersed in water were stable and demonstrated a single population with an average hydrodynamic diameter of 30.6 ± 5.0 nm, as measured with a light-

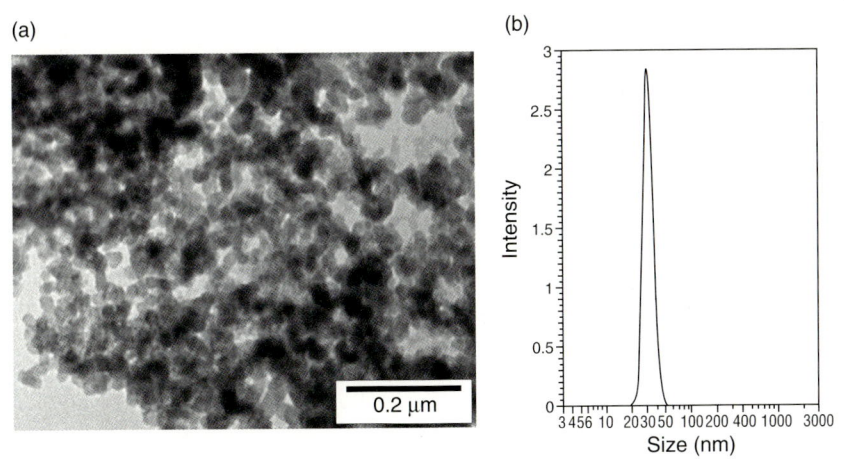

Figure 12.1 Transmission electron microscopy image (a) and size histogram (b) of the P(MAOETIB) nanoparticles.

scattering technique (Figure 12.1b). Thus, an excellent correlation is apparent between the dry and hydrodynamic diameters of the nanoparticles. The P(MAOETIB) nanoparticles were free from traces of the monomer, as was verified using Fourier transform infrared (FTIR) spectroscopy, based on a lack of the C=C double-bond stretching band at ~1623 cm^{-1}. Likewise, ^{1}H NMR (THF-d_8) confirmed a lack of the two peaks of the vinylic protons at 5.61 and 6.16 ppm. It should also be noted that the P(MAOETIB) nanoparticles contain approximately 58% iodine, while the monomer contained 62%, as confirmed by iodine elemental analysis. This difference in the iodine content may be due to the initiator fractions of the polymer, and to adsorption of SDS onto the P(MAOETIB) nanoparticle surfaces. It is also possible that a few aromatic C–I bonds were cleaved during the radical polymerization of MAOETIB, leading to a slight decrease in the iodine content of the polymer relative to the monomer.

12.3.2.1 Effect of the MAOETIB Concentration

The influence of MAOETIB concentration on the size and size distribution of the P(MAOETIB) nanoparticles is shown in Figure 12.2. Clearly, the diameter of the particles is increased in line with increases in the MAOETIB concentration; for example, in the presence of 1%, 6% and 8% MAOETIB, the size of the P(MAOETIB) nanoparticles was increased from 25.9 ± 8.6 to 41.7 ± 7.8 and 46.7 ± 5.4 nm, respectively. It can be assumed that an increase in the initial monomer concentration may increase its concentration within the micelles and, as a consequence, increase the diameter of the formed particles [39]. The same tendency, with regards to the influence of styrene concentration on the size of polystyrene nanoparticles formed by emulsion polymerization, was reported by Martinez [40].

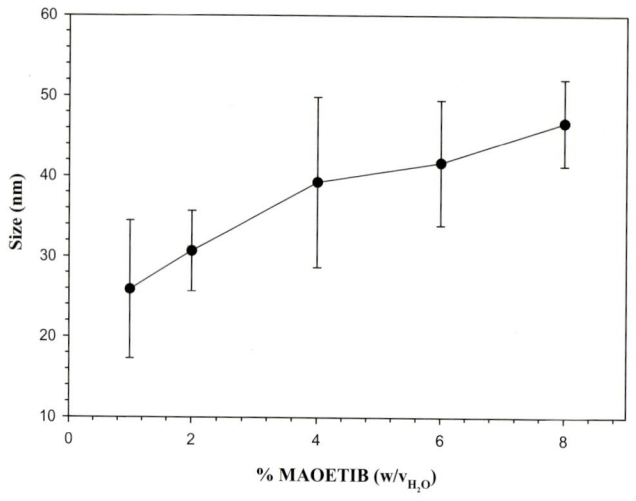

Figure 12.2 Effect of MAOETIB concentration on the size and size distribution of the P(MAOETIB) nanoparticles.

Table 12.1 Effect of the initiator concentration on the size and size distribution of the P(MAOETIB) nanoparticles.[a]

Diameter (nm)	%PPS (w/v_{H2O})
30.6 ± 5.0	0.05
287.4 ± 61.4	0.3
355.8 ± 139.4	1

a The P(MAOETIB) nanoparticles were prepared according to Section 12.3.1, with 5 ml of toluene solution containing different PPS concentrations.

12.3.2.2 Effect of the Initiator Concentration

The effects of PPS on the size and size distribution of the P(MAOETIB) nanoparticles are listed in Table 12.1. Here, the data show that increasing the PPS concentration from 0.05% to 0.3%, and to 1% leads to larger particles with a broader size distribution: from 30.6 ± 5.0 to 287.4 ± 61.4 nm, and to 355.8 ± 139.4 nm, respectively. A similar effect of the initiator concentration on the size of particles was previously reported by Boguslavsky et al. [39]. It can be assumed that increasing the initiator concentration causes an increase in the oligomeric radical concentration, and thus in the number of P(MAOETIB) chains. This may lead to an increase in the number of P(MAOETIB) nanoparticles (more nuclei) and an increase in their size (i.e., more P(MAOETIB) chains will participate in the growing

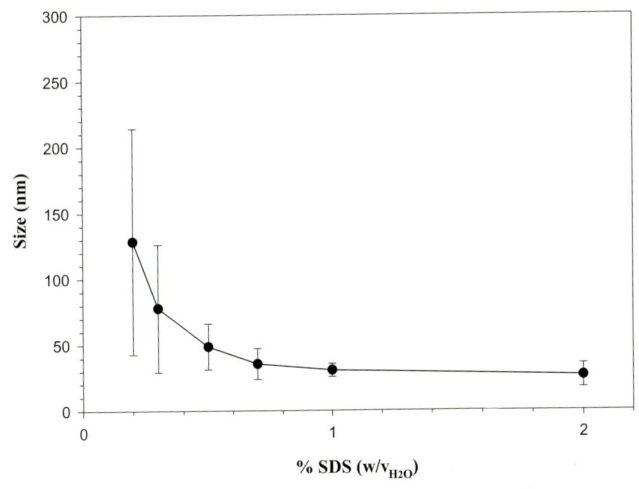

Figure 12.3 Effect of the SDS concentration on the size and size distribution of the P(MAOETIB) nanoparticles.

process). Moreover, a higher initiator concentration will increase the growth rate of the oligomeric chains, thus favoring secondary nucleation during the particle growth stage, and thereby increasing the particle size distribution.

12.3.2.3 Effect of the Surfactant Concentration

The effect of SDS concentration on the size and size distribution of the formed P(MAOETIB) nanoparticles is shown in Figure 12.3. Here, an increase in SDS concentration leads to the formation of smaller particles with a narrow size distribution. For example, raising the SDS concentration from 0.3% to 0.7%, and to 2% (w/v), led to a decrease in the average size and size distribution of the P(MAOETIB) nanoparticles, from 77.9 ± 48.4 to 35.6 ± 11.8 nm, and to 26.3 ± 9.0 nm. A similar effect of surfactant concentration on the diameter and size distribution of particles was previously reported by Martinez et al. [40] and Boguslavsky et al. [39]. In these cases, a higher surfactant concentration resulted in an increasing amount of surfactant molecules adsorbed onto the nuclei and, therefore, to a better protection against growing processes.

12.3.3
In Vitro X-Ray Visibility of the P(MAOETIB) Nanoparticles

The *in vitro* radiopacity of the dried and water-dispersed P(MAOETIB) nanoparticles was demonstrated using a computed tomography (CT) scanner (HeliCAT II, Marconi). Quantization (in Hounsfield units; HU) of the opacification of nanoparticles was carried out using image-processing software (CDP DiagNet v. 5.55).

The X-ray visibility of the P(MAOETIB) nanoparticles is shown in Figure 12.4. The CT image of the dried P(MAOETIB) particles placed in an Eppendorf pipette,

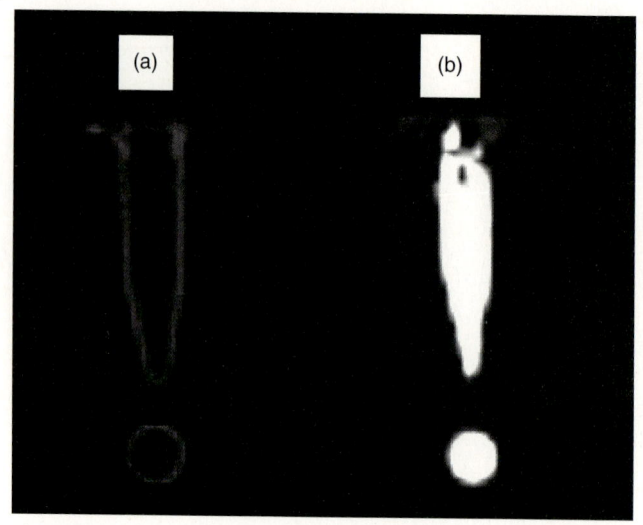

Figure 12.4 (a) X-ray visibility of an empty Eppendorf tip and section; (b) X-ray visibility of an Eppendorf containing dry P(MAOETIB) nanoparticles and section.

Table 12.2 Opacification of the P(MAOETIB) nanoparticles dispersed in saline at different concentrations.[a]

[Nanoparticles] (mg ml^{-1})	Opacification (HU)
0 (saline only)	0
1	16 ± 4
2	42 ± 4
4	115 ± 3
8	185 ± 6
16	362 ± 1

a The P(MAOETIB) nanoparticles of 30.6 ± 5.0 nm were prepared according to Section 12.3.1. HU, Hounsfield units.

while its section (Figure 12.4b) confirmed an excellent radiopaque nature compared to the image of the empty Eppendorf (and its section) (Figure 12.4a), which is almost transparent to X-ray irradiation.

Details of the opacification of P(MAOETIB) nanoparticles dispersed in water at different concentrations are listed in Table 12.2 (data in HU). These data confirmed the significant rise in intensity of the CT signal as the concentration of the P(MAOETIB) particles dispersed in water was increased. Typically, the CT signal intensities of radiopaque nanoparticles dispersed in water at concentrations of 2, 8, and 16 mg ml^{-1} were 42 ± 4, 185 ± 6, and 362 ± 1 HU, respectively.

12.3.4
In Vivo X-Ray Visibility of the P(MAOETIB) Nanoparticles: Preliminary Studies

A dog model was used in order to evaluate the *in vivo* contrast ability of the P(MAOETIB) nanoparticles. For this, a saline dispersion of P(MAOETIB) nanoparticles (30.6 ± 5.0 nm hydrodynamic diameter; concentration 3 mg ml^{-1}; dose level 40 mg kg^{-1}) was administered intravenously (by a slow drip infusion) to a single dog (body weight 20 kg). Subsequently, pelvic, abdominal, and leg CT scans were performed before and at 24 h after administration of the radiopaque nanoparticles. In addition, full blood tests (including a complete blood count and a full chemical analysis) were conducted before, and at 24 h and one month after, injection of the nanoparticles. The dog was observed over a six-month period after injection, to monitor any changes in its health.

The CT images and enhancement (in HU) of the animal's organs (popliteal lymph nodes, liver, kidney and spleen) before and after P(MAOETIB) nanoparticle administration are shown in Figure 12.5. A significant lymph node enhancement, at 24 h after nanoparticle administration was observed, both visually and quantitatively (40 HU before compared to 80 HU after injection). The same trend was seen in the liver, kidneys and spleen, with the CT signal intensity being enhanced from 70 to 100 HU, 30 to 64 HU, and 73 to 147 HU, respectively. Although only one dog was used in this preliminary *in vivo* study, the results confirmed not only the potential contrast properties of the P(MAOETIB) nanoparticles, but also that they could be injected safely via the intravenous route. No change in either blood tests or the animal's general health was observed during the six months after injection. The fact that the iodinated nanoparticles caused a significant enhancement of the lymph nodes at 24 h after injection was not surprising, as nanoparticles of up to ~70 nm diameter are well-known to undergo significant uptake into the lymph nodes, following phagocytosis by the resident macrophages [41].

A contrasting situation arises in metastases, however, which do not possess an intact phagocytosing system. Thus, the consequent inability of nanoparticles to be accumulated in the metastatic lymph nodes [41] provides the basis for using superparamagnetic iron oxide (SPIO) nanoparticles to distinguish between normal and metastatic lymph nodes [41]. Until now, the only imaging modality using SPIO has been MRI, which is not only very costly but also unavailable at most medical centers [41]. As the P(MAOETIB) nanoparticles have very good X-ray visibility, a potential clinical role for them might be in the detection of various pathogenic zones, such as cancer metastases in lymph nodes, by using CT. Clearly, there is a need to expand these *in vivo* studies, using greater numbers of animals in the presence and absence of cancer in specified organs.

12.4
Radiopaqe Iodinated P(MAOETIB–GMA) Copolymeric Nanoparticles

Although, previously, the P(MAOETIB) nanoparticles were relatively easily created, the nanoparticle aqueous dispersion proved to be unstable and tended to

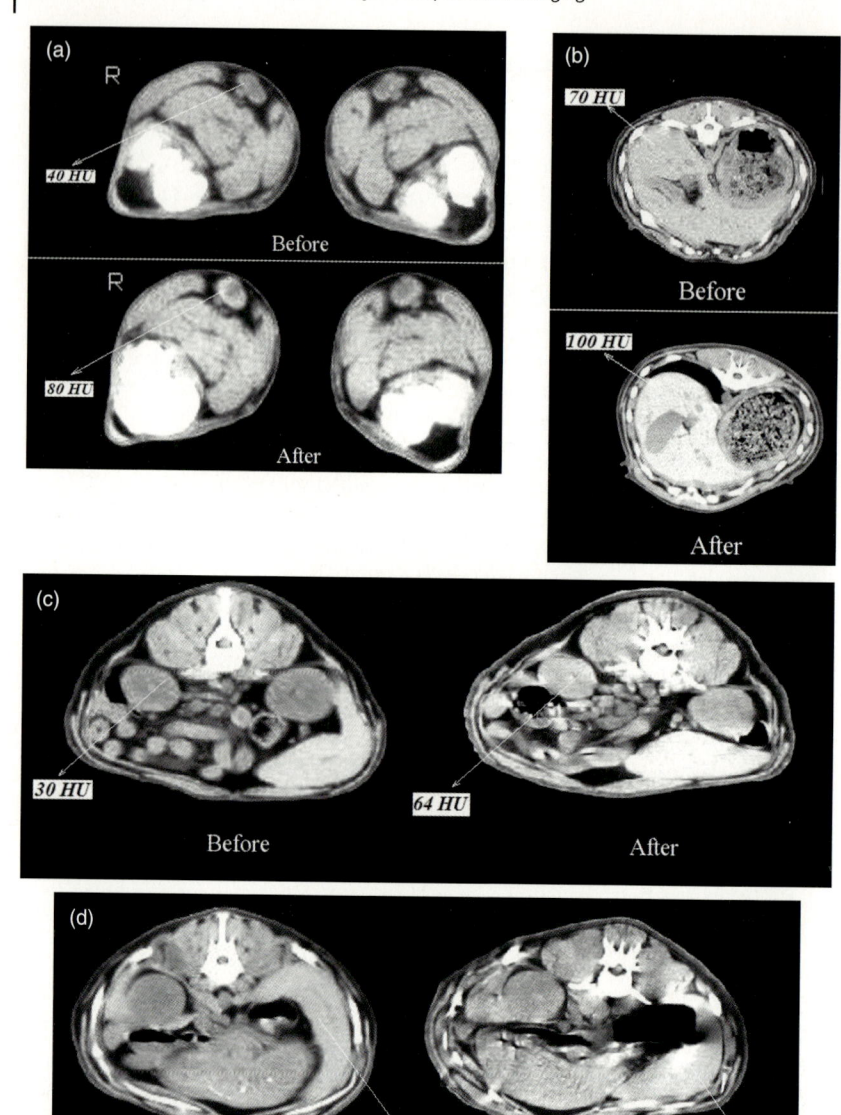

Figure 12.5 Computed tomography images and opacification (in HU) of dog organs before and at 24 h after intravenous administration of the P(MAOETIB) nanoparticles dispersed in saline. (a) Popliteal lymph nodes; (b) Liver; (c) Kidney; (d) Spleen.

agglomerate, particularly at a weight concentration of dispersed nanoparticles above ~0.3% (3 mg ml^{-1}). The agglomeration rate also increased in line with the concentration of nanoparticles in the aqueous phase, preventing their efficient *in vivo* use as contrast agents for medical X-ray imaging. The *in vivo* CT-imaging in a dog was conducted after the intravenous administration of P(MAOETIB) nanoparticles dispersed in saline, at a concentration of 3 mg ml^{-1} (0.3%). Because of this low particle concentration, an injection volume of 266 ml was required in order to observe any radiopacity. Unfortunately, such a high volume would be inconvenient and impractical for clinical use. Moreover, despite the large volume injected, only a minimal change in the X-ray opacity of tissues was observed, while the low concentration of P(MAOETIB) homo-nanoparticles in saline did not permit blood pool imaging.

The following subsections describe the synthesis of iodinated copolymeric nanoparticles of narrow size distribution, via the emulsion copolymerization of the monomers, MAOETIB, and a relatively low concentration of glycidyl methacrylate (GMA). As the surface of the resultant copolymeric nanoparticles is far more hydrophilic than that of the P(MAOETIB) nanoparticles, P(MAOETIB–GMA) nanoparticles would be significantly more stable against agglomeration in aqueous continuous phase.

12.4.1
Synthesis of P(MAOETIB–GMA) Copolymeric Nanoparticles

The P(MAOETIB–GMA) nanoparticles were prepared by emulsion copolymerization of MAOETIB with GMA, based on a procedure similar to that used to prepare the P(MAOETIB) nanoparticles. Briefly, radiopaque copolymeric nanoparticles of 25.5 ± 4.2 nm dry diameter were formed by the introduction into a vial containing 20 ml of a 1% SDS aqueous solution and 0.05% PPS (10 mg), a 5 ml aliquot of a toluene solution containing 396 mg MAOETIB and 4 mg GMA. The mixture was shaken at 73 °C for 12 h, after which the organic phase containing the toluene and excess monomers was separated from the aqueous phase. Excess SDS was removed from the aqueous dispersion by extensive dialysis. The dried radiopaque P(MAOETIB–GMA) nanoparticles were then obtained by lyophilization. The effect of the wt% ratio [MAOETIB]/[GMA], while maintaining the total monomer (GMA + MAOETIB) weight constant (400 mg) on the size and size distribution of the P(MAOETIB–GMA) nanoparticles, was also elucidated.

12.4.2
Influence of the Weight Ratio [MAOETIB]/[GMA] on the Iodine Content and Size and Size Distribution of the Copolymeric Nanoparticles

The influence of the wt% ratio [MAOETIB]/[GMA] on the iodine content and on the hydrodynamic size and size distribution of the P(MAOETIB–GMA) nanoparticles, are shown in Figure 12.6a and b, respectively. In these experiments, the total monomer amount remained constant (0.4 g), while the weight ratio

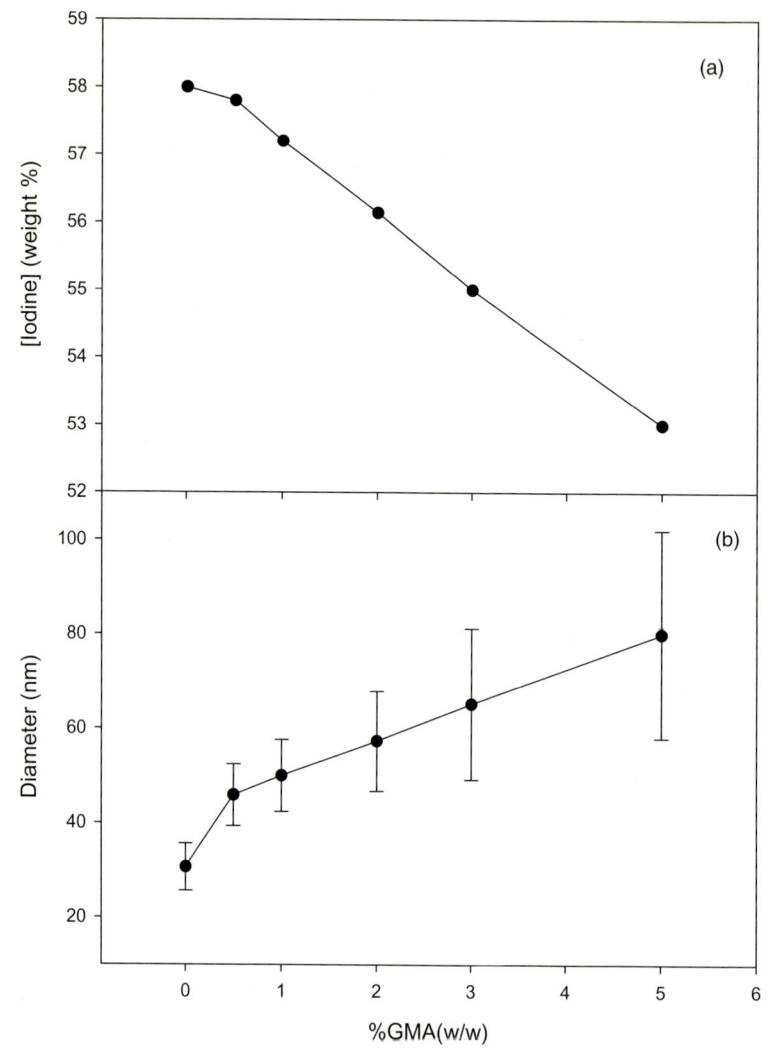

Figure 12.6 Effect of GMA wt% on (a) the iodine content and (b) the hydrodynamic size and size distribution of the P(MAOETIB–GMA) nanoparticles. The P(MAOETIB–GMA) nanoparticles were prepared in the presence of 400 mg total monomer concentration at various [MAOETIB]/[GMA] wt% ratios.

[MAOETIB]/[GMA] was altered. As might be expected, the iodine content of the P(MAOETIB–GMA) nanoparticles was decreases as the GMA wt% increased (Figure 12.6a). For example, in the absence and presence of 0.5%, 1% and 2% GMA, the iodine wt% of the copolymeric nanoparticles was decreased from 58.0% to 57.8%, 57.2%, and 56.1%, respectively. The size and size distribution of the nanoparticles was seen to increase as the GMA wt% increased (Figure 12.6b). For example, in the absence and presence of 0.5%, 1% and 2% GMA, the hydrodynamic diameter of the P(MAOETIB–GMA) nanoparticles was increased from 30.6 ± 5.0 to 45.9 ± 6.5, 50.0 ± 7.6 and 57.3 ± 10.6 nm, respectively. The optimal copolymeric nanoparticles for *in vivo* use as contrast agents for medical X-ray imaging applications are those particles that optimize the ratio between the iodine content on the one hand, and the size and size distribution on the other hand. Thus, it would seem that the most suitable copolymeric nanoparticles are those prepared in the presence of 99.5% MAOETIB and 0.5% GMA. However, when dispersed in water these copolymeric nanoparticles proved not to be stable at nanoparticle concentrations above 2–3%, so that the X-ray radiopacity per unit volume of an aqueous dispersion containing these nanoparticles would most likely be insufficient for *in vivo* use. On the other hand, although the copolymeric nanoparticles prepared in the presence of 99% MAOETIB and 1% GMA contained slightly less iodine (57.8% and 57.2%), and their size and size distribution was slightly higher (45.9 ± 6.5 and 50.0 ± 7.6 nm, respectively), the aqueous dispersion of these nanoparticles would be stable, such that their size and size distribution would be maintained at nanoparticle concentrations of up to 15 wt%. The copolymeric nanoparticles prepared at a wt% ratio of [MAOETIB]/[GMA] less than 1 (99% MAOETIB and 1% GMA) were shown to possess a lower iodine content and a significantly higher size and size distribution (see Figure 12.6b). Thus, the optimal copolymeric nanoparticles selected for further detailed studies were those prepared at a wt% ratio [MAOETIB]/[GMA] of 99/1.

12.4.3
Characterization of the P(MAOETIB-GMA) Nanoparticles Prepared at a Weight Ratio [MAOETIB]/[GMA] of 99/1

A TEM image of dried P(MAOTIB–GMA) nanoparticles, prepared at a wt% ratio [MAOETIB]/[GMA] of 99/1 (according to Section 12.4.1) is shown in Figure 12.7a. These iodinated nanoparticles are spherical and composed of a single population with a dry diameter of 25.5 ± 4.2 nm. In contrast, the P(MAOETIB–GMA) nanoparticles dispersed in water possessed a single population with a hydrodynamic diameter of 50.0 ± 7.6 nm, as measured by light scattering (Figure 12.7b). The major difference between the dry and hydrodynamic diameters of the nanoparticles was most likely due to a significant adsorption of water molecules onto the surface of the nanoparticles. Such surface water adsorption illustrates the hydrophilic nature of the copolymeric P(MAOETIB–GMA) nanoparticles; indeed, it is this hydrophilic surface that allows the nanoparticles to remain stable in an aqueous continuous phase at very high concentrations (up to ca. 15%). On the other hand, the P(MAOETIB) nanoparticles produced via the homopolymerization of MAOETIB (see Section 12.3.1) have average dry and hydrodynamic diameters of 28.9 ± 6.3 and 30.6 ± 5.0 nm, respectively. The slight difference between the dry and hydrodynamic diameters of these homopolymeric nanoparticles, compared to that of the copolymeric nanoparticles, indicates the relative lipophilic nature of the

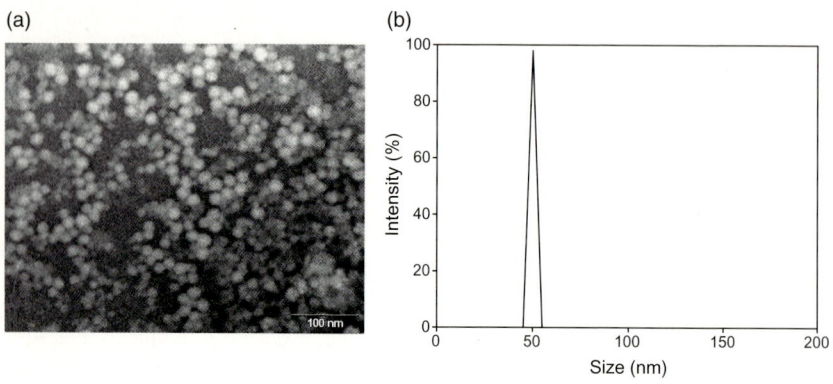

Figure 12.7 Transmission electron microscopy image (a) and size histogram (b) of the P(MAOETIB–GMA) nanoparticles. The P(MAOETIB–GMA) nanoparticles were prepared in the presence of 400 mg total monomer concentration at a [MAOETIB]/[GMA] wt% ratio of 99/1.

P(MAOETIB) nanoparticles, and explains their relative instability in the aqueous continuous phase.

These optimal P(MAOETIB–GMA) nanoparticles were free from traces of the monomer, as verified by ^{1}H NMR (THF-d$_8$) (by a lack of two peaks of the vinylic protons at 5.61 and 6.16 ppm), and by FTIR (by a lack of the C=C double-bond stretching band at about 1623 cm^{-1}). Surprisingly, the FTIR spectrum of the P(MAOETIB–GMA) copolymeric nanoparticles did not show peaks at 845 and 910 cm^{-1}, corresponding to the epoxide vibrational bands of the GMA monomeric units. Consequently, it was assumed that, under experimental conditions, each of the epoxide groups would split open to two hydroxyl groups. Indeed, the FTIR spectrum of the P(MAOETIB–GMA) nanoparticles indicated a clear typical absorption peak of the –OH stretching band at about 3500 cm^{-1}.

The size and size distribution of the P(MAOETIB–GMA) and P(MAOETIB) nanoparticles dispersed in water (7 mg ml^{-1}) and stored at room temperature for 30 days are shown in Figure 12.8. The data show clearly that the size and size distribution of the P(MAOETIB–GMA) nanoparticles (Figure 12.8a) are preserved during the entire measurement period, while those of the P(MAOETIB) nanoparticles (Figure 12.8b) are increased consistently. For example, the hydrodynamic size and size distribution of the P(MAOETIB–GMA) nanoparticles, immediately after synthesis and again after 7 and 30 days, were 50.0 ± 7.6, 50.7 ± 6.0 and 50.9 ± 6.5 nm respectively, whereas those of the P(MAOETIB) nanoparticles were 30.6 ± 5.0, 49.2 ± 21.1 and 185.0 ± 72.2 nm, respectively. These results illustrate the stability towards agglomeration during storage of the copolymeric nanoparticles, and the instability of their homopolymeric counterparts. To illustrate this, two vials which contain either the copolymeric or homopolymeric nanoparticles dispersed in water (7 mg ml^{-1}), and which have been stored at room temperature

12.4 Radiopaqe Iodinated P(MAOETIB–GMA) Copolymeric Nanoparticles | 357

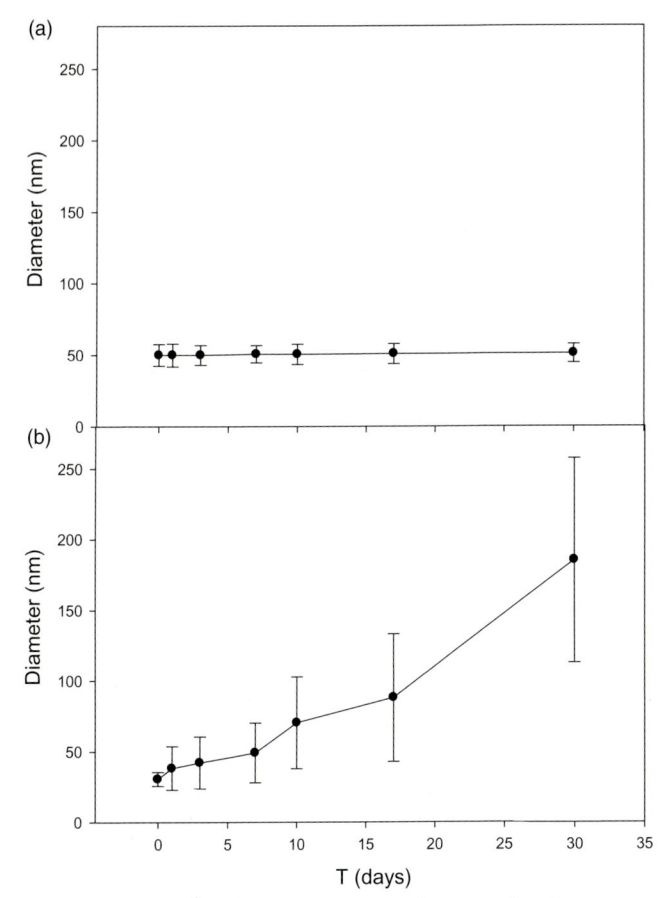

Figure 12.8 Changes with time in the size and size distribution of the P(MAOETIB–GMA) (a) and the P(MAOETIB) (b) nanoparticle dispersions in water (7 mg ml^{-1}). The P(MAOETIB–GMA) nanoparticles were prepared in the presence of 400 mg total monomer concentration at a [MAOETIB]/[GMA] wt% ratio of 99/1. The P(MAOETIB) nanoparticles were prepared in the presence of 400 mg of MAOETIB.

for three months, are shown in Figure 12.9. The P(MAOETIB–GMA) (left) aqueous dispersion is homogeneous, and has the same appearance as immediately after preparation. By contrast, the P(MAOETIB) aqueous dispersion has separated into two phases, an aqueous phase and a white precipitate composed of the agglomerated P(MAOETIB) nanoparticles. It would appear that the remarkable hydrophilic nature of the surface of the copolymeric nanoparticles, achieved by hydrolysis of the surface PGMA, allows them to remain stable in the absence of any surfactant or stabilizer, even at concentrations up to 150 mg ml^{-1}.

In vivo studies require the dispersion of the copolymeric nanoparticles in a physiological aqueous continuous phase. The stability against agglomeration of

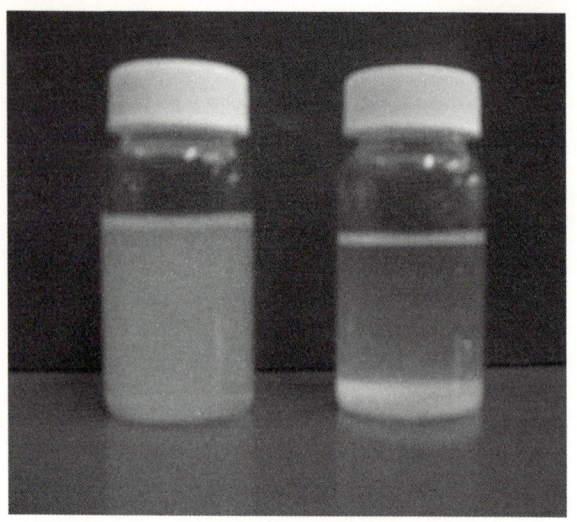

Figure 12.9 Visibility of the P(MAOETIB–GMA) (left) and P(MAOETIB) (right) nanoparticle dispersions in water (7 mg ml^{-1}) after a three-month storage period at room temperature.

the P(MAOETIB–GMA) nanoparticles dispersed in water (7 mg ml^{-1}) containing different concentrations of sodium chloride, is shown in Figure 12.10. Clearly, the size and size distribution of these nanoparticles are increased in line with the NaCl concentration; in this case, in the absence and presence of 0.2%, 0.5% and 0.85% NaCl the sizes were increased from 50.0 ± 7.6 nm to 50.1 ± 7.6, 92.0 ± 32.7 and 276.0 ± 101.2 nm, respectively. These results indicated clearly that the copolymeric nanoparticles were unstable in saline (0.85% NaCl). This effect of salt concentration on the stability of colloidal particles is well known, and may be explained by a salting-out effect of surface-adsorbed water molecules towards solvation of the ions [42]. In contrast, the results confirmed the stability of copolymeric nanoparticles in a 5% dextrose aqueous solution; that is, the size and size distribution of 5%, 8% and 15% copolymeric nanoparticles dispersed in a 5% dextrose aqueous solution were 51.0 ± 6.5, 50.8 ± 7.2 and 50.3 ± 7.4 nm, respectively. A similar size and size distribution were also noted in the absence of dextrose. As an aqueous 5% dextrose solution has a similar osmolarity to blood [43], the *in vivo* animal X-ray imaging trials were conducted using copolymeric nanoparticles (80 mg ml^{-1}) dispersed in a 5% dextrose aqueous continuous phase, as follows.

12.4.4
In Vivo X-Ray Visibility of the P(MAOETIB-GMA) Nanoparticles

Two male Copenhagen rats (body weight ca. 500 g) were used to evaluate the distribution and X-ray visibility of P(MAOETIB–GMA) nanoparticles in the blood,

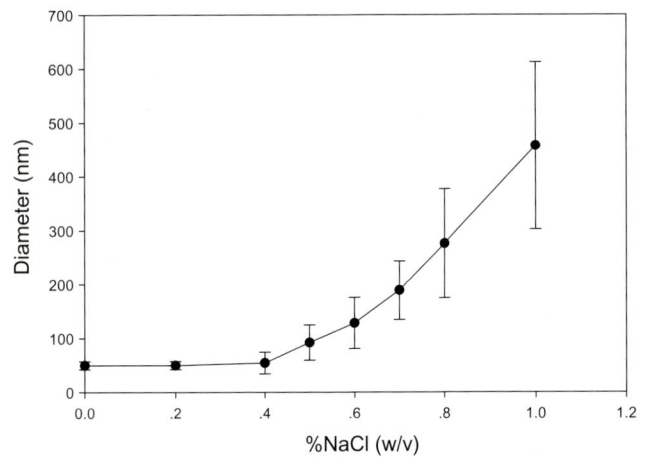

Figure 12.10 Size and size distribution of the P(MAOETIB–GMA) nanoparticles dispersed in water (7 mg ml^{-1}) containing different amounts of sodium chloride. The P(MAOETIB–GMA) nanoparticles were prepared in the presence of 400 mg total monomer concentration at a [MAOETIB]/[GMA] wt% ratio of 99/1.

lymph nodes, liver, and spleen. For this purpose, a 5% dextrose dispersion of the P(MAOETIB–GMA) nanoparticles (80 mg ml^{-1}) was injected (400 mg kg^{-1}) into the tail vein of each rat. Subsequently, CT scans were carried out at 2 and 30 min after administration of the radiopaque nanoparticles.

Similarly, two male Black6 mice (body weight ca. 30 g) with a liver cancer model were used to evaluate the potential of the nanoparticles for liver imaging. For this purpose, the P(MAOETIB–GMA) nanoparticles (80 mg ml^{-1}) were dispersed in 5% dextrose solution, and injected into the tail vein at a dose level of 600 mg kg^{-1}. Subsequently, CT scans were carried out before, and at 3 min and 4 h after, the injection.

The *in vivo* radiopacity of the P(MAOETIB–GMA) nanoparticles dispersed in 5% dextrose solution was performed with a prototype flat-panel detector-based volume CT scanner (Siemens, Erlangen, Germany) [44]. The total scan time was 40 s, with a rotation time of 20 s. A tube voltage of 80 kV and a tube current of 50 mA with continuous radiation were selected. Motion-gating to reduce the effects of respiratory motion was performed, as described elsewhere [45]. The reconstruction field-of-view was 7.6 cm transaxially with a reconstruction matrix of 512 × 512 pixels and an axial slice spacing of 0.2 mm; this resulted in a voxel size of 0.15 × 0.15 × 0.2 mm. A modestly sharp reconstruction kernel (H80s) was used for the image reconstruction. Image post-processing was performed using standard 3-D post-processing steps, for example, volume-rendering.

The rapid intravenous (0.2 ml s^{-1}) injection of the nanoparticles was well tolerated by the animals, with no early signs of intolerance, such as breathing

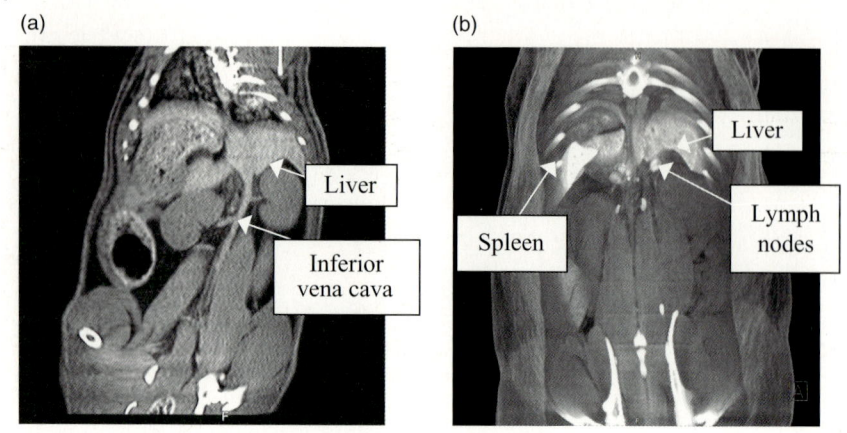

Figure 12.11 Enhancement after intravenous injection into the tail vein of a rat of 2.5 ml of a 5% dextrose dispersion of the P(MAOETIB–GMA) nanoparticles (80 mg ml^{-1}). CT scanning was carried out at 2 min (a) and 30 min (b) after injection, as shown in the coronal reformation (a) and volume-rendering (b) techniques of CT scanning.

interruption, cardiac stress or allergic reaction being witnessed. During a six-month observation period no signs of any reduced state of health were identified, such as inappetence, ruffled fur, doziness, or weight loss.

The CT images of a rat after the intravenous administration of a copolymeric nanoparticle dispersion is shown in Figure 12.11. An enhancement of the blood pool and a slight enhancement of the liver was visible at 2 min (Figure 12.11a), while beyond this time (30 min) a strong uptake of nanoparticles by cells of the reticuloendothelial system (RES), and thus a strong enhancement of the spleen, liver, and lymph nodes, was observed. It should be noted, that in order to achieve the same contrast results in a human patient of body weight 75 kg, the highest concentration in which the nanoparticles would remain stable (150 mg ml^{-1}) should be used, and 200 ml of the nanoparticles dispersion injected intravenously.

CT sections, demonstrating a cancerous liver before, and at 3 min and 4 h after an intravenous injection of the copolymeric nanoparticle dispersion (80 mg ml^{-1}; 300 µl injected) are shown in Figure 12.12. From 3 min after injection (Figure 12.12b), the enhancement was increased gradually such that, after 4 h a strong enhancement of healthy liver tissue was apparent (Figure 12.12c). In contrast, the cancerous liver tissue can be favorably delineated as dark nonenhancing spots. The enhancement of healthy liver tissue was due to a high uptake of the nanoparticles by cells of the RES (e.g., macrophages, Kupffer cells) that reside within the liver [31, 32]. As the cancerous tumor cells displace the macrophages, the X-ray density is not noticeably affected in the diseased areas after particle injection.

In conclusion, P(MAOETIB–GMA) nanoparticles can be used to better diagnose metastatic disease, which particularly holds true for micro-CT systems that require

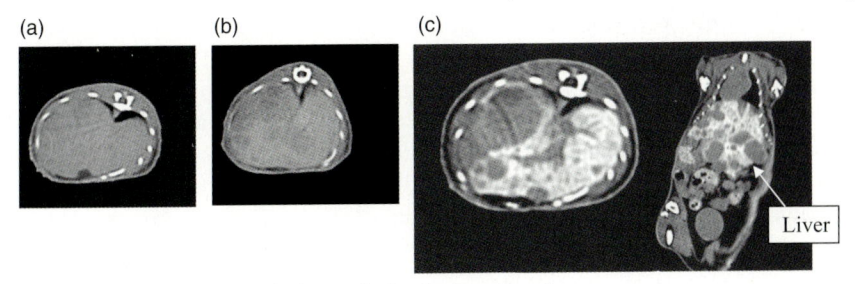

Figure 12.12 Cancerous liver before and after the injection of 300 μl 5% dextrose dispersion of the P(MAOETIB–GMA) nanoparticles (80 mg ml^{-1}) into the tail vein of a mouse. CT scanning was carried out before (a), and at 3 min (b) and 4 h (c) after injection.

longer scanning times and thus a much more persistent labeling of the healthy liver tissue. It should be noted that, in order to achieve the same healthy liver tissue enhancement in a human patient of body weight 75 kg, the highest concentration in which the nanoparticles would remain stable (150 mg ml^{-1}) should be used, and 200 ml of the nanoparticles dispersion injected intravenously.

12.5 Summary

Currently, X-ray-based CT is among the most convenient imaging/diagnostic tools used in hospitals, in terms of its availability, efficiency, and cost. However, most – if not all – of the polymeric nanoparticle contrast agents reported to date are not yet suitable for clinical use, due to problems relating to the safety, biodegradability, and instability (due to agglomeration) of the concentrated radiopaque nanoparticles when in aqueous physiological dispersions.

This chapter has included descriptions of the synthesis of two types of iodinated nanoparticle of narrow size distribution, prepared by heterogeneous polymerization of the monomer MAOETIB, namely P(MAOETIB) nanoparticles prepared by the emulsion polymerization of MAOETIB, and P(MAOETIB–GMA) copolymeric nanoparticles prepared by the emulsion copolymerization of MAOETIB and residual GMA. In addition, γ-Fe$_2$O$_3$/P(MAOETIB) core–shell dual-mode (CT/MRI) nanoparticles were prepared by the seeded emulsion polymerization of MAOETIB in the presence of the γ-Fe$_2$O$_3$ nanoparticles dispersed in an aqueous continuous phase [34]. (Note: Descriptions of these dual-mode nanoparticles were excluded from this chapter, for reasons of limited space.)

These three types of iodinated nanoparticle were planned for use as diagnostic agents for the RES system, notably for the liver, kidneys, and lymph nodes. Currently, the most suitable iodinated nanoparticles for clinical use are the P(MAOETIB–GMA) copolymeric nanoparticles which, with a dry diameter of

25.5 ± 4.2 nm contain a relatively high iodine load (58 wt%), and are also stable in aqueous physiological continuous phases containing up to 15% nanoparticles. Future plans include an extension of the studies with such copolymeric nanoparticles, particularly with regards to their *in vivo* biodistribution and biodegradability, and to monitor their potential role in the diagnosis of lymph node metastases. An equally important area for the future is the development of targeted molecular imaging using radiopaque nanoparticles, although significant improvements will be required in the CT instrumentation if this goal is to be achieved.

References

1 Liversidge, G.G., Cooper, E.R., Shaw, J.M. and Mcintire, G.L. X-ray contrast compositions useful in medical imaging. U.S. Patent 5, 451, 393, 1995.

2 Krause, W. (2002) Liver-specific X-ray contrast agents. *Topics in Current Chemistry*, **222**, 173–200.

3 Hung-Wen, S. and Wen-Chang, C. (2008) Photosensitive high-refractive-index poly(acrylic acid) graft-poly(ethylene glycol methacrylate) nanocrystalline titania hybrid films. *Macromolecular Chemistry and Physics*, **209**, 1778–86.

4 Labella, R., Davy, K.W.M., Lambrechts, P., Van Meerbeek, B. and Vanherle, G. (1998) Monomethacrylate co-monomers for dental resins. *European Journal of Oral Sciences*, **106**, 816–24.

5 Krause, W. and Schneider, P.W. (2002) Chemistry of X-ray contrast agents. *Topics in Current Chemistry*, **222**, 107–50.

6 Vera, D.R. and Mattrey, R.F. (2002) A molecular CT blood pool contrast agent. *Academic Radiology*, **9**, 784–92.

7 Torchilin, V.P. (2002) PEG-based micelles as carriers of contrast agents for different imaging modalities. *Advanced Drug Delivery Reviews*, **54**, 235–52.

8 Torchilin, V.P. (1999) Polychelating amphiphilic polymers to optimize the concentration of contrast agents used in medical diagnostic imaging. *Chemtech*, **29**, 27–34.

9 Torchilin, V.P. (2000) Polymeric contrast agents for medical imaging. *Current Pharmaceutical Biotechnology*, **1**, 183–215.

10 Masahiko, O., Takeshi, Y., Tomohiro, A., Hiroki, U., Tadashi, K. and Tatsuo, K. (2002) Synthesis and properties of radiopaque polymer hydrogels II: copolymers of 2,4,6-triiodophenyl- or N-(3-carboxy-2,4,6-triiodophenyl)-acrylamide and p-styrene sulfonate. *Journal of Molecular Structure*, **602**, 17–28.

11 Horak, D., Metalova, M., Svec, F., Drobnik, J., Katal, J., Borovicka, M., Adamyan, A.A., Voronkova, O.S, and Gumargalieva, K.Z. (1987) Hydrogels in endovascular embolization. III. Radiopaque spherical particles, their preparation and properties. *Biomaterials*, **8**, 142–5.

12 Jayakrishnan, A., Thanoo, B.C., Rathinam, K. and Mohanty, M. (1990) Preparation and evaluation of radiopaque hydrogel microspheres based on PHEMA/iothalamic acid and PHEMA/iopanoic acid as particulate emboli. *Journal of Biomedical Materials Research*, **24**, 993–1004.

13 Mottu, F., Rufenacht, D.A., Laurent, A. and Doelker, E. (2002) Iodine-containing cellulose mixed esters as radiopaque polymers for direct embolization of cerebral aneurysms and arteriovenous malformations. *Biomaterials*, **23**, 121–31.

14 Chithambara, T.B. and Jayakrishnan, A. (1990) Barium sulphate loaded p(HEMA) microspheres as artificial emboli: preparation and properties. *Biomaterials*, **11**, 477–81.

15 Chithambara, T.B. and Jayakrishnan, A. (1991) Tantalum loaded silicone microspheres as particulate emboli. *Journal of Microencapsulation*, **8**, 95–101.

16 Chithambara, T.B., Sunny, M. and Jayakrishnan, A. (1991) Ta-loaded polyurethane microspheres for particulate

embolization: preparation and properties. *Biomaterials*, **12**, 525–8.
17 Horak, D., Metalova, M. and Rypacek, F. (1997) New radiopaque polyHEMA-based hydrogel particles. *Journal of Biomedical Materials Research*, **34**, 183–8.
18 Benzina, A., Kruft, M.A.B., van der Veen, F.H., Bar, F.H., Blezer, R., Lindhout, T. and Koole, L.H. (1996) A versatile three-iodine molecular building block leading to new radiopaque polymeric biomaterials. *Journal of Biomedical Materials Research*, **32**, 459–66.
19 Jayakrishnan, A. and Chithambara, T.B. (1992) Synthesis and polymerization of some iodine-containing monomers for biomedical application. *Journal of Applied Polymer Science*, **44**, 743–8.
20 Kruft, M.A.B., Benzina, A., Bar, F.H., van der Veen, F.H., Bastiaansen, C.W.M. and Blezer, R. (1994) Studies of two new radiopaque polymeric biomaterials. *Journal of Biomedical Materials Research*, **28**, 1259–66.
21 Nirmala, R.J., Juby, P. and Jayakrishnan, A. (2006) Polyurethanes with radiopaque properties. *Biomaterials*, **27**, 160–6.
22 Moszner, N., Saiz, U., Klester, A.M. and Rheinberger, V. (1995) Synthesis and polymerization of hydrophobic iodine-containing methacrylates. *Die Angewandte Makromolekulare Chemie*, **224**, 115–23.
23 Davy, K.W.M. and Anseau, M.R. (1997) X-ray opaque methacrylate polymers for biomedical applications. *Polymer International*, **43**, 143–54.
24 Davy, K.W.M., Anseau, M.R. and Berry, C. (1997) Iodinated methacrylate copolymers as x-ray opaque denture base acrylics. *Journal of Dentistry*, **25**, 499–505.
25 Saralidze, K.S., Aldenhoff, B.J., Knetsch, L.W. and Koole, L.H. (2003) Injectable polymeric microspheres with x-ray visibility. Preparation, properties, and potential utility as new traceable bulking agents. *Biomacromolecules*, **4**, 793–8.
26 Vasquez, B., Ginebra, M.P., Gil, F.G., Planell, J.A., Bravo, A.L. and Roman, J.S. (1999) Radiopaque acrylic cements prepared with a new acrylic derivative of iodo-quinoline. *Biomaterials*, **20**, 2047–53.
27 Emans, P.J., Saralidze, K., Knetsch, L.W., Gijbels, J.J., Kuijer, R. and Koole, L.H. (2005) Development of new injectable bulking agents: biocompatibility of radiopaque polymeric microspheres studies in a mouse model. *Journal of Biomedical Materials Research*, **73A**, 430–6.
28 Jay, M. and Ryo, U.Y. (1991) Biodegradable, low biological toxicity radiographic contrast medium and method of X-ray imaging. U.S. Patent 5, 019, 370.
29 Idee, J.M., Nachman, I., Port, M., Petta, M., Lem, G.L., Greneur, S.L., Dencausse, A., Meyer, D. and Corot, C. (2003) Iodinated contrast media: from non-specific to blood-pool agents. *Topics in Current Chemistry*, **222**, 151–71.
30 Kreuter, J. (2001) Nanoparticulate systems for brain delivery of drugs. *Advanced Drug Delivery Reviews*, **47**, 65–81.
31 Popovtzer, R., Agrawal, A., Kotov, N.A., Popovtzer, A., Balter, J., Carey, E.T. and Kopelman, R. (2008) Targeted gold nanoparticles enable molecular CT imaging of cancer. *Nano Letters*, **8** (12), 4593–6.
32 Galperin, A., Margel, D., Baniel, J., Dank, G., Biton, H. and Margel, S. (2007) Novel radiopaque iodinated polymeric nanoparticles for x-ray imaging uses. *Biomaterials*, **28**, 4461–6.
33 Aviv, H., Bartling, S., Kiessling, F. and Margel, S. (2009) Radiopaque iodinated copolymeric nanoparticles for x-ray imaging applications. *Biomaterials*, **30**, 5610–16.
34 Galperin, A. and Margel, S. (2007) Synthesis and characterization of radiopaque magnetic core-shell nanoparticles for X-ray imaging applications. *Journal of Biomedical Materials Research*, **83B**, 490–8.
35 Galperin, A. and Margel, S. (2006) Synthesis and characterization of radiopaque polystyrene microspheres by dispersion polymerization for x-ray imaging applications. *Journal of Polymer Science*, **44**, 3859–68.
36 Galperin, A. and Margel, S. (2006) Synthesis and characterization of new micrometer-sized polymeric composite particles of narrow size distribution by a single-step swelling of uniform polystyrene template microspheres for X-ray imaging applications. *Biomacromolecules*, **7**, 2650–60.

37 Aviv, H., Bartling, S., Budjan, J. and Margel, S. (2010) Synthesis of novel dual modality (CT/MRI) core- shell macroparticles for embolization purposes. *Biomacromolecules*, **11** (6), 1600–7.

38 Margel, S. and Galperin, A. (2009) New core and core-shell nanoparticles containing iodine for x-ray imaging applications. US pending Patent Application No: 11/887,762. Publication No. US2009/0311192 A1.

39 Boguslavsky, L., Baruch, S. and Margel, S. (2005) Synthesis and characterization of polyacrylonitrile nanoparticles by dispersion/emulsion polymerization process. *Journal of Colloid and Interface Science*, **289**, 71–85.

40 Martinez, A., Gonzalez, C. and Porras, M. (2005) Nano-sized latex particles obtained by emulsion polymerization using an amphiphilic block copolymer as surfactant. *Colloids and Surfaces A: Physicochemical and Engineering Aspects*, **270**, 67–71.

41 Harisinghani, M.G., Barentsz, J., Hahn, P.F., Deserno, W.M., Tabatabaei, S., Hulsbergen van de Kaa, C., De la Rosette, J. and Weissleder, R. (2003) Noninvasive detection of clinically occult lymph-node metastases in prostate cancer. *New England Journal of Medicine*, **348**, 2491–9.

42 Belova, V.V., Voshkin, A.A., Kholkin, A.I. and Payrtman, A.K. (2009) Solvent extraction of some lanthanides from chloride and nitrate solutions by binary extractants. *Hydrometallurgy*, **97**, 198–203.

43 Simon, D.W. (2009) Perioperative fluid and electrolyte balance in children. *Anaesthesia and Intensive Care Medicine*, **10**, 93–7.

44 Gupta, R., Grasruck, M., Suess, C., Bartling, S.H., Schmidt, B. and Stierstorfer, K. (2006) Ultra-high resolution flat-panel volume CT: fundamental principles, design architecture, and system characterization. *European Radiology*, **16**, 1191–205.

45 Bartling, S.H., Dinkel, J., Stiller, W., Grasruck, M., Madisch, I. and Kauczor, H.U. (2008) Intrinsic respiratory gating in small-animal CT. *European Radiology*, **18**, 1375–84.

13
Solid Lipid Nanoparticles to Improve Brain Drug Delivery

Paolo Blasi, Aurélie Schoubben, Stefano Giovagnoli, Carlo Rossi and Maurizio Ricci

13.1
Introduction

In recent years, the problem of drug delivery to the brain has been increasingly investigated due to the spreading of neurodegenerative diseases [1–5]. In fact, population aging, human immunodeficiency virus (HIV) infections, and increased risk factors have dramatically promoted the incidence of pathologies such as Parkinson's (PD) and Alzheimer's disease (AD) [6–9]. To date, a number of bottlenecks have limited drug-development programs [5, 10, 11]. Although the brain structures have been investigated, their complex homeostasis remains largely unknown, and this has limited the understanding of the physiopathology and etiology of central nervous system (CNS) disorders.

One basic issue in brain drug development is that most molecules are unable to cross the blood–brain barrier (BBB), or that they do not reach the brain in therapeutic concentrations [12, 13]. There is, consequently, a strong need for alternative approaches to overcome the BBB issue and, among the different strategies proposed over the years, nanotechnologies appear to show much promise.

In this chapter the most interesting data relating to nanoparticles produced from lipid materials (e.g., triglyceride and wax) with a melting temperature higher than 37 °C will be reported. These particles, which are referred to as solid lipid nanoparticles (SLNs), are intended to provide drug delivery to the brain. Following a brief introduction on the peculiar biological characteristics of the brain and its interfaces with the body, a summary of the different strategies proposed to deliver therapeutics through the BBB will be provided. Following a general overview of the physico-chemical aspects and preparation methods of SLNs, the most relevant data evidencing brain drug accumulation will be reported, and discussed with respect to their mechanism(s) of action. The potential value of SLNs as contrast agents for brain imaging will also be treated. Finally, the toxicity of SLNs will be examined with respect to that of polymeric nanoparticles (NPs).

Nanomaterials for the Life Sciences Vol.10: Polymeric Nanomaterials. Edited by Challa S. S. R. Kumar
Copyright © 2011 WILEY-VCH Verlag GmbH & Co. KGaA, Weinheim
ISBN: 978-3-527-32170-4

13.2
The General Problem of Brain Drug Delivery

13.2.1
Basic Brain Physiology

The CNS consists of the *encephalon* or brain, and the *medulla spinalis* or spinal cord (Figure 13.1). Both regions are well protected against trauma, the brain by the skull bones and the spinal cord by the vertebra, while three membranes – the *meninges* – are positioned in between the brain and the skull bones (Figure 13.1). The brain and spinal cord are continuous with one another at the level of the upper border of the atlas vertebra.

The *dura mater* lies directly under the skull, the *pia mater* lies over the brain, while the *arachnoid* resides in between the former two. The brain is surrounded by the *subarachnoid space*, positioned between the pia mater and arachnoid, in which the *cerebrospinal fluid* (CSF) runs. The CSF, which is produced mainly by the *choroid plexus* (CP), fills the ventricles and the subarachnoid space, and is in direct relationship with the brain interstitial fluid [14].

The brain is one of the most – if not *the* most – well-protected organs in the human body, with both anatomic and physiological barriers providing protection against the external environment. Indeed, specific interfaces tightly regulate the exchanges between the peripheral circulation and the CSF circulatory system. These interfaces include: (i) the CP epithelium (blood/ventricular CSF); (ii) the arachnoid epithelium (blood/subarachnoid CSF); and (iii) the BBB formed by the cerebrovascular endothelium (blood/interstitial fluid of the brain). By tightly regulating both the entrance and clearance of endogenous and exogenous molecules, these barriers provide an excellent homeostasis of the CNS.

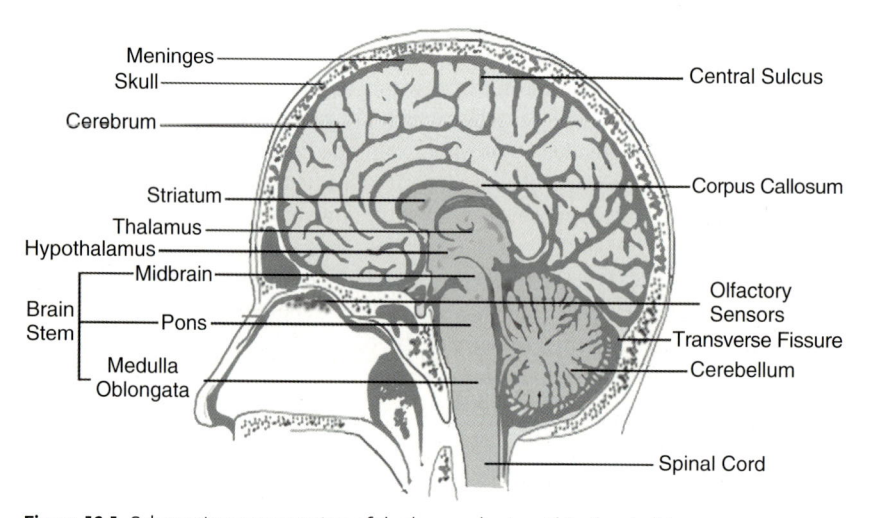

Figure 13.1 Schematic representation of the human brain within the skull bones.

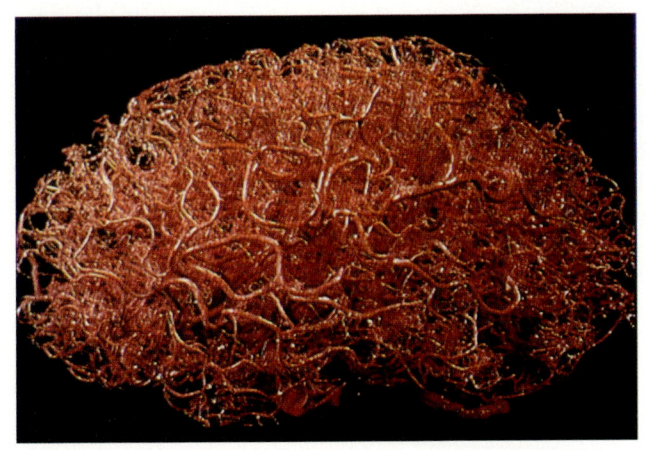

Figure 13.2 Latex calc of the complete brain vasculature, the blood–brain barrier. Zlokovic, B.V. and Apuzzo M.L.J. (1998) Strategies to circumvent vascular barriers of the central nervous system. *Neurosurgery*, **43**, 877–8. Reproduced with permission © Wolters Kluwer Health.

The BBB, which represents the main obstacle against achieving effective CNS pharmacological treatments [4, 12, 13], is constituted by the brain microvasculature that, due to its anatomic and physiologic features, greatly restricts the number of drugs that can enter the brain after systemic administration. The volume occupied by the capillary and endothelial cells respectively is about 1% and 0.1% of the brain volume; however, the brain microvasculature develops a total surface area of ~20 m^2 and a total length of ~600 km (Figure 13.2) [13].

The brain capillaries present no *fenestrae*, a small amount of pinocytosis vesicles and particular tight junctions (TJs) that are also known as the *zonula occludens* [15, 16]. The TJs are structures that form a narrow and continuous seal surrounding each endothelial and epithelial cell at the apical border, the aim being to strictly regulate the movement of molecules through the paracellular pathway. The brain microvasculature TJs show some differences in morphology, composition, and complexity compared to those of both the *epithelia* and peripheral *endothelia*. BBB feature development and maintenance are possible because of the interactions with neurons, astrocytes [17], pericytes [18] and microglia [19], which together form the so-called neurovascular unit [20].[1]

The *astrocytes* have a multitude of functions that are important to brain homeostasis, such as K$^+$ level maintenance, neurotransmitter inactivation, and growth factor and cytokine production and regulation [17]. The *pericytes* seem to be involved in virtually all of the processes of BBB formation, differentiation, and integrity [18].

The presence of transport systems (which generally are divided into three classes of carrier-mediated transport, active efflux transport, and receptor-mediated transport) in the different regions of the BBB further complicates the scenario, by conferring to the barrier a dynamic permeability [13]. Indeed, although these

---

1) The "neurovascular unit" represents a new and more advanced concept of the BBB.

transporters may be used for drug-targeting purposes, they generally hinder drug delivery to the brain. In this regard, P-glycoproteins (P-gps) considerably complicate brain drug targeting [21]. The P-gps, which are ATP-dependent transport proteins common to many tissues, are present at the apical membranes of different types of cell, including brain microvascular endothelial cells. These proteins are aimed at the elimination (by an efflux transport process) of potentially harmful foreign substances that are able to cross the BBB. It has been demonstrated, both *in vitro* and *in vivo*, that the BBB P-gps can prevent the accumulation within the brain of many drugs, such as vincristine, anthracyclines, taxanes, cyclosporine A, digoxin and dexamethasone [21].

13.2.2
Brain Drug-Delivery Strategies

In general, the major parameters that affect drug entry into the brain from the systemic circulation are lipid solubility, molecular mass, and charge. Typically, neutral highly lipophilic molecules with a molecular mass below ~500 Da might freely cross the BBB but, unfortunately, 95–98% of the potential CNS drugs do not possess these physico-chemical characteristics [13]. Consequently, several invasive and noninvasive approaches have been developed to overcome the BBB.

Direct brain drug administration can be achieved by either intracerebral or intraventricular infusion [2, 22], although such procedures are mainly used to treat severe pathologic conditions, such as brain tumors [23]. Unfortunately, however, this route of administration is often poorly effective due to the low brain tissue diffusivity, the high CSF turnover rate, and high risk of infection [13, 24]. The use of implanted drug-delivery systems, biodegradable polymeric microparticles, or stem cells have been proposed to improve the compliance and effectiveness of these treatments [25, 26].

BBB reversible disruption, as pioneered by Professor Edward Neuwelt [27], was one of the earliest techniques to be employed, and involves the injection of a hypertonic solution (e.g., mannitol) into the carotid artery. The resultant osmotic shock causes endothelial cell shrinking such that the BBB remains open for about 30 min, the timeframe used to administer the drug solution [28, 29]. BBB modulation can also be achieved by using inflammation mediators (e.g., leukotrienes and histamine) and vasoactive peptides (e.g., bradykinin and its analog Cereport® (RMP-7)) [26]. BBB disruption can also be provoked by ultrasound and electromagnetic radiation, although contrasting results have been reported for this approach [2].

The chemical modification and development of suitably targeted drug-delivery systems represent alternative approaches [4]. Chemical modification may involve the conjugation of water-soluble drugs to lipophilic molecules, such as fatty acids, acetyl or methyl groups, so as to enhance lipid-mediated transport [26]. In this regard, prodrugs such as levodopa [25], codeine, and heroin [2] may cross the BBB and become locked-in by enzymatic degradation, which in turn produces more-hydrophilic analogues. A similar approach implies the transformation of a drug into a nutrient-like analog capable of exploiting specific transporters [30].

Other drugs, such as doxorubicin and nerve growth factor can also reach the brain when conjugated to proteins, peptides (e.g., insulin), or monoclonal antibodies (e.g., the peptidomimetic antibody OX26) [2, 22, 25].

P-gp inhibition has been proposed as a further strategy to enhance brain drug accumulation [31]. However, as P-gp inhibition also hinders the elimination of toxic compounds, the application of this strategy should be critically addressed.

Finally, colloidal carriers such as polymeric micelles, polymeric NPs, SLNs, liposomes, and nanogels have been used to enhance BBB penetration [32–36]. Among the aforementioned colloids, the SLNs seem very promising, and will be discussed in greater detail later in the following subsections.

13.3
Solid Lipid Nanoparticles for Brain Drug Delivery

13.3.1
General Information

SLNs refer to colloids produced by lipids, namely fatty acids, mono-, di-, and triglycerides, waxes and nonsaponificable lipids, that are solids at body temperature. The SLN story began in 1991 when, independently, different research groups (notably, led by Dr Maria Rosa Gasco at the University of Turin, Italy, and Dr Rainer Helmut Müller at the Free University of Berlin, Germany), reported the production of this novel and alternative drug carrier system [37, 38].

SLNs combine the advantages of liposomes, with respect to the low material toxicity, with that of polymeric NPs, in terms of controlled release [39]. Lipids and waxes, due to their low toxicity and high biocompatibility, present potential advantages over synthetic polymers for the preparation of polymeric NPs. In fact, SLNs have been virtually proposed for all administration routes, namely parenteral [39], oral [40], dermal [41, 42], and ocular [43]. In addition, these carriers may be produced by homogenizing a melted lipid in a hot surfactant solution, using well-known and easily scalable methods such as microfluidization (also referred to as high-pressure homogenization).

In fact, high-pressure homogenization and microemulsion represent the two original methods employed for SLN production [44–46]. Besides these methodologies (which will be described in detail below), other techniques include solvent emulsification–evaporation or –diffusion, water/oil/water (w/o/w) double emulsion, and high-speed stirring and/or sonication [45, 46]. More recently, membrane contactor [47], electrospray [48], and coacervation [49] techniques have also been developed.

Hot high-pressure homogenization involves the dispersal or solubilization of a drug in the melted lipid, and successive transfer into a surfactant aqueous solution at the same temperature in order to form an oil-in-water pre-emulsion, using high-shear mixing. The pre-emulsion is then homogenized at the same temperature, using a high-pressure homogenizer.

In the case of *cold high-pressure homogenization*, a drug dispersion or solution in the melted lipid is cooled down and micronized to obtain a powder that is then dispersed, using high-shear mixing, in a cold aqueous solution containing a stabilizer. This suspension is then processed using high-pressure homogenization at a low temperature. This technique is mostly used with highly temperature-sensitive or water-soluble compounds, as it limits the degradation and increases the practical loading of the drug, respectively [44–46].

The *microemulsion* method implies the preparation of a warm microemulsion with 10% lipid, 15% surfactant, and up to 10% cosurfactant. The emulsion is then poured into an excess of cold water to form the SLN, that is then separated by either ultrafiltration or lyophilization [44–46]. These methods possess several advantages over other techniques, such as the absence of an organic solvent and ultrasonication that may cause contamination of the final product. Among the most recently developed techniques, both *coacervation* [49] and the use of a *membrane contactor* [47] avoid the use of organic solvent and ultrasounds, but their scalability is uncertain. Typically, up to 150 kg of SLN suspension per hour, and a batch size of up to 1.1 liters, can be processed when using high-pressure homogenization and microemulsion, respectively [44, 50]. When compared to SLNs, the polymeric NPs large batch preparation is more complicated [51], although when preparing poly(alkylcyanoacrylate) (PACA) NPs either monomers or preformed polymers can be used [52]. When using monomers, the polymerization must be carried out either in an emulsion or at the interface, leading to the production of nanospheres and nanocapsules, respectively. In both cases, the polymerization is initiated by the hydroxyl ions of water and proceeds by anionic polymerization. With interfacial polymerization, water- or oil-loaded nanocapsules can be produced according to the type of emulsion used (water-in-oil or oil-in-water emulsion systems). When preformed polymers are employed, either nanoprecipitation provoked by the addition of a nonsolvent to the polymeric solution, or conventional emulsion–solvent evaporation techniques are used [52]. This brief description of the preparation of PACA NPs provides evidence for the complexity of their production on an industrial scale. With few exceptions, SLNs have been used to target the brain parenchyma from the systemic circulation, in which case sterilization of the administered material is paramount (issues related to this process will be addressed later). The sterilization of SLNs may be carried out (at least, in theory) by using filtration, γ-irradiation, and vapor sterilization [45]. When the particle sizes are below 0.2 μm, sterile filtration can be considered the method of choice; if this is carried out on the emulsion, then filter occlusion may be avoided or reduced, making the process highly compliant [53]. If the correct filter features are chosen, the methodology will have no effect on the particles' characteristics, such as mean diameter, polydispersity, zeta potential, and physical stability [53].

With regards to *γ-irradiation*, free radicals are often produced that may lead to chemical alterations of the lipid matrix. Physical changes, such as an increased particle size or aggregation, may also occur when heat sterilization is applied. For instance, the autoclaving of SLNs stabilized with Poloxamer 188 led to a particle dimension increase [54]. Although attention must be paid to thermosensitive actives, this technique has been shown compatible with all preparations that are adequately stabilized, or in which the lipid concentration is sufficiently low (2–10%) [55–58]. Finally, when *microwaves* were used successfully to sterilize para-acyl-calix-arene-based SLNs, they caused no physical alterations [59].

In comparison with SLNs, very few reports have been made concerning polymeric NP sterilization. The use of aseptic conditions has also been suggested for PACA NP preparation, with one report proposing the use of γ-irradiation rather than autoclaving or formaldehyde treatment that would lead to particle aggregation [60, 61]. In light of these considerations, for both particulate systems, a preparation conducted under sterile conditions remains the best choice overall of the sterilization methods reported above.

13.3.2
Physico-Chemical Aspects of Lipid Packing and SLN Structure

Lipid amphiphiles tend to self-assemble by entropy-driven hydrophobic interactions. This hydrophobic effect is correlated to a decrease in chemical potential that, for an alkane of n carbon units is [62]:

$$\Delta\mu = 10.2 + 3.7n \text{ kJ mol}^{-1} \tag{13.1}$$

In this way, the lipid monomers minimize the area of aqueous contact, but the entropy of mixing will compete with monomer aggregation. Therefore, the actual formation of aggregates occurs when a certain number of dispersed monomers is reached (for a surfactant, this is termed the critical micellar concentration, CMC) [63]. In this regard, the chemical potential of a N-molecule assembly is given by [62–67]:

$$\mu_N^0 = 2\gamma a_0 + \frac{\gamma}{a}(a-a_0)^2 \tag{13.2}$$

where γ is a positive interfacial energy per unit area, a is the actual area per molecule, and a_0 is an "optimum" area per molecule [63]. This results from the competition between the interfacial energy and electrostatic, steric, van der Waals, and hydration energies that define a particular shape.

A relatively simple approach to aggregate shape assessment is provided by the Israelachvili's packing parameter (P) [65, 68] (Equation 13.3):

$$P = \frac{V}{AL} \tag{13.3}$$

where V is the molar volume of the hydrophobic moiety, A the cross-sectional area of the hydrophilic portion, and L the critical length of the hydrocarbon chain.

According to the P-value and the molecular geometry, conical or truncated conical, cylinder, and reverse conical geometries give rise to reverse micelles, bilayers and direct micelles, respectively [68, 69].

By obeying the same rules, solid lipid aggregates can be interpreted as the result of hydrocarbon chains packing into well-defined structures. The aggregate geometry depends on the hydrocarbon chain properties, such as the number of carbon atoms, the degree of saturation, and the substituents. Generally, hydrocarbon chains have been observed to pack according to two separate subcell classes. The first class is characterized by a tight assembly with a specific chain–chain interaction, whereas in the second class the specific interactions are lost and the chains pack more loosely as a result of partial rotation along the chain axis [70, 71]. In the simplest case of single-chain lipids, specific chain interaction leads to two-dimensional (2-D) rectangular and oblique lattices that can be sorted out into parallel and perpendicular orthorhombic, and parallel triclinic and monoclinic subcells, respectively. Hybrid lattices can occur for complex lipids, such as acylglycerols, in which the interaction of hydrocarbon chains is constrained by the link to a common moiety.

On the other hand, nonspecific chain interactions result in a greater distance between the chains, which allows for the rotation of carbon atoms (Figure 13.3). This so-called "rotator phase" is characteristic of phospholipids [70], and consists of hexagonal and pseudo-hexagonal arrangements with gauche conformers located principally to the distal end of the chains.

Such different subcell arrangements determine lipid polymorphism. In fact, hexagonal, orthorhombic perpendicular and triclinic parallel subcells have been associated with α, β' and β forms, respectively. Although, in the past, lipid polymorph classification has suffered from a rather confusing nomenclature [71, 72], these three polymorphs are recognized as being the three major solid-state organi-

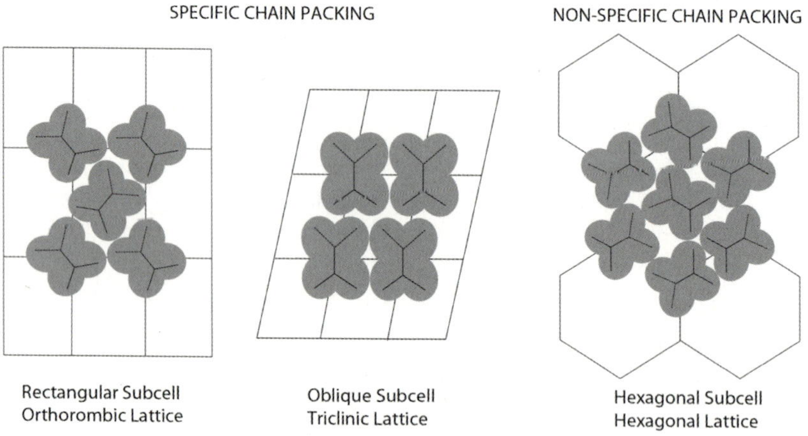

Figure 13.3 Hydrocarbon chain packing in the solid state. Specific chain packing includes orthorhombic and triclinic lattices corresponding to two-dimensional rectangular and oblique subcells. Nonspecific chain packing leads to hexagonal lattice and subcell.

zations of simple and complex lipids, in particular di- and triglycerides [71, 73–76]. Naturally, the α form is the thermodynamically least stable, while β' is metastable and, under certain conditions, is easily converted to β.

A number of sub-forms, which have been classified according to X-ray diffraction patterns and melting temperatures, include β'1, β'2 (also referred to as γ or sub-α), β'3, and so forth, which possess the same chain packing of the β' form but have decreasing melting points. These forms have been recorded for complex *cis*-unsaturated lipids, such as 1-stearoyl-2-oleoyl-*sn*-glycerol [77], 1,3-stearoyl-2-dioleoyl-*sn*-glycerol [78], *N*-, *O*-diacylethanolamines [79], and natural fats such as cocoa butter [80].

Transitions among the different forms can occur by changing the environmental conditions. Moreover, factors such as unsaturation and differences in chain length may also contribute to the melting path. In fact, shorter odd-numbered chains seem to prefer β' orthorhombic crystals, which transit through a pseudo-hexagonal phase (pseudo-α) before melting [70]. On the other hand, longer odd-numbered chains, which possess a higher melting temperature, show an additional true hexagonal packing prior to melt. In turn, even-numbered chains of more than 20 carbons prefer β triclinic lattices that convert directly into hexagonal assemblies (α form) before melting [70, 81, 82].

On the other side, the presence of double bonds prevents mixing with saturated chains and promotes phase segregation, even if the two chains are part of the same molecule [70, 83]. Immiscibility occurs even when two chains differ in length of more than four atoms [70].

In the case of di- and triglycerides, phase segregation occurs with the chains stretching towards opposite directions, thus packing into different layers hinged around the same head group [70, 80, 84]. In triglycerides, segregated trilayers have been recorded [70, 85], the arrangement of which depends on the chain difference and on the binding position on the glycerol group. That is, different chains in position 1 or 2 segregate the two equal chains into a bilayer, and the other chains into an interdigitated monolayer [70]. When all three chains are different, however, the structure assignment becomes troublesome, although crystallization does occur with evidence of specific chain–chain interaction. On the other hand, segregation is not possible in water, as the chains are forced to pack into one phase due to the orientation of the polar group towards the water phase. In this case, chain tilting produces preferentially hexagonal phases, as observed for phospholipids [86].

Lipid packing determines the inner structure of lipid colloidal particles, such as SLNs [44, 68, 69, 87, 88]. In fact, even when shaped into SLN, lipids retain their polymorphism.

The particle size in the colloidal domain can, however, dramatically alter the properties of solid aggregates [69, 89]. In fact, as the size decreases, the surface area and the curvature radius and are seen to increase, together with the surface energy, owing to the presence of a larger fraction of molecules located at or near the particle surface [89].

In this regard, if F_V is the volume fraction of randomly distributed spherical particles of radius r, then the number of molecules per unit volume N_V (the number density) is given by Equation 13.4:

$$N_V = \frac{3F_V}{4\pi r^3} \tag{13.4}$$

Here, N_V can be assumed to be the number of lipid molecules on the SLN surface, and is inversely correlated to the SLN size as it decreases very rapidly as the particle size increases.

As a consequence, the features of SLNs differ from those of bulk lipids [90, 91]. In fact, according to lipid composition and particle size, SLNs can show a rheological behavior that can be predicted by the Krieger's equation (Equation 13.5) [92]:

$$\eta = \eta_0 \left(1 - \frac{\varphi}{p}\right)^{-[\eta]p} \tag{13.5}$$

where η is the viscosity of the dispersion, η_0 is the viscosity of the medium, $[\eta]$ is the intrinsic viscosity, φ is the volume fraction of the dispersed phase, and p is the volume fraction of the dispersed phase at most dense packing. As an example, cetylpalmitate, stearic alcohol and Softisan 138 SLN show about a tenfold increased storage and loss modulus compared to the respective microparticles [91].

In contrast to polymeric NPs, which commonly show a spherical shape, pure triglyceride SLNs from a spherical shape tend to transform into platelet-like structures [93], where the platelets are formed by two or three lipid layers each about 10–18 nm thick (Figure 13.4).

This step-like shape offers a large surface for particle–particle interaction, thus fostering the formation of semisolid network-like structures. Based on this assumption, the SLN intrinsic viscosity $[\eta]$ can be associated to platelet-like rotational ellipsoids under weak shear conditions by the relation of Hinc and Leal (Equation 13.6) [94]:

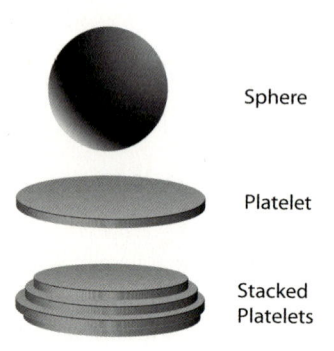

Figure 13.4 Classical spherical shape of nanoparticles compared to characteristic SLN platelet and stacked-platelet structures.

$$[\eta] = \frac{32}{15\pi} \frac{D}{h} \tag{13.6}$$

where h is the platelet thickness, and D the platelet diameter.

Since SLNs are usually obtained by crystallization upon the rapid cooling of hot lipid melt emulsions, lipid polymorphs will influence the SLNs' final shape, properties, and stability.

In fact, bulk triglycerides crystallize upon rapid cooling into the unstable α-form, which can convert to β' and then β forms upon heating and/or storage. Moreover, polymorph conversion is much faster in SLNs compared to bulk materials, and most likely also follows a different path [73–75, 87, 90, 91].

When platelet SLNs are associated only with β polymorphs, the particle surface area will increase as α converts to β, with a greater risk of clumping [74]. In addition, it seems that the lower the lipid melting point, the better the chance of an α to β transformation during storage [73].

The melting point of triglyceride SLNs depletes with particle size following the Gibbs–Thomson equation, either for spherical (curved) or plane surface particles [95, 96] (Equation 13.7):

$$-\frac{T_0 - T}{T_0} \approx \ln \frac{T}{T_0} = -\frac{2\gamma_{sl} V_s}{r \Delta H_{fus}} \tag{13.7}$$

where T is the melting temperature of a particle with radius r, T_0 is the melting temperature of the bulk material at the same external pressure, γ_{sl} is the interfacial tension at the solid–liquid interface, V_s is the specific volume of the solid, and ΔH_{fus} is the specific heat of fusion.

This behavior can impair the stability of SLN formulations by producing supercooled melts [87].

In this regard, drug loading may also affect SLN polymorphism as a result of an intercalation of the drug between the lipid layers [87]. Freshly prepared blank Compritol® 888 SLN showed small amounts of the α-form, which disappeared upon drug entrapment [97]. This increased entropy favored a further melting point depletion, with the possible formation of supercooled melts rather than solid structures [73, 87]. The risk is greater for short-chain lipids, such as trimyristin and trilaurin, which form liquid dispersions [73]. However, a tendency towards supercooling can be recorded even with triglyceride SLNs having chain lengths of up to 18 carbons [87].

Drug loadings from 1% up to 50% (w/w) can cause drastic changes in the lipid structure, and drug expulsion may occur upon storage. The rate of drug expulsion is correlated to the rate of α to β polymorphic transformation.

13.3.3
Surfactants and SLNs for Brain Targeting

The protocols employed for the preparation of SLNs imply the use of a proper surfactant to stabilize the solid dispersion that is formed. The most common stabilizers used for SLN preparation include ionic and nonionic surface active agents, such as phospholipids, Polysorbates (Tweens®), Poloxamers (Pluronics®), derivatized fatty acids, and their combinations [75, 98, 99].

Along with lipid and particle characteristics [100–102], such molecules affect the SLN inner structure and, subsequently, the loading capacity [103, 104]. In particular, tripalmitin SLNs show the stable β-form when nonionic surfactants are employed, whereas ionic surfactants cause the retention of a greater amount of the α-form [100]. As a consequence, the particle shape and morphology can also be modified since, as previously mentioned, a higher β-form retention produces platelet-like particles, whereas a higher amount of the α-form gives rise to particles of different shapes, as a function of size. Large particles (>200 nm) are mostly spherical, while small particles (<100 nm) have a blocky isometric layered shape [100]. The effect of surfactants on the SLN polymorphic behavior suggests that the transition begins mainly from the surface of the particle. Surfactants, such as Poloxamers 407 and 188 or Poloxamine 908, can be used to modify even the SLN degradation rate [105, 106]. The aforementioned effect can be ascribed to difficulties in anchoring the enzyme onto the particle surface, due to surfactant steric hindrance. This provides the possibility of modulating, *in vivo*, the SLN half-life by using different surfactants.

Most of the scalable preparation methods are based on high-pressure homogenization techniques where, in general, a higher homogenization pressure, a greater number of homogenization cycles and a higher process temperature will lead to a reduction in SLN size [102, 107, 108]. Yet, the coupled effect of cavitation forces and high shear, downsizing melt droplets, might favor aggregation. This effect has been observed for Compritol® 888 and Softisan® 142 SLNs and is, of course, susceptible to both lipid and surfactant characteristics [102, 108]. Consequently, an effective surfactant, beyond that needed to stabilize the lipid dispersion, will also functions as a surface modifier [109–111].

Polysorbates – in particular polysorbate 80 (P80) – have shown some predominance over other surfactants for delivering NPs to the brain (see below) [32]. Hypothetically, P80 exploits the monooleic chain to anchor onto the SLN surface, and this may have a major impact on the lipid's physical state. However, P80-coated long-chain triglyceride SLNs have been shown to retain the solid lipid matrix in spite of a slight melting depletion. Nevertheless, certain issues may arise for low-melting lipids, as Softisan® 142 SLN, which may become liquid at 37 °C [53].

13.3.4
Evidence of Brain Drug Accumulation

The first evidence of lipid NP translocation across the BBB dates back 15 years [112], with subsequent reports of different detailed pharmacokinetics studies confirming the great potential of SLNs as carriers for brain drug delivery. In fact, both camptothecin- and doxorubicin-loaded SLNs showed a significant brain drug accumulation after oral and intravenous administration, respectively [113, 114]. A subsequent series of *in vivo* experiments corroborated this first evidence, and demonstrated that a significant brain targeting could indeed be obtained when the SLNs were stabilized with poly(ethylene glycol) (PEG)-containing surfactants [113–116]. In this case, intact stealth[2] and conventional SLNs were visible in rat CSF at 20 min after administration [115].

In one of these investigations, conventional treatment (e.g., doxorubicin solution) provided higher drug concentrations in all tissues but the brain parenchyma, at 30 min after administration. The SLNs, when stabilized by PEG-containing surfactants, were shown to accumulate about fivefold more drug in the brain than did conventional NPs (~10 and ~2 μg g^{-1} tissue, respectively), whereas among animals treated with doxorubicin solution the drug was essentially undetected in the brain parenchyma [116].

Camptothecin was more efficiently delivered to the brain when loaded into SLNs stabilized with poloxamer 188 (P188), than when administered as a solution. Similar results were obtained after both intravenous and oral administration [113, 117]. In particular, after systemic administration the brain maximum concentration (C_{max}) was augmented by 180% [117] while, in the former case, the mean residence time (MRT) and the area under the curve (AUC)/dose ratio were almost four- and tenfold increased with respect to the drug solution [117].

Analogous results were obtained with tobramycin and idarubicin [40, 118]. In these cases, the SLNs were able to enhance tobramycin brain concentrations after either parenteral or duodenal administration. However, a significant accumulation, with respect to tobramycin solution, was achieved only by parenteral administration. A different behavior was seen for idarubicin-loaded SLNs, which furnished significant brain targeting in rats via both routes of administration [40, 118].

In situ brain perfusion was also used to demonstrate SLN uptake over a short time frame [119]. Although P80 demonstrated a better performance in targeting NPs to the brain, Brij® 78-coated SLNs delivered a greater amount of paclitaxel compared to Taxol® (paclitaxel solution in the surfactant Cremophor® ELR : ethanol, 50 : 50, v/v), which is the commercial formulation [120].

In one *in vivo* study performed by the present authors in collaboration with Dr. G. Traina and Prof. M. Brunelli (University of Pisa), 280 nm Compritol® SLNs stabilized by P80 were shown capable of delivering the fluorescent dye Nile Red to different brain structures. In particular, at 24 h after parenteral administration, fluorescence was concentrated to the *corpus callosus* (Figure 13.5) (G. Traina *et al.*, unpublished data). Although the crossing of intact SLNs was not apparent, nor could be confirmed by the sole presence of fluorescent dye in the parenchyma, these results were of great interest, and further investigations are currently under way.

Because of the low SLN encapsulation efficiency generally obtained with hydrophilic compounds, the pro-drug or lipid–drug conjugate (LDC) approach has been proposed [121, 122]. This method has many advantages, including the ability

---

2) "Stealth" refers to a nanoparticle being able to avoid the mononuclear phagocytic system. This is generally achieved by particle surface modification by the adsorption of PEG-containing surfactant or the covalent linkage of PEG chains.

Figure 13.5 Fluorescence microscopy image of a slice (depth 200 μm) of rat brain at 24 h after treatment with fluorescent P80-coated Compritol® SLNs (280 nm). Traina, G., Blasi, P., Schoubben, A., Giovagnoli, S., Rossi, C., Brunelli, M., Ricci, M., unpublished data.

to improve encapsulation efficiency, the increase in biological barrier permeability, and the possibility to shape the LDC as nanoparticulate matter [123]. The latter effect is of particular interest when considering the limited particle uptake generally observed at the BBB surface and the possibility to reduce excipient accumulation and toxicity [32].

Colloidal LDC demonstrated, *in vivo*, the capacity to accumulate on the BBB surface, an effect which may prove to be useful for creating a high drug gradient between the brain capillaries and the parenchyma with the achievement of a higher brain drug concentrations [121]. Analogously to the LDC approach, when drug NPs or nanocrystals are coated with adequate surfactants they may be successfully employed as a brain-targeting strategy [39].

All such evidence of brain targeting by SLNs coated with PEG-containing surfactants may be corroborated by a large amount of data obtained with polymeric NPs, using the same strategy.

Colloidal particles, following intravenous administration, are efficiently opsonized and then cleared from the bloodstream by the phagocytic cells within minutes, mainly distributing into the liver (60–90% of the injected dose), spleen (2–10%), lungs (3–20% and more), and bone marrow (>1%) [124]. One of the first issues to be solved to increase the usefulness of NPs is to avoid opsonization in order to augment the drug's half-life. Thus, the correct modification of the NPs' surface characteristics and size represents the most common strategy to alter the distribution of NPs in the body, and to enhance not only the blood circulation time of the drug but also its deposition in organs other than those of the reticuloendothelial system (RES) [125, 126]. In this respect, different approaches have been proposed to modify the hydrophobic particle surface, such as the physical adsorption of PEG-containing surfactants or surface covalent modifications with PEG derivatives [127–131]. Professor Jörg Kreuter and colleagues, while investigating the effect of different surfactants on NP body distribution, noted that some (especially P80) were able to target NPs to the brain, in significant amounts [132–134]. Indeed, P80 was found to be the most effective surfactant for improving the *in vivo* brain drug uptake of different compounds that normally are unable to penetrate the BBB, such as dalargine [132–134]. The brain-targeting ability of the P80-coated NPs has been demonstrated *in vivo* with a large variety of active compounds, including tubocurarine [135], loperamide [136], doxorubicin [137], nerve growth factor [138], and 5-chloro-7-iodo-8-hydroxyquinoline [139]. Most recently, polysorbate 188 was also seen to be effective for directing NPs to the brain parenchyma [140].

Although evidence that P80-coated NPs can target therapeutic molecules to the brain has been clearly established, the mechanism by which such drug transport across the BBB is achieved has not yet been fully explained, and has led to considerable debate. Subsequently, several different potential mechanisms have been proposed [33], among which the most plausible are as follows [32, 33]:

- The increased NP retention in the brain blood capillaries, and/or their adsorption to the capillary walls, can create a higher concentration gradient, thus enhancing transport across the BBB and leading to brain drug accumulation.
- The endocytosis of NPs by the endothelial cells can permit drug release within these cells, with subsequent drug diffusion into the brain parenchyma.
- The transcytosis of NPs with the bound drug can cause the drug to be released directly into the brain parenchyma.

P80-coated NP endocytosis is believed to be mediated by protein adsorption on the surface, followed by subsequent interaction (triggering the endocytosis) with specific BBB luminal receptors. Specifically, the hypothesis postulates that the adsorption of apolipoprotein E (ApoE) or B onto P80-coated NPs could be responsible for the interaction with the BBB, and the resultant endocytosis [121, 141, 142]. These findings have been supported by recent *in vivo* studies indicating that, following intravenous administration, human serum albumin NPs with covalently bound ApoE are taken up into the cerebral endothelium by an endocytic mechanism, and this is followed by transcytosis into brain parenchyma [141].

As noted above, the surface characteristics of NPs are largely responsible for their fate in the body, although, besides surface hydrophobicity, charge has also a role. In particular, NPs with a slightly positive charge were found to accumulate higher etoposide concentrations (10- and 14-fold) in the brain at 1 and 4 h after administration, when compared to negatively charged particles or the drug solution itself [143]. In another study, cationic SLNs were used to deliver higher clozapine concentrations to the brain with respect to neutral SLNs and drug suspensions [144]. Unfortunately, large amounts of both negatively charged and positively charged NPs caused a clear increase in the cortical cerebrovascular volume, which is known to represent a sign of BBB disruption, with evident limitation of this strategy [145].

13.3.5
SLN Potentiality in Brain Imaging

During recent years, nanotechnology has expanded beyond the borders of the brain targeting field towards the new horizon of brain imaging which, nonetheless, is based on the same principles already applied for drug delivery [146]. In fact, an enhanced NP extravasation into the brain at a tumor site, due to the altered vascular architecture of the tissue, would result in a higher concentration of a local contrast agent, thus aiding the detection process [147]. In addition, the possibility of developing multifunctional nanocarriers, containing both ferromagnetic agents and surface-targeting moieties, would lead to contrast agent-loaded NPs becoming a more selective and potent diagnostic tool [146, 148].

Superparamagnetic iron oxide (SPIO) particles, which are characterized by a large magnetic moment in the presence of a static external magnetic field, have been proposed for applications in magnetic resonance imaging (MRI) since the late 1980s and, indeed, the number of commercially available products continues to expand [149].

From a practical viewpoint, SPIO particles consist of a core of either magnetite (F_3O_4; this is mostly utilized for biomedical applications) or maghemite (γF_2O_3). The core is covered by a hydrophilic coating that prevents agglomeration of the colloidal suspension, and causes the particles to be easily dispersed in biological fluids.

Many coating materials have been proposed for this role, including dextran, carboxydextran, carboxymethylated dextran, starch, PEGylated starch, PEG, arabinogalactan, glycosaminoglycan, organic siloxane, sulfonated styrene-divinylbenzene, amino acids, α-hydroxyacids (e.g., citric, tartaric, gluconic), hydroxamate (arginine hydroxamate), or dimercaptosuccinic acid (DMSA) [150–153].

The nature of the surface coating as well as the geometric arrangement of the coating on the iron oxide surface, determines the final size[3] of the colloidal particles and also play a significant role in the particles' pharmacokinetics and biodistribution.

Multiple components determine the efficacy and affect the application of these agents, including the size of the iron oxide crystals, the charge, the nature of the coating materials, and the hydrodynamic size of the coated particle. In fact, these physico-chemical characteristics affect particle stability, biodistribution, opsonization and metabolism, as well as clearance from the vascular system.

The blood half-lives of the various iron oxide NPs administered to patients can range from 1h to 24–36h [154]. A long blood half-life represents an important feature that renders such systems suitable as complementary or alternative MRI contrast agents when compared to gadolinium-based contrast agents (GBCAs). In

---

[3] According to their hydrodynamic diameter, these nanoparticles can be classified in three groups, namely: standard SPIOs (50–180nm coating included), ultrasmall superparamagnetic iron oxide particles (USPIOs) (10–50nm), and very small superparamagnetic iron oxide particles (VSPIOs) (<10nm).

fact, the enhancing signal image intensity evaluation of intracranial tumors following ferumoxide, ferumoxtran and GBCA infusion, showed a more effective performance of ferumoxtran with respect to ferumoxides.

Another advantage of NPs is their capacity for highly selective molecular tumor targeting.

It is also possible that NPs could be engineered to image certain subpopulations of cells, such as stem cells or endothelial cells. Currently, a variety of iron oxide NPs are under development, and have shown promise as tumor-specific contrast agents [155]. Recently, active targeting to the CNS following a NP surface modification (e.g., the addition of monoclonal antibodies or peptides) has been produced for glioma imaging [156].

On the basis of these findings, NP-based materials may be considered ideal devices to deliver both contrast agents and drugs to brain tumors. Unlike other NPs, the SLNs have not been investigated closely as a tool for MRI and, despite their potential in terms of scale-up, loading, versatile structural properties, easy preparation and biocompatibility, very few reports exist describing contrast agent-loaded SLNs [157, 158].

In this respect, SLNs loaded with colloidal iron oxides (SLN-FeA and SLN-FeB) were compared *in vitro* and *in vivo* to the commercial MRI contrast agent Endorem®. Following parenteral administration to rats, both types of SLN-Fe showed a slower blood clearance and a more prolonged CNS retention time (up to 135 min) compared to Endorem®. The results also showed that, following its inclusion into SLNs, Endorem® would become a new type of contrast agent, that would be taken up by the liver but not cross the BBB. This contrasted with Endorem®-containing SLNs, which showed a clear uptake into the CNS. It would appear, therefore, that the kinetics of SLN-Fe is related to the SLNs rather than to their iron oxide content. These findings have placed SLNs among a list of novel promising MRI contrast agent delivery systems for the CNS.

As a half-way point between imaging and therapy, future SLN applications may well become focused on photodynamic therapy (PDT) [148]. In fact, multifunctional SLNs loaded with a photosensitizer molecule and iron oxide may be used successfully for imaging and PDT, if they were to be adequately coated with a glioma tumor-targeting moiety, and following a better intracranial localization of the tumor.

Finally, the SLNs represent a promising tool in thermal ablation, the application of which, when combined with chemotherapy, may be potentially effective in the treatment of malignant gliomas, but would be strongly limited by its nonfocalized field of action. Recent *in vivo* studies, conducted both in animal glioma models and in selected patients, showed a prolongation of survival and a regression of tumor growth after the injection of magnetic NPs into the tumors, followed by exposure to an alternating magnetic field [159]. It has also been recently shown in animal models that the concomitant application of a brain-targeting magnetic field during the intravenous administration of iron oxide NPs, induced a fivefold increase in tumor exposure to the NPs and a 3.6-fold rise in selective NP accumulation in tumors, compared to the normal brain tissue [160].

All of these findings suggest the potential use of SLNs as neuroimaging and concomitantly therapeutic carriers, thus clearing the way to a selective, localized, and concomitant or subsequent thermotherapy and chemotherapy. Moreover, a prolonged retention time would allow long-lasting observations to be made of tumor behavior, and prove valuable in the noninvasive *in vivo* MRI monitoring of the therapeutic effects produced by SLN-delivered drugs.

13.3.6
Toxicity Issues

As the main aim of these nanocarriers is their administration to humans, they must be biocompatible and biodegradable, and/or at least excreted via renal filtration, while their biodegradation products must be nontoxic. Whilst the initial studies on NP toxicity were carried out on polymeric NPs, and have provided a wealth of data (see below), studies of the toxicity of SLNs or lipid materials have been conducted only recently.

In this section, attention will be focused on the toxicity of PACA NPs, the aim being to provide a comparison with the data available for SLNs. PACA NPs can be degraded via three different pathways, the main process involves esterases that lead to the formation of alkyl alcohol and poly(cyanoacrylic) acid [52]. Yet, all three degradation mechanisms give rise to potentially toxic byproducts, and it is therefore fundamentally important to evaluate their effects on the organism [52]. In the past, PACA NPs have been assayed on several cell lines, including macrophages [161, 162], L929 fibroblasts [163, 164], hepatocytes [165, 166], cancer cells DC3F [167], mesenchymal cells [168], Swiss 3T3 cells [169, 170], and murine cerebral microvessels [171]. Based on the results of these studies, it appeared that PACA NPs were able to activate macrophages (the "respiratory burst") [162] and to cause membrane perforation [161]. Moreover, the NP-based cytotoxicity was correlated to alkyl chain length and particle dimensions, both of which parameters influenced the degradation rate [164]. In fact, NPs obtained from long side-chain polymers showed a slow degradation rate that permitted particle adhesion to the cells and a resultant high concentration of byproducts close to the membrane. In contrast, short side-chain and small particle dimensions led to a rapid degradation and a burst release of degradation products [164]. Hence, whilst both long and short side-chain polymers were capable of causing toxicity, the short-chain polymers proved to be generally more toxic. Alternatively, the repetitive administration of long-chain polymer NPs may lead to "waste disposal" in the brain endothelial cells and parenchyma. From a general standpoint, SLNs should be characterized by a higher acceptability with respect to polymeric NPs. In fact, their components (e.g., mono-, di-, triglycerides, fatty acids or waxes) are degraded and release endogenous byproducts. Despite there being no SLN formulation available currently for endovenous administration, parenteral nutrition, containing fatty acids, has been used clinically since the early 1960s [44]. It is also important to underline that, besides the tolerability of solid lipid components, generally recognized as safe (GRAS),

13.3 Solid Lipid Nanoparticles for Brain Drug Delivery

surfactants used to guarantee SLN stability must be acceptable as well [44]. Whilst SLN toxicity must be investigated in order to understand the potential of this carrier for brain drug delivery, there is unfortunately a lack of studies that have directly compared the toxicity of PACA and lipid NPs. In addition, *in vitro* toxicity studies involving one or other carrier are difficult to compare, since in virtually all cases the cell lines have been different. However, an interesting comparison showed that Compritol® and cetylpalmitate SLN were 20- and 10-fold less toxic (in terms of cell viability), respectively than polyester polymers (approved by the FDA for parenteral use) [172, 173].

The toxicity of SLNs is strongly related to their lipid nature, and to the surfactant used to stabilize the NPs. For example, both Dynasan® 114 and Compritol® SLNs had no influence on HL 60 and murine peritoneal macrophage viability in the range 0.015 to 1.5% [111, 174], whereas Dynasan® 118 and Imwitor® (glycerol monostearate) SLNs showed macrophage viability reductions at 0.1% [174]. This cytotoxicity seemed to be related to the higher amount of stearic acid released upon SLN degradation over time. However, Imwitor® SLN led to a high viability of MCF-7 breast cells when incubated at 0.05% [175]. Dynasan® 114 SLN (up to 0.1%) was also very well tolerated by RAW 264.7 macrophages, with the exception of those stabilized with a cationic surfactant, namely cetylpyridinium chloride, which proved to be toxic in dose-dependent manner [176]. In a recent report which evaluated SLN toxicity on A549 cancer cell lines [177], all of the lipids investigated (i.e., monostearin, stearic acid, glycerol tristearate, Compritol® ATO 888) led to a 50% cell growth inhibition in the range of 0.03 to 0.05%. The toxicity of cetylpalmitate and Compritol® SLNs was also studied *in vivo* after intravenous injection in mice [178]. In this case, no histological abnormalities or acute inflammation were detected with cetylpalmitate SLNs, whereas with Compritol® pathological signs were apparent. In particular, the liver was characterized by cell necrosis, mononuclear cell infiltration and Kupffer cell hyperplasia, and the spleen by lymph follicle destruction and a loss of normal architecture. However, at six weeks after the last treatment the situation was restored in both the liver and spleen, where only lipid droplets could be observed [178]. It should be stressed that the Compritol® dose administered to mice was equivalent to six boluses of 100 g of lipids in a 75 kg human; clearly, for human therapy the administered dose would be much lower, with a potential reduction of organ histological anomalies [178]. When P80-coated cetylpalmitate SLN toxicity was investigated *in vivo* by means of the chorioallantoic membrane assay, neither the SLNs, P80 stabilizing solution nor bulk cetylpalmitate had elicited any inflammation or neoangiogenesis at 24 and 48 h after application to seven-day-old chicken embryos [53, 179].

When considering the above-reported data, SLN appear as excellent candidates for brain delivery, especially as the cytotoxicity of SLN is much lower than that of the PACA NPs. Moreover, the fact that the SLN drug loading is generally higher than that of the PACA NPs becomes an essential point when considering that only 1–2% of the administered dose will reach the brain.

13.4
Concluding Remarks

As anticipated by Richard Feinman in his famous lecture "There's plenty of room at the bottom" [180], nanotechnologies are destined to have a strong impact in present-day life. Clearly, both pharmaceutical and medical areas will benefit from such new technologies, with NPs in their different forms (e.g., dendrimers, liposomes, polymeric NPs, SLNs) being peculiarly involved in drug delivery and targeting and, therefore, having a major role in brain drug delivery. It is becoming increasingly apparent that SLNs will deeply impact on progress in drug delivery and, in particular, of brain drug targeting. Although the SLNs represent quite recently developed drug-delivery systems [181], they have reached the cosmetic product market within 14 years of their conception [182], and surely will soon be approved for topical pharmaceutical applications. Such a rapid breakthrough of the SLN formulations is directly related to the GRAS state of the materials employed, many of which have already been approved for cosmetic and pharmaceutical dermal uses.

Nevertheless, the parenteral use of these colloidal carriers—especially via the intravenous route—will demand stricter rules to match the authorities' requirements. Consequently, despite lipids having been used since the 1960s in parenteral nutrition products [39], the approval of intravenous SLN formulations still seems far away. In fact, as foreseen by one of the "fathers" of SLN, it is likely that a period of more than ten years will be required before approval is gained for the first SLN parenteral formulation (R.H. Müller, personal communication).

As noted in this chapter, whilst the major potential of SLNs for brain drug delivery should speed the process from the bench to the bedside, such a goal can of course only be achieved by establishing the safety and efficacy of SLN-based formulations for brain drug delivery.

References

1 Pardridge, W.M. (2002) Why is the global CNS pharmaceutical market so under-penetrated? *Drug Discovery Today*, 7, 5.

2 Begley, D.J. (2004) Delivery of therapeutic agents to the central nervous system: the problems and the possibilities. *Pharmacology and Therapeutics*, 104, 29–45.

3 Neuwelt, E.A., Abbott, N.J., Abrey, L., Banks, W.A., Blakley, B., Davis, T., Engelhardt, B., Grammas, P., Nedergaard, M., Nutt, J., Pardridge, W., Rosenberg, G.A., Smith, Q. and Drewes, L.R. (2008) Strategies to advance translational research into brain barriers. *Lancet Neurology*, 7, 84–96.

4 Ricci, M., Blasi, P., Giovagnoli, S. and Rossi, C. (2006) Delivering drugs to the central nervous system: a medicinal chemistry or a pharmaceutical technology issue? *Current Medicinal Chemistry*, 13, 1757–75.

5 Su, Y. and Sinko, P.J. (2006) Drug delivery across the blood-brain barrier: why is it difficult? How to measure and improve it? *Expert Opinion on Drug Delivery*, 3, 419–35.

6 World Health Organization (2002) The global burden of disease project.

http://www.who.int (accessed 30 July 2010)

7 Croquelois, A., Assal, G., Annoni, J.M., Staub, F., Gronchi, A., Bruggimann, L., Dieguez, S. and Bogousslavsky, J. (2005) Diseases of the nervous system: patients' aetiological beliefs. *Journal of Neurology, Neurosurgery, and Psychiatry*, **76**, 582–4.

8 Meyerhoff, J.D. (2001) Effects of alcohol and HIV infection on the central nervous system. *Alcohol Research and Health*, **25**, 288–98.

9 Levy, R.M. and Bredesen, D.E. (1988) Central nervous system dysfunction in acquired immunodeficiency syndrome. *Journal of Acquired Immune Deficiency Syndromes*, **1**, 41–64.

10 Alavijeh, M.S., Chishty, M., Qaiser, M.Z. and Palmer, A.M. (2005) Drug metabolism and pharmacokinetics, the blood-brain barrier, and central nervous system drug discovery. *NeuroRx*, **2**, 554–71.

11 Lawrence, R.N. and Richard, S. (2002) Sykes contemplates the future of the Pharma industry. *Drug Discovery Today*, **7**, 645–8.

12 Neuwelt, E. (1989) *Implication of the Blood–Brain Barrier and Its Manipulation*, vol. 1, Basic Science Aspects, Plenum Publishing Corporation, New York, USA.

13 Pardridge, W.M. (2001) *Brain Drug Targeting: the Future of Brain Drug Development*, Cambridge University Press, Cambridge, UK.

14 Noback, C.R., Strominger, N.L. and Demarest, R.J. (1991) *The Human Nervous System: Introduction and Review*, 4th edn, Williams & Wilkins, New York, USA.

15 Ballabh, P., Braun, A. and Nedergaard, M. (2004) The blood-brain barrier: an overview: structure, regulation, and clinical implications. *Neurobiology of Disease*, **16**, 1–13.

16 Lapierre, L.A. (2000) The molecular structure of the tight junction. *Advanced Drug Delivery Reviews*, **41**, 255–64.

17 Abbott, N.J. (2002) Astrocyte-endothelial interactions and blood-brain barrier permeability. *Journal of Anatomy*, **200**, 629–38.

18 Balabanov, R. and Dore-Duffy, P. (1998) Role of the CNS microvascular pericyte in the blood-brain barrier. *Journal of Neuroscience Research*, **53**, 637–44.

19 Hanisch, U.-K. and Kettenmann, H. (2007) Microglia: active sensor and versatile effector cells in the normal and pathologic brain. *Nature Neuroscience*, **10**, 1387–94.

20 Neuwelt, E.A. (2004) Mechanisms of disease: the blood-brain barrier. *Neurosurgery*, **54**, 131–40.

21 Stouch, T.R. and Gudmundsson, O. (2002) Progress in understanding the structure–activity relationships of P-glycoprotein. *Advanced Drug Delivery Reviews*, **54**, 315–28.

22 Groothuis, D.R. (2000) The blood-brain and blood-tumor barriers: a review of strategies for increasing drug delivery. *Neuro-Oncology*, **2**, 45–59.

23 Wang, P.P., Frazier, J. and Brem, H. (2002) Local drug delivery to the brain. *Advanced Drug Delivery Reviews*, **54**, 987–1013.

24 Krewson, C.E., Klarman, M.L. and Saltzman, W.M. (1995) Distribution of nerve growth factor following direct delivery to brain interstitium. *Brain Research*, **680**, 196–206.

25 Garcia-Garcia, E., Andrieux, K., Gil, S. and Couvreur, P. (2005) Colloidal carriers and blood-brain barrier (BBB) translocation: a way to deliver drugs to the brain? *International Journal of Pharmaceutics*, **298**, 274–92.

26 Temsamani, J., Scherrmann, J.M., Rees, A.R. and Kaczorek, M. (2000) Brain drug delivery technologies: novel approaches for transporting therapeutics. *Pharmaceutical Science and Technology Today*, **3**, 155.

27 Neuwelt, E.A., Frenkel, E.P., Diehl, J., Vu, L.H., Rapoport, S. and Hill, S. (1980) Reversible osmotic blood-brain barrier disruption in humans: implications for the chemotherapy of malignant brain tumors. *Neurosurgery*, **7**, 44–52.

28 Neuwelt, E.A., Hill, S.A. and Frenkel, E.P. (1984) Osmotic blood-brain barrier modification and combination chemotherapy: concurrent tumor regression in areas of barrier opening

and progression in brain regions distant to barrier opening. *Neurosurgery*, **15**, 362–6.

29 Neuwelt, E. (1989) *Implication of the Blood–Brain Barrier and Its Manipulation*, vol. 2, Clinical Aspects, Plenum Publishing Corporation, New York, USA.

30 Battaglia, G., La Russa, M., Bruno, V., Arenare, L., Ippolito, R., Copani, A., Bonina, F. and Nicoletti, F. (2000) Systemically administered D-glucose conjugates of 7-chlorokynurenic acid are centrally available and exert anticonvulsant activity in rodents. *Brain Research*, **860**, 149–56.

31 Kabanov, A.V., Batrakova, E.V. and Miller, D.W. (2003) Pluronic® block copolymers as modulators of drug efflux transporter activity in the blood-brain barrier. *Advanced Drug Delivery Reviews*, **55**, 151–64.

32 Blasi, P., Giovagnoli, S., Schoubben, A., Ricci, M. and Rossi, C. (2007) Solid lipid nanoparticles for targeted brain drug delivery. *Advanced Drug Delivery Reviews*, **59**, 454–77.

33 Kreuter, J. (2005) Application of nanoparticles for the delivery of drugs to the brain. *International Congress Series*, **1277**, 85–94.

34 Huwyler, J., Wu, D. and Pardridge, W.M. (1996) Brain drug delivery of small molecules using immunoliposomes. *Proceedings of the National Academy of Sciences of the United States of America*, **93**, 14164–9.

35 Costantino, L., Gandolfi, F., Tosi, G., Rivasi, F., Vandelli, M.A. and Forni, F. (2005) Peptide-derivatized biodegradable nanoparticles able to cross the blood brain barrier. *Journal of Controlled Release*, **108**, 84–96.

36 Aktas, Y., Yemisci, M., Andrieux, K., Gursoy, R.N., Alonso, M.J., Fernandez-Megia, E., Novoa-Carballal, R., Quiñoá, E., Riguera, R., Sargon, M.F., Celik, H.H., Demir, A.S., Hincal, A.A., Dalkara, T., Capan, Y. and Couvreur, P. (2005) Development and brain delivery of chitosan-PEG nanoparticles functionalized with the monoclonal antibody OX26. *Bioconjugate Chemistry*, **16**, 1503–11.

37 Lucks, J.S., Müller, R.H. and Köning, B. (1992) Solid lipid nanoparticles (SLN) – an alternative drug carrier system. *European Journal of Pharmaceutics and Biopharmaceutics*, **38**, 33s.

38 Gasco, M.R. (1993) Method for producing solid lipid microspheres having a narrow size distribution. U.S. Patent 5, 250, 236.

39 Müller, R.H., Radtke, M. and Wissing, S.A. (2004) Solid lipid nanoparticles and nanostructured lipid carriers, in *Encyclopedia of Nanoscience and Nanotechnology* (ed. H.S. Nalwa), American Scientific Publishers, Los Angeles, pp. 43–56.

40 Zara, G.P., Bargoni, A., Cavalli, R., Fundarò, A., Vighetto, D. and Gasco, M.R. (2002) Pharmacokinetics and tissue distribution of idarubicin-loaded solid lipid nanoparticles after duodenal administration to rats. *Journal of Pharmaceutical Sciences*, **91**, 1324–33.

41 Puglia, C., Blasi, P., Rizza, L., Schoubben, A., Bonina, F., Rossi, C. and Ricci, M. (2008) Lipid nanoparticles for prolonged topical delivery: an in vitro and in vivo investigation. *International Journal of Pharmaceutics*, **357**, 295–304.

42 Puglia, C., Filosa, R., Peduto, A., de Caprariis, P., Rizza, L., Bonina, F. and Blasi, P. (2006) Evaluation of alternative strategies to optimize ketorolac transdermal delivery. *AAPS PharmSciTech*, **7**, Article 64.

43 Cavalli, R., Gasco, M.R., Chetoni, P., Burgalassi, S. and Saettone, M.F. (2002) Solid lipid nanoparticles (SLN) as ocular delivery system for tobramycin. *International Journal of Pharmaceutics*, **238**, 241–5.

44 Wissing, S.A., Kaiser, O. and Müller, R.H. (2004) Solid lipid nanoparticles for parenteral drug delivery. *Advanced Drug Delivery Reviews*, **56**, 1257–72.

45 Mehnert, W. and Mäder, K. (2001) Solid lipid nanoparticles – production, characterization and applications. *Advanced Drug Delivery Reviews*, **47**, 165–96.

46 de Mendoza, A.E.-H., Campanero, M.A., Mollinedo, F. and Blanco-Prieto, M.J. (2009) Lipid nanomedicines for

anticancer drug therapy. *Journal of Biomedical Nanotechnology*, **5**, 323–43.

47 El-Harati, A.A., Charcosset, C. and Fessi, H. (2006) Influence of the formulation for solid lipid nanoparticles prepared with a membrane contactor. *Pharmaceutical Development and Technology*, **11**, 153–7.

48 Trotta, M., Cavalli, R., Trotta, C., Bussano, R. and Costa, L. (2009) Electrospray technique for solid lipid-based particle production. *Drug Development and Industrial Pharmacy*, **36** (4), 431–8.

49 Battaglia, L., Gallarate, M., Cavalli, R. and Trotta, M. (2010) Solid lipid nanoparticles produced through a coacervation method. *Journal of Microencapsulation*, **27**, 78–85.

50 Marengo, E., Cavalli, R., Caputo, O., Rodriguez, L. and Gasco, M.R. (2000) Scale-up of the preparation process of solid lipid nanospheres. Part I. *International Journal of Pharmaceutics*, **205**, 3–13.

51 Kaur, I.P., Bhandari, R., Bhandari, S. and Kakkar, V. (2008) Potential of solid lipid nanoparticles in brain targeting. *Journal of Controlled Release*, **127**, 97–109.

52 Vauthier, C., Dubernet, C., Fattal, E., Pinto-Alphandary, H. and Couvreur, P. (2003) Poly(alkylcyanoacrylates) as biodegradable materials for biomedical applications. *Advanced Drug Delivery Reviews*, **55**, 519–48.

53 Blasi, P., Schoubben, A., Giovagnoli, S., Puglia, C., Barberini, L., Cirotto, C., Rossi, C. and Ricci, M. (2009) Colloidal dispersions for targeted brain drug delivery. *European Journal of Pharmaceutical Sciences*, **38 S**, 25–6.

54 Schwarz, C., Mehnert, W., Lucks, J.S. and Müller, R.H. (1994) Solid lipid nanoparticles (SLN) for controlled drug delivery. I. Production, characterization and sterilization. *Journal of Controlled Release*, **30**, 83–96.

55 Müller, R.H., Freitas, C., zur Mühlen, A. and Mehnert, W. (1996) Solid lipid nanoparticles (SLN) for controlled drug delivery. *European Journal of Pharmaceutical Sciences*, **4**, S75–6.

56 Heiati, H., Tawashi, R. and Phillips, N.C. (1998) Drug retention and stability of solid lipid nanoparticles containing azidothymidine palmitate after autoclaving, storage and lyophilisation. *Journal of Microencapsulation*, **15**, 173–84.

57 Cavalli, R., Caputo, O. and Gasco, M.R. (2000) Preparation and characterization of solid lipid nanospheres containing paclitaxel. *European Journal of Pharmaceutical Sciences*, **10**, 305–9.

58 Cavalli, R., Caputo, O., Parlotti, M.E., Trotta, M., Scarnecchia, C. and Gasco, M.R. (1997) Sterilization and freeze-drying of drug-free and drug-loaded solid lipid nanoparticles. *International Journal of Pharmaceutics*, **148**, 47–54.

59 Shahgaldian, P., Da Silva, E., Colemana, A.W., Rather, B. and Zaworotko, M.J. (2003) Para-acyl-calix-arene based solid lipid nanoparticles (SLNs): a detailed study of preparation and stability parameters. *International Journal of Pharmaceutics*, **253**, 23–38.

60 Kreuter, J. (2006) Nanoparticles as drug delivery systems for the brain, in *Drugs and the Pharmaceutical Sciences* (ed. S. Benita), Harwood Academic Publishers, Amsterdam, pp. 689–706.

61 Sommerfeld, P., Schroeder, U. and Sabel, B.A. (1998) Sterilization of unloaded polybutylcyanoacrylate nanoparticles. *International Journal of Pharmaceutics*, **164**, 113–18.

62 Gruner, M., Cullis, P.R., Hope, M.J. and Tilcock, C.P.S. (1985) Lipid polymorphism: the molecular basis of nonbilayer phases. *Annual Review of Biophysics and Biophysical Chemistry*, **14**, 211–38.

63 Israelachvili, J.N., Marcelja, S. and Horn, R.G. (1980) Physical principles of membrane organization. *Quarterly Reviews of Biophysics*, **13**, 121–200.

64 Israelachvili, J.N., Mitchell, D.J. and Ninham, B.W. (1977) Theory of self-assembly of lipid bilayers and vesicles. *Biochimica et Biophysica Acta*, **470**, 185–201.

65 Israelachvili, J.N., Mitchell, D.J. and Ninham, B.W. (1976) Theory of self-assembly of hydrocarbon amphiphiles into micelles and bilayers. *Journal of the Chemical Society, Faraday Transactions*, **272**, 1525–68.

66 Mitchell, D.J. and Ninham, B.W. (1981) Micelles, vesicles and microemulsions. *Journal of the Chemical Society, Faraday Transactions,* **77**, 601–29.

67 Tanford, C. (1980) *The Hydrophobic Effect: Formation of Micelles and Biological Membranes,* 2nd edn, John Wiley & Sons, Inc., New York.

68 Israelachvili, J.N. (1994) The science and applications of emulsions – an overview. *Colloids and Surfaces A – Physicochemical and Engineering Aspects,* **91**, 1–8.

69 Bummer, P.M. (2004) Physical chemical considerations of lipid-based oral drug delivery-solid lipid nanoparticles. *Critical Reviews in Therapeutic Drug Carrier Systems,* **21**, 1–19.

70 Small, D.M. (1984) Lateral chain packing in lipids and membranes. *Journal of Lipid Research,* **25**, 1490–500.

71 Chapman, D. (1962) The polymorphism of glycerides. *Chemical Reviews,* **62**, 433–56.

72 Foubert, I., Dewettink, K., Van de Walle, D., Dijkstra, A.J. and Quinn, P.J. (2007) Physical properties: structural and physical characterisics, in *The Lipid Handbook with CD-ROM,* 3rd edn (eds D. Frank, J.L. Gunstone, A.J. Harwood and A.J. Dijkstra), CRC Press, Boca Raton, FL, pp. 471–509.

73 Bunjes, H., Westesen, K. and Koch, M.H.J. (1996) Crystallization tendency and polymorphic transitions in triglyceride nanoparticles. *International Journal of Pharmaceutics,* **129**, 159–73.

74 Westesen, K. and Siekmann, B. (1997) Investigation of the gel formation of phospholipid-stabilized solid lipid nanoparticles. *International Journal of Pharmaceutics,* **151**, 35–45.

75 Freitas, C. and Müller, R.H. (1999) Correlation between long-term stability of solid lipid nanoparticles (SLN™) and crystallinity of the lipid phase. *European Journal of Pharmaceutics and Biopharmaceutics,* **47**, 125–32.

76 Oh, J.H., Mc Curdy, A.R., Clark, S. and Swanson, B.G. (2002) Characterization and thermal stability of polymorphic forms of synthesized tristearin. *Journal of Food Science,* **67**, 2911–17.

77 Di, L. and Small, D.M. (1993) Physical behavior of the mixed chain diacylglycerol, 1-stearoyl-2-oleoyl-*sn*-glycerol: difficulties in chain packing produced marked polymorphism. *Journal of Lipid Research,* **34**, 1611–23.

78 Kaneko, F., Yano, J. and Sato, K. (1998) Diversity in the fatty-acid conformation and chain packing of cis-unsaturated lipids. *Current Opinion in Structural Biology,* **8**, 417–25.

79 Tarafdar, P.K. and Swamy, M.J. (2009) Polymorphism in "L" shaped lipids: structure of N-, O-diacylethanolamines with mixed acyl chains. *Chemistry and Physics of Lipids,* **162**, 25–33.

80 D'Souza, V., deMan, J.M. and deMan, L. (1990) Short spacings and polymorphic forms of natural and commercial solid fats: a preview. *Journal of the American Oil Chemists' Society,* **67**, 835–43.

81 Da Silva, E., Bresson, S. and Rousseau, D. (2009) Characterization of the three major polymorphic forms and liquid state of tristearin by Raman spectroscopy. *Chemistry and Physics of Lipids,* **157**, 113–19.

82 Sato, K. (2001) Crystallization behaviour of fats and lipids – a review. *Chemical Engineering Science,* **56**, 2255–65.

83 Schmidt, W.F., Mookherji, S. and Crawford, M.A. (2009) Unit cell volume and liquid-phase immiscibility in oleate-stearate lipid mixtures. *Chemistry and Physics of Lipids,* **158**, 10–15.

84 Abes, M. and Narine, S.S. (2007) Crystallization and phase behavior of fatty acid esters of 1,3-propanediol I: pure systems. *Chemistry and Physics of Lipids,* **149**, 14–27.

85 Goto, M., Kodali, D.R., Small, D.M., Honda, K., Kozawa, K. and Uchida, T. (1992) Single crystal structure of a mixed-chain triacylglycerol: 1,2-dipalmitoyl-3-acetyl-*sn*-glycerol. *Proceedings of the National Academy of Sciences of the United States of America,* **89**, 8083–6.

86 Fenske, D.B. and Cullis, P.R. (1995) Lipid polymorphism, in *The Encyclopedia of Nuclear Magnetic Resonance* (eds D.M. Grant and R.K. Harris) John Wiley & Sons, Inc., New York, pp. 2730–5.

87 Westesen, K., Bunjes, H. and Koch, M.H.J. (1997) Physicochemical characterization of lipid nanoparticles

and evaluation of their drug loading capacity and sustained release potential. *Journal of Controlled Release*, **48**, 223–36.

88 Kristl, J., Volk, B., Ahlin, P., Gomba, K. and Šentjur, M. (2003) Interactions of solid lipid nanoparticles with model membranes and leukocytes studied by EPR. *International Journal of Pharmaceutics*, **256**, 133–40.

89 Poole, C.P. and Owens, F.J. (2003) Introduction to physics of the solid state, in *Introduction to Nanotechnology* (eds C.P. Poole and F.J. Owens), John Wiley & Sons, Inc., Hoboken, New Jersey, pp. 8–34.

90 Bunjes, H., Koch, M.H.J. and Westesen, K. (2000) Effect of particle size on colloidal solid triglycerides. *Langmuir*, **16**, 5234–41.

91 Lippacher, A., Müller, R.H. and Mäder, K. (2002) Semisolid SLNTM® dispersions for topical application: influence of formulation and production parameters on viscoelastic properties. *European Journal of Pharmaceutics and Biopharmaceutics*, **53**, 155–60.

92 Krieger, I.M. (1972) Rheology of monodisperse lattices. *Advances in Colloid and Interface Science*, **3**, 111–36.

93 Jores, K., Mehnert, W., Drechsler, M., Bunjes, H., Johann, C. and Mäder, K. (2004) Investigations on the structure of solid lipid nanoparticles (SLN) and oil-loaded solid lipid nanoparticles by photon correlation spectroscopy, field-flow fractionation and transmission electron microscopy. *Journal of Controlled Release*, **95**, 217–27.

94 Hinch, E.J. and Leal, L.G. (1972) The effect of Brownian motion on the rheological properties of a suspension of non-spherical particles. *Journal of Fluid Mechanics*, **52**, 683–712.

95 Siekmann, B. and Westesen, K. (1994) Thermoanalysis of the recrystallization process of melt-homogenized glyceride nanoparticles. *Colloids and Surfaces B*, **3**, 159–75.

96 Jackson, K.L. and McKenna, G.B. (1990) The melting behavior of organic materials confined in porous solids. *Journal of Chemical Physics*, **93**, 9002–11.

97 Souto, E.B., Mehnert, W. and Müller, R.H. (2006) Polymorphic behaviour of Compritol 888 ATO as bulk lipid and as SLN and NLC. *Journal of Microencapsulation*, **23**, 417–33.

98 Schubert, M.A., Harms, M. and Müller-Goymann, C.C. (2006) Structural investigations on lipid nanoparticles containing high amounts of lecithin. *European Journal of Pharmaceutical Sciences*, **27**, 226–36.

99 Schöler, N., Olbrich, C., Tabatt, K., Müller, R.H., Hahn, H. and Liesenfeld, O. (2001) Surfactant, but not the size of solid lipid nanoparticles (SLN) influences viability and cytokine production of macrophages. *International Journal of Pharmaceutics*, **221**, 57–67.

100 Bunjes, H., Koch, M.H.J. and Westesen, K. (2003) Influence of emulsifiers on the crystallization of solid lipid nanoparticles. *Journal of Pharmaceutical Sciences*, **92**, 1509–20.

101 Üner, M., Wissing, S.A., Yener, G. and Müller, R.H. (2004) Influence of surfactants on the physical stability of solid lipid nanoparticle (SLN) formulations. *Pharmazie*, **59**, 331–2.

102 Reddy, L.H. and Murthy, R.S.R. (2005) Etoposide-loaded nanoparticles made from glyceride lipids: formulation, characterization, in vitro drug release, and stability evaluation. *AAPSPharmSciTech*, **6**, E 24.

103 Jenning, V. and Gohla, S. (2000) Comparison of wax and glyceride solid lipid nanoparticles (SLN®). *International Journal of Pharmaceutics*, **196**, 219–22.

104 Schubert, M.A. and Müller-Goymann, C.C. (2005) Characterisation of surface-modified solid lipid nanoparticles (SLN): influence of lecithin and nonionic emulsifier. *European Journal of Pharmaceutics and Biopharmaceutics*, **61**, 77–86.

105 Oldrich, C., Kayser, O. and Müller, R.H. (2002) Lipase degradation of Dynasan 114 and 116 solid lipid nanoparticles (SLN) – effect of surfactants, storage time and crystallinity. *International Journal of Pharmaceutics*, **237**, 119–28.

106 Oldrich, C., Kayser, O. and Müller, R.H. (2002) Enzymatic degradation of Dynasan 114 SLN-effect of surfactants and particle size. *Journal of Nanoparticle Research*, **4**, 121–9.

107 Liedtke, S., Wissing, S., Müller, R.H. and Mäder, K. (2000) Influence of high pressure homogenisation equipment on nanodispersions characteristics. *International Journal of Pharmaceutics*, **196**, 183–5.

108 Blasi, P., Giovagnoli, S., Schoubben, A., Ricci, M. and Rossi, C. (2006) Preparation and characterization of Tween® 80 coated solid lipid nanoparticles for brain drug targeting. Proceeding of the 13th International Pharmaceutical Technology Symposium, pp. 577–8.

109 Weyhers, H., Lück, M., Mehnert, W., Souto, E.B. and Müller, R.H. (2005) Surface modified solid lipid nanoparticles (SLN) analysis of plasma protein adsorption patterns by two-dimensional polyacrylamide gel electrophoresis (2-D PAGE). *International Journal of Molecular Medicine*, **1**, 196–201.

110 Göppert, T.M. and Müller, R.H. (2005) Adsorption kinetics of plasma proteins on solid lipid nanoparticles for drug targeting. *International Journal of Pharmaceutics*, **302**, 172–86.

111 Müller, R.H., Rühl, D., Runge, S., Schulze-Forster, K. and Mehnert, W. (1997) Cytotoxicity of solid lipid nanoparticles as a function of the lipid matrix and the surfactant. *Pharmaceutical Research*, **14**, 458–62.

112 Minagawa, T., Sakanaka, K., Inaba, S.I., Sai, Y., Tamai, I., Suwa, T. and Suji, A. (1996) Blood-brain-barrier transport of lipid microspheres containing linprost, a prostaglandin I2 analogue. *Journal of Pharmacy and Pharmacology*, **48**, 1016–22.

113 Yang, S.C., Zhu, J.B., Lu, Y., Liang, B.W. and Yang, C.Z. (1999) Body distribution of camptothecin solid lipid nanoparticles after oral administration. *Pharmaceutical Research*, **16**, 751–7.

114 Zara, G.P., Cavalli, R., Fundarò, A., Bargoni, A., Caputo, O. and Gasco, M.R. (1999) Pharmacokinetics of doxorubicin incorporated in solid lipid nanospheres (SLN). *Pharmacological Research*, **40**, 281–6.

115 Podio, V., Zara, G.P., Carazzone, M., Cavalli, R. and Gasco, M.R. (2000) Biodistribution of stealth and non-stealth solid lipid nanospheres after intravenous administration to rats. *Journal of Pharmacy and Pharmacology*, **52**, 1057–63.

116 Fundarò, A., Cavalli, R., Bargoni, A., Vighetto, D., Zara, G.P. and Gasco, M.R. (2000) Non-stealth and stealth solid lipid nanoparticles (SLN) carrying doxorubicin: pharmacokinetics and tissue distribution after i.v. administration to rats. *Pharmacological Research*, **42**, 337–43.

117 Yang, S.C., Lu, L.F., Cai, Y., Zhu, J.B., Liang, B.W. and Yang, C.Z. (1999) Body distribution in mice of intravenously injected camptothecin solid lipid nanoparticles and targeting effect on brain. *Journal of Controlled Release*, **59**, 299–307.

118 Bargoni, A., Cavalli, R., Zara, G.P., Fundarò, A., Caputo, O. and Gasco, M.R. (2001) Transmucosal transport of tobramycin incorporated in solid lipid nanoparticles (SLN) after duodenal administration to rats. Part II-tissue distribution. *Pharmacological Research*, **43**, 497–502.

119 Koziara, J.M., Lockman, P.R., Allen, D.D. and Mumper, R.J. (2003) In situ blood-brain barrier transport of nanoparticles. *Pharmaceutical Research*, **20**, 1772–8.

120 Koziara, J.M., Lockman, P.R., Allen, D.D. and Mumper, R.J. (2004) Paclitaxel nanoparticles for the potential treatment of brain tumors. *Journal of Controlled Release*, **99**, 259–69.

121 Gessner, A., Olbrich, C., Schröder, W., Kayser, O. and Müller, R.H. (2001) The role of plasma proteins in brain targeting: species dependent protein adsorption patterns on brain-specific lipid drug conjugate (LDC) nanoparticles. *International Journal of Pharmaceutics*, **214**, 87–91.

122 Olbrich, C., Gessner, A., Kayser, O. and Müller, R.H. (2002) Lipid-drug conjugate (LDC) nanoparticles as novel carrier system for the hydrophilic antitrypanosomal drug diminazenediaceturate. *Journal of Drug Targeting*, **10**, 387–96.

123 Müller, R.H. and Keck, C.M. (2004) Challenges and solutions for the delivery

124 of biotech drugs – a review of drug nanocrystal technology and lipid nanoparticles. *Journal of Biotechnology*, **113**, 151–70.

124 Kreuter, J. (1994) Nanoparticles, in *Colloidal Drug Delivery Systems* (ed. J. Kreuter), Marcel Dekker, New York, pp. 219–342.

125 Kreuter, J. (2001) Nanoparticulate systems for brain delivery of drugs. *Advanced Drug Delivery Reviews*, **47**, 65–81.

126 Leu, D., Manthey, B., Kreuter, J., Speiser, P. and DeLuca, P.P. (1984) Distribution and elimination of coated polymethyl [2–14C] methacrylate nanoparticles after intravenous injection in rats. *Journal of Pharmaceutical Sciences*, **73**, 1433–7.

127 Illum, L. and Davis, S.S. (1983) Effect of non-ionic surfactant poloxamer 338 on the fate and deposition of polystyrene microspheres following intravenous administration. *Journal of Pharmaceutical Sciences*, **72**, 1086–9.

128 Illum, L. and Davis, S.S. (1984) The organ uptake of intravenously administered colloidal particles can be altered using a non-ionic surfactant (poloxamer 338). *FEBS Letters*, **167**, 79–82.

129 Tröster, S.D., Wallis, K.H., Müller, R.H. and Kreuter, J. (1992) Correlation of the surface hydrophobicity of 14C poly(methyl methacrylate) nanoparticles to their body distribution. *Journal of Controlled Release*, **20**, 247–60.

130 Calvo, P., Gouritin, B., Chacun, H., Desmaële, D., D'Angelo, J., Noël, J.P., Georgin, D., Fattal, E., Andreux, J.P. and Couvreur, P. (2001) Long-circulating PEGylated polycyanoacrylate nanoparticles as new drug carrier for brain delivery. *Pharmaceutical Research*, **18**, 1157–66.

131 Gref, R., Minamitake, Y., Peracchia, M.T., Trubetskoy, V., Torchilin, V. and Langer, R. (1994) Biodegradable long-circulating polymeric nanospheres. *Science*, **263**, 1600–3.

132 Kreuter, J., Alyautdin, R.N., Kharkevich, D.A. and Ivanov, A.A. (1995) Passage of peptides through the blood-brain barrier with colloidal polymer particles (nanoparticles). *Brain Research*, **674**, 171–4.

133 Alyautdin, R.N., Gothier, D., Petrov, V., Kharkevich, D. and Kreuter, J. (1995) Analgesic activity of the hexapeptide dalargin adsorbed on the surface of polysorbate 80-coated poly(butylcyanoacrylate) nanoparticles. *European Journal of Pharmaceutics and Biopharmaceutics*, **41**, 44–8.

134 Schröder, U. and Sabel, B.A. (1996) Nanoparticles, a drug carrier system to pass the blood-brain barrier, permit central analgesic effects of intravenous dalargin injections. *Brain Research*, **710**, 121–4.

135 Alyautdin, R.N., Tezikov, E.B., Ramges, P., Kharkevich, D.A., Begley, D.J. and Kreuter, J. (1998) Significant entry of tubocurarine into the brain of rats by adsorption to polysorbate 80-coated polyisobutylcyanoacrylate nanoparticles: an in situ brain perfusion study. *Journal of Microencapsulation*, **15**, 67–74.

136 Alyautdin, R.N., Petrov, V.E., Langer, K., Berthold, A., Kharkevich, D.A. and Kreuter, J. (1997) Delivery of loperamide across the blood-brain barrier with polysorbate 80-coated polybutylcyanoacrylate nanoparticles. *Pharmaceutical Research*, **14**, 325–8.

137 Steiniger, S.C.J., Kreuter, J., Khalansky, A.S., Skidan, I.N., Bobruskin, A.I., Smirnova, Z.S., Severin, S.E., Uhl, R., Kock, M., Geiger, K.D. and Gelperina, S.E. (2004) Chemotherapy of glioblastoma in rats using doxorubicin loaded nanoparticles. *International Journal of Cancer*, **109**, 759–67.

138 Abdel Wahab, B.A., Evgenivetch, P.V. and Alyautdin, R.N. (2005) Brain targeting of nerve growth factor using poly(butylcyanoacrylate) nanoparticles. *The Internet Journal of Pharmacology*, **3** (2).

139 Roney, C., Kulkarni, P., Arora, V., Antich, P., Bonte, F., Wu, A., Mallikarjuana, N.N., Manohar, S., Liang, H.F., Kulkarni, A.R., Sung, H.W., Sairam, M. and Aminabhavi, T.M. (2005) Targeted nanoparticles for drug delivery through the blood-brain barrier for Alzheimer's disease. *Journal of Controlled Release*, **108**, 193–214.

140 Petri, B., Bootz, A., Khalansky, A., Hekmatara, T., Müller, R.H., Uhl, R., Kreuter, J. and Gelperina, S. (2007) Chemotherapy of brain tumour using doxorubicin bound to surfactant-coated poly(butyl cyanoacrylate) nanoparticles: revisiting the role of surfactants. *Journal of Controlled Release*, **117**, 51–8.

141 Zensi, A., Begley, D., Pontikis, C., Legros, C., Mihoreanu, L., Wagner, S., Büchel, C., von Briesen, H. and Kreuter, J. (2009) Albumin nanoparticles targeted with Apo E enter the CNS by transcytosis and are delivered to neurons. *Journal of Controlled Release*, **137**, 78–86.

142 Kreuter, J., Shamenkov, D., Petrov, V., Ramge, P., Cychutek, K., Koch-Brandt, C. and Alyautdin, R. (2002) Apolipoprotein-mediated transport of nanoparticle-bound drugs across the blood-brain barrier. *Journal of Drug Targeting*, **10**, 317–25.

143 Reddy, L.H., Sharma, R.K., Chuttani, K., Mishra, A.K. and Murthy, R.R. (2004) Etoposide-incorporated tripalmitin nanoparticles with different surface charge: formulation, characterization, radiolabeling, and biodistribution studies. *The AAPS Journal*, **6**, Article 23.

144 Manjunath, K. and Venkateswarlu, V. (2005) Pharmacokinetics, tissue distribution and bioavailability of clozapine solid lipid nanoparticles after intravenous and intraduodenal administration. *Journal of Controlled Release*, **107**, 215–28.

145 Reddy, L.H., Sharma, R.K., Chuttani, K., Mishra, A.K. and Murthy, R.S.R. (2005) Influence of administration route on tumor uptake and biodistribution of etoposide loaded solid lipid nanoparticles in Dalton's lymphoma tumor bearing mice. *Journal of Controlled Release*, **105**, 185–98.

146 Weinstein, J.S., Varallyay, C.G., Dosa, E., Gahramanov, S., Hamilton, B., Rooney, W.D., Muldoon, L.L. and Neuwelt, E.A. (2010) Superparamagnetic iron oxide nanoparticles: diagnostic magnetic resonance imaging and potential therapeutic applications in neurooncology and central nervous system inflammatory pathologies, a review. *Journal of Cerebral Blood Flow and Metabolism*, **30**, 15–35.

147 Orringer, D.A., Koo, Y.E., Chen, T., Kopelman, R., Sagher, O. and Philbert, M.A. (2009) Small solutions for big problems: the application of nanoparticles to brain tumor diagnosis and therapy. *Clinical Pharmacology and Therapeutics*, **85**, 531–4.

148 Reddy, G.R., Bhojani, M.S., McConville, P., Moody, J., Moffat, B.A., Hall, D.E., Kim, G., Koo, Y.L., Woolliscroft, M.J., Sugai, J.V., Johnson, T.D., Philbert, M.A., Kopelman, R., Rehemtulla, A. and Ross, B.D. (2006) Vascular targeted nanoparticles for imaging and treatment of brain tumors. *Clinical Cancer Research*, **12**, 6677–86.

149 Stark, D.D., Weissleder, R., Elizondo, G., Hahn, P.F., Saini, S., Todd, L.E., Wittenberg, J. and Ferrucci, J.T. (1988) Superparamagnetic iron oxide: clinical application as a contrast agent for MR imaging of the liver. *Radiology*, **168**, 297–301.

150 Bonnemain, B. (1998) Superparamagnetic agents in magnetic resonance imaging: physicochemical characteristics and clinical applications – a review. *Journal of Drug Targeting*, **6**, 167–74.

151 Neuwelt, E.A., Várallyay, C.G., Manninger, S., Solymosi, D., Haluska, M., Hunt, M.A., Nesbit, G., Stevens, A., Jerosch-Herold, M., Jacobs, P.M. and Hoffman, J.M. (2007) The potential of ferumoxytol nanoparticle magnetic resonance imaging, perfusion, and angiography in central nervous system malignancy: a pilot study. *Neurosurgery*, **60**, 601–12.

152 Laurent, S., Forge, D., Port, M., Roch, A., Robic, C., Vander Elst, L. and Muller, R.N. (2008) Magnetic iron oxide nanoparticles: synthesis, stabilization, vectorization, physicochemical characterizations, and biological applications. *Chemical Reviews*, **108**, 2064–110.

153 Mornet, S., Vasseur, S., Grasset, F. and Duguet, E. (2004) Magnetic nanoparticle design for medical diagnosis and therapy. *Journal of Materials Chemistry*, **14**, 2161–75.

154 Corot, C., Robert, P., Idée, J.M. and Port, M. (2006) Recent advances in iron oxide nanocrystal technology for medical imaging. *Advanced Drug Delivery Reviews*, **58**, 1471–504.

155 Veiseh, O., Sun, C., Gunn, J., Kohler, N., Gabikian, P., Lee, D., Bhattarai, N., Ellenbogen, R., Sze, R., Hallahan, A., Olson, J. and Zhang, M. (2005) Optical and MRI multifunctional nanoprobe for targeting gliomas. *Nano Letters*, **5**, 1003–8.

156 Chen, K., Xie, J., Xu, H., Behera, D., Michalski, M.H., Biswal, S., Wang, A. and Chen, X. (2009) Triblock copolymer coated iron oxide nanoparticle conjugate for tumor integrin targeting. *Biomaterials*, **30**, 6912–19.

157 Peira, E., Marzola, P., Podio, V., Aime, S., Sbarbati, A. and Gasco, M.R. (2003) In vitro and in vivo study of solid lipid nanoparticles loaded with superparamagnetic iron oxide. *Journal of Drug Targeting*, **11**, 19–24.

158 Brioschi, A.M., Calderoni, S., Zara, G.P., Priano, L., Gasco, M.R. and Mauro, A. (2009) Solid lipid nanoparticles for brain tumors therapy: state of the art and novel challenges. *Progress in Brain Research*, **180**, 193–223.

159 Maier-Hauff, K., Rothe, R., Scholz, R., Gneveckow, U., Wust, P., Thiesen, B., Feussner, A., von Deimling, A., Waldoefner, N., Felix, R. and Jordan, A. (2007) Intracranial thermotherapy using magnetic nanoparticles combined with external beam radiotherapy: results of a feasibility study on patients with glioblastoma multiforme. *Journal of Neuro-Oncology*, **81**, 53–60.

160 Chertok, B., David, A.E., Huang, Y. and Yang, V.C. (2007) Glioma selectivity of magnetically targeted nanoparticles: a role of abnormal tumor hydrodynamics. *Journal of Controlled Release*, **122**, 315–23.

161 Kante, B., Couvreur, P., Dubois-Krack, G., De Meester, C., Guiot, P., Roland, M., Mercier, M. and Speiser, S. (1982) Toxicity of polyalkylcyanoacrylate nanoparticles I: free NPs. *Journal of Pharmaceutical Sciences*, **71**, 786–90.

162 Gaspar, R., Préat, V., Opperdoes, F.R. and Roland, M. (1992) Macrophage activation by polymeric nanoparticles of polyalkylcyanoacrylates: activity against intracellular *Leishmania donovani* associated with hydrogen peroxide production. *Pharmaceutical Research*, **9**, 782–7.

163 Huang, C.Y. and Lee, Y.D. (2006) Core-shell type of nanoparticles composed of poly[(*n*-butyl cyanoacrylate)-*co*-(2-octyl cyanoacrylate)] copolymers for drug delivery application: synthesis, characterization and *in vitro* degradation. *International Journal of Pharmaceutics*, **325**, 132–9.

164 Lherm, C., Müller, R.H., Puisieux, F. and Couvreur, P. (1992) Alkylcyanoacrylate drug carriers II. Cytotoxicity of cyanoacrylate nanoparticles with different alky chain length. *International Journal of Pharmaceutics*, **84**, 13–22.

165 Fernandez-Urrusuno, R., Fattal, E., Porquet, D., Feger, J. and Couvreur, P. (1995) Evaluation of liver toxicological effects induced by polyalkylcyanoacrylate nanoparticles. *Toxicology and Applied Pharmacology*, **130**, 272–9.

166 Fernandez-Urrusuno, R., Fattal, E., Feger, J., Couvreur, P. and Thérond, P. (1997) Evaluation of hepatic antioxidant systems after intravenous administration of polymeric nanoparticles. *Biomaterials*, **18**, 511–17.

167 Kubiak, C., Couvreur, P., Manil, L. and Clausse, B. (1989) Increased cytotoxicity of nanoparticle-carried adriamycin in vitro and potentiation by verapamil and amiodarone. *Biomaterials*, **10**, 553–6.

168 Gipps, E.M., Groscurth, P., Kreuter, J. and Speiser, P.P. (1987) The effects of polyalkylcyanoacrylate nanoparticles on human normal and malignant mesenchymal cells in vitro. *International Journal of Pharmaceutics*, **40**, 23–31.

169 Tseng, Y.C., Tabata, Y., Hyon, S.H. and Ikada, Y. (1990) In vitro toxicity of 2-cyanoacrylate polymers by cell culture method. *Journal of Biomedical Materials Research*, **24**, 1355–67.

170 Maaβen, S., Fattal, E., Müller, R.H. and Couvreur, P. (1993) Cell cultures for the assessment of toxicity and uptake of polymeric particulate drug carriers. *STP Pharma Sciences*, **3**, 11–22.

171 Vinters, H.V. and Ho, H.W. (1988) Effects of isobutyl 2-cyanoacrylate polymer on cultured cells derived from murine cerebral microvessels. *Toxicology in Vitro*, **2**, 37–41.

172 Müller, R.H. and Olbrich, C. (1999) Solid lipid nanoparticles (SLN): phagocytic uptake, in vitro cytotoxicity and in vivo biodegradation., 2nd communication. *La Pharmacie Industrielle*, **61**, 564–9.

173 Müller, R.H., Maaβen, S., Weyhers, H., Specht, F. and Lucks, J.S. (1996) Cytotoxicity of magnetide-loaded polylactide, polylactide/glycolide particles and solid lipid nanoparticles. *International Journal of Pharmaceutics*, **138**, 85–94.

174 Schöler, N., Hahn, H., Müller, R.H. and Liesenfeld, O. (2002) Effect of lipid matrix and size of solid lipid nanoparticles (SLN) on the viability and cytokine production of macrophages. *International Journal of Pharmaceutics*, **231**, 167–76.

175 You, J., Wan, F., de Cui, F., Sun, Y., Du, Y.-Z. and Hu, F.Q. (2007) Preparation and characteristic of vinorelbine bitartrate-loaded solid lipid nanoparticles. *International Journal of Pharmaceutics*, **343**, 270–6.

176 Olbrich, C., Schöler, N., Tabatt, K., Kayser, O. and Müller, R.H. (2004) Cytotoxicity studies of Dynasan 114 solid lipid nanoparticles (SLN) on RAW 264.7 macrophages-impact of phagocytosis on viability and cytokine production. *Journal of Pharmacy and Pharmacology*, **56**, 883–91.

177 Yuan, H., Miao, J., Du, Y.-Z., You, J., Hu, F.-Q. and Zeng, S. (2008) Cellular uptake of solid lipid nanoparticles and cytotoxicity of encapsulated paclitaxel in A549 cancer cells. *International Journal of Pharmaceutics*, **348**, 137–45.

178 Weyhers, H., Ehlers, S., Hahn, H., Souto, E.B. and Mueller, R.H. (2006) Solid lipid nanoparticles (SLN)–effects of lipid composition on in vitro degradation and in vivo toxicity. *Pharmazie*, **61**, 539–44.

179 Blasi, P., De Santis, E., Schoubben, A., Giovagnoli, S., Puglia, C., Rossi, C., Barberini, L., Cirotto, C. and Ricci, M. (2009) In vivo biocompatibility studies on lipid nanoparticles. AAPS Annual Meeting and Exposition, November 8–12, 2009 Los Angeles, CA, USA.

180 Feynman, R.P. (1960) There's plenty of room at the bottom. An invitation to enter a new field of physics. *Engineering and Science*, **23** (5), 22–36.

181 Müller, R.H. (2007) Lipid nanoparticles: recent advances. *Advanced Drug Delivery Reviews*, **59**, 375.

182 Pardeike, J., Hommoss, A. and Müller, R.H. (2009) Lipid nanoparticles (SLN, NLC) in cosmetic and pharmaceutical dermal products. *International Journal of Pharmaceutics*, **366**, 170–84.

Part Three
Nanoscaffolds, Nanotubes, and Nanowires

Nanomaterials for the Life Sciences Vol.10: Polymeric Nanomaterials. Edited by Challa S. S. R. Kumar
Copyright © 2011 WILEY-VCH Verlag GmbH & Co. KGaA, Weinheim
ISBN: 978-3-527-32170-4

14
Architectural and Surface Modification of Nanofibrous Scaffolds for Tissue Engineering

Jerani T.S. Pettikiriarachchi, Clare L. Parish, David R. Nisbet and John S. Forsythe

14.1
Introduction

Tissue engineering aims to develop biologically functional scaffolds for the repair, replacement, or regeneration of damaged tissues. As the nanofibrous scaffolds produced by electrospinning technology resemble the fibrillar architecture of the natural extracellular matrix (ECM) [1–4], extensive investigations are currently under way in tissue engineering, with the aim of biologically and chemically optimizing the fibers so as to enhance their bioactivity. In this chapter, the most recent strategies employed for the modification of nanofibrous scaffolds are reviewed. Whilst a plethora of information exists detailing the various methods for optimizing the morphology of electrospun fibers [5–7], very few data exist reviewing chemical modification methods. Consequently, a cumulation of the latest information in both of these areas is provided, which aims to present a concise review of the morphological and chemical methods used to functionalize organic nanofibrous scaffolds. Thus, an overview is provided of the electrospun scaffolds used in tissue engineering, the electrospinning process itself, and various postfabrication treatment strategies used to enhance the biological functionality of the scaffolds.

14.2
Tissue Engineering Scaffolds

Patients suffering from tissue or organ damage, as a result of disease or trauma, mainly rely on drug therapy and surgical transplantation. Unfortunately, cell or organ transplants are often hindered by donor scarcity, site morbidity, and immunorejection, thereby prompting a search for alternate solutions. A cornerstone of tissue engineering is the development of biologically functional scaffolds by employing disciplines from materials science, nanotechnology, biomedicine, biochemistry, and the medical sciences. Such scaffolds are not substitutes, nor do

Nanomaterials for the Life Sciences Vol.10: Polymeric Nanomaterials. Edited by Challa S. S. R. Kumar
Copyright © 2011 WILEY-VCH Verlag GmbH & Co. KGaA, Weinheim
ISBN: 978-3-527-32170-4

they entirely mimic the features of natural tissue; rather, they provide a temporary supporting platform to encourage natural tissue regeneration.

In providing a matrix for cell growth, tissue-engineered scaffolds must reflect some functionality features of the natural scaffolding system found in the body, the ECM. The ECM occupies the space between cells, and is intricately organized to support these cells and their protein networks. It provides structural support and anchorage sites for cells, and also functions to segregate tissue and regulate intercellular communication through the presentation or release of biochemical signals such as adhesion peptides and growth factors [8]. Therefore, a critical requirement for tissue-engineered scaffolds is the integration of optimal architectural and biochemical characteristics to facilitate controlled cellular interactions and responses.

The design of a scaffold, that mimics some key features of the ECM must extend beyond a static replacement, and must interact with the host tissue by eliciting desirable cellular responses. Prerequisite scaffold features for optimum host integration include:

- A high porosity to facilitate cell migration, proliferation, and nutrient and metabolite transportation.
- Biomolecules for scaffold–cell adhesion and interactions.
- Mechanical properties that match the native tissue, meet short-term functional requirements, and do not interfere with long-term function.
- Biocompatibility in regards to inflammation and cytotoxicity.
- Nontoxic degradation byproducts that are bioeliminable or remain inert over time.
- Reduced scarring.
- Three-dimensionality in order to fill the injury site and generate an environment that is capable of interacting with the entire cell surface and neighboring cells.

Although the fabrication of scaffolds that incorporate these features has proven to be complex, the inclusion of such features will be essential to avoid host rejection and ascertain successful functionality. Decellularized ECM scaffolds repopulated with cells obtained from the host can be utilized in the production of biologically functional scaffolds that reduce the possibility of immunorejection. This was exemplified by recent experiments where cadaver organs were recellularized to yield a pumping heart [9] and functional trachea [10, 11]. Unfortunately, this technique requires donor tissues, and may also have a limited functional recovery; in the case of the beating heart, only 2% of the adult heart pumping function was achieved [9].

Research is currently under way to develop an optimal variety of scaffolds that satisfy many of the aforementioned biocompatibility and functional criteria. Various types of scaffold, including as nanofibers, hydrogels, self-assembling peptides, freeze-dried, phase-separated-foams and gas-foams, have been investigated for their tissue regeneration potential. As each scaffold type possesses different advantages and limitations, further research is clearly required before they can be employed in medical applications.

14.3
Nanofibrous Scaffolds

Nanofibrous scaffolds used in tissue engineering have an architectural similarity to that of the ECM found in most tissues. The ECM consists of a hierarchical arrangement of extracellular macromolecules that are produced and secreted locally by cells living within the matrix. Polysaccharide chains called glycosaminoglycans (GAGs) and fibrous proteins such as collagen, elastin, fibronectin and laminin (all of which have structural and cellular adhesive properties) comprise the ECM [3, 4, 12]. The ECM plays a number of important roles in regulating cell morphology, behavior, development, survival, migration, proliferation, and functionality [2, 3, 8]. Thus, by creating a scaffold which is reminiscent of the ECM, it should be possible to facilitate cell growth and tissue reformation at a site of injury.

Technological advances have enabled the fabrication of nanoscaled fibrous scaffolds for a variety of biomedical applications, such as tissue engineering, wound dressing, and drug-release systems that are already used in many nonbiomedical applications such as membranes, sensors, and protective clothing [13]. Today, nanofibrous scaffolds are becoming increasingly popular in the field of tissue engineering, due to the enhanced characteristics that they display. In particular, the tissue engineering of cardiac [14]/vascular grafts [15], bone [16–18], cartilage [19], skin [20–22], and neural systems (both peripheral [23, 24] and central [25, 26] nervous systems) all represent applications where nanofibrous scaffolds are currently being considered as viable treatment strategies.

The characteristics of nanofibers that render them suitable candidates include an inherently large surface area-to-volume ratio, a fibrous morphology, fewer structural defects, and superior mechanical properties compared to other forms of scaffolding [12, 27]. These features render the scaffold more biologically compatible and also enable intimate cellular interactions in a manner similar to the ECM.

Nanofibers can be prepared by using a variety of techniques that include drawing, template synthesis, thermally induced phase separation, self-assembly, and electrospinning. Electrospinning is convenient when manufacturing nanofibrous scaffolds for tissue engineering, due to its versatility across a broad range of polymer systems, as well as its controllability, reproducibility, high productivity, and the ability to mass produce single, continuous fibers. This is practical not only for research purposes but also for scaling-up production, should nanofibrous scaffolds prove to be successful in tissue regeneration.

14.4
Electrospinning

Electrospinning is a rapid method of producing very fine polymer fibers with diameters in the submicron range. It is a top-down nanomanufacturing process

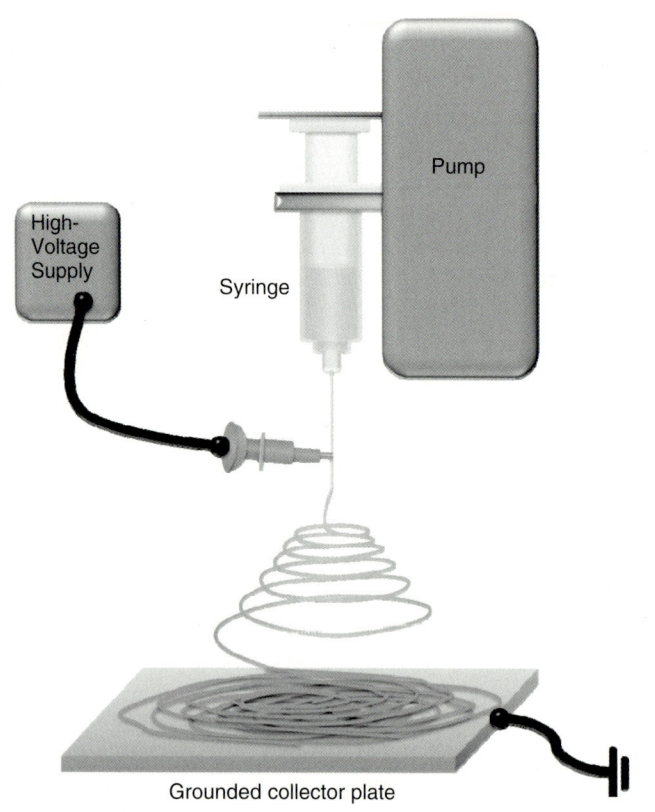

Figure 14.1 Schematic set-up of an electrospinner. Under the influence of a high-voltage electric field, the polymer solution at the syringe tip is stretched to form a nanofiber, which is then collected on a grounded plate to yield a nonwoven nanofibrous scaffold.

that involves a complex combination of polymer science, electronics and fluid mechanics functioning simultaneously to draw out, uniaxially, a viscoelastic polymer solution into a fiber under the influence of an electric field. The fiber formed is collected as a nonwoven mesh to produce a three-dimensional (3D) scaffolding construct.

A typical electrospinner consists of a high-voltage source, a capillary containing the polymer solution, and a grounded collecting device (Figure 14.1). A host of information is available on the process of electrospinning (for reviews, see Refs [5–7]). Briefly, the process involves the application of a high voltage to a capillary containing a polymer solution. The meniscus formed at the capillary tip, due to surface tension, is imbalanced by the introduction of competitive charge repulsion and surface charge contraction, and this results in the formation of a Taylor cone. At a critical voltage, a polymer jet is ejected and accelerates through the electric field, undergoing a whipping motion as it elongates, and thins due to electrostatic repulsion. As the jet travels towards the grounded collector, the solvent evaporates, leaving behind a solid polymeric fiber that eventually builds up to form a nonwoven fibrous mesh. The fiber diameters typically range from several nanometers up to 1 μm [28]. An important point here is that the collector regulates the fiber orientation; for example, a plate may produce randomly oriented fibers, while a rotating mandrel may produce aligned fibers.

The fundamental mechanism of electrospinning has only been characterized qualitatively, although quantitative theories have been proposed [29]. Notably, recent progress in the field of electrospinning has led to the development of alternate electrospinner set-ups such as the Nanospider™ and bubble electrospinning, which utilize different polymer feed systems and allow for the industrial-scale production of fibrous mats. Whilst the underlying mechanisms of these processes are identical, the Nanospider™ [30] uses a rotating cylinder that is partially immersed in the polymer solution to initiate Taylor cones, whereas bubble electrospinning [31] relies on the formation of a bubble at the polymer solution surface.

The processing–property relationships in electrospinning have been investigated in several studies, under different conditions. The processing parameters that regulate the fiber traits can be classified as ambient parameters, solution properties, and processing variables. The common factors that underpin fiber traits and their effects are listed in Table 14.1.

14.5
Cellular Interactions with Polymeric Nanofibers

The ideal tissue engineering scaffold should have a surface that is biologically compatible for cell attachment, growth, proliferation, and migration but, unfortunately, an electrospun scaffold alone is insufficient to satisfy these criteria. Although the nanofibrous nature of electrospun scaffolds resembles certain aspects of the ECM, and provides a large surface area for cellular interactions, various modifications are often necessary to enhance the biological recognition of the scaffolds by cells. Whilst such scaffold modifications may be either topographical or chemical in nature, they must act to capture the most important biological aspects of the different tissues, each of which possesses a unique ECM.

Cells respond to a variety of extracellular biochemical and physical cues in their immediate environment and, as a consequence, create a feedback system to regulate their function and survival. Commonly, cells respond to biochemical environmental cues via membrane-bound receptors (or ion channels), while tactile cues are detected by filopodia (finger-like protrusions) such as those seen on the navigating growth cones of neurons [47, 48]. When activated, these "sensors" initiate a cascade of intracellular signaling processes to transmit the signal and trigger a response.

Importantly, haptotactic, biochemical, and mechanical cues can be incorporated into scaffolds to manipulate cellular responses and, in the context of tissue damage, to promote regeneration. Cell migration, differentiation and function within a

Table 14.1 Ambient, solution, and processing parameters that affect the structure of electrospun fibers.

Parameter type	Parameter	Effect	Reference(s)
Ambient	Temperature	Increasing temperature decreases solution viscosity, producing smaller fiber diameters	[5]
	Humidity	Increasing humidity induced pore formation on fibers	[32]
Solution	Viscosity/Concentration	Low solution viscosity/concentration produced beads on fibers. Increasing viscosity/concentration produced more uniform fibers and increased the fiber diameter	[3, 29, 33–37]
	Conductivity	Increasing conductivity minimized fiber bead formation and yields smaller fiber diameters	[29, 38, 39]
	Surface tension	Variable effects on fibers	[29, 34, 38]
	Polymer molecular weight	Increasing molecular weight decreased fiber beading and droplet formation	[14, 29, 40]
Processing	Flow rate	High flow rates produce larger fibers that are still wet upon reaching the collector	[3, 14, 29, 34, 36–38]
	Voltage	High voltages induce fiber beading, but ambiguous relationship	[3, 14, 36, 37, 41–43]
	Working distance	There is a critical working distance for collecting dry fibers; variations induce beading	[29, 32, 33, 37, 38, 44]
	Collector geometry and kinetics	Plates collect randomly oriented fibers while conductive frames and mandrels collect aligned fibers. Yarns and braided fibers can also be obtained	[23, 36, 45, 46]

tissue is regulated by topographical and biochemical cues, and by gradients specific to that tissue. Similarly, temporally and spatially regulated proteins regulate the growth and guidance of cellular processes such as the axons of nerves. In addition, cells are capable of transducing the mechanical [49] and surface energy [50] properties of their surroundings. A cell's ability to sense topographical cues provides nanofibrous scaffolds with a capacity to orient and guide cellular processes through contact guidance mechanisms [6, 51, 52]. Furthermore, these scaffolds can be imbibed with a range of biochemical cues and mechanical properties reminiscent of those occurring *in vivo* [53, 54].

Cells are highly sensitive to changes in surface energy, chemistry, and mechanical properties. Hence, it is necessary to develop an understanding of the key environmental factors that affect the development, survival and function of various cells, and to incorporate these features into electrospun nanofibrous scaffolds. As the requirements for each cell type will vary, it is not feasible to fabricate a non-specific scaffold for all tissue engineering applications; rather, scaffolds must be customized with the essential haptotactic, biochemical, and mechanical features for the specific tissue that is to be regenerated.

One important factor that pertains to the biological compatibility of an electrospun scaffold is the nature of the polymer utilized. Both, synthetic and naturally derived polymers can be electrospun to form nanofibers. Naturally derived polymers, such as collagen, laminin and fibronectin, possess an arginine-glycine-aspartic acid (RGD) sequence that is recognized by integrin cell surface receptors and mediates cell–cell and cell–ECM adhesion [4]. This minimizes any need for the artificial incorporation of proteins and their derivates to enhance cell adhesion and growth. However, natural polymers also possess inferior mechanical properties compared to biodegradable synthetic organic polymers, which are mechanically superior and readily scalable [55, 56]. Some polymers also need to be crosslinked in order to retain their fibrous form [20, 57, 58], while others may have a low thermostability. In order to circumvent these drawbacks, both natural and synthetic organic polymers have been blended to produce spinnable fibers with improved properties for tissue engineering applications [24, 56, 59]. However, despite the enhanced biological compatibility of nanofibrous scaffolds, further modifications may be required for cellular recognition.

Polymeric electrospun fibers may lack the appropriate chemical and topographical surface properties, such as chemical composition, hydrophilicity, crystallinity, crosslink density, and roughness. Thus, electrospun scaffolds should be biologically compatibilized to create an environment that is favorable for cell growth and will enhance scaffold–cell interactions. In general, the key to fabricating any scaffold for tissue engineering lies in the creation of a noncytotoxic microenvironment containing, at minimum, the basic biochemical and physical cues for the survival of a specific cell type. The main approaches currently under investigation to improve the properties of nanofibrous scaffolds are the optimization of fiber architecture and surface properties.

14.6
Optimizing Fiber and Scaffold Architecture

Fiber architecture is predominantly governed by the ambient, solvent, and processing conditions employed during electrospinning, and refers to the thickness, orientation, and structure of the polymer strands produced. The fiber architecture has a capacity to elicit directed cellular responses, and its optimization can enhance these responses.

14.6.1
Fiber Diameter

Fiber diameter is an important parameter that underpins several scaffold architectural properties, in addition to cellular interactions and penetration. The fiber diameter of electrospun scaffolds is governed by parameters such as the polymer constituents comprising the solution, its concentration, the applied voltage, the working distance (the distance between the needle tip and collector), and the plate size [33, 36]. The effect of these variables on fiber diameter is detailed in Table 14.1. It should be noted that the inverse effects of these parameter adjustments also apply.

Regulation of the fiber diameter is critical in tissue engineering, as it determines the scaffold architectural features (e.g., fiber density, porosity, pore size, and inter-fiber spacing), and hence controls cellular infiltration and proliferation [20, 26]. From a tissue engineering perspective, fibers of larger diameter increase the porosity and pore size such that cells are more readily able to penetrate the scaffold when the pore size exceeds the diameter of the cell body [14]; this is also in accordance with Eichhorn's theory [60]. However, there is some evidence that thicker fibers may induce a greater foreign body reaction when implanted [61, 62]. In particular, fibers of 1–6 µm diameter produced a thinner fibrotic capsule than did fibers with 6–16 µm diameters [62] after subcutaneous implantation. However, the optimal fiber diameter required is likely to be tissue-specific. Controlling the fiber diameter during electrospinning remains a challenging task, as there is a broad fiber diameter variability and this poses a drawback to the technique. Although a method whereby positively charged amiphiphile octadecyl rhodamine or octadecyltrimethylammmonium bromide doping of the polymer solution has been proposed as a means of achieving smaller fiber diameters with limited variability, it is not commonly used, despite having been shown not to cause any cytotoxic effects [63].

Despite fiber diameter variability in electrospun scaffolds, a normal distribution with a mean fiber diameter is achievable. Both, fiber uniformity and diameter will have significant effects on cell adhesion and growth kinetics, such that a critical fiber diameter exists for the optimized growth of some specific cell phenotypes [64]. The critical fiber diameter is also influenced by the fiber texture; for example, the rate of proliferation of mesenchymal stem cells (MSCs) is independent of the fiber diameter on smooth fibers, but rough fibers with a critical diameter of 10 µm will enhance cell adhesion and proliferation [65]. Variations in cell adhesion to fibers of differing diameters may be related to changes in integrin receptor expression at the cell surface. In adult human dermal fibroblast cells, integrin receptor distribution occurs predominantly along the perimeter of the cells when they are cultured on poly(methyl methacrylate) (PMMA) fibers with diameters exceeding a critical value of 970 nm, whereas a more uniform receptor distribution was achieved on smaller fibers [22]. Thus, not only fiber diameter but also fiber variability and texture, impact synergistically on cell growth and tissue regeneration.

14.6.2
Fiber Orientation

Fiber alignment is considered an important concept to attain organized cell growth and tissue formation for tissue engineering applications. Haptotactic cues in the immediate environment can direct cell growth, especially through topographical cues [51]. Electrospun nanofibrous scaffolds can also present topographical cues on a hierarchy of levels, from fiber texture to fiber orientation. The orientation of fibers is believed to direct cell growth, such that an oriented tissue formation can ultimately be achieved.

The alignment of electrospun fibers is predominantly customized by varying parameters associated with the collecting device, including its shape and speed. Conventionally, parallel electrodes and fast-spinning mandrels have been employed in aligning fibers, whereas collector plates and slow-spinning mandrels have been used in the production of randomly oriented fibers. The influence of collection disk rotation speed, needle size, needle tip shape and syringe pump flow rate has been studied in the creation of highly aligned fibers with minimal crossing points [46]. Further investigations into designing methods to improve fiber alignment have also received attention. For instance, the use of parallel plates with bifurcating (angled) edges as a collector device will improve alignment [66]. Whereas, in some instances, it may be desirable to physically direct the morphology of cells, achieving this by means of fiber alignment has some drawbacks. Fiber alignment presents a dichotomy in that, although it facilitates directed cell growth, it also reduces scaffold porosity, which in turn reduces the flow of nutrients and waste to and from the cells, and also hinders cell penetration into the scaffold. In this sense, such a scaffold could be perceived by cells to be a textured two-dimensional (2-D) surface instead of a highly tailored three-dimensional (3-D) scaffold. Thus, it may be beneficial for a scaffold to only be partially aligned in order to offer contact guidance cues while retaining its three-dimensionality.

Aligned fibrous scaffolds have been considered viable options in a number of tissue engineering applications. Neural tissue engineering, in particular, benefits from aligned nanofibers, as the growth cone located at the axon tip in neurons responds to tactile cues encouraging directed neurite outgrowth [23, 46, 52]. Similarly, aligned nanofibers influence the cell morphology and organization of MSCs [67], and direct their differentiation [68]. This alludes to the possibility of developing organized cellular tissues such as cartilage or muscle on nanofibrous scaffolds with optimized architectures. Fibroblasts also exhibit a parallel growth tactile response to aligned nanofibers [22], although the structure of the human dermis does not dictate a necessity for directional cell growth. Even though cell growth parallel to fiber direction is noted here, the nature of the response to tactile cues may be cell-specific. A perpendicular contact guidance of neural cells has also been reported [69]. Perpendicular contact guidance has also been observed *in vivo* following the implantation of aligned nanofibrous scaffolds in the caudate putamen of adult rat brains [70] (Figure 14.2).

14.6.3
Core–Shell Fibers

Core–shell-structured polymeric nanofibers have been attracting research interest due to their unique architecture and properties. The processing method of these

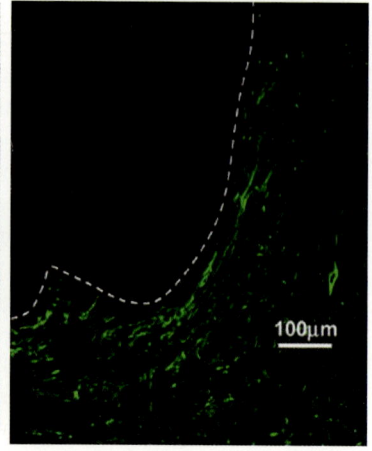

Figure 14.2 Neural cells on a polycaprolactone (PCL) scaffold exhibit contact guidance parallel to the nanofibers. (a) Scanning electron microscopy image with host-derived neural cells in the caudate putamen exhibiting perpendicular contact guidance on a partially aligned fibrous scaffold *in vivo*. Reproduced with permission from Ref. [52]; © 2008, Elsevier; (b) Confocal image of stained neurites traversing the implant. The dotted white line indicates the interface between the scaffold and endogenous tissue. Reproduced with permission from Ref. [70]; © 2009, Elsevier.

fibers enables synthetic and natural polymers to be combined, thereby achieving a scaffold with mechanical and biological compatibility. It also allows for materials that are normally not able to be electrospun to be spun in the presence of a "spinnable" shell material. Core–shell-structured nanofibers can also be employed in some tissue engineering applications, as vessels for drug loading, where the internal fiber compartment contains drugs that are progressively released in an attempt to attain sustained release profiles [71] and modulate cell function.

Several methods are available by which core–shell nanofibers can be electrospun. Electrospun nanofibers can be used as a template for chemical vapor deposition (CVD) or surface-initiated atom transfer radical polymerization (ATRP) to yield the shell component [72]. Recently, it has become more versatile to produce core–shell nanofibers via coaxial electrospinning, where single [73] and multichannel [74] fibers can be produced with various functional components. An exemplary coaxial two-capillary spinneret is depicted in Figure 14.3a. This technique has also been used to "mold" nanofibers by the coaxial spinning of a poly(glycerol sebacate)–polylactide core and polylactide shell, which was subsequently removed upon curing the core [75]. In addition, a novel strategy whereby emulsions [71] and homogeneous solutions are electrospun to form core–shell nanofibers due to phase separation events has also been developed [72]. A transmission electron microscopy (TEM) image of a core–shell fiber is depicted in Figure 14.3b.

The dual functionality of core–shell fibers in providing physical scaffold support, while being capable of releasing biomolecules such as drugs and proteins, makes them a versatile candidate for tissue engineering. In particular, a compromise

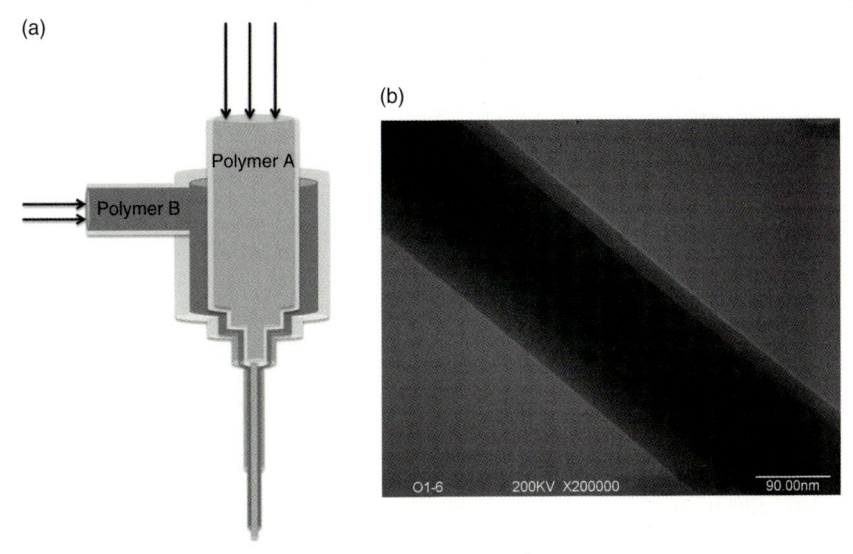

Figure 14.3 Core–shell nanofibers. (a) A schematic example of a coaxial, two-capillary spinneret for the production of core–shell nanofibres; (b) Transmission electron microscopy image of a poly(ethylene oxide)/chitosan core–shell nanofiber. Panel (b) reproduced with permission from Ref. [72]; © 2009, American Chemical Society.

between mechanical properties and surface functionality is attainable by the coaxial spinning of fibers with a synthetic polymer core and a natural polymer shell [72]. Further to enhancing the biological compatibility of the construct, core–shell fibers can promote cell adhesion if natural polymers containing cell-anchoring peptide sequences such as RGD are utilized as the shell. Tissue regeneration can also be facilitated through the encapsulation of biomolecules within the core of core–shell fibrous scaffolds that are gradually released over time. Sustained release profiles have been displayed by core–shell fibers containing nerve growth factor (NGF) [76] and bovine serum albumin (BSA) [71]. This provides a consistent trophic supply to cells growing in the scaffold, as indicated by a promotion of cell proliferation [71]. However, the mechanism of protein release from the core–shell fibers remains unclear.

14.6.4
Pore Size

Although high porosity is an inherent advantage of electrospun scaffolds, a lack of control over the pore size remains an obstacle affecting their utility in tissue engineering applications. As a result of the random deposition of fibers on top of each other, electrospun scaffolds typically have low pore sizes of approximately 5 µm [77]. This size is adequate for a limited number of cell types, such as adult brain-derived stem cells [26], neural cells [70] and MSCs [78] to infiltrate the scaffold, but is not so for others. Consequently, for some cell phenotypes, there is a poor cell infiltration, limited substance diffusivity, and a lack of vascularization.

Enlarging the pore size of electrospun scaffolds can be achieved by the incorporation of leachable template particles. In one such approach, a cryogenic electrospinning technique was used to generate ice crystals as pore templates by subsequent liquefaction and extraction, with minimal toxic effects [2, 77]. This technique can enhance the 3-D porosity of the scaffold, but the scaffolds produced will lack uniformity in pore size due to difficulties associated with controlling the growth of the ice crystals. A more controlled approach is to utilize template salt particles (such as sodium chloride or calcium carbonate), where the micro- and nanosized particles produce scaffolds with secondary nano/micro porosity [79]. The use of sodium chloride particles has negligible cytotoxic affects, as observed in the adhesion and proliferation of chondrocytes [80]. Another technique based on the leachant concept involves spinning layers of polyurethane (PU) and polyethylene oxide (PEO), the latter of which is extracted via polymer dissolution to enhance scaffold porosity [15]. This technique promoted the growth and penetration of smooth muscle cells (Figure 14.4).

Physical or chemical modification of the collector device is another approach by which the pore size of electrospun scaffolds can be enhanced. Through the design of a cylindrical frame comprised of metal struts as the collector, a poly(D,L-lactide-co-glycolide) and collagen scaffold with 85–90% porosity and 90–130 μm pore sizes was created [81]. In a novel approach, a collector with an enamel coating containing montmorillonite clay particles that modulated the electrical properties (i.e., charge distribution and density) was capable of regulating the fiber deposition pattern [82]. This deposition method provided some control over the pore sizes through patterning of the enamel coating, and also provides control over fiber orientation.

14.6.5
Layering Fibers

The layering of fibers during electrospinning has been used to achieve a variety of outcomes, including an increased pore size, an enhanced three-dimensionality, replication of the ECM organization, and the tuning of mechanical properties. Creating an electrospun scaffold that has sufficiently large pores to facilitate cell infiltration remains an issue to be addressed for many applications. The porous structure of electrospun scaffolds can be enlarged through the layering of polymers that dissolve in different solvents (such as PU and PEO), and dissolving one type of fiber to increase the pore size at intervals throughout the scaffold [15]. These layered constructs have demonstrated a higher bioactivity and cell penetration compared to a monolayered scaffold. A comparison of the cell proliferation responses to scaffolds produced by this layering technique and a solution–mixture technique produced similar responses, although both scaffolds performed better than conventional electrospun scaffolds [83]. In a similar fashion, bilayered nano-

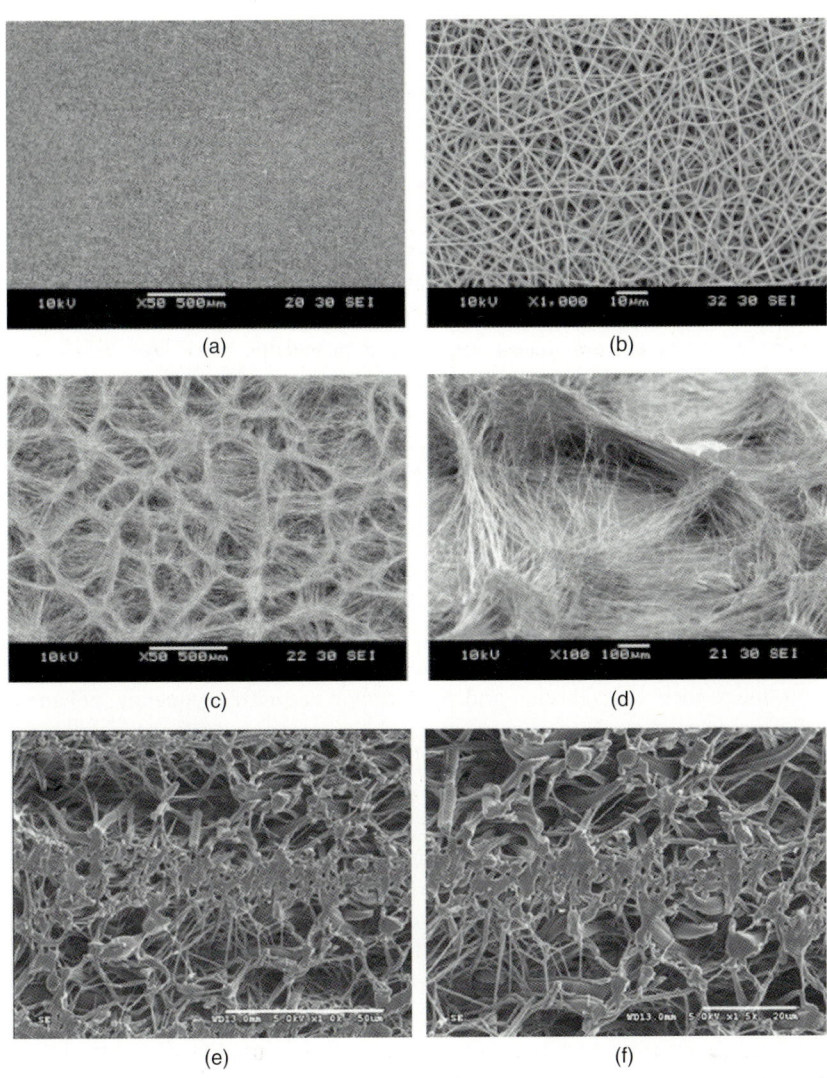

Figure 14.4 Scanning electron microscopy images comparing (a, b) the scaffold architecture of conventional electrospun scaffolds, (c, d) the cryogenic electrospun scaffold with enhanced porosity, and (e, f) the layered PEO leached PU scaffold. Panels (a–d) reproduced with permission from Ref. [2]; © 2009, John Wiley & Sons, Inc.; panels (e) and (f) reproduced with permission from Ref. [15]; © 2009, Koninklijke Brill NV.

and micro-fiber scaffolds of poly(ε-caprolactone) (PCL) have been used to generate variations in pore size, while retaining porosity within the scaffold [84]. In one study, increasing the layer thickness of the scaffold hindered rat marrow stromal cell infiltration under both static and flow perfusion culture conditions, indicating a limitation with this technique. In contrast, human embryonic stem cells (HESCs) and myoblasts were able to readily infiltrate a layered scaffold prepared using a hydrospinning technique (electrospinning into a water bath and progressively stacking layers) when compared to fibers collected on a plate [85]. The hydrospun scaffold was 99% porous with pore diameters of over 100 μm, thus creating a 3-D environment in which cells were able to penetrate.

The main function of layered scaffolding techniques is to accommodate cell penetration, proliferation, and migration. Mimicking the structure of the ECM may not only enhance penetration by cells in the vicinity, but it may also help mimic the load-bearing capacity of the tissue. An electrospun scaffold with an angle-ply, multilamellar architecture that replicated the anatomic form of the annulus fibrosis of spinal cord disks, had mechanical parity with native tissue [78]. Similarly, a tubular composite comprising a polylactide outer layer and a silk fibroin–gelatin inner layer possessed appropriate biomechanical properties, provided a favorable environment for cell growth, and did not induce macrophage or lymphocyte activation after subcutaneous implantation, indicating good candidature for blood vessel tissue engineering [86].

14.7
Optimizing the Fiber Surface

The surfaces of nanofibrous scaffolds typically require post-fabrication treatment to optimize their cell adhesion and proliferation capacity. Frequently, polymers that possess good mechanical properties lack the necessary surface properties for biological applications. Therefore, in order to avoid compromising the scaffolds' mechanical integrity by using polymers with inferior properties, it is convenient to surface-treat the nanofibers to attain the desired surface properties. Surface modification alters the chemical and/or physical properties by modifying the atomic or molecular structure at the surface, altering the surface topography, or by coating with a different material. Irrespective of the method employed for the surface treatment, there are inherent limitations in efficiency. As electrospun fibers are nanoscale, the thickness or depth of the surface treatment must also be in the range of a few nanometers. In addition, the 3-D nature of the scaffold makes it difficult to achieve a homogeneous coverage of all fiber surfaces and treatment penetration in thicker scaffolds. Despite these drawbacks, the surface treatment of fibers outweighs the alternative loss in fiber mechanical strength that would result from selecting polymers with desirable surface characteristics. The most common fiber surface-modification methods that have been, and still are, employed include polymer blending, chemical treatment, biomolecule attachment, coating, graft polymerization, and plasma treatment.

14.7.1
Polymer Blending

Although not strictly a post-treatment strategy, polymer blending can be used as a surface-treatment strategy whereby two (or more) materials are physically blended together to attain a new biocomposite with superior surface and/or mechanical properties. The process requires some miscibility between the two materials, but does not require chemical bonding, which makes it one of the most easily used methods for the introduction of functional molecules into electrospun fibers. Blending is typically performed to modify the material properties of the parent polymer in terms of biological recognition, hydrophilicity, tensile strength, and degradation [23]. Hence, this technique modifies not only the surface properties of the fibers but also their bulk properties.

While blending may improve the cell compatibility of synthetic nanofibers, it can also cause a deterioration of the mechanical properties [56]. Altering the polymer content in a mixture also impacts upon its degradation. For example, an increase in the gelatin content of poly(L-lactide-*co*-ε-caprolactone) (PLLA–PCL) nanofibers increases the degradation rate due to hydrolytic cleavage, partial enzymatic degradation, and an increase in hydrophilicity [87]. The blending of poly(D-lysine) (PDL) in poly(ethylene glycol) (PEG) alters the mode of degradation from the surface to the bulk of the scaffold, in addition to optimizing the surface properties [21]. Furthermore, "hairy" nanofibers composed of polymer–peptide conjugates of PEO-[serine-glutamic acid-glutamic acid] blended in PEO are formed by exploiting the electric field present during electrospinning, which drives the polarizable conjugate to the fiber surface [88].

Blending is also a convenient method by which scaffolds with biomimetic properties can be prepared. For instance, the incorporation of hydroxyapatite (HA) particles in electrospun nanofibers creates scaffolds with some similarities in structure to bone, but also acts to strengthen the scaffold. The ECM of natural bone contains nanosized HA crystals between the collagen molecules [89]. Reflecting this, a HA- dispersed PLLA nanofibrous scaffold has been synthesized where the Ca^{2+} of the HA particles interact with the acid end groups of PLLA to promote strong surface bonding, high tensile properties, and also superior osteoblast adhesion and growth [16, 90]. Furthermore, electrically conductive blends of PLLA–PCL and polyaniline (PANI) nanofibers that modulate the induction of myoblasts into myotubes, without external electrical stimulation, have been designed for skeletal tissue engineering applications [91].

14.7.2
Coating Fibers

Coating is a simple means of functionalizing a fibrous scaffold. In this physical approach, a material containing the desired functional properties is used to coat the surface of the polymer fibers although, as a result, the coatings may be prone to detachment and the technique can lack reproducibility and controllability. Nevertheless, coating remains the most convenient method of functionalizing electrospun polymeric scaffolds, without chemical reactivity.

In general, coatings are used to enhance the performance of nanofibers, especially to enhance cell–scaffold interactions. For instance, both PCL and poly(lactic-*co*-glycolic acid) (PLGA) electrospun scaffolds have been coated with polypyrrole to enhance electrical conduction for signal transduction, particularly in the context of neural tissue engineering [92, 93]. Polypyrrole is capable of electrically stimulating neurons [94, 95]; however, as it belongs to the rigid-rod host family, polypyrrole is not suitable for electrospinning, which makes coating an attractive alternative.

PCL fibrous scaffolds have also been coated with gelatin via layer-by-layer (LbL) deposition using poly(styrene sulfonate) sodium salt as a negatively charged polyelectrolyte and calcium phosphate to promote preosteoblast cell growth, on an otherwise hydrophobic unfavorable surface, for bone tissue engineering [96]. Subsequently, electrostatic self-assembly was used to deposit poly(diallyldimethy lammonium chloride) and poly(sodium 4-styrene sulfonate) to form a layered coating on an electrospun PCL which promoted fibroblast cell growth [97].

14.7.3
Chemical Modification of Fibers

Chemical modification methods introduce or create reactive chemical species at the fiber surfaces, in order to improve the chemical and physical properties. Typically, chemical modifications involve nucleophilic or electrophilic reactions at susceptible functional groups on the surface. The chemical surface treatment of polymer fibers is frequently used to enhance the hydrophilicity of scaffold surfaces and, in turn to improve the adhesion, proliferation, and differentiation of cells.

The partial hydrolysis [26, 98] and aminolysation [99] of polyester scaffolds has commonly been used to modify surface hydrophilicity, or to introduce new functional groups. During oxidation, random chemical scission of ester linkages along the polymer backbone yields carboxylic and hydroxyl groups, while the aminolysation of esters involves nucleophilic attack of the amine at the carbonyl group. The hydrolysis of PLLA scaffolds in an aqueous solution of sodium hydroxide promotes calcium ion binding to enhance HA mineralization for bone tissue engineering applications [100]. Another chemical treatment proposed to promote calcium phosphate crystallization on electrospun scaffolds is an alternate immersion of the scaffold in solutions containing calcium ions and phosphate ions [17, 18]. This technique promotes the precipitation of calcium phosphate nuclei that can trap calcium ions around the fibers, in order for bone calcium phosphate to grow. Chemical surface modification is, therefore, an efficient method of functionalizing the surface of electrospun scaffolds, especially as the scaffold integrity is retained and solvents can permeate the scaffold, thus ensuring the functionalization of deeply embedded fibers. However, careful selection of the hydrolysis agents and treatment duration are needed.

The chemical surface modification of electrospun scaffolds is a common step for the attachment of biologically functional molecules. Some polymers used for electrospinning scaffolds lack inherent functional groups capable of reacting with biomolecules, and so require chemical conditioning to generate the necessary surface functionality. The bioconjugation of proteins to synthetic polymers commonly occurs through amine coupling of lysine amino acid residues and sulfhydryl coupling of cysteine residues [101]. The chemical introduction of amino groups to the surface of nanofibers prepares them for the attachment of various biomolecules such as brain-derived neurotrophic factor (BDNF), collagen, chitosan, and peptides via a coupling agent [25, 27, 102]. In some instances, a covalently bound spacer arm may be beneficial when tethering biomolecules. These short-chain spacers extend the reactive group away from the scaffold surface to minimize steric hindrance, and often produce a more active complex. Biomolecule conjugates utilized in the functionalization of nanofibrous scaffolds are discussed in Section 14.7.6.

14.7.4
Polymer Grafting onto Surfaces

Graft coupling (or "grafting-to") commonly involves a condensation reaction between a reactive group on the polymer surface and the graft polymer chain. In contrast, graft polymerization (or "grafting-from") relies on surface initiation for polymerization to occur, and also results in a higher graft density. Surface initiation can be attained by chemical modification or irradiation via plasma discharge, ultraviolet (UV) light, gamma rays, and electron beams.

The efficacy of polymer grafting lies in the ability to modify the polymer surface to possess different properties, the controllability and reproducibility of the technique, and the long-term stability of covalently grafted chains. In essence, grafting addresses the issues related to physical surface-modification techniques such as blending and coating. Grafting is an efficient post fabrication surface-modification technique that imbibes electrospun fibers with nanotopographical and chemical cues that influence cell behavior through the generation of "bottle-brush"- like fibers. Graft coupling has been employed in the surface functionalization of polyethersulfone nanofibers where poly(acrylic acid) polymer chains were grafted via UV photopolymerization [103]. Similarly, solvent-resistant nanofibers that are thermoresponsive have been prepared through click chemistry. In this case, block copolymers of poly(4-vinylbenzyl chloride) and poly(glycidyl methacrylate) were electrospun and exposed to sodium azide in order to generate azido surface groups that would undergo the click reaction with the acetylene group of alkyene-terminated poly(N-isopropylacrylamide) [104]. Conversely, graft polymerization has also been used in the fabrication of bottle-brush fibers. The surface-initiated ATRP of lead dimethacrylate and acidolysis by H_2S was employed in the production of poly(methylmethacrylate-co-vinylbenzyl chloride) and PbS hybrid nanofibers [105]. An advantage of grafting, by either coupling or polymerization, is that a stable nonhydrolyzable link is formed between the nanofiber surface and the preformed polymer, which ensures longevity of the functional component.

The grafting of natural molecules onto organic substrates has become a popular method for functionalizing nanofibers to enhance cell adhesion and the biological compatibility of scaffolds. For example, PLLA nanofibers have been rendered cell-adhesive through grafting of gelatin complexed with N-maleic acyl-chitosan [106]. A noteworthy feature of this scaffold was that human umbilical vein endothelial cells (HUVECs) cultured on it retained metabolic activity after exposure to shear stress, thus highlighting its potential for vascular tissue engineering applications. Similarly, the biological functionality of gelatin has popularized its use as a graft agent; indeed, it has been grafted onto various scaffolds using different graft methods. For example, gelatin has been grafted on to PLLA scaffolds by aminolysis of the PLLA fibers and subsequent glutaraldehyde coupling [107]. A more complex method of grafting gelatin onto polyethylene terephthalate (PET) nanofibers involved fiber treatment with formaldehyde to yield hydroxyl groups on the surface, followed by graft polymerization of methacrylic acid initiated by Ce(IV). Following this, gelatin was grafted on using a carbodiimide coupling agent [108]. The grafting of gelatin improves the adhesion and spreading of cells on scaffolds.

14.7.5
Plasma Treatment of Fibers

The surface of polymers, such as nanofibrous scaffolds, can be tailored for optimal hydrophilicity, topography, and functional groups by plasma surface treatment. Plasma treatment can be employed as a technique for attaining a clean surface, promoting chemical adhesion, tailoring surface energy, and improving biocompatibility. A beneficial characteristic of plasma treatment is that it is surface-limited (ca. 500–1000 in depth) [109]. This dry chemical method represents an attractive modification technique in the field of tissue engineering, as it also avoids residual toxicity complications generated by aqueous solutions in wet chemical processes.

Plasma – which commonly is referred to as the "fourth state of matter" – is a partially or wholly ionized gas consisting of electrons, photons, ions, free radicals, atoms, and molecules with a neutral net charge. There are two types of plasma: hot or high-temperature plasma, and cold or low-temperature plasma, denoting the temperatures at which the ionized gas is generated [110]. Low-temperature plasma is used in surface-modification applications under vacuum conditions, where low pressures minimize the collision frequency between particles within the plasma, and the reaction occurs at low temperatures [109]. During the process of plasma surface modification, an evacuated reaction chamber is filled with a low-pressure gas that is subsequently energized by a form of energy (radiofrequency, direct or indirect current, and microwaves) to create a glow discharge plasma [110]. The energetic species in the ionized gas bombard the substrate surface and transfer their energy to it. Energy dissipation then occurs through physical and chemical processes in the form of etching/degradation, deposition/grafting, crosslinking, and functionalization [109, 110]. Unique reactions can be promoted at the substrate surface by the choice of gas reagent. For example, reagents such as oxygen, nitrogen, water and air are used to remove organics, leaving functional oxygen-containing groups that enhance the hydrophilicity and improve adhesive bonding [111]. Conversely, reagents such as argon, helium and nitrogen

Figure 14.5 Schematic overview of the plasma treatment process. The plasma treatment of a polymer substrate exposes its surface to a UV glow discharge in the presence of an excited gas species which collide with it, resulting in one of several outcomes: crosslinking of the substrate polymer chains; functionalization of the polymer surface; deposition (grafting) of the species onto the surface; or etching (degradation) of the substrate.

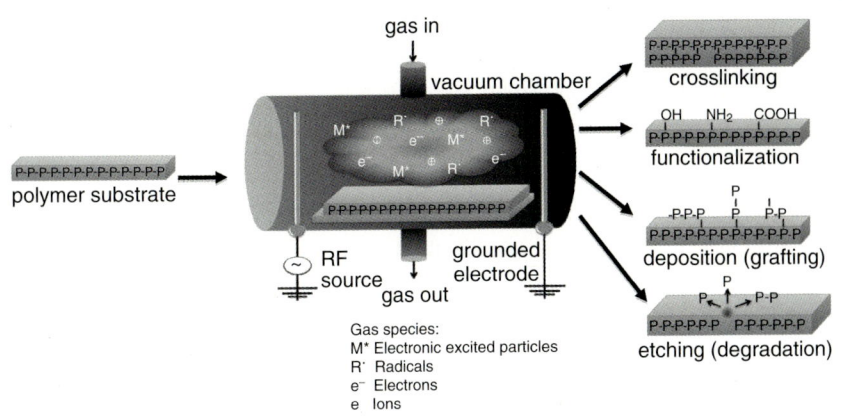

remove organics from oxidation-sensitive surfaces; the resultant surface modification merely spans several atomic layers in depth. A schematic overview of a plasma reactor is depicted in Figure 14.5.

In tissue engineering, plasma treatment has been employed predominantly to enhance cell interactions with the fiber surface. Plasma treatment changes the hydrophilicity and topography of fibers [112], which in turn can either promote or prohibit cell attachment. Within the plasma treatment process there are numerous variable parameters; therefore, findings are specific to the set parameters being tested. For instance, a comparison between argon and oxygen gas plasma treatment of PCL on cell growth found that oxygen treatment enhanced the adhesion and proliferation of fibroblasts, chondrocytes and osteoblasts [112], whereas an argon plasma treatment of silk fibroin facilitated chondrocyte proliferation and increased synthesis of the ECM protein GAG [113]. In addition to enhancing cell adhesion and proliferation, the surface-oxidative properties of plasma treatment can also be utilized for polymer grafting. Both, acrylic acid [114] and collagen [115], have been covalently immobilized onto nanofibrous scaffolds surface-modified by using plasma polymerization, to endorse the growth of fibroblast and somatic stem cells, respectively. However, it has also been suggested that the plasma treatment of PCL nanofibers alone surpasses the cellular adhesion properties attained via collagen-grafted PCL scaffolds [116].

The mutual benefits of plasma treatment promoting cell adhesion and biomolecule grafting renders this technique versatile for functionalizing the surface of electrospun scaffolds. However, the limited penetration depth of plasma through the micropores of electrospun scaffolds means that concealed subsurface fibers may not be treated. This may be detrimental for tissue engineering applications, where cell penetration is required, and for this reason it has at times become preferential to use wet chemical surface-treatment methods. Nonetheless, plasma treatment provides a convenient batch-process method for introducing surface functionality with minimal toxicity effects.

14.7.6
Biomolecule Attachment

The presence of appropriate biomolecules at the site of injury is critical to achieving tissue regeneration (here, a biomolecule refers to biologically significant ligands, including peptides and proteins). *In vivo*, biomolecules act to direct the adhesion, proliferation, migration and differentiation of cells through the regulation of metabolic pathways and gene expression. Biomolecules can be present in either soluble or immobilized forms, and influence cells through interactions with intracellular proteins or cell membrane receptors, respectively. The type of molecule present, its structure and quantity, govern cellular function [117].

The attachment of biomolecules onto scaffold surfaces introduces a degree of biorecognition. Biomolecules have recently been covalently tethered to electrospun scaffolds, as the solvents used in the manufacture of synthetic organic scaffolds are detrimental to most biological molecules, thereby rendering any form of pre-manufacture blending ineffective. The tethering of biomolecules onto nanofibrous scaffolds is advantageous, as the mechanical strength of the polymer can be preserved while achieving improved biological properties. Furthermore, tethered biomolecules have a prolonged location-specific lifespan, and cannot be phagocytosed by cells (unlike soluble biomolecules) [118]. Consequently, a sustained activation of signaling pathways can be utilized to study the influence of particular biomolecules on cells, leading to the development of customized tissue-engineered scaffolds.

Despite these merits, the retention of bioactivity by biomolecules, once they have been tethered onto scaffolds, remains an issue. For example, many ligands have high conformation and orientation specificities that are imperative to bioactivity. Although the attachment of biomolecules onto surfaces presumes reactivity between particular chemical groups, it is likely that a number of ligands are incorrectly oriented, and that this may negatively influence its functionality relative to the soluble form [119].

The details of some biomolecules that have been attached to electrospun scaffolds are listed in Table 14.2. In particular, proteins such as collagen, laminin and its derivatives have received attention due to their ECM origins and cell-anchoring features. It is believed that, by incorporating essential ECM proteins, the scaffold biocompatibility can be enhanced by various methods, including facilitating cell adhesion, morphology, migration and proliferation, as well as by altering the surface hydrophilicity, thus promoting tissue regeneration [115, 123–125]. Current investigations are also examining the effects of biologically relevant ligand/protein gradients [54, 121, 128] on electrospun scaffolds. Such gradients are important in ensuring an appropriate spatial positioning within a tissue (including cell migration and differentiation) and axonal growth (in the context of neural cells). Thus,

Table 14.2 Summary of biomolecules tethered onto electrospun scaffolds.

Biomolecule	Electrospun polymer	Tissue engineering application	Reference
Laminin (and its derivatives)	Poly(ε-caprolactone)	Peripheral nerve regeneration	[54]
		Adipose-derived stem cells delivery	[120]
	PLLA	Skin and bone	[27]
		Peripheral nerve regeneration	[1]
Fibronectin	Polylactide	Adult liver	[121]
	Silk	Skeletal tissue regeneration	[67]
Collagen	Polyurethanesulfone	Stem cell infiltration	[115]
		Neural tissue regeneration	[122]
	Poly(methyl methacrylate and acrylic acid) (PMMAAA)	Skin tissue	[123]
	Poly(ε-caprolactone)	Skin tissue and bone	[27]
	Poly(acrylonitrile-co-acrylic acid)	Liver	[117]
	PGA, PLLA, PLGA	Skin	[114]
Selectin	PLG/collagen	Bone tissue	[124]
Eggshell membrane proteins	Poly(ε-caprolactone)	Membrane barriers	[125]
Chitosan	Poly(ε-caprolactone)	Skin and bone	[27]
Heparin	Poly(ε-caprolactone)	Nonspecific	[126]
BDNF	Poly(ε-caprolactone)	Neural tissue regeneration	[25]
HA	PLGA/collagen	Bone regeneration	[55]
Bone morphogenetic protein-2	Chitosan	Bone regeneration	[127]

PGA, poly(glycolic acid).

electrospun scaffolds with an in-built cell growth directionality can be fabricated for specific applications.

14.8
Challenges with Fibrous Scaffolds in Tissue Engineering

Although nanofibrous scaffolds offer exciting prospects for tissue engineering applications, several limitations restrict their current use. The major problems encountered by electrospun nanofibrous scaffolds are associated with accommodating cell growth (adhesion, proliferation, migration and differentiation), scaffold surface functionalization, mechanical properties, and biological compatibility.

The primary challenges associated with electrospun nanofibrous scaffolds, that are currently being resolved, include:

- *Limited pore dimensions:* During electrospinning, fibers tend to form dense fiber bundles, especially when aligned; this leads to the production of small pores that limit cell infiltration.

- *Limited three-dimensionality:* Electrospun scaffolds have limited thicknesses, which may be sufficient to achieve some cell penetration, but they remain incapable of filling large cavities in damaged tissues. Furthermore, an increase in the thickness of nanofibrous scaffolds may decrease nutrient and waste diffusion, adversely affecting cell infiltration.

- *Limited surface functionality:* In addition to having the appropriate surface topography and chemical composition for cell adherence, the fibers must also interact with specific cell types through growth and differentiation factors, depending on the application. It may also be necessary to have functional gradients to induce directional growth of cells and, in turn, of tissues.

- *Toxicity of residual solvents:* This includes the solvents used to dissolve the polymers, as well as those used in surface functionalization processes.

- *Degradation:* Although the necessity of this feature is debatable for some applications [129], it is imperative that the degradation of scaffolds generates nontoxic byproducts, prevents inflammation, and that preferably they are bioeliminable. Furthermore, the mechanical properties of the scaffold must continue to offer structural support to tissues during this process.

- *Mechanical compatibility:* A mismatch in mechanical properties between the scaffold and surrounding tissue can result in tissue collapse and/or microtearing, and exacerbate the foreign body response.

- *Biocompatibility:* The final scaffold must not only be noncytotoxic to the selective tissue type, but also be compatible with other cell types such as blood cells, which infiltrate the scaffold and undergo angiogenesis. Neither must the scaffold elicit a chronic foreign body response.

Many of these primary issues hinder cell penetration. However, these challenges are not unique to electrospun scaffolds—they are also encountered when employing other types of scaffolding, such as hydrogels, self-assembling peptide scaffolds, freeze-dried scaffolds, and phase-separated scaffolds. Whilst the design of a scaffold that can overcome these biological and morphological hurdles remains a challenging task, it is crucial to ensure maximal benefits for *in vivo* applications. By overcoming these challenges, nanofibrous scaffolds would promote cell attachment, proliferation and migration, while simultaneously providing mechanically compatible structural support.

In addition, there are wider challenges relating to the use of nanofibrous scaffolds (and other types of scaffolding) in tissue engineering that are noteworthy. Engineered scaffolds for tissue regeneration face production challenges such as mass production, reproducibility, reliability, costs, and government regulations. There are also consumer challenges that involve societal acceptance and clinical practicality, which must be addressed. While these appear to be long-term challenges, they remain worthy of consideration when biologically functionalizing nanofibrous scaffolds for tissue engineering.

14.9
Summary

Nanofibrous scaffolds possess many structural features that can be tuned to optimize their compatibility with cells. The diameter and orientation of nanofibers in an electrospun scaffold can influence cell organization, proliferation, migration, and differentiation, as well as direct their outgrowths via haptotactic cues. The correct interfiber distance can allow cell infiltration into the scaffold, proliferation, the transport of nutrients and metabolites and, in some cases, also facilitate contact guidance cues. To some extent, these features can be controlled by a refinement of the processing conditions. Other features such as the fiber structure and layering, can also be used to tailor a scaffold. Most organic polymers produce surfaces that are incompatible for cell adhesion; consequently, core–shell-structured nanofibers have been fabricated through homogeneous solutions and emulsions to create nanofibers with natural polymer surface coatings that are favorable for cell adhesion. Moreover, this method retains the mechanical integrity of the scaffold while introducing biological functionality. The layering of nanofibers can also facilitate cellular infiltration and growth, in addition to enhancing the three-dimensionality of the construct and tailoring of their mechanical properties.

Overall, many structural and chemical methods have been developed that can be used to optimize nanofibers for tissue engineering applications. Whilst each method has its advantages and disadvantages, the functionality imbibed to the nanofibers will underpin which material is the most suitable for any given application. Functionalizing the scaffold to enhance biological compatibility represents a preliminary step towards deploying these constructs *in vivo*.

14.10
Future Perspectives

Although, in this chapter, much emphasis has been placed on the different biological functionalization techniques that can be used to enhance the capacity of nanofibrous scaffolds to interact with cells, this by no means makes the scaffold biologically compatible. Current investigations into nanofibrous scaffolds involve the optimization of cell culture techniques, using *in vitro* conditions. Yet, the *in vivo* conditions differ significantly, with complicating factors such as heterogeneous cell populations and organization, immune responses, and vascularization. While these factors need not be addressed immediately, there are other factors that should be addressed at the current "design stage" in the manufacture of nanofibrous scaffolds. Morphological optimization at the nano-, micro- and macroscales, and scaffold mechanical properties, each play important roles in controlling cellular responses and host integration. The conformation of a scaffold determines its ability to fill lesions, to form seamless boundaries with surrounding tissues, and to govern cell penetration. The compatibility of mechanical properties is another requisite for scaffolds designed for *in vivo* use, in order to render them load-bearing and/or flexible (where applicable), and to minimize tissue tearing due to moduli mismatches. Thus, architectural and mechanical optimization are equally important as surface modification in the design and fabrication of nanofibrous scaffolds.

The increased interest in tissue engineering during recent years may serve as an indicator that scaffolds tailored for various tissues could approach clinical testing sometime in the near future, and may soon be followed by "off-the-shelf" surgically implantable constructs. At present, however, further studies are required on the tailoring of scaffold functionalities for various tissue types, and their *in vitro* assessment. Subsequently, *in vivo* assessments and redesigning to rectify any problems must be conducted before the successful production of nanofibrous scaffolds for tissue engineering applications.

References

1 Koh, H., Yong, T., Chan, C. and Ramakrishna, S. (2008) Enhancement of neurite outgrowth using nano-structured scaffolds coupled with laminin. *Biomaterials*, **29** (26), 3574–82.

2 Leong, M., Rasheed, M., Lim, T. and Chian, K. (2009) In vitro cell infiltration and in vivo cell infiltration and vascularization in a fibrous, highly porous poly(D,L-lactide) scaffold fabricated by cryogenic electrospinning technique. *Journal of Biomedical Materials Research Part A*, **91A** (1), 231–40.

3 Han, D. and Gouma, P.I. (2006) Electrospun bioscaffolds that mimic the topology of extracellular matrix. *Nanomedicine*, **2** (1), 37–41.

4 Rho, K.S., Jeong, L., Lee, G., Seo, B.M., Park, Y.J., Hong, S.D. et al. (2006) Electrospinning of collagen nanofibers: effects on the behavior of normal human keratinocytes and early-stage wound healing. *Biomaterials*, **27** (8), 1452–61.

5 Pham, Q.P., Sharma, U. and Mikos, A.G. (2006) Electrospinning of polymeric nanofibers for tissue engineering applications: a review. *Tissue Engineering*, **12** (5), 1197–211.

6 Sill, T.J. and Recum, H.Av. (2008) Electrospinning: applications in drug delivery and tissue engineering. *Biomaterials*, **29**, 1989–2006.

7 Teo, W.E. and Ramakrishna, S. (2006) A review on electrospinning design and nanofibre assemblies. *Nanotechnology*, **17** (14), R89–106.

8 Sell, S., Barnes, C., Smith, M., McClure, M., Madurantakam, P., Grant, J. et al. (2007) Extracellular matrix regenerated: tissue engineering via electrospun biomimetic nanofibers. *Polymer International*, **56** (11), 1349–60.

9. Ott, H.C., Matthiesen, T.S., Goh, S.K., Black, L.D., Kren, S.M., Netoff, T.I. et al. (2008) Perfusion-decellularized matrix: using nature's platform to engineer a bioartificial heart. *Nature Medicine*, **14** (2), 213–21.
10. Macchiarini, P., Jungebluth, P., Go, T., Asnaghi, M.A., Rees, L.E., Cogan, T.A. et al. (2008) Clinical transplantation of a tissue-engineered airway. *Lancet*, **372** (9655), 2023–30.
11. Hollander, A., Macchiarini, P., Gordijn, B. and Birchall, M. (2009) The first stem cell-based tissue-engineered organ replacement: implications for regenerative medicine and society. *Regenerative Medicine*, **4** (2), 147–8.
12. Park, H., Cannizzaro, C., Vunjak-Novakovic, G., Langer, R., Vacanti, C.A. and Farokhzad, O.C. (2007) Nanofabrication and microfabrication of functional materials for tissue engineering. *Tissue Engineering*, **13** (8), 1867–77.
13. Agarwal, S., Greiner, A. and Wendorff, J. (2009) Electrospinning of manmade and biopolymer nanofibers-progress in techniques, materials, and applications. *Advanced Functional Materials*, **19** (18), 2863–79.
14. Balguid, A., Mol, A., van Marion, M., Bank, R., Bouten, C. and Baaijens, F. (2009) Tailoring fiber diameter in electrospun poly(epsilon-caprolactone) scaffolds for optimal cellular infiltration in cardiovascular tissue engineering. *Tissue Engineering: Part A*, **15** (2), 437–44.
15. Shin, J., Lee, Y., Heo, S., Park, S., Kim, S. and Kim, Y. (2009) Manufacturing of multi-layered nanofibrous structures composed of polyurethane and poly(ethylene oxide) as potential blood vessel scaffolds. *Journal of Biomaterials Science Polymer Edition*, **20** (5-6), 757–71.
16. Sui, G., Yang, X.P., Mei, F., Hu, X.Y., Chen, G.Q., Deng, X.L. et al. (2007) Poly-L-lactic acid/hydroxyapatite hybrid membrane for bone tissue regeneration. *Journal of Biomedical Materials Research Part A*, **82A** (2), 445–54.
17. Araujo, J., Martins, A., Leonor, I., Pinho, E., Reis, R. and Neves, N. (2008) Surface controlled biomimetic coating of polycaprolactone nanofiber meshes to be used as bone extracellular matrix analogues. *Journal of Biomaterials Science Polymer Edition*, **19** (10), 1261–78.
18. Hayashi, S., Ohkawa, K., Yamamoto, H., Yamaguchi, M., Kimoto, S. and Kurata, S. (2009) Calcium phosphate crystallization on electrospun cellulose non-woven fabrics containing synthetic phosphorylated polypeptides. *Macromolecular Materials and Engineering*, **294** (5), 315–22.
19. Li, X., Xie, J., Lipner, J., Yuan, X., Thomopoulos, S. and Xia, Y. (2009) Nanofiber scaffolds with gradations in mineral content for mimicking the tendon-to-bone insertion site. *Nano Letters*, **9** (7), 2763–8.
20. Powell, H. and Boyce, S. (2008) Fiber density of electrospun gelatin scaffolds regulates morphogenesis of dermal-epidermal skin substitutes. *Journal of Biomedical Materials Research Part A*, **84A** (4), 1078–86.
21. Cui, W., Zhu, X., Yang, Y., Li, X. and Jin, Y. (2009) Evaluation of electrospun fibrous scaffolds of poly(DL-lactide) and poly(ethylene glycol) for skin tissue engineering. *Materials Science and Engineering C, Biomimetic Materials, Sensors and Systems*, **29** (6), 1869–76.
22. Liu, Y., Ji, Y., Ghosh, K., Clark, R., Huang, L. and Rafailovichz, M. (2009) Effects of fiber orientation and diameter on the behavior of human dermal fibroblasts on electrospun PMMA scaffolds. *Journal of Biomedical Materials Research Part A*, **90A** (4), 1092–106.
23. Gupta, D., Venugopal, J., Prabhakaran, M., Dev, V., Low, S. and Choon, A. (2009) Aligned and random nanofibrous substrate for the in vitro culture of Schwann cells for neural tissue engineering. *Acta Biomaterialia*, **5** (7), 2560–9.
24. Prabhakaran, M., Venugopal, J., Chyan, T.T., Hai, L., Chan, C. and Lim, A. (2008) Electrospun biocomposite nanofibrous scaffolds for neural tissue engineering. *Tissue Engineering: Part A*, **14** (11), 1787–97.
25. Horne, M.K., Nisbet, D.R., Forsythe, J.S. and Parish, C. (2010) Three dimensional nanofibrous scaffolds incorporating

26 Nisbet, D.R., Pattanawong, S., Ritchie, N.E., Shen, W., Finkelstein, D.I., Horne, M.K. et al. (2007) Interaction of embryonic cortical neurons on nanofibrous scaffolds for neural tissue engineering. *Journal of Neural Engineering*, **4** (2), 35–41.

immobilized BDNF promote proliferation and differentiation of cortical neural stem cells. *Stem Cells and Development*, **June**, 843–52.

27 Mattanavee, W., Suwantong, O., Puthong, S., Bunaprasert, T., Hoven, V.P. and Supaphol, P. (2009) Immobilization of biomolecules on the surface of electrospun polycaprolactone fibrous scaffolds for tissue engineering. *ACS Applied Materials and Interfaces*, **1** (5), 1076–85.

28 Wang, Y.K., Yong, T. and Ramakrishna, S. (2005) Nanofibres and their influence on cells for tissue regeneration. *Australian Journal of Chemistry*, **58** (10), 704–12.

29 Beglou, M.J. and Haghi, A.K. (2008) Electrospun biodegradable and biocompatible natural nanofibers: a detailed review. *Cellulose Chemistry and Technology*, **42** (9-10), 441–62.

30 Kostakova, E., Meszaros, L. and Gregr, J. (2009) Composite nanofibers produced by modified needleless electrospinning. *Materials Letters*, **63** (28), 2419–22.

31 Yang, R., He, J., Xu, L. and Yu, J. (2009) Bubble-electrospinning for fabricating nanofibers. *Polymer*, **50** (24), 5846–50.

32 Ki, C.S., Baek, D.H., Gang, K.D., Lee, K.H., Um, I.C. and Park, Y.H. (2005) Characterization of gelatin nanofiber prepared from gelatin-formic acid solution. *Polymer*, **46** (14), 5094–102.

33 Zhao, L., He, C., Gao, Y., Cen, L., Cui, L. and Cao, Y. (2008) Preparation and cytocompatibility of PLGA scaffolds with controllable fiber morphology and diameter using electrospinning method. *Journal of Biomedical Materials Research Part B, Applied Biomaterials*, **87B** (1), 26–34.

34 Zuo, W.W., Zhu, M.F., Yang, W., Yu, H., Chen, Y.M. and Zhang, Y. (2005) Experimental study on relationship between jet instability and formation of beaded fibers during electrospinning. *Polymer Engineering and Science*, **45** (5), 704–9.

35 Tan, A., Ifkovits, J., Baker, B., Brey, D., Mauck, R. and Burdick, J. (2008) Electrospinning of photocrosslinked and degradable fibrous scaffolds. *Journal of Biomedical Materials Research Part A*, **87A** (4), 1034–43.

36 Beachley, V. and Wen, X. (2009) Effect of electrospinning parameters on the nanofiber diameter and length. *Materials Science and Engineering C, Biomimetic Materials, Sensors and Systems*, **29** (3), 663–8.

37 Chen, Z., Wei, B., Mo, X. and Cui, F. (2009) Diameter control of electrospun chitosan-collagen fibers. *Journal of Polymer Science Part B, Polymer Physics*, **47** (19), 1949–55.

38 Zhang, C.X., Yuan, X.Y., Wu, L.L., Han, Y. and Sheng, J. (2005) Study on morphology of electrospun poly(vinyl alcohol) mats. *European Polymer Journal*, **41** (3), 423–32.

39 Kwon, I.K., Kidoaki, S. and Matsuda, T. (2005) Electrospun nano- to microfiber fabrics made of biodegradable copolyesters: structural characteristics, mechanical properties and cell adhesion potential. *Biomaterials*, **26** (18), 3929–39.

40 Gupta, P., Elkins, C., Long, T.E. and Wilkes, G.L. (2005) Electrospinning of linear homopolymers of poly(methyl methacrylate): exploring relationships between fiber formation, viscosity, molecular weight and concentration in a good solvent. *Polymer*, **46** (13), 4799–810.

41 Kim, H.S., Kim, K., Jin, H.J. and Chin, I.J. (2005) Morphological characterization of electrospun nano-fibrous membranes of biodegradable poly(L-lactide) and poly(lactide-*co*-glycolide). *Macromolecular Symposia*, **224**, 145–54.

42 Jarusuwannapoom, T., Hongroijanawiwat, W., Jitjaicham, S., Wannatong, L., Nithitanakul, M., Pattamaprom, C. et al. (2005) Effect of solvents on electro-spinnability of polystyrene solutions and morphological appearance of resulting electrospun polystyrene fibers. *European Polymer Journal*, **41** (3), 409–21.

43 Meechaisue, C., Dubin, R., Supaphol, P., Hoven, V.P. and Kohn, J. (2006) Electrospun mat of tyrosine-derived polycarbonate fibers for potential use as tissue scaffolding material. *Journal of Biomaterials Science, Polymer Edition*, **17** (9), 1039–56.

44 Geng, X.Y., Kwon, O.H. and Jang, J.H. (2005) Electrospinning of chitosan dissolved in concentrated acetic acid solution. *Biomaterials*, **26** (27), 5427–32.

45 Dalton, P.D., Klee, D. and Moller, M. (2005) Electrospinning with dual collection rings. *Polymer*, **46** (3), 611–14.

46 Wang, H., Mullins, M., Cregg, J., Hurtado, A., Oudega, M. and Trombley, M. (2009) Creation of highly aligned electrospun poly-L-lactic acid fibers for nerve regeneration applications. *Journal of Neural Engineering*, **6** (1), article 016001.

47 Canty, A.J. and Murphy, M. (2008) Molecular mechanisms of axon guidance in the developing corticospinal tract. *Progress in Neurobiology*, **85** (2), 214–35.

48 Blizzard, C.A., Haas, M.A., Vickers, J.C. and Dickson, T.C. (2007) Cellular dynamics underlying regeneration of damaged axons differs from initial axon development. *European Journal of Neuroscience*, **26** (5), 1100–8.

49 Engler, A.J., Sen, S., Sweeney, H.L. and Discher, D.E. (2006) Matrix elasticity directs stem cell lineage specification. *Cell*, **126** (4), 677–89.

50 Nisbet, D.R., Pattanawong, S., Nunan, J., Shen, W., Horne, M.K., Finkelstein, D.I. et al. (2006) The effect of surface hydrophilicity on the behavior of embryonic cortical neurons. *Journal of Colloid and Interface Science*, **299** (2), 647–55.

51 Corey, J.M., Lin, D.Y., Mycek, K.B., Chen, Q., Samuel, S., Feldman, E.L. et al. (2007) Aligned electrospun nanofibers specify the direction of dorsal root ganglia neurite growth. *Journal of Biomedical Materials Research Part A*, **83A** (3), 636–45.

52 Ghasemi-Mobarakeh, L., Prabhakaran, M., Morshed, M., Nasr-Esfahani, M. and Ramakrishna, S. (2008) Electrospun poly(epsilon-caprolactone)/gelatin nanofibrous scaffolds for nerve tissue engineering. *Biomaterials*, **29** (34), 4532–9.

53 Li, G. and Hoffman-Kim, D. (2008) Tissue-engineered platforms of axon guidance. *Tissue Engineering Part B: Reviews*, **14** (1), 33–51.

54 Valmikinathan, C., Wang, J., Smiriglio, S., Golwala, N. and Yu, X. (2009) Magnetically induced protein gradients on electrospun nanofibers. *Combinatorial Chemistry and High-Throughput Screening*, **12** (7), 656–63.

55 Ngiam, M., Liao, S., Patil, A., Cheng, Z., Chan, C. and Ramakrishna, S. (2009) The fabrication of nano-hydroxyapatite on PLGA and PLGA/collagen nanofibrous composite scaffolds and their effects in osteoblastic behavior for bone tissue engineering. *Bone*, **45** (1), 4–16.

56 Jose, M., Thomas, V., Dean, D. and Nyairo, E. (2009) Fabrication and characterization of aligned nanofibrous PLGA/collagen blends as bone tissue scaffolds. *Polymer*, **50** (15), 3778–85.

57 Sisson, K., Zhang, C., Farach-Carson, M., Chase, D. and Rabolt, J. (2009) Evaluation of cross-linking methods for electrospun gelatin on cell growth and viability. *Biomacromolecules*, **10** (7), 1675–80.

58 Schiffman, J., Stulga, L. and Schauer, C. (2009) Chitin and chitosan: transformations due to the electrospinning process. *Polymer Engineering and Science*, **49** (10), 1918–28.

59 Bajgai, M., Aryal, S., Bhattarai, S., Bahadur, K., Kim, K. and Kim, H. (2008) Poly(epsilon-caprolactone) grafted dextran biodegradable electrospun matrix: a novel scaffold for tissue engineering. *Journal of Applied Polymer Science*, **108** (3), 1447–54.

60 Eichhorn, S.J. and Sampson, W.W. (2005) Statistical geometry of pores and statistics of porous nanofibrous assemblies. *Journal of the Royal Society, Interface*, **2** (4), 309–18.

61 Smith, M.J., Smith, D.C., White, K.L. and Bowlin, G.L. (2007) Immune response testing of electrospun polymers: an important consideration in the evaluation of biomaterials. *Journal of Engineered Fibers and Fabrics*, **2** (2), 41–7.

62 Sanders, J.E., Cassisi, D.V., Neumann, T., Golledge, S.L., Zachariah, S.G., Ratner, B.D. et al. (2003) Relative influence of polymer fiber diameter and surface charge on fibrous capsule thickness and vessel density for single-fiber implants. *Journal of Biomedical Materials Research. Part A: Early View*, **65** (4), 462–7.

63 Lin, K., Chua, K., Christopherson, G., Lim, S. and Mao, H. (2007) Reducing electrospun nanofiber diameter and variability using cationic amphiphiles. *Polymer*, **48** (21), 6384–94.

64 Chen, M., Patra, P., Warner, S. and Bhowmick, S. (2007) Role of fiber diameter in adhesion and proliferation of NIH 3T3 fibroblast on electrospun polycaprolactone scaffolds. *Tissue Engineering*, **13** (3), 579–87.

65 Moroni, L., Licht, R., de Boer, J., de Wijn, J. and van Blitterswijk, C. (2006) Fiber diameter and texture of electrospun PEOT/PBT scaffolds influence human mesenchymal stem cell proliferation and morphology, and the release of incorporated compounds. *Biomaterials*, **27** (28), 4911–22.

66 Secasanu, V., Giardina, C. and Wang, Y. (2009) A novel electrospinning target to improve the yield of uniaxially aligned fibers. *Biotechnology Progress*, **25** (4), 1169–75.

67 Meinel, A., Kubow, K., Klotzsch, E., Garcia-Fuentes, M., Smith, M. and Vogel, V. (2009) Optimization strategies for electrospun silk fibroin tissue engineering scaffolds. *Biomaterials*, **30** (17), 3058–67.

68 Wise, J., Yarin, A., Megaridis, C. and Cho, M. (2009) Chondrogenic differentiation of human mesenchymal stem cells on oriented nanofibrous scaffolds: engineering the superficial zone of articular cartilage. *Tissue Engineering: Part A*, **15** (4), 913–21.

69 Gomez, N., Lu, Y., Chen, S. and Schmidt, C. (2007) Immobilized nerve growth factor and microtopography have distinct effects on polarization versus axon elongation in hippocampal cells in culture. *Biomaterials*, **28** (2), 271–84.

70 Nisbet, D., Rodda, A., Horne, M., Forsythe, J. and Finkelstein, D. (2009) Neurite infiltration and cellular response to electrospun polycaprolactone scaffolds implanted into the brain. *Biomaterials*, **30** (27), 4573–80.

71 Yang, Y., Li, X., Cui, W., Zhou, S., Tan, R. and Wang, C. (2008) Structural stability and release profiles of proteins from core-shell poly (DL-lactide) ultrafine fibers prepared by emulsion electrospinning. *Journal of Biomedical Materials Research Part A*, **86A** (2), 374–85.

72 Zhang, J., Yang, D., Xu, F., Zhang, Z., Yin, R. and Nie, J. (2009) Electrospun core-shell structure nanofibers from homogeneous solution of poly(ethylene oxide)/chitosan. *Macromolecules*, **42** (14), 5278–84.

73 Dror, Y., Salalha, W., Avrahami, R., Zussman, E., Yarin, A.L., Dersch, R. et al. (2007) One-step production of polymeric microtubes by co-electrospinning. *Small*, **3** (6), 1064–73.

74 Zhao, Y., Cao, X. and Jiang, L. (2007) Bio-mimic multichannel microtubes by a facile method. *Journal of the American Chemical Society*, **129** (4), 764–5.

75 Yi, F. and Lavan, D. (2008) Poly(glycerol sebacate) nanofiber scaffolds by core/shell electrospinning. *Macromolecular Bioscience*, **8** (9), 803–6.

76 Yan, S., Xiaoqiang, L., Lianjiang, T., Chen, H. and Xiumei, M. (2009) Poly(L-lactide-*co*-epsilon-caprolactone) electrospun nanofibers for encapsulating and sustained releasing proteins. *Polymer*, **50**, 4212–19.

77 Lannutti, J., Reneker, D., Ma, T., Tomasko, D. and Farson, D.F. (2007) Electrospinning for tissue engineering scaffolds. *Materials Science and Engineering C – Biomimetic and Supramolecular Systems*, **27** (3), 504–9.

78 Nerurkar, N.L., Baker, B.M., Sen, S., Wible, E.E., Elliott, D.M. and Mauck, R.L. (2009) Nanofibrous biologic laminates replicate the form and function of the annulus fibrosus. *Nature Materials*, **8** (12), 986–92.

79 Wang, Y., Wang, B., Wang, G., Yin, T. and Yu, Q. (2009) A novel method for preparing electrospun fibers with nano-/micro-scale porous structures. *Polymer Bulletin*, **63** (2), 259–65.

80. Kim, T., Chung, H. and Park, T. (2008) Macroporous and nanofibrous hyaluronic acid/collagen hybrid scaffold fabricated by concurrent electrospinning and deposition/leaching of salt particles. *Acta Biomaterialia*, **4** (6), 1611–19.
81. Yang, Y., Zhu, X., Cui, W., Li, X. and Jin, Y. (2009) Electrospun composite mats of poly[(D,L-lactide)-*co*-glycolide] and collagen with high porosity as potential scaffolds for skin tissue engineering. *Macromolecular Materials and Engineering*, **294** (9), 611–19.
82. Zucchelli, A., Fabiani, D., Gualandi, C. and Focarete, M. (2009) An innovative and versatile approach to design highly porous, patterned, nanofibrous polymeric materials. *Journal of Materials Science*, **44** (18), 4969–75.
83. Kim, G., Park, J. and Park, S. (2007) Surface-treated and multilayered poly(epsilon-caprolactone) nanofiber webs exhibiting enhanced hydrophilicity. *Journal of Polymer Science Part B, Polymer Physics*, **45** (15), 2038–45.
84. Pham, Q., Sharma, U. and Mikos, A. (2006) Electrospun poly(epsilon-caprolactone) microfiber and multilayer nanofiber/microfiber scaffolds: characterization of scaffolds and measurement of cellular infiltration. *Biomacromolecules*, **7** (10), 2796–805.
85. Tzezana, R., Zussman, E. and Levenberg, S. (2008) A layered ultraporous scaffold for tissue engineering, created via a hydrospinning method. *Tissue Engineering Part C: Methods*, **14** (4), 281–8.
86. Wang, S., Zhang, Y., Wang, H., Yin, G. and Dong, Z. (2009) Fabrication and properties of the electrospun polylactide/silk fibroin-gelatin composite tubular scaffold. *Biomacromolecules*, **10** (8), 2240–4.
87. Jeong, S., Lee, A., Lee, Y. and Shin, H. (2008) Electrospun gelatin/poly(L-lactide-*co*-epsilon-caprolactone) nanofibers for mechanically functional tissue-engineering scaffolds. *Journal of Biomaterials Science Polymer Edition*, **19** (3), 339–57.
88. Sun, X.Y., Shankar, R., Borner, H.G., Ghosh, T.K. and Spontak, R.J. (2007) Field-driven biofunctionalization of polymer fiber surfaces during electrospinning. *Advanced Materials*, **19** (1), 87–91.
89. Jang, J., Castano, O. and Kim, H. (2009) Electrospun materials as potential platforms for bone tissue engineering. *Advanced Drug Delivery Reviews*, **61** (12), 1065–83.
90. Deng, X.L., Sui, G., Zhao, M.L., Chen, G.Q. and Yang, X.P. (2007) Poly(L-lactic acid)/hydroxyapatite hybrid nanofibrous scaffolds prepared by electrospinning. *Journal of Biomaterials Science, Polymer Edition*, **18** (1), 117–30.
91. Jun, I., Jeong, S. and Shin, H. (2009) The stimulation of myoblast differentiation by electrically conductive sub-micron fibers. *Biomaterials*, **30** (11), 2038–47.
92. Xie, J., MacEwan, M., Willerth, S., Li, X., Moran, D. and Sakiyama-Elbert, S. (2009) Conductive core-sheath nanofibers and their potential application in neural tissue engineering. *Advanced Functional Materials*, **19** (14), 2312–18.
93. Lee, J., Bashur, C., Goldstein, A. and Schmidt, C. (2009) Polypyrrole-coated electrospun PLGA nanofibers for neural tissue applications. *Biomaterials*, **30** (26), 4325–35.
94. Guimard, N.K., Gomez, N. and Schmidt, C.E. (2007) Conducting polymers in biomedical engineering. *Progress in Polymer Science*, **32**, 876–921.
95. Ateh, D.D., Navsaria, H.A. and Vadgama, P. (2006) Polypyrrole-based conducting polymers and interactions with biological tissues. *Journal of the Royal Society, Interface*, **3** (11), 741–52.
96. Li, X., Xie, J., Yuan, X. and Xia, Y. (2008) Coating electrospun poly(epsilon-caprolactone) fibers with gelatin and calcium phosphate and their use as biomimetic scaffolds for bone tissue engineering. *Langmuir*, **24** (24), 14145–50.
97. Dubas, S., Kittitheeranun, P., Rangkupan, R., Sanchavanakit, N. and Potiyaraj, P. (2009) Coating of polyelectrolyte multilayer thin films on nanofibrous scaffolds to improve cell adhesion. *Journal of Applied Polymer Science*, **114** (3), 1574–9.

98 Chen, F., Lee, C.N. and Teoh, S.H. (2007) Nanofibrous modification on ultra-thin poly(epsilon-caprolactone) membrane via electrospinning. *Materials Science and Engineering C – Biomimetic and Supramolecular Systems*, **27** (2), 325–32.

99 Nisbet, D., Yu, L., Zahir, T., Forsythe, J. and Shoichet, M. (2008) Characterization of neural stem cells on electrospun poly(epsilon-caprolactone) submicron scaffolds: evaluating their potential in neural tissue engineering. *Journal of Biomaterials Science, Polymer Edition*, **19** (5), 623–34.

100 Chen, J., Chu, B. and Hsiao, B.S. (2006) Mineralization of hydroxyapatite in electrospun nanofibrous poly(L-lactic acid) scaffolds. *Journal of Biomedical Materials Research. Part A: Early View*, **79** (2), 307–17.

101 Hermanson, G.T. (1996) *Bioconjugate Techniques*, Elsevier, USA.

102 Grafahrend, D., Calvet, J., Klinkhammer, K., Salber, J., Dalton, P. and Moller, M. (2008) Control of protein adsorption on functionalized electrospun fibers. *Biotechnology and Bioengineering*, **101** (3), 609–21.

103 Chua, K.N., Chai, C., Lee, P.C., Tang, Y.N., Ramakrishna, S., Leong, K.W. et al. (2006) Surface-aminated electrospun nanofibers enhance adhesion and expansion of human umbilical cord blood hematopoietic stem/progenitor cells. *Biomaterials*, **27** (36), 6043–51.

104 Fu, G.D., Xu, L.Q., Yao, F., Zhang, K., Wang, X.F., Zhu, M.F. et al. (2009) Smart nanofibers from combined living radical polymerization, "click chemistry" and electrospinning. *ACS Applied Materials and Interfaces*, **1** (2), 239–43.

105 Ye, J., Chen, Y.W., Zhou, W.H., Wang, X.F., Guo, Z.P. and Hu, Y.H. (2009) Preparation of Polymer@PbS hybrid nanofibers by surface-initiated atom transfer radical polymerization and acidolysis by H_2S. *Materials Letters*, **63** (16), 1425–7.

106 Zhu, A., Zhao, F. and Ma, T. (2009) Photo-initiated grafting of gelatin/N-maleic acyl-chitosan to enhance endothelial cell adhesion, proliferation and function on PLA surface. *Acta Biomaterialia*, **5** (6), 2033–44.

107 Cui, W., Li, X., Chen, J., Zhou, S. and Weng, J. (2008) In situ growth kinetics of hydroxyapatite on electrospun poly(DL-lactide) fibers with gelatin grafted. *Crystal Growth and Design*, **8** (12), 4576–82.

108 Ma, Z., Kotaki, M., Yong, T., He, W. and Ramakrishna, S. (2005) Surface engineering of electrospun polyethylene terephthalate (PET) nanofibers towards development of a new material for blood vessel engineering. *Biomaterials*, **26** (15), 2527–36.

109 Kuzuya, M., Sasai, Y., Yamauchi, Y. and Kondo, S.I. (2008) Pharmaceutical and biomedical engineering by plasma techniques. *Journal of Photopolymer Science and Technology*, **21** (6), 785–98.

110 Vasita, R., Shanmugam, K. and Katti, D.S. (2008) Improved biomaterials for tissue engineering applications: surface modification of polymers. *Current Topics in Medicinal Chemistry*, **8** (4), 341–53.

111 Yoo, H.S., Kim, T.G. and Park, T.G. (2009) Surface-functionalized electrospun nanofibers for tissue engineering and drug delivery. *Advanced Drug Delivery Reviews*, **61** (12), 1033–42.

112 Martins, A., Pinho, E., Faria, S., Pashkuleva, I., Marques, A. and Reis, R. (2009) Surface modification of electrospun polycaprolactone nanofiber meshes by plasma treatment to enhance biological performance. *Small*, **5** (10), 1195–206.

113 Baek, H., Park, Y., Ki, C., Park, J. and Rah, D. (2008) Enhanced chondrogenic responses of articular chondrocytes onto porous silk fibroin scaffolds treated with microwave-induced argon plasma. *Surface and Coatings Technology*, **202** (22-23), 5794–7.

114 Park, K., Ju, Y.M., Son, J.S., Ahn, K.D. and Han, D.K. (2007) Surface modification of biodegradable electrospun nanofiber scaffolds and their interaction with fibroblasts. *Journal of Biomaterials Science, Polymer Edition*, **18** (4), 369–82.

115 Shabani, I., Haddadi-Asl, V., Seyedjafari, E., Babaeijandaghi, F. and Soleimani, M.

(2009) Improved infiltration of stem cells on electrospun nanofibers. *Biochemical and Biophysical Research Communications*, **382** (1), 129–33.
116. Prabhakaran, M., Venugopal, J., Chan, C. and Ramakrishna, S. (2008) Surface modified electrospun nanofibrous scaffolds for nerve tissue engineering. *Nanotechnology*, **19** (45), article 455102.
117. Wang, Z.G., Wan, L.S. and Xu, Z.K. (2009) Immobilization of catalase on electrospun nanofibrous membranes modified with bovine serum albumin or collagen: coupling site-dependent activity and protein-dependent stability. *Soft Matter*, **5** (21), 4161–8.
118. Choi, J. and Yoo, H. (2007) Electrospun nanofibers surface-modified with fluorescent proteins. *Journal of Bioactive and Compatible Polymers*, **22** (5), 508–24.
119. Doran, M.R., Markway, B.D., Aird, I.A., Rowlands, A.S., George, P.A., Nielsen, L.K. *et al.* (2009) Surface-bound stem cell factor and the promotion of hematopoietic cell expansion. *Biomaterials*, **30** (25), 4047–52.
120. Santiago, L.Y., Nowak, R.W., Rubin, J.P. and Marra, K.G. (2006) Peptide-surface modification of poly(caprolactone) with laminin-derived sequences for adipose-derived stem cell applications. *Biomaterials*, **27** (15), 2962–9.
121. Woodrow, K., Wood, M., Saucier-Sawyer, J., Solbrig, C. and Saltzman, W. (2009) Biodegradable meshes printed with extracellular matrix proteins support micropatterned hepatocyte cultures. *Tissue Engineering: Part A*, **15** (5), 1169–79.
122. Li, W., Guo, Y., Wang, H., Shi, D., Liang, C. and Ye, Z. (2008) Electrospun nanofibers immobilized with collagen for neural stem cells culture. *Journal of Materials Science Materials in Medicine*, **19** (2), 847–54.
123. Duan, Y., Wang, Z., Yan, W., Wang, S., Zhang, S. and Jia, J. (2007) Preparation of collagen-coated electrospun nanofibers by remote plasma treatment and their biological properties. *Journal of Biomaterials Science, Polymer Edition*, **18** (9), 1153–64.
124. Ma, K., Chan, C., Liao, S., Hwang, W., Feng, Q. and Ramakrishna, S. (2008) Electrospun nanofiber scaffolds for rapid and rich capture of bone marrow-derived hematopoietic stem cells. *Biomaterials*, **29** (13), 2096–103.
125. Jia, J., Duan, Y., Yu, J. and Lu, J. (2008) Preparation and immobilization of soluble eggshell membrane protein on the electrospun nanofibers to enhance cell adhesion and growth. *Journal of Biomedical Materials Research Part A*, **86A** (2), 364–73.
126. Lu, Y., Jiang, H., Tu, K. and Wang, L. (2009) Mild immobilization of diverse macromolecular bioactive agents onto multifunctional fibrous membranes prepared by coaxial electrospinning. *Acta Biomaterialia*, **5** (5), 1562–74.
127. Park, Y.J., Kim, K.H., Lee, J.Y., Ku, Y., Lee, S.N., Min, B.M. *et al.* (2006) Immobilization of bone morphogenetic protein-2 on nanofibrous chitosan membrane for enhanced guided bone regeneration. *Biotechnology and Applied Biochemistry*, **43**, 17–24.
128. Moore, K., Macsween, M. and Shoichet, M. (2006) Immobilized concentration gradients of neurotrophic factors guide neurite outgrowth of primary neurons in macroporous scaffolds. *Tissue Engineering*, **12** (2), 267–78.
129. Nisbet, D.R., Bourne, J.A. and Forsythe, J.S. (2010) A commentary on neural tissue engineering in the central nervous system – Interfacing a lesion, in *Biomaterials Developments and Applications* (ed. H. Bourg and A. Lisle), Nova Science Publishers, pp. 453–463.

15
Controlling the Shape of Organic Nanostructures: Fabrication and Properties

Rabih O. Al-Kaysi and Christopher J. Bardeen

15.1
Introduction

During recent years, the majority of investigations in the field of nanotechnology have focused on the fabrication of nanostructures composed of metallic or inorganic components. Only recently have systematic investigations begun into the potential use of organic nanostructures. If the question were to be asked why inorganic and metallic elements have become the leading materials in nanotechnology, one explanation would lie in the differences between the chemical bondings that hold organic molecules, inorganic ions, or metal atoms together. An inorganic nanostructure is held together by strong covalent or ionic bonds; thus, uniform inorganic nanowires with controllable dimensions can be grown via chemical vapor deposition (CVD) methods on a surface containing inorganic nanoparticle seeds. This method is generally used for most inorganic materials that can be either vaporized or synthesized under reduced pressure. The growth of inorganic nanostructures, such as quantum dots (QDs) can also be size-controlled by using solution-based chemical capping methods. Hence, the fabrication of inorganic nanostructures is guided by the strong ionic or metallic bonds which glue the building blocks of the nanostructure together. These strong interatomic or interionic forces facilitate the production of different shapes of nanoparticles, such as nanoprisms [1], nanocubes [2], nanopyramids [3], and spindle shapes [4], that allow the subsequent exploration of shape-dependent chemistry. Moreover, the drive towards ever-smaller electronic circuits and processors has compelled research groups to focus on creating nanostructures from conducting and semiconducting materials that typically are either inorganic or metallic in nature. Finally, these types of material are robust with respect to the electron microscopy techniques that are widely used to characterize nanostructures. Soft materials, such as organic molecules composed of carbon and hydrogen atoms, are much more prone to electron beam-induced charging and damage. Thus, even if an organic nanostructure can be successfully created, it may not be easy to characterize using standard electron microscopy methods.

Nanomaterials for the Life Sciences Vol. 10: Polymeric Nanomaterials. Edited by Challa S. S. R. Kumar
Copyright © 2011 WILEY-VCH Verlag GmbH & Co. KGaA, Weinheim
ISBN: 978-3-527-32170-4

Covalently bonded organic nanostructures such as carbon nanotubes (CNTs) [5], graphene sheets [6, 7], fullerenes [8], and nanodiamonds [9], comprise a class of carbon-based nanostructures that are robust and have undergone extensive investigation. These materials are typically produced via a high-temperature electric discharge, where the strong covalent bond acts as a template for seeding further C–C bonds. Unfortunately, in many cases a mixture of structures results (an example being the case of CNTs), and a formidable task ensues to separate the different types of nanostructure [10, 11]. In recent years, these different types of carbon allotrope have been the subject of much research effort, especially with regards to their novel electronic properties. Moreover, as this research has attracted widespread attention, such covalent nanostructures will not form the subject of this chapter. Likewise, no attention will be paid to the study of single polymer molecules, which may be thought of as covalent organic nanowires.

Instead, attention will be focused on nanostructures composed of many organic molecules which are held together by noncovalent forces such as van der Waals, dipole–dipole, or hydrogen bonds. These forces are typically one to two orders of magnitude weaker than the bonding interactions that hold inorganic covalent or ionic solids together. The weak intermolecular interactions in these organic solids have two important consequences:

- These "soft" materials are more difficult to characterize, as they are more easily damaged or deformed by standard methods such as high-voltage electron microscopy.

- Controlling the growth of these materials (either by using covalent capping groups or by size-selected catalytic growth – two strategies commonly used in the inorganic field) is not feasible as there is no covalent "handle" on the assembly of stable molecules.

Consequently, the controlled growth of noncovalent organic nanostructures presents a series of unique challenges to the materials chemist.

In this chapter, the various techniques used to prepare organic nanostructures with well-defined sizes and shapes are summarized, and some examples given. The dominant theme will be the quest to prepare organic nanostructures, and to identify the packing properties of the molecules within the nanostructure, whether single crystals, polycrystalline, or amorphous. In lacking the covalent bonding of inorganic materials to drive crystal formation, organic molecules may have several energetically similar options when assembled into a solid. These issues are highlighted in a discussion of the methods used to prepare organic nanostructures, such as mechanical milling, vacuum sublimation, supramolecular self-assembly, reprecipitation, and emulsification. Attention is also focused on the use of hard templates, such as anodic aluminum oxide (AAO), to fix the size and shape of one-dimensional (1-D) nanostructures made from organic molecules, including nanowires, nanotubes or nanorods. To date, AAO templates appears to offer the simplest method for fabricating 1-D organic nanostructures. The novel properties observed in some organic nanostructures will also be discussed, and some applications for organic nanostructures highlighted. Clearly, novel applications provide motivation for the further development of synthetic methods in this field. Finally, the current challenges in the field will be summarized, and suggestions made for possible future research.

15.2
Milling, Soft-Templating, and Other Methods for Preparing Organic Nanostructures

15.2.1
Mechanical Milling

Perhaps the simplest way to reduce the size of organic bulk crystals is to use a mortar and pestle; indeed, many chemists have experiences grinding an organic compound with KBr to prepare pellets for infrared measurements. Whilst, in general, manual grinding can produce only crystallites with micron dimensions, mechanical mills and attrition devices are available that incorporate smooth metallic spheres that can be shaken with the material to be ground at very high speed and amplitude [12]. Unfortunately, mechanical milling is not without its side effects on the crystal structure of the milled material, with unwanted polymorphs and polycrystalline inorganic nanoparticles being produced during mechanical milling due to the high stresses impacting on the crystalline phase. The mechanical milling of organic materials typically yields particle sizes on the order of 1 μm and above [13], as well as a broad particle size distribution and polycrystallinity. Despite these problems, mechanical milling represents one of the few options available for creating organic hybrid and cocrystal particles [14–16]. Mechanochemistry effects can also lead to bond breaking and chemical transformation of the starting material [17]. Finally, the absence of a solvent from a milled batch of organic nanoparticles permits the latter to aggregate together via noncovalent intermolecular forces; this is similar to what happens in reprecipitated organic nanoparticles when they fall out of aqueous suspension. It may also be possible to grind the organic material in the presence of a surfactant, which acts as a coating and prevents the nanoparticles from sticking together.

15.2.2
Vapor Growth

Many attempts to grow 1-D nanostructures via vapor deposition (or sublimation) techniques failed to yield homogeneous nanowires, but rather led to the creation of platelets, whiskers, or any other shape predetermined by the nature of the crystal packing of the organic material [18]. Usually, the resultant samples contained a broad size distribution of crystallites, as the growth proceeded in a statistical manner. There are some exceptions where organic nanostructures can be grown using vapor deposition or via a high-temperature solid-phase reaction. For example, 1-D organic nanorods (average diameter 40 nm, and several microns long) were synthesized by using a metal nanoparticle to catalyze nanorod growth at elevated temperatures [19, 20]. Several groups have shown that the vacuum evaporation of a variety of conjugated molecules onto single crystal surfaces can result in arrays of nanowires [21–29]. The deposition of a solution of a molecule of interest onto a surface, followed by rapid solvent evaporation, has also been shown to produce nanostructures for some molecules [30, 31]. In all these cases, however, obtaining a high yield of uniform nanostructures was seen to depend on a somewhat fortuitous combination of having the correct molecule and the correct surface, and consequently such a method would be difficult to generalize. A more general approach would be to allow the particles to form directly in the vapor phase, and to measure their properties while suspended in a vacuum. In fact, this method has been used extensively to characterize organic aerosols for atmospheric sciences, and has also been adapted to examine the size-scaling of material properties [32].

15.2.3
Supramolecular Self-Assembly

One-dimensional organic nanostructures have been grown by employing the principles of supramolecular chemistry and noncovalent interactions [33]. An organic aromatic molecule designed to have both hydrophobic and hydrophilic ends can be precipitated from water to form very long, self-assembled nanowires [34, 35]. Such nanowires can also be created via $\pi-\pi$ interaction and the stacking of small-molecule semiconductors and conducting polymers [36–38]. The molecular packing within these nanowires often differs from the crystal structure of the bulk material. Another example of supramolecular self-assembly was the observation that intramolecular charge-transfer molecules containing both electron-donating and -accepting moieties, can be reprecipitated from aqueous solution to form nanoribbons, nanotubes, or even nanowires [39, 40]. Unfortunately, this method is limited to very few molecules, as the slightest variation in chemical structure yields different, and unpredictable, supramolecular shapes. The supramolecular assembly approach relies on the design of molecules with functional groups that enhance packing. By its very nature, the molecular structure determines the shape of the resulting nanostructure, which in turn makes it difficult to generate different shapes using the same molecule. A second issue here is that many supramolecular assemblies are relatively disordered in solution, and lack the long-range order inherent in solid-state crystalline materials.

15.2.4
Reprecipitation

Whilst self-assembly is facilitated by certain design features of the molecules, most molecules are able to self-assemble in the sense that they crystallize. Consequently, the more general method of reprecipitation has been widely used to create organic nanoparticles [13, 41–51]. For this, a dilute solution (~0.001 M) of an organic com-

15.2 Milling, Soft-Templating, and Other Methods for Preparing Organic Nanostructures

pound in a water-miscible organic solvent [e.g., ethanol or tetrahydrofuran (THF)] is injected rapidly in a large volume of stirring water. A rapid precipitation of the water-insoluble organic compound results in a suspension of particles in the water, the size of which can be tuned to a degree by varying the concentration of the injected organic solution and the temperature of the liquid [52]. This method can be applied to a wide variety of organic compounds that are very poorly soluble in water. An added benefit is that the suspension of organic nanoparticles is easy to handle, which facilitates measurement of their photophysical properties using standard methods such as absorption and fluorescence for liquid solutions. The same principle can be applied to any material where the solute has a low aqueous solubility, including polymers [53].

Whilst many reports have been made describing the use of reprecipitation to form spherical (zero-dimensional) organic nanoparticles, a variety of 1-D shapes (rods, tubes, ribbons) have also been produced in this way [54, 55]. In many cases, a degree of control over the size or morphology of the nanostructures can be obtained simply by varying the preparation conditions, and this in turn provides a means of studying their size-dependent optical properties [44, 56–61]. In most cases, the physical origin of these effects is unclear, as conjugated organic molecules would not be expected to be capable of supporting large-radius Bohr excitons that would exhibit true quantum confinement effects [62, 63].

Subsequently, the reprecipitation method has been altered in many ways. For example, the addition of a surfactant can lead to the reprecipitated organic nanoparticles being stabilized, together with a degree of size control [64, 65]. The direct condensation of organic compounds in the gaseous state in a surfactant/water solution can also be used to produce organic nanoparticle suspensions that are stable and can be concentrated without aggregation [66]. Unfortunately, the reprecipitation process may be disturbed if a probe sonicator is used to grind the reprecipitated organic nanoparticles in order to generate smaller particle sizes and a more uniform distribution [67]. Laser absorption may also lead to an effective milling action on organic microcrystals suspended in a liquid, leading to nanocrystals [68]. Finally, the mixing of two solvents through a microporous membrane ("membrane-mixing") has been recognized as a way to utilize the different solubilities when manufacturing perylene nanoparticles [50].

Unfortunately, the reprecipitation method has several disadvantages in its use, the primary problem being that the nanoparticle suspension is unstable and the particles will eventually aggregate to form larger-sized chunks that settle out of solution. In some cases, the nanoparticle suspension may be very stable due to a high surface zeta-charge, and the reprecipitated particles will tend to be spherical in shape with a variable size distribution. However, it is equally likely that the nanoparticles will have different shapes and tend to be polycrystalline. One frequent problem with reprecipitation is that only minute quantities of organic nanoparticles are required to be suspended in large quantities of water, to minimize aggregation. Whilst the method does not generally yield monocrystalline dispersions, there are exceptions where the reprecipitated particles are crystalline, but micron-sized [69].

15.2.5
Electrospinning

One very versatile method for preparing polymer nanowires is that of electrospinning, where an electrostatic field is used to draw a polymer solution rapidly through a narrow nozzle. The success of the technique is based on the unique properties of polymers; for example, when a polymer is soluble in organic solvents, electrospinning can produce ultra-long polymer nanowires with thicknesses that depend on the applied potential [70–73]. Recently, electrospinning has been used to grow nanowires from a variety of common polymers such as polypyrrole and polystyrene. In fact, polymer nanowires with diameters as small as 10 nm have been obtained by the careful control of parameters such as the electric field strength, the polymer concentration, and the temperature. Unfortunately, electrospinning suffers from one problem in that it results in long, tangled masses of nanowires, rather than oriented or dispersed samples; however, the nanowires can often be shortened by the application of ultrasound.

15.2.6
Emulsification

Perhaps the most commercially successful method used to date to prepare organic nanoparticles is the use of stable emulsions to create polymer nanospheres with tunable diameters and narrow size distributions. This method has been applied successfully to a variety of polymers, including poly(methyl methacrylate) (PMMA) [74] and conjugated polymers [75–77]. For stable polymers, the nanoparticle surface can be functionalized with a variety of chemically active pendant groups.

The size of the nanospheres created is determined by the reaction temperature and the amount of initiator injected, with sizes ranging from 5 nm to many microns. This method has also been successfully extended to the synthesis of nanoparticles composed of small molecules [78, 79]. One disadvantage of the emulsion method – and, indeed, of any method that relies on surfactants – is that the resulting structures will never consist of a single component, but rather will possess a coating of the surfactant molecule. For some applications – particularly electronic transport measurements – this contaminant may have a major influence on the measured properties. The nanoparticles produced in this manner are also often amorphous or of low crystallinity.

15.3
Hard-Templating Methods for Preparing Organic Nanostructures

The idea of using a structured template to direct crystal growth is not new, with porous Vycor glass having in the past been used as a medium for growing organic crystals and for their isolation [80]. More recently, Ibanez and coworkers showed that porous sol–gel glasses could be used as a medium in which different types of molecular solids can be grown [81, 82]. Since, in such porous glasses, there always exists a wide distribution of pore sizes and shapes, it became clear that the field of nanomaterials would benefit greatly from a wide availability of templates of uniform shape and size. In fact, such templates became freely available about twenty years ago, in the form of porous AAO, and have subsequently become the basis for the creation of nanowires and nanotubes composed of a huge variety of materials [83, 84].

15.3.1
AAO Templates

Metals such as titanium, zirconium, and aluminum form ordered and evenly spaced metal oxide nanopores (in the case of Al) or nanotubes (in the case of Ti) on the surface of the metal when anodized in an aqueous acidic solution. Previously, AAO membranes have been prepared via a two-step anodization of pure aluminum (99.999%) in an acidic solution [85–87]. In this case, the aluminum is first annealed in a vacuum oven at ~600 °C and then thoroughly degreased and electropolished. Several factors will determine the size, separation, and length of the honeycomb-structured nanopores created: typically, the time of anodization will determine the thickness of the membrane, while the voltage and type of electrolyte will determine the pore size and separation. In fact, the pore size can be custom-fabricated from less than 7 nm to more than 300 nm, and all sizes in between. Varying the potential during the anodization process will result in novel nanostructured AAO templates with a regular periodic hierarchal branching [88]. Likewise, AAO templates with cone-shaped nanopores were successfully fabricated on small substrates [89]. Although, today, it has become common practice to prepare custom AAO templates in the laboratory, Whatman Inc. can provide commercial AAO membrane filters (Anodisc-13) with an average pore diameter of 200 nm, a 60 μm-thick membrane, and a pore density on the order of 10^9 pores per cm^2. (It should be noted that, for Whatman templates, the quoted pore diameters of 100 nm and 20 nm exist only for the first 2 μm of the top layer of the membrane; for the remaining length, the pores are about 200 nm in diameter.) Nonetheless, these readily available and relatively inexpensive templates are excellent for producing 1-D nanostructures with a relatively narrow diameter distribution (±30 nm). If required, AAO membranes with smaller and more uniform pore diameters and template thicknesses are available from specialized companies such as Synkera (www.synkera.com), but these are more expensive. However, unlike the Whatman templates the Synkera templates are transparent and more rigid.

Before organic materials are grown in them, the templates are usually ultrasonicated with solvents of varying polarity (water, acetone, ethanol, THF, chloroform) to remove any industrial contaminants and to unclog the pores. The alumina surface of the nanopores can be activated by drying the template in a vacuum oven at a temperature close to 240 °C. Further washing of the template with a high-purity solvent, such as THF, by placing it over a glass frit and applying suction, is usually sufficient to remove organic contaminants.

15.3.2
Template-Assisted Synthesis of Polymer Nanowires, Nanotubes, and Nanorods

With several reviews having been produced on the use of AAO templates to synthesize polymer nanowires or nanotubes [90, 91], it is considered "easy" to create 1-D polymer nanostructures that are flexible and can withstand the rigor of fabrication. Previously, 1-D polymer nanostructures were synthesized using a template wetting technique, whereby a solution of the polymer was passed through the AAO membrane. Capillary action then forced the polymer solution to spread evenly on the inside pores of the AAO template. As the solvent evaporated, a layer of polymer was deposited on the inside walls of the AAO pores, with the thickness of the nanotube depending on the concentration of the polymer solution. Typically, the more concentrated the solution, the thicker was the nanotube wall deposited. Unfortunately, this method is limited to polymers that are soluble in organic solvents, and to polymer solutions that have a low viscosity that allows them to flow through the AAO nanopores. The first attempt to fabricate 1-D polymer nanostructures was made by Martin and coworkers, who prepared polymer nanotubes by running polymer solutions through AAO templates [92]. Subsequently, other groups have used this technique to prepare polymer nanotubes from a variety of polymers [93–95]. In this case, the polymer nanotube's external diameter was equal to that of the AAO nanopore diameter, while the length was equal to the thickness of the template. The nanotubular morphology was confirmed using electron microscopy. Similar techniques based on wetting template methods were used to synthesize arrays of polymer nanotubes made from conductive polymers, ferroelectric polymers [96], semiconducting polymers, nonpolar polymers [97], poly(vinyl pyrrolidone) [98], and magnetic molecularly imprinted polymers [99]. Insulating polymers such as polyamide 66 were used to make nanotubes for coating Pt nanowires, thus forming nanocables [100–102].

An alternative method, in which the AAO template is wetted with a molten polymer, is limited to polymers that have a low melting point or are able to flow through nanopores. If an amorphous polymer is used, it must be heated well above its glass transition temperature, or above the melting point in the case of a partially crystalline polymer. Polymer nanotubes made from polytetrafluoroethylene (PTFE), polystyrene, or PMMA were successfully fabricated following the above-mentioned method. The nanopore size and period of wetting determined whether a nanotube or a nanowire was obtained. For a 200 nm AAO template, brief wetting resulted in polymer nanotubes with wall thickness between 10 and 20 nm, while extended wetting yielded nanowires. For smaller-sized AAO nanopores, wetting the template with molten polymer for brief periods of time produced nanowires [103, 104]. Therefore, nanotube or nanowire formation is a function of the AAO template pore size, and the smaller the nanopore size the greater the chances are for nanowire formation. Confinement effects on block copolymers introduced via the melt method can lead to different mesoscale ordering [105].

In situ polymerization offers yet another method of forming polymer nanowires or nanotubes [106]. As an example, Martin and coworkers prepared polyacetylene fibers by chemically reacting acetylene precursors directly inside the template channels [107]. Simple heating can lead to the creation of conjugated polymer nanotubes or nanorods within the templates [108]. An interesting example of this is the formation of graphene nanorods by the carbonization of molecular precursors introduced into AAO pores [109]. Polypyrrole nanotubes with a 60 nm external diameter were synthesized via an alternating current electropolymerization of a pyrrole solution [110]. It was also revealed that the formation of nanotubes or nanowires through electropolymerization depends on the potential as well as the monomer concentration. This method has also been used to produce oligopyrene nanowires [111]. The monomer 3,4-ethylenedioxythiophene has been electropolymerized into poly(3,4-ethylenedioxythiophene) (PEDOT) nanowires and nanotubes; in this case, the formation of PEDOT nanotubes were grown at a potential lower than 1.4 V (versus SCE), regardless of the initial concentration of the monomer [112].

15.3.3
Template-Assisted Synthesis of Small-Molecule Nanowires, Nanotubes, and Nanorods

15.3.3.1 Solution-Based Template Wetting Method
When an AAO template is dipped in a solution of the small molecule, the capillary action of the alumina nanopores will draw in the solution rapidly. Then, when the solvent evaporates, the solute is deposited on the nanopore walls because the high surface energy of the pores is stronger than the intermolecular forces; thus, a nanotube is formed. This approach works not only for organic molecules but also for organosilanes such as hexaphenylsilole [113] and organometallic complexes such as the (porphryinato)(phthalocyaninato)europium complex [114]. Several small molecules were utilized to fabricate nanotubes or nanowires via solution-based template wetting. This process has been used to make nanotubes composed of C_{60} [115] and perylene [116], while inorganic–organic hybrid core–shell structures can also be produced using this approach. When pyrene nanotubes (100 nm diameter) were synthesized by the dip-dry method, the formed nanotubes had an evenly deposited thick pyrene coating. These nanotubes could serve as a coating for an inorganic material, such as AgI, that can be deposited inside them [117]. Nanotubes and nanowires composed of chiral quinidine molecules have also been investigated [118].

15.3.3.2 Melting–Recrystallization Template Wetting Method
If, in some cases, small organic molecules have very low solubility in organic solvents, the alternative to solution-based template wetting is to melt the organic molecule and to wet the AAO template with the molten material. This method can also be extended to soluble organic small molecules. A molten sample of dibenzoylmethane (DBM), a small organic molecule with a relatively low melting point (80 °C), was prepared and a 200 nm-pore diameter AAO template then dipped into the molten compound, such that the organic liquid immediately wetted the

nanopore walls. It was observed that, the longer the immersion time, the thicker the deposited organic layer on the nanopore walls; ultimately, a tube wall thickness from 20 to 70 nm was observed with increasing immersion time from 10 to 120 min, respectively. However, complete filling of the nanopores was prevented due to the strong adhesive forces between the AAO walls and the organic molecules; in fact, these forces proved to be stronger than the cohesive forces between the molecules themselves. An X-ray diffraction (XRD) analysis of the DBM nanotubes inside the template revealed that the DBM molecules had arranged themselves preferentially along the (121) plane, while the fluorescence emission from the nanotubes was seen to be 4.6-fold greater than that from the bulk [119]. A similar method has been used to form nanotubes of pentacene, an important polycyclic aromatic hydrocarbon used in organic semiconductor technology [120].

15.3.3.3 Sublimation Method

Another means of introducing small molecules inside AAO templates is by sublimation, whereby the sublimed small molecules percolate through the nanochannels, where they condense. Previously, this method was used to prepare pyrene and 1,4-bis[2-(5-phenyloxazolyl)]benzene (POPOP) nanowires (20, 70 and 130 nm in diameter and several micrometers long) [121]. These nanowires were most likely polycrystalline in nature, although no such evidence was presented. The fluorescence spectra of the nanowires indicated a dependence on the diameter of the nanowire. Thus, the 20 nm-diameter nanowires showed monomer-like fluorescence, most likely due to an inability of the molecules at the surface to form pairs. The ratio of unpaired molecules at the surface was greater than those paired in the middle of the nanowire. Large-diameter nanowires showed fluorescence similar to that of bulk crystals. Although, unfortunately, these authors did not elaborate on the experimental conditions for preparing these nanowires, the present authors' experience suggests that subliming small molecules (e.g., anthracene, naphthalene, POPOP or pyrene) close to the AAO template results in an immediate clogging of the AAO template pores and the growth of crystals on the surface. It is possible that the aforementioned authors used heated AAO templates, or sophisticated vacuum techniques. Notably, the sublimation method is limited to small organic molecules that are sufficiently stable to be sublimed, whereas most small organic molecules either melt or decompose at elevated temperatures.

15.3.3.4 Electrophoretic Deposition Method

Whilst both solution and melting recrystallization methods yield nanotubes and nanowires, depending on the processing conditions, electrophoretic deposition represents an alternative approach to preparing small-molecule nanowires or nanotubes. When one face of an AAO template was coated with a thin film of Au and affixed to an anode, the bottom side of the pores becomes positively charged, and will attract small organic molecules with a high electron density. An example of this procedure is when a 0.1 M solution of pyrene was prepared in an electrolytic solution of boron trifluoride diethyl etherate (BFEE). When the coated AAO template was used as a working electrode, the pyrene units were electro-oligomerized, thus forming a continuous nanowire; in this way, nanowires of oligopyrene that were 200 nm thick and 60 µm long (average number of repeat units = 6) were synthesized. The length of the nanowire could be controlled by the time of electrodeposition. The resulting nanowires were polycrystalline and exhibited a multicolored fluorescence emission, depending on the degree of conjugation between the pyrene units in each oligomer [111]. Electrophoretic deposition was also used to prepare charge-transfer organic salt nanotubes of bis(ethylenedithio)tetrathiafulvalene (BEDT-TTF) [91], and also nanowires composed of a perylenediimide derivative. In the latter case the length of the nanowires was proportional to the electrophoretic deposition time, while photoconductivity measurements on the aligned wires showed an order of magnitude more photosensitivity than that for cast films [122].

15.3.3.5 Solvent-annealing Method

Among the different 1-D small-molecule organic nanostructures prepared, nanotubular structures are recognized as being thermodynamically favored over a fully packed nanowire, due to the strong surface area energy of the AAO nanopores. These 1-D nanostructures tend to be polycrystalline, amorphous, or a mixture of the two. The polycrystalline nature of the organic nanostructures contrasts with that of their inorganic counterparts, since the latter are highly crystalline and, in many cases, the whole 1-D structure is a single crystal. Thus, a worthwhile goal might be to identify a general method for preparing a highly crystalline organic 1-D nanostructure using AAO templates. It is recognized that to deposit organic molecules inside the AAO pores is a straightforward procedure; likewise, it is known that, given the opportunity, the molecules inside the pores would arrange themselves preferentially into a low-energy crystalline packing motif. Thus, the problem is how to provide molecules with an opportunity to escape from their higher-energy amorphous or polycrystalline packing arrangement, and to reorder as a single crystal. Yet, this is simply the process of annealing a crystal, which is achieved using elevated temperatures and is facilitated by the presence of solvent molecules that can help to disassemble (dissolve) the initially formed solid.

The general procedure of this is outlined in Figure 15.1a [123]. In practical terms, an AAO template (Whatman-13, 200 nm, 25 mm diameter) has sufficient pores to accommodate a maximum of 6–9 mg of organic material. The AAO membrane is suspended using a specially designed Teflon holder that holds the AAO template from the edge, inside a standard 16 oz (ca. 500 ml) jar. About 0.2 ml of a concentrated solution of the organic small molecule (100–200 mg ml^{-1} solvent) is placed on the surface of the template (in the present authors' laboratory, THF was most often used as it is a good solvent for most organic small molecules). Argon gas was introduced into the jar at a rate of 1 bubble every 8–10 s. Room-temperature solvent vapor annealing causes the deposited organic layer on the surface of the AAO template to dissolve in the solvent vapors, while capillary action inside the template prevents the solution from dripping. The organic compound slowly crystallizes in the nanopores as the solvent evaporates (~48 h). The annealed

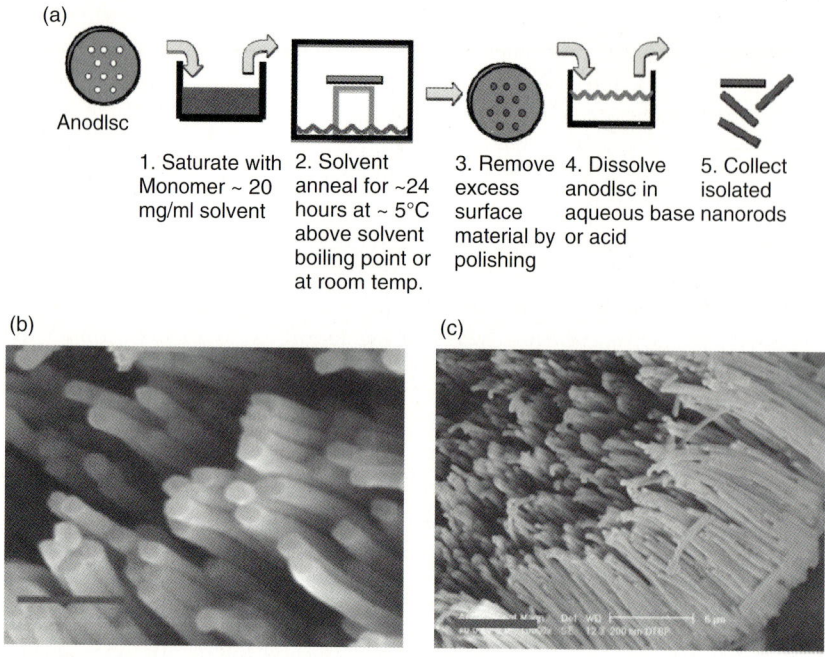

Figure 15.1 (a) Steps for fabricating solvent-annealed molecular crystal nanorods; (b) Scanning electron microscopy images of molecular crystal nanorods made from a 2,7-di-*tert*-butylpyrene (scale bar = 500 nm); (c) SEM of the same material (scale bar = 5 μm). Reproduced with permission from Ref. [123]; © 2006, Royal Society of Chemistry.

AAO template surface was polished using 2000-grit sandpaper to remove excess organic solid on the template surfaces. Finally, the template was etched out using phosphoric or aqueous sodium hydroxide, liberating the nanorods. It should be noted that this general procedure can vary depending on the molecule, and in many cases an even simpler experimental set-up will suffice. To prepare 9-anthracene carboxylic acid (9AC) nanorods, for example, the loaded template can be suspended in a bell-jar that is inverted on a glass plate, with a solvent-soaked paper tissue (Kimwipe) placed at the bottom. Slow evaporation of the solvent (in this case, THF) at room temperature over the course of 24h is usually sufficient to grow crystalline nanorods of this molecule.

Several small organic molecules were used to prepare highly crystalline organic nanowires (which are usually referred to as nanorods); 2,7-di-*tert*-butylpyrene was used to prepare the first sample of crystalline nanorods (Figure 15.1b,c). A modified solvent-annealing procedure was used where, instead of room-temperature solvent annealing, the jar/AAO template were placed in an oven for two days. Again, the solvent used was THF (boiling point 65 °C), and the oven temperature was set at 3 °C above the boiling point of THF. This procedure yielded highly

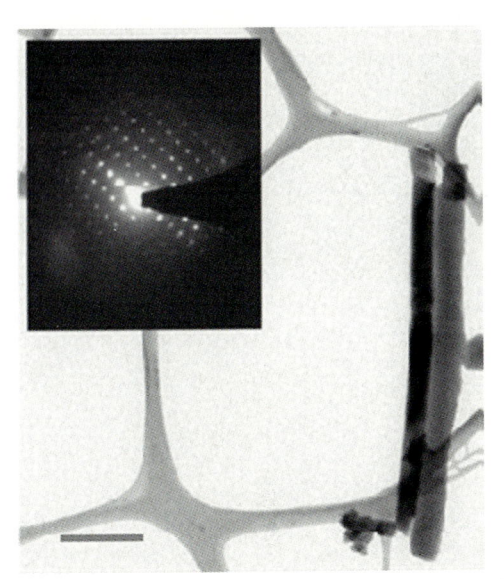

Figure 15.2 Transmission electron microscopy dark-field image of 2,7-di-tert-butylpyrene nanorods. The lack of grain boundaries that would lead to a speckled appearance of the rods is an indication of their high crystallinity. Scale bar = 500 nm. Inset: Electron diffraction pattern of the nanorod showing well-defined pattern with no diffuse halo that would indicate amorphous or polycrystalline regions. Reproduced with permission from Ref. [123]; © 2006, Royal Society of Chemistry.

crystalline nanorods. The evidence was established by well-defined spots in the electron diffraction pattern from a single nanorod, and by dark-field experiments that revealed the nanorod to be single crystal with over a minimum of 15 μm length (Figure 15.2).

Other compounds used to prepare nanorods included m-terphenyl and 9-cyanoanthracene. In the case of m-terphenyl, a dramatic effect of solvent annealing was apparent from "before" and "after" scanning electron microscopy (SEM) images (see Figure 15.3). After depositing the m-terphenyl into the AAO pores, the typical case appears where the molecules have formed open-ended tubes with irregular surface features; however, after solvent annealing, solid nanorods with smooth, uniform surfaces are visible. The annealing and crystal growth process most likely proceeds in a hierarchical fashion, where the initially formed microcrystallites dissolve and then reform more oriented crystals, which then fuse together. Evidence of this process was seen in nanorods examined before the annealing has been completed, where small, disk-like structures, intermediate between the polycrystalline starting material and the single crystal, were observed [123].

The ability to grow high-quality, single-crystal organic nanorods has resulted in the observation of novel phenomena, perhaps the most dramatic of which is the ability of these nanoscale crystals to survive chemical transformations that would

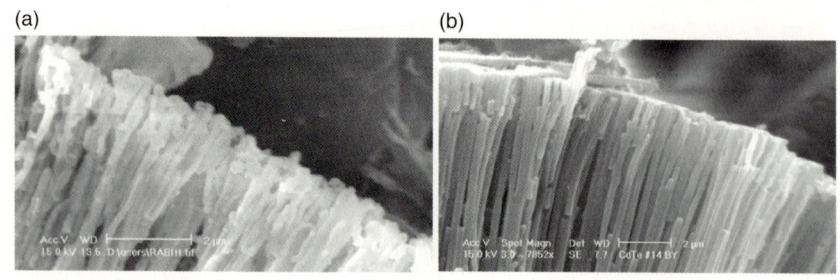

Figure 15.3 Scanning electron microscopy images of (a) *m*-terphenyl nanotubes before annealing, and (b) crystalline nanorods after solvent annealing in THF. Reproduced with permission from Ref. [123]; © 2006, Royal Society of Chemistry.

destroy larger crystals. This is vividly illustrated for 1-D organic nanostructures composed of photoreactive anthracene ester derivatives. Highly crystalline nanorods of anthracene-9-carboxylic acid *tert*-butyl ester (9TBAE) [124], prepared using the solvent-annealing method, exhibit unique photomechanical properties. When large micron-sized crystals of 9TBAE were irradiated with ultraviolet (UV) light (365 nm), the molecules photodimerized via a [4+4] photocycloaddition to form the anti- or "head-to-tail" photodimer. The photochemical reaction produced internal stress that resulted in crystal disintegration. However, when 200 nm nanorods of 9TBAE were irradiated, a clean crystal–crystal transformation was observed with an anisotropic expansion along the long axis. The single-crystal transformation was confirmed using transmission electron microscopy (TEM) and by the electron diffraction pattern that showed discrete spots instead of diffuse concentric circles, typical of polycrystalline nanorods. Whilst irradiation results in the fragmentation of bulk crystals, the 200 nm-diameter nanorods proved to be very robust under a variety of irradiation conditions, even after undergoing a very large (15%) expansion upon exposure to 365 nm light, as measured using atomic force microscopy (AFM) (Figure 15.4) and fluorescence microscopy.

The ability to photoreact crystalline nanorods has opened up a novel route to the production of crystalline polymer nanorods, as demonstrated by Al-Kaysi *et al.* [125]. When the photopolymerization of a photoactive monomer was achieved inside the AAO templates, the first step was to load the AAO template with a concentrated solution of the monomer (in this case, 9AC-ME). The solvent was then slowly evaporated at elevated temperatures, to yield the highly crystalline monomer nanorods. The A–A-type monomer units were aligned in an alternating face-to-face sandwich packing at a distance less than 4 . Subsequent irradiation of the loaded AAO template with UV light caused photopolymerization of the monomer to give regioregular poly(9AC-ME). The polymer nanorods retained the crystalline order of the monomer to yield a very rigid, highly insoluble, highly crystalline polymer. In this way, polymer nanorods were synthesized with diameters ranging from 35 to 200 nm (Figure 15.5).

15.3 Hard-Templating Methods for Preparing Organic Nanostructures | 443

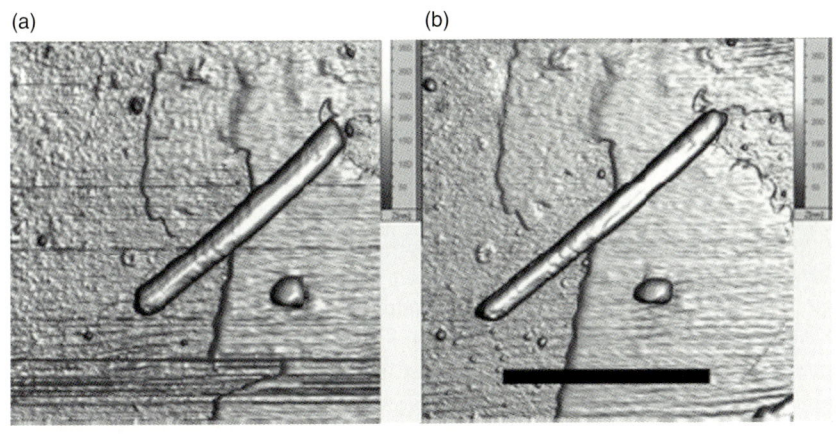

(a)   (b)

Figure 15.4 Atomic force microscopy scan of a single 200 nm-diameter 9-TBAE nanorod. (a) Before illumination; (b) After 365 nm illumination. The rod expands by 15%. Scale bar = 4 μm. Reproduced with permission from Ref. [124]; © 2006, American Chemical Society.

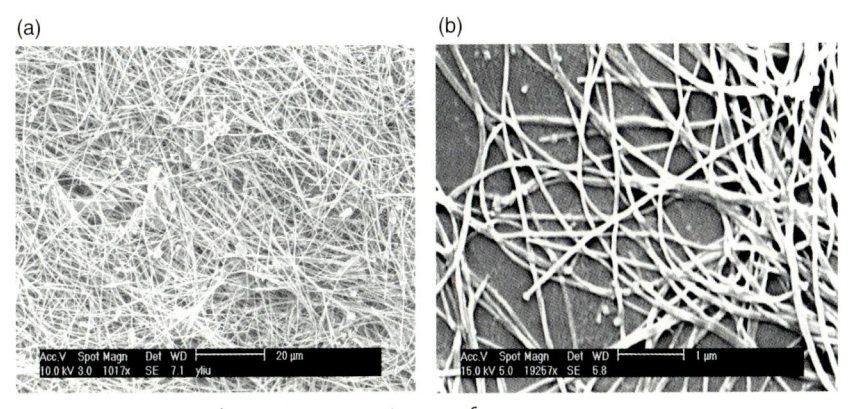

(a)   (b)

Figure 15.5 Scanning electron microscopy images of crystalline polymer nanorods. (a) 200 nm-diameter poly(9AC-ME) nanowires; (b) 35 nm-diameter poly(9AC-ME) nanowires. Reproduced with permission from Ref. [125]; © 2007, American Chemical Society.

Organic nanorods are able not only to survive photochemical transformations, but also to undergo bimolecular chemical reactions, without losing their original shape or crystallinity. For example, when 9-methyl anthracene (9MA) nanorods of 200 nm diameter were fabricated via a solvent-annealing method, the nanorods were retained inside the template, which was suspended in an aqueous suspension of 1,2,4,5-tetracyanobenzene (TCNB). After several days, a solid-state diffusion of the TCNB molecules inside the 9MA nanorods was observed. Following removal of the template, the nanorods were seen to have changed into an intimate 1:1 charge-transfer salt of 9MA:TCNB. The charge-transfer nanorods were single crystals, as confirmed by TEM and electron diffraction experiments. A similar diffusion experiment on micron-sized crystals of 9MA resulted in disintegration of that crystal and formation of the polycrystalline charge transfer (CT) complex [126].

It should be mentioned that the solvent-annealing method can be used also to produce crystalline nanorods in chemically modified AAO pores. Indeed, it has been found that crystalline 9AC rods can be grown in untreated alumina, in alumina treated with a surfactant (e.g., sodium dodecylsulfate), and in pores that have been coated with a sol–gel silica layer. The procedure for preparing organic nanorods with an ultrathin silica coating is similar to that for regular solvent-annealed nanorods, producing hybrid organic–inorganic core–shell structures which retain the orientation and crystallinity of the uncoated nanorods [127].

15.4
Applications of Noncovalent Organic Nanostructures

15.4.1
Electronic and Optical Devices

Covalent organic nanostructures (graphene and CNTs) have been incorporated into a variety of electronic devices, including field-effect transistors (FETs), photovoltaic cells, and chemical sensors. Whilst this large and rapidly expanding field has been the subject of several recent reviews, the use of noncovalent organic nanostructures in similar applications has received much less attention. One clear obstacle to combining noncovalent solids with other materials within an electronic device is their propensity to dissolve. Mixing CNTS together with a conjugated polymer is straightforward to accomplish, as the covalent bonds keep both nanostructured materials intact as they are mixed. However, following the same procedure with a molecular crystal nanowire would most likely result in the loss of the nanowire morphology, as its constituent molecules dissolve in the organic solvent.

Despite the challenges in processing, organic nanostructures have been used as components in both optical and electronic devices. Isolated organic nanoparticles can exhibit interesting properties in aqueous suspensions, including photocatalysis [128], anisotropic fluorescent probes [129], and optical switching [48]. Suspended nanoparticles can also be used as test-beds for solid-state chemical reactions that are difficult to probe in bulk crystals [130, 131]. Crystalline small-molecule and polymeric nanowires have both been used as optical waveguides, and have exhibited lasing action under optical pumping conditions [132–135]. One-dimensional organic nanostructures have also been used as field-emission electrodes due to their large aspect ratios [136], while the observation of electroluminescence from oriented needle crystals [137] indicates that these structures may have applications in low-cost displays. By utilizing advances in nanofabrication, many groups have attached electric contacts to both individual and arrays of nanowires to measure their electrical properties [29, 38, 138–142]. For example, Briseno et al. have fabricated FETs from individual molecular crystal nanowires, and measured carrier mobilities similar to those observed in bulk crystals of the same molecule [143, 144].

In the area of photovoltaics, the most common approach to producing high-efficiency organic solar cells is the bulk heterojunction (BHJ) architecture. In this approach, the nanoscale mixing of two different organic components – an electron donor and an electron acceptor – shortens the mean free path of optically excited excitons and enhances the photoinduced generation of electron-hole pairs within the device [145]. BHJ solar cells are usually composed of a mixture of a conjugated polymer donor (e.g., polythiophene) and a small-molecule acceptor (e.g., C_{60}) [146]. Control of the nanoscale mixing is regarded as the key to developing higher efficiency and more reliable BHJ materials, and this is driving many research groups to investigate how to create and characterize organic nanostructures within bicomponent systems. In fact, several groups are pursuing the creation of BHJ layers composed entirely of small molecules [147, 148], although in this case the problem of how to control the mixing is even more challenging than in the polymeric materials. It is certainly possible that insights gained from the control of small-molecule nanostructures may help to control the nanoscale mixing of different small molecules in the BHJ architecture.

15.4.2
Chemical Sensing

The high surface area:volume ratio of organic nanostructures makes them especially useful for sensing applications that, eventually, will be used for detecting explosives, drugs, and environmental toxins. Although the subject of sensors is very broad, and beyond the scope of this chapter, a brief discussion will be provided of sensors based on noncovalent organic nanostructures, the fluorescence of which is turned on or off, depending on the material detected. Again, examples of organic nanostructures as sensors are more numerous in the area of covalent organic nanostructures, such as CNTs [149]. In the field of noncovalent assemblies, there are many examples of using polymer nanowires to detect various analytes [150, 151]. In cases of electrical readout, the analyte triggers a change in the carrier mobility, and thus the resistance. Polymer nanowires have been used both for gas sensing [152–154] and for the detection of biological analytes in aqueous media [155, 156]. For sensors that generate an optical readout (most commonly fluorescence), the binding of an analyte changes the amount of fluorescence emitted by the material. An example of a nanostructured polymeric sensor based on fluorescence was recently provided by Lee et al., who employed a solution-based fabrication of polymer nanotubes to prepare PMMA nanotubes doped with tiny amounts of perylene as the fluorescent material. For this, the perylene was added to a 5% PMMA solution in acetone, and the PMMA solution then passed through an AAO membrane to form perylene-doped PMMA nanotubes. With the perylene/

PMMA nanotubes still inside the AAO template, an aqueous solution of disodium fluorescein (DSF) was passed through. When the AAO was excited with UV light, which was absorbed by the perylene and not by the DSF, the perylene fluorescence was quenched while the DSF fluorescence evolved. This was indicative of an efficient Förster-type energy transfer from the perylene (donor) inside the PMMA nanotube to the DSF (acceptor) in solution [157].

Since molecular crystals tend to have charge mobilities that are several orders of magnitude larger than their polymeric counterparts, sensors based on these materials would have a much larger dynamic range [158]. However, the use of molecular crystal nanostructures as sensors remains a largely unexplored area. An example of a fluorescence pH sensor based on small-molecule nanostructures was described by Zhao et al., who grew vertically aligned single-crystal organic nanowires made from 1,5-diaminoanthraquinone over a glass substrate, using a sublimation technique. The nanowires produced had diameters ranging from 80 to 500 nm, and a length of several microns, depending on the evaporation temperature. The red fluorescence of these nanowires could be quenched when exposed to minute amounts of acid (HCl), and regenerated when exposed to ammonia gas [22].

15.4.3
Photomechanical Actuation

The manipulation of nanometer- and micrometer-scale objects is a central goal in many areas of science, from physics to biology. Photons are ideal for controlling motion as they can penetrate deep into three-dimensional (3-D) samples such as biological media, and do not require any physical contact between the controller and the object. To convert the photons into mechanical work, small-scale photomechanical actuators are required. The development of photomechanical actuators has centered on the use of photochemically active chromophores (e.g., azobenzene and spiropyran), embedded in polymers and liquid crystal elastomers [159, 160]. Recently, a new type of organic photomechanical material, based on single-component photochemically active molecular crystals, has been demonstrated [161–167]. The present authors' group has investigated the behavior of molecular crystalline nanorods composed of anthracene derivatives that undergo a [4+4] photodimerization reaction. In particular, the photoinduced response seen in the anthracene ester rods suggests that these materials could be viable candidates for photomechanical actuator devices. In this type of reaction, a single anthracene absorbs a photon and reacts with a neighboring molecule to form a head-to-tail bridged dimer [124, 125, 168]. The nanoscale dimensions of rods composed of these anthracene derivatives permit them to undergo crystal-to-crystal photoreactions that would cause a macroscopic crystal to disintegrate. This robustness appears to be a general feature of organic nanocrystals [169, 170]. In the case of 9AC, the head-to-head photodimer dissociates spontaneously within a few minutes, resetting the system (Scheme 15.1). Thus, this reversible photochemical solid-state system could form the basis of a photomechanical actuator.

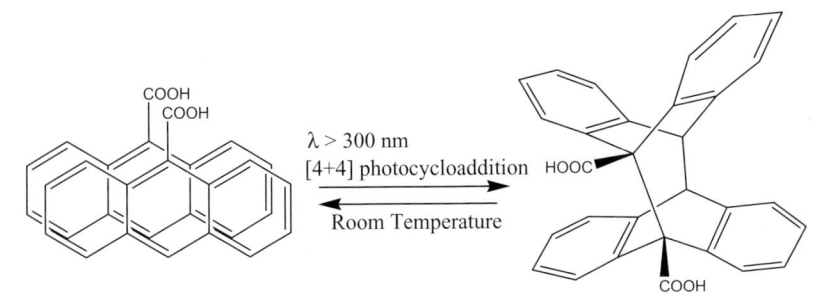

15.4 Applications of Noncovalent Organic Nanostructures | 447

Scheme 15.1 Photodimerization of 9AC in the solid state. Note that the carboxylic acid groups are on the same sides of the dimer, leading to a thermally unstable head-to-head photodimer.

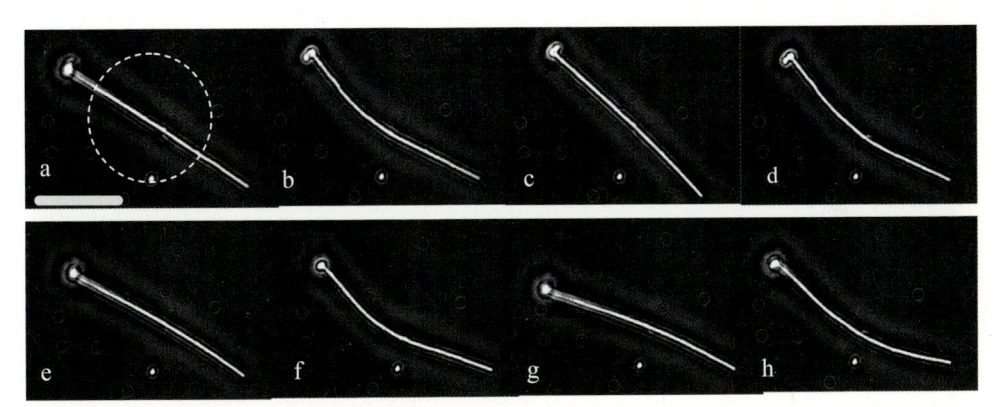

Figure 15.6 Phase-contrast microscopy image of a single 200 nm-diameter 9AC nanorod bending (365 nm light on) and unbending (light off) when the center segment is illuminated. The sequence a–h follows four bending-unbending cycles. Scale bar = 20.7 μm. Reproduced with permission from Ref. [168]; © 2007, Wiley-VCH.

Although the 9AC rods did not undergo a large expansion (as did 9TBAE), it was possible to induce bending by irradiating a local segment of a single nanorod. The nanorod regains its original structure, and the mechanical motion can be recycled several times (Figure 15.6). The initial demonstrations of reversible photomechanical molecular crystals provided evidence that they could be useful as small-scale photoactuators. However, to assess whether the nanorod photoresponse could be harnessed to generate controlled, microscale displacements, several aspects of their photomechanical response must be understood in more detail. Hence, an assessment was made as to whether it was possible to control the magnitude, rate and orientation of the photoinduced deformations in the nanorods [171]. In order to precisely control the location and magnitude of the bend, two-photon excitation (2PE) was used to localize the region of reacted molecules in 3-D space. 2PE allowed advantage to be taken of the superior penetration characteristics of the near-infrared (NIR) opposed to the UV light used in previous studies. By using this method, it was possible to demonstrate the controlled bending of both 200 nm- and 35 nm-diameter nanorods, which are the smallest photomechanical structures yet demonstrated experimentally. It was also shown that the sequential bending of longer segments could, in some cases, lead to an actual translation of entire rods. By measuring the bend angles as a function of irradiation time and intensity, both the rate and magnitude of deformation were shown to be consistent with a simple kinetic model, where the bend angle is linearly proportional to the fraction of dimerized molecules. This quantitative understanding provides a foundation for both engineering the nanoscale photomechanical response, and for a more detailed chemical understanding of the phenomenon.

15.5
Future Challenges and Outlook

Today, although organic nanotechnology is still at its early stages of development, there remain several challenges in the field. Yet, success in overcoming these challenges will determine the future outlook for the field which is, undoubtedly, very bright. One important challenge concerns the identification of the potential applications of these materials. In this chapter, attention has been centered on applications in diverse areas such as drug delivery, optoelectronic devices, chemical sensing, and photomechanical actuation, all of which represent promising avenues for further investigation. However, there may be other applications beyond these three areas, and the ultimate capability of organic nanostructures in such applications is also unknown. Thus, the practical question of the potential applications of organic nanostructures is intimately tied to more fundamental scientific questions regarding their properties. Clearly, a better understanding is needed of the size effect of organic nanostructures on the mechanical, chemical, electrical conductivity and photophysical properties when compared to the bulk, in order to parallel the understanding of inorganic nanostructures. A second major challenge is the development of new methods to control the size, shape, and crystallinity of organic nanostructures. Most nanostructures prepared to date have consisted of quasi zero-dimensional (nanospheres) or 1-D (nanowires) structures that comprise only a small subset of potential geometric shapes. With few exceptions, organic compounds cannot be efficiently fashioned into shaped nanostructures without the aid of a template. It has been seen that AAO templates with variable pore sizes may serve as a cheap platform for producing highly crystalline cylindrical nanostructures; the question remains as to whether it is possible to move beyond cylindrical organic nanostructures? By using AAO templates with a hierarchal branching, it may be possible to fabricate organic nanostructures with branches. An example would be the mesoscale dendrimer made from one or more highly crystalline organic compounds. However, there may be limits to the templating approach, and it remains an open question as to whether a new approach, perhaps based on self-assembly, could be developed for the preparation of more complicated shapes.

Overall, the field of organic nanostructures is wide open, combining aspects of materials science, organic chemistry, and physical chemistry. The drive towards environment-friendly energy solutions and technology will eventually lead to further applications for organic nanostructures that tend to be biodegradable and less toxic than inorganic or heavy-metal nanostructures.

Acknowledgments

C.J.B. acknowledges the support of the National Science Foundation through grants CHE-0719039 and DMR-0907310. R. O. Al-Kaysi acknowledges the support of KSAU-HS/KAIMRC through grant RC08/093.

References

1 Wang, D. and Song, C. (2005) *Journal of Physical Chemistry B*, **109**, 12697–700.
2 Galush, W.J., Shelby, S.A., Mulvihill, M.J., Tao, A., Yang, P. and Groves, J.T. (2009) *Nano Letters*, **9**, 2077–82.
3 Tian, Y., Liu, H., Zhao, G. and Tatsuma, T. (2006) *Journal of Physical Chemistry B*, **110** (46), 23478–81.
4 Shen, H.X., Xua, M.M., Yana, X., Yao, J.L., Hana, S.Y. and Gu, R.A. (2010) *Colloids and Surfaces A: Physicochemical and Engineering Aspects*, **353** (2–3), 204–9.
5 Tasis, D., Tagmatarchis, N., Bianco, A. and Prato, M. (2006) *Chemical Reviews*, **106** (3), 1105–13.
6 Geim, A.K. and Novoselov, K.S. (2007) *Nature Materials*, **6** (3), 183–91.
7 Fuhrer, M.S. (2009) *Nature*, **459**, 1037.
8 Kroto, H.W., Heath, J.R., O'Brien, S.C., Curl, R.F. and Smalley, R.E. (1985) *Nature*, **318**, 162–3.
9 Boudou, J.-P., Curmi, P.A., Jelezko, F., Wrachtrup, J., Aubert, P., Sennour, M., Balasubramanian, G., Reuter, R., Thorel, A. and Gaffet, E. (2009) *Nanotechnology*, **20** (23), 235602.
10 Ménard-Moyon, C., Izard, N., Doris, E. and Mioskowski, C. (2006) *Journal of the American Chemical Society*, **128** (20), 6552–3.
11 Krupke, R. and Hennrich, F. (2005) *Advanced Engineering Materials*, **7** (3), 111–16.
12 Castro, C.L.D. and Mitchell, B.S. (2002) Nanoparticles from mechanical attrition, in *Synthesis, Functionalization and Surface Treatment of Nanoparticles* (ed. E.B.M.-I. Baraton), American Scientific Publishers, pp. 1–15.
13 Horn, D. and Rieger, J. (2001) *Angewandte Chemie International Edition*, **40**, 4330–61.
14 Kuroda, R., Imai, Y. and Tajima, N. (2002) *Chemical Communications*, (23), 2848–9.
15 Braga, D., Maini, L., Polito, M., Mirolo, L. and Grepioni, F. (2003) *Chemistry–A European Journal*, **9**, 4362–70.
16 Hayashi, K., Morii, H., Iwasaki, K., Horie, S., Horiishi, N. and Ichimura, K. (2007) *Journal of Materials Chemistry*, **17**, 527–30.

17 Kaupp, G. (2009) *Crystal Engineering Communications*, **11**, 388–403.

18 Sears, G.W. (1963) *Journal of Chemical Physics*, **39**, 2846–7.

19 Liu, H., Li, Y., Xiao, S., Gan, H., Jiu, T., Li, H., Jiang, L., Zhu, D., Yu, D., Xiang, B. and Chen, Y. (2003) *Journal of the American Chemical Society*, **125**, 10794–5.

20 Liu, H., Li, Y., Xiao, S., Li, H., Jiang, L., Zhu, D., Xiang, B., Chen, Y. and Yu, D. (2004) *Journal of Physical Chemistry B*, **108**, 7744–7.

21 Debe, M.K. and Drube, A.R. (1995) *Journal of Vacuum Science Technology B*, **13**, 1236–41.

22 Zhao, Y.S., Wu, J. and Huang, J. (2009) *Journal of the American Chemical Society*, **131**, 3158–9.

23 Wang, Z.-C., Xiao, W.-C., Ding, X.-L., Ma, Y.-P., Xue, W. and He, S.-G. (2008) *Nanotechnology*, **19**, 505703/1–5.

24 Balzer, F. and Rubahn, H.G. (2002) *Surface Science*, **507–510**, 588–92.

25 Ichikawa, M., Hibino, R., Inoue, M., Haritani, T., Hotta, S., Koyama, T. and Taniguchi, Y. (2003) *Advanced Materials*, **15**, 213–17.

26 Yanagi, H., Morikawa, T., Hotta, S. and Yase, K. (2001) *Advanced Materials*, **13**, 313–17.

27 Balzer, F. and Rubahn, H.G. (2001) *Applied Physics Letters*, **79**, 3860–2.

28 Andreev, A., Matt, G., Brabec, C.J., Sitter, H., Badt, D., Seyringer, H. and Sariciftci, N.S. (2000) *Advanced Materials*, **12**, 629–33.

29 Sun, Y., Tan, L., Jiang, S., Qian, H., Wang, Z., Yan, D., Di, C., Wang, Y., Wu, W., Yu, G., Yan, S., Wang, C., Hu, W., Liu, Y. and Zhu, D. (2007) *Journal of the American Chemical Society*, **129**, 1882–3.

30 Wakayama, Y., Mitsui, T., Onodera, T., Oikawa, H. and Nakanishi, H. (2006) *Chemical Physics Letters*, **417**, 503–8.

31 Jiang, L., Fu, Y., Li, H. and Hu, W. (2008) *Journal of the American Chemical Society*, **130** (12), 3937–41.

32 Firanescu, G., Hermsdorf, D., Ueberschaer, R. and Signorell, R. (2006) *Physical Chemistry Chemical Physics*, **8**, 4149–65.

33 Hoeben, F.J.M., Jonkheijm, P., Meijer, E.W. and Schenning, A.P.H.J. (2005) *Chemical Reviews*, **105**, 1491–546.

34 Ryu, J.-H., Hong, D.-J. and Lee, M. (2008) *Chemical Communications*, (9), 1043–54.

35 Balakrishnan, K., Datar, A., Oitker, R., Chen, H., Zuo, J. and Zang, L. (2005) *Journal of the American Chemical Society*, **127**, 10496–7.

36 Loi, S., Wiesler, U.M., Butt, H.J. and Mullen, K. (2000) *Chemical Communications*, (13), 1169–70.

37 Yamamoto, Y., Fukushima, T., Suna, Y., Ishii, N., Saeki, A., Seki, S., Tagawa, S., Taniguchi, M., Kawai, T. and Aida, T. (2006) *Science*, **314**, 1761–4.

38 Durkut, M., Mas-torrent, M., Hadley, P., Jonkheijm, P., Schenning, A.P.H., Meijer, E.W., George, S. and Ajayaghosh, A. (2006) *Journal of Chemical Physics*, **124**, 154704/1–6.

39 Zhang, X., Zhang, X., Wang, B., Zhang, C., Chang, J.C., Lee, C.S. and Lee, S.-T. (2008) *Journal of Physical Chemistry C*, **112**, 16264–8.

40 Zhang, X., Zhang, X., Shi, W., Meng, X., Lee, C. and Lee, S. (2007) *Angewandte Chemie International Edition*, **46**, 1525–8.

41 Kasai, H., Nalwa, H.S., Oikawa, H., Okada, S., Matsuda, H., Minami, N., Kakuta, A., Ono, K., Mukoh, A. and Nakanishi, H. (1992) *Japanese Journal of Applied Physics*, **31**, L1132–4.

42 Nalwa, H.S., Kasai, H., Okada, S., Oikawa, H., Matsuda, H., Kakuta, A., Mukoh, A. and Nakanishi, H. (1993) *Advanced Materials*, **5**, 758–60.

43 Al-Kaysi, R.O., Muller, A.M., Ahn, T.S., Lee, S. and Bardeen, C.J. (2005) *Langmuir*, **21** (17), 7990–4.

44 Kasai, H., Kamatani, H., Okada, S., Oikawa, H., Matsuda, H. and Nakanishi, H. (1996) *Japanese Journal of Applied Physics*, **35**, L221–3.

45 Keuren, E.V., Georgieva, E. and Adrian, J. (2001) *Nano Letters*, **1** (3), 141–4.

46 Keuren, E.V., Georgieva, E. and Durst, M. (2003) *Journal of Dispersion Science and Technology*, **24** (5), 721–9.

47 Ji, X., Ma, Y., Cao, Y., Zhang, X., Xie, R., Fu, H., Xiao, D. and Yao, J. (2001) *Dyes and Pigments*, **51**, 87–91.

48 An, B.K., Kwon, S.-K., Jung, S.-D. and Park, S.Y. (2002) *Journal of the American Chemical Society*, **124**, 14410–15.

49 Ou, Z., Yao, H. and Kimura, K. (2006) *Chemistry Letters*, **35**, 782–3.

50 Jia, Z., Xiao, D., Yang, W., Ma, Y., Yao, J. and Liu, Z. (2004) *Journal of Membrane Science*, **241**, 387–92.

51 Latterini, L., Roscini, C., Carlotti, B., Aloisi, G.G. and Elisei, F. (2006) *Physica Status Solidi A*, **203** (6), 1470–5.

52 Katagi, H., Kasai, H., Okada, S., Oikawa, H., Komatsu, K., Matsuda, H., Liu, Z. and Nakanishi, H. (1996) *Japanese Journal of Applied Physics*, **35**, L1364–6.

53 Kook, S.K. and Kopelman, R. (1992) *Journal of Physical Chemistry*, **96**, 10672–6.

54 Tian, Z., Chen, Y., Yang, W., Yao, J., Zhu, L. and Shuai, Z. (2004) *Angewandte Chemie International Edition*, **43**, 4060–3.

55 Zhao, Y.S., Yang, W., Xiao, D., Sheng, X., Yang, X., Shuai, Z., Luo, Y. and Yao, J. (2005) *Chemistry of Materials*, **17**, 6430–5.

56 Bisht, P.B., Fukuda, K. and Hirayama, S. (1997) *Journal of Physical Chemistry B*, **101**, 8054–8.

57 Fu, H.-B. and Yao, J.-N. (2001) *Journal of the American Chemical Society*, **123**, 1434–9.

58 Onodera, T., Kasai, H., Okada, S., Oikawa, H., Mizuno, K., Fujitsuka, M., Ito, O. and Nakanishi, H. (2002) *Optical Materials*, **21**, 595–8.

59 Oikawa, H., Mitsui, T., Onodera, T., Kasai, H., Nakanishi, H. and Sekiguchi, T. (2003) *Japanese Journal of Applied Physics*, **42**, L111–13.

60 Wang, F., Han, M.Y., Mya, K.Y., Wang, Y. and Lai, Y.H. (2005) *Journal of the American Chemical Society*, **127**, 10350–5.

61 Zhang, X., Shang, X., Shi, W., Meng, X., Lee, C. and Lee, S. (2005) *Journal of Physical Chemistry B*, **109**, 18777–80.

62 Matsui, A.H., Mizuno, K., Nishi, O., Matsushima, Y., Shimizu, M., Goto, T. and Takeshima, M. (1995) *Chemical Physics*, **194**, 167–74.

63 Shen, Z. and Forrest, S.R. (1997) *Physical Review B*, **55**, 10578–92.

64 Ogura, T., Tanoura, M., Tatsuhara, K. and Hiraki, A. (1994) *Bulletin of the Chemical Society of Japan*, **67**, 3143–9.

65 Destree, C., Ghijsen, J. and Nagy, J.B. (2007) *Langmuir*, **23** (4), 1965–73.

66 Kostler, S., Rudorfer, A., Haase, A., Satzinger, V., Jakopic, G. and Ribitsch, V. (2009) *Advanced Materials*, **21**, 1–6.

67 Kang, P., Chen, C., Hao, L., Zhu, C., Hu, Y. and Chen, Z. (2004) *Materials Research Bulletin*, **39**, 545–51.

68 Asahi, T., Sugiyama, T. and Masuhara, H. (2008) *Accounts of Chemical Research*, **41**, 1790–8.

69 Abyan, M., Caro, D. and Fery-Forgues, S. (2009) *Langmuir*, **25**, 1651–8.

70 Reneker, D.H. and Chun, I. (1996) *Nanotechnology*, **7**, 216–23.

71 Huang, Z.-M., Zhangb, Y.-Z., Kotakic, M. and Ramakrishna, S. (2003) *Composites Science and Technology*, **63** (15), 2223–53.

72 Bognitzki, M., Czado, W., Frese, T., Schaper, A., Hellwig, M., Steinhart, M., Greiner, A. and Wendorff, J.H. (2001) *Advanced Materials*, **13**, 70–2.

73 Kakade, M.V., Givens, S., Garner, K., Lee, K.H., Chase, D.B. and Rabolt, J.F. (2007) *Journal of the American Chemical Society*, **129**, 2777–82.

74 D'Amato, R., Venditti, I., Russo, M.V. and Falconieri, M. (2006) *Journal of Applied Polymer Science*, **102**, 4493–9.

75 Lal, M., Kumar, N.D., Joshi, M.P. and Prasad, P.N. (1998) *Chemistry of Materials*, **10**, 1065–8.

76 Landfester, K., Montenegro, R., Scherf, U., Guntner, R., Asawapirom, U., Patil, S., Neher, D. and Kietzke, T. (2002) *Advanced Materials*, **14**, 651–5.

77 Piok, T., Gadermaier, C., Wenzl, F.P., Patil, S., Montenegro, R., Landfester, K., Lanzani, G., Cerullo, G., Scherf, U. and List, E.J.W. (2004) *Chemical Physics Letters*, **389**, 7–13.

78 Debuigne, F., Jeunieau, L., Wiame, M. and Nagy, J.B. (2000) *Langmuir*, **16**, 7605–11.

79 Kwon, E., Oikawa, H., Kasai, H. and Nakanishi, H. (2007) *Crystal Growth and Design*, **7**, 600–2.

80 Jackson, C.L. and McKenna, G.B. (1990) *Journal of Chemical Physics*, **93**, 9002–11.

81 Ibanez, A., Maximov, S., Guiu, A., Chaillout, C. and Baldeck, P.L. (1998) *Advanced Materials*, **10**, 1540–3.

82 Botzung-Appert, E., Monnier, V., Duong, T.H., Pansu, R. and Ibanez, A. (2004) *Chemistry of Materials*, **16**, 1609–11.

83 Martin, C.R. (1994) *Science*, **266**, 1961–6.
84 Thomas, A., Goettmann, F. and Antonietti, M. (2008) *Chemistry of Materials*, **20**, 738–55.
85 Asoh, H., Nishio, K., Nakao, M., Tamamura, T. and Masudaa, H. (2001) *Journal of the Electrochemical Society*, **148** (4), B152–6.
86 Schneider, J.J., Engstler, J., Budna, K.P., Teichert, C. and Franzka, S. (2005) *European Journal of Inorganic Chemistry*, **2005** (12), 2352–9.
87 Lee, W., Scholz, R. and Gosele, U. (2008) *Nano Letters*, **8**, 2155–60.
88 Meng, G., Jung, Y.J., Cao, A., Vajtai, R. and Ajayan, P.M. (2005) *Proceedings of the National Academy of Sciences of the United States of America*, **102** (20), 7074–8.
89 Yoo, B.-Y., Hendricks, R.K., Ozkanb, M. and Myunga, N.V. (2006) *Electrochimica Acta*, **51**, 3543–50.
90 Bae, C., Yoo, H., Kim, S., Lee, K., Jiyoung, K., Sung, M.M. and Shin, H. (2008) *Chemistry of Materials*, **20** (3), 756–67.
91 Ji, H.-X., Hu, J.-S., Tang, Q.-X., Hu, W.-P., Song, W.-G. and Wan, L.-J. (2006) *Advanced Materials*, **18**, 2753–7.
92 Cepak, V.M. and Martin, C.R. (1999) *Chemistry of Materials*, **11**, 1363–7.
93 Song, G., She, X., Fu, Z. and Li, J. (2004) *Journal of Materials Research*, **19** (11), 3324–8.
94 Zheng, R.K., Yang, Y., Wang, Y., Wang, J., Chan, H.L.W., Choy, C.L., Jin, C.G. and Li, X.G. (2005) *Chemical Communications*, (11), 1447–9.
95 Oh, H.J., Jeong, Y.S., Kwon, S.H., Heo, C.H., Ki, B.S. and Chi, C.S. (2007) *Diffusion and Defect Data Part B, Solid State Phenomena*, **124/126**, 1109–12.
96 Wang, C.-C., Shen, Q.-D., Tang, S.-C., Wu, Q., Bao, H.-M., Yang, C.-Z. and Jiang, X.-Q. (2008) *Macromolecular Rapid Communications*, **29** (9), 724–8.
97 She, X., Song, G., Li, J., Han, P., Yang, S., Wang, S. and Zhi, P. (2006) *Polymer Journal*, **38** (7), 639–42.
98 Qiao, J., Zhang, X., Meng, X., Zhou, S., Wu, S. and Lee, S.-T. (2005) *Nanotechnology*, **16**, 433–6.
99 Li, Y., Yin, X.-F., Chen, F.-R., Yang, H.-H., Zhuang, Z.-X. and Wang, X.-R. (2006) *Macromolecules*, **39** (13), 4497–9.
100 She, X.L., Song, G.J., Peng, Z., Li, J.J., Lim, C.T., Tan, E.P.S., Lv, L. and Zhao, X.S. (2007) *Polymer Journal*, **39** (10), 1025–9.
101 Li, J.-J., Song, G.-J., She, X.-L., Han, P., Peng, Z. and Chen, D. (2006) *Polymer Journal*, **38** (6), 554–8.
102 She, X., Song, G., Li, J., Han, P., Yang, S. and Peng, Z. (2006) *Journal of Materials Research*, **21** (5), 1209–14.
103 Zhang, M., Dobriyal, P., Chen, J.-T. and Russell, T.P. (2006) *Nano Letters*, **6** (5), 1075–9.
104 Steinhart, M., Wendorff, J.H. and Wehrspohn, R.B. (2003) *ChemPhysChem*, **4**, 1171–6.
105 Xiang, H., Shin, K., Kim, T., Moon, S.I., McCarthy, T.J. and Russell, T.P. (2004) *Macromolecules*, **37**, 5660–4.
106 Yu, B., Gao, Y. and Li, H. (2004) *Journal of Applied Polymer Science*, **91**, 425–30.
107 Liang, W. and Martin, C.R. (1990) *Journal of the American Chemical Society*, **112**, 9666–8.
108 Kim, K. and Jin, J.I. (2001) *Nano Letters*, **1**, 631–6.
109 Zhi, L., Wu, J., Li, J., Kolb, U. and Mullen, K. (2005) *Angewandte Chemie International Edition*, **44**, 2120–3.
110 Liu, L., Zhao, C., Zhao, Y., Jia, N., Zhou, Q., Yan, M. and Jiang, Z. (2005) *European Polymer Journal*, **41**, 2117–21.
111 Qu, L. and Shi, G. (2004) *Chemical Communications*, (24), 2800–1.
112 Xiao, R., Cho, S.I., Liu, R. and Lee, S.B. (2007) *Journal of the American Chemical Society*, **129** (14), 4483–9.
113 Heng, L., Zhai, J., Qin, A., Zhang, Y., Dong, Y., Tang, B.Z. and Jiang, L. (2007) *ChemPhysChem*, **8** (10), 1513–18.
114 Zhao, L., Yang, W., Luo, Y., Zhai, T., Zhang, G. and Yao, J. (2005) *Chemistry–A European Journal*, **11**, 3773–8.
115 Liu, H., Li, Y., Jiang, L., Luo, H., Xiao, S., Fang, H., Li, H., Zhu, D., Yu, D., Xu, J. and Xiang, B. (2002) *Journal of the American Chemical Society*, **124**, 13370–1.
116 Zhao, L., Yang, W., Ma, Y., Yao, J., Lib, Y. and Liub, H. (2003) *Chemical Communications*, (19), 2442–3.

117 Zhang, X., Ju, W., Gu, M., Meng, X., Shi, W., Zhang, X. and Lee, S. (2005) *Chemical Communications*, (33), 4202–4.

118 Gan, H., Liu, H., Li, Y., Liu, Y., Lu, F., Jiu, T. and Zhu, D. (2004) *Chemical Physics Letters*, **399**, 130–4.

119 Zhao, L., Yang, W., Zhang, G., Zhai, T. and Yao, J. (2003) *Chemical Physics Letters*, **379**, 479–83.

120 Barrett, C., Iacopino, D., O'Carroll, D., DeMarzi, G., Tanner, D.A., Quinn, A.J. and Redmond, G. (2007) *Chemistry of Materials*, **19** (3), 338–40.

121 Lee, J.-K., Koh, W.-K., Chaeb, W.-S. and Kim, Y.-R. (2002) *Chemical Communications*, (2), 138–9.

122 Bai, R., Ouyang, M., Zhou, R.-J., Shi, M.-M., Wang, M. and Chen, H.-Z. (2008) *Nanotechnology*, **19**, 055604/1–6.

123 Al-Kaysi, R.O. and Bardeen, C.J. (2006) *Chemical Communications*, **11**, 1224–6.

124 Al-Kaysi, R.O., Muller, A.M. and Bardeen, C.J. (2006) *Journal of the American Chemical Society*, **128** (50), 15938–9.

125 Al-Kaysi, R.O., Dillon, R.J., Kaiser, J.M., Mueller, L.J., Guirado, G. and Bardeen, C.J. (2007) *Macromolecules*, **40**, 9040–4.

126 Al-Kaysi, R.O., Müller, A.M., Frisbee, R.J. and Bardeen, C.J. (2009) *Journal of Crystal Growth and Design*, **9** (4), 1780–5.

127 Al-Kaysi, R.O., Dillon, R.J., Zhu, L. and Bardeen, C.J. (2008) *Journal of Colloid and Interface Science*, **327**, 102–7.

128 Kim, H.Y., Bjorklund, T.G., Lim, S.-H. and Bardeen, C.J. (2003) *Langmuir*, **19**, 3941–6.

129 Hosaka, N., Obata, M., Suzuki, M., Saiki, T., Takeda, K. and Kuwata-Gonokami, M. (2008) *Applied Physics Letters*, **92**, 113305/1–3.

130 Chin, K.K., Natarajan, A., Gard, M.N., Capmpos, L.M., Shepherd, H., Johansson, E. and Garcia-Garibay, M.A. (2007) *Chemical Communications*, (41), 4266–8.

131 Lebedeva, N.V., Tarasov, V.F., Resendiz, M.J.E., Garcia-Garibay, M.A., White, R.C. and Forbes, M.D.E. (2010) *Journal of American Chemical Society*, **132**, 82–4.

132 Yanagi, H. and Morikawa, T. (1999) *Applied Physics Letters*, **75**, 187–9.

133 Balzer, F., Bordo, V.G., Simonsen, A.C. and Rubahn, H.G. (2003) *Physical Review B*, **67**, 115408/1–4508/8.

134 Quochi, F., Cordella, F., Mura, A., Bongiovanni, G., Balzer, F. and Rubahn, H.G. (2006) *Applied Physics Letters*, **88**, 041106/1–3.

135 O'Carroll, D., Lieberwirth, I. and Redmond, G. (2007) *Nature Nanotechnology*, **2**, 180–4.

136 Cui, S., Liu, H., Gan, L., Li, Y. and Zhu, D. (2008) *Advanced Materials*, **20**, 2918–25.

137 Yanagi, H., Morikawa, T. and Hotta, S. (2002) *Applied Physics Letters*, **81**, 1512–14.

138 Martin, C.R. (1995) *Accounts of Chemical Research*, **28**, 61–8.

139 Guo, Y., Tang, Q., Liu, H., Zhang, Y., Li, Y., Hu, W., Wang, S. and Zhu, D. (2008) *Journal of the American Chemical Society*, **130**, 9198–9.

140 Chung, J.W., Yang, H., Singh, B., Moon, H., An, B.K., Lee, S.Y. and Park, S.Y. (2009) *Journal of Materials Chemistry*, **19**, 5920–5.

141 Dong, H., Jiang, S., Jiang, L., Liu, Y., Li, H., Hu, W., Wang, E., Yan, S., Wei, Z., Xu, W. and Gong, X. (2009) *Journal of the American Chemical Society*, **131**, 17315–20.

142 Tang, Q., Tong, Y., Hu, W., Wan, Q. and Bjoernholm, T. (2009) *Advanced Materials*, **21**, 4234–7.

143 Briseno, A.L., Mannsfeld, S.C.B., Lu, X., Xiong, Y., Jenekhe, S.A., Bao, Z. and Xia, Y. (2007) *Nano Letters*, **7** (3), 668–75.

144 Briseno, A.L., Mannsfeld, S.C.B., Reese, C., Hancock, J.M., Xiong, Y., Jenekhe, S.A., Bao, Z. and Xia, Y. (2007) *Nano Letters*, **7** (9), 2847–53.

145 Yu, G., Gao, J., Hummelen, J.C., Wudl, F. and Heeger, A.J. (1995) *Science*, **270**, 1789–91.

146 Gunes, S., Neugebauer, H. and Saricifti, N.S. (2007) *Chemical Reviews*, **107**, 1324–38.

147 Gebeyehu, D., Maennig, B., Drechsel, J., Leo, K. and Pfeiffer, M. (2003) *Solar Energy Materials and Solar Cells*, **79**, 81–92.

148 Peumans, P., Uchida, S. and Forrest, S.R. (2003) *Nature*, **425**, 158–62.

149 Liu, Z., Tabakman, S., Welsher, K. and Dai, H. (2009) *Nano Research*, **2**, 85–120.

150 Cao, Y., Kovalev, A.E., Xiao, R., Kim, J., Mayer, T.S. and Mallouk, T.E. (2008) *Nano Letters*, **8**, 2757–61.

151 Antohe, V.A., Radu, A., Matefi-Tempfli, M., Attout, A., Yunus, S., Bertrand, P., Dutu, C.A., Vlad, A., Melinte, S., Matefi-Tempfli, S. and Piraux, L. (2009) *Applied Physics Letters*, **94**, 073118/1–3.

152 Ma, X., Li, G., Wang, M., Cheng, Y., Bai, R. and Chen, H. (2006) *Chemistry–A European Journal*, **12**, 3254–60.

153 Dan, Y., Cao, Y., Mallouk, T.E., Evoy, S. and Johnson, A.T.C. (2009) *Nanotechnology*, **20**, 434014/1–4.

154 Shirsat, M.D., Bangar, M.A., Deshusses, M.A., Myung, N.V. and Mulchandani, A. (2009) *Applied Physics Letters*, **94**, 083502/1–3.

155 Wanekaya, A.K., Chen, W., Myung, N.V. and Mulchandani, A. (2006) *Electroanalysis*, **18**, 533–50.

156 Bangar, M.A., Shirale, D.J., Chen, W., Myung, N.V. and Mulchandani, A. (2009) *Analytical Chemistry*, **81**, 2168–75.

157 Lee, S., Muller, A.M., Al-Kaysi, R. and Bardeen, C.J. (2006) *Nano Letters*, **6** (7), 1420–4.

158 Wright, J.D. (1995) *Molecular Crystals*, 2nd edn, Cambridge University Press, New York.

159 Ikeda, T., Mamiya, J. and Yu, Y. (2007) *Angewandte Chemie International Edition*, **46**, 506–28.

160 Finkelmann, H., Nishikawa, E., Pereira, G.G. and Warner, M. (2001) *Physical Review Letters*, **87**, 015501/1–4.

161 Irie, M., Kobatake, S. and Horichi, M. (2001) *Science*, **291** (5509), 1769–72.

162 Kobatake, S., Takami, S., Muto, H., Ishikawa, T. and Irie, M. (2007) *Nature*, **446**, 778–81.

163 Colombier, I., Spagnoli, S., Corval, A., Baldeck, P.L., Giraud, M., Leaustic, A., Yu, P. and Irie, M. (2007) *Journal of Chemical Physics*, **126**, 011101/1–3.

164 Uchida, K., Sukata, S., Matsuzawa, Y., Akazawa, M., Jong, J.J.D., Katsonis, N., Kojima, Y., Nakamura, S., Areephong, J., Meetsma, A. and Feringa, B.L. (2008) *Chemical Communications*, (3), 326–8.

165 Lange, C.W., Foldeaki, M., Nevodchikov, V.I., Cherkasov, V.K., Abakumov, G.A. and Pierpont, C.G. (1992) *Journal of the American Chemical Society*, **114**, 4220–2.

166 Flannigan, D.J., Lobastov, V.A. and Zewail, A.H. (2007) *Angewandte Chemie International Edition*, **46**, 9206–10.

167 Flannigan, D.J., Samartzis, P.C., Yurtsever, A. and Zewail, A.H. (2009) *Nano Letters*, **9**, 875–81.

168 Al-Kaysi, R.O. and Bardeen, C.J. (2007) *Advanced Materials*, **19**, 1276–80.

169 Bucar, D.K. and MacGillivray, L.R. (2007) *Journal of the American Chemical Society*, **129**, 32–3.

170 Takahashi, S., Miura, H., Kasai, H., Okada, S., Oikawa, H. and Nakanishi, H. (2002) *Journal of the American Chemical Society*, **124**, 10944–5.

171 Good, J.T., Burdett, J.J. and Bardeen, C.J. (2009) *Small*, **5** (24), 2902–9.

16
Conducting Polymer Nanowires and Their Biomedical Applications
Robert Lee and Adam K. Wanekaya

16.1
Introduction

Conducting polymers (CPs) are polymers with spatially extended π-bonding systems due to alternating single and double carbon–carbon bonds. They normally are comprised of C, H, and simple heteroatoms such as N, O, and S. The unique intrinsic conductivity of these organic materials arises from π-conjugation originating from the overlap of π-electrons. The applications of conducting polymers in biosensing is mainly due to changes in their electrical properties resulting from the depletion or accumulation of charge carriers on exposure to a targeted analyte. Therefore, the specific surface area of conducting polymers is crucial in determining their sensitivity in sensing applications.

The CP nanowires form part of the one-dimensional (1-D) nanostructured materials that include material such as carbon nanotubes (CNTs), metal nanowires, metal oxide nanowires, nanotubes, nanosprings, and nanobelts. Theoretically, these materials should have increased sensitivities compared to the bulk materials, due to their small size, high aspect ratios, high surface-to-volume ratios, and unusual target-binding properties. Target-binding properties involving nanowires have significant effects on their physical and chemical properties, thus providing a mode of signal transduction not necessarily available within bulk structures of the same material. The ability to fabricate nanowires of different aspect ratios allows the fine-tuning of their electrical and optical properties that are critical to the function and integration of high-density nanoscale arrays. Also, nanowires are extremely attractive for nanoelectronics because they can function both as sensing devices and as the wires that access them. Further, the direct conversion of chemical, biological, or physical information into an electronic signal takes advantage of the existing microeletronic technology leading to miniaturized sensor devices. Because the electrical properties of nanowires are strongly influenced by minor perturbations, they provide an efficient avenue for the much desired rapid, label-free and direct detection of various targets (Figure 16.1). The advantages of label-free detection include:

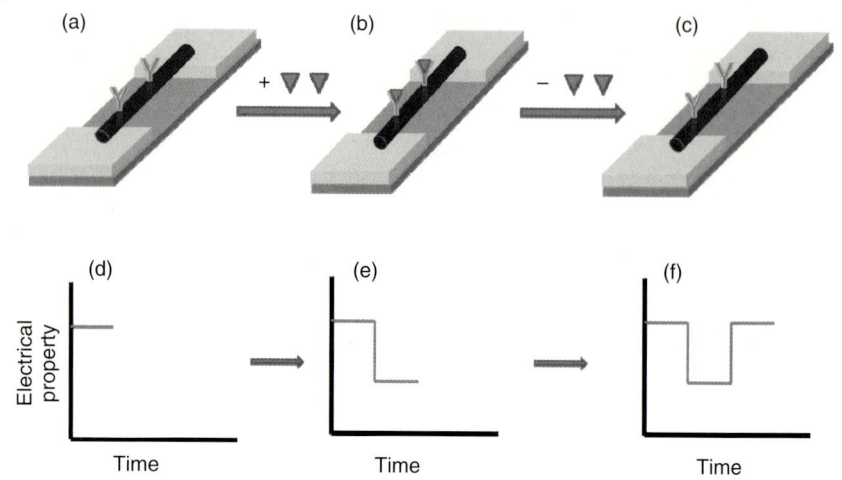

Figure 16.1 Schematic of nanowire-based electrical detection.
(a) Nanowire modified with a biomolecule; (b) Specific
binding of targets; (c) The target unbinds;
(d–f) Corresponding changes in electrical property of the
nanowire with time as a result of charge depletion/addition
due to target binding/unbinding.

- A simple homogeneous assay format, without separation and washing
- Increased sensitivity
- Rapid near-real-time response.

These advantages have the potential to impact biological research, biomedical, bioterrorism, and environmental applications.

Indeed, 1-D nanostructured materials such as CNTs and silicon nanowires have demonstrated profound performance in device fabrication in general, and label-free detection in particular [1–7]. However, CP nanowires have unique advantages over other 1-D nanostructured materials, in that they can be modified before, during, and after polymerization; this makes them very adaptable materials as optimal conditions can be used for each step. The conductivities of CPs can be modulated up to 15 orders of magnitude by changing the dopant and/or dopant/monomer ratios, while the conductivities of CP nanowires can be modulated by controlling their oxidation state. Further, the single-segment and multisegmented CP nanowires can be easily fabricated under ambient conditions with benign reagents, using template-directed methods [8–15]. In this chapter, details of the fabrication, modification, alignment/assembly and biomedical applications of CP nanowires and CP nanotubes related to the most significant research conducted during the past three years will be discussed. (Note: It is advised that readers consult prior publications to obtain information regarding CP nanostructures [16–19].)

16.2
Fabrication of Conducting Polymer Nanowires

Although, in general, a good fabrication technique for CP nanowires should enable the simultaneous control of morphology, dimensions, and properties, template-directed electrochemical deposition is the preferred technique for fabricating single-segment and multisegmented CP nanowires. In the first step of the process, a thin film of a conductive layer is evaporated onto one face of the template; this is followed by the sequential electrodeposition of a sacrificial metal layer, material A, material B, material C, and so on (see Figure 16.2). Following deposition of the desired components, the nanowires are released by dissolving the sacrificial layer and the template, with the length of the nanowires being determined by the current density and the deposition time, and their diameter determined by the pore diameter of the template. Chemical polymerization can similarly be achieved by simply immersing the template into a solution of the monomer and an oxidizing agent, and aligned bundles of CP nanowires have been generated *in situ* in this fashion [20]. The presence of an appropriate initiator, as well as careful tuning of the reaction conditions, were found to be crucial parameters for producing aligned nanowires.

Recently, the fabrication of nanowires by the electropolymerization of monomers within nanoscale channel arrays connecting two electrodes has proved to be very attractive [21, 22]. As the nanoscale channels have built-in electrical contacts that become interconnects to the array components, this procedure avoids the harsh, and often cumbersome, template dissolution processes that are necessary in conventional template-directed fabrication methods. Further, any manipulation, alignment and electrical-contacting procedures that are required in the conventional-template directed methods are no longer necessary.

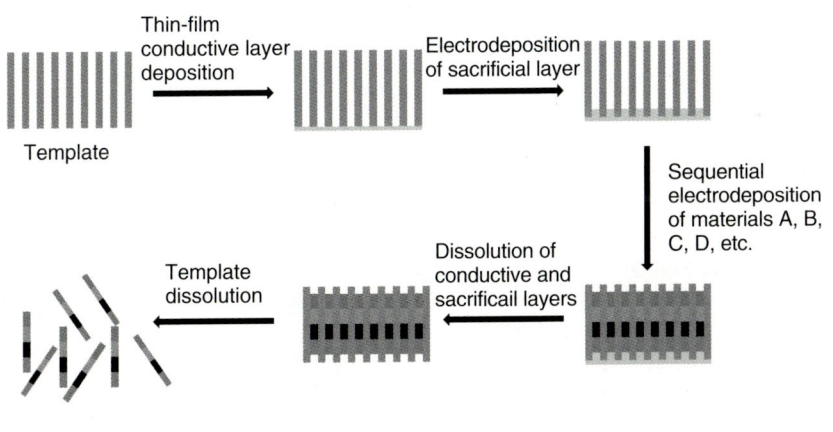

Figure 16.2 Schematic of template-based fabrication of conducting polymer nanowires.

DNA has also been used as a template in the fabrication of polyaniline nanowires. Recently, He and coworkers fabricated polyaniline nanowires by stretching, aligning, and immobilizing double-stranded λ-DNA on a thermally oxidized silicon chip, using a molecular combing method [23]. The DNA templates were then incubated in a protonated aniline monomer solution to emulsify and organize the aniline along the DNA chains. Finally, the aligned monomers were polymerized enzymatically by adding horseradish peroxidase (HRP) and H_2O_2 successively to form polyaniline/DNA nanowires.

Other techniques for CP nanowire fabrication include electrospinning [24–26], lithography [27–29], template-directed vapor deposition polymerization [30], and mechanical stretching [31].

16.3
Surface Modification of Conducting Polymer Nanowires

Although a significant amount of research on the modification of CPs has been based on thin films, some of the techniques should also be applicable to the modification of CP nanowires. In principle, such modifications can be carried out in four ways, namely before, during and after the polymerization process, with the fourth procedure being an entrapment technique where the biomolecule is immobilized during the electrochemical polymerization process. The modification of CP nanowires is necessary for imparting biocompatibility and recognition/binding properties. The interface between the biomolecules and the CP nanowires is critical; for optimal bioactivity and sensitivity, the surface chemistry should ensure the proper attachment, orientation, and accessibility of the biomolecule. Consequently, a number of different modification schemes have been used to attach the various biomolecules to the CP nanowire surfaces.

More recently, the CP nanowires have been modified with enzymes, amino acids, antibodies, nucleic acids, and even with aptamers, with modification following polymerization having been proved as the most popular approach. In this case, an appropriate functional group on the CP nanowire binds covalently to another functional group on the targeted biomolecule. This requires the synthesis of CP nanowires with reactive entities that can be used as anchoring points. Different functional groups, such as carboxylic and amine, have been incorporated successfully into the CP nanowires in this fashion. For example, the presence of carboxylic groups on the CP nanowire surfaces has enabled the direct linking of enzymes [32], amino acids [8], aptamers [33], antibodies, and other proteins [9, 34]. In this case, the carboxylic group of the CP nanowires binds covalently with the primary amine groups in the biomolecules through the carbodiimide (1-ethyl-3-[3-dimethylaminopropyl]carbodiimide hydrochloride; EDC)-mediated (Figure 16.3) [8, 9] or 4-(4,6-dimethoxy-1,3,5-triazin-2-yl)-4-methylmorpholinium chloride-mediated amidation/condensation reactions (Figure 16.4) [32–34]. Alternatively, EDC can be used to activate the carboxylic groups of the biomolecules prior to linkage with the amine groups of the CP nanowires [35]. Glutaradehyde has also

16.3 Surface Modification of Conducting Polymer Nanowires

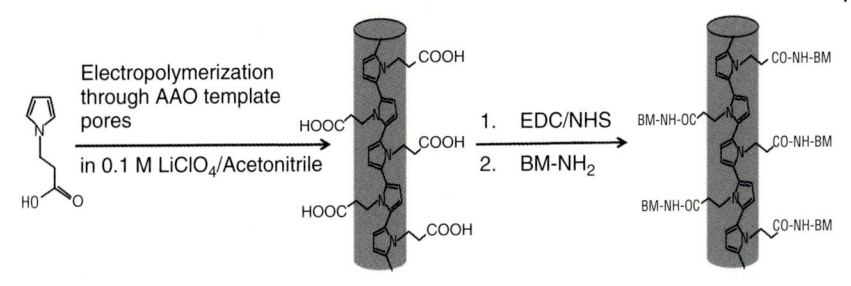

Figure 16.3 Reaction scheme for the EDC-mediated modification of carboxylated CP nanowires. Adapted from Refs [8, 16]. BM, biomolecule; EDC, 1-ethyl-3-(3-dimethylaminopropyl) carbodiimide hydrochloride; NHS, N-hydroxysuccinimide.

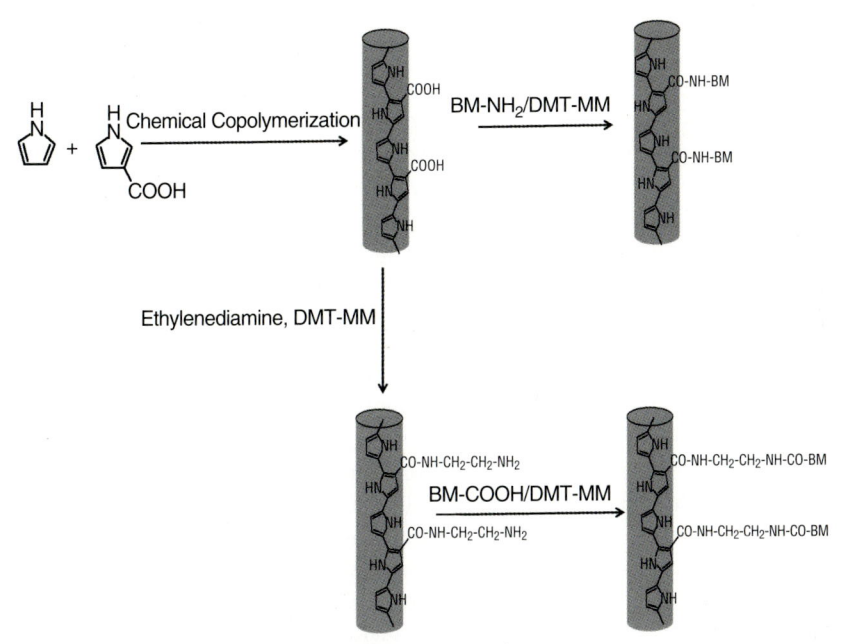

Figure 16.4 Reaction scheme for the DMT-MM-mediated modification of carboxylated CP nanowires. Adapted from Refs [32–34]. BM, biomolecule; DMT-MM, 4-(4,6-dimethoxy-1,3,5-triazin-2-yl)-4-methylmorpholinium chloride).

been used as linker between the amine group of CP nanowires and the carboxyl group of the biomolecules [35]. In this particular application, EDC was found to be a better linker than glutaraldehyde for linking the cancer antigen-125 (CA-125) to the CP nanowires.

Chelator CP nanotubes have been successfully modified with histidine-tagged proteins via copper coordination. In this case, the chelator, nitrilotriacetic acid

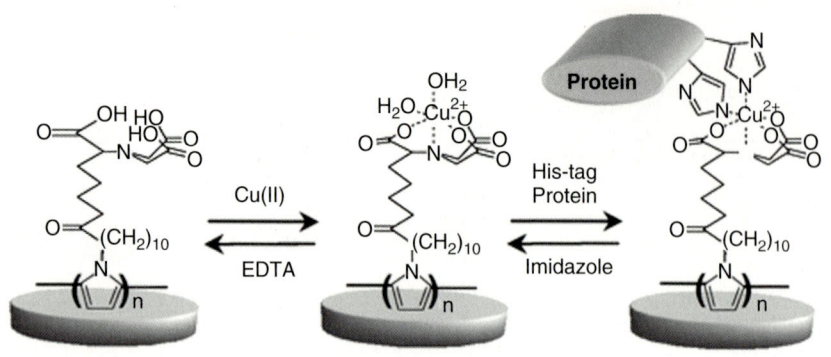

Figure 16.5 Schematic representation of the immobilization of histidine-tagged biomolecules on conducting polymers. Reproduced from Ref. [36]; © 2005, American Chemical Society.

(NTA), was incorporated into the monomer before polymerization, which resulted in the fabrication of NTA-functionalized CP nanotubes; the latter were then exposed to Cu^{2+}, resulting in their complexation. Finally, exposure to histidine-tagged proteins resulted in an immobilization of the nanotubes through their coordination with Cu^{2+} (Figure 16.5) [37]. The major advantage of the modification after polymerization procedure is the opportunity to use optimal conditions for each step, in order to obtain desirable CP nanowire conductivities and biomolecule attachments.

The second technique modification technique involves entrapment of the biomolecule within the CP nanowire matrix. In this case, the monomer is polymerized in a solution containing an appropriate dopant and the targeted biomolecule. Avidin has been used to modify polypyrrole (Ppy) nanowires through this procedure in a single step within lithographically defined nanoscale channels connecting two electrodes [22]. Similarly, avidin-modified segmented gold–Ppy–gold nanowires have been fabricated using a template-directed electrochemical deposition [38]. The main advantages of the entrapment technique are that: (i) the biomolecule undergoes no chemical reaction that could affect its bioactivity; and (ii) specific reactive functional groups are not required on the monomer. The potential disadvantages associated with this procedure include the possibility of reduced accessibility, catalytic activity, and flexibility of the biomolecule as a result of random immobilization and steric hindrances with the CP nanowire matrix.

The emerging technique of imprinting CP nanowires with biomolecules has shown much promise [39, 40]. In this procedure, the imprint molecule is immobilized on the pore walls of silica-coated nanoporous alumina membranes, followed by the template-directed CP nanowire deposition (Figure 16.6). The alumina membranes are then dissolved, leaving the CP nanowires with biological molecule binding sites on the surface. This technique may represent a good alternative to the immobilization of bioreceptors for affinity sensor development.

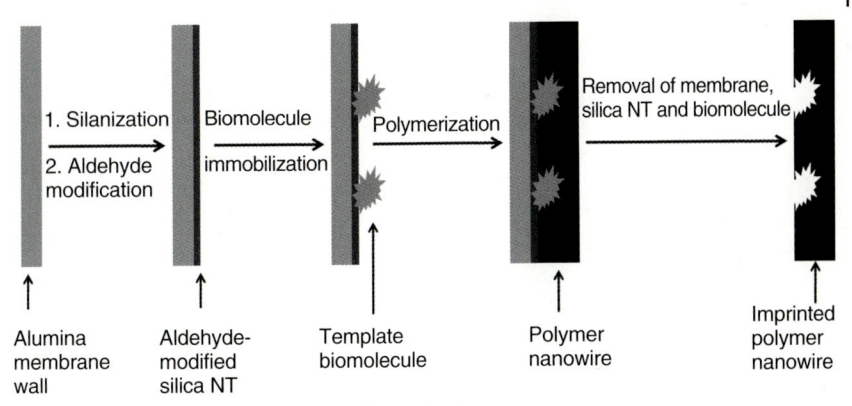

Figure 16.6 Schematic representation of biomolecule imprinting process on a polymer nanowire. Adapted from Ref. [39].

16.4
Assembly/Alignment of Conducting Polymer Nanowires

Whereas, considerable progress has been made in the synthesis of CP nanowires, device fabrication remains somewhat of a challenge due to lack of efficient processes for the assembly and alignment of materials with the desired connectivity into useful architectures and practical functional devices. Therefore, the assembly of CP nanowires into practical and functional devices is sometimes necessary, depending on the method used for their fabrication. Those CP nanowires fabricated in predefined nanochannels do not require any alignment, whereas those fabricated by template-directed methods require some type of assembly/alignment. A good alignment technique should be affordable, fast, defect-tolerant, and also compatible with a variety of materials in various dimensions.

Recently, *magnetic alignment* has been the preferred technique for assembling CP nanowires into functional devices. For this, ferromagnetic elements such as Ni and Co have been electrodeposited with CP nanowires to fabricate multisegmented nanowires with magnetic properties. For example, Au–Ppy–Ni–Au nanowires have been fabricated by their sequential electrodeposition in anodic alumina oxide (AAO) templates (Figure 16.7).

Magnetized multisegmented nanowires exhibit an interesting behavior when suspended in a fluid. Under the influence of an external magnetic field, the magnetized segment controls the orientation of the entire multisegmented nanowire, by allowing only two possible orientations, rather like a magnetic dipole. This property has been used to align nanowires onto gold-coated Ni electrodes, thus resulting in the fabrication of functional devices [13]. The ferromagnetic Ni electrodes were essential for the controlled alignment of the nanowires.

Other research groups have used assembled CP nanowires with active carboxyl groups onto electrode surfaces that had been treated with (3-aminopropyl)

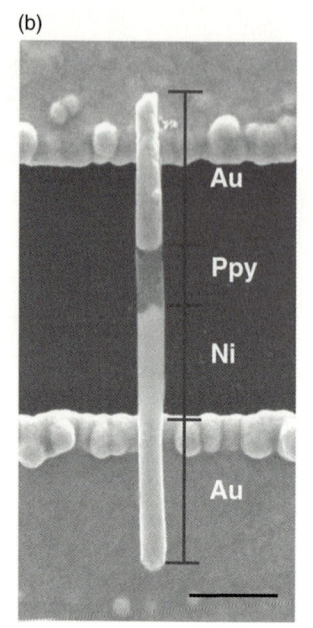

Figure 16.7 (a) Schematic of magnetic alignment; (b) Scanning electron microscopy image of single multisegmented nanowires aligned across electrodes, using the magnetic technique. Scale bar=1 μm. Reproduced from Ref. [13]; © 2009, Wiley Interscience.

trimethoxysilane (APS). The latter compound has primary amino groups that react with the CP nanowire carboxyl groups in the presence of 4-(4,6-dimethoxy-1,3,5-triazin-2-yl)-4-methylmorpholinium chloride (Figure 16.8) [32–34, 41].

16.5
Biomedical Applications of Conducting Polymer Nanowires

16.5.1
Sensing and Detection

16.5.1.1 Proteins and Disease Markers
Most biomedical applications of CP nanowires have involved the biosensing of proteins. For example, poly(pyrrolepropylic) (polyPPA) nanowires were modified with anti-human serum albumin (anti-HSA) utilized in the detection of human serum albumin (HSA), a molecule indicative of incipient renal disease. The electrical detection was based on the selective binding of HSA onto the immobilized anti-HAS, whereby the binding event altered the electrical properties of the CP

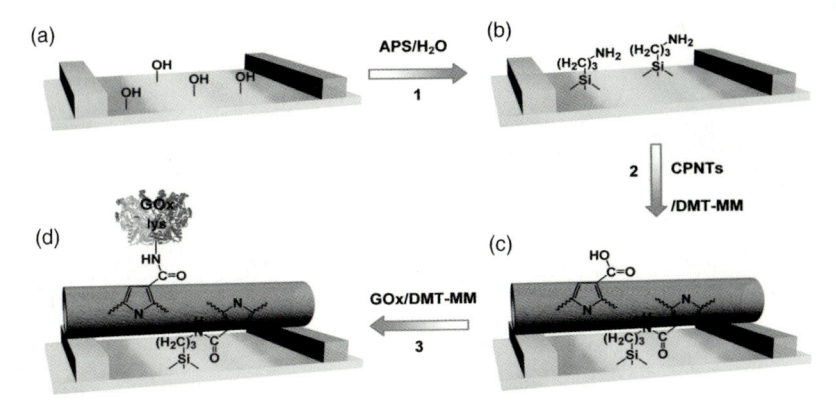

Reaction 1. Hydrolysis: $H_2N(CH_2)_3Si(OCH_3)_3 + 3H_2O \rightarrow H_2N(CH_2)_3Si(OH)_3 + 3CH_3OH$
Condensation: $H_2N(CH_2)_3Si(OH)_3 + 3OH\text{-substrate} \rightarrow H_2N(CH_2)_3Si(O)_3\text{-substrate}$

Reaction 2. Condensation: $\text{pyrrole-COOH} + H_2N(CH_2)_3Si(O)_3\text{-substrate} \rightarrow$
$\text{pyrrole-CONH}(CH_2)_3Si(O)_3\text{-substrate}$

Reaction 3. Condensation: $\text{pyrrole-COOH} + H_2N\text{-lys-GOx} \rightarrow \text{pyrrole-CONH-lys-GOx}$

Figure 16.8 Schematic of nanowires assembly and functionalization of CP nanotubes. (a) Microelectrode substrate; (b) Aminosilane-treated substrate; (c) Nanotube alignment; (d) Nanotube functionalization. Reproduced from Ref. [32]. APS, (3-aminopropyl)trimethoxysilane; DMM-MM, 4-(4,6-dimethoxy-1,3,5-triazin-2-yl)-4-methylmorpholinium chloride; GOx, glucose oxidase.

nanowires, thus enabling real-time detection in a field-effect transistor (FET) format (Figure 16.9) [9]. In this case, pyrrolepropylic acid (PPA) was synthesized and electropolymerized on AAO templates prior to being modified with anti-HSA. Other proteins, such as bovine serum albumin (BSA), had a negligible effect on the electrical properties of the nanowires. Notably, this device enable the detection of HSA down to a concentration of 50 nM.

In a similar, recently described approach, the CA-125 antigen was detected using a single Ppy nanowire chemiresistive immunosensor that had been modified with anti-CA-125 antibodies [35]. For this, the nanowire was anchored across two gold electrodes by selectively electrodepositing gold onto the electrodes. The CA-125 antigen is detected in many ovarian cancer cells, and can be used as a marker to monitor the treatment or recurrence of ovarian cancer. In this case, the nanowire immunosensor exhibited a detection limit of 1 U ml^{-1} CA-125, and a linear dynamic range of up to 1000 U ml^{-1} (Figure 16.10).

Aptamer-modified CP nanotubes have been used to detect thrombin, a coagulation protein. In this case, the CP nanotubes were covalently immobilized and assembled onto microelectrode substrates that had been treated with APS. The thrombin aptamer was modified with a primary aliphatic amino linker, and then covalently bound to the carboxylated CP nanotube in the presence of

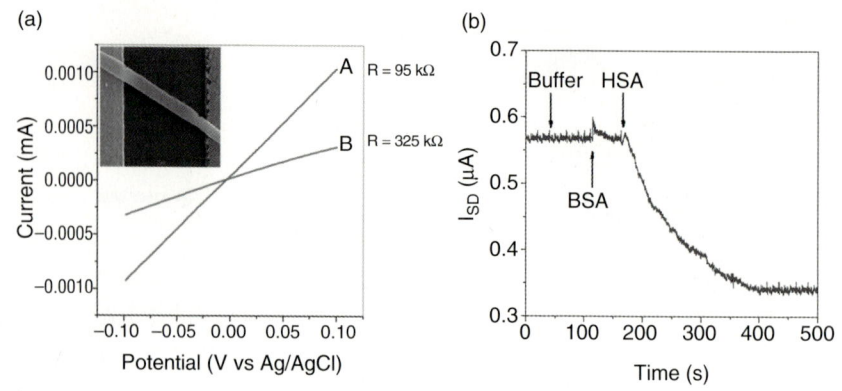

Figure 16.9 (a) Current versus voltage (I–V) profile recorded across anti HAS-modifies nanowires before (A) and after (B) binding with HAS; (b) Real-time dynamic current response on sequential exposure of buffer, BSA, and HSA. Reproduced from Ref. [9]; © 2009, Springer.

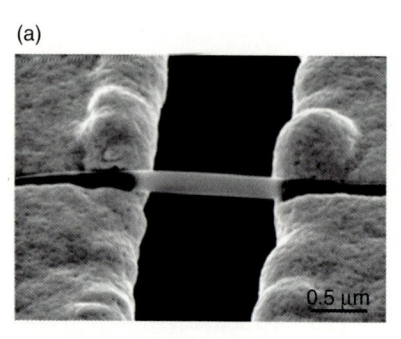

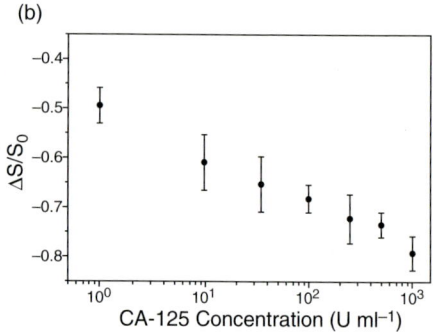

Figure 16.10 (a) Scanning electron microscopy image of a maskless electrodeposited single polypyrrole (PPy) nanowire device, showing selective gold electrodeposition on the electrodes resulting in a partial anchoring of the nanowire; (b) Calibration curve in terms of normalized conductance change of a single Ppy nanowire biosensor for CA-125 detection in spiked human blood plasma. Reproduced from Ref. [35]; © 2009, American Chemical Society.

4-(4,6-dimethoxy-1,3,5-triazin-2-yl)-4-methylmorpholinium as the condensing agent [33]. The FET-configured aptasensor was able to detect thrombin down to 50 nM, a level that was reasonably comparable to that detected with CNT-based FET biosensors [42].

16.5.1.2 Detection of Bacteria

Polyaniline (PANI) nanowires have been employed in the detection of *Bacillus* species, by using an electrical detection combined with lateral flow technology [43]. Although the basic technology that underlies lateral flow systems was first

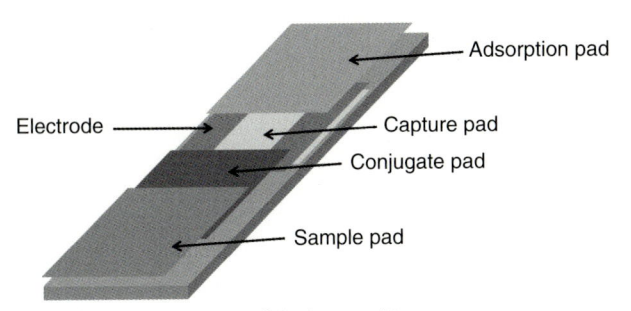

Figure 16.11 Components of the bacteria biosensor.

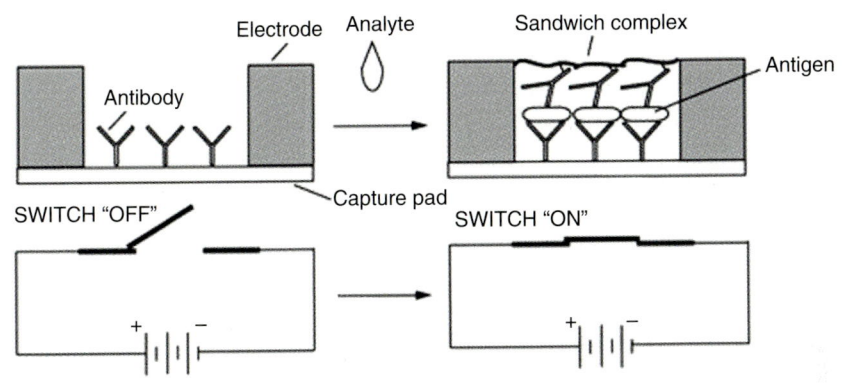

Figure 16.12 Upper row: Schematic of the capture pad before and after analyte application. Bottom row: Schematic of the corresponding electrical circuit formed before and after analyte application.

described during the 1960s, the first commercial application was the home pregnancy test strip, launched in 1988.

For bacterial sensing, the biosensor consisted of four membranous pads; a sample application pad, a conjugate pad, a capture pad, and the absorption pad. Silver electrodes were constructed along the sides of the capture pad (Figure 16.11). The conjugate pad consisted of PANI nanowires conjugated to *Bacillus* antibodies, while the capture pad consisted of permanent *Bacillus* capture antibodies that had been immobilized using glutaraldehyde chemistry. When a liquid sample containing the target analyte was applied to the sample application pad, the pad was rehydrated and mixed with the PANI nanowires conjugated to the *Bacillus* antibodies; this resulted in a binding of the two species. The target–PANI-nanowire–antibody conjugate mixture then flowed by capillary action, passing over the capture pad where the capture antibody became bound to the target to form a sandwich complex. The PANI-nanowire then completed the circuit by making contact with the two electrodes, and the electrical resistance was recorded (Figure 16.12). The biosensor sensitivity in pure cultures of *Bacillus cereus* was found to be 10^1 to 10^2 colony-forming units (CFU) ml^{-1} [43].

Similar detections of *Klebsiella pneumoniae, Pseudomonas aeruginosa, Escherichia coli* and *Enterococcus faecalis* have also been made using an "on–off"-type PANI-nanowire sensor. In this case, attachment of the bacteria onto the PANI-nanowires caused a local modification of the nanowire electrical conductivity, rendering them electrically inhomogeneous [44]. These defects in the nanowires enabled an easy flow of charge carriers, thus switching on the sensor, which was insensitive below a threshold number of bacteria. Despite similarities in the switching effects, the time profile, retention time and half-width of the signals that could be measured were specific for each bacterium, because of the unique interactions with the PANI-nanowires.

16.5.1.3 Detection of Small Molecules

Small molecules, such as glucose, have also been detected using CP nanotubes, as demonstrated by Jang and coworkers [32]. This involved the covalent modification of CP nanotubes with glucose oxidase (GOx). Challenging the biosensor with glucose resulted in the production of H_2O_2 that altered the charge transport property of the CP nanotubes, leading to an increase in source–drain current (I_{SD}) of the conducting polymers. (Figure 16.13). In another approach, GOx was immobilized in a Ppy-nanowire–platinum nanoparticle composite on a glassy carbon electrode, and used for the amperometric detection of H_2O_2 from glucose oxidation [45].

16.5.1.4 Detection of Heavy-Metal Ions and Pesticides

Previously, CP nanowires have been used for the detection of environmentally important species such as heavy-metals and organophosphates. In the first case, Cu^{2+} was detected at 0.6 ppt levels using a chemiresistive biosensor that was based on a single Ppy-NTA nanotube. The chelator, NTA, was incorporated into the monomer before polymerization; this resulted in NTA-functionalized CP nanotubes that were then exposed to Cu^{2+}, resulting in their sensitive detection [37]. However, the selectivity of the sensor was limited by the presence of competing metals such as Ni^{2+} or Cd^{2+}, which also complexed with NTA.

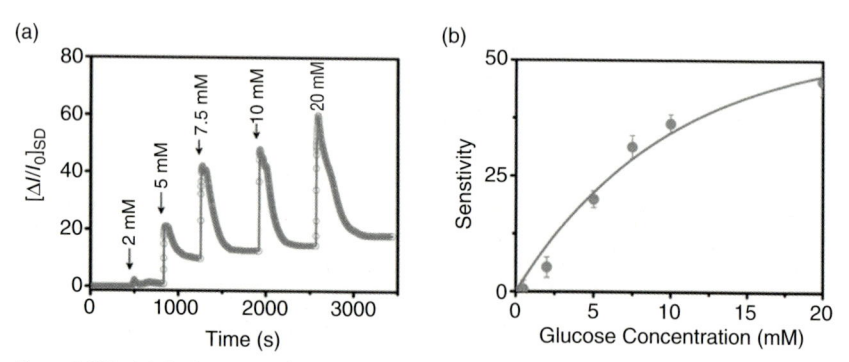

Figure 16.13 (a) Real-time I_{SD} change upon consecutive addition of 2 to 20 nM glucose (VSD = −10 mV); (b) The corresponding calibration curve. VSD, source-drain potential.

An organophosphate, methyl parathion, was detected by the immobilization of acetylcholinesterase (AChE) onto gold nanoparticle–Ppy-nanowire composites on glassy carbon electrodes. Inhibition of AChE activity by the organosphosphate was measured by monitoring the interaction of AChE with the substrate acetylthiocholine, that produces electroactive thiocholine. In this case, a too little inhibition resulted in extensive thiocholine production, and *vice versa* [46].

16.5.2
Drug Delivery and DNA Carriers

The unique actuating properties of conducting polymers make them interesting materials for investigation as potential drug-release platforms. Recently, polyethylenedioxythiopene (PEDOT) nanotubes were investigated for controlled drug release [47], whereby PEDOT was first electropolymerized on top of electrospun polylactide-*co*-glycolide) (PLGA) nanofibers (ca. 100 nm diameter) that had been loaded with dexamethasone, an anti-inflammatory drug. The subsequent degradation of PLGA, a biodegradable polymer, led to the release of dexamethasone that could be controlled by the expansion or contraction of the PEDOT nanotubes, caused by their electrical stimulation (Figure 16.14). In this case, the drug release was believed to be related directly to the force and duration of contraction, with more forceful and longer contractions causing the release of more drug. The electrical stimulation causes electrons to be injected into the PEDOT nanotube, such that the positive charges in the polymer are compensated. In order to maintain an overall charge neutrality, counterions are expelled towards the solution; this causes the nanotube to contract, forcing the drug out from the nanotube ends (Figure 16.14). In contrast, the PEDOT around the PLGA slows down the drug release.

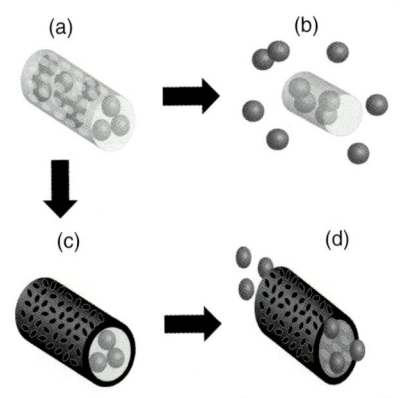

Figure 16.14 Schematic of PEDOT nanotube-based controlled drug release.
(a) Dexamethasone-loaded PLGA;
(b) Degradation of PLGA leading to drug release; (c) Electrochemical deposition of PEDOT around PLGA fibers; (d) Drug release is slowed; (e) PEDOT nanotubes in neutral electrical condition; (f) Electrical stimulation controls release of the drug. Reproduced from Ref. [47]; © 2006, Wiley Interscience.

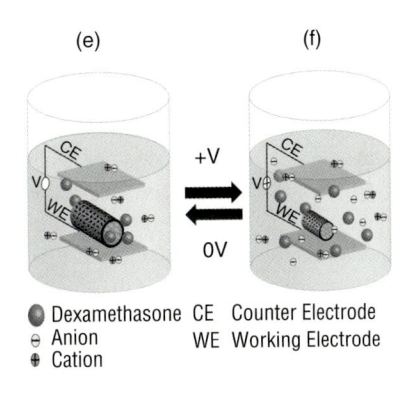

Carboxylated CP nanotubes have been applied as molecular probes and DNA carriers, by attaching them to amine-functionalized silica nanoparticles and subsequently conjugating them with pyreneacetic acid (a photoluminescent molecule) and immobilizing single-stranded DNA (ssDNA) on the surface [30]. Likewise, the attachment of CP nanotubes to pyreneacetic acid and BRAC 1 (a tumor suppressor gene, the mutation of which increases susceptibility to breast cancer) was also demonstrated [30], indicating the potential of CP nanotubes for *in vivo* gene delivery and expression.

16.6
Summary and Future Perspectives

There is no doubt that CP nanowires have recently emerged as excellent 1-D nanomaterials for fabricating state-of-the-art sensor devices, notably as they exhibit certain characteristics that render them truly adaptable. For example, they can be fabricated under ambient conditions, using benign reagents, and they can also be modified before, during, and after polymerization, which provides them with a robust nature such that optimal conditions can be used for each step. The conductivities of CPs can be modulated up to 15 orders of magnitude, by changing the dopant and/or dopant/monomer ratios, while those of CP nanowires can be modulated by controlling their oxidation state. Furthermore, the CP nanowires can also function at ambient or low operating temperatures.

Although tremendous advances have been made in the fabrication of CP nanowires, CP nanotubes, and related structures, several problems remain associated with the large-scale fabrication of these materials with well-controlled and consistent dimensions, morphology, and phase purity. In the case of template-fabricated nanowires, the main challenge involves improving the development of methods for the integration, assembly and alignment of these materials into functional device architectures. Likewise, the mechanical properties of these materials will become an important issue that must be addressed if they are to achieve commercial success. Whether these materials can withstand long-term storage under variable environmental conditions, mechanical stresses such as stretching, and bending and compression, or whether they would need to be mechanically or chemically reinforced remains to be determined. Moreover, if such changes were necessary, how might they affect the performance of the materials? Clearly, such questions need to be addressed before the materials achieve commercial success.

Whereas, great progress has been made recently in the fabrication, modification, and assembly of these materials into functional devices, it is clear that further research is required in this area during the coming years.

References

1 Zheng, G., Patolsky, F., Cui, Y., Wang, W.U. and Lieber, C.M. (2005) Multiplexed electrical detection of cancer markers with nanowire

1. sensor arrays. *Nature Biotechnology*, **12**, 1294.
2. Li, C. et al. (2005) Complementary detection of prostate-specific antigen using in2o3 nanowires and carbon nanotubes. *Journal of the American Chemical Society*, **127** (36), 12484–5.
3. Cui, Y. et al. (2001) Nanowire nanosensors for highly sensitive and selective detection of biological and chemical species. *Science*, **293** (5533), 1289–92.
4. Wang, W., Chen, C., Lin, K., Fang, Y. and Lieber, C.M. (2005) Label-free detection of small-molecule–protein interactions by using nanowire nanosensors. *Proceedings of the National Academy of Sciences of the United States of America*, **102**, 3208–12.
5. Patolsky, F., Zheng, G., Hayden, O., Lakadamyali, M., Zhuang, X. and Lieber, C.M. (2004) Electrical detection of single viruses. *Proceedings of the National Academy of Sciences of the United States of America*, **101**, 14017–22.
6. Star, A., Tu, E., Niemann, J., Gabriel, J.P., Joiner, S. and Valcke, C. (2006) Label-free detection of DNA hybridization using carbon nanotube network field-effect transistors. *Proceedings of the National Academy of Sciences of the United States of America*, **103**, 921–6.
7. Chen, R.J., Bangsaruntip, S., Drouvalakis, K.A., Wong, N., Kam, S., Shim, M., Li, Y., Kim, W., Utz, P.J. and Dai, H. (2003) Noncovalent functionalization of carbon nanotubes for highly specific electronic biosensors. *Proceedings of the National Academy of Sciences of the United States of America*, **100**, 4984–9.
8. Sagar, T. et al. (2010) Rapid and efficient removal of heavy metal ions from aqueous media using cysteine-modified polymer nanowires. *Journal of Applied Polymer Science*, **116** (1), 308–13.
9. Tolani, S. et al. (2009) Towards biosensors based on conducting polymer nanowires. *Analytical and Bioanalytical Chemistry*, **393** (4), 1225–31.
10. Martin, C.R. (1994) Nanomaterials: a membrane-based synthetic approach. *Science*, **266** (5193), 1961–6.
11. Martin, C.R. (1995) Template synthesis of electronically conductive polymer nanostructures. *Accounts of Chemical Research*, **28** (2), 61–8.
12. Samantha, A.M. et al. (2007) Metal/conducting-polymer composite nanowires. *Small*, **3** (2), 239–43.
13. Mangesh, A.B. et al. (2009) Magnetically assembled multisegmented nanowires and their applications. *Electroanalysis*, **21** (1), 61–7.
14. Park, S., Chung, S.-W. and Mirkin, C.A. (2004) Hybrid organic–inorganic, rod-shaped nanoresistors and diodes. *Journal of the American Chemical Society*, **126** (38), 11772–3.
15. Sarah, J.H. et al. (2006) Multisegmented one-dimensional nanorods prepared by hard-template synthetic methods. *Angewandte Chemie International Edition*, **45** (17), 2672–92.
16. Wanekaya, A.K., Chen, W., Myung, N.V. and Mulchandani, A. (2007) Conducting polymer nanowire-based biosensors, in *Handbook of Biosensors and Biochips* (eds R.S. Marks, D.C. Cullen, I. Karube, C.R. Lowe and H.H. Weetal), John Wiley & Sons, Ltd., pp. 831–42.
17. Wanekaya, A.K., Chen, W., Myung, N.V. and Mulchandani, A. (2007) Conducting polymer nanowire-based BioFET for label-free detection, in *Smart Biosensor Technology* (eds G.K. Knopf and A.S. Bassi), CRC Press, Boca Raton, FL, pp. 133–49.
18. Adam, K.W. et al. (2006) Nanowire-based electrochemical biosensors. *Electroanalysis*, **18** (6), 533–50.
19. Cho, S.I. and Lee, S.B. (2008) Fast electrochemistry of conductive polymer nanotubes: synthesis, mechanism, and application. *Accounts of Chemical Research*, **41** (6), 699–707.
20. Wang, Y., Tran, H.D. and Kaner, R.B. (2009) Template-free growth of aligned bundles of conducting polymer nanowires. *Journal of Physical Chemistry C*, **113** (24), 10346–9.
21. Ramanathan, K. et al. (2004) Individually addressable conducting polymer nanowires array. *Nano Letters*, **4** (7), 1237–9.
22. Ramanathan, K. et al. (2005) Bioaffinity sensing using biologically functionalized conducting-polymer nanowire. *Journal of the American Chemical Society*, **127** (2), 496–7.

23 Ma, Y. et al. (2004) Polyaniline nanowires on Si surfaces fabricated with DNA templates. *Journal of the American Chemical Society*, **126** (22), 7097–101.

24 Jun, K. and Craighead, H.G. (2003) Fabrication of oriented polymeric nanofibers on planar surfaces by electrospinning. *Applied Physics Letters*, **83** (2), 371–3.

25 Kameoka, J. et al. (2002) An electrospray ionization source for integration with microfluidics. *Analytical Chemistry*, **74** (22), 5897–901.

26 Liu, H. et al. (2004) Polymeric nanowire chemical sensor. *Nano Letters*, **4** (4), 671–5.

27 Kim, S.R. et al. (2002) Fabrication of polymeric substrates with well-defined nanometer-scale topography and tailored surface chemistry. *Advanced Materials*, **14** (20), 1468–72.

28 Dong, B. et al. (2005) Patterning of conducting polymers based on a random copolymer strategy: toward the facile fabrication of nanosensors exclusively based on polymers. *Advanced Materials*, **17** (22), 2736–41.

29 Maynor, B.W. et al. (2001) Direct-writing of polymer nanostructures: poly(thiophene) nanowires on semiconducting and insulating surfaces. *Journal of the American Chemical Society*, **124** (4), 522–3.

30 Jang, J., Ko, S. and Kim, Y. (2006) Dual-functionalized polymer nanotubes as substrates for molecular-probe and DNA-carrier applications. *Advanced Functional Materials*, **16** (6), 754–9.

31 He, H.X., Li, C.Z. and Tao, N.J. (2001) Conductance of polymer nanowires fabricated by a combined electrodeposition and mechanical break junction method. *Applied Physics Letters*, **78** (6), 811–13.

32 Yoon, H., Ko, S. and Jang, J. (2008) Field-effect-transistor sensor based on enzyme-functionalized polypyrrole nanotubes for glucose detection. *Journal of Physical Chemistry B*, **112** (32), 9992–7.

33 Hyeonseok, Y. et al. (2008) A novel sensor platform based on aptamer-conjugated polypyrrole nanotubes for label-free electrochemical protein detection. *ChemBioChem*, **9** (4), 634–41.

34 Hyeonseok, Y. et al. (2009) Polypyrrole nanotubes conjugated with human olfactory receptors: high-performance transducers for FET-type bioelectronic noses. *Angewandte Chemie International Edition*, **48** (15), 2755–8.

35 Bangar, M.A. et al. (2009) Single conducting polymer nanowire chemiresistive label-free immunosensor for cancer biomarker. *Analytical Chemistry*, **81** (6), 2168–75.

36 Haddour, N., Cosnier, S. and Gondran, C. (2005) Electrogeneration of a poly(pyrrole)-NTA chelator film for a reversible oriented immobilization of histidine-tagged proteins. *Journal of the American Chemical Society*, **127** (16), 5752–3.

37 Aravinda, C.L. et al. (2009) Label-free detection of cupric ions and histidine-tagged proteins using single poly(pyrrole)-NTA chelator conducting polymer nanotube chemiresistive sensor. *Biosensors and Bioelectronics*, **24** (5), 1451–5.

38 Hernandez, R.M. et al. (2004) Template fabrication of protein-functionalized gold–polypyrrole–gold segmented nanowires. *Chemistry of Materials*, **16** (18), 3431–8.

39 Li, Y. et al. (2005) Protein recognition via surface molecularly imprinted polymer nanowires. *Analytical Chemistry*, **78** (1), 317–20.

40 Yang, H.-H. et al. (2005) Surface molecularly imprinted nanowires for biorecognition. *Journal of the American Chemical Society*, **127** (5), 1378–9.

41 Yoon, H. and Jang, J. (2008) A field-effect-transistor sensor based on polypyrrole nanotubes coupled with heparin for thrombin detection. *Molecular Crystals and Liquid Crystals*, **491**, 21–31.

42 So, H.-M. et al. (2005) Single-walled carbon nanotube biosensors using aptamers as molecular recognition elements. *Journal of the American Chemical Society*, **127** (34), 11906–7.

43 Pal, S., Alocilja, E.C. and Downes, F.P. (2007) Nanowire labeled direct-charge transfer biosensor for detecting *Bacillus* species. *Biosensors and Bioelectronics*, **22** (9–10), 2329–36.

44 Langer, J.J. et al. (2009) New "ON-OFF"-type nanobiodetector. *Biosensors and Bioelectronics*, **24** (9), 2947–9.

45 Li, J. and Lin, X. (2007) Glucose biosensor based on immobilization of glucose oxidase in poly(o-aminophenol) film on polypyrrole-Pt nanocomposite modified glassy carbon electrode. *Biosensors and Bioelectronics*, **22** (12), 2898–905.

46 Gong, J., Wang, L. and Zhang, L. (2009) Electrochemical biosensing of methyl parathion pesticide based on acetylcholinesterase immobilized onto Au-polypyrrole interlaced network-like nanocomposite. *Biosensors and Bioelectronics*, **24** (7), 2285–8.

47 Abidian, M.R., Kim, D.H. and Martin, D.C. (2006) Conducting-polymer nanotubes for controlled drug release. *Advanced Materials*, **18** (4), 405–9.

17
Organic Nanowires and Nanotubes for Biomedical Applications
Keunsoo Jeong and Chong Rae Park

17.1
Introduction

For many years, one-dimensional (1-D) nanostructures such as nanowires (NWs) and nanotubes (NTs) have fascinated materials scientists by their peculiar morphology and properties in a variety of applications, ranging from electronics to biomedicine [1–6]. In biomedical applications especially, the 1-D nanostructures of organic molecules possess certain advantages over conventional micelle-type spherical structures of both natural [7–11] and synthetic molecules [12–14]. An example of this is that conducting NWs and NTs are each more effective than spherical structures for the delivery of electrical signals for electrical biosensors, due to the free passage of electrons that they provide.

In the past, many different techniques have been developed to fabricate organic NWs and NTs. Notably, the self-assembly of organic molecules is one of the most widely adopted processes used to fabricate organic NWs and NTs [15–21]. In fact, as many 1-D biomolecular nanostructures are driven by the self-assembly process [15–17], the majority of synthetic molecules [18–21] – if correctly designed – can also form similar nanostructures via the self-assembly process. As an alternative route for the fabrication of various organic NWs and NTs, the template-based has certain advantages in that the diameter, density, and length of the NWs and NTs can be relatively easily controlled, if the template itself is prepared in controlled manner. Unfortunately, it is not so easy to produce appropriate templates with pore channels of a desired diameter, length, and surface chemistry. Moreover, another negative aspect of this approach that the template itself must be completely removed without causing damage to the integrity of the NWs or NTs that have been produced.

The fabrication of 1-D nanostructures based on carbon nanotubes (CNTs) represents yet another preparative class for NTs. Indeed, CNT-based NTs for the biomedical applications can be obtained relatively easily by functionalizing CNTs via either the covalent attachment of functional groups, or by physical attachment via $\pi-\pi$ interactions between CNTs and mediating molecules with functional groups.

Many different types of NW and NT (including CNTs), have shown great potential in a wide variety of biomedical applications, including optical imaging, electrical sensing, and cancer therapy. The unique functions and properties of nanostructures of this type offers new opportunities to identify possible applications for organic NWs, NTs and CNTs. In order to demonstrate the current state of the art of this topic, and to further the current understanding of organic NWs and NTs, some recent advances in the fabrication and possible biomedical application of these materials are outlined, together with suggestions as to their future development.

17.2
Fabrication of Organic Nanowires and/or Nanotubes

17.2.1
Self-Assembly Processes

Both organic NWs and NTs, when fabricated via the self-assembly of building block molecules, may have reactive functional groups on their surface, and so may be used as carriers following their conjugation with materials such as drugs and imaging materials. The self-assembly process has been suggested as one of the most effective methods to fabricate NWs and NTs of either low- or high-molecular-weight organic molecules. However, in order to achieve successful self-assembly, the organic molecules must include a molecular structure that gives rise to relativistic intermolecular interactions. Examples of these include hydrophobic and hydrophilic moieties in the case of surfactants [22–24], rigid and flexible or long and short units in the case of block copolymers [12–14, 18], and functional or polar groups that ensure hydrogen bonding and/or dipole–dipole interaction in the case of biomolecules such as proteins and nucleotides [15–17].

The rod–coil molecules consisting of aromatic rod and flexible polyethylene oxide (PEO) [25–27], when dissolved in water, self-assemble into discrete spherical micelles with a uniform diameter of about 10 nm [25]. This spherical micellar structure slowly changes to cylindrical wires over about seven days [26], indicating that cylindrical wires represent a thermodynamically stable structure, which is consistent with rod–coil molecules based on a long rod length. Cylindrical nanostructures with a uniform diameter of about 10 nm and lengths of up to several hundred nanometers are shown in Figure 17.1.

In another approach, the amphiphilic rigid macrocycle with hydrophilic dendritic chains demonstrates a unique example of tubules in aqueous solution, as shown in Figure 17.2 [28]. In this case, the rigid macrocyclic segments stack directly on top of each other to form a tubular aggregate that is composed of a hydrophobic, stiff interior with a hydrophobic internal cavity and a hydrophilic, flexible exterior.

A gemini-shaped hexabenzocoronene amphiphile was also reported as an organic building block to form NTs, the graphitic wall of which is densely covered

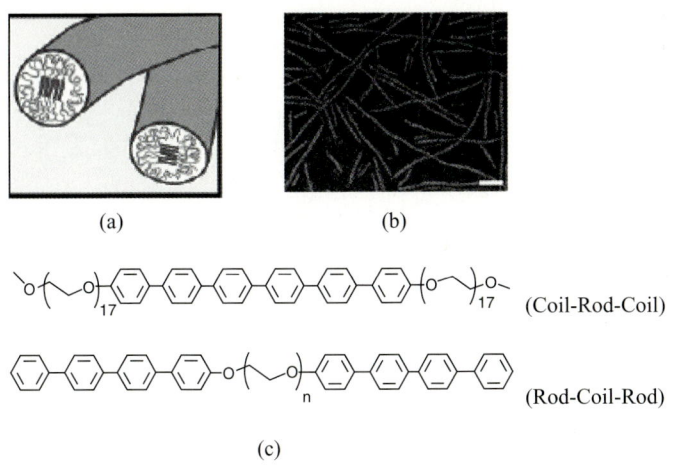

Figure 17.1 (a) Schematic illustration and (b) transmission electron microscopy image of NWs resulting from self-assembly of (c) rigid-coil type molecules. Reproduced with permission from Ref. [26]; © 2005, American Chemical Society.

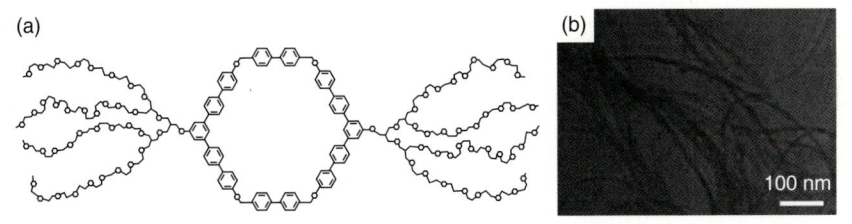

Figure 17.2 (a) Molecular structure and (b) transmission electron microscopy image of a nanotube based on macrocyclic molecule with hydrophilic tails. Reproduced with permission from Ref. [28]; © 2005, The Royal Society of Chemistry.

by positively charged molecules (see Figure 17.3) [29, 30]. The NTs may be dispersed well in aqueous media, with further functionalization of the surface of NTs being made possible by the interaction of cations on the surface with anion guests (Figure 17.3b). These unique features of the NTs may be useful when serving as carriers of anionic biomolecules, including DNA.

Polycyclic aromatic hydrocarbons (PAHs) may also serve as building block molecules to form NWs and NTs via self-assembly, induced by π–π interactions between the PAH molecules. In fact, the functionalization of PAHs improves both their aggregation behavior and their potential for biomedical applications. The introduction of heteroatoms into PAHs affects their intermolecular interactions when they are fabricated into NWs and/or NTs. Notably, a series of amphiphilic centrally charged PAHs exhibited variable types of 1-D nanostructure with regards

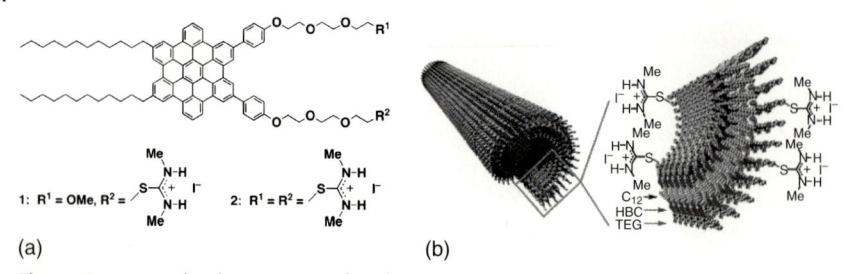

Figure 17.3 (a) Molecular structure of isothiouronium ion-appended hexa-*peri*-hexabenzocoronene; (b) Representation of the graphitic nanotube from hexa-*peri*-hexabenzocoronene. Reproduced with permission from Ref. [29]; © 2007, American Chemical Society.

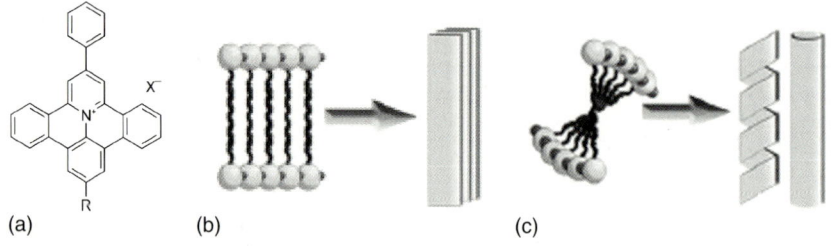

Figure 17.4 (a) Polycyclic aromatic hydrocarbons (PAHs) and schematic representation of self-assembly of PAHs to (b) nanoribbons, (c) helices, and tubes. Reproduced with permission from Ref. [31]; © 2007, Wiley-VCH Verlag GmbH & Co. KGaA.

to the length of the alkyl chains and the counterions of the charged PAHs (Figure 17.4) [31].

It is widely accepted that both natural and synthetic polypeptides form a nanostructure of the type associated with β-sheet peptides [32–34]. When polypeptide units are branched to a coil-type polymer (e.g., PEO), the polypeptide branch units form β-sheet-structured strands, whereby the side groups of the amino acids in the peptide chains orient alternately to opposite sides of the sheet. In an aqueous medium, the PEO with peptide branches will self-assemble into nanofibrils in which the β-sheet peptides are organized such that each β-strand runs perpendicular to the fibril axis, as shown in Figure 17.5. However, when the hydrophilic polymer (such as PEO) is substituted with a hydrophobic polymer, the β-sheet units self-assemble into a bilayered structure.

17.2.2
Template-Based Synthetic Processes

Typically, conducting polymer (CP)-based NWs and NTs have been prepared using a hard template synthesis method, and used as biosensors [35–43]. In order to

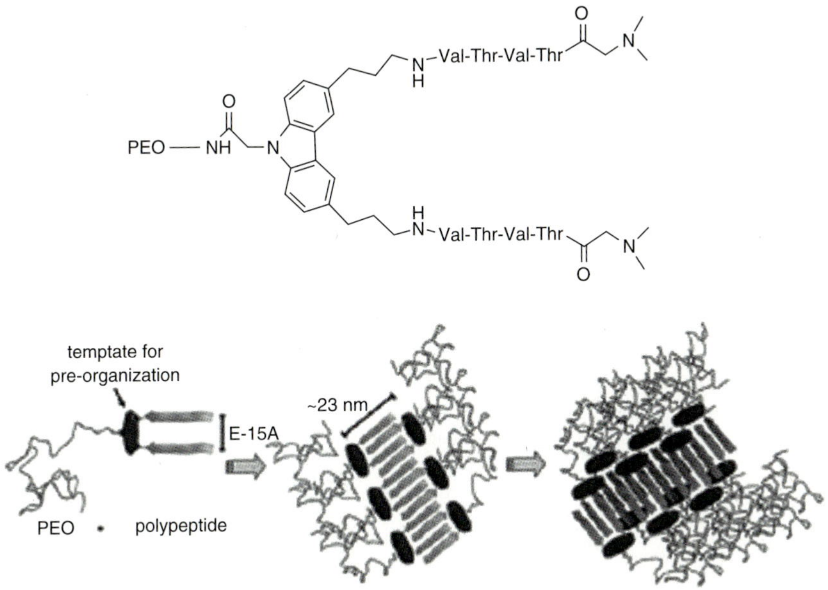

Figure 17.5 Artificially designed PEO-coated β-sheet nanostructure. Reproduced with permission from Ref. [34]; © 2008, The Royal Society of Chemistry.

prepare CP-NWs and CP-NTs, the hard templates – which usually take the form of anodized aluminum oxide (AAO) or polymer-based membranes – are first immersed in a solution of monomer, after which the initiator is added to start the polymerization. During this procedure, the CP was synthesized within the pores of the templates, and the CP-NWs and CP-NTs could subsequently be obtained following removal of the templates. As an example, composite NWs of Au–polypyrrole (Au-PPy) presenting various diameters were efficiently prepared, as shown schematically in Figure 17.6 [36]. Similarly, polythiophene (PT) and poly(3-methylthiophene) (P3MT) -based NTs and NWs, with diameters in the range of 100–200 nm, were synthesized by using an AAO template [37].

The NWs of gold/polyaniline (Au/PANI) composites, with an average diameter of 50–60 nm, were prepared by reacting the aniline monomer with chlorauric acid (HAuCl$_4$), followed by a self-assembly process in the presence of d-camphor-10-sulfonic acid (CSA), that acts as both a dopant and surfactant [44]. Subsequently, the formation probability and size of the Au/PANI NW were found to depend on the molar ratio of aniline to HAuCl$_4$ and the concentration of CSA, respectively. The directly measured conductivity of a single Au/PANI was 77.2 S cm^{-1}. Hollow PANI NWs, with an average diameter of 50–60 nm, were also obtained by dissolving the Au core of the Au/PANI NWs.

In another approach, self-assembled surfactants formed 1-D nanostructures which proved capable of playing the role of soft template for the preparation of

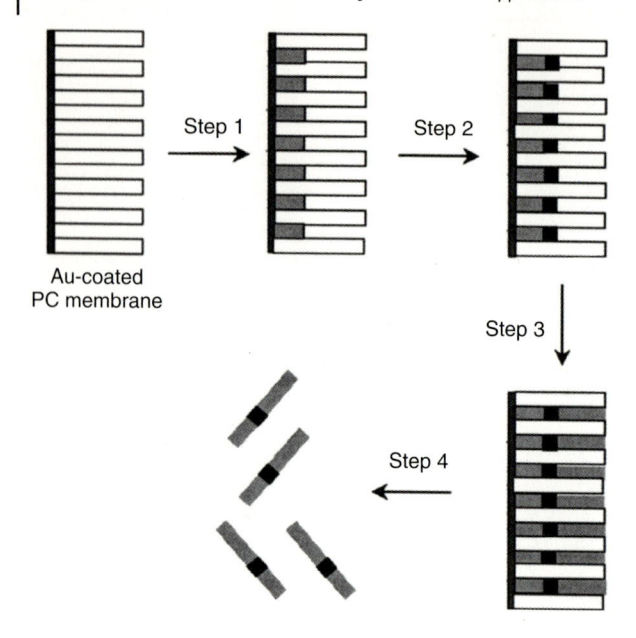

Figure 17.6 Schematic illustration of the fabrication of formation of Au–PPy–Au NWs, using a hard template. Reproduced with permission from Ref. [36]; © 2005, The Electrochemical Society.

NWs and NTs [45–47]. Consequently, wire-, ribbon-, and sphere-like nanostructures of PPy were synthesized using various self-assembled surfactants (anionic, cationic, or nonionic surfactants) with oxidizing agents [45]. Similarly, a surfactant containing a carbon source moiety was presynthesized and used as both a soft template and a carbon source. On the basis of this structure-directing agent, uniform carbon NWs with a diameter less than 1 nm were developed by using a confined self-assembly, as shown in Figure 17.7 [46]. Both, PANI and PPy NWs were also prepared via a hierarchical assembly process, using the rod-like tobacco mosaic virus as a template [47]. The resultant NWs could also be used as biosensors.

17.2.3
Nanotubes Based on Modified CNTs

Today, single-walled carbon nanotubes (SWNTs) and multi-walled carbon nanotubes (MWNTs) are typically prepared via by chemical vapor deposition (CVD) [48, 49]. As yet, the practical applications of CNTs have been limited because of their poor solubility, especially with regards to water-solubility for biomedical applica-

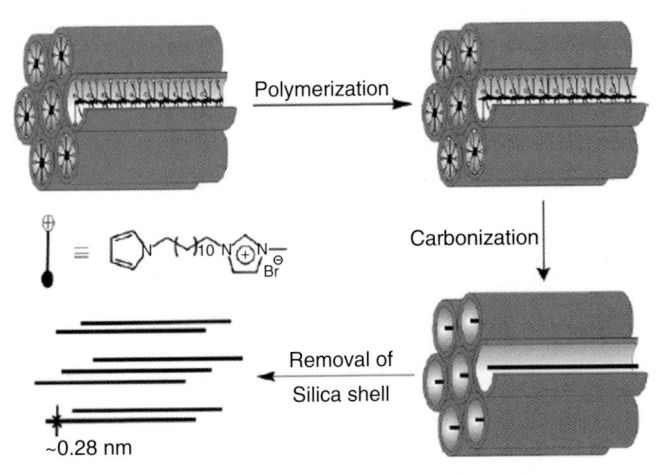

Figure 17.7 Preparation of carbon NWs via the self-assembly of surfactants containing a carbon source. Reproduced with permission from Ref. [46]; © 2007, American Chemical Society.

tions. Consequently, the surface of the CNTs must be modified so as to improve their dispersion or solubility in aqueous media. Such modification of the CNT surfaces can be achieved by either the covalent [50–56] or noncovalent [57–62] attachment of hydrophilic molecules onto the side walls of the CNTs.

Zhao et al. [50] prepared water-soluble SWNTs by the covalent conjugation of hydrophilic polymers such as poly(aminobenzene sulfonic acid) (PABS) and poly(ethylene glycol) (PEG), as shown in Figure 17.8. In fact, the carboxyl groups on the surface of the CNTs can be introduced by the purification and oxidation of CNTs [63], and then used to form the covalent conjugation between CNTs and other molecules, such as oligomeric polyimide (PI) and octadecylamine (ODA).

In order to avoid problematic issues arising from the covalent functionalization of CNTs (i.e., damage of the CNTs and subsequent loss of electrical conductivity, etc.), the hydrophilic molecules are attached noncovalently to the surfaces of the CNTs. Typically, amphiphilic surfactants are used to introduce water-solubility to CNTs; for example, poly(ethylene oxide)–phospholipid (PEO–PL), a polymeric surfactant, has been widely used for the surface modification of CNTs. In this case, the PL groups bind to the surfaces of the CNTs by hydrophobic interactions, while the PEO chains protrude into the water [59–62]. Nakayama et al. [58] reported a noncovalent functionalization of SWNTs by fluorescein–polyethylene oxide (fluor–PEO), whereby the aromatic fluorescein group binds to the surfaces of the SWNTs via π–π stacking. The dispersion of fluor–PEO-functionalized SWNT proved to be stable in water, and showed no aggregation.

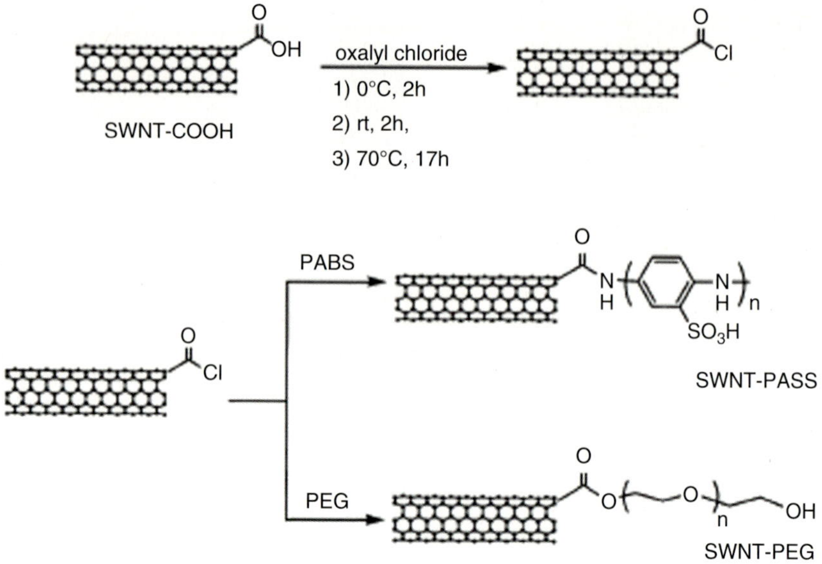

Figure 17.8 Synthesis of water-soluble SWNT–PABS and SWNT–PEG. Reproduced with permission from Ref. [50]; © 2005, American Chemical Society.

17.3
Biomedical Applications of Nanowires and/or Nanotubes

17.3.1
Biosensors

Recently, electrochemical sensors have shown great promise for a wide range of biomedical or environmental applications. Notably, CP NWs and NTs can be used in a variety of electrochemical biosensing techniques for various biomolecules, such as proteins, nucleic acids, viruses, and glucose [64–68]. An example of this was a single conducting PPy NW of 200 nm thickness, 500 nm width, and 3 μm length, which demonstrated a pH-sensing capability [69] based on the direct proportionality between the conductivity of the PPy NW and pH. Subsequently, Kaner et al. [70] developed PANI NW thin-film sensors and compared these to conventional PANI sensors, by exploring their various response mechanisms such as acid doping (HCl), swelling ($CHCl_3$), and conformational changes in the PANI backbone (CH_3OH). Typically, a rapid reduction in resistance was observed within a short period of time when PANI was exposed to HCl, which caused protonation of the imine nitrogen. The charge created by the protonation was balanced by the resulting negatively charged chloride, while the change in conductivity was brought about by the formation of polarons (radical cations) that traveled on the PANI backbone. The mechanism involved in sensing the swelling behavior was that chloroform molecules are relatively small, and can diffuse efficiently into the polymeric matrix; this causes the structure to expand, in association with a decrease in the conductivity of the film. In case of conformational changes in the PANI backbone, methanol interacts with the nitrogen atoms of PANI, leading to an expansion of the PANI chains into a linear form, which in turn decreases the resistance of the film. When PANI was electrospun into nanofibers, with the aid of PEO, those nanofibers of 100–500 nm diameter also exhibited a sensing capability for NH_3 gas [71]. This was based on a measurable change in conductance that arose from the deprotonation of PANI , due to the diffusion of NH_3 into the PANI nanofibers. In contrast, exposure to N_2 did not cause any change in the conductivity of the nanofibers. Notably, the response time of the sensor was greatly affected by the diameter of the nanofibers, and indicated a diffusion-dependent process. Tseng et al. [72] demonstrated real-time electronic sensing in both the gas and in the solution media, by using an array of PANI nanoframework–electrode junctions. The same device could also be used for pH-sensing.

Those NTs based on functionalized CNTs have been extensively investigated as possible biomedical sensors. Indeed, Star et al. [73] reported the creation of CNT network field-effect transistors (CNTNFETs) that could selectively detect DNA immobilization and hybridization. Likewise, CNTNFETs with immobilized synthetic oligonucleotides have been shown to specifically recognize target DNA sequences. Such charge-based DNA detection is possible because there is a strong effect of the DNA counterions on the electronic response of the CNTNFETs.

A composite of chitosan (CHIT) filled with platinum nanowires (PtNW) and SWNTs (denoted PtNW–SWNT–CHIT) was prepared by, first, an electrodeposition of Pt in AAO templates, followed by dispersal of the PtNW together with SWNTs in CHIT solution [74]. The PtNW–SWNT–CHIT film-modified electrode demonstrated a linear response range with excellent sensitivity, as well as a good repeatability and stability for glucose biosensing purposes.

The possible use of an MWNT film as a sensor was also suggested [75]. In this case, the MWNT films were prepared via a solution/filtration method and bonded directly onto specimens, using a nonconductive adhesive, as shown in Figure 17.9.

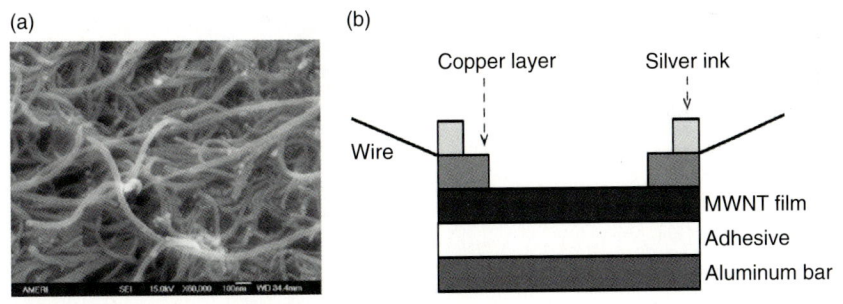

Figure 17.9 Scanning electron microcopy image of a MWNT film; (b) Schematic illustration of a MWNT film sensor. Reprinted with permission from Ref. [75]; © 2008, IOP Publishing Ltd.

The finding that the change in resistance of the MWNT film was proportional to the applied stain indicated that the film might potentially be used for structural health monitoring and vibration control applications.

17.3.2
Cancer Therapy

17.3.2.1 CNTs for Photothermal Therapy

The destruction of tumors using hyperthermia has been under investigation for some time, mainly because thermal therapeutics are minimally (or even non-) invasive, relatively simple to perform, and have the potential to treat tumors in situations where surgical resection is not feasible. Unfortunately, as the simple heating techniques used cannot distinguish between the tumor lesion and healthy tissues, the activating energy source will inevitably penetrate also the healthy tissues.

Many investigations have been conducted on the treatment of tumors with hyperthermia, employing deep-penetrating near-infrared (NIR) lasers with or without a heat-releasing agent [76, 77]. Initially, CNTs were suggested for use as a heat release agent in photothermal therapy, under NIR irradiation; the procedure for the NIR induction of photothermal therapy is shown in Figure 17.10 [77]. Subsequently, Kam et al. showed that continuous NIR radiation could cause tumor destruction, based on the local heating of SWNTs, due to their strong optical absorbance in the NIR region. Selective tumor destruction could also be achieved by the functionalization of SWNTs with targeting moieties [76]. Likewise, it was shown that solid malignant tumors could be completely destroyed, without harmful side effects, by an injection of SWNTs followed by NIR irradiation, whereas in mice treated with SWNTs, but not irradiated, the tumors continued to grow [77]. In both of these studies the surfaces of the SWNTs were modified by conjugation with PEG, while the lengths of the SWNTs were cut to within a few hundred nanometers in order to improve their water-solubility.

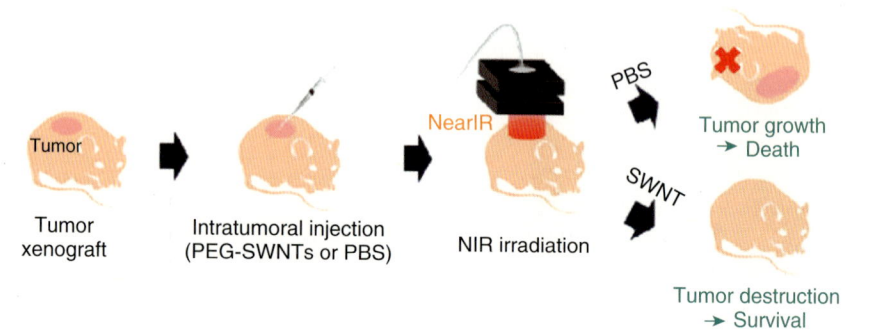

Figure 17.10 Schematic representation of the SWNT-mediated photothermal treatment of tumors in mice. Reproduced with permission from Ref. [77]; © 2009, American Chemical Society.

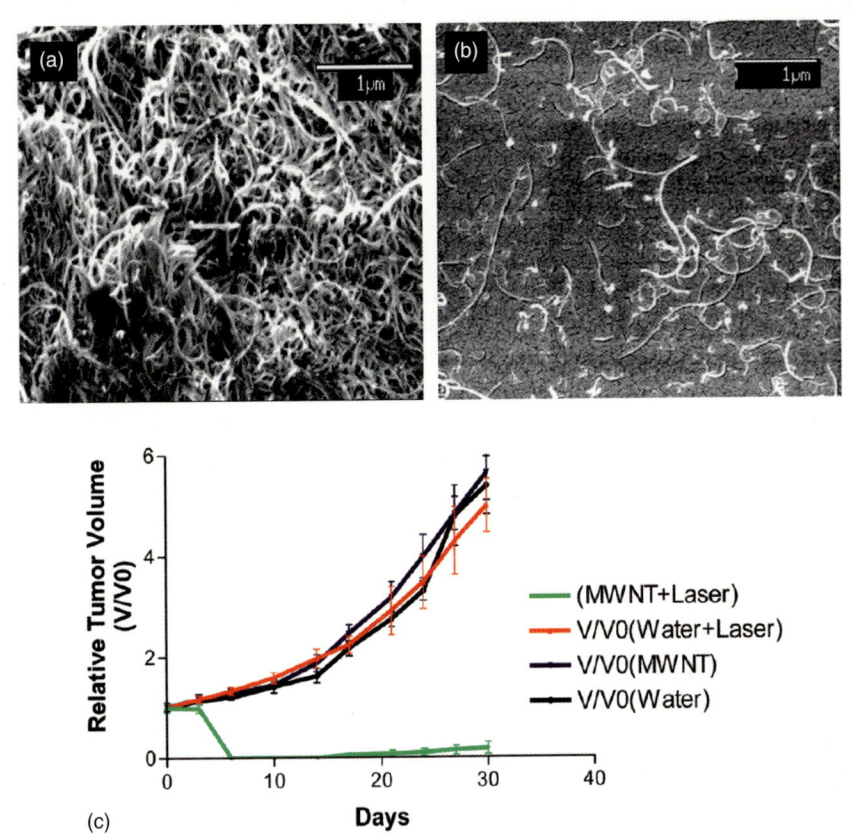

Figure 17.11 Scanning electron microscopy images.
(a) Non-DNA-encased MWNTs; (b) DNA-encased MWNTs;
(c) Change in relative volume of the four tumor groups under NIR irradiation. Reproduced with permission from Ref. [78]; © 2009, American Chemical Society.

Recently, MWNTs were also investigated as heat-release agents [78] (see Figure 17.11). In this case, the surfaces of the MWNTs were modified by the noncovalent attachment of DNA to improve their water-solubility. The DNA–MWNTs were then used to safely destroy malignant tumors *in vivo*. Upon NIR irradiation of the DNA–MWNTs, heat is generated, there being a linear correlation between the irradiation time and the laser power. The growth rate of tumors treated with DNA–MWNT injections was clearly distinguished from that of untreated, control tumors. Moreover, the nonmalignant tissues displayed no long-term damage after receiving the same treatment.

17.3.2.2 CNTs for Radiofrequency Ablation

Radiofrequency ablation (RFA) can be defined as the destruction of malignant cells by using an alternating current (AC) at an electromagnetic frequency that falls into the range characteristic of radiofrequency (RF) [79]. The mechanisms of tissue heating in RFA are based on the conversion of electrical energy into thermal energy. During the RFA procedure, an electrode is first inserted into the target tissue; the RF current then flows from the generator through the electrode into the tissue, so as to form an entire electric circuit. Under a certain voltage as the potential energy, the poorest conductors in the circuit are biological tissues, with a higher impedance. As the ions of the tissue attempt to follow the change in direction of the AC, ionic agitation occurs, and this results in frictional heat of the tissue (i.e., resistive or ohmic heating). When the temperature increases to a certain level (normally >70 °C), the tumor cells will be destroyed.

Recently, SWNTs have also been used as heat-release agent under a RF field [80]. In this case, the SWNTs were functionalized with a polymer based on poly(phenylene ethylene) (PPE). The PPE-functionalized SWNTs were exposed to a 13.56 MHz RF field, which induced an efficient heating of aqueous suspensions of SWNTs (see Figure 17.12). This phenomenon was applied to produce a selective thermal destruction *in vitro* of cancer cells that contained internalized SWNTs. Typically, the SWNT-treated tumors showed complete necrosis, whereas the control tumors survived.

17.3.3
Optical Bioimaging

As the surface areas of NWs and NTs are relatively high, it is possible to load large amounts of biomaterials onto their surfaces. Moreover, if the NWs and NTs are water-soluble and their size is sufficiently small, they may be used as a carrier of

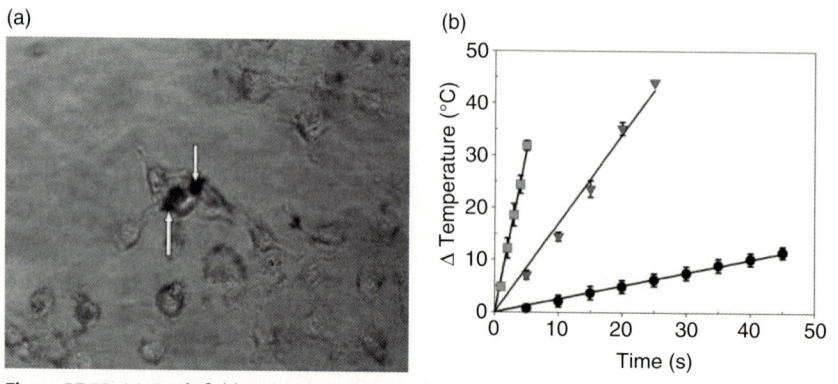

Figure 17.12 (a) Brightfield microscopic image demonstrating the intracellular collections of SWNTs (arrows) in Hep3B human hepatocellular carcinoma cells; (b) Temperature change of aqueous SWNT suspensions under an RF field. •, no SWNTs; ▼, SWNT suspensions of 50 mg l^{-1}; □, SWNT suspension of 250 mg l^{-1}. Reproduced with permission from Ref. [80]; © 2007, Wiley-VCH Verlag GmbH & Co. KGaA.

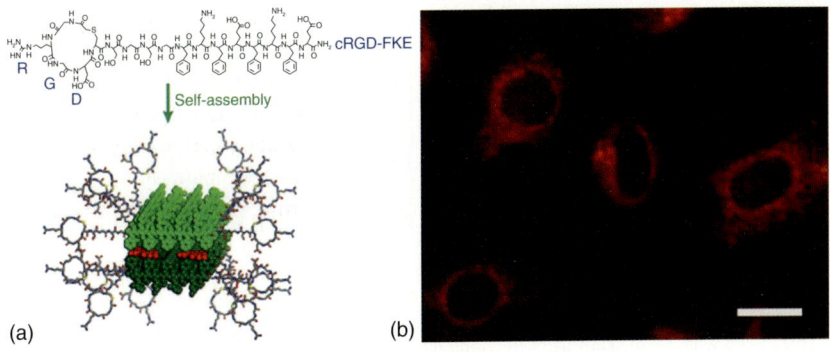

Figure 17.13 (a) Illustration of β-sheet-like nanoribbon structure by the self-assembly of cRGD-containing supramolecules; (b) Optical image of the Nile red delivered intracellularly into HeLa cells, as loaded onto cRGD units. Scale bar = 50 μm. Reproduced with permission from Ref. [81]; © 2008, The Royal Society of Chemistry.

biomaterials. For example, a cyclic arginine-glycine-aspartic acid (cRGD)-coated nanoribbon structure formed via the self-assembly of supramolecular building blocks, including a cRGD peptide, could be used as a carrier of Nile red dye into the cells, which would be valuable for optical bioimaging (Figure 17.13) [81].

In certain cases, color-tuned fluorescent organic NWs and NTs were prepared via the self-assembly of organic building blocks [21]; in fact, the NWs and NTs could serve as imaging materials themselves if they were water-soluble. Massuyeau et al. [82] prepared photoluminescent nanofibers of poly-(p-phenylene-vinylene) (PPV), using a template of polycarbonate nanoporous membrane, and observed unique features such as a blue-shifted emission with a higher quantum yield. This effect was attributed to the cancellation of interchain interactions, consistent with nanoscale tubular structures formed from weakly interacting and short polymer chain segments. These NTs open up perspectives for tunable photoluminescence properties in the blue spectral range, and for biochemical applications.

Raman spectroscopy is a well-established bioanalytical tool that demonstrates many advantages, including an excellent sensitivity to small structural and chemical changes, a high spatial resolution, and resistance to autofluorescence and photobleaching [83]. Although Raman spectroscopy has been used occasionally to image biological processes within living cells and excised tissues, the inherently weak intensity has limited its application. However, Zavaleta *et al.* reported the use of Raman spectroscopy to visualize cancer cells, with SWNTs as the diagnostic imaging contrast agent [84, 85]. In this case, optimized noninvasive Raman microscopy was used to evaluate tumor targeting and the localization of RGD–SWNTs in mice. In addition, SWNTs conjugated with cRGD peptides were used as a contrast agent for photoacoustic imaging under NIR irradiation [86]. The photoacoustic imaging of living subjects offers a higher spatial resolution, and also

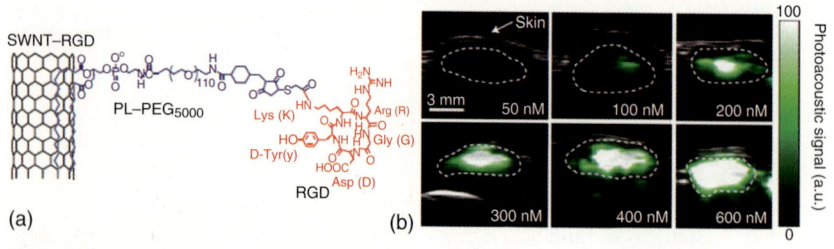

Figure 17.14 (a) Molecular structure of SWNTs modified with cRGD; (b) Vertical slice in the 3-D photoacoustic image of mice injected with SWNTs at concentrations of 50 to 600 nM. Reproduced with permission from Ref. [86]; © 2008, Macmillan Publishers Ltd.

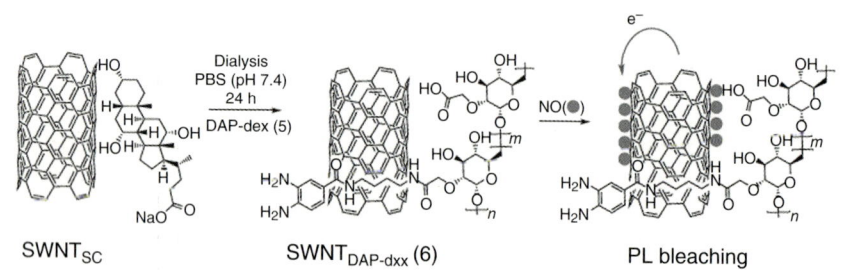

Figure 17.15 Preparation of SWNT–DAP–dex, and mechanism for NIR fluorescence quenching by NO. Reproduced with permission from Ref. [87]; © 2009, Macmillan Publishers Ltd.

allows deeper tissues to be imaged compared to most optical imaging techniques. Unfortunately, as many diseases do not exhibit a natural photoacoustic contrast, it is necessary to administer a photoacoustic contrast agent. The intravenous administration of these targeted NTs to mice bearing tumors resulted in an eight-fold greater photoacoustic signal in the tumor than in mice injected with nontargeted CNTs (see Figure 17.14).

Recently, the optical detection of gas molecule-mediated SWNTs was reported [87]. For this, 3,4-diaminophenyl-functionalized dextran (DAP-dex) was first synthesized and wrapped with SWNTs that would impart a rapid and selective fluorescence detection of nitric oxide (NO), a messenger of biological signaling (see Figure 17.15). The NIR fluorescence of SWNT-DAP-dex is immediately and directly quenched by NO, but not by other reactive nitrogen and oxygen species. Moreover, such quenching was reversible, and shown to be caused by electron transfer from the top of the valence band of the SWNT to the lowest unoccupied molecular orbital (LUMO) of NO. The resultant optical sensor proved to be capable of a real-time and spatially resolved detection of NO, produced by stimulating NO synthase in macrophages.

17.4
Summary

In this chapter, recent advances in the fabrication and biomedical applications of organic NWs and NTs (including CNTs) have been described. It is remarkable that, within a very short period of time, these 1-D nanostructures have surpassed the capabilities of traditional imaging, delivery, and sensing devices.

To date, organic NWs and NTs have been prepared typically via the self-assembly of organic building blocks, the template synthesis of polymeric molecules and, occasionally, via an electrospinning process. The surface-functionalized CNTs, which are water-soluble, have also been explored for their biomedical applications.

The electrical conductivity of organic NWs and NTs has been utilized for conductivity-based biosensing that is capable of achieving a multiplexed detection, without probe labeling, but which is impossible with currently available technology. In the case of optical imaging, 1-D nanostructures can–unlike previous micron-scale formulations–penetrate endothelial barriers to reach specific sites, and also visualize targeted sites by absorbing the excitation source, including NIR radiation and radiofrequency. CNTs have also been applied to cancer therapy via hyperthermia, due to their inherent thermal properties.

Despite the remarkable recent advances that have been made with organic NWs and NTs, the fabrication methods and surface functionalization of these 1-D nanostructures will require continuous improvement if their application in the field of biomedicine is to be expanded. For example, their poor water-solubility remains problematic, and this limits their biomedical applications. For CNTs, the water-solubility might be improved to some extent, by conducting various surface modifications and also by controlling their lengths. Unfortunately, however, these techniques may not always be applicable to organic NWs and NTs, due to their structural instability in aqueous media, and it is therefore necessary to consider the surface functionality of 1-D nanostructures from the design stage of building block molecules. Furthermore, the photoluminescence and conductivity of the organic NWs and NTs might also need to be improved in order to widen their area of application. From the standpoint of their practical application in various areas of biomedicine, the incorporation of these 1-D nanostructures into routine functional integrated devices remains a clear challenge.

Whilst these 1-D nanostructures, particularly in the field of biomedicine, continue to offer unlimited research opportunities, they are clearly awaiting a massive and intensive interdisciplinary collaborative research program in order to progress.

References

1 Portney, N.G. and Ozkan, M. (2006) Nano-oncology: drug delivery, imaging, and sensing. *Analytical and Bioanalytical Chemistry*, **384**, 620–30.

2 Jain, K.K. (2008) Recent advances in nanooncology. *Technology in Cancer Research and Treatment*, **7**, 1–13.

3 Gordon, A.T., Lutz, G.E., Boninger, M.L. and Cooper, R.A. (2007) Introduction to nanotechnology: potential applications in physical medicine and rehabilitation. *American Journal of Physical Medicine and Rehabilitation*, **86**, 225–41.

4 Singh, R., Pantarotto, D., Lacerda, L., Pastorin, G., Klumpp, C., Prato, M., Bianco, A. and Kostarelos, K. (2006) Tissue biodistribution and blood clearance rates of intravenously administered carbon nanotube radiotracers. *Proceedings of the National Academy of Sciences of the United States of America*, **103**, 3357–62.

5 Smart, S.K., Cassady, A.I., Lu, G.Q. and Martin, D.J. (2006) The biocompatibility of carbon nanotubes. *Carbon*, **44**, 1034–47.

6 Pham, Q.P., Sharma, U. and Mikos, A.G. (2006) Electrospinning of polymeric nanofibers for tissue engineering applications: a review. *Tissue Engineering*, **12**, 1197–211.

7 Jeong, K., Lee, W., Cha, J., Park, C.R., Cho, Y.W. and Kwon, I.C. (2008) Regioselective succinylation and gelation behavior of glycol chitosan. *Macromolecular Research*, **16**, 57–61.

8 Park, J.H., Cho, Y.W., Son, Y.J., Kim, K., Chung, H., Jeong, S.Y., Choi, K., Park, C.R., Park, R.W., Kim, I.S. and Kwon, I.C. (2006) Preparation and characterization of self-assembled nanoparticles based on glycol chitosan bearing adriamycin. *Colloid and Polymer Science*, **284**, 763–70.

9 Cha, C., Lee, W.B., Cho, Y.W., Ah, C.H., Kwon, I.C. and Park, C.R. (2006) Preparation and characterization of cisplatin-incorporated chitosan hydrogels, microparticles, and nanoparticles. *Macromolecular Research*, **14** (5), 573–8.

10 Son, Y.J., Jang, J.S., Cho, Y.W., Chung, H., Park, R.W., Kwon, I.C., Kim, I.S., Park, J.Y., Seo, S.S., Park, C.R. and Jeong, S.Y. (2003) Biodistribution and anti-tumor efficacy of doxorubicin loaded glycol-chitosan nanoaggregates by EPR effect. *Journal of Controlled Release*, **91**, 135–45.

11 Kim, Y.H., Gihm, S.H., Park, C.R., Lee, K.Y., Kim, T.W., Kwon, I.C., Chung, H. and Jeong, S.Y. (2001) Structural characteristics of size-controlled self-aggregates of deoxycholic acid-modified chitosan and their application as a DNA delivery carrier. *Bioconjugate Chemistry*, **12**, 932–8.

12 He, Y., Li, Z., Simone, P. and Lodge, T.P. (2006) Self-assembly of block copolymer micelles in an ionic liquid. *Journal of the American Chemical Society*, **128**, 2745–50.

13 Qin, S., Geng, Y., Discher, D.E. and Yang, S. (2006) Temperature-controlled assembly and release from polymer vesicles of poly(ethylene oxide)-block-poly(N-isopropylacrylamide). *Advanced Materials*, **18**, 2905–9.

14 Ghoroghchian, P.P., Li, G., Levine, D.H., Davis, K.P. and Bates, F.S. (2006) Bioresorbable vesicles formed through spontaneous self-assembly of amphiphilic poly(ethylene oxide)-block-polycaprolactone. *Macromolecules*, **39**, 1673–5.

15 Clausen, C.H., Jensen, J., Castillo, J., Dimaki, M. and Svendsen, W.E. (2008) Qualitative mapping of structurally different dipeptide nanotubes. *Nano Letters*, **8**, 4066–9.

16 Li, Y., Dong, M., Otzen, D.E., Yao, Y., Liu, B., Besenbacher, F. and Mamdouh, W. (2009) Influence of tunable external stimuli on the self-assembly of guanosine supramolecular nanostructures studied by atomic force microscope. *Langmuir*, **25**, 13432–7.

17 Miranda, F.F., Iwasaki, K., Akashi, S., Sumitomo, K., Kobayashi, M., Yamashita, I., Tame, J.R.H. and Heddle, J.G. (2009) A self-assembled protein nanotube with high aspect ratio. *Small*, **5**, 2077–84.

18 Bae, C., Shin, H. and Moon, J. (2007) Facile route to aligned one-dimensional arrays of colloidal nanoparticles. *Chemistry of Materials*, **19**, 1531–3.

19 Guérin, G., Raez, J., Wang, X.-S., Manners, I. and Winnik, M.A. (2006) Polyferrocenylsilane block copolymers: nanotubes and nanowires through self-assembly. *Progress in Colloid and Polymer Science*, **132**, 152–60.

20 Kim, S.W., Cho, H.G. and Park, C.R. (2009) Catalyst-free and template-free preparation of semi-cylindrical carbon nanoribbons. *Carbon*, **47**, 2391–5.

21. Ah, B.K., Gihm, S.H., Chung, J.W., Park, C.R., Kwon, S.K. and Park, S.Y. (2009) Color-tuned highly fluorescent organic nanowires/nanofabrics: easy massive fabrication and molecular structural origin. *Journal of the American Chemical Society*, **131**, 3950–7.
22. Sagisaka, M., Hino, M., Nakanishi, Y., Inui, Y., Kawaguchi, T., Tsuchiya, K., Sakai, H., Abe, M. and Yoshizawa, A. (2009) Self-assembly of double-tail anionic surfactant having cyanobiphenyl terminal groups in water. *Langmuir*, **25**, 10230–6.
23. Arai, N., Yasuoka, K. and Zeng, X.C. (2008) Self-assembly of surfactants and polymorphic transition in nanotubes. *Journal of the American Chemical Society*, **130**, 7916–20.
24. Yuan, Z., Yin, Z., Sun, S. and Hao, J. (2008) Densely stacked multilamellar and oligovesicular vesicles, bilayer cylinders, and tubes joining with vesicles of a salt-free catanionic extractant and surfactant system. *Journal of Physical Chemistry B*, **112**, 1414–19.
25. Ryu, J.H., Jang, C.J., Yoo, Y.S., Lim, S.G. and Lee, M. (2005) Supramolecular reactor in aqueous environment: aromatic cross Suzuki coupling reaction at room temperature. *Journal of Organic Chemistry*, **70**, 8956–62.
26. Ryu, J.H. and Lee, M. (2005) Transformation of isotropic fluid to nematic gel triggered by dynamic bridging of supramolecular nanocylinders. *Journal of the American Chemical Society*, **127**, 14170–1.
27. Yang, W.Y., Lee, E. and Lee, M. (2006) Tubular organization with coiled ribbon from amphiphilic rigid-flexible macrocycle. *Journal of the American Chemical Society*, **128**, 3484–5.
28. Ryu, J.H., Oh, N.K. and Lee, M. (2005) Tubular assembly of amphiphilic rigid macrocycle with flexible dendrons. *Chemical Communications*, **41**, 1770–2.
29. Zhang, G., Jin, W., Fukushima, T., Kosaka, A., Ishii, N. and Aida, T. (2007) Formation of water-dispersible nanotubular graphitic assembly decorated with isothiouronium ion groups and its supramolecular functionalization. *Journal of the American Chemical Society*, **129**, 719–22.
30. Jin, W., Fukushima, T., Kosaka, A., Niki, M., Ishii, N. and Aida, T. (2005) Controlled self-assembly triggered by olefin metathesis: cross-linked graphitic nanotubes from an amphiphilic hexa-peri-hexabenzocoronene. *Journal of the American Chemical Society*, **127**, 8284–5.
31. Wu, D., Zhi, L., Bodwell, G.J., Cui, G., Tsao, N. and Mullen, K. (2007) Self-assembly of positively charged discotic PAHs: from nanofibers to nanotubes. *Angewandte Chemie International Edition*, **46**, 5417–20.
32. Cherny, I. and Gazit, E. (2008) Amyloids: not only pathological agents but also ordered nanomaterials. *Angewandte Chemie International Edition*, **47**, 4062–9.
33. Lim, Y.B. and Lee, M. (2008) Nanostructures of beta-Sheet peptide: steps towards bioactive functional materials. *Journal of Materials Chemistry*, **18**, 723–7.
34. Eckhardt, I., Groenewolt, M., Krauseb, E. and Börner, H. G. (2005) Rational design of oligopeptide organizers for the formation of poly(ethylene oxide) nanofibers. *Chemical Communications*, **41**, 2814–16
35. Zhang, X., Zhang, J., Song, W. and Liu, Z. (2006) Controllable synthesis of conducting polypyrrole nanostructures. *Journal of Physical Chemistry B*, **110**, 1158–65.
36. Reynesa, O. and Demoustier-Champagnez, S. (2005) Template electrochemical growth of polypyrrole and gold-polypyrrole-gold nanowire arrays. *Journal of the Electrochemical Society*, **152**, D130–5.
37. Park, D.H., Kim, B.H., Jang, M.K., Bae, K.Y., Lee, S.J. and Joo, J. (2005) Synthesis and characterization of polythiophene and poly (3-methylthiophene) nanotubes and nanowires. *Synthetic Metals*, **153**, 341–4.
38. Cheng, F.L., Zhang, M.L. and Wang, H. (2005) Fabrication of polypyrrole nanowire and nanotube arrays. *Sensors*, **5**, 245–9.
39. Joo, J., Kim, B.H., Park, D.H., Kim, H.S., Seo, D.S., Shim, J.H., Lee, S.J., Ryu, K.S., Kim, K., Jin, J.I., Lee, T.J. and Lee, C.J. (2005) Fabrication and applications of

conducting polymer nanotube, nanowire, nanohole, and double wall nanotube. *Synthetic Metals*, **153**, 313–16.

40 Wang, J., Dai, J. and Yarlagadda, T. (2005) Carbon nanotube–conducting-polymer composite nanowires. *Langmuir*, **21**, 9–12.

41 Zheng, R.K., Chan, H.L.W. and Choy, C.L. (2005) A simple template-based hot-press method for the fabrication of metal and polymer nanowires and nanotubes. *Nanotechnology*, **16**, 1928–34.

42 Zheng, R.K., Yang, Y., Wang, Y., Wang, J., Chan, H.L.W., Choy, C.L., Jin, C.G. and Li, X.G. (2005) A simple and convenient route to prepare poly(vinylidene fluoride trifluoroethylene) copolymer nanowires and nanotubes. *Chemical Communications*, **41**, 1447–9.

43 Dougherty, S. and Liang, J. (2009) Core–shell polymer nanorods by a two-step template wetting process. *Nanotechnology*, **20**, 295301.

44 Huang, K., Zhang, Y., Long, Y., Yuan, J., Han, D., Wang, Z., Niu, L. and Chen, Z. (2006) Preparation of highly conductive, self-assembled gold/polyaniline nanocables and polyaniline nanotubes. *Chemistry–A European Journal*, **12**, 5314–19.

45 Baron, R., Willner, B. and Willner, I. (2007) Biomolecule–nanoparticle hybrids as functional units for nanobiotechnology. *Chemical Communications*, **43**, 323–32.

46 Zhang, W., Cui, J., Tao, C., Lin, C., Wu, Y. and Li, G. (2009) Confined self-assembly approach to produce ultrathin carbon nanofibers. *Langmuir*, **25**, 8235–9.

47 Niu, Z., Liu, J., Lee, L., Bruckman, M., Zhao, D., Koley, G. and Wang, Q. (2007) Biological templated synthesis of water-soluble conductive polymeric nanowires. *Nano Letters*, **7**, 3729–33.

48 Baughman, R., Zakhidov, A. and Heer, W. (2002) Carbon nanotubes—the route toward applications. *Science*, **297**, 787–92.

49 Rao, C.N.R., Satishkumar, B.C., Govindaraj, A. and Nath, M. (2001) Nanotubes. *ChemPhysChem*, **2**, 78–105.

50 Zhao, B., Hu, H., Yu, A., Perea, D. and Haddon, R. (2005) Synthesis and characterization of water soluble single-walled carbon nanotube graft copolymers. *Journal of the American Chemical Society*, **127**, 8197–203.

51 Lin, Y., Zhou, B., Martin, R., Henbest, K., Harruff, B., Riggs, J., Guo, Z., Allard, L. and Sun, Y. (2005) Visible luminescence of carbon nanotubes and dependence on functionalization. *Journal of Physical Chemistry B*, **109**, 14779–82.

52 Wang, X., Liu, H., Jin, Y. and Chen, C. (2006) Polymer-functionalized multiwalled carbon nanotubes as lithium intercalation hosts. *Journal of Physical Chemistry B*, **110**, 10236–40.

53 Liu, Y., Du, Z., Li, Y., Zhang, C., Li, C., Yang, X. and Li, H. (2006) Surface covalent encapsulation of multiwalled carbon nanotubes with poly(acryloyl chloride) grafted poly(ethylene glycol). *Journal of Polymer Science: Part A, General Papers*, **44**, 6880–7.

54 Chattopadhyay, J., Cortez, F., Chakraborty, S., Slater, N. and Billups, W. (2006) Synthesis of water-soluble PEGylated single-walled carbon nanotubes. *Chemistry of Materials*, **18**, 5864–8.

55 Yang, S., Fernando, S., Liu, J., Wang, J., Sun, H., Liu, Y., Chen, M., Huang, Y., Wang, X., Wang, H. and Sun, Y. (2008) Covalently PEGylated carbon nanotubes with stealth character in vivo. *Small*, **4**, 940–4.

56 Yan, L., Poon, Y., Chan-Park, M., Chen, Y. and Zhang, Q. (2008) Individually dispersing single-walled carbon nanotubes with novel neutral pH water-soluble chitosan derivatives. *Journal of Physical Chemistry C*, **112**, 7579–87.

57 Nakayama, N., Bangsaruntip, S., Sun, X., Welsher, K. and Dai, H. (2007) Noncovalent functionalization of carbon nanotubes by fluorescein-polyethylene glycol: supramolecular conjugates with pH-dependent absorbance and fluorescence. *Journal of the American Chemical Society*, **129**, 2448–9.

58 Liu, Z., Cai, W., He, L., Nakayama, N., Chen, K., Sun, X., Chen, X. and Dai, H. (2007) In vivo biodistribution and highly efficient tumour targeting of carbon nanotubes in mice. *Nature Nanotechnology*, **2**, 47–52.

59 Liu, Z., Sun, X., Nakayama, N. and Dai, H. (2007) Supramolecular chemistry on water-soluble carbon nanotubes for drug loading and delivery. *ACS Nano*, **1**, 50–6.

60 Liu, Z., Chen, K., Davis, C., Sherlock, S., Cao, Q., Chen, X. and Dai, H. (2008) Drug delivery with carbon nanotubes for in vivo cancer treatment. *Cancer Research*, **68**, 6652–60.

61 Welsher, K., Liu, Z., Daranciang, D. and Dai, H. (2008) Selective probing and imaging of cells with single walled carbon nanotubes as near-infrared fluorescent molecules. *Nano Letters*, **8**, 586–90.

62 Park, C., Lee, S., Lee, J.H., Lim, J., Lee, S.C., Park, M., Lee, S.S., Kim, J., Park, C.R. and Kim, C. (2007) Controlled assembly of carbon nanotubes encapsulated with amphiphilic block copolymer. *Carbon*, **45**, 2072–8.

63 Cho, H.G., Kim, S.W., Lim, H.J., Yun, C.H., Lee, H.S. and Park, C.R. (2009) A simple and highly effective process for the purification of single-walled carbon nanotubes synthesized with arc-discharge. *Carbon*, **47**, 3544–9.

64 Huang, X. and Choi, Y. (2007) Chemical sensors based on nanostructured materials. *Sensors and Actuators B*, **122**, 659–71.

65 Chen, P., Shen, G. and Zhou, C. (2008) Chemical sensors and electronic noses based on 1-D metal oxide nanostructures. *IEEE Transactions on Nanotechnology*, **7**, 668–82.

66 He, B., Morrow, T. and Keating, C. (2008) Nanowire sensors for multiplexed detection of biomolecules. *Current Opinion in Chemical Biology*, **12**, 522–8.

67 Roy, S. and Gao, Z. (2009) Nanostructure-based electrical biosensors. *Nano Today*, **4**, 318–34.

68 Cheng, M., Cuda, G., Bunimovich, Y., Gaspari, M., Heath, J., Hill, H., Mirkin, C., Nijdam, A., Terracciano, R., Thundat, T. and Ferrari, M. (2006) Nanotechnologies for biomolecular detection and medical diagnostics. *Current Opinion in Chemical Biology*, **10**, 11–19.

69 Yun, M., Myung, N., Vasquez, R., Lee, C., Menke, E. and Penner, R. (2004) Electrochemically grown wires for individually addressable sensor arrays. *Nano Letters*, **4**, 419–22.

70 Virji, S., Huang, J., Kaner, R. and Weiller, B. (2004) Polyaniline nanofiber gas sensors: examination of response mechanisms. *Nano Letters*, **4**, 491–6.

71 Liu, H., Kameoka, J., Czaplewski, D. and Craighead, H. (2004) Polymeric nanowire chemical sensor. *Nano Letters*, **4**, 671–5.

72 Wang, J., Chan, S., Carlson, R., Luo, Y., Ge, G., Ries, R., Heath, J. and Tseng, H. (2004) Electrochemically fabricated polyaniline nanoframework electrode junctions that function as resistive sensors. *Nano Letters*, **4**, 1693–7.

73 Star, A., Tu, E., Niemann, J., Gabriel, J., Joiner, C. and Valcke, C. (2006) Label-free detection of DNA hybridization using carbon nanotube network field-effect transistors. *Proceedings of the National Academy of Sciences of the United States of America*, **103**, 921–6.

74 Qu, F., Yang, M., Shen, G. and Yu, R. (2007) Electrochemical biosensing utilizing synergic action of carbon nanotubes and platinum nanowires prepared by template synthesis. *Biosensors and Bioelectronics*, **22**, 1749–55.

75 Li, X., Levy, C. and Elaadil, L. (2008) Multiwalled carbon nanotube film for strain sensing. *Nanotechnology*, **19**, 045501.

76 Kam, N., O'Connell, M., Wisdom, J. and Dai, H. (2005) Carbon nanotubes as multifunctional biological transporters and near-infrared agents for selective cancer cell destruction. *Proceedings of the National Academy of Sciences of the United States of America*, **102**, 11600–5.

77 Moon, H.K., Lee, S.H. and Choi, H.C. (2009) In vivo near-infrared mediated tumor destruction by photothermal effect of carbon nanotubes. *ACS Nano*, **3**, 3707–13.

78 Ghosh, S., Dutta, S., Gomes, E., Carroll, D., D'Agostino, R., Olson, J., Guthold, M. and Gmeiner, W. (2009) Increased heating efficiency and selective thermal ablation of malignant tissue with DNA-encased multiwalled carbon nanotubes. *ACS Nano*, **3**, 2667–73.

79 Ni, Y., Mulier, S., Miao, Y., Michel, L. and Marchal, G. (2005) *Abdominal Imaging*, **30**, 381–400.

80 Gannon, C., Cherukuri, P., Yakobson, B., Cognet, L., Kanzius, J., Kittrell, C., Weisman, R., Pasquali, M., Schmidt, H.,

Smalley, R. and Curley, S. (2007) Carbon nanotube-enhanced thermal destruction of cancer cells in a noninvasive radiofrequency field. *Cancer*, **110**, 2654–65.

81 Lim, Y.B., Kwon, O.J., Lee, E., Kim, P.H., Yun, C.O. and Lee, M. (2008) A cyclic RGD-coated peptide nanoribbon as a selective intracellular nanocarrier. *Organic and Biomolecular Chemistry*, **6**, 1944–8.

82 Massuyeau, F., Duvail, J., Athalin, H., Lorcy, J., Lefrant, S., Wéry, J. and Faulques, E. (2009) Elaboration of conjugated polymer nanowires and nanotubes for tunable photoluminescence properties. *Nanotechnology*, **20**, 155701.

83 Ryder, A.G. (2005) Surface enhanced Raman scattering for narcotic detection and applications to chemical biology. *Current Opinion in Chemical Biology*, **9**, 489–93.

84 Keren, S., Zavaleta, C., Cheng, Z., Zerda, A., Gheysens, O. and Gambhir, S. (2008) Noninvasive molecular imaging of small living subjects using Raman spectroscopy. *Proceedings of the National Academy of Sciences of the United States of America*, **105**, 5844–9.

85 Zavaleta, C., Zerda, A., Liu, Z., Keren, S., Cheng, Z., Schipper, M., Chen, X., Dai, H. and Gambhir, S. (2008) Noninvasive Raman spectroscopy in living mice for evaluation of tumor targeting with carbon nanotubes. *Nano Letters*, **8**, 2800–5.

86 Zerda, A., Zavaleta, C., Keren, S., Vaithilingam, S., Bodapati, S., Liu, Z., Levi, J., Smith, B., Ma, T., Oralkan, O., Cheng, Z., Chen, X., Dai, H., Khuri-Yakub, B. and Gambhir, S. (2008) Carbon nanotubes as photoacoustic molecular imaging agents in living mice. *Nature Nanotechnology*, **3**, 557–62.

87 Kim, J.H., Heller, D., Jin, H., Barone, P., Song, C., Zhang, J., Trudel, L., Wogan, G., Tannenbaum, S. and Strano, M. (2009) The rational design of nitric oxide selectivity in single-walled carbon nanotube near-infrared fluorescence sensors for biological detection. *Nature Chemistry*, **1**, 473–81.

18
Rosette Nanotubes for Targeted Drug Delivery

Sarabjeet Singh Suri, Hicham Fenniri and Baljit Singh

18.1
Introduction

Nanotechnology was originally defined as "... the creation of useful materials, devices, and systems used to manipulate matter that are small scale ranging between 1 and 100 nm" (nano.cancer.gov). During the past ten years, as potential nanotechnological applications have expanded exponentially in multiple directions – including the medical sciences – the definition of nanotechnology has been broadened. Recently, Theis and coworkers suggested the addition of two points to the original definition of nanotechnology, namely that: (i) the nanodevice must exhibit properties that only arise because of the nanoscale dimensions; and (ii) the peculiar behavior of the nanodevice must be predictable through the construction of appropriate mathematical models [1].

Both, carbon and metallic nanotubes have been extensively studied for a wide variety of applications due to their mechanical stability, conductance, and large surface areas. Whilst these nanotubes continue to attract intense attention, especially for drug-delivery purposes, they are associated with some serious drawbacks. For example, both moisture and oxygen can affect the charge and structure of the carbon/metallic nanotubes. Moreover, the nanotubes pose challenges in their solubility, in the reproducibility of their precise structural properties, and the medical applications associated with such properties. During recent years, many attempts have been made to deploy carbon and gold nanoparticles for the delivery of peptides or proteins of interest to intracellular targets. For example, various nuclear localization signal peptides have been conjugated to gold nanoparticles, either through a secondary protein such as bovine serum albumin (BSA) [2] or through a thioalkyl triazole linker [3]. Unfortunately, these conjugates are large, branched polypeptides that are difficult to characterize and can also cause oxidative stress due to their metal content. In addition, the solubility, distribution, bioavailability and safety of such nanoparticles are questionable. Consequently, protein- or peptide-based nanotubes offer an attractive alternative for targeted drug delivery, live cell imaging, surgery, and implants. Moreover, during recent years several notable steps have been taken in this direction.

Nanomaterials for the Life Sciences Vol.10: Polymeric Nanomaterials. Edited by Challa S. S. R. Kumar
Copyright © 2011 WILEY-VCH Verlag GmbH & Co. KGaA, Weinheim
ISBN: 978-3-527-32170-4

In this chapter, information is provided related to the latest updates of biofriendly nanotubes, including the novel rosette nanotubes employed for drug delivery purposes, with special reference being made to receptor-mediated delivery.

18.2
Peptide-Based Nanotubes

The molecular self-assembly of proteins represents the main bottom-up approach for the affordable large-scale production of nanomaterials. Self-assembling peptide-based nanotubes prepared from cyclic peptides have structural and functional properties that are suitable for a variety of applications in the biological, medical, and materials sciences. Proteins and natural–synthetic peptides represent the most versatile molecular building blocks, due to their extensive biochemical, conformational and functional diversity, and their precise regulatory mechanisms. These peptides also offer a unique specificity of interactions with other proteins or biomolecules that might be explored for targeted drug delivery, using bionanotubes. Due to the high complexity of proteins, a thorough understanding of the physical and chemical properties that initiate and control their self-assembling properties is required. Within this context, simple and short peptides offer much more avenues to understand the synthesis, nature, properties and functions of bionanotubes. For example, cationic dipeptides derived from phenyl-alanine (Phe) such as, NH_2–Phe–Phe–NH_2, will self-assemble into nanotubes at neutral pH and rearrange into spherical structures that are approximately 100 nm in diameter [4]. These nanotubes will convert spontaneously into vesicles, and are endocytosed by the cells – a property which provides potential applications in both gene and drug delivery. Peptide-based nanotubes hold tremendous potential as an efficient targeted drug-delivery system, because it is now possible to design molecular recognition motif(s) into a synthetic peptide, in order to deliver, specifically, a wide range of substances inside the cellular environment in a ligand/site-specific pattern. The cargoes also include the delivery of water-insoluble molecules and large biological molecules for medical and cosmetic applications.

Next in molecular complexity are the longer synthetic linear peptide nanotubes of approximately 2–3 nm length, with a hydrophilic head of one or two charged amino acids and a hydrophobic tail of more than four consecutive hydrophobic amino acids, for example, V6D, A6D, G8DD, and KV6. Upon dissolution in water, these peptides form a complex network of cationic or anionic open-ended bionanotubes stabilized by their hydrophobic peptides. Unlike conventional surfactants, peptide surfactants are packed by H–H bonding, although some peptide surfactants also display typical β-sheet structure imposed on a well-extended peptide backbone.

By far the most extensive data on cyclic peptide nanotubes are based on the molecules designed by Hartgerink and colleagues [5]. Cyclic peptides with an even number of alternating D- and L- amino acids stack upon each other through an extensive complex intermolecular hydrogen bonding, to form long cylindrical nanostructures with an anti-parallel β-sheet structure. The outer surface of these nanotubes is defined by all of the amino acid side chains, and thus can be controlled by peptide engineering, or by the covalent attachment of polymers or peptide sequences with specific targeting sequences, producing polymer shells around the nanotube core. The ability to engineer the outer surface properties of a peptide nanotube has revolutionized the area of medical nanotechnology, and led to the development of new antimicrobial and cytotoxic agents, to the controlled and targeted release of drugs, and to new artificial ion channels that are controlled by peptide design. Today, self-assembling rosette nanotubes based on a wide variety of cyclic peptides have been developed.

18.3
Self-Assembling Rosette Nanotubes

18.3.1
Self-Assembly Peptides

Similar to many other landmark discoveries, the self-assembling peptides were discovered accidentally by Zhang and coworkers, as the peptide EAK16 from yeast [6, 7]. In this case, a protein known as Zuotin was identified that had an interesting repetitive 16-residue peptide sequence motif n-AEAEAKAKAEAEAKAK-c (EAK16-II). The four ionic self-complementary peptides, namely EAK16-II, RDA16-I, RAD16-II, and EAK-I (yeast Zuotin) form stable β-sheet structures in water and undergo spontaneous assembly to form nanofiber scaffolds [6]. These discoveries led to the design of peptide-based nanotubes that have generally very small dimensions, and can be delivered to specific cells or tissues by the addition of specific motifs in the peptide. This opportunity to manipulate biological materials offers the target-specific delivery of water-insoluble molecules for medical, diagnostic, and cosmetic applications.

18.3.2
G^C Motif Self-Assembly: Novel Helical Rosette Nanotubes

18.3.2.1 Novelty
In recent years, the intracellular targeting of proteins of interest has been attempted by using carbon and gold nanoparticles as delivery vehicles. Some of the conjugates – such as those where the signal peptide is linked to gold nanoparticles through a secondary protein – contain large, branched polypeptides and may induce oxidative stress due to their metal content [2, 3]. Furthermore, their solubility and bioavailability are also questionable. Recently, the Fenniri group has made major advances to overcome these disadvantages through the synthesis of guanine–cytosine (G^C) motif self-assembling, biologically inspired, metal-free helical rosette nanotubes (RNTs) that are water-soluble upon synthesis, and biocompatible by design [8, 9].

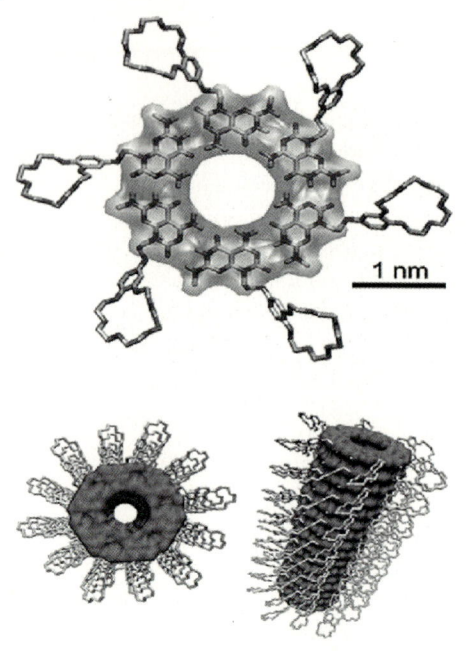

Figure 18.1 Hierarchical self-assembly in to a six-membered supermacrocycle rosette (upper), and resulting nanotube, top (lower left) and side (lower right) views models. Each crown ether site within the assembled structure provides 328 of open space for binding of a molecular guest. Reproduced with permission from Ref. [8]; © 2002, ACS Publications.

18.3.2.2 G^C Motif Self-Assembly Process

The RNTs are obtained through the spontaneous self-assembly of a synthetic heterobicyclic G^C motif featuring the Watson–Crick H-bond donor/acceptor arrays of guanine and cytosine (Figure 18.1) [9]. The hierarchical self-assembly of the G^C motif into six-membered rosettes maintained by 18 hydrogen bonds is followed by the stacking of rosettes to form nanotubes approximately 3.5 nm in diameter with an inner diameter of 1.1 nm. The outer diameter varies depending on substitutions on the G^C motif, which are expressed on the outer surface of the resulting nanotubes. The length of the tube is regulated through a variety of factors, including temperature (Figure 18.2) [10].

18.3.2.3 Built-In Strategy for Manipulating the Properties of RNTs

The RNTs represent a promising class of nanomaterial due to their synthetic accessibility and amenability to chemical functionalization. For instance, RNTs with different surface groups displaying chiroptical [8] and hierarchical [11] tunability, high thermal stability [12], and entropy-driven self-assembly behavior [9] in aqueous or polar media have been reported. In principle, any functional group that is covalently conjugated to the G^C motif is ultimately expressed on the outer

18.3 Self-Assembling Rosette Nanotubes

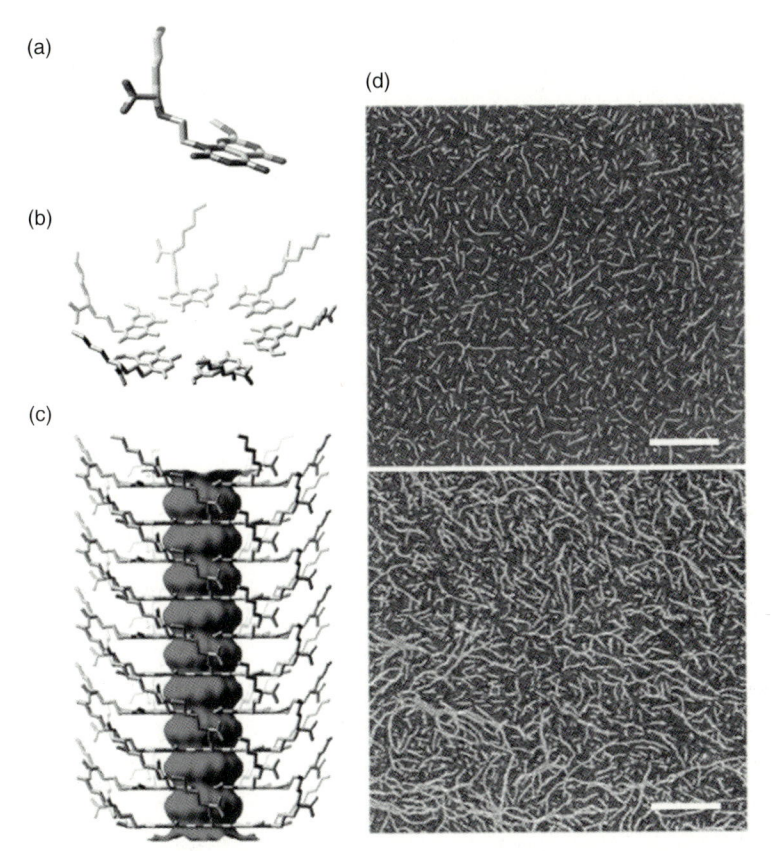

Figure 18.2 Module (a) self-assembles into a hexameric rosette (b), which then self-organizes into a K RNT (c), as shown in the negatively stained transmission electron microscopy images (d). The upper panel in (d) shows unheated K RNT, while the lower panel shows heated K RNT. The red color indicates the central core of K RNT. Scale bars = 250 nm. Reproduced with permission from Ref. [10]; © 2009, National Academy of Sciences.

surface of the RNT, thereby offering a robust "built-in" strategy for manipulating the physical and biological properties of the RNTs [12]. The RNTs undergo an entropically driven self-assembly process, whereby the nanotubes grow extensively upon heating. By using this G^C motif six-membered rosettes technology, K (lysine) RNT, RGDSK RNT, and RGDSK K (1, 5, and 10% molar ratio, respectively) RNT have been synthesized (Figure 18.3). The synthesis, characterization and biological functions of these nanotubes are described below.

The K1 G^C base derivative was coassembled with different concentrations of the RGDSK G^C base derivative (both as the trifluoroacetate salts), and the resultant hybrid RNT characterized by ^{1}H and ^{13}C NMR, high-resolution mass spectrometry, and elemental analysis. The molecular models of KRNT and RGDSK RNT are shown in Figure 18.4. K RNT was obtained by attaching a lysine molecule to

K-RNT RGDSK-RNT RGDSK/K-RNT
(1:10)

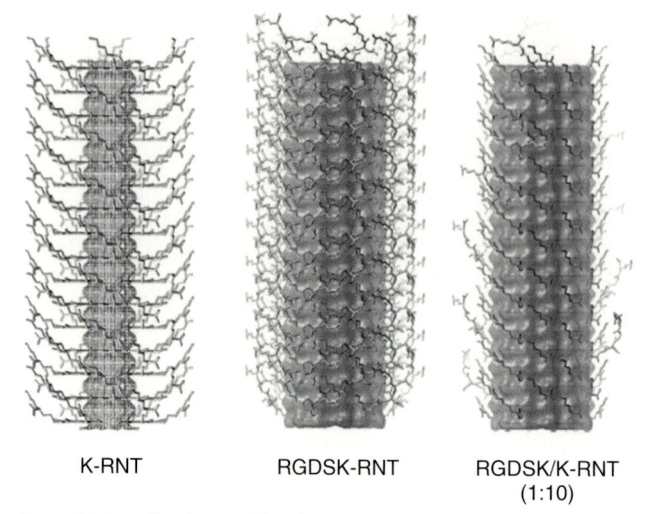

Figure 18.3 Molecular models of K RNT, RGDSK RNT and hybrid nanotube RGDSK/K RNT. The latter is composed of RGDSK RNT and K RNT in a molar ratio of 1:10. *Note*: The colors are used only to distinguish between the different RNTs. Reproduced with permission from Ref. [13]; © 2009, Elsevier.

a G^C motif through H–H bonding shared by a NH_2 group. Two G^C motif, each with an attached lysine molecule, were joined together and underwent self-assembly to yield a K RNT (Figure 18.4a, b). This was confirmed using NMR spectroscopy, circular dichroism spectroscopy, variable-temperature UV/visible melting studies, dynamic light scattering, tapping-mode atomic force microscopy (AFM), and transmission electron microscopy (TEM). The molecular formula of K G^C derivative was $C_{15}H_{24}N_8O_4$, and the exact mass and molecular weight were 380.19 and 380.4 Da, respectively. In agreement with the calculated average diameter of 3.5 nm, TEM images of K RNT featured a diameter of 4.0 ± 0.3 nm [12]. RGDSK RNT (Figure 18.4c, d) was also characterized using ^1H and ^{13}C NMR, high-resolution mass spectrometry and elemental analysis. The molecular formula of the RGDSK G^C derivative was $C_{30}H_{49}N_{15}O_{11}$, and the exact mass and molecular weight were 795.37 and 795.8 Da, respectively.

18.3.3
Biological Functions of RNTs

Helical RNTs have significant potential as an efficient drug-delivery system [12–14]. The most important features of these RNTs are their biofriendly nature and their ability to tag a variety of peptides for targeted delivery. The biological functions of these RNTs have been characterized in both *in vitro* and *in vivo* models. Recently, the intratracheal administration of K-RNT to C57BL/6 mice was shown to induce a very low level of transient lung inflammation at a high dose (50 μg per

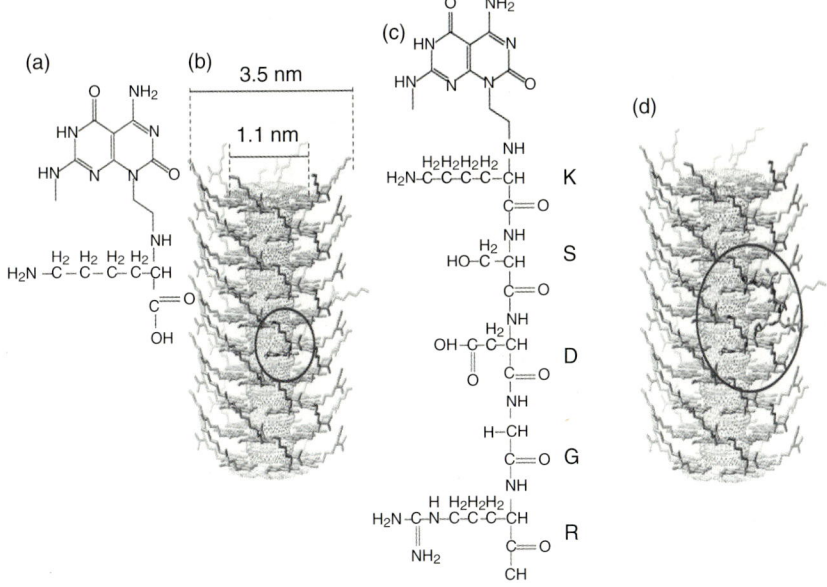

Figure 18.4 Molecular models of K1 (a, b) and RGDSK RNTs (c, d). K RNT was obtained by attaching a lysine molecule to the G^C motif through H–H bonding shared by a NH$_2$ group. Two G^C motif, each with an attached lysine molecule, are joined together and undergo self-assembly to yield a K RNT. The molecular formula of the K G^C derivative was C$_{15}$H$_{24}$N$_8$O$_4$, and the exact mass and molecular weight were 380.19 and 380.4 Da, respectively (Figure 18.3a, b). In agreement with the calculated average diameter of 3.5 nm, TEM images of K RNT featured a diameter of 4.0 ± 0.3 nm [12]. RGDSK RNT (Figure 18.3c, d) was also characterized by ^{1}H and ^{13}C NMR, high-resolution mass spectrometry, and elemental analysis. The molecular formula of the RGDSK G^C derivative was C$_{30}$H$_{49}$N$_{15}$O$_{11}$, and the exact mass and molecular weight were 795.37 and 795.8 Da, respectively.

mouse), whereas the lower doses did not induce any visible signs of inflammation [15]. The *in vitro* data obtained showed that the K RNT neither reduced cell viability nor induced the expression of inflammatory molecules in human bronchial epithelial adenocarcinoma Calu-3 cells [16]. K RNT was also found to interface well with macrophages, which are important immune cells, without activating them [10]. Taken together, these data showed that K RNT does not induce any obvious signs of toxicity, either *in vivo* and *in vitro*.

Attention has also been focused on elucidating the molecular signals induced by candidate RNTs, because the molecular signals underlie the phenotypic changes observed in the cells. In this case, an evaluation was made of the cell-signaling events caused by lysine-functionalized RNTs (K RNT) coassembled with Arg-Gly-Asp-Ser-Lys functionalized RNTs (RGDSK RNTs) for the induction of inflammation and apoptosis in human adenocarcinoma Calu-3 cells [13]. When coassembled in a molar ratio of 1:10 µM, these composite RNTs (referred to as RGDSK/K RNTs) rapidly induced the phosphorylation of P38 mitogen-activated protein kinases (MAPKs) within 2 min. The MAPKs are known to play a key role in regulating signal transduction and various cellular functions [17]. In particular, P38 MAPK, one of the three distinct families of MAPKs, becomes activated by the dual phosphorylation of tyrosine and threonine at the TGY motif, and is known to be involved in the regulation of both inflammation [18–21] and apoptosis [22–24]. It was observed that, when the Calu-3 cells were treated with higher concentrations of RGDSK/K RNTs (>10:100 μM), a P38 MAPK-dependent increase in secretion of tumor necrosis factor-alpha (TNF-α) resulted [13]. The RGDSK/K RNTs also induced a concentration- and P38 MAPK-dependent increase in caspase-3 activity and DNA fragmentation in Calu-3 cells. The pro-apoptotic properties of these RNTs were also confirmed using a complementary DNA (cDNA) microarray that showed an overexpression of pro-apoptotic genes, namely BCL2, caspase-3, Bcl-2 homologous antagonist/killer (BAK) protein (BAK1), cell death-inducing DFFA-like effector b (CIDEB), tumor protein p53 binding protein 2 (TP53BP2), FAS, TNF, and Fas ligand (FASLG). These results suggested that the RNTs could be used as a drug to induce apoptosis in cancer cells, or as a versatile platform to deliver a variety of biologically active molecules for cancer therapy.

Lung diseases, which are accompanied by inflammation, continue to induce significant mortality and morbidity in humans and animals [25, 26]. Inflammation in the lung and other organs is characterized by the migration of activated neutrophils, which are critical to combat bacterial infections. However, the migration of too-many activated neutrophils can cause excessive tissue damage, leading to morbidity and mortality [27, 28]. Thus, an investigation was made of the use of RNT in modulating the behavior of neutrophils to regulate inflammation in various organs, including the lungs. Specifically, the potential for RGDSK/K RNT tubes to interfere with the migration of neutrophils was examined. The data acquired showed that RGDSK/K RNT inhibited neutrophil migration in response to chemotactic molecules by interfering with the phosphorylation processes. Interestingly, in contrast to the apoptotic actions of the RNT on adenocarcinoma cells, the RGDSK/K RNT did not induce apoptosis in either normal or activated neutrophils [63].

In order to realize the biomedical potential of RNTs, there is a critical need to understand the mechanisms of uptake, internalization, and fate of RNTs within the cells. It is also important to investigate the fate of RNTs in the low-pH and enzyme-rich milieu of lysosomes. To address this issue, a twin-base hybrid RNT (TBL) was synthesized that was composed of RGDSK and fluoroscein isothiocyanate (FITC) which provided an ability to track the hybrids *in vitro* in human differentiated macrophages cells (S.S. Suri, A. Myles, H. Fenniri and B. Singh, unpublished observations). Subsequent scanning electron microscopy (SEM) studies confirmed the nanoscale dimensions of the RGDSK-TBL RNTs, while confocal microscopy demonstrated the colocalization of FITC/RGDSK-TBL RNT and integrin $\alpha_v\beta_3$-Cy5 on the macrophages within 2 min of incubation at 37 °C. The RNTs were also seen to colocalize with EEA1, which is a marker for early endosomes. Inhibition of the colocalization of the FITC/RGDSK-TBL RNT with the integrin, in addition to translocation into endosomes at 37 °C, confirmed the presence of an energy-dependent, receptor-mediated process. The incubation of differentiated macrophages with RGDSK-TBL RNT (>5 µM) resulted in almost 100% apoptosis within 12 h. It was suggested that the receptor-mediated endocytosis of FITC/RGDSK-TBL RNT might be exploited for the rapid and efficient delivery of tagged RNTs into the cells.

Metallic stents, when inserted into partially clogged arteries to promote normal blood flow and prevent myocardial infarction, do not interact closely with the arterial endothelial cells. Invariably, the latter cells will fail to adhere to the stent, which may become loose and be dislodged. Recently acquired data have demonstrated enhanced endothelial cell functions on RNT-coated titanium vascular stents [29], whereby a low concentration (0.1 mg ml^{-1}) of K RNT-coated titanium caused increases in endothelial cell densities by 37% and 52% when compared to uncoated titanium, after 4 h and three days, respectively [29]. Such excellent cytocompatibility of RNTs suggests the need for further molecular characterization of these novel nanomaterials in vascular stent applications.

Today, bone fractures, osteoporosis, and bone cancer represent a major health problem, especially among the rapidly aging populations of the Western hemisphere. The design of biomimetic bone tissue engineering materials that could restore and improve damaged bone tissues provides an exciting opportunity to improve traditional orthopedic implants. The RNTs have attractive properties that improve osteoblast (bone-forming cell) functions, including adhesion when coated onto a traditional implant material such as titanium [30, 31], or embedded in hydrogels [32, 33]. Such properties have been attributed to the nanoscale, biologically inspired features and the rich lysine or peptide moieties of RNTs. The temperature-dependent self-assembling properties of K RNT alter the protein-dependence of osteoblast adhesion in this system [30]. Recent evidence has shown that a biomimetic orthopedic hydrogel nanocomposite based on the self-assembly property of RNTs, the osteoconductive properties of nanocrystalline hydroxyapatite (HA), and the biocompatible properties of hydrogels, specifically poly(2-hydroxyethyl methacrylate) (pHEMA), caused a very significant increase (up to 2-4-fold) in osteoblast adhesion when compared to hydrogel controls [32]. The RNTs also stimulated HA nucleation and mineralization along their main axis, in a manner reminiscent of the HA/collagen assembly pattern in natural bone. These exciting properties of biomimetic nanocrystalline HA/RNT hydrogel composites make them promising candidates for the further study of bone tissue-engineering applications.

Recently, attention has been focused on RGDSK/K RNT, because of the ability for RGD peptides to target and modulate the actions of the integrin $\alpha_v\beta_3$, a cell-surface receptor for extracellular matrix (ECM) ligands such as vitronectin, fibronectin, and fibrinogen. The $\alpha_v\beta_3$ integrin is either absent or expressed at low levels in normal cells, but its expression is increased in breast cancer and melanoma [34, 35]. The integrin $\alpha_v\beta_3$ is a widely recognized target for the development of molecular probes for imaging angiogenesis, and for cancer therapy [36]. Wagner and colleagues covalently coupled DI17E6, a monoclonal antibody against the $\alpha_v\beta_3$ integrin, to human serum albumin nanoparticles to target the $\alpha_v\beta_3$

integrin-positive melanoma cells [37]. Moreover, an increased cytotoxicity of doxorubicin-loaded DI17E6 nanoparticles was demonstrated in melanoma cells positive for integrin $\alpha_v\beta_3$. Recently, the possible use was reported of RGD-tagged RNTs to target integrin-overexpressing lung adenocarcinoma cells, which led to the induction of apoptosis in these cells [13]. Based on this proof of principle, the aim is to further characterize the use of RGDSK/K RNT in targeting inflammatory and tumor cells expressing integrin $\alpha_v\beta_3$.

18.4
Stability Issues

In order to fully explore the potential of peptide-based nanofabrication technology for medical applications, it is necessary to investigate the stability, integrity, and functionality of peptide nanomers. Unlike naturally occurring proteins and peptides, bionanotubes are generally much more stable under extreme conditions. For example, bionanotubes synthesized from β-amyloid and diphenyl glycine peptides showed thermal stability up to 200 °C, as well as a high chemical stability which makes them suitable for potential use in nanoelectromechanics, functional nanodevices, and medical applications such as bone repair surgeries [38, 39]. The pH of the solvent is also an important factor controlling the stability of bionanotubes, because of the need to maintain the charge of amino acids in order to avoid the aggregation of nanotubes into large clumps and to withstand the wide range of pH-values encountered in normal and inflamed organs. Nanotubes prepared from diphenylalanine peptide are highly stable both in aqueous and dry conditions [38, 40]. Generally, alanine and valine produce more homogeneous and stable nanotubes than do glycine and leucine, whereas for cationic peptides lysine or histidine are preferred over arginine, possibly due to steric effects. Glycine and aspartic acid have also attracted interest because these amino acids are believed to have been present in the prebiotic environment of the early Earth.

18.5
Nanomaterials for Receptor-Mediated Targeting

Currently, there is an accumulating wealth of data to show that nanoparticles are inherently suitable for targeting cell-surface receptors for the delivery of drugs and genes. Such receptor targeting has the advantages of speed, efficiency, and specificity. Recently acquired data on the targeting of integrin $\alpha_v\beta_3$ by RNTs have shown that the RNTs are highly amenable to configuration to target other types of receptor that regulate the transformation and activation of cells in cancers and inflammatory diseases. Some examples of when nanoparticles have been used and RNT may be configured to achieve an even higher efficiency of drug and gene delivery are provided in the following subsections.

18.5.1
Human Epidermal Growth Factor Receptor (EGFR)

The EGFR is a member of the kinase receptors [41], and is overexpressed in breast cancer, ovarian cancer, prostate cancer, bladder cancer, glioblastoma, non-small cell lung cancer, and head and neck cancer. The EGFR has also been found to play a significant role in the progression of several human malignancies [42]. Magadala and colleagues fabricated EGFR-targeting peptide-gelatin-based nanoparticles for the encapsulation of reporter plasmid DNA, and examines the potential for its delivery and transfection in human pancreatic cancer cells [43]. Recently, EGFR-conjugated immunoliposomes and bionanocapsules have also been used for targeted drug delivery to brain tumors [44]. Nanoparticles prepared from polymeric poly(lactide-co-glycolide) (PLGA), loaded with rapamycin and conjugated anti-EGFR antibodies, caused apoptosis and cell death in breast cancer cells [45]. Likewise, human serum albumin nanoparticles conjugated to trastuzumab to target human EGFR-2-overexpressing cells were used to inhibit polo-like kinase-1 (Plk-1), which promotes cell proliferation, and to induce mitotic chaos, upregulate apoptosis, and inhibit tumor growth [46–49].

18.5.2
Vasoactive Pituitary Adenylate Cyclase (VPAC)-Activating Peptide Receptors

Vasoactive intestinal peptide (VIP), a 28-mer peptide, is a member of the secretin glucagon superfamily, and has a widespread distribution in the gastrointestinal, neuronal, endocrine, and immune systems. VIP binds to the G-protein-coupled VPAC receptors (viz., VPAC-1 and VPAC-2) which, under rapid recycling, leads to an efficient internalization into the cytoplasm [50]. The high-affinity receptors for VIP are overexpressed in a wide range of human tumors, including pituitary adenoma, gastrinoma, medullary thyroid carcinoma, lung cancer, pancreatic tumor, colorectal carcinoma, ovarian carcinoma, prostate carcinoma, urinary bladder carcinoma, and breast carcinoma [51]. Although initially, the use of VIP for tumor cell targeting was interrupted due to an enzymatic degradation of the peptides, protamine-based nanoparticles ("proticles") have been developed which afford protection against such enzymatic degradation while triggering the release and internalization of the conjugate by human tumors [51]. These studies have opened possibility for proticles to be used as a potential peptide-mediated targeting for both *in vivo* and clinical applications.

18.5.3
Transferrin Receptor (TfR)

Transferrin is a well-studied ligand for tumor targeting and gene delivery [52]. Transferrin binds to the TfRs on the cell surface, followed by a rapid receptor-mediated endocytosis. The TfR has been exploited by a specific and efficient delivery of doxorubicin with apotransferrin protein nanoparticles [53]. Nanotubes to target TfRs have also been used to deliver drug and to transfect genes across the blood–brain barrier into the brain [54].

18.5.4
Folate Receptor (FR)

At least four isoforms of the FR are known, namely α, β, γ/γ', and δ [55]. FR-α is one of the most promising and most investigated epithelial cancer markers that are known to be overexpressed on many human cancer cell surfaces, but is rarely found on normal cell surfaces. Recently, the importance of FR-β as a cellular target has been recognized in myelogenous leukemia and chronic inflammatory diseases. The FRs are N-glycosylated proteins with a high binding affinity to folate. Liposomal oligodeoxyribonucleotide has been targeted to FRs to deliver the anti-sense EGFR gene, and to inhibit cell growth and EGFR expression [56, 57]. Recently, a folate–poly(ethylene glycol) (PEG)–poly(ethylene imine) (PEI) conjugate has been used efficiently for the delivery of small interfering RNA (siRNA) [58]. Cowpea mosaic virus (CPMV), a well-characterized nanoparticle, has also been used for targeting FRs in cancer cells [59]. Compared to viral carriers, nonviral gene delivery systems have shown good biocompatibility and safety, but low transfection efficiencies. Recently, modified folate–chitosan and cholesterol derivatives were used to prepare charge-changing solid lipid nanoparticles capable of enhancing reporter gene expression by targeting FR-expressing cells [60]. Recently, folate-conjugated N-trimethyl chitosan was used to shuttle FITC-conjugated bovine serum albumin (BSA) to a cancer cell line (SKOV3 cells) that overexpressed the FR [61].

18.6
Ethical Issues and Future Directions

Although, similar to any newly emerging field of science, nanomedicine is presently at an experimental or even "primitive" stage, it offers new horizons for improving human health. Clearly, there is a need to develop rigorous safety standards for the use of nanoproducts, including those intended for medical purposes, and for this purpose the ethical, legal, and social implications of nanotechnologies have been elaborated in order to develop frameworks aimed at minimizing any prospective negative effects of the nanomaterials cleared for clinical use [62]. Some of the important issues related to the use of nanotubes in the medical field include toxicity, immunogenicity, biocompatibility, stability, pharmacokinetics (e.g., absorption, metabolism, distribution, and excretion), pharmacodynamics, privacy, and integrity. The atomic dimensions of the nanomaterials impart on them special characteristics, and thus special precautions must be followed during their handling and use. Blessed with extremely small dimensions, nanoparticles can enter the cells of organisms, avoid natural defenses, and be transported to organs and tissues in a target-specific manner. Yet, these qualities may prove to be harmful to humans if used without thorough investigation. In particular, it is the surface-to-mass ratio of nanomaterials that provides the greatest cause for concern. A check-list that might be followed before a nanobased drug can be released for biomedical applications might be as follows:

- Monitor any effects on DNA, and complete a thorough investigation of the transcription of genes.
- Check the distribution of any drug–shuttle nanocomplexes in the body.
- Monitor the metabolism and toxic effects of the nanoshuttle.
- Check the stability of the .nanoparticles
- Monitor the time release of the drug.
- Monitor any adverse side effects.
- Check any interference with cellular pathways.
- Monitor the biological half-life of the nanoshuttle.

18.7
Conclusions

The rapid developments in the creation and characterization of new nanomaterials for biomedical applications continue to inspire hope for the treatment and prevention of infectious diseases. The ability of novel nanoparticles to target specific receptors both selectively and sensitively, ultimately to deliver biological and chemical materials directly into the cells, will surely lead to the development of many "smart bullets." Within this context, the discovery of novel RNTs, which are not only organic but also biologically inspired and highly nimble, lends credibility to the claims that nanomedicine may fulfill its promised potential.

References

1 Theis, T., Parr, D., Binks, P., Ying, J., Drexler, K.E., Schepers, E., Mullis, K., Bai, C., Boland, J.J., Langer, R., Dobson, P., Rao, C.N. and Ferrari, M. (2006) nan'o.tech.nol'o.gy n. *Nature Nanotechnology*, **1**, 8–10.

2 Tkachenko, A.G., Xie, H., Liu, Y., Coleman, D., Ryan, J., Glomm, W.R., Shipton, M.K., Franzen, S. and Feldheim, D.L. (2004) Cellular trajectories of peptide-modified gold particle complexes: comparison of nuclear localization signals and peptide transduction domains. *Bioconjugate Chemistry*, **15**, 482–90.

3 Oyelere, A.K., Chen, P.C., Huang, X., El-Sayed, I.H. and El-Sayed, M.A. (2007) Peptide-conjugated gold nanorods for nuclear targeting. *Bioconjugate Chemistry*, **18**, 1490–7.

4 Yan, X., He, Q., Wang, K., Duan, L., Cui, Y. and Li, J. (2007) Transition of cationic dipeptide nanotubes into vesicles and oligonucleotide delivery. *Angewandte Chemie International Edition*, **46**, 2431–4.

5 Hartgerink, J.D., Beniash, E. and Stupp, S.I. (2001) Self-assembly and mineralization of peptide-amphiphile nanofibers. *Science*, **294**, 1684–8.

6 Nagai, Y., Unsworth, L.D., Koutsopoulos, S. and Zhang, S. (2006) Slow release of molecules in self-assembling peptide nanofiber scaffold. *Journal of Controlled Release*, **115**, 18–25.

7 Zhang, S., Lockshin, C., Herbert, A., Winter, E. and Rich, A. (1992) Zuotin, a putative Z-DNA binding protein in *Saccharomyces cerevisiae*. *The EMBO Journal*, **11**, 3787–96.

8 Fenniri, H., Deng, B.L. and Ribbe, A.E. (2002) Helical rosette nanotubes with tunable chiroptical properties. *Journal of the American Chemical Society*, **124**, 11064–72.

9 Fenniri, H., Deng, B.L., Ribbe, A.E., Hallenga, K., Jacob, J. and Thiyagarajan, P. (2002) Entropically driven self-assembly of multichannel rosette nanotubes. *Proceedings of the National Academy of Sciences of the United States of America*, **99** (Suppl. 2), 6487–92.

10 Journeay, W.S., Suri, S.S., Moralez, J.G., Fenniri, H. and Singh, B. (2009) Macrophage inflammatory response to self-assembling rosette nanotubes. *Small*, **5**, 1446–52.

11 Tikhomirov, G., Yamazaki, T., Kovalenko, A. and Fenniri, H. (2008) Hierarchical self-assembly of organic prolate nanospheroids from hydrophobic rosette nanotubes. *Langmuir*, **24**, 4447–50.

12 Moralez, J.G., Raez, J., Yamazaki, T., Motkuri, R.K., Kovalenko, A. and Fenniri, H. (2005) Helical rosette nanotubes with tunable stability and hierarchy. *Journal of the American Chemical Society*, **127**, 8307–9.

13 Suri, S.S., Rakotondradany, F., Myles, A.J., Fenniri, H. and Singh, B. (2009) The role of RGD-tagged helical rosette nanotubes in the induction of inflammation and apoptosis in human lung adenocarcinoma cells through the P38 MAPK pathway. *Biomaterials*, **30**, 3084–90.

14 Suri, S.S., Fenniri, H. and Singh, B. (2007) Nanotechnology-based drug delivery systems. *Journal of Occupational Medicine and Toxicology*, **2**, 16.

15 Journeay, W.S., Suri, S.S., Moralez, J.G., Fenniri, H. and Singh, B. (2008) Rosette nanotubes show low acute pulmonary toxicity in vivo. *International Journal of Nanomedicine*, **3**, 373–83.

16 Journeay, W.S., Suri, S.S., Moralez, J.G., Fenniri, H. and Singh, B. (2008) Low inflammatory activation by self-assembling Rosette nanotubes in human Calu-3 pulmonary epithelial cells. *Small*, **4**, 817–23.

17 Zhang, W. and Liu, H.T. (2002) MAPK signal pathways in the regulation of cell proliferation in mammalian cells. *Cell Research*, **12**, 9–18.

18 Kumar, S., Boehm, J. and Lee, J.C. (2003) p38 MAP kinases: key signalling molecules as therapeutic targets for inflammatory diseases. *Nature Reviews Drug Discovery*, **2**, 717–26.

19 Schindler, J.F., Monahan, J.B. and Smith, W.G. (2007) p38 pathway kinases as anti-inflammatory drug targets. *Journal of Dental Research*, **86**, 800–11.

20 Strassburger, M., Braun, H. and Reymann, K.G. (2008) Anti-inflammatory treatment with the p38 mitogen-activated protein kinase inhibitor SB239063 is neuroprotective, decreases the number of activated microglia and facilitates neurogenesis in oxygen-glucose-deprived hippocampal slice cultures. *European Journal of Pharmacology*, **592**, 55–61.

21 Temming, K., Lacombe, M., van der Hoeven, P., Prakash, J., Gonzalo, T., Dijkers, E.C., Orfi, L., Keri, G., Poelstra, K., Molema, G. and Kok, R.J. (2006) Delivery of the p38 MAPkinase inhibitor SB202190 to angiogenic endothelial cells: development of novel RGD-equipped and PEGylated drug-albumin conjugates using platinum(II)-based drug linker technology. *Bioconjugate Chemistry*, **17**, 1246–55.

22 Reddy, K.B., Nabha, S.M. and Atanaskova, N. (2003) Role of MAP kinase in tumor progression and invasion. *Cancer and Metastasis Reviews*, **22**, 395–403.

23 Sun, Y. and Sinicrope, F.A. (2005) Selective inhibitors of MEK1/ERK44/42 and p38 mitogen-activated protein kinases potentiate apoptosis induction by sulindac sulfide in human colon carcinoma cells. *Molecular Cancer Therapeutics*, **4**, 51–9.

24 Wang, X.B., Gao, H.Y., Hou, B.L., Huang, J., Xi, R.G. and Wu, L.J. (2007) Nanoparticle realgar powders induce apoptosis in U937 cells through caspase MAPK and mitochondrial pathways. *Archives of Pharmacal Research*, **30**, 653–8.

25 Benjamim, C.F., Hogaboam, C.M. and Kunkel, S.L. (2004) The chronic consequences of severe sepsis. *Journal of Leukocyte Biology*, **75**, 408–12.

26 Frevert, C.W. and Warner, A.E. (1999) Respiratory distress resulting from acute

lung injury in the veterinary patient. *Journal of Veterinary Internal Medicine*, **6**, 154–65.
27. Lee, W.L. and Downey, G.P. (2001) Neutrophil activation and acute lung injury. *Current Opinion in Critical Care*, **7**, 1–7.
28. Zemans, R.L., Colgan, S.P. and Downey, G.P. (2009) Transepithelial migration of neutrophils: mechanisms and implications for acute lung injury. *American Journal of Respiratory Cell and Molecular Biology*, **40**, 519–35.
29. Fine, E., Zhang, L., Fenniri, H. and Webster, T.J. (2009) Enhanced endothelial cell functions on rosette nanotube-coated titanium vascular stents. *International Journal of Nanomedicine*, **4**, 91–7.
30. Chun, A.L., Moralez, J.G., Webster, T.J. and Fenniri, H. (2005) Helical rosette nanotubes: a biomimetic coating for orthopedics? *Biomaterials*, **26**, 7304–9.
31. Sato, M. and Webster, T.J. (2004) Nanobiotechnology: implications for the future of nanotechnology in orthopedic applications. *Expert Review of Medical Devices*, **1**, 105–14.
32. Zhang, L., Rakotondradany, F., Myles, A.J., Fenniri, H. and Webster, T.J. (2009) Arginine-glycine-aspartic acid modified rosette nanotube-hydrogel composites for bone tissue engineering. *Biomaterials*, **30**, 1309–20.
33. Zhang, L., Ramsaywack, S., Fenniri, H. and Webster, T.J. (2008) Enhanced osteoblast adhesion on self-assembled nanostructured hydrogel scaffolds. *Tissue Engineering Part A*, **14**, 1353–64.
34. Ruoslahti, E. (2002) Specialization of tumour vasculature. *Nature Reviews Cancer*, **2**, 83–90.
35. Singh, B., Fu, C. and Bhattacharya, J. (2000) Vascular expression of the $\alpha_v\beta_3$ integrin in lung and other organs. *American Journal of Physiology*, **278**, L217–26.
36. Lim, E.H., Danthi, N., Bednarski, M. and Li, K.C. (2005) A review: integrin alphavbeta3-targeted molecular imaging and therapy in angiogenesis. *Nanomedicine*, **1**, 110–14.
37. Wagner, S., Rothweiler, F., Anhorn, M.G., Sauer, D., Riemann, I., Weiss, E.C., Katsen-Globa, A., Michaelis, M., Cinatl, J., Jr, Schwartz, D., Kreuter, J., von Briesen, H. and Langer, K. (2010) Enhanced drug targeting by attachment of an anti-alphav integrin antibody to doxorubicin loaded human serum albumin nanoparticles. *Biomaterials*, **31**, 2388–98.
38. Adler-Abramovich, L., Reches, M., Sedman, V.L., Allen, S., Tendler, S.J. and Gazit, E. (2006) Thermal and chemical stability of diphenylalanine peptide nanotubes: implications for nanotechnological applications. *Langmuir*, **22**, 1313–20.
39. Reches, M. and Gazit, E. (2006) Designed aromatic homo-dipeptides: formation of ordered nanostructures and potential nanotechnological applications. *Physical Biology*, **3**, S10–19.
40. Ryu, J. and Park, C.B. (2009) Synthesis of diphenylalanine/polyaniline core/shell conducting nanowires by peptide self-assembly. *Angewandte Chemie International Edition*, **48**, 4820–3.
41. Doebele, R.C., Oton, A.B., Peled, N., Camidge, D.R. and Bunn, P.A., Jr (2010) New strategies to overcome limitations of reversible EGFR tyrosine kinase inhibitor therapy in non-small cell lung cancer. *Lung Cancer*, **69** (1), 1–12.
42. Rogers, S.J., Harrington, K.J., Rhys-Evans, P., O-Charoenrat, P. and Eccles, S.A. (2005) Biological significance of c-erbB family oncogenes in head and neck cancer. *Cancer and Metastasis Reviews*, **24**, 47–69.
43. Magadala, P. and Amiji, M. (2008) Epidermal growth factor receptor-targeted gelatin-based engineered nanocarriers for DNA delivery and transfection in human pancreatic cancer cells. *AAPS Journal*, **10**, 565–76.
44. Feng, B., Tomizawa, K., Michiue, H., Miyatake, S., Han, X.J., Fujimura, A., Seno, M., Kirihata, M. and Matsui, H. (2009) Delivery of sodium borocaptate to glioma cells using immunoliposome conjugated with anti-EGFR antibodies by ZZ-His. *Biomaterials*, **30**, 1746–55.
45. Acharya, S., Dilnawaz, F. and Sahoo, S.K. (2009) Targeted epidermal growth factor receptor nanoparticle bioconjugates for breast cancer therapy. *Biomaterials*, **30**, 5737–50.
46. Gumireddy, K., Reddy, M.V., Cosenza, S.C., Boominathan, R., Baker, S.J., Papathi, N., Jiang, J., Holland, J. and

Reddy, E.P. (2005) ON01910, a non-ATP-competitive small molecule inhibitor of Plk1, is a potent anticancer agent. *Cancer Cell*, **7**, 275–86.

47 Liu, X., Lei, M. and Erikson, R.L. (2006) Normal cells, but not cancer cells, survive severe Plk1 depletion. *Molecular and Cellular Biology*, **26**, 2093–108.

48 Lowery, D.M., Clauser, K.R., Hjerrild, M., Lim, D., Alexander, J., Kishi, K., Ong, S.E., Gammeltoft, S., Carr, S.A. and Yaffe, M.B. (2007) Proteomic screen defines the Polo-box domain interactome and identifies Rock2 as a Plk1 substrate. *The EMBO Journal*, **26**, 2262–73.

49 Steinhauser, I., Langer, K., Strebhardt, K. and Spankuch, B. (2009) Uptake of plasmid-loaded nanoparticles in breast cancer cells and effect on Plk1 expression. *Journal of Drug Targeting*, **17**, 627–37.

50 Ou, X., Tan, T., He, L., Li, Y., Li, J. and Kuang, A. (2005) Antitumor effects of radioiodinated antisense oligonuclide mediated by VIP receptor. *Cancer Gene Therapy*, **12**, 313–20.

51 Ortner, A., Wernig, K., Kaisler, R., Edetsberger, M., Hajos, F., Kohler, G., Mosgoeller, W. and Zimmer, A. (2010) VPAC receptor-mediated tumor cell targeting by protamine-based nanoparticles. *Journal of Drug Targeting*, **18** (6), 457–67.

52 Misra, S., Hascall, V.C., De Giovanni, C., Markwald, R.R. and Ghatak, S. (2009) Delivery of CD44 shRNA/nanoparticles within cancer cells: perturbation of hyaluronan/CD44v6 interactions and reduction in adenoma growth in Apc Min/+ MICE. *Journal of Biological Chemistry*, **284**, 12432–46.

53 Krishna, A.D., Mandraju, R.K., Kishore, G. and Kondapi, A.K. (2009) An efficient targeted drug delivery through apotransferrin loaded nanoparticles. *PLoS One*, **4**, e7240.

54 Xia, C.F., Boado, R.J., Zhang, Y., Chu, C. and Pardridge, W.M. (2008) Intravenous glial-derived neurotrophic factor gene therapy of experimental Parkinson's disease with Trojan horse liposomes and a tyrosine hydroxylase promoter. *Journal of Gene Medicine*, **10**, 306–15.

55 Zhao, X., Li, H. and Lee, R.J. (2008) Targeted drug delivery via folate receptors. *Expert Opinion on Drug Delivery*, **5**, 309–19.

56 Leamon, C.P., Cooper, S.R. and Hardee, G.E. (2003) Folate-liposome-mediated antisense oligodeoxynucleotide targeting to cancer cells: evaluation in vitro and in vivo. *Bioconjugate Chemistry*, **14**, 738–47.

57 Wang, S., Lee, R.J., Cauchon, G., Gorenstein, D.G. and Low, P.S. (1995) Delivery of antisense oligodeoxyribonucleotides against the human epidermal growth factor receptor into cultured KB cells with liposomes conjugated to folate via polyethylene glycol. *Proceedings of the National Academy of Sciences of the United States of America*, **92**, 3318–22.

58 Kim, S.H., Mok, H., Jeong, J.H., Kim, S.W. and Park, T.G. (2006) Comparative evaluation of target-specific GFP gene silencing efficiencies for antisense ODN, synthetic siRNA, and siRNA plasmid complexed with PEI-PEG-FOL conjugate. *Bioconjugate Chemistry*, **17**, 241–4.

59 Destito, G., Yeh, R., Rae, C.S., Finn, M.G. and Manchester, M. (2007) Folic acid-mediated targeting of cowpea mosaic virus particles to tumor cells. *Chemistry and Biology*, **14**, 1152–62.

60 Liu, Z., Zhong, Z., Peng, G., Wang, S., Du, X., Yan, D., Zhang, Z., He, Q. and Liu, J. (2009) Folate receptor mediated intracellular gene delivery using the charge changing solid lipid nanoparticles. *Drug Delivery*, **16**, 341–7.

61 Zheng, Y., Cai, Z., Song, X., Chen, Q., Bi, Y., Li, Y. and Hou, S. (2009) Preparation and characterization of folate conjugated N-trimethyl chitosan nanoparticles as protein carrier targeting folate receptor: in vitro studies. *Journal of Drug Targeting*, **17** (4), 1–10.

62 Spagnolo, A.G. and Daloiso, V. (2009) Outlining ethical issues in nanotechnologies. *Bioethics*, **23**, 394–402.

63 Le, M.H.A., Suri, S.S., Felaniaina, R., Fenniri, H. and Singh, B. (2010) Rosette nanotubes inhibit bovine neutrophil chemotaxis. *Veterinary Research*, **41**, 75.

Index

a

accumulation
- brain drug   376ff.
- LDC   378
- nanocarrier   123, 139, 292
- organic nanoparticle   223, 273
- poly(ethylene glycol) (PEG)   124
- tissue   273
- tumor   140

aggregation
- carboxy-functionalized NPs   328
- copolymer   243f.
- in vivo   344
- micellar   75
- monomer   371
- organic nanoparticle   223, 225
- poly(divinyl benzene) (poly-DVB)   35
- polymer   35, 237
- polymersomes   163
- radiopaque iodinated P(MAOETIB–GMA)   357f.
- solid lipid nanoparticles (SLNs)   370ff.

atomic force microscopy (AFM)
- lipid-polymer films and coatings   278
- molecular crystal nanorod   442f.
- nanogel   49f.
- nanotube forest   111
- rosette nanotubes (RNTs)   498

b

biocompatibility
- alginate   298
- carbon nanotubes (CNTs)   76, 108
- chitosan   298
- core–shell magnetic NPs   293
- nanofibrous scaffolds   398
- poly(ethylene glycol) (PEG)   298
- polymersomes   164, 229
- polysaccharide-derived particles   203
- quantum dots (QDs)   127
- scaffolds   418

biodegradable
- cellular uptake   323f.
- core–shell polymeric NPs   288ff.
- nanogels   47
- polyester   321ff.
- poly(ethylene glycol) (PEG)   124
- poly(n-butylcyanacrylate) (PBCA)   320f.
- SRP   74

biofouling   267
blood–brain barrier (BBB)   165, 321, 365ff.
- reversible disruption   368

bond
- acetal   47
- coordination   42ff.
- covalent   47, 430
- disulfide   4, 47
- ester   47
- hydrogen   4, 42f., 61, 430
- intra-gel   6
- ionic   42f.
- labile   4ff.
- noncovalent   42, 444ff.
- nonreactive   5
- permanent   4f., 8
- ph-cleavable   132
- reformation   7
- reversible   7
- rupture   7f., 12f.
- stiffness   7, 10, 12
- thiol   4

bottom-up approach, see synthesis
brain imaging, see medical imaging
bulk heterojunction (BHJ) architecture   445

c

cancer thearpy
- chemotherapy   139, 298
- photodynamic therapy (PDT)   300f., 381
- photothermal therapy   482f.

- polymeric-based core–shell NPs 300ff.
- radiofrequency ablation 483f.
- solid lipid nanoparticles (SLNs) 366ff.
- thermotherapy 382

Cauchy–Green strain tensor 16, 20

cell
- adhesion 82, 105
- differentiation 329ff.
- penetration 136f., 169

cellular
- therapeutics 314ff.
- uptake 73, 312, 315, 317ff.

central nervous system (CNS)
- disorder 365
- targeting 381

chelator 138f., 459f.

chemotherapy, see cancer therapy

chromatography
- dye affinity 86
- gel-permeation (GPC) 43, 45, 50, 263
- high-performance liquid (HPLC) 263
- thin-layer (TLC) 263

circular dichroism spectroscopy 498

click chemistry 32, 77

cluster
- grape-like 46
- nanogel 12, 19

coating
- fibers 411f.
- lipid– 269ff.
- nanotubes 108
- pinhole-free 70
- PNIPAAm 79
- polyelectrolyte 297
- polysorbate 80– 376ff.
- stimulus-responsive polymer (SRP) 73

complexes
- anionic polymer– cationic polymer 43f.
- anionic polymer–metal ion 43f.
- avidin 165
- drug–polymer 298
- gadolinium 316

computational fluid dynamics (CFD) simulation 200f.

computed tomography (CT) 127, 345, 349ff.

confocal microscopy 166, 294

conjugates
- CNTs–PNIPAAm 76
- ligand–PEG–PE 128f.
- lipid–drug (LDC) 377f.
- oligonucleotide (ODN)–PEG 132
- PEG–lipid 125
- pro drug 377

- SRPs–QDs 73f.
- SWNTs–paclitaxel (PTX) 480
- transferrin–PEG–liposomes 129

contact angle, see wettability

continuum elasticity theory 5

contrast agents, see medical imaging

copolymers
- crosslinking methods 40f.
- diblock 34, 38, 43, 50, 65, 158
- graft 44, 158
- multiblock 158ff.
- radiopaque iodinated P(MAOETIB-GMA) 351ff.
- random block 158
- star 36f., 158
- tetrablock 160
- triblock 40, 62, 74, 158

core–shell biomedical applications 285, 293ff.
- bioimaging 293f.
- biosensing in Diabetis Mellitus 295f.
- glucose monitoring 295f.
- labeling 293f.

core–shell nanostructures
- colloids 290, 292
- functionalized 298ff.
- multifunctional 286, 293f.
- nanocapsules 286, 288, 316f., 334, 370
- nanofibers 405ff.
- nanogels 31, 35, 46
- nanospheres 286, 370
- quantum dots 286
- polymeric NPs 286ff.
- surface-properties 292
- synthesis 289

critical micellar concentration (CMC), see micelle

critical micelle temperature (CMT), see micelle

critical solution temperature
- lower (LCST) 35, 41, 46, 61ff.
- upper (UCST) 61ff.

crosslinking agent
- divinyl benzene (DVB) 28, 32, 34f., 37
- methylenebis(acrylamide) (BIS) 28, 30, 33ff.
- methyl methacrylate (MMA) 32

crosslinking
- bifunctional 29
- covalent 28, 75, 291, 316
- density 39
- dual 4f., 7ff.
- intermolecular 46
- ionically 291
- linear chains 40f.

– micelles 36
– nanocomplexes 41
– nanoparticles 41
– noncovalent 28, 42
– photothermal 39
– side-chain 39
cytoplasmic localization 327
cytotoxicity
– QDs 295
– SLN 383

d

Damkohler number 236, 243
defects 4, 466
deformation
– bulk 17, 22
– elastic 16
– gel 6, 8ff.
– -gradient tensor 16
– modes 22
– plastic 13
– tensile 4, 8f., 13f.
detection
– labeling 293f.
– -free 455f.
– real time 463f.
dialysis 41, 199
dispersion
– mono– 286
– nano– 199, 245
– poly– 232, 273
– precipitation–redispersion 199
– stabilization 223
divinyl benzene (DVB), see crosslinking agent
DNA
– complexation 204
– encapsulated 174
– intracellular delivery 135f.
drug delivery for the brain 366ff.
– solid lipid nanoparticles (SLNs) 369ff.
– strategies 368f.
drug delivery systems (DDSs)
– bifunctional core–shell nanostructures 296ff.
– controlled (CDDS) 77f., 203, 344
– dendrimers 127
– encapsulated PLGA–PEG NPs 245f.
– green 46
– in vitro release 290
– liposome-mediated 65, 76, 233f.
– long-circulating liposomes 124ff.
– longevity 122ff.
– magnetically-sensitive 133f., 136
– microfluidic-related organic NP 244ff.

– multifunctional nanocarrier 121ff.
– nanofibrous scaffolds 399
– nanowire conducting polymer (CP) 467f.
– nonliposomal long-circulating 126f.
– organic NPs 222, 226f., 244ff.
– pH-sensitive systems 130ff.
– polymersomes 76, 170
– polysaccharide-based nanoparticulate 287, 291, 296
– quantum dots (QDs) 127
– redox-sensitive 135f.
– stimuli-responsive release 130f., 246ff.
– targetability 122ff.
– temperature-sensitive 133f., 136
– theranostics 138ff.
– ultrasound-sensitive 134
drug expulsion 375
dynamic light scattering (DLS)
– lipopolymers 273
– rosette nanotubes (RNTs) 498

e

ellipsometry
– lipid-polymer films and coatings 278
emulsification, see synthesis
emulsion
– double 230, 269, 369
– inverse mini– 314
– micro– 195, 370
– mini– 29f., 195, 313ff.
– oil-in-oil 195
– oil-in-water 29, 33, 195, 197, 370
– organic NPs 228ff.
– water-in-oil 29, 269, 370
– water-in-oil-in-water 269, 369
– single 159, 269
endocytosis of NPs 331f., 379
endocytotic pathway 312, 331f.
enhanced permeation and retention (EPR) effect 122f., 301
enzymatic degradation 368
extracellular matrix (ECM) 79, 397ff.
extravasation 123

f

field-effect transistors (FETs)
– CNT-based biosensors 464, 481
– covalent organic nanostructures 444
– nanowire CP 463
– peptide nanostructures 111
finite element approximation
– deformation of nanogels 16, 18
flow
– cytometry 323

– -driven microcapsules 5
fluorescence-activated cell sorting
  (FACS) 315, 319, 325
fluorescence microscopy
– molecular crystal nanorod 442
– near infrared (NIR) 169, 486
– organic NPs 206
– QDs 300
– self-assembled peptides 113, 116
– solid lipid nanoparticles (SLNs) 378
– thioflavin 113, 116
fluorescence recovery after photobleaching
  (FRAP)
– lipid-polymer films and coatings 278
Food and Drug Administration (FDA)
– cetylpalmitate SLN 383
– PEGylated QDs 287
– polyester polymers 383
– poly(ethylene oxide)–poly(propylene oxide)
  (PEO–PPO) 158
– poly(N-isopropylacrylamide)
  (PNIPAAm) 61
Fourier transform (FT) 302
– fast (FFT) filtering 166
– infrared spectroscopy (FTIR) 263, 347
functionalization 66
– nanoparticles (NPs) 72f., 129f., 312f., 324ff.
– ORMOCER 113ff.
– scaffolds 397, 418
– surface 312f., 324ff.

g

gadolinium-based contrast agents
  (GBCAs) 316f., 335, 380f.
gamma-scintigraphy, *see* medical imaging
gene therapy
– contrast agents 314
– polysomes 174f.
– organic nanoparticles 204
generally recognized as save (GRAS) 382
grafting
– density 76, 86
– -from 69, 413
– poly(ethylene glycol) (PEG) 70
– polymer onto surface 413f.
– -to 69f., 413

h

half-life
– circulation 265, 267, 272, 275, 298
– gadolinium-based contrast agents
  (GBCAs) 380f.
– lipopolymer hybrid NPs 265

heterojunction, *see* bulk heterojunction
Hookean spring 6, 8
hydrophilic
– conformation 66, 161
– monomer 35
– poly(ethylene glycol) (PEG) 124
– polymer 38f., 64, 124, 163
– poly(N-isopropylacrylamide)
  (PNIPAM) 31
hydrodynamic
– flow focusing 231ff.
– radius 49, 65, 346, 351, 355
hydrogel 60ff.
– cationic 64
– chitosan 64
– layer 292
– PDMAEMA 64f.
– salt-responsive 86
– swelling–deselling 64, 72
hydrophobic
– monomers 33
– polymer 35, 38f., 65, 106f., 371

i

immobilization
– CP nanowire 460
– enzyme 62
– spiropyran 35
immunoresponse 292
interaction
– cell–cell 403
– cell–ECM 403
– cell–nanofiber 401ff.
– cell–nanoparticle 248, 317ff.
– cell–scaffold 403
– cellular 311
– chain–chain 372f.
– drug–carrier 123
– enhanced repulsive 124
– enzyme–hydrogel 302
– Hookean spring 6, 8
– host–guest 204
– hydrophobic 42, 44f., 106, 371
– inter-gel 6, 18
– intermolecular 46
– intramolecular 64
– nanoparticle–organelle 249
– π–π stacking 77, 432, 475, 479
– particle–particle 374
– potential 6f., 34
interface
– solid/liquid 169, 375
– water/oil 317
interfacial

– energy 371
– film 197
– polyaddition 317
– tension 167, 375
interferometric reflectance spectroscopy 72
intracellular delivery
– cell-penetrating peptide
 (CPP)-mediated 136f.
– of drug nanocarriers 122, 135ff.
– targeting 130ff.
irradiation
– electron beam 41, 413, 429
– γ– 370, 413
– quantum-ray 41
– ultraviolet (UV) 39, 41ff.
Israelachvili's packing parameter 371

k

kinetics
– enzyme 83
– growth 329
– homogeneous competitive 244
– organic NP formation 244, 253
– uptake 312, 320, 334
Krieger's equation 374

l

labeling, see detection
lab-on-a-chip technology 204, 221f.
laser scanning microscopy (LSM) 315
– confocal (cLSM) 323, 327f., 334
– subcellular localization 327f., 334
lattice spring model (LSM), see model
Levi–Civita tensor 20
lipid-coated, see coating
lipid-polymer nanomaterials, see
 lipopolymers
lipopolymers
– applications, see nanostructure application
– coatings 260, 275, 277ff.
– conjugates 260f.
– films 260, 277ff.
– glyolipids 262
– hybrid NPs 260, 264ff.
– packing 371ff.
– phospholipids 262
– properties 263
– synthesis 261f., 266ff.
– synthetic 260f.
– tether 278
liposomes
– large multilamellar vesicles (LMV) 266
– large unilamellar vesicles (LUV) 266
– long-circulating 124ff.
– PEGylated 137f., 267f.
– polymer-decorated 60
– small unilamellar vesicles (SUV) 266
– stimulus-responsive polymers
 (SRPs) 75f., 267
lithography
– electron-beam 193, 251
– holographic 193
– nanoimprint (NIL) 193f., 251, 460f.
– particle replication in nonwetting
 templates (PRINT) 193f.
– photo– 193
– step– and flash imprint (S-FIL) 194
– X-ray 193

m

macrogels 30
magic bullets 226
magnetic resonance image (MRI), see
 medical imaging
marker
– bio– 293
– chemical 5
– damage 5
– sensing of desease 462f.
mass spectrometry (MS)
– nanotubes 109
– lipopolymers 263
mean residence time (MRT) 377
medical imaging
– brain imaging 380ff.
– contrast agents 138f., 311, 314, 316f.
– gamma-scintigraphy 138
– in vitro 140
– in vivo 139f.
– magnetic resonance image (MRI) 137f.,
 169f., 300, 311, 314f.
– ultrasonography 138
membrane
– alumina 71
– aluminum oxide (AAO) 435
– bilayer 267, 278
– conformations 159ff.
– contactors 200
– fluidity 161
– fusion 162f.
– lipid 277f.
– mixers 200
– permeability 85f., 161
– polymersomes 159ff.
– porous 60, 71f., 85
– stability 170
– thickness 161
mesenchymal stem cells (MSCs)

– clathrin-mediated uptake 334
– intracellular uptake 326f.
– proliferation rate 404
methylenebis(acrylamide) (BIS), *see* crosslinking agent
methyl methacrylate (MMA), *see* crosslinking agent
micelle
– critical micellar concentration (CMC) 74, 132, 223
– critical micelle temperature (CMT) 77
– crosslinked 36ff.
– flower-like 40
– formation 36
– ph-sensitivity 132
– polyion complex 43
– polymeric (PM) 237
– spherical 167
– unimeric 75
microcapsule 4
microelectrode array (MEA) 111, 114
microfluidic
– batch synthesis of organic NPs 222ff.
– channels 234f., 247, 249
microfluidic devices 204f., 221f., 227ff.
– aggression time 236f.
– continuous-flow 224
– flow rate ratio (FRR) 234, 241
– flow velocity 235f.
– heat transfer 240f.
– kinetic control 242f.
– microfluidic dimensions 235f.
– micromixing 238f., 243
– mixing time 235ff.
– operating parameters 234ff.
– reactors 228ff.
– residence time distribution (RTD) 239
– solvent-resistant 250
– thermal control 240
microfluidic
– high-throughput processes 249, 251f.
– synthesis of organic nanoparticles 227ff.
– synthetic operations 238ff.
– T-junctions 229, 231
microgel 28, 36ff.
model
– atomistic 5
– Bell 7
– gel lattice spring model (gLSM) 15f., 18
– lattice spring model (LSM) 5ff.
– seven-node 6
molecular weights
– high-molecular weight polymers 161, 168, 188
– low-molecular weight compounds 42, 168, 295
– multifunctional low-molecular weight initiators 37
– nanogels 31
– ultra-high-molecular weight compound 302
– weight-average 49

n

nanobelts 455
nanocapsules, *see* core–shell nanostructures
nanocarrier, *see* drug delivery systems (DDSs)
nanofiber
– biomolecule attachment 416f.
– chemical modification 412f.
– coating 411f.
– fiber diameter 403f.
– fiber orientation 404f., 408ff.
– fiber surface optimization 410ff.
– plasma treatment of 414ff.
– polycaprolactone (PCL) 406, 409, 412
– polymer grafting 413f.
– tissue engineering 397ff.
nanogel
– anionic 31
– dumbbell 37
– capsules 34
– cationic 31, 33
– chemical 28ff.
– cholesterol-bearing pullulan (CHP) 44ff.
– degradable chemical 47f.
– dynamics 7
– hairy 32f., 37, 50
– hexagonal particle 18f., 21ff.
– hollow 34, 39
– Janus 35, 160
– mechanical stability 13f.
– network 6, 8f., 86
– N-isopropylacrylamide (NIPAM) 31, 34f., 47f., 133
– physical 28, 42ff.
– polyacrylamide (PAM) 34
– poly(N-isopropylacrylamide) (PNIPAAm) 30ff.
– poly(vinyl caprolactam) (PVCL) 63
– poly(vinylimidazole) (PVI) 34
– poly(vinyl methyl ether) (PMVE) 41, 6163
– poly(2-vinylpyridine) (P2VP) 34
– poly(4-vinylpyridine) (P2VP) 35
– self-healing 3ff.
– size 28, 30, 34

– star-like 36ff.
– synthesis 28ff.
– three-dimensional (3-D) 18
– tripatite 37
– two-dimensional (2-D) 8, 18
– viscoelastic 21ff.
nanointerfaces 59ff.
nanoparticles (NPs)
– dual-reporter 314
– encapsulated 4, 60
– Fe_3O_4 32
– functionalization 72f., 129f., 312f., 324ff.
– gadolinium 316f., 335, 380
– gold (AuNPs) 34, 67, 72f., 129, 345
– iodinated 344f.
– kinetically frozen 236
– magnetic 73, 108, 126, 140, 247, 314
– MAOETIB 346ff.
– PACA 382f.
– PEGylated 129f.
– polyester 321ff.
– poly(n-butylcyanacrylate) (PBCA) 320f.
– polysaccharide-based 287, 291
– radiopaque 343ff.
– silica 34, 72
– soft 188
– solid organic 188ff.
– sterilization 370f.
– stimulus-responsive polymer (SRP) 72ff.
– superparamagnetic iron oxides (SPIOs) 133f., 136f., 140, 170, 311, 314ff.
– ultrasmall superparamagnetic iron oxides (USPIOs) 380
– very small superparamagnetic iron oxides (VSPIOs) 380
nanoprecipitation, see synthesis
nanorods
– solvent-annealed molecular crystal 440ff.
– stimulus-responsive polymer (SRP) 72
nanoshell, see coating
nanospheres, see core–shell nanostructures
nanostructure applications
– chemical sensing 445f.
– core–shell biomedical applications 285, 293ff.
– lipopolymers 263f., 275f., 279
– nanowire conducting polymer applications 464f., 480ff.
– organic nanoparticle applications 207f.
– nanotubes 464, 480ff.
– photomechanical actuation 446ff.
– polymersomes biomedical applications 161, 169ff.
– stimulus-responsive polymer (SRP) application 77ff.
nanosprings 455
nanostructure
– fabrication, see synthesis
– one-dimensional (1-D) 430f., 433, 439, 455
– size-controlled growth 429f.
– three-dimensional (3-D) 18, 114f., 405, 418
– two-dimensional (2-D) 8, 18, 405
– zero-dimensional 433
nanotubes
– applications, see nanostructure applications
– carbon (CNTs) 76, 107, 430, 455
– charge-transfer organic salt 439
– chelator CP 459f.
– dibenzoylmethane (DBM) 437f.
– fabrication 474ff.
– horizontal alignment 108
– multi-walled carbon (MWNTs) 107, 478, 484
– PEGylated 127, 129
– peptide-based 494ff.
– poly(3,4-ethylenedioxythiophene) (PEDOT) 437, 467
– polymeric-based 300
– rosette (RNTs) 493ff.
– single-walled (SWNTs) 478ff.
– stimulus-responsive polymer (SRP) 72, 76f.
– vertical alignment 107
nanowire conducting polymer applications
– biosensors 462f., 480
– chemiresistive immunosensor 463
– detection 464ff.
– sensing of desease markers 462f.
nanowire conducting polymer
– assembly/alignment 461, 463
– carboxylated 459
– fabrication 457f., 474ff.
– magnetic alignment 461f.
– segmented 457, 460f.
– surface modification 458f.
nanowires
– polyaniline (PANI) 464f., 481
– poly(3,4-ethylenedioxythiophene) (PEDOT) 437
– POPOP 438
– supramolecular self-assembly 430, 431
neo-Hookean compressible material 17
nuclear magnetic resonance spectroscopy (NMR) 263

– radiopaque iodinated P(MAOETIB) NPs 347
– rosette nanotubes (RNTs) 497ff.

o

opacification 350, 352
oponization 124, 290, 292, 378
optical
– bioimaging 484ff.
– microscopy 229
organic nanoparticle
– applications, see nanostructure applications
– dendrimers 188f.
– fullerenes 188f.
– growth 200, 224, 243f., 252
– macromolecules 188f.
– monodisperse 229
– nucleation 200, 224, 243f., 252
– preparation, see synthesis
– properties 226
– size-dependent effects 189, 223, 225
osmotic pressure 344

p

patterning
– inkjet technology 107, 109, 111
– laser-induced forward transfer (LIFT) 111ff.
– three-dimensional (3-D) nanostructures 114ff.
PEGylated, see poly(ethylene glycol)
peptides, see self-assembled peptide nanostructures
pharmaceutical nanocarriers, see drug delivery systems (DDSs)
pharmacological
– activity 344
– inhibitors 312
phase
– separation 165f.
– transition temperature 271
phospholipids, see self-assembled
photodynamic therapy (PDT), see cancer therapy
photo-Fenton reaction, see reaction
photoisomerization 66ff.
photothermal therapy, see cancer therapy
plastic, elongation, see structural rearrangement
polyanions 64ff.
polycations 62, 64
polydispersity 69
polyelectrolyte 203f., 291, 297
poly(ethylene glycol) (PEG) 40

– -block-polyions 43
– cell adhesion inhibitor 82
– -containing monomers 33
– grafting 70
– PEGylated nanocarriers 125ff.
– sterical protecting 124, 128
polyisoprene (PI) 318f., 323
polymer
– amphiphilic 44, 124, 126, 139, 157ff.
– blending 410f.
– brushes 60, 128, 161
– conducting (CP), see nanowires
– core-crosslinked star 36ff.
– cushion 277f.
– degradation 322
– density 7
– longevity 123f.
– star 36ff.
polymerization
– anionic 370
– arm-first method 36f.
– atom transfer radical (ATRP) 31ff.
– cationic 36f.
– controlled radical (CRP) 69, 75
– core-first method 37f.
– dispersion 35
– electron-beam 71
– heterogeneous 345
– in liposomes 34
– in situ 40, 436
– in templates 29ff.
– inverse miniemulsion 29, 33
– living 36ff.
– macroinitiator-type 39
– macromer-type 40
– micelle crosslinking 38ff.
– miniemulsion 29f., 195
– nitroxide-mediated radical (NMRP) 32, 37, 69
– on inorganic NPs 34
– photo– 442
– photoiniferter-mediated 69
– plasma 70f.
– precipitation 29f., 33, 35
– reversible addition-fragmentation chain transfer (RAFT) precipitation 35f., 41f., 46ff.
– ring-opening (ROP) 36f., 69, 263
– surfactant-free emulsion (SFEP) 29f., 32ff.
– suspension 29f.
– two-photon 114
– two-step 32
polymersomes biomedical applications

– artificial cells and organelles 173
– cancer thearpy 172
– controlled release 161
– delivery vectors 172
– gene therapy 174f.
– medical imaging 169ff.
– nanoreactors 172f.
– near-infrared (NIR)-polymersomes 169
polymersomes 76, 157ff.
– chemistry 158ff.
– membrane 159ff.
– PEO-based 161, 163ff.
– porous 171
– preparation 167ff.
– properties 160f., 167
– proteo– 171
– responsive 163f., 167
– size 162f.
– surface chemistry 164ff.
polyplexes 132
polystyrene (PS) 32, 40, 314, 318, 320, 323, 347
– latex particles 126
– negativley 329
  – tissue culture (TCPS) 71, 81
polyzwitterions 65f.
protein
– adhesion 79
– adsorption 80, 292
– –cell junction 79
– denaturization 46

q

quantum confinement 189
quantum dots (QDs)
– encapsulation 73, 141, 170
– PEGylated 127, 130, 287
– size-controlled 429
– stimulus-responsive polymer (SRP) 73
– synthesis 243, 429
quartz crystal microbalance (QCM) 68

r

radiofrequency ablation, see cancer therapy
radiopaque iodinated P(MAOETIB)
  NPs 346ff.
– characterization 346f.
– effect of initiator concentration 348f.
– effect of MAOETIB concentration 347f.
– effect of surfactant concentration 349
Raman spectroscopy 486
reaction
– conjugation 261
– diffusion-limited homogeneous 243

– photo-Fenton 41
– polyaddition 195
– polycondensation 195
– side-chain 41
receptor
– folate (FR) 130, 504
– human epidermal growth factor (EGFR) 503f.
– transferrin (TfR) 129, 503f.
– vasoactive pituitary adenylate cyclase (VPAC)-activating peptide 503
repair-and-go strategy 5
reticuloendothelial system (RES) 123, 264, 267, 273, 287, 344, 360, 378
Reynolds number 233
RNA
– delivery carrier 44
– small interfering (siRNA) 33, 44, 276
Runge–Kutta algorithm 7
rupture 9ff.

s

saturation parameter 13f.
scaffolds
– core–shell fibers 405ff.
– fabrication 399ff.
– fiber diameter 403f.
– fiber orientation 404f., 408ff.
– fiber surface optimization 410ff.
– functionalization 397, 418
– mechanical properties 398, 418
– polycaprolactone (PCL) 406, 409, 412, 415
– porosity 398, 407f., 418
– structure 401ff.
– three-dimensional (3-D) 405, 418
– tissue engineering 397ff.
– two-dimensional (2-D) 405
scanning electron microscopy (SEM)
– cold-field emission gun high-resolution (CFEG-HRSEM) 107
– molecular crystal nanorod 440, 442f.
– nanotube forest 107ff.
– peptide microelectrode array (MEA) 114
  – polycaprolactone (PCL) scaffold 406
– segmented CP nanowire 462
self-assembled
– amphiphilic lipids 259, 265, 271, 371
– block copolymers 38, 230, 243, 265
– G^C motif 495ff.
– layer-by-layer 286, 297, 412
– lipid core–polymer shell hybrid NPs 298
– nanotubes 474f., 495ff.
– nanowires 474ff.

self-assembled peptide nanostructures
- controlled positioning 105ff.
- dielectrophoresis 110f.
- direct transfer 112f.
- fibrils 106, 115ff.
- horizontal alignment 106ff.
- inkjet technology 107, 109, 111
- laser-induced forward transfer (LIFT) 111ff.
- nanotube 106f., 494ff.
- nonlinear lithography 114ff.
- three-dimensional (3-D) nanostructures 114ff.
- transactivating–transduction (TAT) derivative 369
- vapor deposition methods 107ff.
- vertical alignment 106f., 110
self-assembled
- phospholipids 157f.
- polysaccharide-based nanoparticles 291
- supramolecular 430, 432
side groups
- amino 324, 327, 333
- carboxylic 324ff.
- charged 325ff.
sodium dodecyl sulfate (SDS), see surfactant
sodium dodecylbenzene sulfonate (SDBS), see surfactant
sol–gel, see synthesis
solid lipid nanoparticles (SLNs) 234, 246, 365ff.
- brain drug accumulation 376ff.
- brain targeting 375f.
- crystallization 375
- melting temperature 365, 373, 375
- polymorphism 372f., 375
- size 373f.
- stability 375f.
- sterilization 370f.
- structure 371ff.
- synthesis 369ff.
solubility
- CNTs 478f.
- drugs 368
- lipopolymer hybrid nanoparticles 265
- monomer 33
- organic nanoparticles 245
- poly(ethylene glycol) (PEG) 124
- polymer 33, 124
stealth 165, 267, 292, 377
stimulus-responsive polymer (SRP) application
- control of biointerfacial interaction 79ff.
- controlled drug delivery 77ff.

- micofluidic valves 82ff.
- molecular separation 85f.
stimulus-responsive polymer
- based systems 288
- dual– 67ff.
- multi– 67ff.
stimulus-responsive polymer nanointerfaces 59f., 69ff.
- covalent routes 69f.
- grafting-from techniques 69
- grafting-to techniques 69ff.
- nanoparticles 72ff.
- noncovalent routes 70f.
stimulus-responsive polymer
- photoresponsive 66f.
- ph-responsive 62ff.
- thermoresponsive 60ff.
structural rearrangement 9, 11ff.
subcellular
- distribution of charged NPs 327f.
- localization 312, 315
substrates
- silicon 71, 107
- stimulus-responsive polymer-functionalized 71f.
- tissue culture polystyrene (TCPS) 71
surface area to volume ratio 445
surface
- charge 288, 300, 325f.
- crack 9
- functonalization 312f., 324ff.
- lotus-type 108
- wettability 64, 71
surface-enhanced resonant Raman scattering (SERS) 252
surfactant
- brain targeting 375f.
- -free 29
- -initiated 30
- polysorbate 80 (P80) 376ff.
- sodium bis(2-ethylhexyl) sulfosuccinate (AOT) 33
- sodium dodecyl sulfate (SDS) 30, 35
- sodium dodecylbenzene sulfonate (SDBS) 30, 32
- sorbitan monooleate (Span-80) 33
- through polymerization 30
synthesis
- batch synthesis 222f.
- bottom-up approach 29f., 108, 190, 194ff.
- coacervation 369
- chemical vapor deposition (CVD) 406, 429, 478
- comminution method 223

– cyclical freeze-thaw 266
– electrophoretic deposition 110f., 438
– electrospinning 289, 369, 397, 399ff.
– emulsification 289, 434
– emulsion 29, 33, 194f., 197, 229ff.
– ether injection 266
– extrusion 266
– freeze-drying 197f.
– hard-templating methods 434ff.
– high-pressure homogenization 268f., 369f., 376
– ionic gelation 289, 298
– lithographic methods 193f., 458, 460f.
– mechanical milling 190f., 431
– mechanical stretching 458
– mechano-chemical 289, 431
– melting–recrystallization 437f.
– membrane-mixing 433
– microfluidic devices 204f., 221ff.
– monomer MAOETIB 345f.
– nanoprecipitation 230f., 271f.
– precipitation by polyelectrolyte complex formation 203f.
– radiopaque iodinated P(MAOETIB–GMA) 343f., 353
– reactive ion etching (RIE) 109
– reprecipitation 198f., 432f.
– reverse-phase evaporation 266
– sol–gel 72, 76, 289, 298, 434
– solvent-annealing method 439ff.
– sonication 266, 269, 369
– star-like nanogel 36, 40
– sublimation 438
– supercritical fluid techniques 202f.
– supersaturation 231
– template-assisted synthesis of nanostructures 436f., 476ff.
– thin-film hydration method 266, 269
– top-down approach 29, 190f., 222, 399
– vapor condensation-based methods 206f.
– vapor growth 431

t

targeting
– active 292
– brain 375f.
– intracellular 130ff.
– ligand 128f.
– magnetic drug (MDT) 122, 134
– passive 122, 292
– receptor-mediated 502ff.
template
– aluminum oxide (AAO) 430, 435ff.
– -assisted synthesis of nanostructures, see synthesis
– biomolecule 460f.
– -directed fabrication 457f., 460
– pores 435ff.
– Whatman 435
theranostics, see drug delivery systems
thermosensitivity
– nanogels 30f.
– stimulus-responsive polymers 60ff.
thermotherapy, see cancer therapy
time-to-flight secondary ion mass spectrometry (ToF-SIMS) 80
– PNIPAAm coatings 80
tissue engineering 79, 397ff.
tissue regeneration 416
top-down approach, see synthesis
toxicity
– contrast agent 344
– drugs 265
– PACA NPs 382f.
– poly(ethylene glycol) (PEG) 124
– quantum dots (QDs) 170
– solid lipid nanoparticles (SLNs) 365, 369, 382f.
– transfection agents 325
transcytosis of NPs 379
transfection agents 137, 315, 319, 324ff.
transformation
– crystal–crystal 442
– nanogels–microgels 46
– structural 48
transition
– hydrogels 61
– non-reversible 163
– sol–gel 28, 47, 60
– temperature 61
transmission electron microscopy (TEM)
– cellular uptake of NPs 323f.
– core–shell nanostructures 299, 406f.
– lipopolymers 270, 273f.
– molecular crystal nanorod 441f., 444
– nanotubes 108, 475
– polymersomes 166
– radiopaque iodinated P(MAOETIB–GMA) 355f.
– radiopaque iodinated P(MAOETIB) NPs 346f.
– subcellular location 316, 327f.

u

ultrasonography, see medical imaging
ultraviolet (UV) light 39, 41ff.
uptake

– cellular   73, 312, 315, 317ff.
– concentration-dependence   319
– density–uptake relationship   312, 326
– dynamic raft-dependent   332
– influence of polymer   318ff.
– influence of transfection agents   319, 324ff.
– kinetics   312, 320
– lipid raft-dependent   332
– size dependency   328f.
– solid lipid nanoparticles (SLNs)   377, 379

v

vesicles, *see* polymersomes
viscoelastic behavior   5, 15ff.

w

Weibull statistical analysis   9ff.
wettability
– contact angle   64, 108
– controlled   67
– polymer interfaces   66
– surface   64
Wolff rearrangement   68
wound dressing   302, 399

x

X-ray absorption spectroscopy (XRS)   242
X-ray diffraction   438, 442
X-ray imaging   344f., 353, 358ff.
– in vitro visibility of MAOETIB   349f.
– in vivo visibility of MAOETIB   351, 358ff.
X-ray scattering   224

y

yield stress   12
Young's moduli   105

z

zwitterionic
– headgoups   262f.
– poly–   65f.